Greek Alphabet

Alpha	A	α	Iota	I	ι	Rho	P	ρ
Beta	B	β	Kappa	K	κ	Sigma	Σ	σ
Gamma	Γ	γ	Lambda	Λ	λ	Tau	T	τ
Delta	Δ	δ	Mu	M	μ	Upsilon	Υ	υ
Epsilon	E	ε	Nu	N	ν	Phi	Φ	ϕ
Zeta	Z	ζ	Xi	Ξ	ξ	Chi	X	χ
Eta	H	η	Omicron	O	o	Psi	Ψ	ψ
Theta	Θ	θ	Pi	Π	π	Omega	Ω	ω

Conversion Table for Units

Length		
meter (SI unit)	m	
centimeter	cm	$= 10^{-2}$ m
ångström	Å	$= 10^{-10}$ m
micron	μ	$= 10^{-6}$ m
Volume		
cubic meter (SI unit)	m^3	
liter	L	$= $ dm$^3 = 10^{-3}$ m^3
Mass		
kilogram (SI unit)	kg	
gram	g	$= 10^{-3}$ kg
Energy		
joule (SI unit)	J	
erg	erg	$= 10^{-7}$ J
rydberg	Ry	$= 2.179\,87 \times 10^{-18}$ J
electron volt	eV	$= 1.602\,18 \times 10^{-19}$ J
inverse centimeter	cm^{-1}	$= 1.986\,45 \times 10^{-23}$ J
calorie (thermochemical)	Cal	$= 4.184$ J
liter atmosphere	l atm	$= 101.325$ J
Pressure		
pascal (SI unit)	Pa	
atmosphere	atm	$= 101325$ Pa
bar	bar	$= 10^5$ Pa
torr	Torr	$= 133.322$ Pa
pounds per square inch	psi	$= 6.894\,757 \times 10^3$ Pa
Power		
watt (SI unit)	W	
horsepower	hp	$= 745.7$ W
Angle		
radian (SI unit)	rad	
degree	°	$= \dfrac{2\pi}{360}$ rad $= \left(\dfrac{1}{57.295\,78}\right)$ rad
Electrical dipole moment		
C m (SI unit)		
debye	D	$= 3.335\,64 \times 10^{-30}$ C m

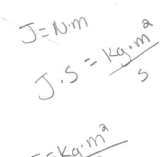

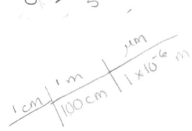

$$N = \frac{kg \cdot m}{s^2}$$

$$J = N \cdot m$$

$$J \cdot s = \frac{kg \cdot m^2}{s}$$

$$J = \frac{kg \cdot m^2}{s^2}$$

$$\frac{1\,cm}{} \cdot \frac{1\,m}{100\,cm} \cdot \frac{\mu m}{1 \times 10^{-6}\,m}$$

The Chemistry Place for *Physical Chemistry*

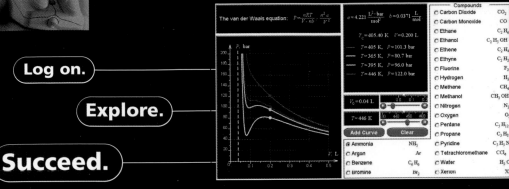

Log on.

Explore.

Succeed.

The Chemistry Place for *Physical Chemistry* features a suite of more than 55 applet-based tutorials developed by the authors for those chapters that have the most challenging math components. A web icon in the margin of the text directs you to specific online exercises that complement the textbook discussion and that encourage you to reason qualitatively about chemical processes, experiment quantitatively, and think critically. Each activity includes a printable worksheet for lab or homework assignments.

Online Interactive Applets and Animations

- Ideal Gas under Constant Pressure or Volume
- Reversible Isothermal Compression of an Ideal Gas
- Reversible Isobaric Compression and Expansion of an Ideal Gas
- Isochoric Heating and Cooling of an Ideal Gas
- Reversible Cyclic Processes
- Reversible Adiabatic Heating and Cooling of an Ideal Gas
- The Reversible Carnot Cycle
- Van der Walls Equation of State
- Comparison between Ideal Gas Law and van der Waals Equation
- Compression Factor and Molar Volume
- Fugacity and the Fugacity Coefficient
- Blackbody Radiation
- Diffraction of Light
- Diffraction from Double Slit
- Transverse, Longitudinal, and Surface Waves
- Interference of Two Traveling Waves
- Interference of Two Standing Waves
- Expanding Functions in Fourier Series
- The Classical Particle in a Box
- Energy Levels for the Particle in a Box
- Probability of Finding the Particle in a Given Interval
- Eigenfunctions for the Two-Dimensional Box
- Acceptable Wave Functions for the Particle in a Box
- Expectation Values for E, p, and x for a Superposition Wave Function
- Energy Eigenfunctions and Eigenvalues for a Finite Depth Box
- Tunneling through a Barrier

- The Heisenberg Uncertainty Principle
- Wave Packets and the Uncertainty Principle
- Expanding the Total Energy Eigenfunctions in Eigenfunctions of the Momentum Operator
- The Classical Harmonic Oscillator
- Energy Levels and Eigenfunctions for the Harmonic Oscillator
- Probability of Finding the Oscillator in the Classically Forbidden Region
- Energy Levels and Emission Spectra
- The Morse Potential
- Normal Modes for H_2O, CO_2, NH_3, and Formaldehyde
- Rotational Spectroscopy of Diatomic Molecules
- Rotational-Vibrational Spectroscopy of Diatomic Molecules
- Simulation of a Laser
- The Molecular Orbital Energy Diagram
- Behavior of the Partition Function
- Variation of q with Temperature
- Vibrational Heat Capacity Simulation
- Benchmark Values for Gas Particle Speed Distributions
- Fick's Second Law
- Sequential Reaction Kinetics
- Branching Reaction Kinetics
- The Lindemann Mechanism
- Michaelis-Menten Enzyme Kinetics
- The Langmuir Isotherm
- Three Dimensional Molecule Visualizations: Ammonia, Carbon Dioxide, Formaldehyde, and Water
- Graphing Routine Tool

System requirements:
- Windows OS 98; NT4; 2000; XP (250 MHz)
- Macintosh OS 9.2; 10.2; 10.3 (233 MHz)
- 64 MB RAM installed
- 1024x768 screen resolution
- 56K modem internet connection or higher
- Browsers for Windows: Internet Explorer 6.0; Netscape 7.1
- Browsers for Macintosh: Internet Explorer 5.2; Netscape 7.1; Safari 1.2
- Plug-ins: Flash 7.0

For technical support
please visit
http://www.aw-bc.com/techsupport/
and complete the appropriate online form.

How to log on to www.aw-bc.com/chemplace:

1. Go to www.aw-bc.com/chemplace
2. Click on your textbook's cover.
3. Click "Register."
4. Scratch off the silver foil coating below to reveal your pre-assigned access code.
5. Enter your pre-assigned access code exactly as it appears below.

6. Complete the online registration form to create your own personal user Login Name and Password.
7. Once your personal Login Name and Password are confirmed by email, go back to www.aw-bc.com/chemplace, click on your textbook's cover, type in your new Login Name and Password, and click "Log In."

Your Access Code is:

More options for your physical chemistry course

Physical Chemistry by Engel and Reid is a flexible text that accommodates your approach to teaching. Choose the complete text for "Thermo-first" courses or a split version of the text, organized to support "Quantum-first" courses. **The choice is yours.**

ALL CHAPTERS

Physical Chemistry

Thomas Engel • Philip Reid

Physical Chemistry
by Thomas Engel, *University of Washington* and Philip Reid, *University of Washington*
©2006 • 970 pp • Casebound
ISBN 0-8053-3842-X

TABLE OF CONTENTS

CHAPTERS 12–29

Quantum Chemistry & Spectroscopy

Thomas Engel

Quantum Chemistry & Spectroscopy
by Thomas Engel
©2006 • 446 pp • Casebound • ISBN 0-8053-3843-8

CHAPTERS 1–11 and 30–37

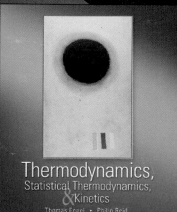

Thermodynamics, Statistical Thermodynamics, & Kinetics

Thomas Engel • Philip Reid

Thermodynamics, Statistical Thermodynamics, & Kinetics
by Thomas Engel and Philip Reid
©2006 • 524 pp • Casebound • ISBN 0-8053-3844

Thermodynamics, Statistical Thermodynamics, and Kinetics

Thomas Engel
University of Washington

Philip Reid
University of Washington

PEARSON

Benjamin Cummings

San Francisco Boston New York
Cape Town Hong Kong London Madrid Mexico City
Montreal Munich Paris Singapore Sydney Tokyo Toronto

Publisher: Jim Smith

Marketing Manager: Scott Dustan

Project Editors: Katie Conley and Lisa Leung

Editorial Assistant: Cinnamon Hearst

Media Producer: Claire Masson

Production Supervisor: Shannon Tozier

Production Editor: Lori Dalberg, Carlisle Publishers Services

Composition: Carlisle Communications, Ltd

Illustrators: Imagineering Media Services, Inc.

Manufacturing Buyer: Michael Early

Text Designer: Carolyn Deacy, Carolyn Deacy Design

Cover Designer: Studio Montage

Text Printer and Binder: RR Donnelley and Sons, Willard

Cover Printer: Phoenix Color

Cover Credit: Adolph Gottlieb, "Vert" 1964, oil on canvas, $60 \times 36''$ © Adolph and Esther
Gottlieb Foundation/Licensed by VAGA, New York, NY

Credits: Some end-of-chapter problems appearing in this book were originally printed in *Physical
Chemistry,* 3rd edition, by Joseph H. Noggle, copyright ©1996 by Joseph H. Noggle, Ph.D., Inc.,
and in *Physical Chemistry,* 3rd edition, by Gilbert W. Castellan, copyright © 1981 by Addison-
Wesley Publishing Company, Inc. Reprinted by permission of Pearson Education, Inc.

Library of Congress Cataloging-in-Publishing Data

Engel, Thomas, 1942-
   Physical chemistry / Thomas Engel, Philip J. Reid.
     p. cm.
   Includes bibliographical references and index.
   ISBN 0-8053-3842-X
  1. Chemistry, Physical and theoretical--Textbooks. I. Reid, Philip J. II. Title.

QD453.3.E54 2005
541--dc22

                          2004029329

Thermodynamics, Statistical Thermodynamics, and Kinetics ISBN 0-8053-3844-6

PEARSON

Benjamin
Cummings

1 2 3 4 5 6 7 8 9 10-DOW-09 08 07 06
www.aw-bc.com

This book is dedicated to my parents, Walter and Juliane, who were my first teachers, and to my cherished family, Esther and Alex, with whom I am still learning.
—Thomas Engel

This book is dedicated to my family—Carolyn, Sierra, and Samantha—for their patience, support, and understanding.
—Philip Reid

About the Authors

Thomas Engel has taught chemistry for more than 20 years at the University of Washington, where he is currently Professor of Chemistry and Associate Chair for the Undergraduate Program. Professor Engel received his bachelor's and master's degrees in chemistry from the Johns Hopkins University, and his Ph.D. in chemistry from the University of Chicago. He then spent 11 years as a researcher in Germany and Switzerland, in which time he received the Dr. rer. nat. habil. degree from the Ludwig Maximilians University in Munich. In 1980, he left the IBM research laboratory in Zurich to become a faculty member at the University of Washington.

Professor Engel's research interests are in the area of surface chemistry, and he has published more than 80 articles and book chapters in this field. He has received the Surface Chemistry or Colloids Award from the American Chemical Society and a Senior Humboldt Research Award from the Alexander von Humboldt Foundation, which has allowed him to establish collaborations with researchers in Germany. He is currently working together with European manufacturers of catalytic converters to improve their performance for diesel engines.

Philip Reid has taught chemistry at the University of Washington since he joined the chemistry faculty in 1995. Professor Reid received his bachelor's degree from the University of Puget Sound in 1986, and his Ph.D. in chemistry from the University of California at Berkeley in 1992. He performed postdoctoral research at the University of Minnesota, Twin Cities, campus before moving to Washington.

Professor Reid's research interests are in the areas of atmospheric chemistry, condensed-phase reaction dynamics, and nonlinear optical materials. He has published more than 70 articles in these fields. Professor Reid is the recipient of a CAREER award from the National Science Foundation, is a Cottrell Scholar of the Research Corporation, and is a Sloan fellow.

Preface

This book grew out of many years of teaching physical chemistry and searching for a thermodynamics textbook that would be accessible to students and demonstrate that thermodynamics is modern, vital, and evolving. This book is intended for undergraduate students majoring in chemistry, biochemistry, and chemical engineering as well as many students majoring in the atmospheric sciences and the biological sciences. The following objectives outline the distinctive features of this book.

- **Focus on pedagogy and logical development of topics.**   In our experience, students understand the main concepts of thermodynamics and statistical mechanics only when the logic behind these concepts is clearly illustrated and explored. In this text, we describe in detail the central ideas that frame thermodynamics and statistical mechanics and the application of these ideas. The goal is to build a solid foundation of understanding that allows the student to understand a variety of topics at a fundamental level, rather than cover a wide variety of topics in modest detail.

- **Illustrate the relevance of thermodynamics to the chemistry of today.** Thermodynamics is perhaps the central area of science that one needs to understand the nature of life on this planet. However, many students struggle to connect the thermodynamics that they learn in books to the world around them. To address this issue, example problems and specific topics are tied together to help the student develop this connection. For example, fuel cells, refrigerators, heat pumps, and real engines are discussed in connection with the second law. Mechanistic aspects of the catalytic synthesis of ammonia are included in discussing equilibrium in reaction mixtures, and atomic scale electrochemistry is discussed to illustrate the relevance of thermodynamics to the scientific issues of today. Our approach to thermodynamics draws on the excellent discussion by Gilbert W. Castellan in *Physical Chemistry,* 3rd edition.

- **Develop statistical mechanics from the "ground up."**   In many physical chemistry texts, a detailed development of statistical mechanics is abandoned in favor of presenting a few central ideas. This text takes the opposite approach, and statistical mechanics is developed starting with probability theory and basic ideas of thermodynamics. The development here largely follows the approach of Leonard Nash as outlined in his text *Elements of Statistical Thermodynamics.* In this approach, basic probability theory is explored and applied to collections of atomic or molecular units. This simple starting point is expanded until the Boltzmann distribution is derived, the canonical ensemble defined, and statistical thermodynamics described in detail. Statistical mechanics provides the critical connection between the macroscopic perspective of thermodynamics and the microscopic perspective of quantum mechanics, and the elegance of this connection is given ample time to develop in the student's mind.

- **Explore modern issues in kinetics.**   We cover the development of chemical kinetics, beginning with a proven pedagogical approach that results in a basic understanding of kinetic concepts. However, the text goes beyond this standard treatment to explore modern approaches in this important area of chemistry. For example, a discussion of single-molecule spectroscopy is presented, and the general area of catalysis is further motivated by discussions of enzyme activity and surface adsorption. The section on chemical kinetics will provide the student with a firm understanding of the general concepts in this field and with the ability to explore more complex ideas.

- **Use web-based simulations to illustrate the concepts being explored and avoid a math overload.** Thermodynamics, statistical mechanics, and kinetics can be mathematically intensive areas to study, and the mathematics involved can distract the student from "seeing" the underlying concepts. To circumvent this problem, web-based simulations have been incorporated as end-of-chapter problems throughout the book so that the student can focus on the science and avoid a math overload. The behavior of thermodynamic behavior molecular ensembles, velocity distribution functions, and kinetic behavior are illustrated through interactive tutorials. These web-based simulations can also be used by instructors during lecture. More than 20 such web-based problems are available on the course website. The course website also includes a graphing routine with a curve-fitting capability, which allows students to print and submit graphical data.

- **Show that learning problem-solving skills is an essential part of physical chemistry.** Many example problems are worked through in each chapter. The end-of-chapter problems cover a range of difficulties suitable for students at all levels. Conceptual questions at the end of each chapter ensure that students practice thinking as quantum chemists.

- **Use color to make learning physical chemistry more interesting.** Four-color images are used to display atomic and molecular orbitals both quantitatively and attractively as well as to make complex images such as the symmetry elements of ethene understandable.

This text contains more material than can be covered in a one-quarter or one-semester course, and this is entirely intentional. Effective use of the text does not require one to proceed sequentially through the chapters or to include all sections. Some topics are discussed in supplemental sections that can be omitted if they are not viewed as essential to the course. Also, many sections are self-contained so that they can be readily omitted if they do not serve the needs of the instructor. For example, many modern physical chemistry courses do not include transport theory, and this chapter can be omitted without impacting student understanding of subsequent chapters. The text is constructed to be flexible to your needs, not the other way around. We welcome the comments of both students and instructors on how the material was used and on how the presentation can be improved. Please contact us at pchem@chem.washington.edu.

Thomas Engel
University of Washington

Philip Reid
University of Washington

Acknowledgments

Many individuals have helped us to get this text into its current form. Students have provided us with feedback directly and through the questions they have asked, which has helped us to understand how they learn. Many of our colleagues, including Gary Drobny, Bill Reinhardt, and especially Mickey Schurr, have been invaluable in critically reading individual chapters and have been generous with their advice and insight, as has Kendrick Shaw, who read all of the chapters in his role as accuracy checker. We are also fortunate to have access to some end-of-chapter problems that were originally presented in *Physical*

Chemistry, 3rd edition, by Joseph H. Noggle, and in *Physical Chemistry,* 3rd edition, by Gilbert W. Castellan. The reviewers, who are listed separately, have made many suggestions for improvement, for which we are very grateful. All those involved in the production process have helped to make this book a reality through their efforts. Special thanks are due to our publisher Jim Smith, who convinced us to take on this task, and to the project editors Lisa Leung, who guided the manuscript through the review and development processes, and Katie Conley, who sheparded the text, art, and supplements through to production.

Reviewers

Ludwik Adamowicz
University of Arizona
Joseph BelBruno
Dartmouth College
Eric Bittner
University of Houston
Juliana Boerio-Goates
Brigham Young University
Alexandre Brolo
University of Victoria
Alexander Burin
Tulane University
Laurie Butler
University of Chicago
Ronald Christensen
Bowdoin College
Jeffrey Cina
University of Oregon
Robert Continetti
University of California, San Diego
Susan Crawford
California State University, Sacramento
H. Floyd Davis
Cornell University
Jimmie Doll
Brown University
D. James Donaldson
University of Toronto
Robert Donnelly
Auburn University
Doug Doren
University of Delaware
Bogdan Dragnea
Indiana University
Cecil Dybowski
University of Delaware
Patrick Fleming
San Jose State University
Rigoberto Hernandez
Georgia Institute of Technology
Ming-Ju Huang
Jackson State University
George Kaminski
Central Michigan University

Katherine Kantardjieff
California State University, Fullerton
Chul-Hyun Kim
California State University, Hayward
Keith Kuwata
Macalester College
Kimberly Lawler-Sagarin
Elmhurst College
Katja Lindenberg
University of California, San Diego
John Lowe
Penn State University
Peter Lykos
Illinois Institute of Technology
David Micha
University of Florida
Daniel Neumark
University of California, Berkeley
Simon North
Texas A&M University
Maria Pacheco
Buffalo State College
Robert Pecora
Stanford University
Jacob Petrich
Iowa State University
David Ritter
Southeast Missouri State University
George Schatz
Northwestern University
Robert Schurko
University of Windsor
Roseanne J. Sension
University of Michigan
Alexa Serfis
Saint Louis University
Bradley Stone
San Jose State University
Michael Tubergen
Kent State University
Tom Tuttle
Brandeis University
James Valentini
Columbia University

Michael Wagner
George Washington University
Robert Walker
University of Maryland
Gary Washington
United States Military Academy, West Point

Charles Watkins
University of Alabama at Birmingham
Rand Watson
Texas A&M University
Mark Young
University of Iowa

Problem Solvers

Sergiy Bubin
University of Arizona
Ming-Ju Huang
Jackson State University
George Kaminski
Central Michigan University
Benjamin Killian
University of Florida
Craig Martens
University of California, Irvine
Matthew Nee
University of California, Berkeley

David Ritter
Southeast Missouri State University
Marc Roussel
University of Lethbridge
Ken Rousslang
University of Puget Sound
John Watts
Jackson State University
Jia Zhou
University of California, Berkeley

Brief Contents

Contents

CHAPTER 1

Fundamental Concepts of Thermodynamics

Thermodynamics provides a description of matter on a macroscopic scale. In this approach, matter is described in terms of bulk properties such as pressure, density, volume, and temperature. The basic concepts employed in thermodynamics, such as system, surroundings, intensive and extensive variables, adiabatic and diathermal walls, equilibrium, temperature, and thermometry, are discussed in this chapter. The usefulness of equations of state, which relate the state variables of pressure, volume, and temperature, is also discussed for real and ideal gases. ∎

1.1 What Is Thermodynamics and Why Is It Useful?

Thermodynamics is the branch of science that describes the behavior of matter and the transformation between different forms of energy on a **macroscopic scale,** or the human scale and larger. Thermodynamics describes a system of interest in terms of its bulk properties. Only a few such variables are needed to describe the system, and the variables are generally directly accessible through measurements. A thermodynamic description of matter does not make reference to its structure and behavior at the microscopic level. For example, 1 mol of gaseous water at a sufficiently low density is completely described by two of the three **macroscopic variables** of pressure, volume, and temperature. By contrast, the **microscopic scale** refers to dimensions on the order of the size of molecules. At the microscopic level, water is a dipolar triatomic molecule, H_2O, with a bond angle of $104.5°$ that forms a network of hydrogen bonds.

In this book, you will learn first about thermodynamics and then about statistical thermodynamics. **Statistical thermodynamics** (Chapters 13 through 15) uses atomic and molecular properties to calculate the macroscopic properties of matter. For example, statistical thermodynamics can show that liquid water is the stable form of aggregation at a pressure of 1 bar and a temperature of 90°C, whereas gaseous water is the stable form at 1 bar and 110°C. Using statistical thermodynamics, macroscopic properties are calculated from underlying molecular properties.

Given that the microscopic nature of matter is becoming increasingly well understood using theories such as quantum mechanics, why is a macroscopic science like thermodynamics relevant today? The usefulness of thermodynamics can be illustrated by describing four applications of thermodynamics:

- You have built a plant to synthesize NH_3 gas from N_2 and H_2. You find that the yield is insufficient to make the process profitable and decide to try to improve the NH_3 output by changing the temperature and/or the pressure. However, you do not know whether to increase or decrease the values of these variables. As will be shown in Chapter 6, the ammonia yield will be higher at equilibrium if the temperature is decreased and the pressure is increased.

- You wish to use methanol to power a car. One engineer provides a design for an internal combustion engine that will burn methanol efficiently according to the reaction $CH_3OH\ (l) + 2O_2\ (g) \rightarrow CO_2\ (g) + 2H_2O\ (l)$. A second engineer designs an electrochemical fuel cell that carries out the same reaction. He claims that the vehicle will travel much farther if powered by the fuel cell than by the internal combustion engine. As will be shown in Chapter 5, this assertion is correct, and an estimate of the relative efficiencies of the two propulsion systems can be made.

- You are asked to design a new battery that will be used to power a hybrid car. Because the voltage required by the driving motors is much higher than can be generated in a single electrochemical cell, many cells must be connected in series. Because the space for the battery is limited, as few cells as possible should be used. You are given a list of possible cell reactions and told to determine the number of cells needed to generate the required voltage. As you will learn in Chapter 11, this problem can be solved using tabulated values of thermodynamic functions.

- Your attempts to synthesize a new and potentially very marketable compound have consistently led to yields that make it unprofitable to begin production. A supervisor suggests a major effort to make the compound by first synthesizing a catalyst that promotes the reaction. How can you decide if this effort is worth the required investment? As will be shown in Chapter 6, the maximum yield expected under equilibrium conditions should be calculated first. If this yield is insufficient, a catalyst is useless.

1.2 Basic Definitions Needed to Describe Thermodynamic Systems

A thermodynamic **system** consists of all the materials involved in the process under study. This material could be the contents of an open beaker containing reagents, the electrolyte solution within an electrochemical cell, or the contents of a cylinder and movable piston assembly in an engine. In thermodynamics, the rest of the universe is referred to as the **surroundings.** If a system can exchange matter with the surroundings, it is called an **open system;** if not, it is a **closed system.** Both open and closed systems can exchange energy with the surroundings. Systems that can exchange neither matter nor energy with the surroundings are called **isolated systems.**

The interface between the system and its surroundings is called the **boundary.** The boundaries determine if energy and mass can be transferred between the system and the surroundings and lead to the distinction between open, closed, and isolated systems. Consider the Earth's oceans as a system, with the rest of the universe being the surroundings. The system–surroundings boundary consists of the solid–liquid interface between the continents and the ocean floor and the water–air interface at the ocean surface. For an open beaker in which the system is the contents, the boundary surface is just inside the inner wall of the beaker, and passes across the open top of the beaker. In this case, energy can be exchanged freely between the system and surroundings through the side and bottom walls, and both matter and energy can be exchanged between the system and surroundings through the open top boundary. The portion of the boundary formed by the beaker in the previous example is called a **wall.** Walls are always boundaries, but a boundary need not be a wall. Walls can be **rigid** or **movable** and **permeable** or **nonpermeable.** An example of a movable wall is the surface of a balloon.

The exchange of energy and matter across the boundary between system and surroundings is central to the important concept of **equilibrium.** The system and surroundings can be in equilibrium with respect to one or more of several different **system variables** such as pressure (P), temperature (T), and concentration. **Thermodynamic equilibrium** refers to a condition in which equilibrium exists with respect to P, T, and concentration. A system can be in equilibrium with its surroundings with respect to a given variable. Equilibrium exists only if that variable does not change with time, and if it has the same value in all parts of the system and surroundings. For example, the interior of a soap bubble[1] (the system) and the surroundings (the room) are in equilibrium with respect to P because the movable wall (the bubble) can reach a position where P on both sides of the wall is the same, and because P has the same value throughout the system and surroundings. Equilibrium with respect to concentration exists only if transport of all species across the boundary in both directions is possible. If the boundary is a movable wall that is not permeable to all species, equilibrium can exist with respect to P, but not with respect to concentration. Because N_2 and O_2 cannot diffuse through the (idealized) bubble, the system and surroundings are in equilibrium with respect to P, but not to concentration. Equilibrium with respect to temperature is a special case that is discussed next.

Temperature is an abstract quantity that is only measured indirectly, for example, by measuring the volume of mercury confined to a narrow capillary, the electromotive force generated at the junction of two dissimilar metals, or the electrical resistance of a platinum wire. At the microscopic level, temperature is related to the mean kinetic energy of molecules. Although each of us has a sense of a "temperature scale" based on the qualitative descriptors *hot* and *cold,* we need a more quantitative and transferable measure of temperature that is not grounded in individual experience. To make this discussion more concrete, we consider a dilute gas under conditions in which the ideal gas law of Equation (1.1) describes the relationship among P, T, and the molar density $\rho = n/V$ with sufficient accuracy:

$$P = \rho RT \tag{1.1}$$

The ideal gas law is discussed further in Section 1.4.

In thermodynamics, **temperature** is the property of a system that determines if the system is in thermal equilibrium with other systems or the surroundings. Equation (1.1) can be rewritten as follows:

$$T = \frac{P}{\rho R} \tag{1.2}$$

This equation shows that for ideal gas systems having the same molar density, a pressure gauge can be used to compare the systems and determine which of T_1 or T_2 is greater. **Thermal equilibrium** between systems exists if $P_1 = P_2$ for gaseous systems with the same molar density.

We use the concepts of temperature and thermal equilibrium to characterize the walls between a system and its surroundings. Consider the two systems with rigid walls shown in Figure 1.1a. Each system has the same molar density and is equipped with a pressure gauge. If we bring the two systems into direct contact, two limiting behaviors are observed. If neither pressure gauge changes, as in Figure 1.1b, we refer to the walls as being **adiabatic.** Because $P_1 \neq P_2$, the systems are not in thermal equilibrium and, therefore, have different temperatures. An example of a system surrounded by adiabatic

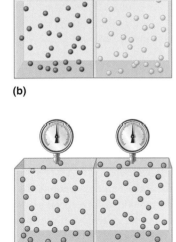

(a)

(b)

(c)

FIGURE 1.1
(a) Two separated systems with rigid walls and the same molar density have different temperatures. (b) The two systems are brought together so that the walls are in intimate contact. Even after a long time has passed, the pressure in each system is unchanged. (c) As in part (b), the two systems are brought together so that the walls are in intimate contact. After a sufficient time has passed, the pressures are equal.

---

[1] For this example, the surface tension of the bubble is assumed to be so small that it can be set equal to zero. This is in keeping with the thermodynamic tradition of weightless pistons and frictionless pulleys.

walls is coffee in a Styrofoam cup with a Styrofoam lid.[2] Experience shows that it is not possible to bring two systems enclosed by adiabatic walls into thermal equilibrium by bringing them into contact, because adiabatic walls insulate against the transfer of "heat." If you push a Styrofoam cup containing hot coffee against one containing ice water, they will not reach the same temperature. Rely on your experience at this point regarding the meaning of heat; a thermodynamic definition will be given in Chapter 2.

The second limiting case is shown in Figure 1.1c. In bringing the systems into intimate contact, both pressures change and reach the same value after some time. We conclude that the systems have the same temperature, $T_1 = T_2$, and say that they are in thermal equilibrium. We refer to the walls as being **diathermal.** Two systems in contact separated by diathermal walls reach thermal equilibrium because diathermal walls conduct heat. Hot coffee stored in a copper cup is an example of a system surrounded by diathermal walls. Because the walls are diathermal, the coffee will quickly reach room temperature.

The **zeroth law of thermodynamics** generalizes the experiment illustrated in Figure 1.1 and asserts the existence of an objective temperature that can be used to define the condition of thermal equilibrium. The formal statement of this law is as follows:

> Two systems that are separately in thermal equilibrium with a third system are also in thermal equilibrium with one another.

There are four laws of thermodynamics, all of which are generalizations from experience rather than mathematical theorems. They have been rigorously tested in more than a century of experimentation, and no violations of these laws have been found. The unfortunate name assigned to the "zeroth" law is due to the fact that it was formulated after the first law of thermodynamics, but logically precedes it. The zeroth law tells us that we can determine if two systems are in thermal equilibrium without bringing them into contact. Imagine the third system to be a thermometer, which is defined more precisely in the next section. The third system can be used to compare the temperatures of the other two systems; if they have the same temperature, they will be in thermal equilibrium if placed in contact.

1.3 Thermometry

The discussion of thermal equilibrium required only that a device exist, called a **thermometer,** that can measure relative hotness or coldness. However, scientific work requires a quantitative temperature scale. For any useful thermometer, the empirical temperature, t, must be a single-valued, continuous, and monotonic function of some thermometric system property designated by x. Examples of thermometric properties are the volume of a liquid, the electrical resistance of a metal or semiconductor, and the electromotive force generated at the junction of two dissimilar metals. The simplest case that one can imagine is if the empirical temperature, t, is linearly related to the value of the thermometric property, x:

$$t(x) = a + bx \qquad (1.3)$$

Equation (1.4) defines a **temperature scale** in terms of a specific thermometric property, once the constants a and b are fixed. The constant a determines the zero of the temperature scale because $t(0) = a$, and the constant b determines the size of a unit of temperature, called a degree.

One of the first practical thermometers was the mercury-in-glass thermometer. It utilizes the thermometric property that the volume of mercury increases monotonically over the temperature range in which it is in the liquid state (between $-38.8°$ and $356.7°C$). In 1745, Carolus Linnaeus gave this thermometer a standardized scale by arbitrarily assigning the values 0 and 100 to the freezing and boiling points of water, respectively. This chosen interval was divided into 100 equal degrees, and the same size

---

[2] In this discussion, Styrofoam is assumed to be a perfect insulator.

degree is used outside of the interval. Because there are 100 degrees between the two calibration points, it is called the **centigrade scale.**

The centigrade scale has been superseded by the **Celsius scale,** which is in widespread use today. The Celsius scale (denoted in units of °C) is similar to the centigrade scale. However, rather than being determined by two fixed points, the Celsius scale is determined by one fixed reference point at which ice, liquid water, and gaseous water are in equilibrium. This point is called the triple point (see Section 8.2) and is assigned the value 0.01°C. On the Celsius scale, the boiling point of water at a pressure of 1 atmosphere is 99.975°C. The size of the degree is chosen to be the same as on the centigrade scale.

Although the Celsius scale is used widely throughout the world today, the numerical values for this temperature scale are completely arbitrary. It would be preferable to have a temperature scale derived directly from physical principles. There is such a scale, called the **thermodynamic temperature scale** or **absolute temperature scale.** For such a scale, the temperature is independent of the substance used in the thermometer, and the constant a in Equation (1.4) is zero. The **gas thermometer** is a practical thermometer with which the absolute temperature can be measured. The thermometric property is the temperature dependence of P for a dilute gas at constant V. The gas thermometer provides the international standard for thermometry at very low temperatures. At intermediate temperatures, the electrical resistance of platinum wire is the standard, and at higher temperatures, the radiated energy emitted from glowing silver is the standard.

How is the gas thermometer used to measure the thermodynamic temperature? Measurements carried out by Robert Boyle in the 19th century demonstrated that the pressure exerted by a fixed amount of gas at constant V varies linearly with temperature on the Celsius scale as shown in Figure 1.2. At the time of Boyle's experiments, temperatures below $-30°C$ were not attainable in the laboratory. However, the P versus T data can be extrapolated to the limiting T value at which $P \to 0$. It is found that these straight lines obtained for different values of V intersect at a common point on the T axis that lies near $-273°C$.

The data show that at constant V, the thermometric property P varies with temperature as

$$P = c + dt \tag{1.4}$$

where t is the temperature on the Celsius scale, and c and d are experimentally obtained proportionality constants.

Figure 1.2 shows that all lines intersect at a single point, even for different gases. This suggests a unique reference point for temperature, rather than the two reference points used in constructing the centigrade scale. The value zero is given to the temperature at which $P \to 0$. However, this choice is not sufficient to define the temperature scale, because the size of the degree is undefined. By convention, the size of the degree on the absolute temperature scale is set equal to the size of the degree on the Celsius scale, because the Celsius scale was in widespread use at the time the absolute temperature scale was formulated. With these two choices, the absolute and Celsius temperature scales are related by Equation (1.5). The scale measured by the ideal gas thermometer is the absolute temperature scale used in thermodynamics. The unit of temperature on this scale is called the **kelvin,** abbreviated K (without a degree sign):

$$T(\text{K}) = T(°\text{C}) + 273.15 \tag{1.5}$$

Using the triple point of water as a reference, the absolute temperature, $T(\text{K})$, measured by an ideal gas thermometer is given by

$$T(\text{K}) = 273.16 \frac{P}{P_{tp}} \tag{1.6}$$

where P_{tp} is the pressure corresponding to the triple point of water. On this scale, the volume of an ideal gas is directly proportional to its temperature; if the temperature is reduced to half of its initial value, V is also reduced to half of its initial value. In practice, deviations from the ideal gas law that occur for real gases must be taken into account when using a gas

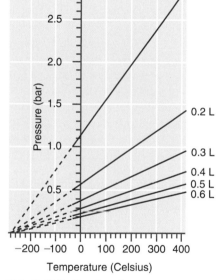

FIGURE 1.2
The pressure exerted by 5.00×10^{-3} mol of a dilute gas is shown as a function of the temperature measured on the Celsius scale for different fixed volumes. The dashed portion indicates that the data are extrapolated to lower temperatures than could be achieved experimentally by Boyle.

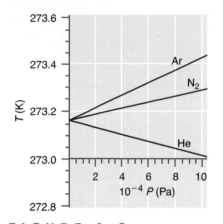

FIGURE 1.3

The temperature measured in a gas thermometer defined by Equation (1.6) is independent of the gas used only in the limit that $P \to 0$.

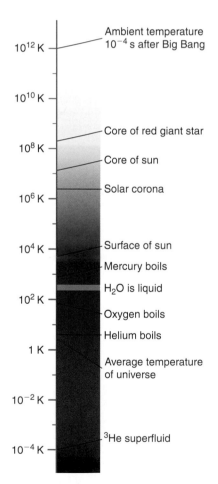

FIGURE 1.4

The absolute temperature is shown on a logarithmic scale together with the temperature of a number of physical phenomena.

thermometer. If data were obtained from a gas thermometer using He, Ar, and N_2 for a temperature very near T_{tp}, they would exhibit the behavior shown in Figure 1.3. We see that the temperature only becomes independent of P and of the gas used in the thermometer if the data are extrapolated to zero pressure. It is in this limit that the gas thermometer provides a measure of the thermodynamic temperature. Because 1 bar = 10^5 Pa, gas-independent T values are only obtained below $P \sim 0.01$ bar. The absolute temperature is shown in Figure 1.4 on a logarithmic scale together with associated physical phenomena.

1.4 Equations of State and the Ideal Gas Law

Macroscopic models in which the system is described by a set of variables are based on experience. It is particularly useful to formulate an **equation of state,** which relates the state variables. Using the absolute temperature scale, it is possible to obtain an equation of state for an ideal gas from experiments. If the pressure of He is measured as a function of the volume for different values of temperature, the set of nonintersecting hyperbolas shown in Figure 1.5 is obtained. The curves in this figure can be quantitatively fit by the functional form

$$PV = \alpha T \tag{1.7}$$

where T is the absolute temperature as defined by Equation (1.6), allowing α to be determined, which is found to be directly proportional to the mass of gas used. It is useful to separate out this dependence by writing $\alpha = nR$, where n is the number of moles of the gas, and R is a constant that is independent of the size of the system. The result is the ideal gas equation of state:

$$PV = NkT = nRT \tag{1.8}$$

where the proportionality constants k and R are called the **Boltzmann constant** and the **ideal gas constant,** respectively; N is the number of molecules; and n is the number of moles of the gas. The equation of state given in Equation (1.8) is known as the **ideal gas law.** Because the four variables are related through the equation of state, any three of these variables is sufficient to completely describe the ideal gas. Note that the total number of moles—not the number of moles of the individual gas components of a gas mixture—appears in the ideal gas law.

Of these four variables, P and T are independent of the amount of gas, whereas V and n are proportional to the amount of gas. A variable that is independent of the size of the system (for example, P and T) is referred to as an **intensive variable,** and one that is proportional to the size of the system (for example, V) is referred to as an **extensive variable.** Equation (1.8) can be written in terms of intensive variables exclusively:

$$P = \rho RT \tag{1.8a}$$

For a fixed number of moles, the ideal gas equation of state has only two independent intensive variables: any two of P, T, and ρ.

For an ideal gas mixture

$$PV = \sum_i n_i RT \tag{1.9}$$

because the gas molecules do not interact with one another. Equation (1.9) can be rewritten in the form

$$P = \sum_i \frac{n_i RT}{V} = \sum_i P_i = P_1 + P_2 + P_3 + \dots \tag{1.10}$$

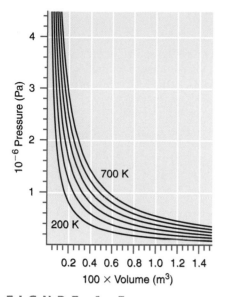

FIGURE 1.5

Illustration of the relationship between pressure and volume of 0.010 mol of He for fixed values of temperature, which differ by 100 K.

In Equation (1.10), P_i is the **partial pressure** of each gas. This equation states that each ideal gas exerts a pressure that is independent of the other gases in the mixture. We also have

$$\frac{P_i}{P} = \frac{\dfrac{n_i RT}{V}}{\displaystyle\sum_i \dfrac{n_i RT}{V}} = \frac{\dfrac{n_i RT}{V}}{\dfrac{nRT}{V}} = \frac{n_i}{n} = x_i \tag{1.11}$$

which relates the partial pressure of a component in the mixture, P_i, with its **mole fraction**, $x_i = n_i/n$, and the total pressure, P.

In the SI system of units, pressure is measured in Pascal (Pa) units, where 1 Pa = 1 N/m². The volume is measured in cubic meters, and the temperature is measured in kelvin. However, other units of pressure are frequently used, and these units are related to the Pascal as indicated in Table 1.1. In this table, numbers that are not exact have been given to five significant figures. The other commonly used unit of volume is the liter (L), where $1 m^3 = 10^3 L$ and $1 L = 1 dm^3 = 10^{-3} m^3$.

In the SI system, the constant R that appears in the ideal gas law has the value $8.314 J K^{-1} mol^{-1}$, where the joule (J) is the unit of energy in the SI system. To simplify calculations for other units of pressure and volume, values of the constant R with different combinations of units are given in Table 1.2.

EXAMPLE PROBLEM 1.1

Starting out on a trip into the mountains, you inflate the tires on your automobile to a recommended pressure of 3.21×10^5 Pa on a day when the temperature is $-5.00°C$. You drive to the beach, where the temperature is $28.0°C$. Assume that the volume of the tire has increased by 3%. What is the final pressure in the tires? The manufacturer of the tire recommends that you not exceed the recommended pressure by more than 10%. Has this limit been exceeded?

Solution

Because the number of moles is constant,

$$\frac{P_i V_i}{T_i} = \frac{P_f V_f}{T_f}; \quad P_f = \frac{P_i V_i T_f}{V_f T_i};$$

$$P_f = \frac{P_i V_i T_f}{V_f T_i} = 3.21\times10^5 Pa \times \frac{V_i}{1.03 V_i} \times \frac{(273.15+28.0)}{(273.15-5.00)} = 3.50\times10^5 Pa$$

This pressure is within 10% of the recommended pressure.

TABLE 1.1

Units of Pressure and Conversion Factors

Unit of Pressure	Symbol	Numerical Value
Pascal	Pa	$1 N m^{-2} = 1 kg m s^{-2}$
Atmosphere	atm	1 atm = 101,325 Pa (exactly)
Bar	bar	$1 bar = 10^5 Pa$
Torr or millimeters of Hg	Torr	1 Torr = 101,325/760 = 133.32 Pa
Pounds per square inch	psi	1 psi = 6,894.8 Pa

TABLE 1.2

The Ideal Gas Constant, R, in Various Units

$R = 8.314 \text{ J K}^{-1} \text{mol}^{-1}$

$R = 8.314 \text{ Pa m}^3 \text{ K}^{-1} \text{mol}^{-1}$

$R = 8.314 \times 10^{-2} \text{ L bar K}^{-1} \text{mol}^{-1}$

$R = 8.206 \times 10^{-2} \text{ L atm K}^{-1} \text{mol}^{-1}$

$R = 62.36 \text{ L Torr K}^{-1} \text{mol}^{-1}$

EXAMPLE PROBLEM 1.2

Consider the composite system, which is held at 298 K, shown in the following figure. Assuming ideal gas behavior, calculate the total pressure, and the partial pressure of each component if the barriers separating the compartments are removed. Assume that the volume of the barriers is negligible.

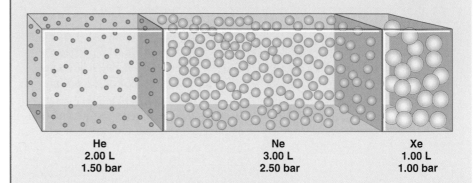

He	Ne	Xe
2.00 L	3.00 L	1.00 L
1.50 bar	2.50 bar	1.00 bar

Solution

The number of moles of He, Ne, and Xe is given by

$$n_{He} = \frac{PV}{RT} = \frac{1.50 \text{ bar} \times 2.00 \text{ L}}{8.314 \times 10^{-2} \text{ L bar K}^{-1} \text{mol}^{-1} \times 298 \text{ K}} = 0.121 \text{ mol}$$

$$n_{Ne} = \frac{PV}{RT} = \frac{2.50 \text{ bar} \times 3.00 \text{ L}}{8.314 \times 10^{-2} \text{ L bar K}^{-1} \text{mol}^{-1} \times 298 \text{ K}} = 0.303 \text{ mol}$$

$$n_{Xe} = \frac{PV}{RT} = \frac{1.00 \text{ bar} \times 1.00 \text{ L}}{8.314 \times 10^{-2} \text{ L bar K}^{-1} \text{mol}^{-1} \times 298 \text{ K}} = 0.0403 \text{ mol}$$

$$n = n_{He} + n_{Ne} + n_{Xe} = 0.464$$

The mole fractions are

$$x_{He} = \frac{n_{He}}{n} = \frac{0.121}{0.464} = 0.261$$

$$x_{Ne} = \frac{n_{Ne}}{n} = \frac{0.303}{0.464} = 0.653$$

$$x_{Xe} = \frac{n_{Xe}}{n} = \frac{0.0403}{0.464} = 0.0860$$

The total pressure is given by

$$P = \frac{(n_{He} + n_{Ne} + n_{Xe})RT}{V} = \frac{0.464 \text{ mol} \times 8.3145 \times 10^{-2} \text{ L bar K}^{-1} \text{mol}^{-1} \times 298 \text{ K}}{6.00 \text{ L}}$$

$$= 1.92 \text{ bar}$$

The partial pressures are given by

$$P_{He} = x_{He}P = 0.261 \times 1.92 \text{ bar} = 0.501 \text{ bar}$$

$$P_{Ne} = x_{Ne}P = 0.653 \times 1.92 \text{ bar} = 1.25 \text{ bar}$$

$$P_{Xe} = x_{Xe}P = 0.0860 \times 1.92 \text{ bar} = 0.165 \text{ bar}$$

1.5 A Brief Introduction to Real Gases

The ideal gas law provides a first look at the usefulness of describing a system in terms of macroscopic parameters. However, we should also emphasize the downside of not taking the microscopic nature of the system into account. For example, the ideal gas law only holds for gases at low densities. Experiments show that Equation (1.8) is accurate to higher values of pressure and lower values of temperature for He than for NH_3. Why is this the case? Real gases will be discussed in detail in Chapter 7. However, because we need to take nonideal gas behavior into account in Chapters 2 through 6, we introduce an equation of state that is valid to higher densities in this section.

An ideal gas is described by two assumptions: the atoms or molecules of an ideal gas do not interact with one another, and the atoms or molecules can be treated as point masses. These assumptions have a limited range of validity, which can be discussed using the potential energy function typical for a real gas, as shown in Figure 1.6. This figure shows the potential energy of interaction of two gas molecules as a function of the distance between them. The intermolecular potential can be divided into regions in which the potential energy is essentially zero ($r > r_{transition}$), negative (attractive interaction) ($r_{transition} > r > r_{V=0}$), and positive (repulsive interaction) ($r < r_{V=0}$). The distance $r_{transition}$ is not uniquely defined and depends on the energy of the molecule. It can be estimated from the relation $|V(r_{transition})| \approx kT$.

As the density is increased from very low values, molecules approach one another to within a few molecular diameters and experience a long-range attractive van der Waals force due to time-fluctuating dipole moments in each molecule. This strength of the attractive interaction is proportional to the polarizability of the electron charge in a molecule and is, therefore, substance dependent. In the attractive region, P is lower than that calculated using the ideal gas law. This is the case because the attractive interaction brings the atoms or molecules closer than they would be if they did not interact. At sufficiently high densities, the atoms or molecules experience a short-range repulsive interaction due to the overlap of the electron charge distributions. Because of this interaction, P is higher than that calculated using the ideal gas law. We see that for a real gas, P can be either greater or less than the ideal gas value. Note that the potential becomes repulsive for a value of r greater than zero. As a consequence, the volume of a gas well above its boiling temperature approaches a finite limiting value as P increases. By contrast, the ideal gas law predicts that $V \to 0$ as $P \to \infty$.

Given the potential energy function depicted in Figure 1.6, under what conditions is the ideal gas equation of state valid? A real gas behaves ideally only at low densities for which $r > r_{transition}$, and the value of $r_{transition}$ is substance dependent. Real gases will be discussed in much greater detail in Chapter 7. However, at this point we introduce a real gas equation of state, because it will be used in the next few chapters. The **van der Waals equation of state** takes both the finite size of molecules and the attractive potential into account. It has the form

$$P = \frac{nRT}{V - nb} - \frac{n^2 a}{V^2} \tag{1.12}$$

This equation of state has two parameters that are substance dependent and must be experimentally determined. The parameters a and b take the finite size of the molecules and the strength of the attractive interaction into account, respectively. (Values of a and b for selected gases are listed in Table 7.4.) The van der Waals equation of state is more accurate in calculating the relationship between P, V, and T for gases than the ideal gas law because a and b have been optimized using experimental results. However, there are other more accurate equations of state that are valid over a wider range than the van der Waals equation. Such equations of state include up to 16 adjustable substance-dependent parameters.

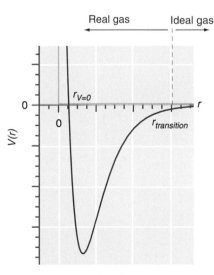

FIGURE 1.6

The potential energy of interaction of two molecules or atoms is shown as a function of their separation, r. The yellow curve shows the potential energy function for an ideal gas. The dashed blue line indicates an approximate r value below which a more nearly exact equation of state than the ideal gas law should be used. $V(r) = 0$ at $r = r_{V=0}$ and as $r \to \infty$.

EXAMPLE PROBLEM 1.3

Van der Waals parameters are generally tabulated with either of two sets of units:

a: Pa m^6 mol^{-2} or bar dm^6 mol^{-2}

b: m^3 mol^{-1} or dm^3 mol^{-1}

Determine the conversion factor to convert one system of units to the other. Note that 1 dm^3 = 10^{-3} m^3 = 1 L.

Solution

$$\text{Pa m}^6 \text{ mol}^{-2} \times \frac{\text{bar}}{10^5 \text{ Pa}} \times \frac{10^6 \text{ dm}^6}{\text{m}^6} = 10 \text{ bar dm}^6 \text{ mol}^{-2}$$

$$\text{m}^3 \text{ mol}^{-1} \times \frac{10^3 \text{ dm}^3}{\text{m}^3} = 10^3 \text{ dm}^3 \text{ mol}^{-1}$$

In Example Problem 1.4, a comparison is made of the molar volume for N_2 calculated at low and high pressures, using the ideal gas and van der Waals equations of state.

EXAMPLE PROBLEM 1.4

a. Calculate the pressure exerted by N_2 at 300 K for molar volumes of 250 and 0.100 L using the ideal gas and the van der Waals equations of state. The values of parameters a and b for N_2 are 1.370 bar dm^6 mol^{-2} and 0.0387 dm^3 mol^{-1}, respectively.

b. Compare the results of your calculations at the two pressures. If P calculated using the van der Waals equation of state is greater than those calculated with the ideal gas law, we can conclude that the repulsive interaction of the N_2 molecules outweighs the attractive interaction for the calculated value of the density. A similar statement can be made regarding the attractive interaction. Is the attractive or repulsive interaction greater for N_2 at 300 K and V_m = 0.100 L?

Solution

a. The pressures calculated from the ideal gas equation of state are

$$P = \frac{nRT}{V} = \frac{1 \text{ mol} \times 8.314 \times 10^{-2} \text{ L bar mol}^{-1}\text{K}^{-1} \times 300 \text{ K}}{250 \text{ L}} = 9.98 \times 10^{-2} \text{ bar}$$

$$P = \frac{nRT}{V} = \frac{1 \text{ mol} \times 8.314 \times 10^{-2} \text{ L bar mol}^{-1}\text{K}^{-1} \times 300 \text{ K}}{0.100 \text{ L}} = 249 \text{ bar}$$

The pressures calculated from the van der Waals equation of state are

$$P = \frac{nRT}{V - nb} - \frac{n^2 a}{V^2}$$

$$= \frac{1 \text{ mol} \times 8.314 \times 10^{-2} \text{ L bar mol}^{-1}\text{K}^{-1} \times 300 \text{ K}}{250 \text{ L} - 1 \text{ mol} \times 0.0387 \text{ dm}^3 \text{ mol}^{-1}} - \frac{(1 \text{ mol})^2 \times 1.370 \text{ bar dm}^6 \text{ mol}^{-2}}{(250 \text{ L})^2}$$

$$= 9.98 \times 10^{-2} \text{ bar}$$

$$P = \frac{1 \text{ mol} \times 8.314 \times 10^{-2} \text{ L bar mol}^{-1}\text{K}^{-1} \times 300 \text{ K}}{0.1 \text{ L} - 1 \text{ mol} \times 0.0387 \text{ dm}^3 \text{ mol}^{-1}} - \frac{(1 \text{ mol})^2 \times 1.370 \text{ bar dm}^6 \text{ mol}^{-2}}{(0.1 \text{ L})^2}$$

$$= 270 \text{ bar}$$

b. Note that the result is identical with that for the ideal gas law for V_m = 250 L, and that the result calculated for V_m = 0.100 L deviates from the ideal gas law result. Because $P^{real} > P^{ideal}$, we conclude that the repulsive interaction is more important than the attractive interaction for this specific value of molar volume and temperature.

Vocabulary

absolute temperature scale	intensive variable	surroundings
adiabatic	isolated system	system
Boltzmann constant	kelvin scale	system variables
boundary	macroscopic scale	temperature
Celsius scale	macroscopic variables	temperature scale
centigrade scale	microscopic scale	thermal equilibrium
closed system	mole fraction	thermodynamic equilibrium
diathermal	movable wall	thermodynamic temperature scale
equation of state	nonpermeable wall	thermometer
equilibrium	open system	van der Waals equation of state
extensive variable	partial pressure	wall
gas thermometer	permeable wall	zeroth law of thermodynamics
ideal gas constant	rigid wall	
ideal gas law	statistical thermodynamics	

Questions on Concepts

Q1.1 The location of the boundary between the system and the surroundings is a choice that must be made by the thermodynamicist. Consider a beaker of boiling water in an airtight room. Is the system open or closed if you place the boundary just outside the liquid water? Is the system open or closed if you place the boundary just inside the walls of the room?

Q1.2 Real walls are never totally adiabatic. Order the following walls in increasing order with respect to their being diathermal: 1-cm-thick concrete, 1-cm-thick vacuum, 1-cm-thick copper, 1-cm-thick cork.

Q1.3 Why is the possibility of exchange of matter or energy appropriate to the variable of interest a necessary condition for equilibrium between two systems?

Q1.4 At sufficiently high temperatures, the van der Waals equation has the form $P \approx RT/(V_m - b)$. Note that the attractive part of the potential has no influence in this expression. Justify this behavior using the potential energy diagram of Figure 1.6.

Q1.5 The parameter a in the van der Waals equation is greater for H_2O than for He. What does this say about the form of the potential function in Figure 1.6 for the two gases?

Problems

Problem numbers in **red** indicate that the solution to the problem is given in the *Student's Solutions Manual*.

P1.1 A sealed flask with a capacity of 1.00 dm³ contains 5.00 g of ethane. The flask is so weak that it will burst if the pressure exceeds 1.00×10^6 Pa. At what temperature will the pressure of the gas exceed the bursting temperature?

P1.2 Consider a gas mixture in a 2.00-dm³ flask at 27.0°C. For each of the following mixtures, calculate the partial pressure of each gas, the total pressure, and the composition of the mixture in mole percent:

a. 1.00 g H_2 and 1.00 g O_2

b. 1.00 g N_2 and 1.00 g O_2

c. 1.00 g CH_4 and 1.00 g NH_3

P1.3 Suppose that you measured the product PV of 1 mol of a dilute gas and found that $PV = 22.98$ L atm at 0°C and 31.18 L atm at 100°C. Assume that the ideal gas law is valid, with $T = t(°C) + a$, and that the value of R is not known. Determine R and a from the measurements provided.

P1.4 A compressed cylinder of gas contains 1.50×10^3 g of N_2 gas at a pressure of 2.00×10^7 Pa and a temperature of 17.1°C. What volume of gas has been released into the atmosphere if the final pressure in the cylinder is 1.80×10^5 Pa? Assume ideal behavior and that the gas temperature is unchanged.

P1.5 A balloon filled with 10.50 L of Ar at 18.0°C and 1 atm rises to a height in the atmosphere where the pressure is 248 Torr and the temperature is −30.5°C. What is the final volume of the balloon?

P1.6   Consider a 20.0-L sample of moist air at 60°C and 1 atm in which the partial pressure of water vapor is 0.120 atm. Assume that dry air has the composition 78.0 mole percent N_2, 21.0 mole percent O_2, and 1.00 mole percent Ar.

a. What are the mole percentages of each of the gases in the sample?

b. The percent relative humidity is defined as %RH $= P_{H_2O}/P_{H_2O}^*$ where P_{H_2O} is the partial pressure of water in the sample and $P_{H_2O}^* = 0.197$ atm is the equilibrium vapor pressure of water at 60°C. The gas is compressed at 60°C until the relative humidity is 100%. What volume does the mixture contain now?

c. What fraction of the water will be condensed if the total pressure of the mixture is isothermally increased to 200 atm?

P1.7   A mixture of 2.50×10^{-3} g of O_2, 3.51×10^{-3} mol of N_2, and 4.67×10^{20} molecules of CO are placed into a vessel of volume 3.50 L at 5.20°C.

a. Calculate the total pressure in the vessel.

b. Calculate the mole fractions and partial pressures of each gas.

P1.8   Liquid N_2 has a density of 875.4 kg m^{-3} at its normal boiling point. What volume does a balloon occupy at 18.5°C and a pressure of 1.00 atm if 2.00×10^{-3} L of liquid N_2 is injected into it?

P1.9   A rigid vessel of volume 0.500 m^3 containing H_2 at 20.5°C and a pressure of 611×10^3 Pa is connected to a second rigid vessel of volume 0.750 m^3 containing Ar at 31.2°C at a pressure of 433×10^3 Pa. A valve separating the two vessels is opened and both are cooled to a temperature of 14.5°C. What is the final pressure in the vessels?

P1.10   A sample of propane (C_3H_8) is placed in a closed vessel together with an amount of O_2 that is 3.00 times the amount needed to completely oxidize the propane to CO_2 and H_2O at constant temperature. Calculate the mole fraction of each component in the resulting mixture after oxidation assuming that the H_2O is present as a gas.

P1.11   A glass bulb of volume 0.136 L contains 0.7031 g of gas at 759.0 Torr and 99.5°C. What is the molar mass of the gas?

P1.12   The total pressure of a mixture of oxygen and hydrogen is 1.00 atm. The mixture is ignited and the water is removed. The remaining gas is pure hydrogen and exerts a pressure of 0.400 atm when measured at the same values of T and V as the original mixture. What was the composition of the original mixture in mole percent?

P1.13   A gas sample is known to be a mixture of ethane and butane. A bulb having a 200.0-cm^3 capacity is filled with the gas to a pressure of 100.0×10^3 Pa at 20.0°C. If the weight of the gas in the bulb is 0.3846 g, what is the mole percent of butane in the mixture?

P1.14   When Julius Caesar expired, his last exhalation had a volume of 500 cm^3 and contained 1.00 mole percent argon. Assume that $T = 300$ K and $P = 1.00$ atm at the location of his demise. Assume further that T and P currently have the same values throughout the Earth's atmosphere. If all of his exhaled CO_2 molecules are now uniformly distributed throughout the atmosphere (which for our calculation is taken to have a thickness of 1.00 km), how many inhalations of 500 cm^3 must we make to inhale one of the Ar molecules exhaled in Caesar's last breath? Assume the radius of the Earth to be 6.37×10^6 m. [Hint: Calculate the number of Ar atoms in the atmosphere in the simplified geometry of a plane of area equal to that of the Earth's surface and a height equal to the thickness of the atmosphere. See Problem 1.15 for the dependence of the barometric pressure on the height above the Earth's surface.

P1.15   The barometric pressure falls off with height above sea level in the Earth's atmosphere as $P_i = P_i^0 e^{-M_i g z/RT}$ where P_i is the partial pressure at the height z, P_i^0 is the partial pressure of component i at sea level, g is the acceleration of gravity, R is the gas constant, and T is the absolute temperature. Consider an atmosphere that has the composition $x_{N_2} = 0.600$ and $x_{CO_2} = 0.400$ and that $T = 300$ K. Near sea level, the total pressure is 1.00 bar. Calculate the mole fractions of the two components at a height of 50.0 km. Why is the composition different from its value at sea level?

P1.16   Assume that air has a mean molar mass of 28.9 g mol^{-1} and that the atmosphere has a uniform temperature of 25.0°C. Calculate the barometric pressure at Denver, for which $z = 1600$ m. Use the information contained in Problem P1.15.

P1.17   Calculate the pressure exerted by Ar for a molar volume of 1.42 L at 300 K using the van der Waals equation of state. The van der Waals parameters a and b for Ar are 1.355 bar dm^6 mol^{-2} and 0.0320 dm^3 mol^{-1}, respectively. Is the attractive or repulsive portion of the potential dominant under these conditions?

P1.18   Calculate the pressure exerted by benzene for a molar volume of 1.42 L at 790 K using the Redlich-Kwong equation of state:

$$P = \frac{RT}{V_m - b} - \frac{a}{\sqrt{T}}\frac{1}{V_m(V_m + b)} = \frac{nRT}{V - nb} - \frac{n^2 a}{\sqrt{T}}\frac{1}{V(V + nb)}$$

The Redlich-Kwong parameters a and b for benzene are 452.0 bar dm^6 mol^{-2} K$^{1/2}$ and 0.08271 dm^3 mol^{-1}, respectively. Is the attractive or repulsive portion of the potential dominant under these conditions?

P1.19   Devise a temperature scale, abbreviated G, for which the magnitude of the ideal gas constant is 1.00 J G^{-1} mol^{-1}.

P1.20   A mixture of oxygen and hydrogen is analyzed by passing it over hot copper oxide and through a drying tube. Hydrogen reduces the CuO according to the reaction CuO + $H_2 \rightarrow$ Cu + H_2O, and oxygen reoxidizes the copper formed according to Cu + 1/2 $O_2 \rightarrow$ CuO. At 25°C and 750 Torr, 100.0 cm^3 of the mixture yields 84.5 cm^3 of dry oxygen measured at 25°C and 750 Torr after passage over CuO and the drying agent. What is the original composition of the mixture?

CHAPTER 2

Heat, Work, Internal Energy, Enthalpy, and the First Law of Thermodynamics

In this chapter, the internal energy, U, is introduced. The first law of thermodynamics relates ΔU to the heat (q) and work (w) that flows across the boundary between the system and the surroundings. Other important concepts introduced include the heat capacity, the difference between state and path functions, and reversible versus irreversible processes. The enthalpy, H, is introduced as a form of energy that can be directly measured by the heat flow in a constant pressure process. We show how ΔU, ΔH, q, and w can be calculated for processes involving ideal gases. ■

2.1 The Internal Energy and the First Law of Thermodynamics

In this section, we focus on the change in energy of the system and surroundings during a thermodynamic process such as an expansion or compression of a gas. In thermodynamics, we are interested in the internal energy of the system, as opposed to the energy associated with the system relative to a particular frame of reference. For example, a spinning container of gas has a kinetic energy relative to a stationary observer. However, the internal energy of the gas is defined relative to a coordinate system fixed on the container. Viewed at a microscopic level, the internal energy can take on a number of forms such as

- the kinetic energy of the molecules;
- the potential energy of the constituents of the system; for example, a crystal consisting of dipolar molecules will experience a change in its potential energy as an electric field is applied to the system;
- the internal energy stored in the form of molecular vibrations and rotations; and
- the internal energy stored in the form of chemical bonds that can be released through a chemical reaction.

The total of all these forms of energy for the system of interest is given the symbol U and is called the **internal energy.**

The **first law of thermodynamics** is based on our experience that energy can be neither created nor destroyed, if both the system and the surroundings are taken into account. This law can be formulated in a number of equivalent forms. Our initial formulation of this law is stated as follows:

> The internal energy, U, of an isolated system is constant.

This form of the first law looks uninteresting, because it suggests that nothing happens in an isolated system. How can the first law tell us anything about thermodynamic processes such as chemical reactions? When changes in U occur in a system in contact with its surroundings, ΔU_{total} is given by

$$\Delta U_{total} = \Delta U_{system} + \Delta U_{surroundings} = 0 \tag{2.1}$$

Therefore, the first law becomes

$$\Delta U_{system} = -\Delta U_{surroundings} \tag{2.2}$$

For any decrease of U_{system}, $U_{surroundings}$ must increase by exactly the same amount. For example, if a gas (the system) is cooled, and the surroundings is also a gas, the temperature of the surroundings must increase.

How can the energy of a system be changed? There are many ways to alter U, several of which are discussed in this chapter. Experience has shown that all changes in a closed system in which no chemical reactions or phase changes occur can be classified as heat, work, or a combination of both. Therefore, the internal energy of such a system can only be changed by the flow of heat or work across the boundary between the system and surroundings. For example, U for a gas can be increased by heating it in a flame or by doing compression work on it. This important recognition leads to a second and more useful formulation of the first law:

$$\Delta U = q + w \tag{2.3}$$

where q and w designate heat and work, respectively. We use ΔU without a subscript to indicate the change in internal energy of the system. What do we mean by heat and work? In the following two sections, we define these important concepts and distinguish between them.

The symbol Δ is used to indicate a change that occurs as a result of an arbitrary process. The simplest processes are those in which only one of P, V, or T changes. A constant temperature process is referred to as **isothermal,** and the corresponding terms for constants P and V are **isobaric** and **isochoric,** respectively.

2.2 Work

In this and the next section, we discuss the two ways in which the internal energy of a system can be changed. **Work** in thermodynamics is defined as any quantity of energy that "flows" across the boundary between the system and surroundings that can be used to change the height of a mass in the surroundings. An example is shown in Figure 2.1. We define the system as the gas inside the adiabatic cylinder and piston. Everything else shown in the figure is in the surroundings. As the gas is compressed, the height of the mass in the surroundings is lowered and the initial and final volumes are defined by the mechanical stops indicated in the figure.

Consider the system and surroundings before and after the process shown in Figure 2.1, and note that the height of the mass in the surroundings has changed. It is this change that distinguishes work from heat. Work has several important characteristics:

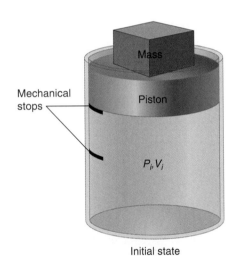

Mechanical stops

Mass

Piston

P_i, V_i

Initial state

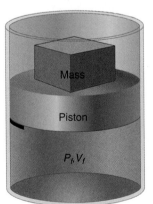

Mass

Piston

P_f, V_f

Final state

FIGURE 2.1

A system is shown in which compression work is being done on a gas. The walls are adiabatic.

- Work is transitory in that it only appears during a change in state of the system and surroundings. Only energy, and not work, is associated with the initial and final states of the systems.

- The net effect of work is to change U of the system and surroundings in accordance with the first law. If the only change in the surroundings is that a mass has been raised or lowered, work has flowed between the system and the surroundings.

- The quantity of work can be calculated from the change in potential energy of the mass, $E_{potential} = mgh$, where g is the gravitational acceleration and h is the change in the height of the mass, m.

- The sign convention for work is as follows. If the height of the mass in the surroundings is lowered, w is positive; if the height is raised, w is negative. In short, $w > 0$ if $\Delta U > 0$. It is common usage to say that if w is positive, work is done on the system by the surroundings. If w is negative, work is done by the system on the surroundings.

How much work is done in the process shown in Figure 2.1? Using a definition from physics, work is done when an object subject to a force, **F,** is moved through a distance, **dl,** according to the path integral

$$w = \int \mathbf{F} \cdot d\mathbf{l} \tag{2.4}$$

Using the definition of pressure as the force per unit area, the work done in moving the mass is given by

$$w = \int \mathbf{F} \cdot d\mathbf{l} = -\iint P_{external} \, dA \, dl = -\int P_{external} \, dV \tag{2.5}$$

The minus sign appears because of our sign convention for work. Note that the pressure that appears in this expression is the external pressure, $P_{external}$, which need not equal the system pressure, P.

An example of another important kind of work, namely, electrical work, is shown in Figure 2.2 in which the contents of the cylinder is the system. Electrical current flows through a conductive aqueous solution and water undergoes electrolysis to produce H_2 and O_2 gas. The current is produced by a generator, like that used to power a light on a bicycle through the mechanical work of pedaling. As current flows, the mass that drives the generator is lowered. In this case, the surroundings do the electrical work on the system. As a result, some of the liquid water is transformed to H_2 and O_2. From electrostatics, the work done in transporting a charge, q, through an electrical potential difference, ϕ, is

$$w_{electrical} = q\phi \tag{2.6}$$

For a constant current, I, that flows for a time, t, $q = It$. Therefore,

$$w_{electrical} = I\phi t \tag{2.7}$$

The system also does work on the surroundings through the increase in the volume of the gas phase at constant pressure. The total work done is

$$w = w_{PV} + w_{electrical} = I\phi t - \int P_{external} \, dV = I\phi t - P_{external} \int dV = I\phi t - P_i(V_f - V_i) \tag{2.8}$$

Other forms of work include the work of expanding a surface, such as a soap bubble, against the surface tension. Table 2.1 shows the expressions for work for four different cases. Each of these different types of work poses a requirement on the walls separating the system and surroundings. To be able to carry out the first three types of work, the walls must be movable, whereas for electrical work, they must be conductive. Several examples of work calculations are given in Example Problem 2.1.

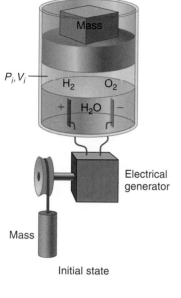

P_i, V_i — H_2 O_2
$+$ H_2O $-$

Mass

Electrical generator

Mass

Initial state

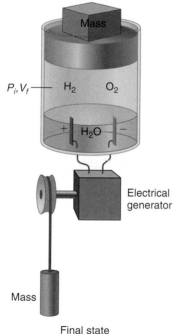

P_i, V_f — H_2 O_2
$+$ H_2O $-$

Mass

Electrical generator

Mass

Final state

F I G U R E   2 . 2
Current produced by a generator is used to electrolyze water and thereby do work on the system as shown by the lowered mass linked to the generator. The gas produced in this process does P–V work on the surroundings, as shown by the raised mass on the piston.

TABLE 2.1

Types of Work

Types of Work	Variables	Equation for Work	Conventional Units
Volume expansion	Pressure (P), volume (V)	$w = -\int P_{external}\, dV$	Pa m^3 = J
Stretching	Tension (γ), length (l)	$w = -\int \gamma\, dl$	N m = J
Surface expansion	Surface tension (γ), area (σ)	$w = -\iint \gamma\, d\sigma$	(N m^{-1}) (m^2) = J
Electrical	Electrical potential (ϕ), electrical charge (q)	$w = \int \phi\, dq$	V C = J

EXAMPLE PROBLEM 2.1

a. Calculate the work involved in expanding 20.0 L of an ideal gas to a final volume of 85.0 L against a constant external pressure of 2.50 bar.

b. A water bubble is expanded from a radius of 1.00 cm to a radius of 3.25 cm. The surface tension of water is 71.99 N m^{-1}. How much work is done in the process?

c. A current of 3.20 A is passed through a heating coil for 30.0 s. The electrical potential across the resistor is 14.5 V. Calculate the work done on the coil.

Solution

a. $w = -\int P_{external}\, dV = -P_{external}(V_f - V_i)$

$$= -2.50\ \text{bar} \times \frac{10^5\ \text{Pa}}{\text{bar}} \times (85.0\ \text{L} - 20.0\ \text{L}) \times \frac{10^{-3}\ \text{m}^3}{\text{L}} = -16.3\ \text{kJ}$$

b. A factor of 2 is included in the calculation below because a bubble has an inner and an outer surface:

$$w = -\iint \gamma\, d\sigma = 2\gamma 4\pi(r_f^2 - r_i^2)$$

$$= -8\pi \times 71.99\ \text{N m}^{-1}(3.25^2\ \text{cm}^2 - 1.00^2\ \text{cm}^2) \times \frac{10^{-4}\ \text{m}^2}{\text{cm}^2}$$

$$= -1.73\ \text{J}$$

c. $w = \int \phi\, dq = \phi q = I\phi t = 14.5\ \text{V} \times 3.20\ \text{A} \times 30.0\ \text{s} = 1.39\ \text{kJ}$

2.3 Heat

Heat[1] is defined in thermodynamics as the quantity of energy that flows across the boundary between the system and surroundings because of a temperature difference be-

---

[1] Heat is perhaps the most misused term in thermodynamics as discussed by Robert Romer [*American Journal of Physics*, 69 (2001), 107–109]. In common usage, it is incorrectly referred to as a substance as in the phrase "Close the door; you're letting the heat out!" An equally inappropriate term is heat capacity (discussed in Section 2.4), because it implies that materials have the capacity to hold heat, rather than the capacity to store energy. We use the terms *heat flow* or *heat transfer* to emphasize the transitory nature of heat. However, you should not think of heat as a fluid or a substance.

tween the system and the surroundings. Just as for work, several important characteristics of heat are of importance:

- Heat is transitory, in that it only appears during a change in state of the system and surroundings. Only energy, and not heat, is associated with the initial and final states of the system and the surroundings.

- The net effect of heat is to change the internal energy of the system and surroundings in accordance with the first law. If the only change in the surroundings is a change in temperature of a reservoir, heat has flowed between system and surroundings. The quantity of heat that has flowed is directly proportional to the change in temperature of the reservoir.

- The sign convention for heat is as follows. If the temperature of the surroundings is lowered, q is positive; if it is raised, q is negative. It is common usage to say that if q is positive, heat is withdrawn from the surroundings and deposited in the system. If q is negative, heat is withdrawn from the system and deposited in the surroundings.

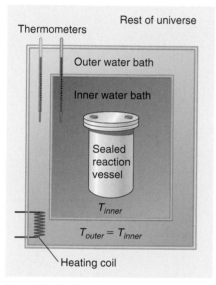

FIGURE 2.3

An isolated composite system is created in which the surroundings to the system of interest are limited in extent. The walls surrounding the inner water bath are rigid.

Defining the surroundings as the rest of the universe is impractical, because it is not realistic to search through the whole universe to see if a mass has been raised or lowered and if the temperature of a reservoir has changed. Experience shows that in general only those parts of the universe close to the system interact with the system. Experiments can be constructed to ensure that this is the case, as shown in Figure 2.3. Imagine that you are interested in an exothermic chemical reaction that is carried out in a rigid sealed container with diathermal walls. You define the system as consisting solely of the reactant and product mixture. The vessel containing the system is immersed in an inner water bath separated from an outer water bath by a container with rigid diathermal walls. During the reaction, heat flows out of the system ($q < 0$), and the temperature of the inner water bath increases to T_f. Using an electrical heater, the temperature of the outer water bath is increased so that at all times, $T_{outer} = T_{inner}$. Because of this condition, no heat flows across the boundary between the two water baths, and because the container enclosing the inner water bath is rigid, no work flows across this boundary. Therefore, $\Delta U = q + w = 0 + 0 = 0$ *for the composite system* made up of the inner water bath and everything within it. Therefore, this composite system is an isolated system that does not interact with the rest of the universe. To determine q and w for the reactant and product mixture, we need to examine only the composite system and can disregard the rest of the universe.

To emphasize the distinction between q and w, and the relationship between q, w, and ΔU, we discuss the two processes shown in Figure 2.4. They are carried out in an isolated system, divided into two subsystems, I and II. In both cases, system I consists solely of the liquid in the beaker, and everything else including the rigid adiabatic walls is in system II. We assume that the temperature of the liquid is well below its boiling point and that its vapor pressure is negligibly small. This ensures that no liquid is vaporized in the process, so that both systems are closed. System II can be viewed as the surroundings for system I and vice versa.

In Figure 2.4a, a Bunsen burner fueled by a propane canister is used to heat the liquid (system I). The boundary between systems I and II is the surface that encloses the liquid, and heat can flow between systems I and II all across this wall. It is observed that the temperature of the liquid increases in the process. From general chemistry, we know that the internal energy of a monatomic gas increases linearly with T. This result can be generalized to state that U is a monotonically increasing function of T for a uniform single-phase system of constant composition. Therefore, because $\Delta T_I > 0$, $\Delta U_I > 0$.

We next consider the changes in system II. From the first law, $\Delta U_{II} = -\Delta U_I < 0$. Observe that the masses of propane and oxygen have decreased, and that the masses of H_2O and CO_2 have increased as a result of the combustion of propane. None of the weights in system II have been lowered as a result of the process, and the volume of both systems has remained constant. We conclude that no work was done by system II

(a)

(b)

FIGURE 2.4
Two systems, I and II, are enclosed in a rigid adiabatic enclosure. System I consists solely of the liquid in the beaker for each case. System II consists of everything else in the enclosure. **(a)** Process in which the liquid is heated using a flame. **(b)** Heating using a resistive coil through which an electrical current flows.

on system I, that is, $w = 0$. Therefore, if $\Delta U_I = q + w > 0$, $q > 0$. Note that because the wall separating systems I and II is diathermal, $T_I = T_{II}$ and therefore $\Delta T_{II} > 0$. If $\Delta T_{II} > 0$, why is $\Delta U_{II} < 0$? This decrease can be attributed to the difference of the internal energy stored in the bonds of the combustion products CO_2 and H_2O compared with the reactants C_3H_8 and O_2. The internal energy U is not a monotonic function of T for this system because the chemical composition changes in the process.

We now consider Figure 2.4b. In this case, the boundary between systems I and II lies just inside the inner wall of the beaker, passes across the open top of the beaker, and just outside of the surface of the heating coil. Note that the heating coil is entirely in system II. Heat can flow between systems I and II all across the boundary surface. Upon letting the mass in system II fall, electricity flows through the heating coil. It is our experience that the temperature of the liquid (system I) will increase. We again conclude that $\Delta U_I > 0$. However, it is not clear what values q and w have.

To resolve this issue, consider the changes in system II of this composite system. From the first law, $\Delta U_{II} = - \Delta U_I < 0$. We see that a mass has been lowered in system II. Can it be concluded that work has been done by system II on system I? The lowering of the mass in system II shows that work is being done, *but it is being done only on the heating coil, which is also in system II. This is the case because the current flow never crosses the boundary between the systems.* Therefore, $w = 0$, where w is the work done on system I. However, $\Delta U_I > 0$, so that if $w = 0$, $q > 0$. The conclusion is that the increase in U_I is due to heat flow from system II to system I caused by the difference between the temperature of the heating filament and the liquid. Note that because $\Delta U_I + \Delta U_{II} = 0$, the heat flow can be calculated from the electrical work, $q = w = I\phi t$.

Why is $\Delta U_{II} < 0$ if $\Delta T_{II} > 0$? The value of U_{II} is the sum of the internal energy of all parts of system II. The value of T_{II} increases, which increases the thermal part of U_{II}. However, the potential energy of system II decreases because the mass is lowered in the process. The net effect of these two opposing changes is that $\Delta U_{II} < 0$.

These examples show that the distinction between heat and work must be made carefully with a clear knowledge of the position and nature of the boundary between the system and the surroundings.

EXAMPLE PROBLEM 2.2

A heating coil is immersed in a 100-g sample of H_2O liquid at 100°C in an open insulated beaker on a laboratory bench at 1 bar pressure. In this process, 10% of the liquid is converted to the gaseous form at a pressure of 1 bar. A current of 2.00 A flows through the heater from a 12.0-V battery for 1.00×10^3 s to effect the transformation. The densities of liquid and gaseous water under these conditions are 997 and 0.590 kg m^{-3}, respectively.

a. It is often useful to replace a real process by a model that exhibits the important features of the process. Design a model system and surroundings, like those shown in Figures 2.1 and 2.2, that would allow you to measure the heat and work associated with this transformation. For the model system, define the system and surroundings as well as the boundary between them.

b. Calculate q and w for the process.

c. How can you define the system for the open insulated beaker on the laboratory bench such that the work is properly described?

Solution

a. The model system is shown in the following figure. The cylinder walls and the piston form adiabatic walls. The external pressure is held constant by a suitable weight.

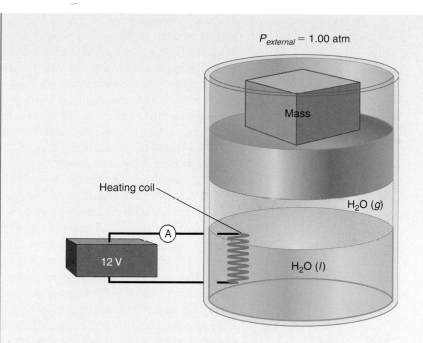

b. In the system shown, the heat input to the liquid water can be equated with the work done on the heating coil. Therefore,

$$q = I\phi t = 2.00 \text{ A} \times 12.0 \text{ V} \times 1.00 \times 10^3 \text{ s} = 24.0 \text{ kJ}$$

As the liquid is vaporized, the volume of the system increases at a constant external pressure. Therefore, the work done by the system on the surroundings is

$$w = P_{external}(V_f - V_i) = 10^5 \text{ Pa} \times \left(\frac{10.0 \times 10^{-3} \text{ kg}}{0.590 \text{ kg m}^{-3}} + \frac{90.0 \times 10^{-3} \text{ kg}}{997 \text{ kg m}^{-3}} - \frac{100.0 \times 10^{-3} \text{ kg}}{997 \text{ kg m}^{-3}} \right)$$

$$= 1.70 \text{ kJ}$$

Note that the electrical work done on the heating coil is much larger than the *P-V* work done in the expansion.

c. Define the system as the liquid in the beaker and the volume containing only molecules of H_2O in the gas phase. This volume will consist of disconnected volume elements dispersed in the air above the laboratory bench.

2.4 Heat Capacity

The process shown in Figure 2.4 provides a way to quantify heat flow in terms of the easily measured electrical work done on the heating coil, $w = I\phi t$. The response of a single-phase system of constant composition to heat input is an increase in T. This is not the case if the system undergoes a phase change, such as the vaporization of a liquid. The temperature of an equilibrium mixture of liquid and gas at the boiling point remains constant as heat flows into the system. However, the mass of the gas phase increases at the expense of the liquid phase.

The response of the system to heat flow is described by a very important thermodynamic property called the **heat capacity.** The heat capacity is a material-dependent property defined by the relation

$$C = \lim_{\Delta T \to 0} \frac{q}{T_f - T_i} = \frac{dq}{dT} \tag{2.9}$$

where C is in the SI unit of J K^{-1}. It is an extensive quantity that, for example, doubles as the mass of the system is doubled. Often, the molar heat capacity, C_m, is used in calculations and it is an intensive quantity with the units of J K^{-1} mol^{-1}. Experimentally, the heat capacity of fluids is measured by immersing a heating coil in the fluid and equating the electrical work done on the coil with the heat flow into the sample. For solids, the heating coil is wrapped around the solid. In both cases, the experimental results must be corrected for heat losses to the surroundings. The significance of the notation dq for an incremental amount of heat is explained in the next section.

The value of the heat capacity depends on the experimental conditions under which it is determined. The most common conditions are constant volume or constant pressure, for which the heat capacity is denoted C_V and C_P, respectively. Values of $C_{P,m}$ at 298.15 K for pure substances are tabulated in Tables 2.2 and 2.3 (see Appendix A, Data Tables), and formulas for calculating $C_{P,m}$ at other temperatures for gases and solids are listed in Tables 2.4 and 2.5, respectively (see Appendix A, Data Tables).

An example of how $C_{P,m}$ depends on T is illustrated in Figure 2.5 for Cl_2. To make the functional form of $C_{P,m}(T)$ understandable, we briefly discuss the relative magnitudes of $C_{P,m}$ in the solid, liquid, and gaseous phases using a microscopic model. A solid can be thought of as a set of interconnected harmonic oscillators, and heat uptake leads to the excitations of the collective vibrations of the solid. At very low temperatures, these vibrations cannot be activated, because the spacing of the vibrational energy levels is large compared to kT. As a consequence, energy cannot be taken up by the solid. Hence, $C_{P,m}$ approaches zero as T approaches zero. For the solid, $C_{P,m}$ rises rapidly with T because the thermal energy available as T increases is sufficient to activate the vibrations of the solid. The heat capacity increases discontinuously as the solid melts to form a liquid. This is the case because the liquid retains all the local vibrational modes of the solid, and more modes with low frequencies become available upon melting. Therefore, the heat capacity of the liquid is greater than that of the solid. As the liquid vaporizes, the local vibrational modes present in the liquid are converted to translations that cannot take up as much energy as vibrations. Therefore, $C_{P,m}$ decreases discontinuously at the vaporization temperature. The heat capacity in the gaseous state increases slowly with temperature as the vibrational modes of the molecule are activated. These changes in $C_{P,m}$ can be calculated for a specific substance using a microscopic model and statistical thermodynamics, as will be discussed in detail in Chapter 15.

Once the heat capacity of a variety of different substances has been determined, we have a convenient way to quantify heat flow. For example, at constant pressure, the heat flow between the system and surroundings can be written as

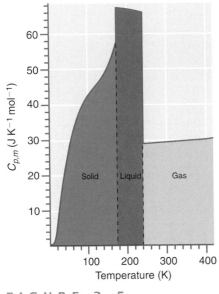

FIGURE 2.5
The variation of $C_{P,m}$ with temperature is shown for Cl_2.

$$q_P = \int_{T_{sys,i}}^{T_{sys,f}} C_P^{system}(T)\,dT = -\int_{T_{surr,i}}^{T_{surr,f}} C_P^{surroundings}(T)\,dT \qquad (2.10)$$

By measuring the temperature change of a thermal reservoir in the surroundings at constant pressure, q_P can be determined. In Equation (2.10), the heat flow at constant pressure has been expressed both from the perspective of the system and from the perspective of the surroundings. A similar equation can be written for a constant volume process. Water is a convenient choice of material for a heat bath in experiments because C_P is nearly constant at the value 4.18 J g^{-1} K^{-1} or 75.3 J mol^{-1} K^{-1} over the range from 0° to 100°C.

EXAMPLE PROBLEM 2.3

The volume of a system consisting of an ideal gas decreases at constant pressure. As a result, the temperature of a 1.50-kg water bath in the surroundings increases by 14.2°C. Calculate q_P for the system.

Solution

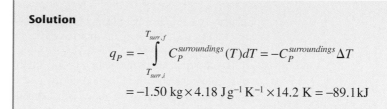

$$q_P = -\int_{T_{surr,i}}^{T_{surr,f}} C_P^{surroundings}(T)dT = -C_P^{surroundings}\Delta T$$

$$= -1.50\ kg \times 4.18\ J\,g^{-1}\,K^{-1} \times 14.2\ K = -89.1\,kJ$$

How are C_P and C_V related for a gas? Consider the processes shown in Figure 2.6 in which a fixed amount of heat flows from the surroundings into a gas. In the constant pressure process, the gas expands as its temperature increases. Therefore, the system does work on the surroundings. As a consequence, not all the heat flow into the system can be used to increase ΔU. No such work occurs for the corresponding constant volume process, and all the heat flow into the system can be used to increase ΔU. Therefore, $dT_P < dT_V$ for the same heat flow dq. For this reason, $C_P > C_V$ for gases.

The same argument applies to liquids and solids as long as V increases with T. Nearly all substances follow this behavior, although notable exceptions occur such as liquid water between 4° and 0°C for which the volume increases as T decreases. However, because ΔV upon heating is much smaller than for a gas, the difference between C_P and C_V for a liquid or solid is much smaller than for a gas.

The preceding remarks about the difference between C_P and C_V have been qualitative in nature. However, the following quantitative relationship, which will be proved in Chapter 3, holds for an ideal gas:

$$C_P - C_V = nR \quad \text{or} \quad C_{P,m} - C_{V,m} = R \tag{2.11}$$

FIGURE 2.6
Not all the heat flow into the system can be used to increase ΔU in a constant pressure process, because the system does work on the surroundings as it expands. However, no work is done for constant volume heating.

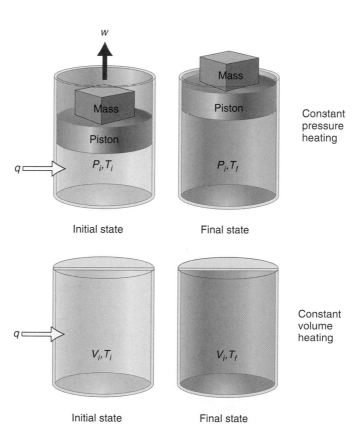

2.5 State Functions and Path Functions

An alternate statement of the first law is that ΔU is independent of the path between the initial and final states and depends only on the initial and final states. We make this statement plausible for the kinetic energy, and the argument can be extended to the other forms of energy listed in Section 2.1. Consider a single molecule in the system. Imagine that the molecule of mass m initially has the speed v_1. We now change its speed incrementally following the sequence $v_1 \rightarrow v_2 \rightarrow v_3 \rightarrow v_4$. The change in the kinetic energy along this sequence is given by

$$\Delta E_{kinetic} = \left(\frac{1}{2} m(v_2)^2 - \frac{1}{2} m(v_1)^2 \right) + \left(\frac{1}{2} m(v_3)^2 - \frac{1}{2} m(v_2)^2 \right) \qquad (2.12)$$

$$+ \left(\frac{1}{2} m(v_4)^2 - \frac{1}{2} m(v_3)^2 \right)$$

$$= \left(\frac{1}{2} m(v_4)^2 - \frac{1}{2} m(v_1)^2 \right)$$

Even though v_2 and v_3 can take on any arbitrary values, they still do not influence the result. We conclude that the change in the kinetic energy depends only on the initial and final speed and that it is independent of the path between these values. Our conclusion remains the same if we increase the number of speed increments in the interval between v_1 and v_2 to an arbitrarily large number. Because this conclusion holds for all molecules in the system, it also holds for ΔU.

This example supports the conclusion that ΔU depends only on the final and initial states and not on the path connecting these states. Any function that satisfies this condition is called a **state function,** because it depends only on the state of the system and not the path taken to reach the state. This property can be expressed in a mathematical form. Any state function, for example, U, must satisfy the equation

$$\Delta U = \int_i^f dU = U_f - U_i \qquad (2.13)$$

This equation states that in order for ΔU to depend only on the initial and final states characterized here by i and f, the value of the integral must be independent of the path. If this is the case, U can be expressed as an infinitesimal quantity, dU, that when integrated, depends only on the initial and final states. The quantity dU is called an **exact differential.** We defer a discussion of exact differentials to Chapter 3.

It is useful to define a cyclic integral, denoted by the symbol $\oint$, as applying to a **cyclic path** such that the initial and final states are identical. For U or any other state function,

$$\oint dU = U_f - U_f = 0 \qquad (2.14)$$

because the initial and final states are the same in a cyclic process.

The state of a single-phase system at fixed composition is characterized by any two of the three variables P, T, and V. The same is true of U. Therefore, for a system of fixed mass, U can be written in any of the three forms $U(V,T)$, $U(P,T)$, or $U(P,V)$. Imagine that a gas of fixed mass characterized by V_1 and T_1 is confined in a piston and cylinder system that is isolated from the surroundings. There is a thermal reservoir in the surroundings at a temperature $T_3 > T_1$. We first compress the system from an initial volume V_1 to a final volume V_2 using a constant external pressure $P_{external}$. The work is given by

$$w = -\int_{V_i}^{V_f} P_{external}\, dV = -P_{external} \int_{V_i}^{V_f} dV = -P_{external}(V_f - V_i) = -P_{external}\Delta V \qquad (2.15)$$

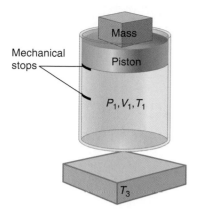

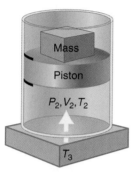

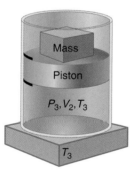

FIGURE 2.7
A system consisting of an ideal gas is contained in a piston and cylinder assembly. The gas in the initial state V_1, T_1 is compressed to an intermediate state, whereby the temperature increases to the value T_2. It is then brought into contact with a thermal reservoir at T_3, leading to a further rise in temperature. The final state is V_2, T_3.

Because the height of the mass in the surroundings is lower after the compression (see Figure 2.7), w is positive and U increases. Because the system consists of a uniform single phase, U is a monotonic function of T, and T also increases. The change in volume ΔV has been chosen such that the temperature of the system T_2 after the compression satisfies the inequality $T_1 < T_2 < T_3$.

We next let an additional amount of heat q flow between the system and surroundings at constant V by bringing the system into contact with the reservoir at temperature T_3. The final values of T and V after these two steps are T_3 and V_2.

This two-step process is repeated for different values of the mass. In each case the system is in the same final state characterized by the variables V_2 and T_3. The sequence of steps that takes the system from the initial state V_1, T_1 to the final state V_2, T_3 is referred to as a **path.** By changing the mass, a set of different paths is generated, all of which originate from the state V_1, T_1, and end in the state V_2, T_3. According to the first law, ΔU for this two-step process is

$$\Delta U = U(T_3, V_2) - U(T_1, V_1) = q + w \tag{2.16}$$

Because ΔU is a state function, its value for the two-step process just described is the same for each of the different values of the mass.

Are q and w also state functions? For this process,

$$w = -P_{external} \Delta V \tag{2.17}$$

and $P_{external}$ is different for each value of the mass, or for each path. Therefore, w is also different for each path; one can take one path from V_1, T_1 to V_2, T_3 and a different path from V_2, T_3 back to V_1, T_1. Because the work is different along these paths, the cyclic integral of work is not equal to zero. Therefore, w is not a state function.

Using the first law to calculate q for each of the paths, we obtain the result

$$q = \Delta U - w = \Delta U + P_{external} \Delta V \tag{2.18}$$

Because ΔU is the same for each path, and w is different for each path, we conclude that q is also different for each path. Just as for work, the cyclic integral of heat is not equal to zero. Therefore, neither q nor w are state functions, and they are called **path functions.**

Because both q and w are path functions, there are no exact differentials for work and heat. Incremental amounts of these quantities are denoted by $đq$ and $đw$, rather than dq and dw, to emphasize the fact that incremental amounts of work and heat are not exact differentials. This is an important result that can be expressed mathematically.

$$\Delta q \neq \int_i^f đq \neq q_f - q_i \quad \text{and} \quad \Delta w \neq \int_i^f đw \neq w_f - w_i \tag{2.19}$$

In fact, there are no such quantities as $\Delta q, q_f, q_i$ and $\Delta w, w_f, w_i$. This is an important result. One cannot refer to the work or heat possessed by a system, because these concepts have no meaning. After a process involving the transfer of heat and work between the system and surroundings is completed, the system and surroundings possess internal energy, but neither possesses heat or work.

Although the preceding discussion may appear pedantic on first reading, it is important to use the terms *work* and *heat* in a way that reflects the fact that they are not state functions. Examples of systems of interest to us are batteries, fuel cells, refrigerators, and internal combustion engines. In each case, the utility of these systems is that work and/or heat flows between the system and surroundings. For example, in a refrigerator, electrical energy is used to extract heat from the inside of the device and to release it in the surroundings. One can speak of the refrigerator as having the capacity to extract heat, but it would be wrong to speak of it as having heat. In the internal combustion engine, chemical energy contained in the bonds of the fuel molecules and in O_2 is released in forming CO_2 and H_2O. This change in internal energy can be used to rotate the wheels of the vehicle, thereby inducing a flow of work between the vehicle and the surroundings. One can refer to the capability of the engine to do work, but it would be incorrect to refer to the vehicle or the engine as containing or having work.

2.6 Equilibrium, Change, and Reversibility

Thermodynamics can only be applied to systems in internal equilibrium, and a requirement for equilibrium is that the overall rate of change of all processes such as diffusion or chemical reaction be zero. How do we reconcile these statements with our calculations of q, w, and ΔU associated with processes in which there is a macroscopic change in the system? To answer this question, it is important to distinguish between the system and surroundings each being in internal equilibrium, and the system and surroundings being in equilibrium with one another.

We first discuss the issue of internal equilibrium. Consider a system made up of an ideal gas, which satisfies the equation of state, $P = nRT/V$. All combinations of P, V, and T consistent with this equation of state form a surface in P–V–T space as shown in Figure 2.8. All points on the surface correspond to equilibrium states of the system. Points that are not on the surface do not correspond to any physically realizable state of the system. Nonequilibrium situations cannot be represented on such a plot, because P, V, and T do not have unique values for a system that is not in internal equilibrium.

Next, consider a process in which the system changes from an initial state characterized by P_i, V_i, and T_i to a final state characterized by P_f, V_f, and T_f as shown in Figure 2.8. If the rate of change of the macroscopic variables is negligibly small, the system passes through a succession of states of internal equilibrium as it goes from the initial to the final state. Such a process is called a **quasi-static process,** and internal equilibrium is maintained in this process. If the rate of change is sufficiently large, the rates of diffusion and intermolecular collisions may not be high enough to maintain the system exactly in a state of internal equilibrium. However, thermodynamic calculations for such a process are still valid as long as it is meaningful to assign a single value of the macroscopic variables P, V, T, and concentration to the system undergoing change. For rapid changes, local fluctuations may occur in the values of the macroscopic variables throughout the system. If these fluctuations are small, quantities such as work and heat can still be calculated using mean values of the macroscopic variables. The same considerations hold for the surroundings.

The concept of a quasi-static process allows one to visualize a process in which the system undergoes a major change in terms of a directed path along a sequence of states in which the system and surroundings are in internal equilibrium. We next distinguish between two very important classes of quasi-static processes, namely, reversible and irreversible processes. It is useful to consider the mechanical system shown in Figure 2.9 when discussing reversible and irreversible processes. Because the two masses have the same value, the net force acting on one end of the wire is zero, and the masses will not move. If an additional mass is placed on either of the two masses, the system is no longer in mechanical equilibrium and the masses will move. In the limit that the incremental mass approaches zero, the velocity at which the initial masses move approaches zero. In this case, one refers to the process as being **reversible,** meaning that the direction of the process can be reversed by placing the infinitesimal mass on the other side of the pulley.

Reversibility in a chemical system can be illustrated by liquid water in equilibrium with gaseous water surrounded by a thermal reservoir. The system and surroundings are both at temperature T. An infinitesimally small increase in T results in a small increase of the amount of water in the gaseous phase, and a small decrease in the liquid phase. An equally small decrease in the temperature has the opposite effect. Therefore, fluctuations in T give rise to corresponding fluctuations in the composition of the system. If an infinitesimal opposing change in the variable that drives the process (temperature in this case) causes a reversal in the direction of the process, the process is reversible.

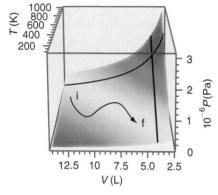

FIGURE 2.8

All combinations of pressure, volume, and temperature consistent with 1 mol of an ideal gas lie on the colored surface. All combinations of pressure and volume consistent with $T = 800$ K and all combinations of pressure and temperature consistent with a volume of 4.0 L are shown as black curves that lie in the P–V–T surface. The third curve corresponds to a path between an initial state i and a final state f that is neither a constant temperature nor a constant volume path.

If an infinitesimal change in the driving variable does not change the direction of the process, one says that the process is **irreversible.** For example, if a large stepwise temperature increase is induced in the system using a heat pulse, the amount of water in the gas phase increases abruptly. In this case, the composition of the system cannot be returned to its initial value by an infinitesimal temperature decrease. This relationship is characteristic of an irreversible process. Although any process that takes place at a rapid rate in the real world is irreversible, real processes can approach reversibility in the appropriate limit. For example, by small variations in the electrical potential in an electrochemical cell, the conversion of reactants to products can be carried out nearly reversibly.

2.7 Comparing Work for Reversible and Irreversible Processes

We concluded in Section 2.5 that w is not a state function and that the work associated with a process is path dependent. This statement can be put on a quantitative footing by comparing the work associated with the reversible and irreversible expansion and compression of an ideal gas. This process is discussed next and illustrated in Figure 2.10.

Consider the following irreversible process, meaning that the internal and external pressures are not equal. A quantity of an ideal gas is confined in a cylinder with a weightless movable piston. The walls of the system are diathermal, allowing heat to flow between the system and surroundings. Therefore, the process is isothermal at the temperature of the surroundings, T. The system is initially defined by the variables T, P_1, and V_1. The position of the piston is determined by $P_{external} = P_1$, which can be changed by adding or removing weights from the piston. Because they are moved horizontally, no work is done in adding or removing the weights. The gas is first expanded at constant temperature by decreasing $P_{external}$ abruptly to the value P_2 (weights are removed), where $P_2 < P_1$. A sufficient amount of heat flows into the system through the diathermal walls to keep the temperature at the constant value T. The system is now in the state defined by T, P_2, and V_2, where $V_2 > V_1$. The system is then returned to its original state in an **isothermal process** by increasing $P_{external}$ abruptly to its original value P_1 (weights are added). Heat flows out of the system into the surroundings in this step. The system has been restored to its original state and, because this is a cyclic process, $\Delta U = 0$. Are q_{total} and w_{total} also zero for the cyclic process? The total work associated with this cyclic process is given by the sum of the work for each individual step:

$$w_{total} = \sum_i -P_{external,i}\Delta V_i = w_{expansion} + w_{compression} = -P_2(V_2 - V_1) - P_1(V_1 - V_2) \quad (2.20)$$

$$= -(P_2 - P_1) \times (V_2 - V_1) > 0 \quad \text{because } P_2 < P_1 \quad \text{and } V_2 > V_1$$

The relationship between P and V for the process under consideration is shown graphically in Figure 2.10, in what is called an **indicator diagram.** An indicator diagram is useful because the work done in the expansion and contraction steps can be evaluated from the appropriate area in the figure, which is equivalent to evaluating the integral $w = -\int P_{external} \, dV$. Note that the work done in the expansion is negative because $\Delta V > 0$, and that done in the compression is positive because $\Delta V < 0$. Because $P_2 < P_1$, the magnitude of the work done in the compression process is more than that done in the expansion process and $w_{total} > 0$. What can one say about q_{total}? The first law states that because $\Delta U = q_{total} + w_{total} = 0$, $q_{total} < 0$.

The same cyclical process is carried out in a reversible cycle. A necessary condition for reversibility is that $P = P_{external}$ at every step of the cycle. This means that P changes during the expansion and compression steps. The work associated with the expansion is

$$w_{expansion} = -\int P_{external} \, dV = -\int P \, dV = -nRT \int \frac{dV}{V} = -nRT \ln \frac{V_2}{V_1} \quad (2.21)$$

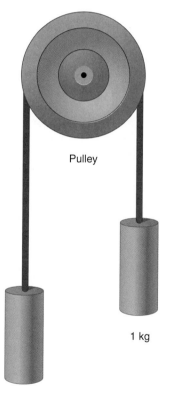

Pulley

1 kg

1 kg

FIGURE 2.9
Two masses of exactly 1 kg each are connected by a wire of zero mass running over a frictionless pulley. The system is in mechanical equilibrium and the masses are stationary.

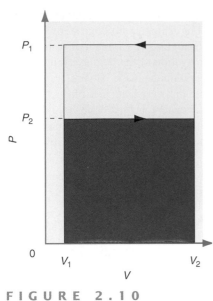

FIGURE 2.10

The work for each step and the total work can be obtained from an indicator diagram. For the compression step, w is given by the total area in red and yellow; for the expansion step, w is given by the red area. The arrows indicate the direction of change in V in the two steps. The sign of w is opposite for these two processes. The total work in the cycle is the yellow area.

This work is shown schematically as the red area in the indicator diagram of Figure 2.11.

If this process is reversed and the compression work is calculated, the following result is obtained:

$$w_{compression} = -nRT \ln \frac{V_1}{V_2} \tag{2.22}$$

It is seen that the magnitude of the work in the forward and reverse processes is equal. The total work done in this cyclical process is given by

$$w = w_{expansion} + w_{compression} = -nRT \ln \frac{V_2}{V_1} - nRT \ln \frac{V_1}{V_2} \tag{2.23}$$

$$= -nRT \ln \frac{V_2}{V_1} + nRT \ln \frac{V_2}{V_1} = 0$$

Therefore, the work done in a reversible isothermal cycle is zero. Because $\Delta U = q + w$ is a state function, $q = -w = 0$ for this reversible isothermal process. Looking at the heights of the weights in the surroundings at the end of the process, we find that they are the same as at the beginning of the process. As will be shown in Chapter 5, q and w are not equal to zero for a reversible cyclical process if T is not constant.

EXAMPLE PROBLEM 2.4

In this example, 2.00 mol of an ideal gas undergoes isothermal expansion along three different paths: (1) reversible expansion from $P_i = 25.0$ bar and $V_i = 4.50$ L to $P_f = 4.50$ bar; (2) a single-step expansion against a constant external pressure of 4.50 bar, and (3) a two-step expansion consisting initially of an expansion against a constant external pressure of 11.0 bar until $P = P_{external}$, followed by an expansion against a constant external pressure of 4.50 bar until $P = P_{external}$.

Calculate the work for each of these processes. For which of the irreversible processes is the magnitude of the work greater?

Solution

The processes are depicted in the following indicator diagram:

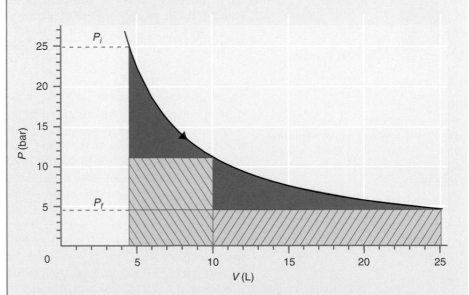

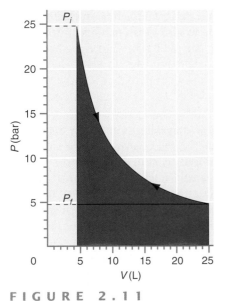

FIGURE 2.11

Indicator diagram for a reversible process. Unlike Figure 2.10, the areas under the P–V curves are the same in the forward and reverse directions.

We first calculate the constant temperature at which the process is carried out, the final volume, and the intermediate volume in the two-step expansion:

$$T = \frac{P_i V_i}{nR} = \frac{25.0 \text{ bar} \times 4.50 \text{ L}}{8.314 \times 10^{-2} \text{ L bar mol}^{-1} \text{ K}^{-1} \times 2.00 \text{ mol}} = 677 \text{ K}$$

$$V_f = \frac{nRT}{P_f} = \frac{8.314 \times 10^{-2} \text{ L bar mol}^{-1} \text{ K}^{-1} \times 2.00 \text{ mol} \times 677 \text{ K}}{4.50 \text{ bar}} = 25.0 \text{ L}$$

$$V_{\text{int}} = \frac{nRT}{P_{\text{int}}} = \frac{8.314 \times 10^{-2} \text{ L bar mol}^{-1} \text{ K}^{-1} \times 2.00 \text{ mol} \times 677 \text{ K}}{11.0 \text{ bar}} = 10.2 \text{ L}$$

The work of the reversible process is given by

$$w = -nRT_1 \ln \frac{V_f}{V_i}$$

$$= -2.00 \text{ mol} \times 8.314 \text{ J mol}^{-1} \text{ K}^{-1} \times 677 \text{ K} \times \ln \frac{25.0 \text{ L}}{4.50 \text{ L}} = -19.3 \times 10^3 \text{ J}$$

We next calculate the work of the single-step and two-step irreversible processes:

$$w_{single} = -P_{external} \Delta V = -4.50 \text{ bar} \times \frac{10^5 \text{ Pa}}{\text{bar}} \times (25.00 \text{ L} - 4.50 \text{ L}) \times \frac{10^{-3} \text{ m}^3}{\text{L}}$$

$$= -9.23 \times 10^3 \text{ J}$$

$$w_{two\text{-}step} = -P_{external} \Delta V = -11.0 \text{ bar} \times \frac{10^5 \text{ Pa}}{\text{bar}} \times (10.2 \text{ L} - 4.50 \text{ L}) \times \frac{10^{-3} \text{ m}^3}{\text{L}}$$

$$- 4.50 \text{ bar} \times \frac{10^5 \text{ Pa}}{\text{bar}} \times (25.00 \text{ L} - 10.2 \text{ L}) \times \frac{10^{-3} \text{ m}^3}{\text{L}}$$

$$= -12.9 \times 10^3 \text{ J}$$

The magnitude of the work is greater for the two-step process than for the single-step process, but less than that for the reversible process.

Example Problem 2.4 suggests that the magnitude of the work for the multistep expansion process increases with the number of steps. This is indeed the case, as shown in Figure 2.12. Imagine that the number of steps n increases indefinitely. As n increases, the pressure difference $P_{external} - P$ for each individual step decreases. In the limit that $n \rightarrow \infty$, the pressure difference $P_{external} - P \rightarrow 0$, and the total area of the rectangles in the indicator diagram approaches the area under the reversible curve. Therefore, the irreversible process becomes reversible and the value of the work equals that of the reversible process.

By contrast, the magnitude of the irreversible compression work exceeds that of the reversible process for finite values of n and becomes equal to that of the reversible process as $n \rightarrow \infty$. The difference between the expansion and compression processes results from the requirement that $P_{external} < P$ at the beginning of each expansion step, whereas $P_{external} > P$ at the beginning of each compression step.

On the basis of these calculations for the reversible and irreversible cycles, we introduce another criterion to distinguish between reversible and irreversible processes. Suppose that a system undergoes a change through one or a number of individual steps, and that the system is restored to its initial state by following the same steps in reverse order. The system is restored to its initial state because the process is cyclical. If the surroundings are also returned to their original state (all masses at the same height and all reservoirs at their original temperatures), the process is reversible. If the surroundings are not restored to their original state, the process is irreversible.

One is often interested in extracting work from a system. For example, it is the expansion of the fuel–air mixture in an automobile engine upon ignition that provides the torque that eventually drives the wheels. Is the capacity to do work similar for reversible and irreversible processes? This question is answered using the indicator

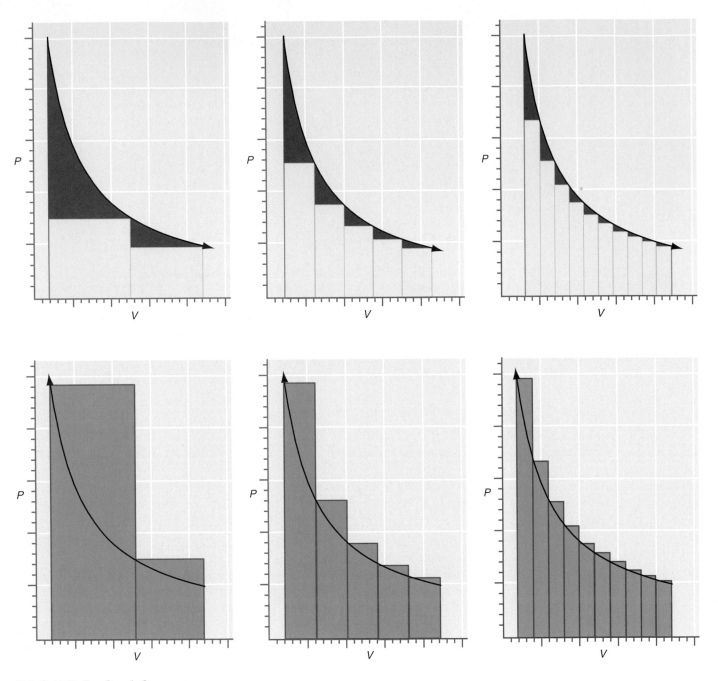

FIGURE 2.12

The work done in an expansion (red area) is compared with the work done in a multistep series of irreversible expansion processes at constant pressure (yellow area) in the top panel. The bottom panel shows analogous results for the compression, where the area under the black curve is the reversible compression work. Note that the total work done in the irreversible expansion and compression processes approaches that of the reversible process as the number of steps becomes large.

diagrams of Figures 2.10 and 2.11 for the specific case of isothermal expansion work, noting that the work can be calculated from the area under the *P-V* curve. We compare the work for expansion from V_1 to V_2 in a single stage at constant pressure to that for the reversible case. For the single-stage expansion, the constant external pressure is given by $P_{external} = nRT/V_2$. However, if the expansion is carried out reversibly, the system pressure is always greater than this value. By comparing the areas in the indicator diagrams of Figure 2.12, it is seen that

$$\left| w_{reversible} \right| \geq \left| w_{irreversible} \right| \qquad (2.24)$$

By contrast, for the compression step,

$$\left| w_{reversible} \right| \leq \left| w_{irreversible} \right| \qquad (2.25)$$

The reversible work is the lower bound for the compression work and the upper bound for the expansion work. This result for the expansion work can be generalized to an important statement that holds for all forms of work: *The maximum work that can be extracted from a process between the same initial and final states is that obtained under reversible conditions.*

Although the preceding statement is true, it suggests that it would be optimal to operate an automobile engine under conditions in which the pressure inside the cylinders differs only infinitesimally from the external atmospheric pressure. This is clearly not possible. A practical engine must generate torque on the drive shaft, and this can only occur if the cylinder pressure is appreciably greater than the external pressure. Similarly, a battery is only useful if one can extract a sizable rather than an infinitesimal current. To operate such devices under useful irreversible conditions, the work output is less than the theoretically possible limit set by the reversible process.

2.8 Determining ΔU and Introducing Enthalpy, a New State Function

How can the ΔU for a thermodynamic process be measured? This will be the topic of Chapter 4, in which calorimetry is discussed. However, this topic is briefly discussed here in order to enable you to carry out calculations on ideal gas systems in the end-of-chapter problems. The first law states that $\Delta U = q + w$. Imagine that the process is carried out under constant volume conditions and that nonexpansion work is not possible. Because under these conditions, $w = -\int P_{external} dV = 0$,

$$\Delta U = q_V \tag{2.26}$$

Equation (2.26) states that ΔU can be experimentally determined by measuring the heat flow between the system and surroundings in a constant volume process.

What does the first law look like under reversible and constant pressure conditions? We write

$$dU = dq_P - P_{external} dV = dq_P - P\,dV \tag{2.27}$$

and integrate this expression between the initial and final states:

$$\int_i^f dU = U_f - U_i = \int dq_P - \int P\,dV = q_P - (P_f V_f - P_i V_i) \tag{2.28}$$

Note that in order to evaluate the integral involving P, we must know the functional relationship $P(V)$, which in this case is $P_i = P_f = P$. Rearranging the last equation, we obtain

$$(U_f + P_f V_f) - (U_i + P_i V_i) = q_P \tag{2.29}$$

Because P, V, and U are all state functions, $U + PV$ is a state function. This new state function is called **enthalpy** and is given the symbol H. [A more rigorous demonstration that H is a state function can be given by invoking the first law for U and applying the criterion of Equation (3.5) to the product PV.]

$$H \equiv U + PV \tag{2.30}$$

As is the case for U, H has the units of energy, and it is an extensive property. As shown in Equation (2.29), ΔH for a process involving only P–V work can be determined by measuring the heat flow between the system and surroundings at constant pressure:

$$\Delta H = q_P \tag{2.31}$$

This equation is the constant pressure analogue of Equation (2.26). Because chemical reactions are much more frequently carried out at constant P than constant V, the energy change measured experimentally by monitoring the heat flow is ΔH rather than ΔU.

2.9 Calculating q, w, ΔU, and ΔH for Processes Involving Ideal Gases

In this section, we discuss how ΔU and ΔH, as well as q and w, can be calculated from the initial and final state variables if the path between the initial and final state is known. The problems at the end of this chapter ask you to calculate q, w, ΔU, and ΔH for simple and multistep processes. Because an equation of state is often needed to carry out such calculations, the system will generally be an ideal gas. Using an ideal gas as a surrogate for more complex systems has the significant advantage that the mathematics is simplified, allowing one to concentrate on the process rather than the manipulation of equations and the evaluation of integrals.

What does one need to know to calculate ΔU? The following discussion is restricted to processes that do not involve chemical reactions or changes in phase. Because U is a state function, ΔU is independent of the path between the initial and final states. To describe a fixed amount of an ideal gas (i.e., n is constant), the values of two of the variables P, V, and T must be known. Is this also true for ΔU for processes involving ideal gases? To answer this question, consider the expansion of an ideal gas from an initial state V_1, T_1 to a final state V_2, T_2. We first assume that U is a function of both V and T. Is this assumption correct? Because ideal gas atoms or molecules do not interact with one another, U will not depend on the distance between the atoms or molecules. Therefore, U is not a function of V, and we conclude that ΔU must be a function of T only for an ideal gas, $\Delta U = \Delta U(T)$.

We also know that for a temperature range over which C_V is constant,

$$\Delta U = q_V = C_V(T_f - T_i) \tag{2.32}$$

2.1 Ideal Gas under Constant Pressure or Volume

Is this equation only valid for constant V? Because U is a function of T only for an ideal gas, Equation (2.32) is also valid for processes involving ideal gases in which V is not constant. Therefore, if one knows C_V, T_1, and T_2, ΔU can be calculated, regardless of the path between the initial and final states.

How many variables are required to define ΔH for an ideal gas? We write

$$\Delta H = \Delta U(T) + \Delta(PV) = \Delta U(T) + \Delta(nRT) = \Delta H(T) \tag{2.33}$$

It is seen that ΔH is also a function of T only for an ideal gas. In analogy to Equation (2.32),

$$\Delta H = q_P = C_P(T_f - T_i) \tag{2.34}$$

Because ΔH is a function of T only for an ideal gas, Equation (2.34) holds for all processes involving ideal gases, whether P is constant or not as long as it is reasonable to assume that C_P is constant. Therefore, if the initial and final temperatures are known or can be calculated, and if C_V and C_P are known, ΔU and ΔH can be calculated *regardless of the path* for processes involving ideal gases using Equations (2.32) and (2.34), as long as no chemical reactions or phase changes occur. Because U and H are state functions, the previous statement is true for both reversible and irreversible processes. Recall that for an ideal gas $C_P - C_V = nR$, so that if one of C_V and C_P is known, the other can be readily determined.

We next note that the first law links q, w, and ΔU. If any two of these quantities can be calculated, the first law can be used to calculate the third. In calculating work, often only expansion work takes place. In this case one always proceeds from the equation

$$w = -\int P_{external}\, dV \tag{2.35}$$

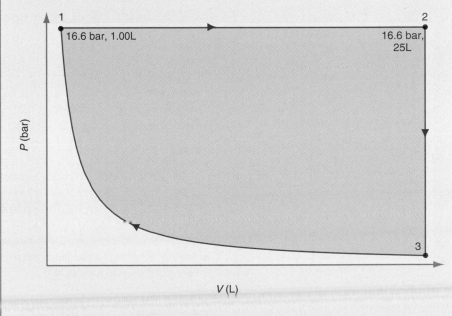
This integral can only be evaluated if the functional relationship between $P_{external}$ and V is known. A frequently encountered case is $P_{external}$ = constant, such that

$$w = -P_{external}(V_f - V_i) \tag{2.36}$$

Because $P_{external} \neq P$, the work considered in Equation (2.36) is for an irreversible process.

A second frequently encountered case is that the system and external pressure differ only by an infinitesimal amount. In this case, it is sufficiently accurate to write $P_{external} = P$, and the process is reversible:

$$w = -\int \frac{nRT}{V} dV \tag{2.37}$$

This integral can only be evaluated if T is known as a function of V. The most commonly encountered case is an isothermal process, in which T is constant. As was seen in Section 2.2, for this case

$$w = -nRT \int \frac{dV}{V} = -nRT \ln \frac{V_f}{V_i} \tag{2.38}$$

In solving thermodynamic problems, it is very helpful to understand the process thoroughly before starting the calculation, because it is often possible to obtain the value of one or more of q, w, and ΔU and ΔH without a calculation. For example, $\Delta U = \Delta H = 0$ for an isothermal process because ΔU and ΔH depend only on T. For an adiabatic process, $q = 0$ by definition. If only expansion work is possible, $w = 0$ for a constant volume process. These guidelines are illustrated in the following two example problems.

EXAMPLE PROBLEM 2.5

A system containing 2.50 mol of an ideal gas for which $C_{V,m}$ = 20.79 J mol^{-1} K^{-1} is taken through the cycle in the following diagram in the direction indicated by the arrows. The curved path corresponds to $PV = nRT$, where $T = T_1 = T_3$.

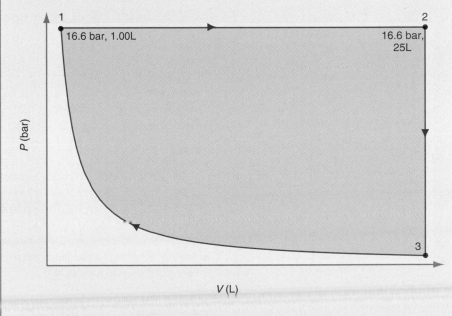

a. Calculate q, w, and ΔU and ΔH for each segment and for the cycle.

b. Calculate q, w, and ΔU and ΔH for each segment and for the cycle in which the direction of each process is reversed.

Solution

We begin by asking whether we can evaluate q, w, ΔU, or ΔH for any of the segments without any calculations. Because the path between states 1 and 3 is isothermal, ΔU and ΔH are zero for this segment. Therefore, from the first law, $q_{3\rightarrow1} = -w_{3\rightarrow1}$. For this reason, we only need to calculate one of these two quantities. Because $\Delta V = 0$ along the path between states 2 and 3, $w_{2\rightarrow3} = 0$. Therefore, $\Delta U_{2\rightarrow3} = q_{2\rightarrow3}$. Again, we only need to calculate one of these two quantities. Because the total process is cyclic, the change in any state function is zero. Therefore, $\Delta U = \Delta H = 0$ for the cycle, no matter which direction is chosen. We now deal with each segment individually.

Segment 1 → 2

The values of n, P_1 and V_1, and P_2 and V_2 are known. Therefore, T_1 and T_2 can be calculated using the ideal gas law. We use these temperatures to calculate ΔU as follows:

$$\Delta U_{1\rightarrow2} = nC_{V,m}(T_2 - T_1) = \frac{nC_{V,m}}{nR}(P_2V_2 - P_1V_1)$$

$$= \frac{2.50\,\text{mol} \times 20.79\,\text{J mol}^{-1}\,\text{K}^{-1}}{2.50\,\text{mol} \times 0.08314\,\text{L bar K}^{-1}\,\text{mol}^{-1}} \times (16.6\,\text{bar} \times 25.0\,\text{L} - 16.6\,\text{bar} \times 1.00\,\text{L})$$

$$= 99.6\,\text{kJ}$$

The process takes place at constant pressure, so

$$w = -P_{external}(V_2 - V_1) = -16.6\,\text{bar} \times \frac{10^5\,\text{N m}^{-2}}{\text{bar}} \times (25.0 \times 10^{-3}\,\text{m}^3 - 1.00 \times 10^{-3}\,\text{m}^3)$$

$$= -39.8\,\text{kJ}$$

Using the first law

$$q = \Delta U - w = 99.6\,\text{kJ} + 39.8\,\text{kJ} \approx 139.4\,\text{kJ}$$

We next calculate T_2, and then $\Delta H_{1\rightarrow2}$:

$$T_2 = \frac{P_2V_2}{nR} = \frac{16.6\,\text{bar} \times 25.0\,\text{L}}{2.50\,\text{mol} \times 0.08314\,\text{L bar K}^{-1}\,\text{mol}^{-1}} = 2.00 \times 10^3\,\text{K}$$

We next calculate $T_3 = T_1$

$$T_1 = \frac{P_1V_1}{nR} = \frac{16.6\,\text{bar} \times 1.00\,\text{L}}{2.50\,\text{mol} \times 0.08314\,\text{L bar mol}^{-1}\,\text{K}^{-1}} = 79.9\,\text{K}$$

$$\Delta H_{1\rightarrow2} = \Delta U_{1\rightarrow2} + \Delta(PV) = \Delta U_{1\rightarrow2} + nR(T_2 - T_1)$$

$$= 99.6 \times 10^3\,\text{J} + 2.5\,\text{mol} \times 8.314\,\text{J mol}^{-1}\,\text{K}^{-1} \times (2000\,\text{K} - 79.9\,\text{K}) = 139.4\,\text{kJ}$$

Segment 2 → 3

As noted above, $w = 0$, and

$$\Delta U_{2\rightarrow3} = q_{2\rightarrow3} = C_V(T_3 - T_2)$$

$$= 2.50\,\text{mol} \times 20.79\,\text{J mol}^{-1}\text{K}^{-1}(79.9\,\text{K} - 2000\,\text{K})$$

$$= -99.6\,\text{kJ}$$

The numerical result is equal in magnitude, but opposite in sign to $\Delta U_{1\rightarrow2}$ because $T_3 = T_1$. For the same reason, $\Delta H_{2\rightarrow3} = -\Delta H_{1\rightarrow2}$.

Segment 3 → 1

For this segment, $\Delta U_{3\rightarrow1} = 0$ and $\Delta H_{3\rightarrow1} = 0$ as noted earlier and $w_{3\rightarrow1} = -q_{3\rightarrow1}$. Because this is a reversible isothermal compression,

$$w_{3\to1} = -nRT \ln \frac{V_1}{V_3} = -2.50 \text{ mol} \times 8.314 \text{ J mol}^{-1}\text{K}^{-1} \times 79.9 \text{ K} \times \ln \frac{1.00 \times 10^{-3}\text{m}^3}{25.0 \times 10^{-3}\text{m}^3}$$

$$= 5.35 \text{ kJ}$$

The results for the individual segments and for the cycle in the indicated direction are given in the following table. If the cycle is traversed in the reverse fashion, the magnitudes of all quantities in the table remain the same, but all signs change.

Path	q (kJ)	w (kJ)	ΔU (kJ)	ΔH (kJ)
1→2	139.4	−39.8	99.6	139.4
2→3	−99.6	0	99.6	−139.4
3→1	−5.35	5.35	0	0
Cycle	34.5	−34.5	0	0

EXAMPLE PROBLEM 2.6

In this example, 2.50 mol of an ideal gas with $C_{V,m} = 12.47$ J mol^{-1} K^{-1} is expanded adiabatically against a constant external pressure of 1.00 bar. The initial temperature and pressure of the gas are 325 K and 2.50 bar, respectively. The final pressure is 1.25 bar. Calculate the final temperature, q, w, ΔU, and ΔH.

Solution

Because the process is adiabatic, $q = 0$, and $\Delta U = w$. Therefore,

$$\Delta U = nC_{v,m}(T_f - T_i) = -P_{external}(V_f - V_i)$$

Using the ideal gas law,

$$nC_{v,m}(T_f - T_i) = -nRP_{external}\left(\frac{T_f}{P_f} - \frac{T_i}{P_i}\right)$$

$$T_f\left(nC_{v,m} + \frac{nRP_{external}}{P_f}\right) = T_i\left(nC_{v,m} + \frac{nRP_{external}}{P_i}\right)$$

$$T_f = T_i \frac{\left(C_{v,m} + \dfrac{RP_{external}}{P_i}\right)}{\left(C_{v,m} + \dfrac{RP_{external}}{P_f}\right)}$$

$$= 325 \text{ K} \times \left(\frac{12.47 \text{ J mol}^{-1}\text{ K}^{-1} + \dfrac{8.314 \text{ J mol}^{-1}\text{ K}^{-1} \times 1.00 \text{ bar}}{2.50 \text{ bar}}}{12.47 \text{ J mol}^{-1}\text{ K}^{-1} + \dfrac{8.314 \text{ J mol}^{-1}\text{ K}^{-1} \times 1.00 \text{ bar}}{1.25 \text{ bar}}}\right) = 268 \text{ K}$$

We calculate $\Delta U = w$ from

$$\Delta U = nC_{V,m}(T_f - T_i) = 2.5 \text{ mol} \times 12.47 \text{ J mol}^{-1}\text{K}^{-1} \times (268 \text{ K} - 325 \text{ K}) = -1.78 \text{ kJ}$$

Because the temperature falls in the expansion, the internal energy decreases:

$$\Delta H = \Delta U + \Delta(PV) = \Delta U + nR(T_2 - T_1)$$

$$= -1.78 \times 10^3 \text{ J} + 2.5 \text{ mol} \times 8.314 \text{ J mol}^{-1}\text{K}^{-1} \times (268 \text{ K} - 325 \text{ K}) = -2.96 \text{ kJ}$$

2.10 The Reversible Adiabatic Expansion and Compression of an Ideal Gas

• • •
WWW
• • •

2.6 Reversible Adiabatic Heating and Cooling of an Ideal Gas

The adiabatic expansion and compression of gases is an important meteorological process. For example, the cooling of a cloud as it moves upward in the atmosphere can be modeled as an adiabatic process because the heat transfer between the cloud and the rest of the atmosphere is slow on the timescale of its upward motion.

Consider the adiabatic expansion of an ideal gas. Because $q = 0$, the first law takes the form

$$\Delta U = w \quad \text{or} \quad C_V dT = -P_{external} dV \tag{2.39}$$

For a reversible adiabatic process, $P = P_{external}$, and

$$C_V dT = -nRT \frac{dV}{V} \quad \text{or, equivalently,} \quad C_V \frac{dT}{T} = -nR \frac{dV}{V} \tag{2.40}$$

Integrating both sides of this equation between the initial and final states,

$$\int_{T_i}^{T_f} C_V \frac{dT}{T} = -nR \int_{V_i}^{V_f} \frac{dV}{V} \tag{2.41}$$

If C_V is constant over the temperature interval $T_f - T_i$, then

$$C_V \ln \frac{T_f}{T_i} = -nR \ln \frac{V_f}{V_i} \tag{2.42}$$

Because $C_P - C_V = nR$ for an ideal gas, Equation (2.42) can be written in the form

$$\ln\left(\frac{T_f}{T_i}\right) = -(\gamma - 1) \ln\left(\frac{V_f}{V_i}\right) \quad \text{or, equivalently,} \quad \frac{T_f}{T_i} = \left(\frac{V_f}{V_i}\right)^{1-\gamma} \tag{2.43}$$

where $\gamma = C_{P,m}/C_{V,m}$. Substituting $T_f/T_i = P_f V_f/P_i V$ in the previous equation, we obtain

$$P_i V_i^{\gamma} = P_f V_f^{\gamma} \tag{2.44}$$

for the adiabatic reversible expansion or compression of an ideal gas. Note that our derivation is only applicable to a reversible process, because we have assumed that $P = P_{external}$.

Reversible adiabatic compression of a gas leads to heating, and reversible adiabatic expansion leads to cooling, as can be concluded from Figure 2.13. Two systems containing 1 mol of N_2 gas have the same volume at $P = 1$ atm. Under isothermal conditions, heat flows out of the system as it is compressed to $P > 1$ atm, and heat flows into the system as it is expanded to $P < 1$ atm to keep T constant. No heat flows into or out of the system under adiabatic conditions. Note in Figure 2.13 that in a reversible adiabatic compression originating at 1 atm, $P_{adiabatic} > P_{isothermal}$ for all $P > 1$ atm. Therefore, this value of P must correspond to a value of T for which $T > T_{isothermal}$. Similarly, in a reversible adiabatic expansion originating at 1 atm, $P_{adiabatic} < P_{isothermal}$ for all $P < 1$ atm. Therefore, this value of P must correspond to a value of T for which $T < T_{isothermal}$.

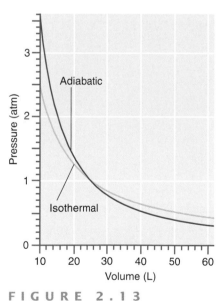

FIGURE 2.13
Two systems containing 1 mol of N_2 have the same P and V values at 1 atm. The red curve corresponds to reversible expansion and compression about $P = 1$ atm under adiabatic conditions. The yellow curve corresponds to reversible expansion and compression about $P = 1$ atm under isothermal conditions.

EXAMPLE PROBLEM 2.7

A cloud mass moving across the ocean at an altitude of 2000 m encounters a coastal mountain range. As it rises to a height of 3500 m to pass over the mountains, it undergoes an adiabatic expansion. The pressure at 2000 and

3500 m is 0.802 and 0.602 atm, respectively. If the initial temperature of the cloud mass is 288 K, what is the cloud temperature as it passes over the mountains? Assume that $C_{P,m}$ for air is 28.86 J K^{-1} mol^{-1} and that air obeys the ideal gas law. If you are on the mountain, should you expect rain or snow?

Solution

$$\ln\left(\frac{T_f}{T_i}\right) = -(\gamma-1)\ln\left(\frac{V_f}{V_i}\right) = -(\gamma-1)\ln\left(\frac{T_f}{T_i}\frac{P_i}{P_f}\right) = -(\gamma-1)\ln\left(\frac{T_f}{T_i}\right) - (\gamma-1)\ln\left(\frac{P_i}{P_f}\right)$$

$$= -\frac{(\gamma-1)}{\gamma}\ln\left(\frac{P_i}{P_f}\right) = -\frac{\left(\dfrac{C_{P,m}}{C_{P,m}-R}-1\right)}{\dfrac{C_{P,m}}{C_{P,m}-R}}\ln\left(\frac{P_i}{P_f}\right)$$

$$= -\frac{\left(\dfrac{28.86 \text{ J K}^{-1}\text{ mol}^{-1}}{28.86 \text{ J K}^{-1}\text{ mol}^{-1} - 8.314 \text{ J K}^{-1}\text{ mol}^{-1}}-1\right)}{\dfrac{28.86 \text{ J K}^{-1}\text{ mol}^{-1}}{28.86 \text{ J K}^{-1}\text{ mol}^{-1} - 8.314 \text{ J K}^{-1}\text{ mol}^{-1}}} \times \ln\left(\frac{0.802 \text{ atm}}{0.602 \text{ atm}}\right) = -0.0826$$

$$T_f = 0.9207 T_i = 265 \text{ K}$$

You can expect snow.

Vocabulary

cyclic path	internal energy	path function
enthalpy	irreversible process	quasi-static process
exact differential	isobaric	reversible process
first law of thermodynamics	isochoric	state function
heat	isothermal	work
heat capacity	isothermal process	
indicator diagram	path	

Questions on Concepts

Q2.1 Electrical current is passed through a resistor immersed in a liquid in an adiabatic container. The temperature of the liquid is varied by 1°C. The system consists solely of the liquid. Does heat or work flow across the boundary between the system and surroundings? Justify your answer.

Q2.2 Explain how a mass of water in the surroundings can be used to determine q for a process. Calculate q if the temperature of a 1.00-kg water bath in the surroundings increases by 1.25°C. Assume that the surroundings are at a constant pressure.

Q2.3 Explain the relationship between the terms *exact differential* and *state function*.

Q2.4 Why is it incorrect to speak of the heat or work associated with a system?

Q2.5 Two ideal gas systems undergo reversible expansion starting from the same P and V. At the end of the expansion, the two systems have the same volume. The pressure in the system that has undergone adiabatic expansion is lower than in the system that has undergone isothermal expansion. Explain this result without using equations.

Q2.6 A cup of water at 278 K (the system) is placed in a microwave oven and the oven is turned on for 1 minute during which it begins to boil. Which of q, w, and ΔU are positive, negative, or zero?

Q2.7 What is wrong with the following statement?: *Because the well-insulated house stored a lot of heat, the temperature didn't fall much when the furnace failed.* Rewrite the sentence to convey the same information in a correct way.

Q2.8 What is wrong with the following statement?: *Burns caused by steam at 100°C can be more severe than those caused by water at 100°C because steam contains more heat than water.* Rewrite the sentence to convey the same information in a correct way.

Q2.9 Describe how reversible and irreversible expansions differ by discussing the degree to which equilibrium is maintained between the system and the surroundings.

Q2.10 A chemical reaction occurs in a constant volume enclosure separated from the surroundings by diathermal walls. Can you say whether the temperature of the surroundings increases, decreases, or remains the same in this process? Explain.

Problems

Problem numbers in **red** indicate that the solution to the problem is given in the *Student's Solutions Manual*.

P2.1 3.00 moles of an ideal gas at 27.0°C expands isothermally from an initial volume of 20.0 dm³ to a final volume of 60.0 dm³. Calculate w for this process (a) for expansion against a constant external pressure of 1.00×10^5 Pa and (b) for a reversible expansion.

P2.2 3.00 moles of an ideal gas are compressed isothermally from 60.0 to 20.0 L using a constant external pressure of 5.00 atm. Calculate q, w, ΔU, and ΔH.

P2.3 A system consisting of 57.5 g of liquid water at 298 K is heated using an immersion heater at a constant pressure of 1.00 bar. If a current of 1.50 A passes through the 10.0-ohm resistor for 150 s, what is the final temperature of the water? The heat capacity for water can be found in Appendix A.

P2.4 For 1 mol of an ideal gas, $P_{external} = P = 200 \times 10^3$ Pa. The temperature is changed from 100°C to 25.0°C, and $C_{V,m} = 3/2R$. Calculate q, w, ΔU, and ΔH.

P2.5 Consider the isothermal expansion of 5.25 mol of an ideal gas at 450 K from an initial pressure of 15.0 bar to a final pressure of 3.50 bar. Describe the process that will result in the greatest amount of work being done by the system with $P_{external} \geq 3.50$ bar and calculate w. Describe the process that will result in the least amount of work being done by the system with $P_{external} \geq 3.50$ bar and calculate w. What is the least amount of work done without restrictions on the external pressure?

P2.6 Calculate ΔH and ΔU for the transformation of 1 mol of an ideal gas from 27.0°C and 1.00 atm to 327°C and 17.0 atm if

$$C_{P,m} = 20.9 + 0.042\frac{T}{K} \text{ in units of J K}^{-1}\text{mol}^{-1}$$

P2.7 Calculate w for the adiabatic expansion of 1 mol of an ideal gas at an initial pressure of 2.00 bar from an initial temperature of 450 K to a final temperature of 300 K. Write an expression for the work done in the isothermal reversible expansion of the gas at 300 K from an initial pressure of 2.00 bar. What value of the final pressure would give the same value of w as the first part of this problem? Assume that $C_{P,m} = 5/2R$.

P2.8 In the adiabatic expansion of 1 mol of an ideal gas from an initial temperature of 25.0°C, the work done on the surroundings is 1200 J. If $C_{V,m} = 3/2R$, calculate q, w, ΔU, and ΔH.

P2.9 An ideal gas undergoes an expansion from the initial state described by P_i, V_i, T to a final state described by P_f, V_f, T in (a) a process at the constant external pressure P_f and (b) in a reversible process. Derive expressions for the largest mass that can be lifted through a height h in the surroundings in these processes.

P2.10 An automobile tire contains air at 320×10^3 Pa at 20.0°C. The stem valve is removed and the air is allowed to expand adiabatically against the constant external pressure of 100×10^3 Pa until $P = P_{external}$. For air, $C_{V,m} = 5/2R$. Calculate the final temperature. Assume ideal gas behavior.

P2.11 3.50 moles of an ideal gas is expanded from 450 K and an initial pressure of 5.00 bar to a final pressure of 1.00 bar, and $C_{P,m} = 5/2R$. Calculate w for the following two cases:

a. The expansion is isothermal and reversible.

b. The expansion is adiabatic and reversible.

Without resorting to equations, explain why the result to part (b) is greater than or less than the result to part (a).

P2.12 An ideal gas described by $T_i = 300$ K, $P_i = 1.00$ bar, and $V_i = 10.0$ L is heated at constant volume until $P = 10.0$ bar. It then undergoes a reversible isothermal expansion until $P = 1.00$ bar. It is then restored to its original state by the extraction of heat at constant pressure. Depict this closed-cycle process in a P–V diagram. Calculate w for each step and for the total process. What values for w would you calculate if the cycle were traversed in the opposite direction?

P2.13 3.00 moles of an ideal gas with $C_{V,m} = 3/2R$ initially at a temperature $T_i = 298$ K and $P_i = 1.00$ bar is enclosed in an adiabatic piston and cylinder assembly. The gas is compressed by placing a 625-kg mass on the piston of diameter 20.0 cm. Calculate the work done in this process and the distance that the piston travels. Assume that the mass of the piston is negligible.

P2.14 A bottle at 21.0°C contains an ideal gas at a pressure of 126.4×10^3 Pa. The rubber stopper closing the bottle is removed. The gas expands adiabatically against $P_{external} = 101.9 \times 10^3$ Pa, and some gas is expelled from the bottle in the process. When $P = P_{external}$, the stopper is quickly replaced. The gas remaining in the bottle slowly warms up to 21.0°C. What is the final pressure in the bottle for a monatomic gas, for which $C_{V,m} = 3/2R$, and a diatomic gas, for which $C_{V,m} = 5/2R$?

P2.15 A pellet of Zn of mass 10.0 g is dropped into a flask containing dilute H_2SO_4 at a pressure of $P = 1.00$ bar and temperature of $T = 298$ K. What is the reaction that occurs? Calculate w for the process.

P2.16 One mole of an ideal gas for which $C_{V,m} = 20.8$ J K^{-1} mol^{-1} is heated from an initial temperature of 0°C to a final temperature of 275°C at constant volume. Calculate q, w, ΔU, and ΔH for this process.

P2.17 One mole of an ideal gas, for which $C_{V,m} = 3/2R$, initially at 20.0°C and 1.00×10^6 Pa undergoes a two-stage transformation. For each of the stages described in the following list, calculate the final pressure, as well as q, w, ΔU, and ΔH. Also calculate q, w, ΔU, and ΔH for the complete process.

a. The gas is expanded isothermally and reversibly until the volume doubles.

b. Beginning at the end of the first stage, the temperature is raised to 80.0°C at constant volume.

P2.18 One mole of an ideal gas with $C_{V,m} = 3/2R$ initially at 298 K and 1.00×10^5 Pa undergoes a reversible adiabatic compression. At the end of the process, the pressure is 1.00×10^6 Pa. Calculate the final temperature of the gas. Calculate q, w, ΔU, and ΔH for this process.

P2.19 One mole of an ideal gas, for which $C_{V,m} = 3/2R$, is subjected to two successive changes in state: (1) From 25.0°C and 100×10^3 Pa, the gas is expanded isothermally against a constant pressure of 20.0×10^3 Pa to twice the initial volume. (2) At the end of the previous process, the gas is cooled at constant volume from 25.0°C to –25.0°C. Calculate q, w, ΔU, and ΔH for each of the stages. Also calculate q, w, ΔU, and ΔH for the complete process.

P2.20 The temperature of 1 mol of an ideal gas increases from 18.0° to 55.1°C as the gas is compressed adiabatically. Calculate q, w, ΔU, and ΔH for this process assuming that $C_{V,m} = 3/2R$.

P2.21 A 1-mol sample of an ideal gas for which $C_{V,m} = 3/2R$ undergoes the following two-step process: (1) From an initial state of the gas described by $T = 28.0$°C and $P = 2.00 \times 10^4$ Pa, the gas undergoes an isothermal expansion against a constant external pressure of 1.00×10^4 Pa until the volume has doubled. (2) Subsequently, the gas is cooled at constant volume. The temperature falls to –40.5°C. Calculate q, w, ΔU, and ΔH for each step and for the overall process.

P2.22 A cylindrical vessel with rigid adiabatic walls is separated into two parts by a frictionless adiabatic piston.

Each part contains 50.0 L of an ideal monatomic gas with $C_{V,m} = 3/2R$. Initially, $T_i = 298$ K and $P_i = 1.00$ bar in each part. Heat is slowly introduced into the left part using an electrical heater until the piston has moved sufficiently to the right to result in a final pressure $P_f = 7.50$ bar in the right part. Consider the compression of the gas in the right part to be a reversible process.

a. Calculate the work done on the right part in this process and the final temperature in the right part.

b. Calculate the final temperature in the left part and the amount of heat that flowed into this part.

P2.23 A vessel containing 1 mol of an ideal gas with $P_i = 1.00$ bar and $C_{P,m} = 5/2R$ is in thermal contact with a water bath. Treat the vessel, gas, and water bath as being in thermal equilibrium, initially at 298 K, and as separated by adiabatic walls from the rest of the universe. The vessel, gas, and water bath have an average heat capacity of $C_P = 7500$ J K^{-1}. The gas is compressed reversibly to $P_f = 10.5$ bar. What is the temperature of the system after thermal equilibrium has been established?

P2.24 The heat capacity of solid lead oxide is given by

$$C_{P,m} = 44.35 + 1.47 \times 10^{-3}\frac{T}{K} \text{ in units of J K}^{-1}\text{ mol}^{-1}$$

Calculate the change in enthalpy of 1 mol of PbO(s) if it is cooled from 500 to 300 K at constant pressure.

P2.25 Consider the adiabatic expansion of 0.500 mol of an ideal monatomic gas with $C_{V,m} = 3/2R$. The initial state is described by $P = 3.25$ bar and $T = 300$ K.

a. Calculate the final temperature if the gas undergoes a reversible adiabatic expansion to a final pressure of $P = 1.00$ bar.

b. Calculate the final temperature if the same gas undergoes an adiabatic expansion against an external pressure of $P = 1.00$ bar to a final pressure $P = 1.00$ bar.

Explain the difference in your results for parts (a) and (b).

P2.26 An ideal gas undergoes a single-stage expansion against a constant external pressure $P_{external}$ at constant temperature from T, P_i, V_i, to T, P_f, V_f.

a. What is the largest mass m that can be lifted through the height h in this expansion?

b. The system is restored to its initial state in a single-state compression. What is the smallest mass m' that must fall through the height h to restore the system to its initial state?

c. If $h = 10.0$ cm, $P_i = 1.00 \times 10^6$ Pa, $P_f = 0.500 \times 10^6$ Pa, $T = 300$ K, and $n = 1.00$ mol, calculate the values of the masses in parts (a) and (b).

P2.27 Calculate q, w, ΔU, and ΔH if 1.00 mol of an ideal gas with $C_{V,m} = 3/2R$ undergoes a reversible adiabatic expansion from an initial volume $V_i = 5.25$ m^3 to a final volume $V_f = 25.5$ m^3. The initial temperature is 300 K.

P2.28 A nearly flat bicycle tire becomes noticeably warmer after it has been pumped up. Approximate this process as a reversible adiabatic compression. Assume the initial pressure and temperature of the air before it is put in the tire to be P_i = 1.00 bar and T_i = 298 K. The final volume of the air in the tire is V_f = 1.00 L and the final pressure is P_f = 5.00 bar. Calculate the final temperature of the air in the tire. Assume that $C_{V,m}$ = 5/2R

P2.29 One mole of an ideal gas with $C_{V,m}$ = 3/2R is expanded adiabatically against a constant external pressure of 1.00 bar. The initial temperature and pressure are T_i = 300 K

and P_i = 25.0 bar. The final pressure is P_f = 1.00 bar. Calculate q, w, ΔU, and ΔH for the process.

P2.30 One mole of N_2 in a state defined by T_i = 300 K and V_i = 2.50 L undergoes an isothermal reversible expansion until V_f = 23.0 L. Calculate w assuming (a) that the gas is described by the ideal gas law and (b) that the gas is described by the van der Waals equation of state. What is the percent error in using the ideal gas law instead of the van der Waals equation? The van der Waals parameters for N_2 are listed in Table 7.4.

Web-Based Simulations, Animations, and Problems

W2.1 A simulation is carried out in which an ideal gas is heated under constant pressure or constant volume conditions. The quantities ΔV (or ΔP), w, ΔU, and ΔT are determined as a function of the heat input. The heat taken up by the gas under constant P or V is calculated and compared with ΔU and ΔH.

W2.2 The reversible isothermal compression and expansion of an ideal gas is simulated for different values of T. The work, w, is calculated from the T and V values obtained in the simulation. The heat, q, and the number of moles of gas in the system are calculated from the results.

W2.3 The reversible isobaric compression and expansion of an ideal is simulated for different values of pressure gas as heat flows to/from the surroundings. The quantities q, w, and ΔU are calculated from the ΔT and ΔV values obtained in the simulation.

W2.4 The isochoric heating and cooling of an ideal gas is simulated for different values of volume. The number of moles of gas and ΔU are calculated from the constant V value and from the T and P values obtained in the simulation.

W2.5 Reversible cyclic processes are simulated in which the cycle is either rectangular or triangular on a P-V plot. For each segment and for the cycle, ΔU, q, and w are determined. For a given cycle type, the ratio of work done on the surroundings to the heat absorbed from the surroundings is determined for different P and V values.

W2.6 The reversible adiabatic heating and cooling of an ideal gas is simulated for different values of the initial temperature. The quantity $\gamma = C_{P,m}/C_{V,m}$ as well as $C_{P,m}$ and $C_{V,m}$ are determined from the P,V values of the simulation, ΔU and ΔH are calculated from the V, T, and P values obtained in the simulation.

CHAPTER 3

The Importance of State Functions: Internal Energy and Enthalpy

The mathematical properties of state functions are utilized to express the infinitesimal quantities dU and dH as exact differentials. By doing so, expressions can be derived that relate the change of U with T and V and the change in H with T and P to experimentally accessible quantities such as the heat capacity and the coefficient of thermal expansion. Although both U and H are functions of any two of the variables P, V, and T, the dependence of U and H on temperature is generally far greater than the dependence on P or V. As a result, for most processes involving gases, liquids, and solids, U and H can be regarded as functions of T only. An exception to this statement is the cooling on the isenthalpic expansion of real gases, which is commercially used in the liquefaction of N_2, O_2, He, and Ar. ∎

3.1 The Mathematical Properties of State Functions

In Chapter 2 we demonstrated that U and H are state functions and that w and q are path functions. We also discussed how to calculate changes in these quantities for an ideal gas. In this chapter, the path independence of state functions is exploited to derive relationships with which ΔU and ΔH can be calculated as functions of P, V, and T for real gases, liquids, and solids. In doing so, we develop the formal aspects of thermodynamics. It will be seen that the formal structure of thermodynamics provides a powerful aid in linking theory and experiment. However, before these topics are discussed, the mathematical properties of state functions need to be outlined.

The thermodynamic state functions of interest here are defined by two variables from the set P, V, and T. In formulating changes in state functions, we will make extensive use of partial derivatives, which are reviewed in the Math Supplement (Appendix B). The following discussion does not apply to path functions such as w and q because a functional relationship such as Equation (3.1) does not exist for path-dependent functions. Consider 1 mol of an ideal gas for which

$$P = f(V,T) = \frac{RT}{V} \tag{3.1}$$

Note that P can be written as a function of the two variables V and T. The change in P resulting from a change in V or T is proportional to the following **partial derivatives:**

$$\left(\frac{\partial P}{\partial V}\right)_T = \lim_{\Delta V \to 0} \frac{P(V + \Delta V, T) - P(V, T)}{\Delta V} = -\frac{RT}{V^2}$$

$$\left(\frac{\partial P}{\partial T}\right)_V = \lim_{\Delta T \to 0} \frac{P(V, T + \Delta T) - P(V, T)}{\Delta T} = \frac{R}{V} \qquad (3.2)$$

The subscript T in $(\partial P/\partial V)_T$ indicates that T is being held constant in the differentiation with respect to V. The partial derivatives in Equation (3.2) allow one to determine how a function changes when the variables changes. For example, what is the change in P if the values of T and V both change? In this case, P changes to $P + dP$ where

$$dP = \left(\frac{\partial P}{\partial T}\right)_V dT + \left(\frac{\partial P}{\partial V}\right)_T dV \qquad (3.3)$$

Consider the following practical illustration of Equation (3.3). You are on a hill and have determined your altitude above sea level. How much will the altitude (denoted z) change if you move a small distance east (denoted by x) and north (denoted by y)? The change in z as you move east is the slope of the hill in that direction, $(\partial z/\partial x)_y$, multiplied by the distance that you move. A similar expression can be written for the change in altitude as you move north. Therefore, the total change in altitude is the sum of these two changes or

$$dz = \left(\frac{\partial z}{\partial x}\right)_y dx + \left(\frac{\partial z}{\partial y}\right)_x dy$$

Because the slope of the hill is a function of x and y, this expression for dz is only valid for small changes dx and dy. Otherwise, higher order derivatives need to be considered.

Second or higher derivatives with respect to either variable can also be taken. The mixed second partial derivatives are of particular interest. Consider the mixed partial derivatives of P:

$$\left(\frac{\partial}{\partial T}\left(\frac{\partial P}{\partial V}\right)_T\right)_V = \frac{\partial^2 P}{\partial T \partial V} = \left(\frac{\partial\left(\frac{\partial\left[\frac{RT}{V}\right]}{\partial V}\right)_T}{\partial T}\right)_V = \left(\frac{\partial\left[-\frac{RT}{V^2}\right]}{\partial T}\right)_V = -\frac{R}{V^2} \qquad (3.4)$$

$$\left(\frac{\partial}{\partial V}\left(\frac{\partial P}{\partial T}\right)_V\right)_T = \frac{\partial^2 P}{\partial T \partial V} = \left(\frac{\partial\left(\frac{\partial\left[\frac{RT}{V}\right]}{\partial T}\right)_V}{\partial V}\right)_T = \left(\frac{\partial\left[\frac{R}{V}\right]}{\partial V}\right)_T = -\frac{R}{V^2}$$

For all state functions f and for our specific case of P, the order in which the function is differentiated does not affect the outcome. For this reason,

$$\left(\frac{\partial}{\partial T}\left(\frac{\partial f(V, T)}{\partial V}\right)_T\right)_V = \left(\frac{\partial}{\partial V}\left(\frac{\partial f(V, T)}{\partial T}\right)_V\right)_T \qquad (3.5)$$

Because Equation (3.5) is only satisfied by state functions f, it can be used to determine if a function f is a state function. If f is a state function, one can write $\Delta f = \int_i^f df = f_{final} - f_{initial}$. This equation states that f can be expressed as an infinitesimal quantity, df, that when integrated depends only on the initial and final states; df is called an **exact differential.** An example of a state function and its exact differential are U and $dU = dq - P_{external}\, dV$.

EXAMPLE PROBLEM 3.1

a. Calculate

$$\left(\frac{\partial f}{\partial x}\right)_y, \left(\frac{\partial f}{\partial y}\right)_x, \left(\frac{\partial^2 f}{\partial x^2}\right)_y, \left(\frac{\partial^2 f}{\partial y^2}\right)_x, \left(\frac{\partial \left(\frac{\partial f}{\partial x}\right)_y}{\partial y}\right)_x, \text{ and } \left(\frac{\partial \left(\frac{\partial f}{\partial y}\right)_x}{\partial x}\right)_y$$

for the function $f(x, y) = ye^x + xy + x \ln y$.

b. Determine if $f(x,y)$ is a state function of the variables x and y.

c. If $f(x,y)$ is a state function of the variables x and y, what is the total differential df?

Solution

a. $\left(\dfrac{\partial f}{\partial x}\right)_y = ye^x + y + \ln y,$ $\qquad \left(\dfrac{\partial f}{\partial y}\right)_x = e^x + x + \dfrac{x}{y}$

$\left(\dfrac{\partial^2 f}{\partial x^2}\right)_y = ye^x,$ $\qquad\qquad\qquad \left(\dfrac{\partial^2 f}{\partial y^2}\right)_x = -\dfrac{x}{y^2}$

$\left(\dfrac{\partial \left(\frac{\partial f}{\partial x}\right)_y}{\partial y}\right)_x = e^x + 1 + \dfrac{1}{y},$ $\qquad \left(\dfrac{\partial \left(\frac{\partial f}{\partial y}\right)_x}{\partial x}\right)_y = e^x + 1 + \dfrac{1}{y}$

b. Because we have shown that

$$\left(\frac{\partial \left(\frac{\partial f}{\partial x}\right)_y}{\partial y}\right)_x = \left(\frac{\partial \left(\frac{\partial f}{\partial y}\right)_x}{\partial x}\right)_y$$

$f(x,y)$ is a state function of the variables x and y. Note that any well-behaved function that can be expressed in analytical form is a state function.

c. The total differential is given by

$$df = \left(\frac{\partial f}{\partial x}\right)_y dx + \left(\frac{\partial f}{\partial y}\right)_x dy$$

$$= (ye^x + y + \ln y)dx + \left(e^x + x + \frac{x}{y}\right)dy$$

Two other important results from differential calculus will be used frequently. Consider a function, $z = f(x, y)$, which can be rearranged to $x = g(y, z)$ or $y = h(x, z)$. For example, if $P = nRT/V$, then $V = nRT/P$ and $T = PV/nR$. In this case

$$\left(\frac{\partial x}{\partial y}\right)_z = \frac{1}{\left(\frac{\partial y}{\partial x}\right)_z} \qquad\qquad (3.6)$$

The **cyclic rule** will also be used:

$$\left(\frac{\partial x}{\partial y}\right)_z \left(\frac{\partial y}{\partial z}\right)_x \left(\frac{\partial z}{\partial x}\right)_y = -1 \qquad\qquad (3.7)$$

Equations (3.6) and (3.7) can be used to reformulate Equation (3.3):

$$dP = \left(\frac{\partial P}{\partial T}\right)_V dT + \left(\frac{\partial P}{\partial V}\right)_T dV$$

TABLE 3.1

Isothermal Coefficient of Expansion for Solids and Liquids at 298 K

Element	$10^6\,\beta$ (K^{-1})	Element or Compound	$10^4\,\beta$ (K^{-1})
Ag(s)	18.9	Hg(l)	18.1
Al(s)	23.1	CCl$_4$(l)	11.4
Au(s)	14.2	CH$_3$COCH$_3$(l)	14.6
Cu(s)	16.5	CH$_3$OH(l)	14.9
Fe(s)	11.8	C$_2$H$_5$OH(l)	11.2
Mg(s)	24.8	C$_6$H$_5$CH$_3$(l)	10.5
Si(s)	2.6	C$_6$H$_6$(l)	12.1
W(s)	4.5	H$_2$O(l)	2.04
Zn(s)	30.2	H$_2$O(s)	1.66

Sources: Benenson, W., Harris, J. W., Stocker, H., and Lutz, H., *Handbook of Physics, Springer*, New York, 2002; Lide, D. R., Ed., *Handbook of Chemistry and Physics*, 83rd ed., CRC Press, Boca Raton, FL, 2002; Blachnik, R., Ed., *D'Ans Lax Taschenbuch für Chemiker und Physiker*, 4th ed., Springer, Berlin. 1998.

Suppose this expression needs to be evaluated for a specific substance, such as N$_2$ gas. What quantities must be measured in the laboratory in order to obtain numerical values for $(\partial P/\partial T)_V$ and $(\partial P/\partial V)_T$? Using Equations (3.6) and (3.7),

$$\left(\frac{\partial P}{\partial V}\right)_T \left(\frac{\partial V}{\partial T}\right)_P \left(\frac{\partial T}{\partial P}\right)_V = -1 \tag{3.8}$$

$$\left(\frac{\partial P}{\partial T}\right)_V = -\left(\frac{\partial P}{\partial V}\right)_T \left(\frac{\partial V}{\partial T}\right)_P = -\frac{\left(\frac{\partial V}{\partial T}\right)_P}{\left(\frac{\partial V}{\partial P}\right)_T} = \frac{\beta}{\kappa} \quad \text{and}$$

$$\left(\frac{\partial P}{\partial V}\right)_T = -\frac{1}{\kappa V}$$

where β and κ are the readily measured **volumetric thermal expansion coefficient** and the **isothermal compressibility,** respectively, defined by

$$\beta = \frac{1}{V}\left(\frac{\partial V}{\partial T}\right)_P \quad \text{and} \quad \kappa = -\frac{1}{V}\left(\frac{\partial V}{\partial P}\right)_T \tag{3.9}$$

Both $(\partial V/\partial T)_P$ and $(\partial V/\partial P)_T$ can be measured by determining the change in volume of the system when the pressure and temperature are varied, while keeping the second variable constant.

The minus sign in the equation for κ is chosen so that values of the isothermal compressibility are positive. For small changes in T and P, Equations (3.9) can be written in the more compact form $V(T_2) = V(T_1)(1 + \beta[T_2 - T_1])$ and $V(P_2) = V(P_1)(1 - \kappa[P_2 - P_1])$. Values for β and κ for selected solids and liquids are shown in Tables 3.1 and 3.2, respectively.

Equation (3.8) is an example of how seemingly abstract partial derivatives can be directly linked to experimentally determined quantities using the mathematical properties of state functions. Using the definitions of β and κ, Equation (3.3) can be written in the form

$$dP = \frac{\beta}{\kappa} dT - \frac{1}{\kappa V} dV \tag{3.10}$$

TABLE 3.2
Isothermal Compressibility at 298 K

Substance	$10^6\, \kappa\,/\mathrm{bar}^{-1}$	Substance	$10^6\, \kappa\,/\mathrm{bar}^{-1}$
Al(s)	1.33	$Br_2(l)$	64
$SiO_2(s)$	2.57	$C_2H_5OH(l)$	110
Ni(s)	0.513	$C_6H_5OH(l)$	61
$TiO_2(s)$	0.56	$C_6H_6(l)$	94
Na(s)	13.4	$CCl_4(l)$	103
Cu(s)	0.702	$CH_3COCH_3\,(l)$	125
C(graphite)	0.156	$CH_3COCH_3(l)$	126
Mn(s)	0.716	$CH_3OH(l)$	120
Co(s)	0.525	$CS_2(l)$	92.7
Au(s)	0.563	$H_2O(l)$	45.9
Pb(s)	2.37	Hg(l)	3.91
Fe(s)	0.56	$SiCl_4(l)$	165
Ge(s)	1.38	$TiCl_4(l)$	89

Sources: Benenson, W., Harris, J. W., Stocker, H., and Lutz, H., *Handbook of Physics*, Springer, New York, 2002; Lide, D. R., Ed., *Handbook of Chemistry and Physics*, 83rd ed., CRC Press, Boca Raton, FL, 2002; Blachnik, R., Ed., *D'Ans Lax Taschenbuch für Chemiker und Physiker*, 4th ed., Springer, Berlin. 1998.

which can be integrated to give

$$\Delta P = \int_{T_i}^{T_f} \frac{\beta}{\kappa}\,dT - \int_{V_i}^{V_f} \frac{1}{\kappa V}\,dV \approx \frac{\beta}{\kappa}(T_f - T)_i - \frac{1}{\kappa}\ln\frac{V_f}{V_i} \tag{3.11}$$

The second expression in Equation (3.11) holds if ΔT and ΔV are small enough that β and κ are constant over the range of integration. Example Problem 3.2 shows a useful application of this equation.

EXAMPLE PROBLEM 3.2

You have accidentally arrived at the end of the range of an ethanol in glass thermometer so that the entire volume of the glass capillary is filled. By how much will the pressure in the capillary increase if the temperature is increased by another 10.0°C? $\beta_{glass} = 4.00 \times 10^{-4}\,(^\circ\mathrm{C})^{-1}$, $\beta_{ethanol} = 11.2 \times 10^{-4}\,(^\circ\mathrm{C})^{-1}$, and $\kappa_{ethanol} = 11.0 \times 10^{-5}\,(\mathrm{bar})^{-1}$. Do you think that the thermometer will survive your experiment?

Solution
Using Equation (3.11),

$$\Delta P = \int \frac{\beta_{ethanol}}{\kappa}\,dT - \int \frac{1}{\kappa V}\,dV \approx \frac{\beta_{ethanol}}{\kappa}\Delta T - \frac{1}{\kappa}\ln\frac{V_f}{V_i}$$

$$= \frac{\beta_{ethanol}}{\kappa}\Delta T - \frac{1}{\kappa}\ln\frac{V_i(1 + \beta_{glass}\Delta T)}{V_i} \approx \frac{\beta_{ethanol}}{\kappa}\Delta T - \frac{1}{\kappa}\frac{V_i\beta_{glass}\Delta T}{V_i}$$

$$= \frac{(\beta_{ethanol} - \beta_{glass})}{\kappa}\Delta T$$

$$= \frac{(11.2 - 4.00) \times 10^{-4}\,(^\circ\mathrm{C})^{-1}}{11.0 \times 10^{-5}\,(\mathrm{bar})^{-1}} \times 10.0\,^\circ\mathrm{C} = 65.5\ \mathrm{bar}$$

In this calculation, we have used the relations $V(T_2) = V(T_1)(1 + \beta[T_2 - T_1])$ and $\ln(1 + x) \approx x$ if $x \ll 1$.

The glass is unlikely to withstand such a large increase in pressure.

3.2 The Dependence of U on V and T

In this section, the fact that dU is an exact differential is used to establish how U varies with T and V. For a given amount of a pure substance or a mixture of fixed composition, U is determined by any two of the three variables P, V, and T. One could choose other combinations of variables to discuss changes in U. However, the following discussion will demonstrate that it is particularly convenient to choose the variables T and V. Because U is a state function, an infinitesimal change in U can be written as

$$dU = \left(\frac{\partial U}{\partial T} \right)_V dT + \left(\frac{\partial U}{\partial V} \right)_T dV \qquad (3.12)$$

This expression says that if the state variables change from T,V to $T + dT, V + dV$, the change in U, dU can be determined in the following way. We determine the slopes of $U(T,V)$ with respect to T and V and evaluate them at T,V. Next, these slopes are multiplied by the increments dT and dV, respectively, and the two terms are added. As long as dT and dV are infinitesimal quantities, higher order derivatives can be neglected.

How can numerical values for $(\partial U/\partial T)_V$ and $(\partial U/\partial V)_T$ be obtained? In the following, we only consider $P-V$ work. Combining Equation (3.12) and the differential expression of the first law,

$$dq - P_{external} dV = \left(\frac{\partial U}{\partial T} \right)_V dT + \left(\frac{\partial U}{\partial V} \right)_T dV \qquad (3.13)$$

The symbol dq is used for an infinitesimal amount of heat as a reminder that heat is not a state function. We first consider processes at constant volume for which $dV = 0$, so that Equation (3.13) becomes

$$dq_V = \left(\frac{\partial U}{\partial T} \right)_V dT \qquad (3.14)$$

Note that in the previous equation, dq_V is the product of a state function and an exact differential. Therefore, dq_V behaves like a state function, but only because the path (constant V) is specified. The quantity dq is *not* a state function.

Although the quantity $(\partial U/\partial T)_V$ looks very abstract, it can be readily measured. For example, imagine immersing a container with rigid diathermal walls in a water bath, where the contents of the container are the system. A process such as a chemical reaction is carried out in the container and the heat flow to the surroundings is measured. If heat flow, dq_V, occurs, a temperature increase or decrease, dT, is observed in the system and the water bath surroundings. Both of these quantities can be measured. Their ratio, dq_V/dT, is a special form of the heat capacity discussed in Section 2.4:

$$\frac{dq_V}{dT} = \left(\frac{\partial U}{\partial T} \right)_V = C_V \qquad (3.15)$$

where dq_V/dT corresponds to a constant volume path and is called the **heat capacity at constant volume.**

The quantity C_V is extensive and depends on the size of the system, whereas $C_{V,m}$ is an intensive quantity. Experiments show that $C_{V,m}$ has different numerical values for dif-

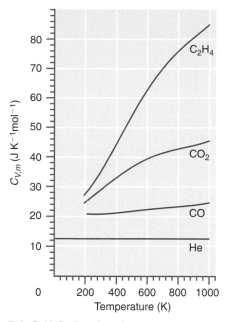

FIGURE 3.1
Molar heat capacities $C_{V,m}$ are shown
for a number of gases. Atoms have only
translational degrees of freedom and,
therefore, have comparatively low
values for $C_{V,m}$ that are independent of
temperature. Molecules with vibrational
degrees of freedom have higher values of
$C_{V,m}$, at temperatures sufficiently high to
activate the vibrations.

ferent substances under the same conditions. Observations show that $C_{V,m}$ is always positive for a single-phase, pure substance or for a mixture of fixed composition, as long as no chemical reactions or phase changes take place in the system. For processes subject to these constraints, U increases monotonically with T. Figure 3.1 shows $C_{V,m}$ as a function of temperature for several gaseous atoms and molecules.

Before continuing with our discussion about how to experimentally determine changes in U with T at constant V for systems of pure substances or for mixtures, let's study heat capacities in a little more detail by looking at them in terms of a microscopic model.

In thermodynamics, the origin for the substance dependence of $C_{V,m}$ is not a matter of inquiry, because thermodynamics is not concerned with the microscopic structure of the system. To obtain numerical results in thermodynamics, system-dependent properties such as $C_{V,m}$ are obtained from experiment or theory. A microscopic model is required to explain why $C_{V,m}$ for a particular substance has its measured value. For example, why is $C_{V,m}$ smaller for gaseous He than for gaseous methanol? To increase the temperature by an amount dT for a system containing helium, the translational energy of the atoms is increased. By contrast, to give the same increase dT in a system containing methanol, the rotational, vibrational, and translational energies of the molecules are all increased simultaneously, because all these degrees of freedom are in equilibrium with one another. Therefore, for a given temperature increment dT, more energy must be added to a mole of methanol than to a mole of He. For this reason, $C_{V,m}$ is larger for gaseous methanol than for gaseous He.

The preceding explanation is valid for high T, but at low T, additional considerations apply. Far less energy is required to excite molecular rotations than molecular vibrations. Therefore, the rotational degrees of freedom contribute to $C_{V,m}$ for molecules even at low temperatures, making $C_{V,m}$ greater for CO than for He at 300 K. However, vibrational degrees of freedom are only excited at higher T. This leads to the additional gradual increase in $C_{V,m}$ observed for CO and CO_2 relative to He for $T > 300$ K. Because the number of vibrational degrees of freedom increases with the number of atoms in the molecule, $C_{V,m}$ is larger for a polyatomic molecule such as C_2H_4 than for CO and CO_2 at a given temperature.

With the definition of C_V, we now have a way to experimentally determine changes in U with T at constant V for systems of pure substances or for mixtures of constant composition in the absence of chemical reactions or phase changes. After C_V has been determined as a function of T as discussed in Section 2.4, the integral is numerically evaluated:

$$\Delta U_V = \int_{T_1}^{T_2} C_V dT = n \int_{T_1}^{T_2} C_{V,m} dT \tag{3.16}$$

Over a limited temperature range, $C_{V,m}$ can often be regarded as a constant. If this is the case, Equation (3.16) simplifies to

$$\Delta U_V = \int_{T_1}^{T_2} C_V dT = C_V \Delta T = n C_{V,m} \Delta T \tag{3.17}$$

which can be written in a different form to explicitly relate q_V and ΔU:

$$\int_i^f dq_V = \int_i^f \left(\frac{\partial U}{\partial T} \right)_V dT \quad \text{or} \quad q_V = \Delta U \tag{3.18}$$

Although dq is not an exact differential, the integral has a unique value if the path is defined, as it is in this case (constant volume). Equation (3.18) shows that ΔU for an arbitrary process in a closed system in which only P–V work occurs can be determined by measuring q under constant volume conditions. As discussed in Chapter 4, the technique of bomb calorimetry uses this approach to determine ΔU for chemical reactions.

Next consider the dependence of U on V at constant T, or $(\partial U/\partial V)_T$. This quantity has the units of $J/m^3 = (J/m)/m^2 = kg\,m\,s^{-2}/m^2 = force/area = pressure$, and is called the **internal pressure.** To explicitly evaluate the internal pressure for different substances, a result will be used that is derived in the discussion of the second law of thermodynamics in Section 5.12:

$$\left(\frac{\partial U}{\partial V}\right)_T = T\left(\frac{\partial P}{\partial T}\right)_V - P \tag{3.19}$$

Using this equation, the total differential of the internal energy can be written as

$$dU = dU_V + dU_T = C_V\, dT + \left[T\left(\frac{\partial P}{\partial T}\right)_V - P\right]dV \tag{3.20}$$

In this equation, the symbols dU_V and dU_T have been used, where the subscript indicates which variable is constant. Equation (3.20) is an important result that applies to systems containing gases, liquids, or solids in a single phase (or mixed phases at a constant composition) if no chemical reactions or phase changes occur. The advantage of writing dU in the form given by Equation (3.20) over that in Equation (3.12) is that $(\partial U/\partial V)_T$ can be evaluated in terms of the system variables P, V, and T and their derivatives, all of which are experimentally accessible.

Once $(\partial U/\partial V)_T$ and $(\partial U/\partial T)_V$ are known, these quantities can be used to determine dU. Because U is a state function, the path taken between the initial and final states is unimportant. Three different paths are shown in Figure 3.2, and dU is the same for these and any other paths connecting V_i,T_i and V_f,T_f. To simplify the calculation, the path chosen consists of two segments, in which only one of the variables changes in a given path segment. An example of such a path is $V_i,T_i \rightarrow V_f,T_i \rightarrow V_f,T_f$. Because T is constant in the first segment,

$$dU = dU_T = \left[T\left(\frac{\partial P}{\partial T}\right)_V - P\right]dV$$

Because V is constant in the second segment, $dU = dU_V = C_V dT$. Finally, the total change in dU is the sum of the changes in the two segments.

3.3 Does the Internal Energy Depend More Strongly on V or T?

Chapter 2 demonstrated that U is a function of T alone for an ideal gas. However, this statement is not true for real gases, liquids, and solids for which the change in U with V must be considered. In this section, we ask if the temperature or the volume dependence of U is most important in determining ΔU for a process of interest. To answer this question, systems consisting of an ideal gas, a real gas, a liquid, and a solid are considered separately. Example Problem 3.3 shows that Equation (3.19) leads to a simple result for a system consisting of an ideal gas.

EXAMPLE PROBLEM 3.3

Evaluate $(\partial U/\partial V)_T$ for an ideal gas and modify Equation (3.20) accordingly for the specific case of an ideal gas.

Solution

$$T\left(\frac{\partial P}{\partial T}\right)_V - P = T\left(\frac{\partial\left[nRT/V\right]}{\partial T}\right)_V - P = \frac{nRT}{V} - P = 0$$

Therefore, $dU = C_V dT$, showing that for an ideal gas, U is a function of T only.

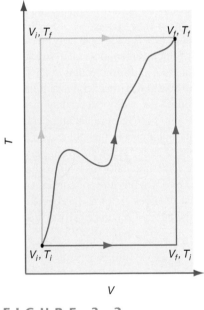

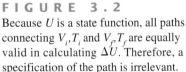

FIGURE 3.2

Because U is a state function, all paths connecting V_i,T_i and V_f,T_f are equally valid in calculating ΔU. Therefore, a specification of the path is irrelevant.

Example Problem 3.3 shows that U is a function of T only for an ideal gas. Specifically, U is not a function of V. This result is understandable in terms of the potential function of Figure 1.6. Because ideal gas molecules do not attract or repel one another, no energy is required to change their average distance of separation (increase or decrease V):

$$\Delta U = \int_{T_i}^{T_f} C_V(T) \, dT \tag{3.21}$$

Recall that because U is a function of T only, Equation (3.21) holds for an ideal gas even if V is not constant.

Next consider the variation of U with T and V for a real gas. The experimental determination of $(\partial U / \partial V)_T$ was carried out by James Joule using an apparatus consisting of two glass flasks separated by a stopcock, all of which were immersed in a water bath. An idealized view of the experiment is shown in Figure 3.3. As a valve between the volumes is opened, a gas initially in volume A expands to completely fill the volume A + B. In interpreting the results of this experiment, it is important to understand where the boundary between the system and surroundings lies. Here, the decision was made to place the system boundary so that it includes all the gas. Initially, the boundary lies totally within V_A, but it moves during the expansion so that it continues to include all gas molecules. With this choice, the volume of the system changes from V_A before the expansion to $V_A + V_B$ after the expansion has taken place.

The first law of thermodynamics [Equation (3.13)] states that

$$dq - P_{external} \, dV = \left(\frac{\partial U}{\partial T}\right)_V dT + \left(\frac{\partial U}{\partial V}\right)_T dV$$

However, all the gas is contained in the system; therefore, $P_{external} = 0$ because a vacuum cannot exert a pressure. Therefore Equation (3.13) becomes

$$dq = \left(\frac{\partial U}{\partial T}\right)_V dT + \left(\frac{\partial U}{\partial V}\right)_T dV \tag{3.22}$$

To within experimental accuracy, Joule found that $dT_{surroundings} = 0$. Because the water bath and the system are in thermal equilibrium, $dT = dT_{surroundings} = 0$. With this observation, Joule concluded that $dq = 0$. Therefore, Equation (3.22) becomes

$$\left(\frac{\partial U}{\partial V}\right)_T dV = 0 \tag{3.23}$$

Because $dV \neq 0$, Joule concluded that $(\partial U / \partial V)_T = 0$. Joule's experiment was not definitive because the experimental sensitivity was limited, as shown in Example Problem 3.4.

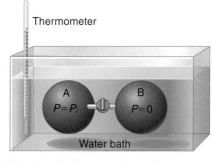

FIGURE 3.3
Schematic depiction of the Joule experiment to determine $(\partial U / \partial V)_T$. Two spherical vessels, A and B, are separated by a valve. Both vessels are immersed in a water bath, the temperature of which is monitored. The initial pressure in each vessel is indicated.

EXAMPLE PROBLEM 3.4

In Joule's experiment to determine $(\partial U / \partial V)_T$, the heat capacities of the gas and the water bath surroundings were related by $C_{surroundings} / C_{system} \approx 1000$. If the precision with which the temperature of the surroundings could be measured is $\pm 0.006°C$, what is the minimum detectable change in the temperature of the gas?

Solution
View the experimental apparatus as two interacting systems in a rigid adiabatic enclosure. The first is the volume within vessels A and B, and the second is the

water bath and the vessels. Because the two interacting systems are isolated from the rest of the universe,

$$q = C_{water\ bath} \Delta T_{water\ bath} + C_{gas} \Delta T_{gas} = 0$$

$$\Delta T_{gas} = -\frac{C_{water\ bath}}{C_{gas}} \Delta T_{water\ bath} = -1000 \times (\pm 0.006°C) = \mp 6°C$$

In this calculation, ΔT_{gas} is the temperature change that the expanded gas undergoes to reach thermal equilibrium with the water bath, which is the negative of the temperature change during the expansion.

Because the minimum detectable value of ΔT_{gas} is rather large, this apparatus is clearly not suited for measuring small changes in the temperature of the gas in an expansion.

More sensitive experiments were carried out by Joule in collaboration with William Thompson (Lord Kelvin). These experiments, which are discussed in Section 3.8, demonstrate that $(\partial U/\partial V)_T$ is small, but nonzero for real gases.

Example Problem 3.3 has shown that $(\partial U/\partial V)_T = 0$ for an ideal gas. We next calculate $(\partial U/\partial V)_T$ and $\Delta U_T = \int_{V_{m,i}}^{V_{m,f}} (\partial U/\partial V)_T dV_m$ for a real gas, in which the van der Waals equation of state is used to describe the gas, as illustrated in Example Problem 3.5.

EXAMPLE PROBLEM 3.5

For a gas described by the van der Waals equation of state, $P = RT/(V_m - b) - a/V_m^2$. Use this equation to complete these tasks:

a. Calculate $(\partial U/\partial V)_T$ using $(\partial U/\partial V)_T = T(\partial P/\partial T)_V - P$.

b. Derive an expression for the change in internal energy, $\Delta U_T = \int_{V_{m,i}}^{V_{m,f}} (\partial U/\partial V)_T dV_m$, in compressing a van der Waals gas from an initial molar volume $V_{m,i}$ to a final molar volume $V_{m,f}$ at constant temperature.

Solution

a. $\displaystyle T\left(\frac{\partial P}{\partial T}\right)_V - P = T\left(\frac{\partial\left[\dfrac{RT}{V_m - b} - \dfrac{a}{V_m^2}\right]}{\partial T}\right)_V - P = \frac{RT}{V_m - b} - P$

$\displaystyle = \frac{RT}{V_m - b} - \frac{RT}{V_m - b} + \frac{a}{V_m^2} = \frac{a}{V_m^2}$

b. $\displaystyle \Delta U_{T,m} = \int_{V_{m,i}}^{V_{m,f}} \left(\frac{\partial U_m}{\partial V}\right)_T dV_m = \int_{V_{m,i}}^{V_{m,f}} \frac{a}{V_m^2} dV_m = a\left(\frac{1}{V_{m,i}} - \frac{1}{V_{m,f}}\right)$

Note that $\Delta U_{T,m}$ is zero if the attractive part of the intermolecular potential is zero.

Example Problem 3.5 demonstrates that $(\partial U/\partial V)_T \neq 0$, and that $\Delta U_{T,m}$ can be calculated if the equation of state of the real gas is known. This allows the relative importance of $\Delta U_{T,m} = \int_{V_{m,i}}^{V_{m,f}} (\partial U_m/\partial V)_T dV_m$ and $\Delta U_{V,m} = \int_{T_i}^{T_f} C_{V,m} dT$ to be determined in a process in which both T and V change, as shown in Example Problem 3.6.

EXAMPLE PROBLEM 3.6

A sample of N_2 gas undergoes a change from an initial state described by $T = 200$ K and $P_i = 5.00$ bar to a final state described by $T = 400$ K and $P_f = 20.0$ bar. Treat N_2 as a van der Waals gas with the parameters $a = 0.137$ Pa m^6 mol^{-2} and $b = 3.87 \times 10^{-5}$ m^3 mol^{-1}. We use the path N_2 (g, $T = 200$ K, $P = 5.00$ bar) → N_2(g, $T = 200$ K, $P = 20.0$ bar) → N_2(g, $T = 400$ K, $P = 20.0$ bar), keeping in mind that all paths will give the same answer for ΔU of the overall process.

a. Calculate $\Delta U_{T,m} = \int_{V_{m,i}}^{V_{m,f}} (\partial U_m / \partial V)_T dV_m$ using the result of Example Problem 3.5. Note that $V_{m,i} = 3.28 \times 10^{-3}$ m^3 mol^{-1} and $V_{m,f} = 7.88 \times 10^{-4}$ m^3 mol^{-1} at 200 K, as calculated using the van der Waals equation of state.

b. Calculate $\Delta U_{V,m} = \int_{T_i}^{T_f} C_{V,m} dT$ using the following relationship for $C_{V,m}$ in this temperature range:

$$\frac{C_{V,m}}{\mathrm{J\,K^{-1}\,mol^{-1}}} = 22.50 - 1.187 \times 10^{-2} \frac{T}{\mathrm{K}} + 2.3968 \times 10^{-5} \frac{T^2}{\mathrm{K^2}} - 1.0176 \times 10^{-8} \frac{T^3}{\mathrm{K^3}}$$

The ratios T^n / K^n ensure that $C_{V,m}$ has the correct dimension.

c. Compare the two contributions to ΔU_m. Can $\Delta U_{T,m}$ be neglected relative to $\Delta U_{V,m}$?

Solution

a. Using the result of Example Problem 3.5,

$$\Delta U_{T,m} = a\left(\frac{1}{V_{m,i}} - \frac{1}{V_{m,f}}\right) = 0.137 \text{ Pa m}^6 \text{ mol}^{-3}$$

$$\times \left(\frac{1}{3.28 \times 10^{-3} \text{ m}^3 \text{ mol}^{-1}} - \frac{1}{7.88 \times 10^{-4} \text{ m}^3 \text{ mol}^{-1}}\right) = -132 \text{ J}$$

b. $$\Delta U_{V,m} = \int_{T_i}^{T_f} C_{V,m} dT = \int_{200}^{400} \left(22.50 - 1.187 \times 10^{-2} \frac{T}{\mathrm{K}} + 2.3968 \times 10^{-5} \frac{T^2}{\mathrm{K^2}} - 1.0176 \times 10^{-8} \frac{T^3}{\mathrm{K^3}} \right)$$

$$\times d\left(\frac{T}{\mathrm{K}}\right) \mathrm{J\ mol^{-1}}$$

$$= (4.50 - 0.712 + 0.447 - 0.0610) \mathrm{kJ\ mol^{-1}} = 4.17 \mathrm{\ kJ\ mol^{-1}}$$

c. $\Delta U_{T,m}$ is 3.1% of $\Delta U_{V,m}$ for this case. In this example, and for most processes, $\Delta U_{T,m}$ can be neglected relative to $\Delta U_{V,m}$ for real gases.

The calculations in Example Problems 3.5 and 3.6 show that to a good approximation $\Delta U_{T,m} = \int_{V_{m,i}}^{V_{m,f}} (\partial U_m / \partial V)_T dV_m \approx 0$ for real gases under most conditions. Therefore, it is sufficiently accurate to consider U as a function of T only [$U = U(T)$] for real gases in processes that do not involve unusually high gas densities.

Having discussed ideal and real gases, what can be said about the relative magnitude of $\Delta U_{T,m} = \int_{V_{m,i}}^{V_{m,f}} (\partial U_m / \partial V)_T dV_m$ and $\Delta U_{V,m} = \int_{T_i}^{T_f} C_{V,m} dT$ for processes involving liquids and solids? From experiments, it is known that the density of liquids and solids varies only slightly with the external pressure over the range in which these two forms of matter are stable. This conclusion is not valid for extremely high pressure conditions such as those in the interior of planets and stars. However, it is safe to say that dV for a solid or liquid is very small in most processes. Therefore,

$$\Delta U_T^{solid, liq} = \int_{V_1}^{V_2} \left(\frac{\partial U}{\partial V}\right)_T dV \approx \left(\frac{\partial U}{\partial V}\right)_T \Delta V \approx 0 \qquad (3.24)$$

because $\Delta V \approx 0$. This result is valid even if $(\partial U / \partial V)_T$ is large.

The conclusion that can be drawn from this section is as follows. Under most conditions encountered by chemists in the laboratory, U can be regarded as a function of T alone for all substances. The following equations gives a good approximation even if V is not constant in the process under consideration:

$$U(T_f, V_f) - U(T_i, V_i) = \Delta U = \int_{T_i}^{T_f} C_V \, dT = n \int_{T_i}^{T_f} C_{V,m} \, dT \qquad (3.25)$$

Note that Equation (3.25) is only applicable to a process in which there is no change in the phase of the system, such as vaporization or fusion, and in which there are no chemical reactions. Changes in U that arise from these processes will be discussed in Chapters 4 and 6.

3.4 The Variation of Enthalpy with Temperature at Constant Pressure

As for U, H can be defined as a function of any two of the three variables P, V, and T. It was convenient to choose U to be a function of T and V because this choice led to the identity $\Delta U = q_V$. Using a similar reasoning, we choose H to be a function of T and P. How does H vary with P and T? The variation of H with T at constant P is discussed next, and a discussion of the variation of H with P at constant T is deferred to Section 3.6.

Consider the constant pressure process shown schematically in Figure 3.4. For this process defined by $P = P_{external}$,

$$dU = dq_P - P \, dV \qquad (3.26)$$

Although the integral of dq is in general path dependent, it has a unique value in this case because the path is specified, namely, $P = P_{external} = $ constant. Integrating both sides of Equation (3.26),

$$\int_i^f dU = \int_i^f dq_P - \int_i^f P \, dV \quad \text{or} \quad U_f - U_i = q_P - P(V_f - V_i) \qquad (3.27)$$

Because $P = P_f = P_i$, this equation can be rewritten as

$$(U_f + P_f V_f) - (U_i + P_i V_i) = q_P \quad \text{or} \quad \boxed{\Delta H = q_P} \qquad (3.28)$$

The preceding equation shows that the value of ΔH can be determined for an arbitrary process at constant P in a closed system in which only P–V work occurs by simply measuring q_P, the heat transferred between the system and surroundings in a constant pressure process. Note the similarity between Equations (3.28) and (3.18). For an arbitrary process in a closed system in which there is no work other than P–V work, $\Delta U = q_V$ if the process takes place at constant V, and $\Delta H = q_P$ if the process takes place at constant P. These two equations are the basis for the fundamental experimental techniques of bomb calorimetry and constant pressure calorimetry discussed in Chapter 4.

A useful application of Equation (3.28) is in experimentally determining the ΔH and ΔU of fusion and vaporization for a given substance. Fusion (solid → liquid) and vaporization (liquid → gas) occur at a constant temperature if the system is held at a constant pressure and heat flows across the system–surroundings boundary. In both of these phase transitions, attractive interactions between the molecules of the system must be overcome. Therefore, $q > 0$ in both cases and $C_P \to \infty$. Because $\Delta H = q_P$, ΔH_{fusion} and $\Delta H_{vaporization}$ can be determined by measuring the heat needed to effect the transition at constant pressure. Because $\Delta H = \Delta U + \Delta(PV)$, at constant P,

$$\Delta U_{vaporization} = \Delta H_{vaporization} - P \Delta V_{vaporization} > 0 \qquad (3.29)$$

The change in volume upon vaporization is $\Delta V_{vaporization} = V_{gas} - V_{liquid} \gg 0$; therefore, $\Delta U_{vaporization} < \Delta H_{vaporization}$. An analogous expression to Equation (3.29) can be written

$P_{external} = P$

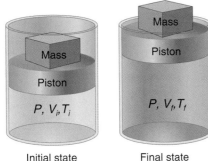

Initial state   Final state

FIGURE 3.4
The initial and final states are shown for an undefined process that takes place at constant pressure.

relating ΔU_{fusion} and ΔH_{fusion}. Note that ΔV_{fusion} is much smaller than $\Delta V_{vaporization}$ and can be either positive or negative. Therefore, $\Delta U_{fusion} \approx \Delta H_{fusion}$. The thermodynamics of fusion and vaporization will be discussed in more detail in Chapter 8.

Because H is a state function, dH is an exact differential, allowing us to link $(\partial H/\partial T)_P$ to a measurable quantity. In analogy to the preceding discussion for dU, dH is written in the form

$$dH = \left(\frac{\partial H}{\partial T}\right)_P dT + \left(\frac{\partial H}{\partial P}\right)_T dP \qquad (3.30)$$

Because $dP = 0$ at constant P, and $dH = \text{đ}q_P$ from Equation (3.28), Equation (3.30) becomes

$$\text{đ}q_P = \left(\frac{\partial H}{\partial T}\right)_P dT \qquad (3.31)$$

Equation (3.31) allows the **heat capacity at constant pressure, C_P,** to be defined in a fashion analogous to C_V in Equation (3.15):

$$C_P \equiv \frac{\text{đ}q_P}{dT} = \left(\frac{\partial H}{\partial T}\right)_P \qquad (3.32)$$

Although this equation looks abstract, C_P is a readily measurable quantity. To measure it, one need only measure the heat flow to or from the surroundings for a constant pressure process together with the resulting temperature change in the limit in which dT and $\text{đ}q$ approach zero and form the ratio $\lim_{dT \to 0} (\text{đ}q/dT)_P$.

As was the case for C_V, C_P is an extensive property of the system and varies from substance to substance. The temperature dependence of C_P must be known in order to calculate the change in H with T. In general, for a constant pressure process in which there is no change in the phase of the system and no chemical reactions,

$$\Delta H_P = \int_{T_i}^{T_f} C_P(T)dT = n\int_{T_i}^{T_f} C_{P,m}(T)dT \qquad (3.33)$$

If the temperature interval is small enough, it can usually be assumed that C_P is constant. In that case,

$$\Delta H_P = C_P \Delta T = nC_{P,m}\Delta T \qquad (3.34)$$

Enthalpy changes that arise from chemical reactions and changes in phase cannot be calculated using Equations (3.33) and (3.34), and the calculation of ΔH for these processes will be discussed in Chapters 4 and 6.

EXAMPLE PROBLEM 3.7

A 143.0-g sample of C(s) in the form of graphite is heated from 300 to 600 K at a constant pressure. Over this temperature range, $C_{P,m}$ has been determined to be

$$\frac{C_{P,m}}{\text{J K}^{-1}\text{mol}^{-1}} = -12.19 + 0.1126\frac{T}{\text{K}} - 1.947\times10^{-4}\frac{T^2}{\text{K}^2} + 1.919\times10^{-7}\frac{T^3}{\text{K}^3}$$

$$- 7.800\times10^{-11}\frac{T^4}{\text{K}^4}$$

Calculate ΔH and q_P. How large is the relative error in ΔH if you neglect the temperature-dependent terms in $C_{P,m}$ and assume that $C_{P,m}$ maintains its value at 300 K throughout the temperature interval?

Solution

$$\Delta H = \frac{m}{M} \int_{T_i}^{T_f} C_{P,m}(T) dT$$

$$= \frac{143.0\,g}{12.00\,g\,mol^{-1}} \frac{J}{mol} \int_{300}^{600} \left(\begin{array}{l} -12.19 + .1126\frac{T}{K} - 1.947 \times 10^{-4} \frac{T^2}{K^2} + 1.919 \\ \times 10^{-7} \frac{T^3}{K^3} - 7.800 \times 10^{-11} \frac{T^4}{K^4} \end{array} \right) d\frac{T}{K}$$

$$= \frac{143.0}{12.00} \times \left[\begin{array}{l} -12.19\frac{T}{K} + 0.0563\frac{T^2}{K^2} - 6.49 \times 10^{-5} \frac{T^3}{K^3} + 4.798 \\ \times 10^{-8} \frac{T^4}{K^4} - 1.56 \times 10^{-11} \frac{T^5}{K^5} \end{array} \right]_{300}^{600} J = 46.85\,kJ$$

From Equation (3.28), $\Delta H = q_P$.

If we had assumed $C_{P,m} = 8.617\,J\,mol^{-1}\,K^{-1}$, which is the calculated value at 300 K, $\Delta H = 143.0\,g/12.00\,g\,mol^{-1} \times 8.617\,J\,K^{-1}\,mol^{-1} \times [600\,K - 300\,K]$ = 30.81 kJ. The relative error is $(30.81\,kJ - 46.85\,kJ)/46.85\,kJ = -34\%$. In this case, it is not reasonable to assume that $C_{P,m}$ is independent of temperature.

3.5 How Are C_P and C_V Related?

To this point, two separate heat capacities, C_P and C_V, have been defined. Are these quantities related? To answer this question, the differential form of the first law is written as

$$dq = C_V\,dT + \left(\frac{\partial U}{\partial V}\right)_T dV + P_{external}\,dV \tag{3.35}$$

Consider a process that proceeds at constant pressure for which $P = P_{external}$. In this case, Equation (3.35) becomes

$$dq_P = C_V\,dT + \left(\frac{\partial U}{\partial V}\right)_T dV + P\,dV \tag{3.36}$$

Because $dq_P = C_P dT$,

$$C_P = C_V + \left(\frac{\partial U}{\partial V}\right)_T \left(\frac{\partial V}{\partial T}\right)_P + P\left(\frac{\partial V}{\partial T}\right)_P = C_V + \left[\left(\frac{\partial U}{\partial V}\right)_T + P\right]\left(\frac{\partial V}{\partial T}\right)_P \tag{3.37}$$

$$= C_V + T\left(\frac{\partial P}{\partial T}\right)_V \left(\frac{\partial V}{\partial T}\right)_P$$

To obtain Equation (3.37), both sides of Equation (3.36) have been divided by dT, and the ratio dV/dT has been converted to a partial derivative at constant P. Equation 3.19 has been used in the last step. Using Equation (3.9) and the cyclic rule, one can simplify Equation (3.37) to

$$C_P = C_V + TV\frac{\beta^2}{\kappa} \quad \text{or} \quad C_{P,m} = C_{V,m} + TV_m\frac{\beta^2}{\kappa} \tag{3.38}$$

Equation (3.38) provides another example of the usefulness of the formal theory of thermodynamics in linking seemingly abstract partial derivatives with experimentally available data. The difference between C_P and C_V can be determined at a given temperature knowing only the molar volume, the coefficient for thermal expansion, and the isothermal compressibility.

Equation (3.38) is next applied to ideal and real gases as well as for liquids and solids, in the absence of phase changes and chemical reactions. Because β and κ are always positive for real and ideal gases, $C_P - C_V > 0$ for these substances. First, $C_P - C_V$ is calculated for an ideal gas, and then it is calculated for liquids and solids. For an ideal gas, $(\partial U / \partial V)_T = 0$ as shown in Example Problem 3.3, and $P(\partial V / \partial T)_P = P(nR/P) = nR$ so that Equation (3.37) becomes

$$C_P - C_V = nR \tag{3.39}$$

This result was stated without derivation in Section 2.4. The partial derivative $(\partial V / \partial T)_P = V\beta$ is much smaller for liquids and solids than for gases. Therefore, generally

$$C_V \gg \left[\left(\frac{\partial U}{\partial V} \right)_T + P \right] \left(\frac{\partial V}{\partial T} \right)_P \tag{3.40}$$

so that $C_P \approx C_V$ for a liquid or a solid. As shown earlier in Example Problem 3.2, it is not feasible to carry out heating experiments for liquids and solids at constant volume because of the large pressure increase that occurs. Therefore, tabulated heat capacity for liquids and solids list $C_{P,m}$ rather than $C_{V,m}$.

3.6 The Variation of Enthalpy with Pressure at Constant Temperature

In the previous section, we learned how H changes with T at constant P. To calculate how H changes as both P and T change, $(\partial H / \partial P)_T$ must be calculated. The partial derivative $(\partial H / \partial P)_T$ is less straightforward to determine in an experiment than $(\partial H / \partial T)_P$. As will be seen, for many processes involving changes in both P and T, $(\partial H / \partial T)_P dT \gg (\partial H / \partial P)_T dP$ and the pressure dependence of H can be neglected relative to its temperature dependence. However, the knowledge that $(\partial H / \partial P)_T$ is not zero is crucial for understanding the operation of a refrigerator and the liquefaction of gases. The following discussion is applicable to gases, liquids, and solids.

Given the definition $H = U + PV$, we begin by writing dH as

$$dH = dU + P\,dV + V\,dP \tag{3.41}$$

Substituting the differential forms of dU and dH,

$$C_P dT + \left(\frac{\partial H}{\partial P} \right)_T dP = C_V\,dT + \left(\frac{\partial U}{\partial V} \right)_T dV + P\,dV + V\,dP \tag{3.42}$$

$$= C_V\,dT + \left[\left(\frac{\partial U}{\partial V} \right)_T + P \right] dV + V\,dP$$

For isothermal processes, $dT = 0$, and Equation (3.42) can be rearranged to

$$\left(\frac{\partial H}{\partial P} \right)_T = \left[\left(\frac{\partial U}{\partial V} \right)_T + P \right] \left(\frac{\partial V}{\partial P} \right)_T + V \tag{3.43}$$

Using Equation (3.19) for $(\partial U / \partial V)_T$,

$$\left(\frac{\partial H}{\partial P} \right)_T = T \left(\frac{\partial P}{\partial T} \right)_V \left(\frac{\partial V}{\partial P} \right)_T + V \tag{3.44}$$

$$= V - T \left(\frac{\partial V}{\partial T} \right)_P$$

The second formulation of Equation (3.44) is obtained through application of the cyclic rule [Equation (3.7)]. This equation is applicable to all systems containing pure substances or mixtures at a fixed composition, provided that no phase changes or chemical reactions take place. The quantity $(\partial H/\partial P)_T$ is evaluated for an ideal gas in Example Problem 3.8.

EXAMPLE PROBLEM 3.8

Evaluate $(\partial H/\partial P)_T$ for an ideal gas.

Solution

$(\partial P/\partial T)_V = (\partial[nRT/V]/\partial T)_V = nR/V$ and $(\partial V/\partial P)_T = RT(d[nRT/P]/dP)_T$ $= -nRT/P^2$ for an ideal gas. Therefore,

$$\left(\frac{\partial H}{\partial P}\right)_T = T\left(\frac{\partial P}{\partial T}\right)_V\left(\frac{\partial V}{\partial P}\right)_T + V = T\frac{nR}{V}\left(-\frac{nRT}{P^2}\right) + V = -\frac{nRT}{P}\frac{nRT}{nRT} + V = 0$$

This result could have been derived directly from the definition $H = U + PV$. For an ideal gas, $U = U(T)$ only and $PV = nRT$. Therefore, $H = H(T)$ only for an ideal gas and $(\partial H/\partial P) = 0$.

Because Example Problem 3.8 shows that H is a function of T only for an ideal gas,

$$\Delta H = \int_{T_i}^{T_f} C_P(T)\,dT = n\int_{T_i}^{T_f} C_{P,m}(T)\,dT \tag{3.45}$$

for an ideal gas. Because H is a function of T only, Equation (3.45) holds for an ideal gas even if P is not constant. This result is also understandable in terms of the potential function of Figure 1.6. Because ideal gas molecules do not attract or repel one another, no energy is required to change their average distance of separation (increase or decrease P).

Equation (3.44) in its general form is next applied to several types of systems. As shown in Example Problem 3.8, $(\partial H/\partial P)_T = 0$ for an ideal gas. For liquids and solids, the first term in Equation (3.44), $T(\partial P/\partial T)_V(\partial V/\partial P)_T$, is usually much smaller than V. This is the case because $(\partial V/\partial P)_T$ is very small, which is consistent with our experience that liquids and solids are difficult to compress. Equation (3.44) establishes that for liquids and solids, $(\partial H/\partial P)_T \approx V$ to a good approximation, and dH can be written as

$$dH \approx C_P\,dT + V\,dP \tag{3.46}$$

for systems that consist only of liquids or solids.

EXAMPLE PROBLEM 3.9

Calculate the change in enthalpy when 124 g of liquid methanol initially at 1.00 bar and 298 K undergoes a change of state to 2.50 bar and 425 K. The density of liquid methanol under these conditions is 0.791 g cm^{-3}, and $C_{P,m}$ for liquid methanol is 81.1 J K^{-1} mol^{-1}.

Solution

Because H is a state function, any path between the initial and final states will give the same ΔH. We choose the path methanol (l, 1.00 bar, 298 K) $\rightarrow$ methanol (l, 1.00 bar, 425 K) $\rightarrow$ methanol (l, 2.50 bar, 425 K). The first step is isothermal, and the second step is isobaric. The total change in H is

$$\Delta H = n\int_{T_i}^{T_f} C_{P,m}dT + \int_{P_i}^{P_f} VdP \approx nC_{P,m}(T_f - T_i) + V(P_f - P_i)$$

$$= 81.1\ \mathrm{J\,K^{-1}\,mol^{-1}} \times \frac{124\ \mathrm{g}}{32.04\ \mathrm{g\,mol^{-1}}} \times (425\ \mathrm{K} - 298\ \mathrm{K})$$

$$+ \frac{124\ \mathrm{g}}{0.791\,\mathrm{g\,cm^{-3}}} \times \frac{10^{-6}\ \mathrm{m^3}}{\mathrm{cm^3}} \times (2.50\ \mathrm{bar} - 1.00\ \mathrm{bar}) \times \frac{10^5\ \mathrm{Pa}}{\mathrm{bar}}$$

$$= 39.9 \times 10^3\ \mathrm{J} + 23.5\ \mathrm{J} \approx 39.9\ \mathrm{kJ}$$

Note that the contribution to ΔH from the change in T is far greater than that from the change in P.

Example Problem 3.9 shows that because molar volumes of liquids and solids are small, H changes much more rapidly with T than with P. Under most conditions, H can be assumed to be a function of T only for solids and liquids. Exceptions to this rule are encountered in geophysical or astrophysical applications, for which extremely large pressure changes can occur.

The conclusion that can be drawn from this section is the following: under most conditions encountered by chemists in the laboratory, H can be regarded as a function of T alone for liquids and solids. It is a good approximation to write

$$H(T_f, P_f) - H(T_i, P_i) = \Delta H = \int_{T_1}^{T_2} C_P dT = n\int_{T_1}^{T_2} C_{P,m} dT \tag{3.47}$$

even if P is not constant in the process under consideration. The dependence of H on P for real gases is discussed in Sections 3.8 and Section 3.9 in connection with the Joule-Thompson experiment.

Note that Equation (3.47) is only applicable to a process in which there is no change in the phase of the system, such as vaporization or fusion, and in which there are no chemical reactions. Changes in H that arise from chemical reactions and changes in phase will be discussed in Chapters 4 and 6.

Having dealt with solids, liquids, and ideal gases, we are left with real gases. For real gases, $(\partial H/\partial P)_T$ and $(\partial U/\partial V)_T$ are small, but still have a considerable effect on the properties of the gases upon expansion or compression. Conventional technology for the liquefaction of gases and for the operation of refrigerators is based on the fact that $(\partial H/\partial P)_T$ and $(\partial U/\partial V)_T$ are not zero for real gases. To derive a formula that will be useful in calculating $(\partial H/\partial P)_T$ for a real gas, the Joule-Thompson experiment is discussed first in the next section.

3.7 The Joule-Thompson Experiment

If the valve on a cylinder of compressed N_2 at 298 K is opened fully, it will become covered with frost, demonstrating that the temperature of the valve is lowered below the freezing point of H_2O. A similar experiment with a cylinder of H_2 leads to a considerable increase in temperature and, potentially, an explosion. How can these effects be understood? To explain them, we discuss the **Joule-Thompson experiment.**

The Joule-Thompson experiment shown in Figure 3.5 can be viewed as an improved version of the Joule experiment because it allows $(\partial U/\partial V)_T$ to be measured

FIGURE 3.5
In the Joule-Thompson experiment, a gas is forced through a porous plug using a piston and cylinder mechanism. The pistons move to maintain a constant pressure in each region. There is an appreciable pressure drop across the plug, and the temperature change of the gas is measured. The upper and lower figures show the initial and final states, respectively. As shown in the text, if the piston and cylinder assembly forms an adiabatic wall between the system (the gases on both sides of the plug) and the surroundings, the expansion is isenthalpic.

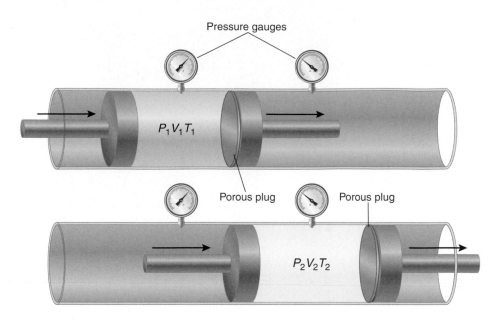

with a much higher sensitivity than in the Joule experiment. In this experiment, gas flows from the high-pressure cylinder on the left to the low-pressure cylinder on the right through a porous plug in an insulated pipe. The pistons move to keep the pressure unchanged in each region until all the gas has been transferred to the region to the right of the porous plug. If N_2 is used in the expansion process ($P_1 > P_2$), it is found that $T_2 < T_1$; in other words, the gas is cooled as it expands. What is the origin of this effect? Consider an amount of gas equal to the initial volume V_1 as it passes through the apparatus from left to right. The total work in this expansion process is the sum of the work performed on each side of the plug separately by the moving pistons:

$$w = w_{left} + w_{right} = -\int_{V_1}^{0} P_1 \, dV - \int_{0}^{V_2} P_2 \, dV = -P_2 V_2 + P_1 V_1 \tag{3.48}$$

Because the pipe is insulated, $q = 0$, and

$$\Delta U = U_2 - U_1 = w = -P_2 V_2 + P_1 V_1 \tag{3.49}$$

This equation can be rearranged to

$$U_2 + P_2 V_2 = U_1 + P_1 V_1 \quad \text{or} \quad H_2 = H_1 \tag{3.50}$$

Note that the enthalpy is constant in the expansion, that is, the expansion is **isenthalpic.** For the conditions of the experiment using N_2, both dT and dP are negative, so $(\partial T/\partial P)_H > 0$. The experimentally determined limiting ratio of ΔT to ΔP at constant enthalpy is known as the **Joule-Thompson coefficient:**

$$\mu_{J-T} = \lim_{\Delta P \to 0} \left(\frac{\Delta T}{\Delta P} \right)_H = \left(\frac{\partial T}{\partial P} \right)_H \tag{3.51}$$

If μ_{J-T} is positive, the conditions are such that the attractive part of the potential dominates, and if μ_{J-T} is negative, the repulsive part of the potential dominates. Using experimentally determined values of μ_{J-T}, $(\partial H/\partial P)_T$ can be calculated. For an isenthalpic process,

$$dH = C_P\, dT + \left(\frac{\partial H}{\partial P}\right)_T dP = 0 \tag{3.52}$$

Dividing through by dP and making the condition $dH = 0$ explicit,

$$C_P \left(\frac{\partial T}{\partial P}\right)_H + \left(\frac{\partial H}{\partial P}\right)_T = 0 \tag{3.53}$$

$$\text{giving} \quad \left(\frac{\partial H}{\partial P}\right)_T = -C_P \mu_{J-T}$$

Equation (3.53) states that $(\partial H/\partial P)_T$ can be calculated using the measurement of material-dependent properties C_P and μ_{J-T}. Because μ_{J-T} is not zero for a real gas, the pressure dependence of H for an expansion or compression process for which the pressure change is large cannot be neglected. Note that $(\partial H/\partial P)_T$ can be positive or negative, depending on the value of μ_{J-T} at the P and T of interest.

If μ_{J-T} is known from experiment, $(\partial U/\partial V)_T$ can be calculated as shown in Example Problem 3.10. This has the advantage that a calculation of $(\partial U/\partial V)_T$ based on measurements of C_P, μ_{J-T}, and the isothermal compressibility κ, is much more accurate than a measurement based on the Joule experiment. Values of μ_{J-T} are shown for selected gases in Table 3.3. Keep in mind that μ_{J-T} is a function of P and ΔP, so the values listed in the table are only valid for a small pressure decrease originating at 1 atm pressure.

TABLE 3.3

Joule-Thompson Coefficients for Selected Substances at 273 K and 1 atm

Gas	μ_{J-T} (K/MPa)
Ar	3.66
C_6H_{14}	−0.39
CH_4	4.38
CO_2	10.9
H_2	−0.34
He	−0.62
N_2	2.15
Ne	−0.30
NH_3	28.2
O_2	2.69

Source: Linstrom, P. J., and Mallard, W. G., Eds., *NIST Chemistry Webbook: NIST Standard Reference Database Number 69*, National Institute of Standards and Technology, Gaithersburg, MD, retrieved from *http://webbook.nist.gov.*

EXAMPLE PROBLEM 3.10

Using Equation (3.43), $(\partial H/\partial P)_T = [(\partial U/\partial V)_T + P](\partial V/\partial P)_T + V$, and Equation (3.19), $(\partial U/\partial V)_T = T(\partial P/\partial T)_V - P$, derive an expression giving $(\partial H/\partial P)_T$ entirely in terms of measurable quantities for a gas.

Solution

$$\left(\frac{\partial H}{\partial P}\right)_T = \left[\left(\frac{\partial U}{\partial V}\right)_T + P\right]\left(\frac{\partial V}{\partial P}\right)_T + V$$

$$= \left[T\left(\frac{\partial P}{\partial T}\right)_V - P + P\right]\left(\frac{\partial V}{\partial P}\right)_T + V - T\left(\frac{\partial P}{\partial T}\right)_V\left(\frac{\partial V}{\partial P}\right)_T + V$$

$$= -T\left(\frac{\partial V}{\partial T}\right)_P + V = -TV\beta + V = V(1 - \beta T)$$

In this equation, β is the volumetric thermal expansion coefficient defined in Equation (3.9).

EXAMPLE PROBLEM 3.11

Using Equation (3.43),

$$\left(\frac{\partial H}{\partial T}\right)_T = \left[\left(\frac{\partial U}{\partial V}\right)_T + P\right]\left(\frac{\partial V}{\partial P}\right)_T + V$$

show that $\mu_{J-T} = 0$ for an ideal gas.

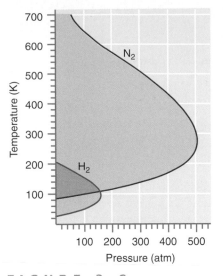

FIGURE 3.6
All along the curves in the figure, $\mu_{J-T} = 0$, and μ_{J-T} is positive to the left of the curves and negative to the right. To experience cooling upon expansion at 100 atm, T must lie between 50 and 150 K for H_2. The corresponding temperatures for N_2 are 100 and 650 K.

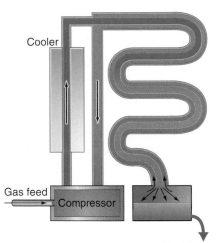

FIGURE 3.7
Schematic depiction of the liquefaction of a gas using an isenthalpic Joule-Thompson expansion. Heat is extracted from the gas exiting from the compressor. It is further cooled in the countercurrent heat exchanger before expanding through a nozzle. Because its temperature is sufficiently low, liquefaction occurs.

Solution

$$\mu_{J-T} = -\frac{1}{C_P}\left(\frac{\partial H}{\partial P}\right)_T = -\frac{1}{C_P}\left[\left(\frac{\partial U}{\partial V}\right)_T\left(\frac{\partial V}{\partial P}\right)_T + P\left(\frac{\partial V}{\partial P}\right)_T + V\right]$$

$$= -\frac{1}{C_P}\left[0 + P\left(\frac{\partial V}{\partial P}\right)_T + V\right]$$

$$= -\frac{1}{C_P}\left[P\left(\frac{\partial [nRT/P]}{\partial P}\right)_T + V\right] = -\frac{1}{C_P}\left[-\frac{nRT}{P} + V\right] = 0$$

In this calculation, we have used the result that $(\partial U/\partial V)_T = 0$ for an ideal gas.

Example Problem 3.11 shows that for an ideal gas, μ_{J-T} is zero. It can be shown that for a van der Waals gas

$$\mu_{J-T} = \frac{1}{C_{P,m}}\left(\frac{2a}{RT} - b\right) \tag{3.54}$$

3.8 Liquefying Gases Using an Isenthalpic Expansion

For real gases, the Joule-Thompson μ_{J-T} can take on either negative or positive values in different regions of P–T space. If μ_{J-T} is positive, a decrease in pressure leads to a cooling of the gas; if it is negative, the expansion of the gas leads to a heating. Figure 3.6 shows the variation of μ_{J-T} with T and P for N_2 and H_2. All along the solid curve, $\mu_{J-T} = 0$. To the left of each curve, μ_{J-T} is positive, and to the right, it is negative. The temperature for which $\mu_{J-T} = 0$ is referred to as the inversion temperature. If the expansion conditions are kept in the region in which μ_{J-T} is positive, ΔT can be made sufficiently large as ΔP decreases in the expansion to liquefy the gas. Note that Equation (3.54) predicts that the inversion temperature for a van der Waals gas is independent of P, which is not in agreement with experiment.

The results in Figure 3.6 are in accord with the observation that a high-pressure ($100 < P < 500$ atm) expansion of N_2 at 300 K leads to cooling and that similar conditions for H_2 lead to heating. To cool H_2 in an expansion, it must first be precooled below 200 K, and the pressure must be less than 160 atm. He and H_2 are heated in an isenthalpic expansion at 300 K for $P < 200$ atm.

The Joule-Thompson effect can be used to liquefy gases such as N_2, as shown in Figure 3.7. The gas at atmospheric pressure is first compressed to a value of 50 to 200 atm, which leads to a substantial increase in its temperature. It is cooled and subsequently passed through a heat exchanger in which the gas temperature decreases to a value within ~50 K of the boiling point. At the exit of the heat exchanger, the gas expands through a nozzle to a final pressure of 1 atm in an isenthalpic expansion. The cooling that occurs because $\mu_{J-T} > 0$ results in liquefaction. The gas that boils away passes back through the heat exchanger in the opposite direction than the gas to be liquefied is passing. The two gas streams are separated, but in good thermal contact. In this process, the gas to be liquefied is effectively precooled, enabling a single-stage expansion to achieve liquefaction.

Vocabulary

cyclic rule

exact differential

heat capacity at constant pressure

heat capacity at constant volume

internal pressure

isenthalpic

isothermal compressibility

Joule-Thompson coefficient

Joule-Thompson experiment

partial derivatives

volumetric thermal expansion coefficient

Questions on Concepts

Q3.1 Why is $C_{P,m}$ a function of temperature for ethane, but not for argon?

Q3.2 Why is $q_v = \Delta U$ only for a constant volume process? Is this formula valid if work other than P–V work is possible?

Q3.3 Refer to Figure 1.6 and explain why $(\partial U/\partial V)_T$ is generally small for a real gas.

Q3.4 Explain without using equations why $(\partial H/\partial P)_T$ is generally small for a real gas.

Q3.5 Why is it reasonable to write $dH \approx C_P dT + V dP$ for a liquid or solid sample?

Q3.6 Why is the equation $\Delta H = \int_{T_i}^{T_f} C_P(T)\, dT = n\int_{T_i}^{T_f} C_{P,m}(T)\, dT$ valid for an ideal gas even if P is not

constant in the process? Is this equation also valid for a real gas? Why or why not?

Q3.7 Heat capacity $C_{P,m}$ is less than $C_{V,m}$ for $H_2O(l)$ between 4° and 5°C. Explain this result.

Q3.8 What is the physical basis for the experimental result that U is a function of V at constant T for a real gas? Under what conditions will U decrease as V increases?

Q3.9 Why does the relation $C_P > C_V$ always hold for a gas? Can $C_P < C_V$ be valid for a liquid?

Q3.10 Can a gas be liquefied through an isoenthalpic expansion if $\mu_{J-T} = 0$?

Problems

Problem numbers in **red** indicate that the solution to the problem is given in the *Student's Solutions Manual*.

P3.1 A differential $dz = f(x,y)dx + g(x,y)dy$ is exact if the integral $\int f(x,y)dx + \int g(x,y)dy$ is independent of the path. Demonstrate that the differential $dz = 2xydx + x^2dy$ is exact by integrating dz along the paths $(1,1) \rightarrow (5,1) \rightarrow (5,5)$ and $(1,1) \rightarrow (3,1) \rightarrow (3,3) \rightarrow (5,3) \rightarrow (5,5)$. The first number in each set of parentheses is the x coordinate, and the second number is the y coordinate.

P3.2 The function $f(x,y)$ is given by $f(x,y) = xy \sin 5x + x^2\sqrt{y} \ln y + 3e^{-2x^2} \cos y$. Determine

$$\left(\frac{\partial f}{\partial x}\right)_y, \left(\frac{\partial f}{\partial y}\right)_x, \left(\frac{\partial^2 f}{\partial x^2}\right)_y, \left(\frac{\partial^2 f}{\partial y^2}\right)_x, \left(\frac{\partial}{\partial y}\left(\frac{\partial f}{\partial x}\right)_y\right)_x$$

and $\left(\frac{\partial}{\partial x}\left(\frac{\partial f}{\partial y}\right)_x\right)_y$ Is $\left(\frac{\partial}{\partial y}\left(\frac{\partial f}{\partial x}\right)_y\right)_x = \left(\frac{\partial}{\partial x}\left(\frac{\partial f}{\partial y}\right)_x\right)_y$?

Obtain an expression for the total differential df.

P3.3 This problem will give you practice in using the cyclic rule. Use the ideal gas law to obtain the three functions $P = f(V,T)$, $V = g(P,T)$, and $T = h(P,V)$. Show that the cyclic rule $(\partial P/\partial V)_T (\partial V/\partial T)_P (\partial T/\partial P)_V = -1$ is obeyed.

P3.4 Using the chain rule for differentiation, show that the isobaric expansion coefficient expressed in terms of density is given by $\beta = -1/\rho (\partial \rho/\partial T)_P$.

P3.5 A vessel is filled completely with liquid water and sealed at 25.0°C and a pressure of 1.00 bar. What is the pressure if the temperature of the system is raised to 60.0°C? Under these conditions, $\beta_{water} = 2.04 \times 10^{-4}\ \text{K}^{-1}$, $\beta_{vessel} = 1.02 \times 10^{-4}\ \text{K}^{-1}$, and $\kappa_{water} = 4.59 \times 10^{-5}\ \text{bar}^{-1}$.

P3.6 Because U is a state function, $(\partial/\partial V (\partial U/\partial T)_V)_T = (\partial/\partial T (\partial U/\partial V)_T)_V$. Using this relationship, show that $(\partial C_V/\partial V)_T = 0$ for an ideal gas.

let $V(T_0, B) \equiv V_0$

P3.7 Because V is a state function, $(\partial/\partial P\,(\partial V/\partial T)_P)_T$ $= (\partial/\partial T\,(\partial V/\partial P)_T)_P$. Using this relationship, show that the isothermal compressibility and isobaric expansion coefficient are related by $(\partial\beta/\partial P)_T = -(\partial\kappa/\partial T)_P$.

P3.8 Integrate the expression $\beta = 1/V\,(\partial V/\partial T)_P$ assuming that β is independent of pressure. By doing so, obtain an expression for V as a function of P and β. derive $V(T, B)$

P3.9 The molar heat capacity $C_{P,m}$ of $SO_2(g)$ is described by the following equation over the range 300 K < 1700 K:

$$\frac{C_{P,m}}{R} = 3.093 + 6.967 \times 10^{-3}\frac{T}{K} - 45.81 \times 10^{-7}\frac{T^2}{K^2}$$

$$+1.035 \times 10^{-9}\frac{T^3}{K^3}$$

In this equation, T is the absolute temperature in Kelvin. The ratios T^n/K^n ensure that $C_{P,m}$ has the correct dimension. Assuming ideal gas behavior, calculate q, w, ΔU, and ΔH if 1 mol of $SO_2(g)$ is heated from 75° to 1350°C at a constant pressure of 1 bar. Explain the sign of w.

P3.10 Starting with the van der Waals equation of state, find an expression for the total differential dP in terms of dV and dT. By calculating the mixed partial derivatives $(\partial/\partial T\,(\partial P/\partial V)_T)_V$ and $(\partial/\partial V\,(\partial P/\partial T)_V)_T$, determine if dP is an exact differential.

P3.11 Obtain an expression for the isothermal compressibility $\kappa = -1/V\,(\partial V/\partial P)_T$ for a van der Waals gas.

P3.12 Regard the enthalpy as a function of T and P. Use the cyclic rule to obtain the expression

$$C_P = -\left(\frac{\partial H}{\partial P}\right)_T\Bigg/\left(\frac{\partial T}{\partial P}\right)_H$$

P3.13 Equation (3.38), $C_P = C_V + TV(\beta^2/\kappa)$, links C_P and C_V with β and κ. Use this equation to evaluate $C_P - C_V$ for an ideal gas.

P3.14 Use $(\partial U/\partial V)_T = (\beta T - \kappa P)/\kappa$ to calculate $(\partial U/\partial V)_T$ for an ideal gas.

P3.15 An 80.0-g piece of gold at 650 K is dropped into 100.0 g of $H_2O(l)$ at 298 K in an insulated container at 1 bar pressure. Calculate the temperature of the system once equilibrium has been reached. Assume that $C_{P,m}$ for Au and H_2O is constant at their values for 298 K throughout the temperature range of interest.

P3.16 A mass of 35.0 g of $H_2O(s)$ at 273 K is dropped into 180.0 g of $H_2O(l)$ at 325 K in an insulated container at 1 bar of pressure. Calculate the temperature of the system once equilibrium has been reached. Assume that $C_{P,m}$ for H_2O is constant at its value for 298 K throughout the temperature range of interest.

P3.17 A mass of 20.0 g of $H_2O(g)$ at 373 K is flowed into 250 g of $H_2O(l)$ at 300 K and 1 atm. Calculate the final temperature of the system once equilibrium has been reached.

Assume that $C_{P,m}$ for H_2O is constant at its values for 298 K throughout the temperature range of interest.

P3.18 Calculate w, q, ΔH, and ΔU for the process in which 1 mol of water undergoes the transition $H_2O(l, 373\text{ K})$ $\rightarrow H_2O(g, 460\text{ K})$ at 1 bar of pressure. The volume of liquid water at 373 K is 1.89×10^{-5} m^3 mol^{-1} and the volume of steam at 373 and 460 K is 3.03 and 3.74×10^{-2} m^3 mol^{-1}, respectively. For steam, $C_{P,m}$ can be considered constant over the temperature interval of interest at 33.58 J mol^{-1} K^{-1}.

P3.19 Because $(\partial H/\partial P)_T = -C_P\mu_{J-T}$, the change in enthalpy of a gas expanded at constant temperature can be calculated. To do so, the functional dependence of μ_{J-T} on P must be known. Treating Ar as a van der Waals gas, calculate ΔH when 1 mol of Ar is expanded from 400 bar to 1.00 bar at 300 K. Assume that μ_{J-T} is independent of pressure and is given by μ_{J-T} $= [(2a/RT) - b]/C_{P,m}$, and $C_{P,m} = 5/2R$ for Ar. What value would ΔH have if the gas exhibited ideal gas behavior?

P3.20 Using the result of Equation (3.8), $(\partial P/\partial T)_V = \beta/\kappa$, express β as a function of κ and V_m for an ideal gas, and β as a function of b, κ, and V_m for a van der Waals gas. predict

P3.21 The Joule coefficient is defined by find $\left(\frac{\partial T}{\partial V}\right)_U C_V$ $(\partial T/\partial V)_U = 1/C_V\,[P - T\,(\partial P/\partial T)_V]$. Calculate the Joule coefficient for an ideal gas and for a van der Waals gas.

P3.22 Use the relation $(\partial U/\partial V)_T = T(\partial P/\partial T)_V - P$ and the cyclic rule to obtain an expression for the internal pressure, $(\partial U/\partial V)_T$, in terms of P, β, T, and κ.

P3.23 Derive the following relation,

$$\left(\frac{\partial U}{\partial V_m}\right)_T = \frac{3a}{2\sqrt{T}V_m(V_m + b)}$$

for the internal pressure of a gas that obeys the Redlich-Kwong equation of state,

$$P = \frac{RT}{V_m - b} - \frac{a}{\sqrt{T}}\frac{1}{V_m(V_m + b)}$$

P3.24 Derive an expression for the internal pressure of a gas that obeys the Bethelot equation of state,

$$P = \frac{RT}{V_m - b} - \frac{a}{TV_m^2}$$

P3.25 For a gas that obeys the equation of state

$$V_m = \frac{RT}{P} + B(T)$$

derive the result

$$\left(\frac{\partial H}{\partial P}\right)_T = B(T) - T\frac{dB(T)}{dT}$$

P3.26 Derive the following expression for calculating the isothermal change in the constant volume heat capacity: $(\partial C_V/\partial V)_T = T(\partial^2 P/\partial T^2)_V$.

P3.27 Use the result of Problem P3.26 to show that $(\partial C_V / \partial V)_T$ for the van der Waals gas is zero.

P3.28 Use the result of Problem P3.26 to derive a formula for $(\partial C_V / \partial V)_T$ for a gas that obeys the Redlich-Kwong equation of state,

$$P = \frac{RT}{V_m - b} - \frac{a}{\sqrt{T}} \frac{1}{V_m(V_m + b)}$$

P3.29 For the equation of state $V_m = RT/P + B(T)$, show that

$$\left(\frac{\partial C_{P,m}}{\partial P} \right)_T = -T \frac{d^2 B(T)}{dT^2}$$

[Hint: Use Equation 3.44 and the property of state functions with respect to the order of differentiation in mixed second derivatives.]

P3.30 Use the relation

$$C_{P,m} - C_{V,m} = T \left(\frac{\partial V_m}{\partial T} \right)_P \left(\frac{\partial P}{\partial T} \right)_V$$

the cyclic rule and the van der Waals equation of state to derive an equation for $C_{P,m} - C_{V,m}$ in terms of V_m and the gas constants R, a, and b.

P3.31 Show that

$$\left(\frac{\partial U}{\partial V} \right)_T = \frac{\beta T - \kappa P}{\kappa}$$

P3.32 Show that the expression $(\partial U / \partial V)_T = T(\partial P / \partial T)_V - P$ can be written in the form

$$\left(\frac{\partial U}{\partial V} \right)_T = T^2 \left(\partial \left[\frac{P}{T} \right] \Big/ \partial T \right)_V = - \left(\partial \left[\frac{P}{T} \right] \Big/ \partial \left[\frac{1}{T} \right] \right)_V$$

CHAPTER 4

Thermochemistry

Thermochemistry is the branch of thermodynamics that investigates the heat flow into or out of a reaction system and deduces the energy stored in chemical bonds. As reactants are converted into products, energy can either be taken up by the system or released to the surroundings. For a reaction that takes place at constant temperature and volume, the heat that flows to the system is equal to ΔU for the reaction. For a reaction that takes place at constant temperature and pressure, the heat that flows to the system is equal to ΔH for the reaction. The enthalpy of formation is defined as the heat flow into the system in a reaction between pure elements that leads to the formation of 1 mol of product. Because H is a state function, the reaction enthalpy can be written as the enthalpies of formation of the products minus those of the reactants. This property allows ΔH and ΔU for a reaction to be calculated for many reactions without carrying out an experiment. ∎

4.1 Energy Stored in Chemical Bonds Is Released or Taken Up in Chemical Reactions

A significant amount of the internal energy or enthalpy of a molecule is stored in the form of chemical bonds. As reactants are transformed to products in a chemical reaction, energy can be released or taken up as bonds are made or broken. For example, consider a reaction in which $N_2(g)$ and $H_2(g)$ dissociate into atoms, and the atoms recombine to form $NH_3(g)$. The enthalpy changes associated with individual steps and with the overall reaction $1/2\ N_2(g) + 3/2\ H_2(g) \rightarrow NH_3(g)$ are shown in Figure 4.1. Note that large enthalpy changes are associated with the individual steps, but that the enthalpy change in the overall reaction is much smaller.

The change in enthalpy or internal energy resulting from chemical reactions appears in the surroundings in the form of a temperature increase or decrease resulting from heat flow and/or in the form of expansion or nonexpansion work. For example, the combustion of gasoline in an automobile engine can be used to do expansion work on the surroundings. Nonexpansion electrical work is possible if the chemical reaction is carried out in an electrochemical cell. In Chapters 6 and 11, the extraction of nonexpansion work from chemical reactions will be discussed. In this chapter, the

Enthalpy changes are shown for individual steps in the overall reaction $1/2\ N_2 + 3/2\ H_2 \rightarrow NH_3$.

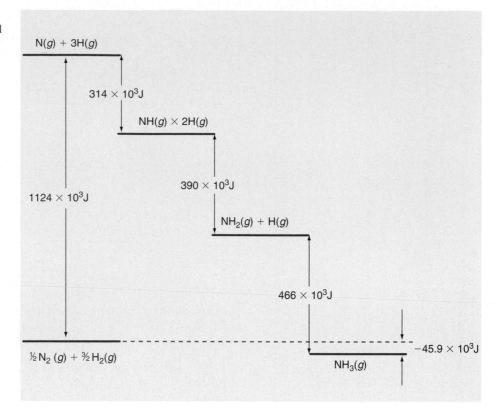

focus is on using measurements of heat flow to determine changes in U and H due to chemical reactions.

4.2 Internal Energy and Enthalpy Changes Associated with Chemical Reactions

In the previous chapters, we discussed how ΔU and ΔH are calculated from work and heat flow between the system and the surroundings for processes that do not involve phase changes or chemical reactions. In this section, this discussion is extended to reaction systems. Because most reactions of interest to chemists are carried out at constant pressure, chemists generally use ΔH rather than ΔU as a measure of the change in energy of the system. However, because ΔH and ΔU are related, one can be calculated from a measurement of the other.

Imagine that a reaction involving a stoichiometric mixture of reactants is carried out in a constant pressure reaction vessel with diathermal walls immersed in a water bath. It is our experience that the temperature of the water bath will either increase or decrease as the reaction proceeds, depending on whether energy is taken up or released in the reaction. If the temperature of the water bath increases, heat flows from the system (the contents of the reaction vessel) to the surroundings (the water bath and the vessel). In this case, we say that the reaction is **exothermic.** If the temperature of the water bath decreases, the heat flows instead from the surroundings to the system, and we say that the reaction is **endothermic.**

The temperature and pressure of the system will also increase or decrease as the reaction proceeds. The total measured values for ΔU (or ΔH) include contributions from two separate components: ΔU for the reaction at constant T and P and ΔU from the change in T and P. To determine ΔU for the reaction, it is necessary to separate these two components. Consider the reaction in Equation (4.1):

$$Fe_3O_4(s) + 4H_2(g) \rightarrow 3Fe(s) + 4H_2O(l) \tag{4.1}$$

This reaction will only proceed at a measurable rate at elevated temperatures. However, as we show later, it is useful to tabulate all values for ΔU and ΔH for reactions at a standard temperature and pressure, generally 298.15 K and 1 bar. The standard state for gases is actually a hypothetical state in which the gas behaves ideally at a pressure of 1 bar. For most reactions, deviations from ideal behavior are very small. Note that the phase (solid, liquid, or gas) for each reactant and product has been specified, because U and H are different for each phase. Because U and H are also functions of T and P, their values must also be specified. Changes in enthalpy and internal energy at the standard pressure of 1 bar are indicated by $\Delta H°$ and $\Delta U°$. The **enthalpy of reaction**, $\Delta H_{reaction}$, at specific values of T and P is defined as the heat withdrawn from the surroundings as the reactants are transformed into products at conditions of constant T and P. It is, therefore, a negative quantity for an exothermic reaction and a positive quantity for an endothermic reaction. How can the reaction enthalpy and internal energy be determined? Because H is a state function, any path can be chosen that proceeds from reactants to products at the standard values of T and P. It is useful to visualize the following imaginary two-step process:

1. The reaction is carried out in an isolated system. The **standard state** values $T = 298.15$ K and $P = 1$ bar are chosen for the initial system conditions. The walls of the constant pressure reaction vessel are switched to an adiabatic condition and the reaction is initiated. At 298.15 K, Fe_3O_4 and Fe are solids, H_2 is a gas, and H_2O is a liquid. The temperature changes from T to T' in the course of the reaction. The reaction is described by

$$(\{Fe_3O_4(s) + 4H_2(g)\}, 298.15 \text{ K}, P = 1 \text{ bar}) \rightarrow (\{3Fe(s) + 4H_2O(l)\}, \tag{4.2}$$
$$T', P = 1 \text{ bar})$$

Because the process in this step is adiabatic, $q_{P,a} = \Delta H_a° = 0$.

2. In the second step, the system comes into thermal equilibrium with the surroundings. The walls of the reaction vessel are switched to a diathermal condition, and heat flow occurs to a very large water bath in the surroundings. The temperature changes from T' back to its initial value of 298.15 K in this process. This process is described by

$$(\{3Fe(s) + 4H_2O(l)\}, T', P = 1 \text{ bar}) \rightarrow (\{3Fe(s) + 4H_2O(l)\}, 298.15 \text{ K}, P = 1 \text{ bar}) \tag{4.3}$$

The heat flow in this step is $\Delta H_b° = q_{P,b}$.

The sum of these two steps is the overall reaction in which the initial and final states are at the same values of P and T, namely, 298.15 K and 1 bar:

$$(\{Fe_3O_4(s) + 4H_2(g)\}, 298.15 \text{ K}, P = 1 \text{ bar}) \rightarrow (\{3Fe(s) + 4H_2O(l)\}, \tag{4.4}$$
$$298.15 \text{ K}, P = 1 \text{ bar})$$

The sum of the ΔH values for the individual steps at 298.15 K, $\Delta H_{reaction}$, is

$$\Delta H_{reaction}° = \Delta H_R° = \Delta H_a° + \Delta H_b° = q_{P,a} + q_{P,b} = 0 + q_{P,b} = q_P \tag{4.5}$$

Note that because H is a state function, it is unimportant if the temperature during the course of the reaction differs from 298.15 K. It is also unimportant if T is higher than 373 K so that H_2O is present as a gas rather than a liquid. If the initial and final temperatures of the system are 298.15 K, then the measured value of q_P is $\Delta H_{reaction}°$ at 298.15 K.

The previous discussion suggests an operational way to determine $\Delta H_{reaction}°$ for a reaction at constant pressure. Because researchers do not have access to an infinitely large water bath, the temperature of the system and surroundings will change in the course of the reaction. The procedure just described, therefore, is modified in the following way.

The reaction is allowed to proceed, and the ΔT that occurs in a finite size water bath in the surroundings that is initially at 298.15 K is measured. If the temperature of the water bath decreases as a result of the reaction, the bath is heated to return it to 298.15 K using an electrical heater. By doing so, we ensure that the initial and final P and T are the same and, therefore, $\Delta H = \Delta H^{\circ}_{reaction}$. How is $\Delta H^{\circ}_{reaction}$ determined for this case? The electrical work done on the heater that restores the temperature to 298.15 K is equal to $\Delta H^{\circ}_{298.15\,K}$, which is equal to $\Delta H^{\circ}_{reaction}$. If the temperature of the water bath increases as a result of the reaction by the amount ΔT, the electrical work done on a heater in the water bath at 298.15 K that increases its temperature by ΔT in a separate experiment is measured. In this case, $\Delta H^{\circ}_{reaction}$ is equal to the negative of the electrical work done on the heater.

Although an experimental method for determining $\Delta H^{\circ}_{reaction}$ has been described, to tabulate the reaction enthalpies for all possible chemical reactions would be a monumental undertaking. Fortunately, $\Delta H^{\circ}_{reaction}$ can be calculated from tabulated enthalpies values for individual reactants and products. This is advantageous because there are far fewer reactants and products than there are reactions among them. Consider $\Delta H^{\circ}_{298.15K}$ for the reaction of Equation (4.1) at $T = 298.15$ K and $P = 1$ bar. These values for P and T are chosen because thermodynamic values are tabulated for these values. However, standard enthalpies for reaction at other values of P and T can be calculated as discussed in Chapters 2 and 3.

$$\Delta H^{\circ}_{298.15\,K} = H^{\circ}_{products} - H^{\circ}_{reactants} \tag{4.6}$$

$$= 3H^{\circ}_m(Fe,s) + 4H^{\circ}_m(H_2O,l) - H^{\circ}_m(Fe_3O_4,s) - 4H^{\circ}_m(H_2,g)$$

The m subscripts refer to molar quantities. Although Equation (4.6) is correct, it does not provide a useful way to calculate $\Delta H^{\circ}_{298.15\,K}$ because there is no experimental way to determine the absolute enthalpies for any element or compound. This is the case because there is no unique reference zero against which individual enthalpies can be measured. For this reason, only ΔH and ΔU, as opposed to H and U, can be determined in an experiment.

Equation (4.6) can be transformed into a more useful form by introducing the enthalpy of formation. The standard **enthalpy of formation,** ΔH°_f, is the enthalpy associated with the reaction in which the only reaction product is 1 mol of the species of interest, and only pure elements in their most stable state of aggregation under the standard state conditions appear as reactants. Note that with this definition, $\Delta H^{\circ}_f = 0$ for an element in its standard state because the reactants and products are identical. The only compounds that are produced or consumed in the reaction $Fe_3O_4(s) + 4H_2(g) \rightarrow 3Fe(s) + 4H_2O(l)$ are $Fe_3O_4(s)$ and $H_2O(l)$. No elements that are not in their standard states appear in the reaction. The formation reactions for the compounds at 298.15 K and 1 bar are

$$H_2(g) + \frac{1}{2}O_2(g) \rightarrow H_2O(l)$$

$$\Delta H^{\circ}_{reaction} = \Delta H^{\circ}_f(H_2O,l) = H^{\circ}_m(H_2O,l) - H^{\circ}_m(H_2,g) - \frac{1}{2}H^{\circ}_m(O_2,g) \tag{4.7}$$

$$3Fe(s) + 2O_2(g) \rightarrow Fe_3O_4(s)$$

$$\Delta H^{\circ}_{reaction} = \Delta H^{\circ}_f(Fe_3O_4,s) = H^{\circ}_m(Fe_3O_4,s) - 3H^{\circ}_m(Fe,s) - 2H^{\circ}_m(O_2,g) \tag{4.8}$$

If Equation (4.6) is rewritten in terms of the enthalpies of formation, a simple equation for the reaction enthalpy is obtained:

$$\Delta H^{\circ}_{reaction} = 4\Delta H^{\circ}_f(H_2O,l) - \Delta H^{\circ}_f(Fe_3O_4,s) \tag{4.9}$$

Note that elements in their standard state do not appear in this equation. This result can be generalized to any chemical transformation

$$v_A A + v_B B + ... \rightarrow v_X X + v_Y Y + ... \tag{4.10}$$

which we write in the form

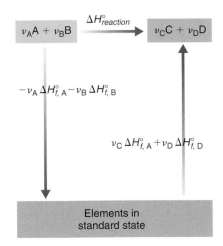

FIGURE 4.2
Equation (4.12) follows from the fact that the enthalpy of both paths is the same because they connect the same initial and final states.

$$\sum_i v_i X_i \qquad (4.11)$$

The X_i refer to all species other than elements in their standard state that appear in the overall equation. The unitless stoichiometric coefficients v_i are positive for products and negative for reactants. The enthalpy change associated with this reaction is

$$\Delta H^\circ_{reaction} = \sum_i v_i \Delta H^\circ_{f,i} \qquad (4.12)$$

The rationale behind Equation (4.12) can also be depicted as shown in Figure 4.2. Two paths are considered between the reactants A and B and the products C and D in the reaction $v_A A + v_B B \rightarrow v_C C + v_D D$. The first of these is a direct path for which $\Delta H^\circ = \Delta H^\circ_{reaction}$. In the second path, A and B are first broken down into their elements, each in its standard state. Subsequently, the elements are combined to form C and D. The enthalpy change for the second route is $\Delta H^\circ = \sum_i v_i \Delta H^\circ_{f,products} - \sum_i v_i \Delta H^\circ_{f,reactants}$. Because H is a state function, the enthalpy change is the same for both paths. This is stated in mathematical form in Equation (4.12).

A specific application of this general equation is given in Equation (4.9). Writing $\Delta H^\circ_{reaction}$ in this form is a great simplification over compiling measured values of reaction enthalpies, because the reaction enthalpies can be calculated using only tabulated values of the formation enthalpies of species other than elements in their standard state. Formation enthalpies for inorganic atoms and molecules are listed in Table 4.1 and formation enthalpies for organic compounds are listed in Table 4.2 (Appendix A, Data Tables).

Another thermochemical convention is introduced at this point in order to be able to calculate enthalpy changes involving electrolyte solutions. The solution reaction that occurs when a salt such as NaCl is dissolved in water is

$$NaCl(s) \rightarrow Na^+(aq) + Cl^-(aq)$$

Because it is not possible to form only positive or negative ions in solution, the measured enthalpy of solution of an electrolyte is the sum of the enthalpies of all anions and cations formed. To be able to tabulate values for enthalpies of formation of individual ions, the enthalpy for the following reaction is set equal to zero at $P = 1$ bar for all temperatures:

$$1/2\, H_2(g) \rightarrow H^+(aq) + e^-(\text{metal electrode})$$

In other words, solution enthalpies of ions are measured relative to that for $H^+(aq)$. The thermodynamics of electrolyte solutions will be discussed in detail in Chapter 10.

As the previous discussion shows, to calculate $\Delta H^\circ_{reaction}$, only the ΔH°_f of the reactant and products are needed. The ΔH°_f are again a *difference* in enthalpy between the compound and its constituent elements, rather than an absolute enthalpy. However, there is a convention that allows absolute enthalpies to be specified using the experimentally determined values of the ΔH°_f of compounds. In this convention, the absolute enthalpy of each pure element in its standard state is set equal to zero. With this convention, the absolute molar enthalpy of any chemical species in its standard state, H°_m, is equal to ΔH°_f for that species. To demonstrate this convention, the reaction in Equation (4.8) is considered:

$$\Delta H^\circ_{reaction} = \Delta H^\circ_f(Fe_3O_4, s) = H^\circ_m(Fe_3O_4, s) - 3H^\circ_m(Fe, s) - 2H^\circ_m(O_2, g) \qquad (4.8)$$

Setting $H^\circ = 0$ for each element in its standard state,

$$\Delta H^\circ_f(Fe_3O_4, s) = H^\circ_m(Fe_3O_4, s) - 3 \times 0 - 2 \times 0 = H^\circ_m(Fe_3O_4, s) \qquad (4.13)$$

Convince yourself that the value of $\Delta H^\circ_{reaction}$ for any reaction involving compounds and elements is unchanged by this convention. In fact, one could choose a different number for the absolute enthalpy of each pure element in its standard state, and it would still not change the value of $\Delta H^\circ_{reaction}$. However, it is much more convenient (and easier to remember) if one sets $H^\circ_m = 0$ for all elements in their standard state. This convention will be used again in Chapter 6 when the chemical potential is discussed.

4.3 Hess's Law Is Based on Enthalpy Being a State Function

As discussed in the previous section, it is extremely useful to have tabulated values of ΔH°_f for chemical compounds at one fixed combination of P and T. With access to these values of ΔH°_f, $\Delta H^\circ_{reaction}$ can be calculated for all reactions among these compounds at the tabulated values of P and T.

But how is ΔH°_f determined? Consider the formation reaction for $C_2H_6(g)$:

$$2C(s) + 3H_2(g) \rightarrow C_2H_6(g) \tag{4.14}$$

Graphite is the standard state of aggregation for carbon at 298.15 K and 1 bar because it is slightly more stable than diamond under these conditions. However, it is unlikely that one would obtain only ethane if the reaction were carried out as written. Given this experimental hindrance, how can ΔH°_f for ethane be determined? To determine ΔH°_f for ethane, we take advantage of the fact that ΔH°_f is path independent. In this context, path independence means that the enthalpy change for any sequence of reactions that sum to the same overall reaction is identical. This statement is known as **Hess's law.** Therefore, one is free to choose any sequence of reactions that leads to the desired outcome. Combustion reactions are well suited for these purposes, because in general they proceed rapidly, go to completion, and produce only a few products. To determine ΔH°_f for ethane, one can carry out the following combustion reactions:

$$C_2H_6(g) + 7/2O_2(g) \rightarrow 2CO_2(g) + 3H_2O(l) \qquad \Delta H^\circ_I \tag{4.15}$$

$$C(s) + O_2(g) \rightarrow CO_2(g) \qquad \Delta H^\circ_{II} \tag{4.16}$$

$$H_2(g) + 1/2\, O_2(g) \rightarrow H_2O(l) \qquad \Delta H^\circ_{III} \tag{4.17}$$

These reactions are combined in the following way to obtain the desired reaction:

$$2 \times [C(s) + O_2(g) \rightarrow CO_2(g)] \qquad 2\Delta H^\circ_{II} \tag{4.18}$$

$$2CO_2(g) + 3H_2O(l) \rightarrow C_2H_6(g) + 7/2O_2(g) \qquad -\Delta H^\circ_I \tag{4.19}$$

$$\underline{3 \times [H_2(g) + 1/2\, O_2(g) \rightarrow H_2O(l)] \qquad 3\, \Delta H^\circ_{III}} \tag{4.20}$$

$$2C(s) + 3H_2(g) \rightarrow C_2H_6(g) \qquad 2\, \Delta H^\circ_{II} - \Delta H^\circ_I + 3\, \Delta H^\circ_{III}$$

We emphasize again that it is not necessary for these reactions to be carried out at 298.15 K. The reaction vessel is immersed in a water bath at 298.15 K and the combustion reaction is initiated. If the temperature in the vessel rises during the course of the reaction, the heat flow that restores the system and surroundings to 298.15 K after completion of the reaction is measured, allowing $\Delta H^\circ_{reaction}$ to be determined at 298.15 K.

Several points should be made about enthalpy changes in relation to balanced overall equations describing chemical reactions. First, because H is an extensive function, multiplying all stoichiometric coefficients with any number changes $\Delta H^\circ_{reaction}$ by the same factor. Therefore, it is important to know which set of stoichiometric coefficients has been assumed if a numerical value of $\Delta H^\circ_{reaction}$ is given. Second, because the units of ΔH°_f for all compounds in the reaction are kJ mol^{-1}, the units of the reaction enthalpy

$\Delta H^\circ_{reaction}$ are also kJ mol^{-1}. One might pose the question "per mole of what?" given that all the stoichiometric coefficients may differ from each other and from one. The answer to this question is per mole of the reaction *as written*. Doubling all the stoichiometric coefficients doubles $\Delta H^\circ_{reaction}$.

EXAMPLE PROBLEM 4.1

The average **bond enthalpy** of the O—H bond in water is defined as one-half of the enthalpy change for the reaction $H_2O(g) \rightarrow 2H(g) + O(g)$. The formation enthalpies, ΔH°_f, for $H(g)$ and $O(g)$ are 218.0 and 249.2 kJ mol^{-1}, respectively, at 298.15 K, and ΔH°_f for $H_2O(g)$ is -241.8 kJ mol^{-1} at the same temperature.

a. Use this information to determine the average bond enthalpy of the O—H bond in water at 298.15 K.

b. Determine the average **bond energy**, ΔU, of the O—H bond in water at 298.15 K. Assume ideal gas behavior.

Solution

a. We consider the sequence

$H_2O(g) \rightarrow H_2(g) + 1/2\,O_2(g)$ $\Delta H^\circ = 241.8$ kJ mol^{-1}

$H_2(g) \rightarrow 2H(g)$ $\Delta H^\circ = 2 \times 218.0$ kJ mol^{-1}

$1/2\,O_2(g) \rightarrow O(g)$ $\Delta H^\circ = 249.2$ kJ mol^{-1}

$H_2O(g) \rightarrow 2H(g) + O(g)$ $\Delta H^\circ = 927.0$ kJ mol^{-1}

This is the enthalpy change associated with breaking both O—H bonds under standard conditions. We conclude that the average bond enthalpy of the O—H bond in water is $\frac{1}{2} \times 927.0\,\text{kJ mol}^{-1} = 463.5\,\text{kJ mol}^{-1}$. We emphasize that this is the average value because the values of ΔH for the transformations $H_2O(g) \rightarrow H(g) + OH(g)$ and $OH(g) \rightarrow O(g) + H(g)$ differ.

b. $\Delta U^\circ = \Delta H^\circ - \Delta(PV) = \Delta H^\circ - \Delta nRT$

$= 927.0\,\text{kJ mol}^{-1} - 2 \times 8.314\,\text{J mol}^{-1}\text{K}^{-1} \times 298.15\,\text{K} - 922.0\,\text{kJ mol}^{-1}$

The average value for ΔU° for the O—H bond in water is $\frac{1}{2} \times 922.0\,\text{kJ mol}^{-1} = 461.0\,\text{kJ mol}^{-1}$. The bond energy and the bond enthalpy are nearly identical.

Example Problem 4.1 shows how bond energies can be calculated from reaction enthalpies. The value of a bond energy is of particular importance for chemists in estimating the thermal stability of a compound as well as its stability with respect to reactions with other molecules. Values of bond energies tabulated in the format of the periodic table together with the electronegativities are shown in Table 4.3 [N. K. Kildahl, *J. Chemical Education*, 72 (1995), 423]. The value of the single bond energy, D_{A-B}, for a combination A—B not listed in the table can be calculated using the empirical relationship due to Linus Pauling in Equation (4.21):

$$D_{A-B} = \sqrt{D_{A-A}D_{B-B}} + 96.5(x_A - x_B)^2 \qquad (4.21)$$

where x_A and x_B are the electronegativity of atoms A and B.

TABLE 4.3

Mean Bond Energies

Selected Bond Energies (kJ/mol)

1	2	13	14	15	16	17	18
H,2.20 432 --- 432 459 565							**He**
Li,0.98 105 --- 243 --- 573	**Be**,1.57 208 --- --- ---,444 632	**B**,2.04 293 --- 389 536,636 613	**C**,2.55 346 602,835 411 358,799 485	**N**,3.04 167 418,942 386 201,607 283	**O**,3.44 142 494 459 142,494 190	**F**,3.98 155 --- 565 --- 155	**Ne**
Na,0.93 72 --- 197 --- 477	**Mg**,1.31 129 --- --- ---,377 513	**Al**,1.61 -- --- 272 --- 583	**Si**,1.90 222 318 318 452,640 565	**P**,2.19 ≈220 ---,481 322 335,544 490	**S**,2.58 240 425 363 ---,523 284	**Cl**,3.16 240 --- 428 218 249	**Ar**
K,0.82 49 --- 180 --- 490	**Sr**,1.00 105 --- --- ---,460 550	**Ga**,1.81 113 --- --- --- ≈469	**Ge**,2.01 188 272 --- --- ≈470	**As**,2.18 146 ---,380 247 301,389 ≈440	**Se**,2.55 172 272 276 --- ≈351	**Br**,2.96 190 --- 362 201 250	**Kr** 50
Rb,0.82 45 --- 163 --- 490	**Sr**,0.95 84 --- --- ---,347 553	**In**,1.78 100 --- --- --- ≈523	**Sn**,1.80 146 --- --- --- ≈450	**Sb**,2.05 121 ---,295 --- ---- ≈420	**Te**,2.10 126 218 238 ---- ≈393	**I**,2.66 149 -- 295 201 278	**Xe** 84 ≈131
Cs,0.79 44 --- 176 --- 502	**Ba**,0.89 44 --- --- 467,561 578	**Tl**,2.04 --- --- --- --- 439	**Pb**,2.33 --- --- --- --- ≈360	**Bi**,2.02 --- ---,192 --- ≈350	**Po**,2.00 116	**At**,2.20	**Rn**

KEY

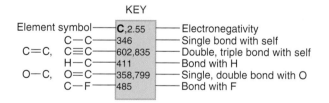

Element symbol	**C**,2.55	Electronegativity
	C—C 346	Single bond with self
C=C, C≡C	602,835	Double, triple bond with self
	H—C 411	Bond with H
O—C, O=C	358,799	Single, double bond with O
	C—F 485	Bond with F

4.4 The Temperature Dependence of Reaction Enthalpies

Suppose that you plan to carry out a reaction that is mildly exothermic at 298.15 K at a higher temperature. Is the reaction endothermic or exothermic at the higher temperature? To answer this question, it is necessary to determine $\Delta H^\circ_{reaction}$ at the elevated temperature. It is assumed that no phase changes occur in the temperature interval of interest. The enthalpy for each reactant and product at temperature T is related to the value at 298.15 K by Equation (4.22), which accounts for the energy supplied in order to heat the substance to the new temperature at constant pressure:

$$H_T^\circ = H_{298.15K}^\circ + \int_{298.15K}^{T} C_P(T')dT' \tag{4.22}$$

The prime in the integral indicates a "dummy variable" that is otherwise identical to the temperature. This notation is needed because T appears in the upper limit of the integral. In Equation (4.22), $H_{298.15\,K}^\circ$ is the absolute enthalpy at 1 bar and 298.15 K. However, because there are no unique values for absolute enthalpies, it is useful to combine similar equations for all reactants and products to obtain the following equation for the reaction enthalpy at temperature T:

$$\Delta H_T^\circ = \Delta H_{298.15K}^\circ + \int_{298.15K}^{T} \Delta C_P(T')dT' \tag{4.23}$$

where

$$\Delta C_p(T') = \sum_i v_i C_{p,i}(T') \tag{4.24}$$

In this equation, the sum is over all reactants and products, *including both elements and compounds*. Elements must be included, because $\Delta H_f^\circ = 0$ only for an element at the standard state temperature of 298.15 K, and $C_{P,i}(T) \neq 0$. A calculation of $\Delta H_{reaction}^\circ$ at an elevated temperature is shown in Example Problem 4.2.

EXAMPLE PROBLEM 4.2

Calculate the enthalpy of formation of HCl(g) at 1450 K and 1 bar pressure given that $\Delta H_f^\circ(\text{HCl}, g) = -92.3 \text{ kJ mol}^{-1}$ at 298.15 K and that

$$C_{P,m}^\circ(\text{H}_2, g) = \left(29.064 - 0.8363 \times 10^{-3}\frac{T}{K} + 20.111 \times 10^{-7}\frac{T^2}{K^2}\right) \text{J K}^{-1}\text{mol}^{-1}$$

$$C_{P,m}^\circ(\text{Cl}_2, g) = \left(31.695 + 10.143 \times 10^{-3}\frac{T}{K} - 40.373 \times 10^{-7}\frac{T^2}{K^2}\right) \text{J K}^{-1}\text{mol}^{-1}$$

$$C_{P,m}^\circ(\text{HCl}, g) = \left(28.165 + 1.809 \times 10^{-3}\frac{T}{K} + 15.464 \times 10^{-7}\frac{T^2}{K^2}\right) \text{J K}^{-1}\text{mol}^{-1}$$

over this temperature range. The ratios T/K and T^2/K^2 appear in these equations in order to have the right units for the heat capacity.

Solution

By definition, the formation reaction is written as

$$1/2 \text{ H}_2(g) + 1/2 \text{ Cl}_2(g) \rightarrow \text{HCl}(g)$$

and

$$\Delta H_{1450K}^\circ = \Delta H_{298.15K}^\circ + \int_{298.15}^{1450} \Delta C_P^\circ(T)dT$$

$$\Delta C_P^\circ(T) = \left[28.165 + 1.809 \times 10^{-3} \frac{T}{K} + 15.464 \times 10^{-7} \frac{T^2}{K^2} \right.$$

$$- \frac{1}{2} \left(29.064 - 0.8363 \times 10^{-3} \frac{T}{K} + 20.111 \times 10^{-7} \frac{T^2}{K^2} \right)$$

$$\left. - \frac{1}{2} \left(31.695 + 10.143 \times 10^{-3} \frac{T}{K} - 40.373 \times 10^{-7} \frac{T^2}{K^2} \right) \right] J\,K^{-1}mol^{-1}$$

$$= \left(-2.215 - 2.844 \times 10^{-3} \frac{T}{K} + 25.595 \times 10^{-7} \frac{T^2}{K^2} \right) J\,K^{-1}mol^{-1}$$

$$\Delta H_{1450K}^\circ = -92.3\,kJ\,mol^{-1} + \int_{298.15}^{1450} \left(-2.215 - 2.844 \times 10^{-3} \frac{T}{K} + 25.595 \times 10^{-7} \frac{T^2}{K^2} \right)$$

$$\times d\frac{T}{K} J\,mol^{-1}$$

$$= -92.3\,kJ\,mol^{-1} - 2.836\,kJ\,mol^{-1} = -95.1\,kJ\,mol^{-1}$$

In this particular case, the change in the reaction enthalpy is not large. This is the case because $\Delta C_P^\circ(T)$ is small, and not because the individual $C_{P,i}^\circ(T)$ are small.

4.5 The Experimental Determination of ΔU and ΔH for Chemical Reactions

For chemical reactions, ΔU and ΔH are generally determined through experiment. In this section, we discuss how these experiments are carried out. If some or all of the reactants or products are volatile, it is necessary to contain the reaction mixture for which ΔU and ΔH are being measured. Such an experiment can be carried out in a **bomb calorimeter,** shown schematically in Figure 4.3. In a bomb calorimeter, the reaction is carried out at constant volume. The motivation for doing so is that if $dV = 0$, $\Delta U = q_V$. Therefore, a measurement of the heat flow provides a direct measurement of $\Delta U_{reaction}^\circ$. Bomb calorimetry is restricted to reaction mixtures containing gases, because it is impractical to carry out chemical reactions at constant volume for systems consisting solely of liquids and solids, as shown in Example Problem 3.2. How are $\Delta U_{reaction}^\circ$ and $\Delta H_{reaction}^\circ$ determined in such an experiment?

The bomb calorimeter is a good illustration of how one can define the system and surroundings to simplify the analysis of an experiment. The system is defined as the contents of a stainless steel thick-walled pressure vessel, the pressure vessel itself, and the inner water bath. Given this definition of the system, the surroundings consist of the container holding the inner water bath, the outer water bath, and the rest of the universe. The outer water bath encloses the inner bath and, through a heating coil, its temperature is always held at the temperature of the inner bath. Therefore, no heat flow will occur through the system and surroundings, and $q = 0$. Because the experiment takes place at constant volume, $w = 0$. Therefore, $\Delta U = 0$. These conditions describe an isolated system of finite size that is not coupled to the rest of the universe. We are only interested in one part of this system, namely, the reaction mixture.

What are the individual components that make up ΔU? Consider the system as consisting of three subsystems: the reactants in the calorimeter, the calorimeter vessel, and the inner water bath. These three subsystems are separated by rigid diathermal walls and are in thermal equilibrium. Energy is redistributed among the subsystems through the following processes. Reactants are converted to products, the energy of the inner water bath changes because of the change in temperature, ΔT, and the same is true of the calorimeter.

FIGURE 4.3

Schematic diagram of a bomb calorimeter. The liquid or solid sample is placed in a cup suspended in the thick-walled steel bomb, which is filled with O_2 gas. The vessel is immersed in an inner water bath and its temperature monitored. The diathermal container is immersed in an outer water bath (not shown) whose temperature is maintained at the same value as the inner bath through a heating coil. By doing so, there is no heat exchange between the inner water bath and the rest of the universe.

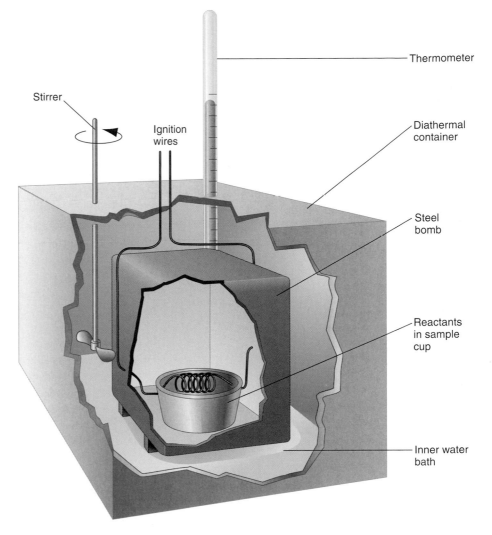

Thermometer

Stirrer

Ignition wires

Diathermal container

Steel bomb

Reactants in sample cup

Inner water bath

$$\Delta U^\circ = \frac{m_s}{M_s}\Delta U^\circ_{reaction,m} + \frac{m_{H_2O}}{M_{H_2O}}C_{H_2O,m}\Delta T + C_{calorimeter}\Delta T = 0 \qquad (4.25)$$

In Equation (4.25), ΔT is the change in the temperature of the inner water bath. The mass of water in the inner bath, m_{H_2O}, its molecular weight, M_{H_2O}, its heat capacity, $C_{H_2O,m}$, and the mass of the sample, m_s, and its molecular weight, M_s, are known. Our interest is in determining $\Delta U^\circ_{combustion,m}$. However, to determine $\Delta U^\circ_{reaction,m}$, the heat capacity of the calorimeter, $C_{calorimeter}$, must first be determined by carrying out a reaction for which $\Delta U^\circ_{reaction,m}$ is already known, as illustrated in Example Problem 4.3. To be more specific, we consider a combustion reaction between a compound and an excess of O_2:

EXAMPLE PROBLEM 4.3

When 0.972 g of cyclohexane undergoes complete combustion in a bomb calorimeter, ΔT of the inner water bath is 2.98°C. For cyclohexane, $\Delta U^\circ_{reaction,m}$ is −3913 kJ mol^{-1}. Given this result, what is the value for $\Delta U^\circ_{reaction,m}$ for the combustion of benzene if ΔT is 2.36°C when 0.857 g of benzene undergoes complete combustion in the same calorimeter? The mass of the water in the inner bath is 1.812×10^3 g, and the $C_{P,m}$ of water is 75.291 J K^{-1} mol^{-1}.

Solution

To calculate the calorimeter constant through the combustion of cyclohexane, we write Equation (4.25) in the following form:

$$C_{calorimeter} = \frac{-\dfrac{m_s}{M_s}\Delta U^\circ_{reaction} - \dfrac{m_{H_2O}}{M_{H_2O}}C_{H_2O,m}\Delta T}{\Delta T}$$

$$= \frac{\dfrac{0.972\,g}{84.16\,g\,mol^{-1}} \times 3913 \times 10^3\,J\,mol^{-1} - \dfrac{1.812\times10^3\,g}{18.02\,g\,mol^{-1}} \times 75.291\,J\,mol^{-1}\,K^{-1} \times 2.98\,^\circ C}{2.98\,^\circ C}$$

$$= 7.59 \times 10^3\,J\,(^\circ C)^{-1}$$

In calculating $\Delta U^\circ_{reaction}$ for benzene, we use the value for $C_{calorimeter}$:

$$\Delta U^\circ_{reaction} = -\frac{M_s}{m_s}\left(\frac{m_{H_2O}}{M_{H_2O}}C_{H_2O,m}\Delta T + C_{calorimeter}\Delta T\right)$$

$$= -\frac{78.12\,g\,mol^{-1}}{0.857\,g} \times \left(\begin{array}{c}\dfrac{1.812\times10^3\,g}{18.02\,g\,mol^{-1}} \times 75.291\,J\,mol^{-1}\,K^{-1} \times 2.36\,^\circ C \\ + \, 7.59\times10^3\,J(^\circ C)^{-1} \times 2.36\,^\circ C\end{array}\right)$$

$$= -3.26 \times 10^6\,J\,mol^{-1}$$

Once ΔU has been determined, ΔH can be determined using the following equation:

$$\Delta H^\circ_{reaction} = \Delta U^\circ_{reaction} + \Delta(PV) \tag{4.26}$$

For reactions involving only solids and liquids, $\Delta U \gg \Delta(PV)$ and $\Delta H \approx \Delta U$. If some of the reactants or products are gases, the small change in the temperature that is measured in a calorimetric experiment can be ignored and $\Delta(PV) = \Delta(nRT) = \Delta nRT$ and

$$\Delta H^\circ_{reaction} = \Delta U^\circ_{reaction} + \Delta nRT \tag{4.27}$$

where Δn is the change in the number of moles of gas in the overall reaction. For the first reaction of Example Problem 4.3,

$$C_6H_{12}(l) + 9O_2(g) \rightarrow 6CO_2(g) + 6H_2O(l) \tag{4.28}$$

and $\Delta n = -3$. Note that at $T = 298.15$ K, the most stable form of cyclohexane and water is a liquid.

$$\Delta H^\circ_{reaction} = \Delta U^\circ_{reaction} - 3RT = -3913 \times 10^3\,kJ\,mol^{-1} - 3 \times 8.314\,J\,K^{-1}mol^{-1} \tag{4.29}$$
$$\times\,298.15\,K$$
$$= -3920 \times 10^3\,J\,mol^{-1}$$

For this reaction, $\Delta H^\circ_{reaction}$ and $\Delta U^\circ_{reaction}$ differ by only 0.2%.

If the reaction under study does not involve gases or highly volatile liquids, there is no need to operate under constant volume conditions. It is preferable to carry out the reaction at constant P, and $\Delta H^\circ_{reaction}$ is directly determined because $\Delta H^\circ_{reaction} = q_P$. A vacuum-insulated vessel with a loosely fitting stopper as shown in Figure 4.4 is adequate for many purposes and can be treated as an isolated composite system. Equation (4.25) takes the following form for constant pressure calorimetry involving the solution of a salt in water:

$$\Delta H^\circ_{reaction} = \frac{m_s}{M_s}\Delta H^\circ_{solution,m} + \frac{m_{H_2O}}{M_{H_2O}}C_{H_2O,m}\Delta T + C_{calorimeter}\Delta T = 0 \tag{4.30}$$

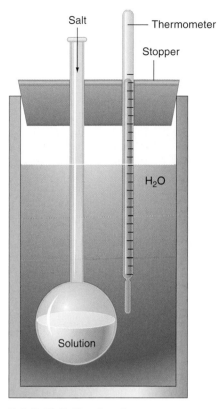

FIGURE 4.4
Schematic diagram of a constant pressure calorimeter suitable for measuring the enthalpy of solution of a salt in a solution.

Labels on figure: Salt, Thermometer, Stopper, H_2O, Solution

Note that in this case the water in the round bottomed reaction vessel is part of the reaction mixture. Because $\Delta(PV)$ is negligibly small for the solution of a salt in a solvent, $\Delta U_{solution} = \Delta H_{solution}$. The solution must be stirred to ensure that equilibrium is attained before ΔT is measured.

EXAMPLE PROBLEM 4.4

The enthalpy of solution for the reaction

$$Na_2SO_4(s) + H_2O(l) \rightarrow 2Na^+(aq) + SO_4{}^{2-}(aq)$$

is determined in a **constant pressure calorimeter.** The calorimeter constant was determined to be 342.5 J K^{-1}. When 1.423 g of Na_2SO_4 is dissolved in 100.34 g of $H_2O(l)$, $\Delta T = 0.037$ K. Calculate ΔH_m for Na_2SO_4 from these data. Compare your result with that calculated using the standard enthalpies of formation in Table 4.1 (Appendix A, Data Tables) and in Chapter 10 in Table 10.1.

Solution

$$\Delta H^\circ_{solution,m} = -\frac{M_s}{m_s}\left(\frac{m_{H_2O}}{M_{H_2O}}C_{H_2O,m}\Delta T + C_{calorimeter}\Delta T\right)$$

$$= -\frac{142.04\ \text{g mol}^{-1}}{1.423\ \text{g}} \times \left(\begin{array}{l}\dfrac{100.34\ \text{g}}{18.02\ \text{g mol}^{-1}} \times 75.3\ \text{J K}^{-1}\text{mol}^{-1} \times 0.037\ \text{K} \\ + 342.5\ \text{J K}^{-1} \times 0.037\ \text{K}\end{array}\right)$$

$$= -2.8 \times 10^3\ \text{J mol}^{-1}$$

We next calculate $\Delta H^\circ_{solution}$ using the data tables.

$$\Delta H^\circ_{solution} = 2\Delta H^\circ_f(Na^+, aq) + \Delta H^\circ_f(SO_4^{2-}, aq) - \Delta H^\circ_f(Na_2SO_4, s)$$
$$= 2 \times (-240.1\ \text{kJ mol}^{-1}) - 909.3\ \text{kJ mol}^{-1} + 1387.1\ \text{kJ mol}^{-1}$$
$$= -2.4\ \text{kJ mol}^{-1}$$

The agreement between the calculated and experimental results is satisfactory.

4.6 Differential Scanning Calorimetry

The constant volume and constant pressure calorimeters described in the preceding section are well suited for measurements on individual samples. Suppose that it is necessary to determine the **enthalpy of fusion** of a dozen related solid materials, each of which is only available in a small quantity. Because experiments are carried out individually, constant volume and constant pressure calorimeters are not well suited to the rapid determination of thermodynamic properties of a series of materials. The technique of choice for these measurements is **differential scanning calorimetry.** The experimental apparatus for such an experiment is shown schematically in Figure 4.5. The word *differential* appears in the name of the technique because the uptake of heat is measured relative to that for a reference material, and *scanning* refers to the fact that the temperature of the sample is varied linearly with time.

The temperature of the cylindrical enclosure is increased linearly with time using a power supply. Heat flows from the enclosure through the disk to the sample because of the temperature gradient generated by the heater. Because all samples and the reference are equidistant from the enclosure, the heat flow to each sample is the same. The reference material is chosen such that its melting point is not in the range of that of the samples.

The measured temperature difference between one sample and the reference, ΔT, is proportional to the difference in the rate of heat uptake of the sample and the reference:

$$\Delta T = \alpha \frac{dq_P}{dt} = \alpha\left(\frac{dq_{reference}}{dt} - \frac{dq_{sample}}{dt}\right) \qquad (4.31)$$

FIGURE 4.5
A differential scanning calorimeter consists of a massive enclosure and lid that are heated to the temperature T using a resistive heater. A support disk in good thermal contact with the enclosure supports multiple samples and a reference material. The temperatures of each of the samples and the reference are measured with a thermocouple.

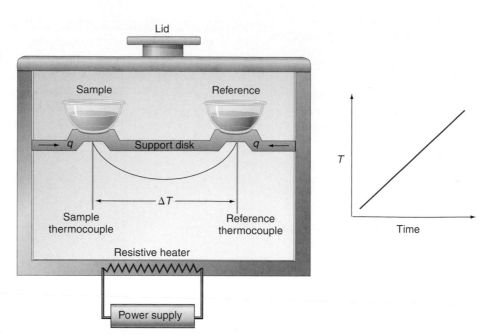

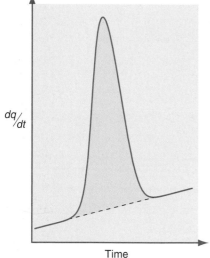

dq/dt

Time

FIGURE 4.6
At temperatures different from the melting temperature, the rate of heat uptake of the sample and the reference are nearly identical. However, at the time corresponding to $T = T_{melting}$, the uptake rate of heat for the sample differs significantly from that of the reference. Kinetic delays due to finite rates of heat conduction broaden the sharp melting transition to finite range.

where α is a calorimeter constant that is determined experimentally. In the temperature range over which no phase transition occurs, $\Delta T \sim 0$. However, as a sample melts, it takes up more heat than the reference and its temperature remains constant while that of the reference increases. The temperature difference between the sample and the reference has the form shown in Figure 4.6 as a function of time. Because the relationship between the time and temperature is known, the time scale can be converted to a temperature scale. Given that the sample has a well-defined melting point, why is the peak in Figure 4.6 broad? The system experiences kinetic delays because the heat flow through the disk is not instantaneous. Therefore, the experimentally determined function dq_P/dt has a finite width and the form of a Gaussian curve. Because the vertical axis is dq_P/dt and the horizontal axis is time, the integral of the area under the curve is

$$\int \left(\frac{dq_P}{dt}\right) dt = q_P = \Delta H \tag{4.32}$$

Therefore, ΔH_{fusion} can be ascertained by determining the area under the curve representing the output data.

The significant advantage of differential scanning calorimetry over constant pressure or constant volume calorimetry is the ability to investigate several samples in parallel and the ability to work with small amounts of sample. As is apparent from Figure 4.5, the apparatus shown is limited to working with solids or to solutions in which phase transitions occur. For example, the technique has been used to determine the temperature range over which proteins undergo the conformational changes associated with denaturation.

Vocabulary

bomb calorimeter	differential scanning calorimetry	enthalpy of reaction
bond energy	endothermic	exothermic
bond enthalpy	enthalpy of formation	Hess's law
constant pressure calorimeter	enthalpy of fusion	standard state

Questions on Concepts

Q4.1 Under what conditions are ΔH and ΔU for a reaction involving gases and/or liquids or solids identical?

Q4.2 If the ΔH_f° for the chemical compounds involved in a reaction are available at a given temperature, how can the reaction enthalpy be calculated at another temperature?

Q4.3 Does the enthalpy of formation of compounds containing a certain element change if the enthalpy of formation of the element under standard state conditions is set equal to 100 kJ mol^{-1} rather than to zero? If it changes, how will it change for the compound A_nB_m if the formation enthalpy of element A is set equal to 100 kJ mol^{-1}?

Q4.4 Is the enthalpy for breaking the first C—H bond in methane equal to the average C—H bond enthalpy in this molecule? Explain your answer.

Q4.5 Why is it valid to add the enthalpies of any sequence of reactions to obtain the enthalpy of the reaction that is the sum of the individual reactions?

Q4.6 The reactants in the reaction $2NO(g) + O_2(g) \rightarrow 2NO_2(g)$ are initially at 298 K. Why is the reaction enthalpy the same if the reaction is (a) constantly kept at 298 K or (b) if the reaction temperature is not controlled and the heat flow to the surroundings is measured only after the temperature of the products is returned to 298 K?

Q4.7 In calculating $\Delta H_{reaction}^\circ$ at 285.15 K, only the ΔH_f° of the compounds that take part in the reactions listed in Tables 4.1 and 4.2 (Appendix A, Data Tables) are needed. Is this statement also true if you want to calculate $\Delta H_{reaction}^\circ$ at 500 K?

Q4.8 What is the point of having an outer water bath in a bomb calorimeter (see Figure 4.3), especially if its temperature is always equal to that of the inner water bath?

Q4.9 What is the advantage of a differential scanning calorimeter over a bomb calorimeter in determining the enthalpy of fusion of a series of samples?

Q4.10 You wish to measure the heat of solution of NaCl in water. Would the calorimetric technique of choice be at constant pressure or constant volume? Why?

Problems

Problem numbers in **red** indicate that the solution to the problem is given in the *Student's Solutions Manual*.

P4.1 Calculate $\Delta H_{reaction}^\circ$ and $\Delta U_{reaction}^\circ$ at 298.15 K for the following reactions:

a. $4NH_3(g) + 6NO(g) \rightarrow 5N_2(g) + 6H_2O(g)$

b. $2NO(g) + O_2(g) \rightarrow 2NO_2(g)$

c. $TiCl_4(l) + 2H_2O(l) \rightarrow TiO_2(s) + 4HCl(g)$

d. $2NaOH(aq) + H_2SO_4(aq) \rightarrow Na_2SO_4(aq) + 2H_2O(l)$
   Assume complete dissociation of NaOH, H_2SO_4, and Na_2SO_4

e. $CH_4(g) + H_2O(g) \rightarrow CO(g) + 3H_2(g)$

f. $CH_3OH(g) + CO(g) \rightarrow CH_3COOH(l)$

P4.2 Calculate $\Delta H_{reaction}^\circ$ and $\Delta U_{reaction}^\circ$ for the oxidation of benzene. Also calculate

$$\frac{\Delta H_{reaction}^\circ - \Delta U_{reaction}^\circ}{\Delta H_{reaction}^\circ} \qquad T = 298K$$

P4.3 Use the tabulated values of the enthalpy of combustion of benzene and the enthalpies of formation of $CO_2(g)$ and $H_2O(l)$ to determine ΔH_f° for benzene.

P4.4 Calculate ΔH_f° for $N_2(g)$ at 650 K using the temperature dependence of the heat capacities from the data tables. How large is the relative error if the molar heat capacity is assumed to be constant at its value of 298.15 K over the temperature interval? *298~650K at P°*
OH for heating N_2 from

P4.5 Several reactions and their standard reaction enthalpies at 25°C are given here:

	$\Delta H_{reaction}^\circ$ (kJ mol^{-1})
$CaC_2(s) + 2H_2O(l) \rightarrow Ca(OH)_2(s)$ + $C_2H_2(g)$	−127.9
$Ca(s) + 1/2\, O_2(g) \rightarrow CaO(s)$	−635.1
$CaO(s) + H_2O(l) \rightarrow Ca(OH)_2(s)$	−65.2

The standard enthalpies of combustion of graphite and $C_2H_2(g)$ are −393.51 and −1299.58 kJ mol^{-1}, respectively. Calculate the standard enthalpy of formation of $CaC_2(s)$ at 25°C.

P4.6 From the following data at 25°C, calculate the standard enthalpy of formation of FeO(s) and of $Fe_2O_3(s)$:

	$\Delta H_{reaction}^\circ$ (kJ mol^{-1})
$Fe_2O_3(s) + 3C(graphite) \rightarrow 2Fe(s)$ + $3CO(g)$	492.6
$FeO(s) + C(graphite) \rightarrow Fe(s) + CO(g)$	155.8
$C(graphite) + O_2(g) \rightarrow CO_2(g)$	−393.51
$CO(g) + 1/2\, O_2(g) \rightarrow CO_2(g)$	−282.98

P4.7 Calculate ΔH_f° for NO(g) at 840 K assuming that the heat capacities of reactants and products are constant over the temperature interval at their values at 298.15 K.

P4.8 Calculate $\Delta H_{reaction}^\circ$ at 650 K for the reaction $4NH_3(g) + 6NO(g) \rightarrow 5N_2(g) + 6H_2O(g)$ using the temperature dependence of the heat capacities from the data tables.

P4.9 From the following data at 298.15 K as well as data in Table 4.1 (Appendix A, Data Tables), calculate the standard enthalpy of formation of $H_2S(g)$ and of $FeS_2(s)$:

$$\Delta H^\circ_{reaction} \text{ (kJ mol}^{-1})$$

$Fe(s) + 2H_2S(g) \rightarrow FeS_2(s) + 2H_2(g)$	-137.0
$H_2S(g) + 3/2\ O_2(g) \rightarrow H_2O(l) + SO_2(g)$	-562.0

P4.10 Calculate the average C—H bond enthalpy in methane using the data tables. Calculate the percent error in equating the average C—H bond energy in Table 4.3 with the bond enthalpy.

P4.11 Use the average bond energies in Table 4.3 to estimate ΔU for the reaction $C_2H_4(g) + H_2(g) \rightarrow C_2H_6(g)$. Also calculate $\Delta U_{reaction}$ from the tabulated values of ΔH°_f for reactant and products (Appendix A, Data Tables). Calculate the percent error in estimating $\Delta U^\circ_{reaction}$ from the average bond energies for this reaction.

P4.12 Calculate the standard enthalpy of formation of $FeS_2(s)$ at 300°C from the following data at 25°C. Assume that the heat capacities are independent of temperature.

Substance	Fe(s)	$FeS_2(s)$	$Fe_2O_3(s)$	S(rhombic)	$SO_2(g)$
ΔH°_f (kJ mol^{-1})			-824.2		-296.81
$C_{P,m}/R$	3.02	7.48		2.72	

You are also given that for the reaction $2FeS_2(s) + 11/2 O_2(g) \rightarrow Fe_2O_3(s) + 4\ SO_2(g)$, $\Delta H^\circ_{reaction} = -1655$ kJ mol^{-1}.

P4.13 At 1000 K, $\Delta H^\circ_{reaction} = -123.77$ kJ mol^{-1} for the reaction $N_2(g) + 3H_2(g) \rightarrow 2NH_3(g)$, with $C_{P,m} = 3.502R$, $3.466R$, and $4.217R$ for $N_2(g)$, $H_2(g)$, and $NH_3(g)$, respectively. Calculate ΔH°_f of $NH_3(g)$ at 300 K from this information. Assume that the heat capacities are independent of temperature.

P4.14 At 298 K, $\Delta H^\circ_{reaction} = 131.28$ kJ mol^{-1} for the reaction $C(graphite) + H_2O(g) \rightarrow CO(g) + H_2(g)$, with $C_{P,m} = 8.53, 33.58, 29.12$, and 28.82 J K^{-1} mol^{-1} for graphite, $H_2O(g)$, $CO(g)$, and $H_2(g)$, respectively. Calculate $\Delta H^\circ_{reaction}$ at 125°C from this information. Assume that the heat capacities are independent of temperature.

P4.15 From the following data, calculate $\Delta H^\circ_{reaction,391.4K}$ for the reaction $CH_3COOH(g) + 2O_2(g) \rightarrow 2H_2O(g) + 2CO_2(g)$:

$$\Delta H^\circ_{reaction} \text{ (kJ mol}^{-1})$$

$CH_3COOH(l) + 2O_2(g) \rightarrow 2H_2O(l) + 2CO_2(g)$	-871.5
$H_2O(l) \rightarrow H_2O(g)$	40.656
$CH_3COOH(l) \rightarrow CH_3COOH(g)$	24.4

Values for $\Delta H^\circ_{reaction}$ for the first two reactions are at 298.15 K, and for the third reaction at 391.4 K.

Substance	$CH_3COOH(l)$	$O_2(g)$	$CO_2(g)$	$H_2O(l)$	$H_2O(g)$
$C_{P,m}/R$	14.9	3.53	4.46	9.055	4.038

P4.16 Consider the reaction $TiO_2(s) + 2\ C(graphite) + 2\ Cl_2(g) \rightarrow 2\ CO(g) + TiCl_4(l)$ for which $\Delta H^\circ_{reaction,298K} = -80$ kJ mol^{-1}. Given the following data at 25°C,

(a) calculate $\Delta H^\circ_{reaction}$ at 135.8°C, the boiling point of $TiCl_4$, and (b) calculate ΔH°_f for $TiCl_4\ (l)$ at 25°C:

Substance	$TiO_2(s)$	$Cl_2(g)$	$C(graphite)$	$CO(g)$	$TiCl_4(l)$
ΔH°_f (kJ mol^{-1})	-945			-110.5	
$C_{P,m}$ (J K^{-1} mol^{-1})	55.06	33.91	8.53	29.12	145.2

Assume that the heat capacities are independent of temperature.

P4.17 Use the following data at 25°C to complete this problem:

$$\Delta H^\circ_{reaction} \text{ (kJ mol}^{-1})$$

$1/2\ H_2(g) + 1/2\ O_2(g) \rightarrow OH(g)$	38.95
$H_2(g) + 1/2\ O_2(g) \rightarrow H_2O(g)$	-241.814
$H_2(g) \rightarrow 2H(g)$	435.994
$O_2(g) \rightarrow 2O(g)$	498.34

Calculate $\Delta H^\circ_{reaction}$ for

a. $OH(g) \rightarrow H(g) + O(g)$

b. $H_2O(g) \rightarrow 2H(g) + O(g)$

c. $H_2O(g) \rightarrow H(g) + OH(g)$

Assuming ideal gas behavior, calculate $\Delta H^\circ_{reaction}$ and $\Delta U^\circ_{reaction}$ for all three reactions.

P4.18 Given the data in Table 4.1 (Appendix A, Data Tables) and the following information, calculate the single bond enthalpies and energies for Si–F, Si–Cl, C–F, N–F, O–F, H–F:

Substance	$SiF_4(g)$	$SiCl_4(g)$	$CF_4(g)$	$NF_3(g)$	$OF_2(g)$	$HF(g)$
ΔH°_f(kJ mol^{-1})	-1614.9	-657.0	-925	-125	-22	-271

P4.19 Given the data in Table 4.3 and the Data Tables, calculate the bond enthalpy and energy of the following:

a. The C–H bond in CH_4

b. The C–C single bond in C_2H_6

c. The C=C double bond in C_2H_4

Use your result from part a to solve parts b and c.

P4.20 A sample of K(s) of mass 2.140 g undergoes combustion in a constant volume calorimeter. The calorimeter constant is 1849 J K^{-1}, and the measured temperature rise in the inner water bath containing 1450 g of water is 2.62 K. Calculate ΔU°_f and ΔH°_f for K_2O.

P4.21 Benzoic acid, 1.35 g, is reacted with oxygen in a constant volume calorimeter to form $H_2O(l)$ and $CO_2(g)$. The mass of the water in the inner bath is 1.240×10^3 g. The temperature of the calorimeter and its contents rises 3.45 K as a result of this reaction. Calculate the calorimeter constant.

P4.22 A sample of $Na_2SO_4(s)$ is dissolved in 225 g of water at 298 K such that the solution is 0.200 molar in Na_2SO_4. A temperature rise of 0.101°C is observed. The calorimeter constant is 330 J K^{-1}. Calculate the enthalpy of solution of Na_2SO_4 in water at this concentration. Compare your result with that calculated using the data in Table 4.1 (Appendix A, Data Tables).

CHAPTER 5

Entropy and the Second and Third Laws of Thermodynamics

Real-world processes have a natural direction of change. Heat flows from hotter bodies to colder bodies, and gases mix rather than separate. Entropy, designated by S, is the state function that predicts the direction of natural, or spontaneous change, and entropy increases for a spontaneous change in an isolated system. For a spontaneous change in a system interacting with its environment, the sum of the entropy of the system and that of the surroundings increases. In this chapter, we introduce entropy, derive the conditions for spontaneity, and show how S varies with the macroscopic variables P, V, and T. ■

5.1 The Universe Has a Natural Direction of Change

To this point, we have discussed q and w, as well as U and H. The first law of thermodynamics states that in any process, the total energy of the universe remains constant. Although the first law requires that a thermodynamic process conserve energy, it does not predict which of several possible processes will occur. Consider the following two examples. A metal rod initially at a uniform temperature could, in principle, undergo a spontaneous transformation in which one end becomes hot and the other end becomes cold without being in conflict with the first law, as long as the total energy of the rod remains constant. However, experience demonstrates that this does not occur. Similarly, an ideal gas that is uniformly distributed in a rigid adiabatic container could undergo a spontaneous transformation such that all of the gas moves to one-half of the container, leaving a vacuum in the other half. Because for an ideal gas, $(\partial U/\partial V)_T = 0$, the energy of the initial and final states is the same. Neither of these transformations violates the first law of thermodynamics—and yet neither occurs.

Experience tells us that there is a natural direction of change in these two processes. A metal rod with a temperature gradient always reaches a uniform temperature at some time after it has been isolated from a heat source. A gas confined to one-half of a container with a vacuum in the other half distributes itself uniformly throughout the container if a valve separating the two parts is opened. The transformations described in the previous paragraph are **unnatural transformations.** The word *unnatural* is used to indicate that in many human lifetimes, these transformations do not occur. By contrast, the reverse processes, in which the temperature gradient along the rod disappears and the gas becomes distributed uniformly throughout the container, are **natural transformations,** also called **spontaneous processes.** A spontaneous process is one that if repeated again and again over the course of a human lifetime leads to the same outcome.

Our experience is sufficient to predict the direction of spontaneous change for the two examples cited, but can the direction of spontaneous change be predicted in less obvious cases? In this chapter, we show that there is a thermodynamic function called entropy that allows one to predict the direction of spontaneous change for a system in a given initial state. Assume that a reaction vessel contains a given number of moles of N_2, H_2, and NH_3 at 600 K and at a total pressure of 280 bar. An iron catalyst is introduced that allows the mixture of gases to equilibrate according to $1/2 N_2 + 3/2 H_2 \rightleftarrows NH_3$. What is the direction of spontaneous change and what are the partial pressures of the three gases at equilibrium? The answer to this question is obtained by calculating the entropy change in the system and the surroundings.

Most students are initially uncomfortable when working with entropy, and there are good reasons for this unease. Entropy is further removed from direct experience than energy, work, or heat. At this stage, it is more important to learn how to calculate changes in entropy than it is to "understand" entropy. In a practical sense, understanding a concept is equivalent to being able to work with that concept. We all know how to work with gravity and can predict the direction in which a mass will move if it is released from our grasp. We can calculate the thrust that a rocket must develop in order to escape from the Earth's gravitational field. However, how many of us could say something sensible if asked what the origin of gravity is? Why do two masses attract one another with an inverse square dependence on distance? The fact that these questions are not easily answered is not a hindrance to living in comfort with gravity. Similarly, if confidence is developed in being able to calculate the change in entropy for processes, the first steps are taken toward "understanding" entropy.

A deeper explanation of entropy will be presented in Chapter 15, on the basis of a microscopic model of matter. Historically, the importance of entropy as a state function that could predict the direction of natural change was known by Carnot in his analysis of steam engines in 1824, nearly 50 years before entropy was understood at a microscopic level by Boltzmann. Even though thermodynamics has no need for a microscopic basis, the underlying microscopic basis of entropy is discussed briefly here to make it less mysterious in the rest of this chapter.

At the microscopic level, matter consists of atoms or molecules that have energetic degrees of freedom (i.e., translational, rotational, vibrational, and electronic), each of which is associated with discrete energy levels that can be calculated using quantum mechanics. Quantum mechanics also characterizes a molecule by a state associated with a set of quantum numbers and a molecular energy. Entropy serves as a measure of the number of quantum states accessible to a macroscopic system at a given energy. Quantitatively, $S = k \ln W$, where W provides a measure of the number of states accessible to the system, and $k = R/N_A$. As demonstrated later in this chapter, the entropy of an isolated system is maximized at equilibrium. Therefore, the approach to equilibrium can be envisioned as a process in which the system achieves the distribution of energy among molecules that corresponds to a maximum value of W and, correspondingly, to a maximum in S.

5.2 Heat Engines and the Second Law of Thermodynamics

The development of entropy used here follows the historical route by which this state function was first introduced. The concept of entropy arose as 19th-century scientists attempted to maximize the work output of engines. In steam engines, heat produced by a wood or coal fire is partially converted to work done on the surroundings by the engine (the system). An automobile engine operates in a cyclical process of fuel intake, compression, ignition and expansion, and exhaust, which occurs several thousand times per minute and is used to perform work on the surroundings. Because the work produced by such engines is

FIGURE 5.1
A schematic depiction of a heat engine is shown. Changes in temperature of the working substance brought about by contacting the cylinder with hot or cold reservoirs generate a linear motion that is mechanically converted to a rotary motion, which is used to do work.

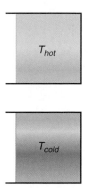
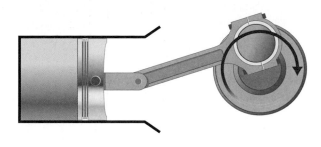

a result of the heat released in a combustion process, they are referred to as **heat engines.** A heat engine is depicted in Figure 5.1. A working substance, which is the system (in this case an ideal gas), is confined in a piston and cylinder assembly with diathermal walls. This assembly can be brought into contact with a hot reservoir at T_{hot} or a cold reservoir at T_{cold}. The expansion or contraction of the gas caused by changes in its temperature drives the piston in or out of the cylinder. This linear motion is converted to circular motion using an eccentric, and the rotary motion is used to do work in the surroundings.

The efficiency of a heat engine is of particular interest. Experience shows that work can be converted to heat with 100% efficiency. Consider an example from calorimetry, discussed in Chapter 4. Electrical work can be done on a resistive heater immersed in a water bath. Observations reveal that the internal energy of the heater is not increased significantly in this process and that the temperature of the water bath increases. One can conclude that all of the electrical work done on the heater has been converted to heat, resulting in an increase in the temperature of the water. What is the maximum theoretical efficiency of the reverse process, the conversion of heat to work? As shown later, it is less than 100%, which limits the efficiency of an automobile engine. *There is a natural asymmetry between converting work to heat and converting heat to work. Thermodynamics provides an explanation for this asymmetry.*

As discussed in Section 2.7, the maximum work output in an isothermal expansion occurs in a reversible process. For this reason, the efficiency of a reversible heat engine is calculated, even though any real engine operates irreversibly, because the efficiency of a reversible engine is an upper bound to the efficiency of a real engine. The purpose of this engine is to convert heat into work by exploiting the spontaneous tendency of heat to flow from a hot reservoir to a cold reservoir. This engine operates in a cycle of reversible expansions and compressions of an ideal gas in a piston and cylinder assembly, and it does work on the surroundings.

This reversible cycle is shown in Figure 5.2 in a P-V diagram. The expansion and compression steps are designed so that the engine returns to its initial state after four steps. Recall that the area within the cycle equals the work done by the engine. Beginning at point a, the first segment is a reversible isothermal expansion in which the gas absorbs heat from the reservoir at T_{hot}, and also does work on the surroundings. In the second segment, the gas expands further, this time adiabatically. Work is also done on the surroundings in this step. At the end of the second segment, the gas has cooled to the temperature T_{cold}. The third segment is an isothermal compression in which the surroundings do work on the system and heat is absorbed by the cold reservoir. In the final segment, the gas is compressed to its initial volume, this time adiabatically. Work is done on the system in this segment, and the temperature rises to its initial value, T_{hot}. In summary, heat is taken up by the engine in the first segment at T_{hot}, and released to the surroundings in the third segment at T_{cold}. Work is done on the surroundings in the first two segments and on the system in the last two segments. An engine is only useful if net work is done on the surroundings, that is, if the magnitude of the work done in the first two steps is greater than the magnitude of the work done in the last two steps. The efficiency of the engine can be calculated by comparing the net work and the heat taken up by the engine from the hot reservoir.

FIGURE 5.2

A reversible Carnot cycle for a sample of an ideal gas working substance is shown on an indicator diagram. The cycle consists of two adiabatic and two isothermal segments. The arrows indicate the direction in which the cycle is traversed. The insets show the volume of gas and the coupling to the reservoirs at the beginning of each successive segment of the cycle. The coloring of the contents of the cylinder indicates the presence of the gas and not its temperature. The volume of the cylinder shown is that at the beginning of the appropriate segment.

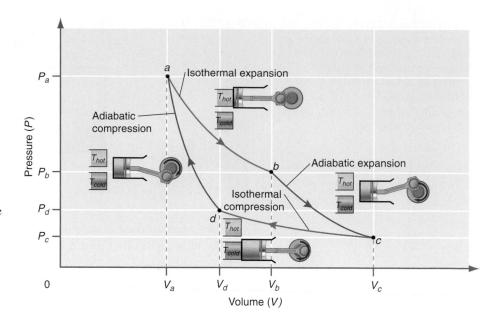

Before carrying out this calculation, the rationale for the design of this reversible cycle should be discussed in more detail. To avoid losing heat to the surroundings at temperatures between T_{hot} and T_{cold}, the adiabatic segments 2 ($b \rightarrow c$) and 4 ($d \rightarrow a$) are used to move the gas between these temperatures. To absorb heat only at T_{hot} and release heat only at T_{cold}, segments 1 ($a \rightarrow b$) and 3 ($c \rightarrow d$) must be isothermal. The reason for using alternating isothermal and adiabatic segments is that no two isotherms at different temperatures intersect, and no two adiabats starting from two different temperatures intersect. Therefore, it is impossible to create a closed cycle of nonzero area in an indicator diagram out of isothermal or adiabatic segments alone. However, net work can be done using alternating adiabatic and isothermal segments. The reversible cycle depicted in Figure 5.2 is called a **Carnot cycle,** after the French engineer who first studied such cycles.

The efficiency of the Carnot cycle can be determined by calculating q, w, and ΔU for each segment of the cycle, assuming that the working substance is an ideal gas. The results are shown first in a qualitative fashion in Table 5.1. The appropriate signs for q and w are indicated. If $\Delta V > 0$, $w < 0$ for the segment, and the corresponding entry for work has a negative sign. For the isothermal segments, q and w have opposite signs from the first law because $\Delta U = 0$ for an isothermal process involving an ideal gas. From Table 5.1, it is seen that

$$w_{cycle} = w_{cd} + w_{da} + w_{ab} + w_{bc} \quad \text{and} \quad q_{cycle} = q_{ab} + q_{cd} \qquad (5.1)$$

Because $\Delta U_{cycle} = 0$,

$$w_{cycle} = -(q_{cd} + q_{ab}) \qquad (5.2)$$

By comparing the areas under the two expansion segments with those under the two compression segments in the indicator diagram in Figure 5.2, you can see that the total

TABLE 5.1

Heat, Work, and ΔU for the Reversible Carnot Cycle

Segment	Initial State	Final State	q	w	ΔU
$a \rightarrow b$	P_a, V_a, T_{hot}	P_b, V_b, T_{hot}	q_{ab} (+)	w_{ab} (−)	$\Delta U_{ab} = 0$
$b \rightarrow c$	P_b, V_b, T_{hot}	P_c, V_c, T_{cold}	0	w_{bc} (−)	$\Delta U_{bc} = w_{bc}$ (−)
$c \rightarrow d$	P_c, V_c, T_{cold}	P_d, V_d, T_{cold}	q_{cd} (−)	w_{cd} (+)	$\Delta U_{cd} = 0$
$d \rightarrow a$	P_d, V_d, T_{cold}	P_a, V_a, T_{hot}	0	w_{da} (+)	$\Delta U_{da} = w_{da}$ (+)
Cycle	P_a, V_a, T_{hot}	P_a, V_a, T_{hot}	$q_{ab} + q_{cd}$ (+)	$w_{ab} + w_{bc} + w_{cd} + w_{da}$ (−)	$\Delta U_{cycle} = 0$

5.1 The Reversible Carnot Cycle

work as seen from the system is negative, meaning that work is done on the surroundings in each cycle. Using this result, we arrive at an important conclusion that relates the heat flow in the two isothermal segments:

$$w_{cycle} < 0, \text{ so that } |q_{ab}| > |q_{cd}| \tag{5.3}$$

In this engine, more heat is withdrawn from the hot reservoir than is deposited in the cold reservoir. It is useful to make a model of this heat engine that indicates the relative magnitude and direction of the heat and work flow, as shown in Figure 5.3a. The figure makes it clear that not all of the heat withdrawn from the higher temperature reservoir is converted to work done by the system (the green disc) on the surroundings.

The efficiency, ε, of the reversible Carnot engine is defined as the ratio of the work output to the heat withdrawn from the hot reservoir. Referring to Table 5.1,

$$\varepsilon = \frac{q_{ab} + q_{cd}}{q_{ab}} = 1 - \frac{|q_{cd}|}{|q_{ab}|} < 1 \text{ because } |q_{ab}| > |q_{cd}|, q_{ab} > 0, \text{ and } q_{cd} < 0 \tag{5.4}$$

Equation (5.4) shows that the efficiency of a heat engine operating in a reversible Carnot cycle is always less than one. Equivalently, not all of the heat withdrawn from the hot reservoir can be converted to work.

These considerations on the efficiency of heat engines led to the Kelvin–Planck formulation of the **second law of thermodynamics:**

> It is impossible for a system to undergo a cyclic process whose sole effects are the flow of heat into the system from a heat reservoir and the performance of an equivalent amount of work by the system on the surroundings.

An alternative, but equivalent, statement of the second law by Clausius is as follows:

> It is impossible for a system to undergo a cyclic process whose sole effects are the flow of heat into the system from a cold reservoir and the flow of an equivalent amount of heat out of the system into a hot reservoir.

Hot reservoir

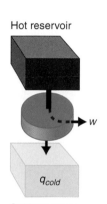

Cold reservoir

(a)

Hot reservoir

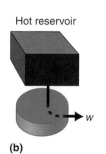

(b)

FIGURE 5.3
(a) A schematic model of the heat engine operating in a reversible Carnot cycle.
(b) The second law of thermodynamics asserts that it is impossible to construct a heat engine that operates using a single heat reservoir and converts the heat withdrawn from the reservoir into work with 100% efficiency as shown. The green disk represents the heat engine.

The second law asserts that the heat engine depicted in Figure 5.3b cannot be constructed. The second law has been put to the test many times by inventors who have claimed that they have invented an engine that has an efficiency of 100%. No such claim has ever been validated. To test the assertion made in this statement of the second law, imagine that such an engine has been invented. We mount it on a boat in Seattle and set off on a journey across the Pacific Ocean. Heat is extracted from the ocean, which is the single heat reservoir, and converted entirely to work in the form of a rapidly rotating propeller. Because the ocean is huge, the decrease in its temperature as a result of withdrawing heat is negligible. By the time we arrive in Japan, not a gram of diesel fuel has been used, because all the heat needed to power the boat has been extracted from the ocean. The money that was saved on fuel is used to set up an office and begin marketing this wonder engine. Does this scenario sound too good to be true? It is. Such an impossible engine is called a **perpetual motion machine of the second kind,** because it violates the second law of thermodynamics. A **perpetual motion machine of the first kind** violates the first law.

The first statement of the second law can be understood using an indicator diagram. For an engine to produce work, the area of the cycle in a P–V diagram must be greater than zero. However, this is impossible in a simple cycle using a single heat reservoir. If $T_{hot} = T_{cold}$ in Figure 5.2, the cycle $a \rightarrow b \rightarrow c \rightarrow d \rightarrow a$ collapses to a line, and the area of the cycle is zero. An arbitrary reversible cycle can be constructed that does not consist

FIGURE 5.4
An arbitrary reversible cycle, indicated by the ellipse, can be approximated to any desired accuracy by a sequence of alternating adiabatic and isothermal segments.

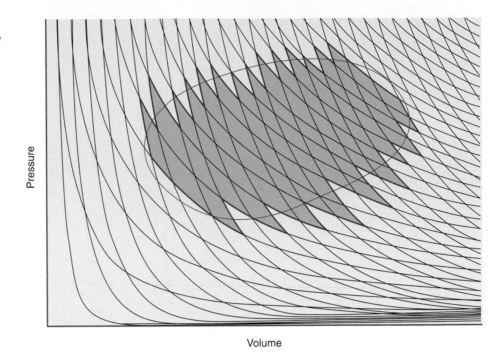

of individual adiabatic and isothermal segments. However, as shown in Figure 5.4, any reversible cycle can be approximated by a succession of adiabatic and isothermal segments, an approximation that becomes exact as the length of each segment approaches zero. It can be shown that the efficiency of such a cycle is also given by Equation (5.9) so that the efficiency of all heat engines operating in any reversible cycle between the same two temperatures, T_{hot} and T_{cold}, is identical.

A more useful form than Equation (5.4) for the efficiency of a reversible heat engine can be derived by using the fact that the working substance in the engine is an ideal gas. Calculating the work flow in each of the four segments of the Carnot cycle using the results of Sections 2.7 and 2.9,

$$w_{ab} = -nRT_{hot} \ln \frac{V_b}{V_a} \qquad w_{ab} < 0 \text{ because } V_b > V_a \tag{5.5}$$

$$w_{bc} = nC_{V,m}(T_{cold} - T_{hot}) \qquad w_{bc} < 0 \text{ because } T_{cold} < T_{hot}$$

$$w_{cd} = -nRT_{cold} \ln \frac{V_d}{V_c} \qquad w_{cd} > 0 \text{ because } V_d < V_c$$

$$w_{da} = nC_{V,m}(T_{hot} - T_{cold}) \qquad w_{da} > 0 \text{ because } T_{hot} > T_{cold}$$

As derived in Section 2.10, the volume and temperature in the reversible adiabatic segments are related by

$$T_{hot}V_b^{\gamma-1} = T_{cold}V_c^{\gamma-1} \quad \text{and} \quad T_{cold}V_d^{\gamma-1} = T_{hot}V_a^{\gamma-1} \tag{5.6}$$

You will show in the end-of-chapter problems that V_c and V_d can be eliminated from the set of Equations (5.5) to yield

$$w_{cycle} = -nR(T_{hot} - T_{cold}) \ln \frac{V_b}{V_a} < 0 \tag{5.7}$$

Because $\Delta U_{a \to b} = 0$, the heat withdrawn from the hot reservoir is

$$q_{ab} = -w = nRT_{hot} \ln \frac{V_b}{V_a} \tag{5.8}$$

and the efficiency of the reversible Carnot heat engine with an ideal gas as the working substance is

$$\varepsilon = \frac{\left|w_{cycle}\right|}{q_{ab}} = \frac{T_{hot} - T_{cold}}{T_{hot}} = 1 - \frac{T_{cold}}{T_{hot}} < 1 \qquad (5.9)$$

The efficiency of this reversible heat engine can approach one only as $T_{hot} \rightarrow \infty$ or $T_{cold} \rightarrow 0$, neither of which can be accomplished in practice. Therefore, heat can never be totally converted to work in a reversible cyclic process. Because w_{cycle} for an engine operating in an irreversible cycle is less than the work attainable in a reversible cycle, $\varepsilon_{irreversible} < \varepsilon_{reversible} < 1$.

EXAMPLE PROBLEM 5.1

Calculate the maximum work that can be done by a reversible heat engine operating between 500 and 200 K if 1000 J is absorbed at 500 K.

Solution

The fraction of the heat that can be converted to work is the same as the fractional fall in the absolute temperature. This is a convenient way to link the efficiency of an engine with the properties of the absolute temperature.

$$\varepsilon = 1 - \frac{T_{cold}}{T_{hot}} = 1 - \frac{200 \text{ K}}{500 \text{ K}} = 0.600$$

$$w = \varepsilon q_{ab} = 0.600 \times 1000 \text{ J} = 600 \text{ J}$$

In this section, only the most important features of heat engines have been discussed. It can also be shown that the efficiency of a reversible heat engine is independent of the working substance. For a more in-depth discussion of heat engines, the interested reader is referred to *Heat and Thermodynamics,* seventh edition, by M. W. Zemansky and R. H. Dittman (McGraw-Hill, 1997). We will return to a discussion of the efficiency of engines when we discuss refrigerators, heat pumps, and real engines in Section 5.11.

5.3 Introducing Entropy

Equating the two formulas for the efficiency of the reversible heat engine given in Equations (5.4) and (5.9),

$$\frac{T_{hot} - T_{cold}}{T_{hot}} = \frac{q_{ab} + q_{cd}}{q_{ab}} \quad \text{or} \quad \frac{q_{ab}}{T_{hot}} + \frac{q_{cd}}{T_{cold}} = 0 \qquad (5.10)$$

The last expression in Equation (5.10) is the sum of the quantity $q_{reversible}/T$ around the Carnot cycle. This result can be generalized to any reversible cycle made up of any number of segments to give the important result stated in Equation (5.11):

$$\oint \frac{dq_{reversible}}{T} = 0 \qquad (5.11)$$

This equation can be regarded as the mathematical statement of the second law. What conclusions can be drawn from Equation (5.11)? Because the cyclic integral of $dq_{reversible}/T$ is zero, this quantity must be the exact differential of a state function. This state function is called the **entropy,** and given the symbol S:

$$dS \equiv \frac{dq_{reversible}}{T} \qquad (5.12)$$

For a macroscopic change,

$$\Delta S = \int \frac{dq_{reversible}}{T} \tag{5.13}$$

Note that whereas $dq_{reversible}$ is not an exact differential, multiplying this quantity by $1/T$ makes the differential exact.

EXAMPLE PROBLEM 5.2

a. Show that the following differential expression is not an exact differential:

$$\frac{RT}{P} dP + R \, dT$$

b. Show that $RTdP + PRdT$, obtained by multiplying the function in part (a) by P, is an exact differential.

Solution

a. For the expression $f(P,T)dP + g(P,T)dT$ to be an exact differential, the condition $\partial f(P,T)/\partial T = \partial g(P,T)/\partial P$ must be satisfied as discussed in Section 3.1. Because

$$\frac{\partial f\left(\dfrac{RT}{P}\right)}{\partial T} = \frac{R}{P} \text{ and } \frac{\partial R}{\partial P} = 0$$

the condition is not fulfilled.

b. Because $\partial(RT)/\partial T = R$ and $\partial(RP)/\partial P = R$, $RTdP + RPdT$ is an exact differential.

Keep in mind that it has only been shown that $\Delta S = \int dq_{reversible}/T$ is a state function. It has not yet been demonstrated that S is a suitable function for measuring the natural direction of change in a process that the system may undergo. We will do so in Section 5.5.

5.4 Calculating Changes in Entropy

The most important thing to remember in doing entropy calculations using Equation (5.13) is that ΔS *must be calculated along a reversible path*. In considering an **irreversible process**, ΔS must be calculated for an equivalent reversible process that proceeds between the same initial and final states corresponding to the irreversible process. Because S is a state function, ΔS is necessarily path independent, *provided that the transformation is between the same initial and final states in both processes*.

Next consider two cases that require no calculation. For any reversible adiabatic process, $q_{reversible} = 0$, so that $\Delta S = \int (dq_{reversible}/T) = 0$. For any cyclic process, $\Delta S = \oint (dq_{reversible}/T) = 0$, because the change in any state function for a cyclic process is zero.

Next consider ΔS for the reversible isothermal expansion or compression of an ideal gas, described by $V_i, T_i \rightarrow V_f, T_i$. Because $\Delta U = 0$ for this case,

$$q_{reversible} = -w_{reversible} = nRT \ln \frac{V_f}{V_i} \text{ and} \tag{5.14}$$

$$\Delta S = \int \frac{dq_{reversible}}{T} = \frac{1}{T} q_{reversible} = nR \ln \frac{V_f}{V_i} \tag{5.15}$$

Note that $\Delta S > 0$ for an expansion ($V_f > V_i$) and $\Delta S < 0$ for a compression ($V_f < V_i$). Although the preceding calculation is for a reversible process, ΔS has exactly the same value for any reversible or irreversible isothermal path that goes between the same initial and final volumes and satisfies the condition $T_f = T_i$. This is the case because S is a state function.

Consider next ΔS for an ideal gas that undergoes a reversible change in T at constant V or P. For a reversible process described by $V_i, T_i \rightarrow V_i, T_f$, $dq_{reversible} = C_V dT$, and

$$\Delta S = \int \frac{dq_{reversible}}{T} = \int \frac{nC_{V,m}dT}{T} \approx nC_{V,m} \ln \frac{T_f}{T_i} \tag{5.16}$$

For a constant pressure process described by $P_i, T_i \rightarrow P_i, T_f$, $dq_{reversible} = C_P dT$ and

$$\Delta S = \int \frac{dq_{reversible}}{T} = \int \frac{nC_{P,m}dT}{T} \approx nC_{P,m} \ln \frac{T_f}{T_i} \tag{5.17}$$

The last expressions in Equations (5.16) and (5.17) are valid if the temperature interval is small enough that the temperature dependence of $C_{V,m}$ and $C_{P,m}$ can be neglected. Again, although ΔS has been calculated for a reversible process, Equations (5.16) and (5.17) hold for any reversible or irreversible process between the same initial and final states for an ideal gas.

The results of the last two calculations can be combined in the following way. Because the macroscopic variables V, T or P, T completely define the state of an ideal gas, any change $V_i, T_i \rightarrow V_f, T_f$ can be separated into two segments, $V_i, T_i \rightarrow V_f, T_i$ and $V_f, T_i \rightarrow V_f, T_f$. A similar statement can be made about P and T. Because S is a state function, ΔS is independent of the path. Therefore, any reversible or irreversible process for an ideal gas described by $V_i, T_i \rightarrow V_f, T_f$ can be treated as consisting of two segments, one of which occurs at constant volume and the other of which occurs at constant temperature. For this two-step process, ΔS is given by

$$\Delta S = nR \ln \frac{V_f}{V_i} + nC_{V,m} \ln \frac{T_f}{T_i} \tag{5.18}$$

Similarly, for any reversible or irreversible process for an ideal gas described by $P_i, T_i \rightarrow P_f, T_f$

$$\Delta S = -nR \ln \frac{P_f}{P_i} + nC_{P,m} \ln \frac{T_f}{T_i} \tag{5.19}$$

In writing Equations (5.18) and (5.19), it has been assumed that the temperature dependence of $C_{V,m}$ and $C_{P,m}$ can be neglected over the temperature range of interest.

EXAMPLE PROBLEM 5.3

Using the equation of state and the relationship between $C_{P,m}$ and $C_{V,m}$ for an ideal gas, show that Equation (5.18) can be transformed into Equation (5.19).

Solution

$$\Delta S = nR \ln \frac{V_f}{V_i} + nC_{V,m} \ln \frac{T_f}{T_i} = nR \ln \frac{T_f P_i}{T_i P_f} + nC_{V,m} \ln \frac{T_f}{T_i}$$

$$= -nR \ln \frac{P_f}{P_i} + n(C_{V,m} + R) \ln \frac{T_f}{T_i} = -nR \ln \frac{P_f}{P_i} + nC_{P,m} \ln \frac{T_f}{T_i}$$

Next consider ΔS for phase changes. Experience shows that a liquid is converted to a gas at a constant boiling temperature through heat input if the process is carried out at constant pressure. Because $q_P = \Delta H$, ΔS for this reversible process is given by

$$\Delta S_{vaporization} = \int \frac{dq_{reversible}}{T} = \frac{q_{reversible}}{T_{vaporization}} = \frac{\Delta H_{vaporization}}{T_{vaporization}} \tag{5.20}$$

Similarly, for the phase change solid $\rightarrow$ liquid,

$$\Delta S_{fusion} = \int \frac{dq_{reversible}}{T} = \frac{q_{reversible}}{T_{fusion}} = \frac{\Delta H_{fusion}}{T_{fusion}} \tag{5.21}$$

Finally, consider ΔS for an arbitrary process involving real gases, solids, and liquids for which β and κ, but not the equation of state, are known. The calculation of ΔS for such substances is described in Supplemental Sections 5.12 and 5.13, in which the properties of S as a state function are fully exploited. The results are stated here. For the system undergoing the change $V_i, T_i \rightarrow V_f, T_f$,

$$\Delta S = \int_{T_i}^{T_f} \frac{C_V}{T} dT + \int_{V_i}^{V_f} \frac{\beta}{\kappa} dV = C_V \ln \frac{T_f}{T_i} + \frac{\beta}{\kappa}(V_f - V_i) \tag{5.22}$$

In deriving the last result, it has been assumed that κ and β are constant over the temperature and volume intervals of interest. Is this the case for solids, liquids, and gases? For the system undergoing a change $P_i, T_i \rightarrow P_f, T_f$,

$$\Delta S = \int_{T_i}^{T_f} \frac{C_P}{T} dT - \int_{P_i}^{P_f} V\beta \, dP \tag{5.23}$$

For a solid or liquid, the last equation can be simplified to

$$\Delta S = C_P \ln \frac{T_f}{T_i} - V\beta(P_f - P_i) \tag{5.24}$$

if V and β are assumed constant over the temperature and pressure intervals of interest. The integral forms of Equations (5.22) and (5.23) are valid for real gases, liquids, and solids. Examples of calculations using these equations are given in Example Problems 5.4 through 5.6. As Example Problem 5.4 shows, Equations (5.22) and (5.23) are also applicable to ideal gases.

EXAMPLE PROBLEM 5.4

One mole of CO gas is transformed from an initial state characterized by $T_i = 320$ K and $V_i = 80.0$ L to a final state characterized by $T_f = 650$ K and $V_f = 120.0$ L. Using Equation (5.22), calculate ΔS for this process. Use the ideal gas values for β and κ. For CO,

$$\frac{C_{V,m}}{J\,mol^{-1}\,K^{-1}} = 31.08 - 0.01452\frac{T}{K} + 3.1415 \times 10^{-5}\frac{T^2}{K^2} - 1.4973 \times 10^{-8}\frac{T^3}{K^3}$$

Solution

For an ideal gas,

$$\beta = \frac{1}{V}\left(\frac{\partial V}{\partial T}\right)_P = \frac{1}{V}\left(\frac{\partial [nRT/P]}{\partial T}\right)_P = \frac{1}{T} \quad \text{and}$$

$$\kappa = -\frac{1}{V}\left(\frac{\partial V}{\partial P}\right)_T = -\frac{1}{V}\left(\frac{\partial[nRT/P]}{\partial P}\right)_T = \frac{1}{P}$$

Consider the following reversible process in order to calculate ΔS. The gas is first heated reversibly from 320 to 650 K at a constant volume of 80.0 L. Subsequently, the gas is reversibly expanded at a constant temperature of 650 K from a volume of 80.0 L to a volume of 120.0 L. The entropy change for this process is obtained using the integral form of Equation (5.22) with the values of β and κ cited earlier. The result is

$$\Delta S = \int_{T_i}^{T_f} \frac{C_V}{T}\, dT + nR \ln \frac{V_f}{V_i}$$

$$\Delta S(\mathrm{J\,K^{-1}mol^{-1}}) = \int_{320}^{650} \frac{\left(31.08 - 0.01452\,\frac{T}{K} + 3.1415\times10^{-5}\,\frac{T^2}{K^2} - 1.4973\times10^{-8}\,\frac{T^3}{K^3}\right)}{\frac{T}{K}}\, d\frac{T}{K}$$

$$+\, 8.314\ \mathrm{J\,K^{-1}} \times \ln\frac{120.0\ \mathrm{L}}{80.0\ \mathrm{L}}$$

$$= 22.024\ \mathrm{J\,K^{-1}mol^{-1}} - 4.792\ \mathrm{J\,K^{-1}mol^{-1}} + 5.027\ \mathrm{J\,K^{-1}mol^{-1}} - 1.207\ \mathrm{J\,K^{-1}mol^{-1}}$$

$$+\, 3.371\ \mathrm{J\,K^{-1}mol^{-1}}$$

$$= 24.42\ \mathrm{J\,K^{-1}mol^{-1}}$$

EXAMPLE PROBLEM 5.5

In this problem, 2.5 mol of CO_2 gas is transformed from an initial state characterized by $T_i = 450$ K and $P_i = 1.35$ bar to a final state characterized by $T_f = 800$ K and $P_f = 3.45$ bar. Using Equation (5.23), calculate ΔS for this process. Assume ideal gas behavior and use the ideal gas value for β. For CO_2,

$$\frac{C_{P,m}}{\mathrm{J\,mol^{-1}K^{-1}}} = 18.86 + 7.937\times10^{-2}\,\frac{T}{K} - 6.7834\times10^{-5}\,\frac{T^2}{K^2} + 2.4426\times10^{-8}\,\frac{T^3}{K^3}$$

Solution

Consider the following reversible process in order to calculate ΔS. The gas is first heated reversibly from 450 to 800 K at a constant pressure of 1.35 bar. Subsequently, the gas is reversibly compressed at a constant temperature of 800 K from a pressure of 1.35 bar to a pressure of 3.45 bar. The entropy change for this process is obtained using Equation (5.23) with the value of $\beta = 1/T$ from Example Problem 5.4.

$$\Delta S(\mathrm{J\,K^{-1}mol^{-1}}) = \int_{T_i}^{T_f} \frac{C_P}{T}\, dT - \int_{P_i}^{P_f} V\beta\, dP = \int_{T_i}^{T_f} \frac{C_P}{T}\, dT - nR\int_{P_i}^{P_f} \frac{dP}{P} = \int_{T_i}^{T_f} \frac{C_P}{T}\, dT - nR\ln\frac{P_f}{P_i}$$

$$= 2.5 \times \int_{450}^{800} \frac{\left(18.86 + 7.937\times10^{-2}\,\frac{T}{K} - 6.7834\times10^{-5}\,\frac{T^2}{K^2} + 2.4426\times10^{-8}\,\frac{T^3}{K^3}\right)}{\frac{T}{K}}\, d\frac{T}{K}$$

$$-2.5\ \mathrm{mol} \times 8.314 \times \ln\frac{3.45\ \mathrm{bar}}{1.35\ \mathrm{bar}}\ \mathrm{J\,K^{-1}mol^{-1}}$$

$$= 27.13\ \mathrm{J\,K^{-1}mol^{-1}} + 69.45\ \mathrm{J\,K^{-1}mol^{-1}} - 37.10\ \mathrm{J\,K^{-1}mol^{-1}} + 8.57\ \mathrm{J\,K^{-1}mol^{-1}}$$

$$-19.50\ \mathrm{J\,K^{-1}mol^{-1}}$$

$$= 48.6\ \mathrm{J\,K^{-1}mol^{-1}}$$

EXAMPLE PROBLEM 5.6

In this problem, 3.00 mol of liquid mercury is transformed from an initial state characterized by $T_i = 300$ K and $P_i = 1.00$ bar to a final state characterized by $T_f = 600$ K and $P_f = 3.00$ bar.

a. Calculate ΔS for this process; $\beta = 18.1 \times 10^{-4}$ K^{-1}, $\rho = 13.54$ g cm^{-3}, and $C_{P,m}$ for Hg(l) = 27.98 J mol^{-1}K^{-1}.

b. What is the ratio of the pressure-dependent term to the temperature-dependent term in ΔS? Explain your result.

Solution

a. Because the volume changes only slightly with temperature and pressure over the range indicated,

$$\Delta S = \int_{T_i}^{T_f} \frac{C_P}{T} dT - \int_{P_i}^{P_f} V\beta\, dP \approx nC_{P,m} \ln\frac{T_f}{T_i} - nV_{m,i}\beta(P_f - P_i)$$

$$= 3.00\,\text{mol} \times 27.98\,\text{J mol}^{-1}\text{K}^{-1} \times \ln\frac{600\,\text{K}}{300\,\text{K}}$$

$$-3.00\,\text{mol} \times \frac{200.59\,\text{g mol}^{-1}}{13.54\,\text{g cm}^{-3} \times \dfrac{10^6\,\text{cm}^3}{\text{m}^3}} \times 18.1 \times 10^{-4}\,\text{K}^{-1} \times 2.00\,\text{bar} \times 10^5\text{Pa bar}^{-1}$$

$$= 58.2\,\text{J K}^{-1} - 1.61 \times 10^{-2}\,\text{J K}^{-1} = 58.2\,\text{J K}^{-1}$$

b. The ratio of the pressure-dependent to the temperature-dependent term is -3×10^{-4}. Because the volume change with pressure is very small, the contribution of the pressure-dependent term is negligible in comparison with the temperature-dependent term.

As Example Problem 5.6 shows, ΔS for a liquid or solid as both P and T change is dominated by the temperature dependence of S. *Unless the change in pressure is very large, ΔS for liquids and solids can be considered to be a function of temperature only.*

5.5 Using Entropy to Calculate the Natural Direction of a Process in an Isolated System

Is entropy useful in predicting the direction of spontaneous change? We now return to the two processes introduced in Section 5.1. The first process concerns the natural direction of change in a metal rod subject to a temperature gradient. Will the gradient become larger or smaller as the system approaches its equilibrium state? To model this process, consider the *isolated* composite system shown in Figure 5.5. Two systems, in the form of metal rods with uniform, but different temperatures $T_1 > T_2$, are brought into thermal contact.

In the following discussion, heat is withdrawn from the left rod; the same reasoning holds if the direction of heat flow is reversed. To calculate ΔS for this irreversible process using the heat flow, one must imagine a reversible process in which the initial and final states are the same as for the irreversible process. In the imaginary reversible process, the rod is coupled to a reservoir whose temperature is lowered very slowly. The temperatures of the rod and the reservoir differ only infinitesimally throughout the process in which an amount of heat, q_P, is withdrawn from the rod. The total change in temperature of the rod, ΔT, is related to q_P by

$$dq_P = C_P dT \quad \text{or} \quad \Delta T = \frac{1}{C_P}\int dq_P = \frac{q_P}{C_P} \tag{5.25}$$

q_P

FIGURE 5.5
Two systems at constant P, each consisting of a metal rod, are placed in thermal contact. The temperatures of the two rods differ by ΔT. The composite system is contained in a rigid adiabatic enclosure (not shown) and is, therefore, an isolated system.

It has been assumed that $\Delta T = T_2 - T_1$ is small enough that C_P is constant over the interval.

Because the path is defined (constant pressure), $\int dq_P$ is independent of how rapidly the heat is withdrawn (the path); it depends only on C_P and ΔT. More formally, because $q_P = \Delta H$ and because H is a state function, q_P is independent of the path between the initial and final states. Therefore, $q_P = q_{reversible}$ if the temperature increment ΔT is identical for the reversible and irreversible processes.

Using this result, the entropy change for this irreversible process in which heat flows from one rod to the other is calculated. Because the composite system is isolated, $q_1 + q_2 = 0$, and $q_1 = -q_2 = q_P$. The entropy change of the composite system is the sum of the entropy changes in each rod:

$$\Delta S = \frac{q_{reversible,1}}{T_1} + \frac{q_{reversible,2}}{T_2} = \frac{q_1}{T_1} + \frac{q_2}{T_2} = q_P \left(\frac{1}{T_1} - \frac{1}{T_2} \right) \qquad (5.26)$$

Because $T_1 > T_2$, the quantity in parentheses is negative. This process has two possible directions:

- If heat flows from the hotter to the colder rod, the temperature gradient will become smaller. In this case, $q_P < 0$ and $dS > 0$.

- If heat flows from the colder to the hotter rod, the temperature gradient will become larger. In this case, $q_P > 0$ and $dS < 0$.

Note that ΔS has the same magnitude, but a different sign, for the two directions of change. Therefore, S appears to be a useful function for measuring the direction of natural change in an isolated system. Experience tells us that the temperature gradient will become less with time. *It can be concluded that the process in which S increases is the direction of natural change in an isolated system.*

Next, consider the second process introduced in Section 5.1 in which an ideal gas spontaneously collapses to half its initial volume without a force acting on it. This process and its reversible analog are shown in Figure 5.6. Recall that U is independent of V for an ideal gas. Because U does not change as V increases, and U is a function of T only for an ideal gas, the temperature remains constant in the irreversible process. Therefore, the spontaneous irreversible process shown in Figure 5.6a is both adiabatic and isothermal and is described by V_i, $T_i \rightarrow 1/2V_i, T_i$. The imaginary reversible process that we use to carry out the calculation of ΔS is shown in Figure 5.6b. In this process, which must have the same initial and final states as the irreversible process, sand is slowly and continuously added to the beaker on the piston to ensure that $P = P_{external}$. The ideal gas undergoes a reversible isothermal transformation described by $V_i, T_i \rightarrow 1/2V_i, T_i$. Because $\Delta U - 0$, $q = -w$. We calculate ΔS for this process:

$$\Delta S = \int \frac{dq_{reversible}}{T} = \frac{q_{reversible}}{T_i} = -\frac{w_{reversible}}{T_i} = nR \ln \frac{\frac{1}{2}V_i}{V_i} = -nR \ln 2 < 0 \qquad (5.27)$$

For the reverse process, in which the gas spontaneously expands so that it occupies twice the volume, the reversible model process is an isothermal expansion for which

$$\Delta S = nR \ln \frac{2V_i}{V_i} = nR \ln 2 > 0 \qquad (5.28)$$

Again, the process with $\Delta S > 0$ is the direction of natural change in this isolated system. The reverse process for which $\Delta S < 0$ is the unnatural direction of change.

The results obtained for isolated systems are generalized in the following statement:

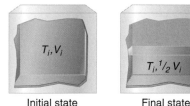

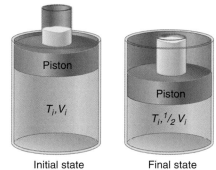

FIGURE 5.6

(a) An irreversible process is shown in which an ideal gas confined in a container with rigid adiabatic walls is spontaneously reduced to half its initial volume. **(b)** A reversible isothermal compression is shown between the same initial and final states as for the irreversible process.

> For any irreversible process in an isolated system, there is a unique direction of spontaneous change: $\Delta S > 0$ for the spontaneous process, $\Delta S < 0$ for the opposite or nonspontaneous direction of change, and $\Delta S = 0$ only for a reversible process. In a quasi-static reversible process, there is no direction of spontaneous change because the system is proceeding along a path, each step of which corresponds to an equilibrium state.

We cannot emphasize too strongly that $\Delta S > 0$ is a criterion for spontaneous change *only* if the system does not exchange energy in the form of heat or work with its surroundings. Note that if any process occurs in the isolated system, it is by definition spontaneous and the entropy increases. Whereas U can neither be created nor destroyed, S for an isolated system can be created ($\Delta S > 0$), but not destroyed.

5.6 The Clausius Inequality

In the previous section, it was shown using two examples that $\Delta S > 0$ provides a criterion to predict the natural direction of change in an isolated system. This result can also be obtained without considering a specific process. Consider the differential form of the first law for a process in which only P–V work is possible:

$$dU = dq - P_{external}\, dV \tag{5.29}$$

Equation (5.29) is valid for both reversible and irreversible processes. If the process is reversible, we can write Equation (5.29) in the following form:

$$dU = dq_{reversible} - P\, dV = T\, dS - P\, dV \tag{5.30}$$

Because U is a state function, dU is independent of the path, and Equation (5.30) holds for both reversible and irreversible processes, as long as there are no phase transitions or chemical reactions, and only P–V work occurs.

To derive the Clausius inequality, we equate the expressions for dU in Equations (5.29) and (5.30):

$$dq_{reversible} - dq = (P - P_{external})dV \tag{5.31}$$

If $P - P_{external} > 0$, the system will spontaneously expand, and $dV > 0$. If $P - P_{external} < 0$, the system will spontaneously contract, and $dV < 0$. In both possible cases, $(P - P_{external})dV > 0$. Therefore, we conclude that

$$dq_{reversible} - dq = TdS - dq \geq 0 \text{ or } TdS \geq dq \tag{5.32}$$

The equality holds only for a reversible process. We rewrite the **Clausius inequality** in Equation (5.32) for an irreversible process in the form

$$dS > \frac{dq}{T} \tag{5.33}$$

However, for an irreversible process in an isolated system, $dq = 0$. *Therefore, we have proved that for any irreversible process in an isolated system, $\Delta S > 0$.*

How can the result from Equations (5.29) and (5.30) that $dU = dq - P_{external}dV$ $= TdS - PdV$ be reconciled with the fact that work and heat are path functions? The answer is that $dw \geq -PdV$ and $dq \leq TdS$, where the equalities hold only for a reversible process. The result $dq + dw = TdS - PdV$ states that the amount by which the work is greater than $-PdV$ and the amount by which the heat is less than TdS in an irreversible process involving only PV work are exactly equal. *Therefore, the differential expression for dU in Equation (5.30) is obeyed for both reversible and irreversible processes.* In Chapter 6, the Clausius inequality is used to generate two new state functions, the Gibbs energy and the Helmholtz energy. These functions allow predictions to be made about the direction of change in processes for which the system interacts with its environment.

The Clausius inequality is next used to evaluate the cyclic integral $\oint dq/T$ for an arbitrary process. Because $dS = dq_{reversible}/T$, the value of the cyclic integral is zero for a reversible process. Consider a process in which the transformation from state 1 to state 2 is reversible, but the transition from state 2 back to state 1 is irreversible:

$$\oint \frac{dq}{T} = \int_1^2 \frac{dq_{reversible}}{T} + \int_2^1 \frac{dq_{irreversible}}{T} \qquad (5.34)$$

The limits of integration on the first integral can be interchanged to obtain

$$\oint \frac{dq}{T} = -\int_2^1 \frac{dq_{reversible}}{T} + \int_2^1 \frac{dq_{irreversible}}{T} \qquad (5.35)$$

Exchanging the limits as written is only valid for a state function. Because $dq_{reversible} > dq_{irreversible}$,

$$\oint \frac{dq}{T} \leq 0 \qquad (5.36)$$

where the equality only holds for a reversible process. Note that the cyclic integral of an exact differential is always zero, but the integrand in Equation (5.36) is only an exact differential for a reversible cycle.

5.7 The Change of Entropy in the Surroundings and $\Delta S_{total} = \Delta S + \Delta S_{surroundings}$

As shown in Section 5.6, the entropy of an isolated system increases in a spontaneous process. Is it true that a process is spontaneous if ΔS for the system is positive? As shown later, this statement is only true for an isolated system. In this section, a criterion for spontaneity is developed that takes into account the entropy change of both the system and the surroundings.

In general, a system interacts only with the part of the universe that is *very close*. Therefore, one can think of the system and the interacting part of the surroundings as forming an interacting composite system that is isolated from the rest of the universe. The part of the surroundings that is relevant for entropy calculations is a thermal reservoir at a fixed temperature, T. The mass of the reservoir is sufficiently large that its temperature is only changed by an infinitesimal amount dT when heat is transferred between the system and the surroundings. Therefore, the surroundings always remain in internal equilibrium during heat transfer.

Next consider the entropy change of the surroundings, whereby the surroundings are at either constant V or constant P. The amount of heat absorbed by the surroundings, $q_{surroundings}$, depends on the process occurring in the system. If the surroundings are at constant V, $q_{surroundings} = \Delta U_{surroundings}$, and if the surroundings are at constant P, $q_{surroundings} = \Delta H_{surroundings}$. Because H and U are state functions, the amount of heat entering the surroundings is independent of the path. In particular, the system and surroundings need not be at the same temperature and q is the same whether the transfer occurs reversibly or irreversibly. Therefore,

$$dS_{surroundings} = \frac{dq_{surroundings}}{T_{surroundings}} \text{ or for a macroscopic change, } \Delta S_{surroundings} = \frac{q_{surroundings}}{T_{surroundings}} \qquad (5.37)$$

Note that the heat that appears in Equation (5.37) is the *actual* heat transferred. By contrast, in calculating ΔS for the system using the heat flow, $dq_{reversible}$ for a *reversible* process that connects the initial and final states of the system must be used, *not* the actual dq for the process. *It is essential to understand this reasoning in order to carry out calculations for ΔS and $\Delta S_{surroundings}$.*

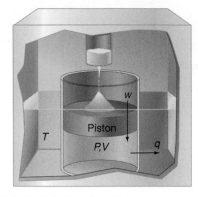

FIGURE 5.7
A sample of an ideal gas (the system) is confined in a piston and cylinder assembly with diathermal walls. The assembly is in contact with a thermal reservoir that holds the temperature at a value of 300 K. Sand falling on the outer surface of the piston increases the external pressure slowly enough to ensure a reversible compression. The directions of work and heat flow are indicated.

This important difference is discussed in calculating the entropy change of the system as opposed to the surroundings with the aid of Figure 5.7. A gas (the system) is enclosed in a piston and cylinder assembly with diathermal walls. The gas is reversibly compressed by an external pressure generated by a stream of sand slowly falling on the external surface of the piston. The piston and cylinder assembly is in contact with a water bath thermal reservoir that keeps the temperature of the gas fixed at the value T. In Example Problem 5.7, ΔS and $\Delta S_{surroundings}$ is calculated for this reversible compression.

EXAMPLE PROBLEM 5.7

One mole of an ideal gas at 300 K is reversibly and isothermally compressed from a volume of 25.0 L to a volume of 10.0 L. Because it is very large, the temperature of the water bath thermal reservoir in the surroundings remains essentially constant at 300 K during the process. Calculate ΔS, $\Delta S_{surroundings}$, and ΔS_{total}.

Solution

Because this is an isothermal process, $\Delta U = 0$, and $q_{reversible} = -w$. From Section 2.7,

$$q_{reversible} = -w = nRT \int_{V_i}^{V_f} \frac{dV}{V} = nRT \ln \frac{V_f}{V_i}$$

$$= 1.00 \text{ mol} \times 8.314 \text{ J mol}^{-1}\text{K}^{-1} \times 300 \text{ K} \times \ln \frac{10.0 \text{ L}}{25.0 \text{ L}} = -2.29 \times 10^3 \text{ J}$$

The entropy change of the system is given by

$$\Delta S = \int \frac{dq_{reversible}}{T} = \frac{q_{reversible}}{T} = \frac{-2.29 \times 10^3 \text{ J}}{300 \text{ K}} = -7.62 \text{ J K}^{-1}$$

The entropy change of the surroundings is given by

$$\Delta S_{surroundings} = \frac{q_{surroundings}}{T} = -\frac{q_{system}}{T} = \frac{2.29 \times 10^3 \text{ J}}{300 \text{ K}} = 7.62 \text{ J K}^{-1}$$

The total change in the entropy is given by

$$\Delta S_{total} = \Delta S + \Delta S_{surroundings} = -7.62 \text{ J K}^{-1} + 7.62 \text{ J K}^{-1} = 0$$

Because the process in Example Problem 5.7 is reversible, there is no direction of spontaneous change and, therefore, $\Delta S_{total} = 0$. In Example Problem 5.8, this calculation is repeated for an irreversible process that goes between the same initial states of the system.

EXAMPLE PROBLEM 5.8

One mole of an ideal gas at 300 K is isothermally compressed by a constant external pressure equal to the final pressure in Example Problem 5.7. At the end of the process, $P = P_{external}$. Because $P \neq P_{external}$ at all but the final state, this process is irreversible. The initial volume is 25.0 L and the final volume is 10.0 L. The temperature of the surroundings is 300 K. Calculate ΔS, $\Delta S_{surroundings}$, and ΔS_{total}.

Solution

We first calculate the external pressure and the initial pressure in the system:

$$P_{external} = \frac{nRT}{V} = \frac{1 \text{ mol} \times 8.314 \text{ J mol}^{-1}\text{K}^{-1} \times 300 \text{ K}}{10.0 \text{ L} \times \frac{1 \text{ m}^3}{10^3 \text{ L}}} = 2.49 \times 10^5 \text{ Pa}$$

$$P_i = \frac{nRT}{V} = \frac{1\ mol \times 8.314\ J\ mol^{-1}K^{-1} \times 300\ K}{25.0\ L \times \dfrac{1\ m^3}{10^3\ L}} = 9.98 \times 10^4\ Pa$$

Because $P_{external} > P_i$, we expect that the direction of spontaneous change will be the compression of the gas to a smaller volume. Because $\Delta U = 0$,

$$q = -w = P_{external}(V_f - V_i) = 2.49 \times 10^5\ Pa \times (10.0 \times 10^{-3}\ m^3 - 25.0 \times 10^{-3}\ m^3)$$

$$= -3.74 \times 10^3\ J$$

The entropy change of the surroundings is given by

$$\Delta S_{surroundings} = \frac{q_{surroundings}}{T} = -\frac{q}{T} = \frac{3.74 \times 10^3\ J}{300\ K} = 12.45\ J\ K^{-1}$$

The entropy change of the system must be calculated on a reversible path and has the value obtained in Example Problem 5.7:

$$\Delta S = \int \frac{dq_{reversible}}{T} = \frac{q_{reversible}}{T} = \frac{-2.29 \times 10^3\ J}{300\ K} = -7.62\ J\ K^{-1}$$

It is seen that $\Delta S < 0$, and $\Delta S_{surroundings} > 0$. The total change in the entropy is given by

$$\Delta S_{total} = \Delta S + \Delta S_{surroundings} = -7.62\ J\ K^{-1} + 12.45\ J\ K^{-1} = 4.83\ J\ K^{-1}$$

The previous calculations lead to the following conclusion: *If the system and the part of the surroundings with which it interacts are viewed as an isolated composite system, the criterion for spontaneous change is $\Delta S_{total} = \Delta S + \Delta S_{surroundings} > 0$.* This statement defines a unique direction of time. A decrease in the entropy of the universe will never be observed, because $\Delta S_{total} \geq 0$. The equality only holds for the universe at equilibrium. However, *any process* that occurs anywhere in the universe is by definition spontaneous and leads to an increase of S_{total}. Because such processes always occur, $\Delta S_{total} > 0$ as time increases. Consider the following consequence of this conclusion: If you view a movie in which two gases in contact with their surroundings are mixed, and the same movie is run backward, you cannot decide which direction corresponds to real time on the basis of the first law. However, using the criterion $\Delta S_{total} \geq 0$, the direction of real time can be established. The English astrophysicist Eddington coined the phrase "entropy is time's arrow" to emphasize this relationship between entropy and time.

Note that a spontaneous process in a system that interacts with its surroundings is not characterized by $\Delta S > 0$, but by $\Delta S_{total} > 0$. The entropy of the system can decrease in a spontaneous process, as long as the entropy of the surroundings increases by a greater amount. In Chapter 6, the spontaneity criterion $\Delta S_{total} = \Delta S + \Delta S_{surroundings} > 0$ will be used to generate two state functions, the Gibbs energy and the Helmholtz energy. These functions allow one to predict the direction of change in processes that interact with their environment using only the changes in state functions of the system.

5.8 Absolute Entropies and the Third Law of Thermodynamics

All elements and many compounds exist in three different states of aggregation. One or more solid phases are the most stable forms at low temperature, and when the temperature is increased to the melting point, a constant temperature transition to the liquid

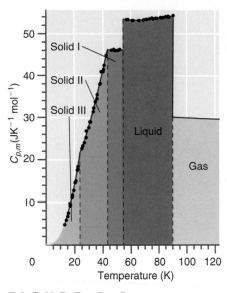

FIGURE 5.8

The experimentally determined heat capacity for O_2 is shown as a function of temperature below 125 K. The dots are data from Giauque and Johnston [*J. American Chemical Society* 51 (1929), 2300]. The red solid lines below 90 K are polynomial fits to these data. The red line above 90 K is a fit to data from the NIST Chemistry Webbook. The yellow line is an extrapolation from 12.73 to 0 K as described in the text. The vertical dashed lines indicate constant temperature-phase transitions, and the most stable phase at a given temperature is indicated in the figure.

phase is observed. After the temperature is increased further, a constant temperature phase transition to a gas is observed at the boiling point. At temperatures higher than the boiling point, the gas is the stable form.

The entropy of an element or a compound is experimentally determined using heat capacity data through the relationship $dq_{reversible,P} = C_P dT$. Just as for the thermochemical data discussed in Chapter 4, entropy values are generally tabulated for a standard temperature of 298.15 K and a standard pressure of 1 bar. We describe such a determination for the entropy of O_2 at 298.15 K, first in a qualitative fashion, and then quantitatively in Example Problem 5.9.

The experimentally determined heat capacity of O_2 is shown in Figure 5.8 as a function of temperature for a pressure of 1 bar. O_2 has three solid phases, and transitions between them are observed at 23.66 and 43.76 K. The solid form that is stable above 43.76 K melts to form a liquid at 54.39 K. The liquid vaporizes to form a gas at 90.20 K. These phase transitions are indicated in Figure 5.8. Experimental measurements of $C_{P,m}$ are available above 12.97 K. Below this temperature, the data are extrapolated to zero kelvin by assuming that in this very low temperature range $C_{P,m}$ varies with temperature as T^3. This extrapolation is based on a model of the vibrational spectrum of a crystalline solid that will be discussed in Chapter 15. The explanation for the dependence of $C_{P,m}$ on T is the same as that presented for Cl_2 in Section 2.4.

Under constant pressure conditions, the molar entropy of the gas can be expressed in terms of the molar heat capacities of the solid, liquid, and gaseous forms and the enthalpies of fusion and vaporization as

$$S_m(T) = S_m(0\,K) + \int_0^{T_f} \frac{C_{P,m}^{solid} dT'}{T'} + \frac{\Delta H_{fusion}}{T_f} + \int_{T_f}^{T_b} \frac{C_{P,m}^{liquid} dT'}{T'} + \frac{\Delta H_{vaporization}}{T_b} \quad (5.38)$$

$$+ \int_{T_b}^{T} \frac{C_{P,m}^{gas} dT'}{T'}$$

If the substance has more than one solid phase, each will give rise to a separate integral. Note that the entropy change associated with the phase transitions solid → liquid and liquid → gas discussed in Section 5.4 must be included in the calculation. To obtain a numerical value for $S_m(T)$, the heat capacity must be known down to zero kelvin, and $S_m(0\,K)$ must also be known.

We first address the issue of the entropy of a solid at zero Kelvin. The **third law of thermodynamics** can be stated in the following form, due to Max Planck:

> The entropy of a pure, perfectly crystalline substance (element or compound) is zero at zero kelvin.

A more detailed discussion of the third law will be presented in Chapter 15. Recall that in a perfectly crystalline atomic (or molecular) solid, the position of each atom is known. Because the individual atoms are indistinguishable, exchanging the positions of two atoms does not lead to a new state. Therefore, a perfect crystalline solid has only one state and $S = k \ln W = k \ln 1 = 0$. The purpose of introducing the third law at this point is that it allows calculations of the absolute entropies of elements and compounds to be carried out for any value of T. To calculate the entropy at a temperature T using Equation (5.38), the $C_{P,m}$ data of Figure 5.8 is graphed in the form $C_{P,m}/T$ as shown in Figure 5.9.

The entropy can be obtained as a function of temperature by numerically integrating the area under the curve in Figure 5.9 and adding the entropy changes associated with phase changes at the transition temperatures. The results for O_2 are shown

FIGURE 5.9

C_p/T as a function of temperature for O_2. The vertical dashed lines indicate constant temperature-phase transitions, and the most stable phase at a given temperature is indicated in the figure.

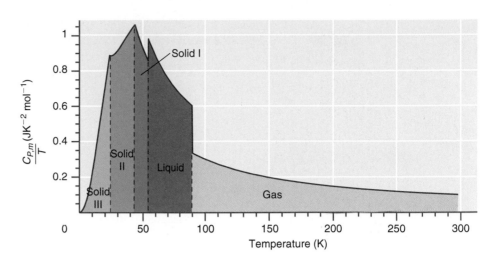

in Figure 5.10. One can also make the following general remarks about the relative magnitudes of the entropy of different substances. Because these remarks will be justified on the basis of a microscopic model, a more detailed discussion will be deferred until Chapter 15.

- Because C_P/T in a single phase region and ΔS for melting and vaporization are always positive, S_m for a given substance is greatest for the gas-phase species. The molar entropies follow the order $S_m^{gas} > S_m^{liquid} > S_m^{solid}$.

- The molar entropy increases with the size of a molecule, because the number of degrees of freedom increases with the number of atoms. A non-linear gas-phase molecule has three translational degrees of freedom, three rotational degrees of freedom, and $3n - 6$ vibrational degrees of freedom. A linear molecule has three translational, two rotational, and $3n - 5$ vibrational degrees of freedom. For a molecule in a liquid, the three translational degrees of freedom are converted to local vibrational modes.

- A solid has only vibrational modes. It can be modeled as a three-dimensional array of coupled harmonic oscillators as shown in Figure 5.11. This solid has a wide spectrum of vibrational frequencies, and solids with a large binding energy have higher frequencies than more weakly bound solids. Because modes with high frequencies are not activated at low temperatures, weakly bound solids have a larger molar entropy than strongly bound solids at low and moderate temperatures.

- The entropy of all substances is a monotonically increasing function of temperature.

FIGURE 5.10

The molar entropy for O_2 is shown as a function of temperature. The vertical dashed lines indicate constant temperature-phase transitions, and the most stable phase at a given temperature is indicated in the figure.

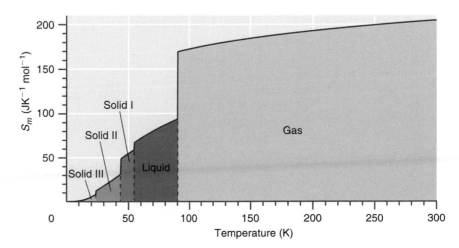

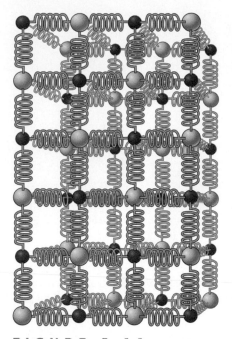

FIGURE 5.11
A useful model of a solid is a three-dimensional array of coupled harmonic oscillators. In solids with a high binding energy, the atoms are coupled by stiff springs.

EXAMPLE PROBLEM 5.9

The heat capacity of O_2 has been measured at 1 atm pressure over the interval 12.97 K $< T <$ 298.15 K. The data have been fit to the following polynomial series in T/K, in order to have a unitless variable:

$0 < T < 12.97$ K:

$$\frac{C_{P,m}(T)}{\text{J mol}^{-1}\,\text{K}^{-1}} = 2.11 \times 10^{-3}\,\frac{T^3}{\text{K}^3}$$

12.97 K $< T < 23.66$ K:

$$\frac{C_{P,m}(T)}{\text{J mol}^{-1}\,\text{K}^{-1}} = -5.666 + 0.6927\,\frac{T}{\text{K}} - 5.191\times10^{-3}\,\frac{T^2}{\text{K}^2} + 9.943\times10^{-4}\,\frac{T^3}{\text{K}^3}$$

23.66 K $< T < 43.76$ K:

$$\frac{C_{P,m}(T)}{\text{J mol}^{-1}\,\text{K}^{-1}} = 31.70 - 2.038\,\frac{T}{\text{K}} + 0.08384\,\frac{T^2}{\text{K}^2} - 6.685\times10^{-4}\,\frac{T^3}{\text{K}^3}$$

43.66 K $< T < 54.39$ K:

$$\frac{C_{P,m}(T)}{\text{J mol}^{-1}\,\text{K}^{-1}} = 46.094$$

54.39 K $< T < 90.20$ K:

$$\frac{C_{P,m}(T)}{\text{J mol}^{-1}\text{K}^{-1}} = 81.268 - 1.1467\,\frac{T}{\text{K}} + 0.01516\,\frac{T^2}{\text{K}^2} - 6.407\times10^{-5}\,\frac{T^3}{\text{K}^3}$$

90.20 K $< T < 298.15$ K:

$$\frac{C_{P,m}(T)}{\text{J mol}^{-1}\,\text{K}^{-1}} = 32.71 - 0.04093\,\frac{T}{\text{K}} + 1.545\times10^{-4}\,\frac{T^2}{\text{K}^2} - 1.819\times10^{-7}\,\frac{T^3}{\text{K}^3}$$

The transition temperatures and the enthalpies for the transitions indicated in Figure 5.8 are as follows:

Solid III $\rightarrow$ solid II	23.66 K	93.8 J mol^{-1}
Solid II $\rightarrow$ solid I	43.76 K	743 J mol^{-1}
Solid I $\rightarrow$ liquid	54.39 K	445 J mol^{-1}
Liquid $\rightarrow$ gas	90.20 K	6815 J mol^{-1}

a. Using these data, calculate S_m° for O_2 at 298.15 K.

b. What are the three largest contributions to S_m°?

Solution

a. $$S_m^\circ(298.15\text{K}) = \int_0^{23.66} \frac{C_{P,m}^{solid,III}\,dT}{T} + \frac{93.8\text{ J}}{23.66\text{ K}} + \int_{23.66}^{43.76} \frac{C_{P,m}^{solid,II}\,dT}{T} + \frac{743\text{ J}}{43.76\text{ K}}$$

$$+ \int_{43.76}^{54.39} \frac{C_{P,m}^{solid,I}\,dT}{T} + \frac{445\text{ J}}{54.39\text{ K}} + \int_{54.39}^{90.20} \frac{C_{P,m}^{liquid}\,dT}{T} + \frac{6815\text{ J}}{90.20\text{ K}}$$

$$+ \int_{90.20}^{298.15} \frac{C_{P,m}^{gas}\,dT}{T}$$

$$= 8.182\text{ J K}^{-1} + 3.964\text{ J K}^{-1} + 19.61\text{ J K}^{-1} + 16.98\text{ J K}^{-1}$$

$$+ 10.13\text{ J K}^{-1} + 8.181\text{ J K}^{-1} + 27.06\text{ J K}^{-1} + 75.59\text{ J K}^{-1}$$

$$+ 35.27\text{ J K}^{-1}$$

$$= 204.9\text{ J mol}^{-1}\,\text{K}^{-1}$$

There is an additional small correction for nonideality of the gas at 1 bar. The currently accepted value is $S_m^{\circ}(298.15\ \text{K}) = 205.152\ \text{J mol}^{-1}\,\text{K}^{-1}$ (Linstrom, P. J., and Mallard, W. G., Eds., *NIST Chemistry Webbook: NIST Standard Reference Database Number 69,* National Institute of Standards and Technology, Gaithersburg, MD, retrieved from *http://webbook.nist.gov.*)

b. The three largest contributions to S_m are ΔS for the vaporization transition, ΔS for the heating of the gas from the boiling temperature to 298.15 K, and ΔS for heating of the liquid from the melting temperature to the boiling point.

The preceding discussion and Example Problem 5.9 show how numerical values of the entropy for a specific substance can be determined at a standard pressure of 1 bar for different values of the temperature. These numerical values can then be used to calculate entropy changes in chemical reactions, as will be shown in Section 5.10.

5.9 Standard States in Entropy Calculations

As discussed in Chapter 4, changes in U and H are calculated using the result that ΔH_f values for pure elements in their standard state at a pressure of 1 bar and a temperature of 298.15 K are zero. For S, the third law provides a natural definition of zero, namely, the crystalline state at zero kelvin. Therefore, the absolute entropy of a compound can be experimentally determined from heat capacity measurements as described earlier. Because S is a function of pressure, tabulated values of entropies refer to a standard pressure of 1 bar. The values of S varies most strongly with P for a gas. From Equation (5.19), for an ideal gas at constant T,

$$\Delta S_m = R \ln \frac{V_f}{V_i} = -R \ln \frac{P_f}{P_i} \tag{5.39}$$

Choosing $P_i = P^{\circ} = 1$ bar,

$$S_m(P) = S_m^{\circ} - R \ln \frac{P(\text{bar})}{P^{\circ}} \tag{5.40}$$

Figure 5.12 shows a plot of the molar entropy of an ideal gas as a function of pressure. It is seen that as $P \to 0$, $S_m \to \infty$. This is a consequence of the fact that as $P \to 0$, $V \to \infty$. As Equation (5.18) shows, the entropy becomes infinite in this limit.

Equation (5.40) provides a way to calculate the entropy of a gas at any pressure. For solids and liquids, S varies so slowly with P, as shown in Section 5.4 and Example Problem 5.6, that the pressure dependence of S can usually be neglected. The value of S for 1 bar of pressure can be used in entropy calculations for pressures that do not differ greatly from 1 bar.

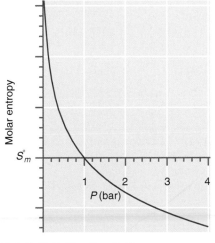

FIGURE 5.12

The molar entropy of an ideal gas is shown as a function of the gas pressure. By definition, at 1 bar, $S_m = S_m^{\circ}$, the standard state molar entropy.

5.10 Entropy Changes in Chemical Reactions

The entropy change in a chemical reaction is a major factor in determining the equilibrium concentration in a reaction mixture. In an analogous fashion to calculating ΔH_R° and ΔU_R° for chemical reactions, ΔS_R° is equal to the difference in the entropies of products and reactions, which can be written as

$$\Delta S_R^{\circ} = \sum_i v_i S_i^{\circ} \tag{5.41}$$

In Equation (5.41), the stoichiometric coefficients v_i are positive for products and negative for reactants. For example, in the reaction

$$Fe_3O_4(s) + 4H_2(g) \rightarrow 3Fe(s) + 4H_2O(l) \qquad (5.42)$$

the entropy change under standard state conditions of 1 bar and 298.15 K is given by

$$\Delta S_{298.15}^{\circ} = 3S_{298.15}^{\circ}(Fe,s) + 4S_{298.15\,K}^{\circ}(H_2O,l) - S_{298.15}^{\circ}(Fe_3O_4,s) - 4S_{298.15}^{\circ}(H_2,g)$$

$$= 3 \times 27.28 \text{ J K}^{-1} \text{mol}^{-1} + 4 \times 69.61 \text{ J K}^{-1} \text{mol}^{-1} - 146.4 \text{ J K}^{-1} \text{mol}^{-1} - 4$$

$$\times 130.684 \text{ J K}^{-1} \text{mol}^{-1}$$

$$= -308.9 \text{ J K}^{-1} \text{mol}^{-1}$$

For this reaction, ΔS is large and negative, primarily because gaseous species are consumed in the reaction, and none are generated. If Δn is the change in the number of moles of gas in the overall reaction, generally ΔS° is positive for $\Delta n > 0$, and negative for $\Delta n < 0$.

Tabulated values of S° are generally available at the standard temperature of 298.15 K, and values for selected elements and compounds are listed in Tables 4.1 and 4.2 (see Appendix A, Data Tables). However, it is often necessary to calculate ΔS° at other temperatures. Such calculations are carried out using the temperature dependence of S discussed in Section 5.4:

$$\Delta S_T^{\circ} = \Delta S_{298.15}^{\circ} + \int_{298.15}^{T} \frac{\Delta C_P^{\circ}}{T'} dT' \qquad (5.43)$$

This equation is only valid if no phase changes occur in the temperature interval between 298.15 K and T. If phase changes occur, the associated entropy changes must be included as they were in Equation (5.38).

EXAMPLE PROBLEM 5.10

The standard entropies of CO, CO_2, and O_2 at 298.15 K are

$$S_{298.15}^{\circ}(CO,g) = 197.67 \text{ J K}^{-1} \text{mol}^{-1}$$
$$S_{298.15}^{\circ}(CO_2,g) = 213.74 \text{ J K}^{-1} \text{mol}^{-1}$$
$$S_{298.15}^{\circ}(O_2,g) = 205.138 \text{ J K}^{-1} \text{mol}^{-1}$$

The temperature dependence of constant pressure heat capacity for of CO, CO_2, and O_2 is given by

$$\frac{C_P^{\circ}(CO,g)}{\text{J K}^{-1} \text{mol}^{-1}} = 31.08 - 1.452 \times 10^{-2} \frac{T}{K} + 3.1415 \times 10^{-5} \frac{T^2}{K^2} - 1.4973 \times 10^{-8} \frac{T^3}{K^3}$$

$$\frac{C_P^{\circ}(CO_2,g)}{\text{J K}^{-1} \text{mol}^{-1}} = 18.86 + 7.937 \times 10^{-2} \frac{T}{K} - 6.7834 \times 10^{-5} \frac{T^2}{K^2} + 2.4426 \times 10^{-8} \frac{T^3}{K^3}$$

$$\frac{C_P^{\circ}(O_2,g)}{\text{J K}^{-1} \text{mol}^{-1}} = 30.81 - 1.187 \times 10^{-2} \frac{T}{K} + 2.3968 \times 10^{-5} \frac{T^2}{K^2}$$

Calculate ΔS° for the reaction $CO(g) + 1/2\, O_2(g) \rightarrow CO_2(g)$ at 475 K.

Solution

$$\frac{\Delta C_P^{\circ}}{\text{J K}^{-1} \text{mol}^{-1}} = \left(18.86 - 31.08 - \frac{1}{2} \times 30.81\right) + \left(7.937 + 1.452 + \frac{1}{2} \times 1.187\right) \times 10^{-2} \frac{T}{K}$$

$$- \left(6.7834 + 3.1415 + \frac{1}{2} \times 2.3968\right) \times 10^{-5} \frac{T^2}{K^2} + (2.4426 + 1.4973)$$

$$\times 10^{-8} \frac{T^3}{K^3}$$

$$= -27.63 + 9.983 \times 10^{-2} \frac{T}{K} - 1.112 \times 10^{-4} \frac{T^2}{K^2} + 3.940 \times 10^{-8} \frac{T^3}{K^3}$$

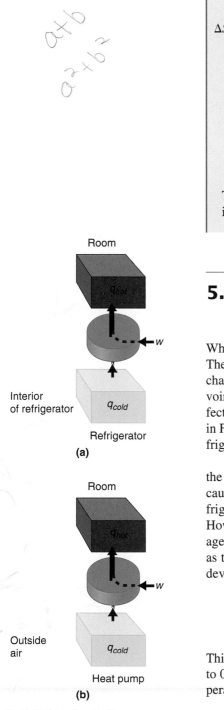

$$\Delta S° = S°_{298.15}(CO_2, g) - S°_{298.15}(CO, g) - \frac{1}{2} \times S°_{298.15}(O_2, g)$$

$$= 213.74 \text{ J K}^{-1} \text{ mol}^{-1} - 197.67 \text{ J K}^{-1} \text{ mol}^{-1} - \frac{1}{2} \times 205.138 \text{ J K}^{-1} \text{ mol}^{-1}$$

$$= -86.50 \text{ J K}^{-1} \text{ mol}^{-1}$$

$$\Delta S°_T = \Delta S°_{298.15} + \int_{298.15}^{T/K} \frac{\Delta C°_p}{T'} \, dT'$$

$$= -86.50 \text{ J K}^{-1} \text{ mol}^{-1}$$

$$+ \int_{298.15}^{475} \frac{\left(-27.63 + 9.983 \times 10^{-2} \frac{T}{K} - 1.112 \times 10^{-4} \frac{T^2}{K^2} + 3.940 \times 10^{-8} \frac{T^3}{K^3}\right)}{\frac{T}{K}} d\frac{T}{K} \text{ J K}^{-1} \text{ mol}^{-1}$$

$$= -86.50 \text{ J K}^{-1} \text{ mol}^{-1} + (-12.866 + 17.654 - 7.605 + 1.0594) \text{ J K}^{-1} \text{ mol}^{-1}$$

$$= -86.50 \text{ J K}^{-1} \text{ mol}^{-1} - 1.757 \text{ J K}^{-1} \text{ mol}^{-1} = -88.26 \text{ J K}^{-1} \text{ mol}^{-1}$$

The value of $\Delta S°$ is negative at both temperatures because the number of moles is reduced in the reaction.

5.11 Refrigerators, Heat Pumps, and Real Engines

What happens if the Carnot cycle in Figure 5.2 is traversed in the opposite direction? The signs of w and q in the individual segments and the signs of the overall w and q are changed. Heat is now withdrawn from the cold reservoir and deposited in the hot reservoir. Because this is not a spontaneous process, work must be done on the system to effect this direction of heat flow. The heat and work flow in such a reverse engine is shown in Figure 5.13. The usefulness of such reverse engines is that they form the basis for refrigerators and heat pumps.

First consider the refrigerator (Figure 5.13a). The interior of a **refrigerator** acts as the system, and the room in which the refrigerator is situated is the hot reservoir. Because more heat is deposited in the room than is withdrawn from the interior of the refrigerator, the overall effect of such a device is to increase the temperature of the room. However, the usefulness of the device is that it provides a cold volume for food storage. The coefficient of performance, η_r, of a reversible Carnot refrigerator is defined as the ratio of the heat withdrawn from the cold reservoir to the work supplied to the device:

$$\eta_r = \frac{q_{cold}}{w} = \frac{q_{cold}}{q_{hot} + q_{cold}} = \frac{T_{cold}}{T_{hot} - T_{cold}} \tag{5.44}$$

This formula shows that as T_{cold} decreases from $0.9\, T_{hot}$ to $0.1\, T_{hot}$, η_r decreases from 9 to 0.1. Equation (5.44) states that if the refrigerator is required to provide a lower temperature, more work is required to extract a given amount of heat.

A household refrigerator typically operates at 255 K in the freezing compartment, and 277 K in the refrigerator section. Using the lower of these temperatures for the cold reservoir and 294 K as the temperature of the hot reservoir (the room), the maximum η_r value is 6.5. This means that for every joule of work done on the system, 6.5 J of heat can be extracted from the contents of the refrigerator. This is the maximum coefficient of performance, and it is only applicable to a refrigerator operating in a reversible Carnot cycle with no dissipative losses. Real-world refrigerators operate in an irreversible cycle, and have dissipative losses associated with friction and turbulence. Therefore, it is

FIGURE 5.13
Reverse Carnot heat engines can be used to induce heat flow from a cold reservoir to hot reservoir with the input of work. The hot reservoir in both **(a)** and **(b)** is a room. The cold reservoir can be configured as the inside of a refrigerator, or outside ambient air.

difficult to achieve η_r values greater than ~1.5 in household refrigerators. This shows the significant loss of efficiency in an irreversible cycle.

Next consider the **heat pump** (Figure 5.13b). A heat pump is used to heat a building by extracting heat from a colder thermal reservoir such as a lake or the ambient air. The maximum coefficient of performance of a heat pump, η_{hp}, is defined as the ratio of the heat pumped into the hot reservoir to the work input to the heat pump:

$$\eta_{hp} = \frac{q_{hot}}{w} = \frac{q_{hot}}{q_{hot} + q_{cold}} = \frac{T_{hot}}{T_{hot} - T_{cold}} \tag{5.45}$$

Assume that T_{hot} = 294 K and T_{cold} = 278 K, typical for a mild winter day. The maximum η_{hp} value is calculated to be 18. Just as for the refrigerator, such high values cannot be attained for real heat pumps operating in an irreversible cycle with dissipative losses. Typical values for commercially available heat pumps lie in the range of 2 to 3.

Heat pumps become less effective as T_{cold} decreases, as shown by Equation (5.45). This decrease is more pronounced in practice, because in order to extract a given amount of heat from air, more air has to be brought into contact with the heat exchanger as T_{cold} decreases. Limitations imposed by the rate of heat transfer come into play. Therefore, heat pumps using ambient air as the cold reservoir are impractical in cold climates. Note that a coefficient of performance of two for a heat pump means that a house can be heated with a heat pump using half the electrical power consumption that would be required to heat the same house using electrical baseboard heaters. This is a significant argument for using heat pumps for residential heating.

It is also instructive to consider refrigerators and heat pumps from an entropic point of view. Transferring an amount of heat, q, from a cold reservoir to a hot reservoir is not a spontaneous process in an isolated system, because

$$\Delta S = q \left(\frac{1}{T_{hot}} - \frac{1}{T_{cold}} \right) < 0 \tag{5.46}$$

However, work can be converted to heat with 100% efficiency. Therefore, the coefficient of performance, η_r, can be calculated by determining the minimum amount of work input required to make ΔS for the withdrawal of q from the cold reservoir, together with the deposition of $q + w$, a spontaneous process.

We next discuss real engines, using the Otto engine, typically used in automobiles, and the diesel engine as examples. The **Otto engine** is the most widely used engine in automobiles. The engine cycle consists of four strokes as shown in Figure 5.14. The intake valve opens as the piston is moving downward, drawing a fuel–air mixture into the cylinder. The intake valve is closed, and the mixture is compressed as the piston moves upward. Just after the piston has reached its highest point, the fuel–air mixture is ignited by a spark plug (not shown), and the rapid heating resulting from the combustion process causes the gas to expand and the pressure to increase. This drives the piston down in a power stroke. Finally, the combustion products are forced out of the cylinder by the upward-moving piston as the exhaust valve is opened. To arrive at a maximum theoretical efficiency for the Otto engine, the reversible Otto cycle shown in Figure 5.15a is analyzed, assuming reversibility.

The reversible Otto cycle begins with the intake stroke $e \rightarrow c$, which is assumed to take place at constant pressure. At this point, the intake valve is closed, and the piston compresses the fuel–air mixture along the adiabatic path $c \rightarrow d$ in the second step. This path can be assumed to be adiabatic because the compression occurs too rapidly to allow much heat to be transferred out of the cylinder. Ignition of the fuel–air mixture takes place at d. The rapid increase in pressure takes place at essentially a constant volume. In this reversible cycle, the combustion is modeled as a quasi-static heat transfer from a se-

FIGURE 5.14
Illustration of the four-stroke cycle of an Otto engine, as explained in the text. The left valve is the intake valve, and the right valve is the exhaust valve.

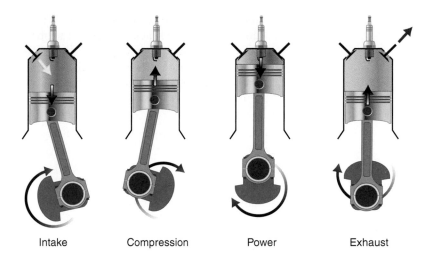

Intake   Compression   Power   Exhaust

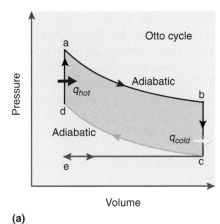

(a)

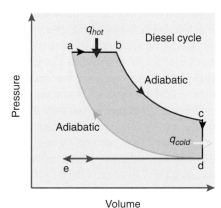

(b)

FIGURE 5.15
Idealized reversible cycles of the (a) Otto and (b) diesel engines.

ries of reservoirs at temperatures ranging from T_d to T_a. The power stroke is modeled as the adiabatic expansion $a \rightarrow b$. At this point, the exhaust valve opens and the gas is expelled. This step is modeled as the constant volume pressure decrease $b \rightarrow c$. The upward movement of the piston expels the remainder of the gas along the line $c \rightarrow e$, after which the cycle begins again.

The efficiency of this reversible cyclic engine can be calculated as follows. Assuming C_V to be constant along the segments $d \rightarrow a$ and $b \rightarrow c$, we write

$$q_{hot} = C_V(T_a - T_d) \quad \text{and} \quad q_{cold} = C_V(T_b - T_c) \tag{5.47}$$

The efficiency is given by

$$\varepsilon = \frac{q_{hot} + q_{cold}}{q_{hot}} = 1 - \frac{q_{cold}}{q_{hot}} = 1 - \left(\frac{T_b - T_c}{T_a - T_d}\right) \tag{5.48}$$

The temperatures and volumes along the reversible adiabatic segments are related by

$$T_c V_c^{\gamma-1} = T_d V_d^{\gamma-1} \quad \text{and} \quad T_b V_c^{\gamma-1} = T_a V_d^{\gamma-1} \tag{5.49}$$

because $V_b = V_c$ and $V_d = V_e$. Recall that $\gamma = C_P / C_V$. T_a and T_b can be eliminated from Equation (5.48) to give

$$\varepsilon = 1 - \frac{T_c}{T_d} \tag{5.50}$$

where T_c and T_d are the temperatures at the beginning and end of the compression stroke $c \rightarrow d$. Temperature T_c is fixed at ~300 K, and the efficiency can be increased only by increasing T_d. This is done by increasing the compression ratio, V_c/V_d. However, if the compression ratio is too high, T_d will be sufficiently high that the fuel ignites before the end of the compression stroke. A reasonable upper limit for T_d is 600 K, and for $T_c = 300$ K, $\varepsilon = 0.50$. This value is an upper limit, because a real engine does not operate in a reversible cycle, and because heat is lost along the $c \rightarrow d$ segment. Achievable efficiencies in Otto engines used in passenger cars lie in the range of 0.20 to 0.30.

In the **diesel engine,** higher compression ratios are possible because only air is let into the cylinder in the intake stroke. The fuel is injected into the cylinder at the end of the compression stroke, thus avoiding spontaneous ignition of the fuel–air mixture during the compression stroke $d \rightarrow a$. Because the compressed air has a high temperature of ~950 K, combustion occurs spontaneously without a spark plug along the constant pressure segment $a \rightarrow b$ after fuel injection. Because the fuel is injected over a time period in which the piston is moving out of the cylinder, this step can be modeled as a constant pressure heat intake. In this segment, it is assumed that heat is absorbed from a series of reservoirs at temperatures between T_a and T_b in a quasi-static process. In the other segments, the same processes occur as described for the reversible Otto cycle.

The heat intake along the segment $a \rightarrow b$ is given by

$$q_{hot} = C_p(T_b - T_a) \tag{5.51}$$

and q_{cold} is given by Equation (5.47). Therefore, the efficiency is given by

$$\varepsilon = 1 - \frac{1}{\gamma}\left(\frac{T_c - T_d}{T_b - T_a}\right) \tag{5.52}$$

Similarly to the treatment of the Otto engine, T_b and T_c can be eliminated from Equation (5.52), and the expression

$$\varepsilon = 1 - \frac{1}{\gamma}\frac{\left(\dfrac{V_b}{V_d}\right)^\gamma - \left(\dfrac{V_a}{V_d}\right)^\gamma}{\left(\dfrac{V_b}{V_d}\right) - \left(\dfrac{V_a}{V_d}\right)} \tag{5.53}$$

can be derived. For typical values $V_b/V_d = 0.2$, $V_a/V_d = 1/15$, $\gamma = 1.5$, and $\varepsilon = 0.64$. The higher efficiency achievable with the diesel cycle in comparison with the Otto cycle is a result of the higher temperature attained in the compression cycle. Real diesel engines used in trucks and passenger cars have efficiencies in the range of 0.30 to 0.35.

The efficiency of widely used engines, the efficiency of the production of electrical power, and the effect of these sectors of human activity on the global climate are matters of current concern. In the 20th century, the world population increased by a factor of 4, and the total electrical work and heat output due to human activity increased by a factor of 16. Because more than 85% of the electrical work and heat output has been generated by burning fossil fuels, the CO_2 concentration in the atmosphere has risen dramatically. At the current rate of increase, it will reach 550 ppm by the end of the 21st century. Modeling studies indicate that this could trigger a global warming equal in magnitude, but opposite in sign, to the last Ice Age.

Two avenues are being pursued to counter this trend: new technologies and conservation. Efforts are under way to identify and optimize technologies that produce heat or electrical work without the emission of greenhouse gases. Technologies for electricity production such as photovoltaic cells, fuel cells based on H_2 combustion, nuclear power, capturing solar power in space and relaying it to the Earth using microwave transmitters, and wind turbines do not produce CO_2 as a by-product. Fuel cells are of particular interest because the work arising from a chemical reaction can be accessed directly, rather than first harvesting the heat from the reaction and subsequently (partially) converting the heat to work. Because a fuel cell converts electrical work to mechanical work, it is not subject to the efficiency limitations imposed on heat engines by the second law. These technologies are in varying states of development along the path to widespread deployment. Lighting is very inefficient, with incandescent lights converting only 2% of the electrical work required to heat the tungsten filament to visible light. The remaining 98% is converted to heat. Fluorescent lighting is a factor of 5 to 6 more efficient.

Until technologies are available that do not affect global warming, large reductions in energy consumption and the emission of greenhouse gases can be realized by conservation. Significant reductions in heat loss can be achieved with more effective building insulation and with window coatings that reduce infrared transmission. Unnecessary heat generation can be reduced by using compact fluorescent lights in place of incandescent lighting. The increased use of commercially available ultraefficient cars that use a factor of 3 less gasoline for a given distance of travel than the current fleet average would substantially reduce CO_2 emissions.

5.12 Using the Fact that S Is a State Function to Determine the Dependence of S on V and T

Section 5.4 showed how the entropy varies with P, V, and T for an ideal gas. In this section, we derive general equations for the dependence of S on V and T that can be applied to solids, liquids, and real gases. We do so by using the property that dS is an exact differential. A similar analysis of S as a function of P and T is carried out in Section 5.13. Consider Equation (5.30), rewritten in the form

$$dS = \frac{1}{T}dU + \frac{P}{T}dV \qquad (5.54)$$

Because $1/T$ and P/T are greater than zero, the entropy of a system increases with the internal energy at constant volume, and increases with the volume at constant internal energy. However, because internal energy is not generally a variable under experimental control, it is more useful to obtain equations for the dependence of dS on V and T.

We first write the total differential dS in terms of the partial derivatives with respect to V and T:

$$dS = \left(\frac{\partial S}{\partial T}\right)_V dT + \left(\frac{\partial S}{\partial V}\right)_T dV \qquad (5.55)$$

To evaluate $(\partial S/\partial T)_V$ and $(\partial S/\partial V)_T$, Equation (5.54) for dS is rewritten in the form

$$dS = \frac{1}{T}\left[C_V dT + \left(\frac{\partial U}{\partial V}\right)_T dV\right] + \frac{P}{T}dV = \frac{C_V}{T}dT + \frac{1}{T}\left[P + \left(\frac{\partial U}{\partial V}\right)_T\right]dV \qquad (5.56)$$

Equating the coefficients of dT and dV in Equations (5.55) and (5.57),

$$\left(\frac{\partial S}{\partial T}\right)_V = \frac{C_V}{T} \quad \text{and} \quad \left(\frac{\partial S}{\partial V}\right)_T = \frac{1}{T}\left[P + \left(\frac{\partial U}{\partial V}\right)_T\right] \qquad (5.57)$$

The temperature dependence of entropy at constant volume can be calculated straightforwardly using the first equality in Equation (5.57):

$$dS = \frac{C_V}{T}dT, \text{ constant } V \qquad (5.58)$$

The expression for $(\partial S/\partial V)_T$ in Equation (5.57) is not in a form that allows for a direct comparison with experiment to be made. To develop a more useful relation, the property is used (see Section 3.1) that because dS is an exact differential, then

$$\left(\frac{\partial}{\partial T}\left(\frac{\partial S}{\partial V}\right)_T\right)_V = \left(\frac{\partial}{\partial V}\left(\frac{\partial S}{\partial T}\right)_V\right)_T \qquad (5.59)$$

Taking the mixed second derivatives of the expressions in Equation (5.57),

$$\left(\frac{\partial}{\partial V}\left(\frac{\partial S}{\partial T}\right)_V\right)_T = \frac{1}{T}\left(\frac{\partial}{\partial V}\left(\frac{\partial U}{\partial T}\right)_V\right)_T \qquad (5.60)$$

$$\left(\frac{\partial}{\partial T}\left(\frac{\partial S}{\partial V}\right)_T\right)_V = \frac{1}{T}\left[\left(\frac{\partial P}{\partial T}\right)_V + \left(\frac{\partial}{\partial T}\left(\frac{\partial U}{\partial V}\right)_T\right)_V\right] - \frac{1}{T^2}\left[P + \left(\frac{\partial U}{\partial V}\right)_T\right]$$

Substituting the expressions for the mixed second derivatives in Equation 5.60 into Equation 5.59, canceling the double mixed derivative of U that appears on both sides of the equation, and simplifying the result, the following equation is obtained:

$$P + \left(\frac{\partial U}{\partial V}\right)_T = T\left(\frac{\partial P}{\partial T}\right)_V \tag{5.61}$$

This equation provides the expression for $(\partial U/\partial V)_T$ that was used without a derivation in Section 3.2. It provides a way to calculate the internal pressure of the system if the equation of state for the substance is known.

Comparing the result in Equation (5.61) with the second equality in Equation (5.57), a practical equation is obtained for the dependence of entropy on volume under constant temperature conditions:

$$\left(\frac{\partial S}{\partial V}\right)_T = \left(\frac{\partial P}{\partial T}\right)_V = -\frac{(\partial V/\partial T)_P}{(\partial V/\partial P)_T} = \frac{\beta}{\kappa} \tag{5.62}$$

where β is the coefficient for thermal expansion at constant pressure, and κ is the isothermal compressibility coefficient. Both of these quantities are readily obtained from experiments. In simplifying this expression, the cyclic rule for partial derivatives, Equation (3.7), has been used.

The result of these considerations is that dS can be expressed in terms of dT and dV as

$$dS = \frac{C_V}{T}dT + \frac{\beta}{\kappa}dV \tag{5.63}$$

Integrating both sides of this equation along a reversible path,

$$\Delta S = \int_{T_i}^{T_f} \frac{C_V}{T}dT + \int_{V_i}^{V_f} \frac{\beta}{\kappa}dV \tag{5.64}$$

This result applies to a single-phase system of a liquid, solid, or gas that undergoes a transformation from the initial result T_i, V_i to T_f, V_f, provided that no phase changes or chemical reactions occur in the system.

SUPPLEMENTAL

5.13 The Dependence of S on T and P

Because chemical transformations are normally carried out at constant pressure rather than constant volume, we need to know how S varies with T and P. The total differential dS is written in the form

$$dS = \left(\frac{\partial S}{\partial T}\right)_P dT + \left(\frac{\partial S}{\partial P}\right)_T dP \tag{5.65}$$

Starting from the relation $U = H - PV$, we write the total differential dU as

$$dU = T\,dS - P\,dV = dH - P\,dV - V\,dP \tag{5.66}$$

This equation can be rearranged to give an expression for dS:

$$dS = \frac{1}{T}dH - \frac{V}{T}dP \tag{5.67}$$

The previous equation is analogous to Equation (5.54), but contains the variable P rather than V.

$$dH = \left(\frac{\partial H}{\partial T}\right)_P dT + \left(\frac{\partial H}{\partial P}\right)_T dP = C_P dT + \left(\frac{\partial H}{\partial P}\right)_T dP \tag{5.68}$$

Substituting this expression for dH into Equation (5.67),

$$dS = \frac{C_P}{T} dT + \frac{1}{T}\left[\left(\frac{\partial H}{\partial P}\right)_T - V\right] dP = \left(\frac{\partial S}{\partial T}\right)_P dT + \left(\frac{\partial S}{\partial P}\right)_T dP \tag{5.69}$$

Because the coefficients of dT and dP must be the same on both sides of Equation (5.69),

$$\left(\frac{\partial S}{\partial T}\right)_P = \frac{C_P}{T} \quad \text{and} \quad \left(\frac{\partial S}{\partial P}\right)_T = \frac{1}{T}\left[\left(\frac{\partial H}{\partial P}\right)_T - V\right] \tag{5.70}$$

The ratio C_P/T is positive for all substances, allowing one to conclude that S is a monotonically increasing function of the temperature.

Just as for $(\partial S/\partial V)_T$ in Section 5.12, the expression for $(\partial S/\partial P)_T$ is not in a form that allows a direct comparison with experimental measurements to be made. Just as in our evaluation of $(\partial S/\partial V)_T$, we equate the mixed second partial derivatives of $(\partial S/\partial T)_P$ and $(\partial S/\partial P)_T$:

$$\left(\frac{\partial}{\partial T}\left(\frac{\partial S}{\partial P}\right)_T\right)_P = \left(\frac{\partial}{\partial P}\left(\frac{\partial S}{\partial T}\right)_P\right)_T \tag{5.71}$$

These mixed partial derivatives can be evaluated using Equation (5.70):

$$\left(\frac{\partial}{\partial P}\left(\frac{\partial S}{\partial T}\right)_P\right)_T = \frac{1}{T}\left(\frac{\partial C_P}{\partial P}\right)_T = \frac{1}{T}\left(\frac{\partial}{\partial P}\left(\frac{\partial H}{\partial T}\right)_P\right)_T \tag{5.72}$$

$$\left(\frac{\partial}{\partial T}\left(\frac{\partial S}{\partial P}\right)_T\right)_P = \frac{1}{T}\left[\left(\frac{\partial}{\partial T}\left(\frac{\partial H}{\partial P}\right)_T\right)_P - \left(\frac{\partial V}{\partial T}\right)_P\right] - \frac{1}{T^2}\left[\left(\frac{\partial H}{\partial P}\right)_T - V\right] \tag{5.73}$$

Equating Equations 5.72 and 5.73,

$$\frac{1}{T}\left(\frac{\partial}{\partial T}\left(\frac{\partial H}{\partial P}\right)_T\right)_P = \frac{1}{T}\left[\left(\frac{\partial}{\partial T}\left(\frac{\partial H}{\partial P}\right)_T\right)_P - \left(\frac{\partial V}{\partial T}\right)_P\right] - \frac{1}{T^2}\left[\left(\frac{\partial H}{\partial P}\right)_T - V\right] \tag{5.74}$$

Simplifying this equation results in

$$\left(\frac{\partial H}{\partial P}\right)_T - V = -T\left(\frac{\partial V}{\partial T}\right)_P \tag{5.75}$$

Using this result and Equation 5.70, the pressure dependence of the entropy at constant temperature can be written in a form that easily allows an experimental determination of this quantity to be made:

$$\left(\frac{\partial S}{\partial P}\right)_T = -\left(\frac{\partial V}{\partial T}\right)_P = -V\beta \tag{5.76}$$

Using these results, the total differential dS can be written in terms of experimentally accessible parameters as

$$dS = \frac{C_P}{T} dT - V\beta\, dP \tag{5.77}$$

Integrating both sides of this equation along a reversible path,

$$\Delta S = \int_{T_i}^{T_f} \frac{C_P}{T} dT - \int_{P_i}^{P_f} V\beta\, dP \tag{5.78}$$

This result applies to a single-phase system of a pure liquid, solid, or gas that undergoes a transformation from the initial result $T_i P_i$ to T_f, P_f, provided that no phase changes or chemical reactions occur in the system.

SUPPLEMENTAL

5.14 The Thermodynamic Temperature Scale

The reversible Carnot cycle provides a basis for the **thermodynamic temperature scale,** a scale that is independent of the choice of a particular thermometric substance. This is the case because all reversible Carnot engines have the same efficiency, regardless of the working substance. The basis for the thermodynamic temperature scale is the fact that the heat withdrawn from a reservoir is a thermometric property. Both on experimental and theoretical grounds, it can be shown that

$$q = a\theta \tag{5.77}$$

where θ is the thermodynamic temperature, and a is an arbitrary scale constant that sets numerical values for the thermodynamic temperature. Using Equations (5.9) and (5.10) for the efficiency of the reversible Carnot engine,

$$\varepsilon = \frac{q_{hot} + q_{cold}}{q_{hot}} = \frac{\theta_{hot} + \theta_{cold}}{\theta_{hot}} \tag{5.78}$$

This equation is the fundamental equation establishing an absolute temperature scale. To this point, we have no numerical values for this scale. Note, however, that $q \to 0$ as $\theta \to 0$, so that there is a natural zero for this temperature scale. Additionally, if we choose one value of θ to be positive, all other values of θ must be greater than zero. Otherwise, we could find conditions under which the heats q_{hot} and q_{cold} have the same sign. This would lead to a perpetual motion machine. Both of these characteristics fit the requirements of an absolute temperature scale.

A numerical scale for the thermodynamic temperature scale can be obtained by assigning the value 273.16 to the θ value corresponding to the triple point of water, and by making the size of a degree equal to the size of a degree on the Celsius scale. With this choice, the thermodynamic temperature scale becomes numerically equal to the absolute temperature scale based on the ideal gas law. However, the thermodynamic temperature scale is the primary scale, because it is independent of the nature of the working substance.

Vocabulary

Carnot cycle

Clausius inequality

diesel engine

entropy

heat engine

heat pump

irreversible process

natural transformations

Otto engine

perpetual motion machine of the first kind

perpetual motion machine of the second kind

refrigerator

second law of thermodynamics

spontaneous process

thermodynamic temperature scale

third law of thermodynamics

unnatural transformations

Questions on Concepts

Q5.1 Which of the following processes is spontaneous?

a. The reversible isothermal expansion of an ideal gas

b. The vaporization of superheated water at 102°C and 1 bar

c. The constant pressure melting of ice at its normal freezing point by the addition of an infinitesimal quantity of heat

d. The adiabatic expansion of a gas into a vacuum

Q5.2 Why are ΔS_{fusion} and $\Delta S_{vaporization}$ always positive?

Q5.3 Why is the efficiency of a Carnot heat engine the upper bound to the efficiency of an internal combustion engine?

Q5.4 The amplitude of a pendulum consisting of a mass on a long wire is initially adjusted to have a very small value. The amplitude is found to decrease slowly with time. Is this process reversible? Would the process be reversible if the amplitude did not decrease with time?

Q5.5 A process involving an ideal gas is carried out in which the temperature changes at constant volume. For a fixed value of ΔT, the mass of the gas is doubled. The process

is repeated with the same initial mass and ΔT is doubled. For which of these processes is ΔS greater? Why?

Q5.6 Under what conditions does the equality $\Delta S = \Delta H / T$ hold?

Q5.7 Under what conditions is $\Delta S < 0$ for a spontaneous process?

Q5.8 Is the equation

$$\Delta S = \int_{T_i}^{T_f} \frac{C_V}{T} dT + \int_{V_i}^{V_f} \frac{\beta}{\kappa} dV = C_V \ln\frac{T_f}{T_i} + \frac{\beta}{\kappa}(V_f - V_i)$$

valid for an ideal gas?

Q5.9 Without using equations, explain why ΔS for a liquid or solid is dominated by the temperature dependence of S as both P and T change.

Q5.10 You are told that $\Delta S = 0$ for a process in which the system is coupled to its surroundings. Can you conclude that the process is reversible? Justify your answer.

Problems

Problem numbers in **red** indicate that the solution to the problem is given in the *Student's Solutions Manual*.

P5.1 Beginning with Equation (5.5), use Equation (5.6) to eliminate V_c and V_d to arrive at the result $w_{cycle} = nR(T_{hot} - T_{cold}) \ln V_b / V_a$.

P5.2 Consider the reversible Carnot cycle shown in Figure 5.2 with 1 mol of an ideal gas with $C_V = 3/2R$ as the working substance. The initial isothermal expansion occurs at the hot reservoir temperature of $T_{hot} = 600°C$ from an initial volume of 3.50 L (V_a) to a volume of 10.0 L (V_b). The system then undergoes an adiabatic expansion until the temperature falls to $T_{cold} = 150°C$. The system then undergoes an isothermal compression and a subsequent adiabatic compression until the initial state described by $T_a = 600°C$ and $V_a = 3.50$ L is reached.

a. Calculate V_c and V_d.

b. Calculate w for each step in the cycle and for the total cycle.

c. Calculate ε and the amount of heat that is extracted from the hot reservoir to do 1.00 kJ of work in the surroundings.

P5.3 Using your results from Problem P5.2, calculate q, ΔU, and ΔH for each step in the cycle and for the total cycle described in Problem P5.2.

P5.4 Using your results from Problems P5.2 and P5.3, calculate $\Delta S, \Delta S_{surroundings}$, and ΔS_{total} for each step in the cycle and for the total cycle described in Problem P5.2.

P5.5 Calculate ΔS if the temperature of 1 mol of an ideal gas with $C_V = 3/2R$ is increased from 150 to 350 K under conditions of (a) constant pressure and (b) constant volume.

P5.6 One mole of N_2 at 20.5°C and 6.00 bar undergoes a transformation to the state described by 145°C and 2.75 bar. Calculate ΔS if

$$\frac{C_{P,m}}{J\,mol^{-1}\,K^{-1}} = 30.81 - 11.87 \times 10^{-3} \frac{T}{K} + 2.3968 \times 10^{-5} \frac{T^2}{K^2}$$

$$- 1.0176 \times 10^{-8} \frac{T^3}{K^3}$$

P5.7 One mole of an ideal gas with $C_V = 3/2R$ undergoes the transformations described in the following list from an initial state described by $T = 300$ K and $P = 1.00$ bar. Calculate q, w, ΔU, ΔH, and ΔS for each process.

a. The gas is heated to 450 K at a constant external pressure of 1.00 bar.

b. The gas is heated to 450 K at a constant volume corresponding to the initial volume.

c. The gas undergoes a reversible isothermal expansion at 300 K until the pressure is half of its initial value.

P5.8 Calculate $\Delta S_{surroundings}$ and ΔS_{total} for each of the processes described in Problem P5.7. Which of the processes is a spontaneous process? The state of the surroundings for each part is as follows:

a. 450 K, 1 bar

b. 450 K, 1 bar

c. 300 K, 0.500 bar

P5.9 At the transition temperature of 95.4°C, the enthalpy of transition from rhombic to monoclinic sulfur is 0.38 kJ mol^{-1}.

a. Calculate the entropy of transition under these conditions.

b. At its melting point, 119°C, the enthalpy of fusion of monoclinic sulfur is 1.23 kJ mol^{-1}. Calculate the entropy of fusion.

c. The values given in parts (a) and (b) are for 1 mol of sulfur; however, in crystalline and liquid sulfur, the molecule is present as S_8. Convert the values of the enthalpy and entropy of fusion in parts (a) and (b) to those appropriate for S_8.

P5.10

a. Calculate ΔS if 1 mol of liquid water is heated from 0° to 100°C under constant pressure if $C_{P,m} = 75.291$ J K^{-1} mol^{-1}.

b. The melting point of water at the pressure of interest is 0°C and the enthalpy of fusion is 6.0095 kJ mol^{-1}. The boiling point is 100°C and the enthalpy of vaporization is 40.6563 kJ mol^{-1}. Calculate ΔS for the transformation $H_2O(s, 0°C) \rightarrow H_2O(g, 100°C)$.

P5.11 One mole of an ideal gas with $C_{V,m} = 5/2R$ undergoes the transformations described in the following list from an initial state described by $T = 250$ K and $P = 1.00$ bar. Calculate q, w, ΔU, ΔH, and ΔS for each process.

a. The gas undergoes a reversible adiabatic expansion until the final pressure is half its initial value.

b. The gas undergoes an adiabatic expansion against a constant external pressure of 0.500 bar until the final pressure is half its initial value.

c. The gas undergoes an expansion against a constant external pressure of zero bar until the final pressure is equal to half of its initial value.

P5.12 The standard entropy of Pb(s) at 298.15 K is 64.80 J K^{-1} mol^{-1}. Assume that the heat capacity of Pb(s) is given by

$$\frac{C_{P,m}(Pb,s)}{J\,mol^{-1}\,K^{-1}} = 22.13 + 0.01172\frac{T}{K} + 1.00\times10^{-5}\frac{T^2}{K^2}$$

The melting point is 327.4°C and the heat of fusion under these conditions is 4770 J mol^{-1}. Assume that the heat capacity of Pb(l) is given by

$$\frac{C_{P,m}(Pb,l)}{J\,K^{-1}\,mol^{-1}} = 32.51 - 0.00301\frac{T}{K}$$

a. Calculate the standard entropy of Pb(l) at 500°C.

b. Calculate ΔH for the transformation Pb(s, 25°C) → Pb(l, 500°C).

P5.13 Between 0°C and 100°C, the heat capacity of Hg(l) is given by

$$\frac{C_{P,m}(Hg,l)}{J\,K^{-1}\,mol^{-1}} = 30.093 - 4.944\times10^{-3}\frac{T}{K}$$

Calculate ΔH and ΔS if 1 mol of Hg(l) is raised in temperature from 0° to 100°C at constant P.

P5.14 One mole of a van der Waals gas at 27°C is expanded isothermally and reversibly from an initial volume of 0.020 m^3 to a final volume of 0.060 m^3. For the van der Waals gas, $(\partial U/\partial V)_T = a/V_m^2$. Assume that $a = 0.556$ Pa m^6 mol^{-2}, and that $b = 64.0\times10^{-6}$ m^3 mol^{-1}. Calculate q, w, ΔU, ΔH, and ΔS for the process.

P5.15 The heat capacity of α-quartz is given by

$$\frac{C_{P,m}(\alpha\text{-quartz},s)}{J\,K^{-1}\,mol^{-1}} = 46.94 + 34.31\times10^{-3}\frac{T}{K} - 11.30\times10^5\frac{T^2}{K^2}$$

The coefficient of thermal expansion is given by $\beta = 0.3530 \times 10^{-4}$ K^{-1} and $V_m = 22.6$ cm^3 mol^{-1}. Calculate ΔS_m for the transformation α–quartz (25.0°C, 1 atm) → α–quartz (225°C, 1000 atm).

P5.16 Calculate $\Delta S_{surroundings}$ and ΔS_{total} for the processes described in parts (a) and (b) of Problem P5.11. Which of the processes is a spontaneous process? The state of the surroundings for each part is as follows:

a. 250 K, 0.500 bar

b. 300 K, 0.500 bar

P5.17 Calculate ΔS, ΔS_{total}, and $\Delta S_{surroundings}$ when the volume of 85.0 g of CO initially at 298 K and 1.00 bar increases by a factor of three in (a) an adiabatic reversible expansion, (b) an expansion against $P_{external} = 0$, and (c) an isothermal reversible expansion. Take $C_{P,m}$ to be constant at the value 29.14 J mol^{-1} K^{-1} and assume ideal gas behavior. State whether each process is spontaneous.

P5.18 One mole of an ideal gas with $C_{V,m} = 3/2R$ is transformed from an initial state $T = 600$ K and $P = 1.00$ bar to a final state $T = 250$ K and $P = 4.50$ bar. Calculate ΔU, ΔH, and ΔS for this process.

P5.19 An ideal gas sample containing 2.50 moles for which $C_{V,m} = 3/2R$ undergoes the following reversible cyclical process from an initial state characterized by $T = 450$ K and $P = 1.00$ bar:

a. It is expanded reversibly and adiabatically until the volume doubles.

b. It is reversibly heated at constant volume until T increases to 450 K.

c. The pressure is increased in an isothermal reversible compression until $P = 1.00$ bar.

Calculate q, w, ΔU, ΔH, ΔS, $\Delta S_{surroundings}$, and ΔS_{total} for each step in the cycle, and for the total cycle. The temperature of the surroundings is 300 K.

P5.20 One mole of $H_2O(l)$ is compressed from a state described by $P = 1.00$ bar and $T = 298$ K to a state described by $P = 800$ bar and $T = 450$ K. In addition, $\beta = 2.07 \times 10^{-4}$ K^{-1} and the density can be assumed to be constant at the value 997 kg m^{-3}. Calculate ΔS for this transformation, assuming that $\kappa = 0$.

P5.21 A 25.0 g mass of ice at 273 K is added to 150.0 g of $H_2O(l)$ at 360 K at constant pressure. Is the final state of the system ice or liquid water? Calculate ΔS for the process. Is the process spontaneous?

P5.22 15.0 g of steam at 373 K is added to 250.0 g of $H_2O(l)$ at 298 K at constant pressure of 1 bar. Is the final state of the system steam or liquid water? Calculate ΔS for the process.

P5.23 The maximum theoretical efficiency of an internal combustion engine is achieved in a reversible Carnot cycle. Assume that the engine is operating in the Otto cycle and that $C_{V,m} = 5/2R$ for the fuel–air mixture initially at 298 K (the temperature of the cold reservoir). The mixture is compressed by a factor of 8.0 in the adiabatic compression step. What is the maximum theoretical efficiency of this engine? How much would the efficiency increase if the compression ratio could be increased to 30? Do you see a problem in doing so?

P5.24 One mole of $H_2O(l)$ is supercooled to $-2.25°C$ at 1 bar pressure. The freezing temperature of water at this pressure is $0.00°C$. The transformation $H_2O(l) \rightarrow H_2O(s)$ is suddenly observed to occur. By calculating ΔS, $\Delta S_{surroundings}$, and ΔS_{total}, verify that this transformation is spontaneous at $-2.25°C$. The heat capacities are given by $C_p(H_2O(l)) = 75.3$ $J\,K^{-1}\,mol^{-1}$ and $C_p(H_2O(s)) = 37.7$ $J\,K^{-1}\,mol^{-1}$, and $\Delta H_{fusion} = 6.008$ kJ mol^{-1} at $0.00°C$. Assume that the surroundings are at $-2.25°C$. [*Hint:* Consider the two pathways at 1 bar: (a) $H_2O(l, -2.25°C) \rightarrow H_2O(s, -2.25°C)$ and (b) $H_2O(l, -2.25°C) \rightarrow H_2O(l, 0.00°C) \rightarrow H_2O(s, 0.00°C) \rightarrow H_2O(s, -2.25°C)$. Because S is a state function, ΔS must be the same for both pathways.]

P5.25 An air conditioner is a refrigerator with the inside of the house acting as the cold reservoir and the outside atmosphere acting as the hot reservoir. Assume that an air conditioner consumes 1.50×10^3 W of electrical power, and that it can be idealized as a reversible Carnot refrigerator. If the coefficient of performance of this device is 2.50, how much heat can be extracted from the house in a 24-hour period?

P5.26 The interior of a refrigerator is typically held at 277 K and the interior of a freezer is typically held at 255 K. If the room temperature is 294 K, by what factor is it more expensive to extract the same amount of heat from the freezer than from the refrigerator? Assume that the theoretical limit for the performance of a reversible refrigerator is valid.

P5.27 The Chalk Point, Maryland, generating station supplies electrical power to the Washington, D.C., area. Units 1 and 2 have a gross generating capacity of 710 MW (megawatt). The steam pressure is 25×10^6 Pa, and the superheater outlet temperature (T_h) is $540°C$. The condensate temperature (T_c) is $30.0°C$.

a. What is the efficiency of a reversible Carnot engine operating under these conditions?

b. If the efficiency of the boiler is 91.2%, the overall efficiency of the turbine, which includes the Carnot efficiency and its mechanical efficiency, is 46.7%, and the efficiency of the generator is 98.4%, what is the efficiency of the total generating unit? (Another 5% needs to be subtracted for other plant losses.)

c. One of the coal burning units produces 355 MW. How many metric tons (1 metric ton = 1×10^6 g) of coal per hour are required to operate this unit at its peak output if the enthalpy of combustion of coal is 29.0×10^3 kJ kg^{-1}?

P5.28 The mean solar flux at the Earth's surface is ~4.00 J cm^{-2} min^{-1}. In a nonfocusing solar collector, the temperature can reach a value of $90.0°C$. A heat engine is operated using the collector as the hot reservoir and a cold reservoir at 298 K. Calculate the area of the collector needed to produce one horsepower (1 hp = 746 watts). Assume that the engine operates at the maximum Carnot efficiency.

P5.29 A refrigerator is operated by a 0.25-hp (1 hp = 746 watts) motor. If the interior is to be maintained at $-20°C$ and the room temperature is $35°C$, what is the maximum heat leak (in watts) that can be tolerated? Assume that the coefficient of performance is 75% of the maximum theoretical value.

P5.30 An electrical motor is used to operate a Carnot refrigerator with an interior temperature of $0°C$. Liquid water at $0°C$ is placed into the refrigerator and transformed to ice at $0°C$. If the room temperature is $20°C$, what mass of ice can be produced in one minute by a 0.25-hp motor that is running continuously? Assume that the refrigerator is perfectly insulated and operates at the maximum theoretical efficiency.

P5.31 Calculate $\Delta S°$ for the reaction $H_2(g) + Cl_2(g) \rightarrow 2HCl(g)$ at 650 K. Omit terms in the temperature-dependent heat capacities higher than T^2/K^2.

Web-Based Simulations, Animations, and Problems

W5.1   The reversible Carnot cycle is simulated with adjustable values of T_{hot} and T_{cold} and ΔU, q, and w are determined for each segment and for the cycle. The efficiency is also determined for the cycle.

Chemical Equilibrium

In the previous chapter, criteria for the spontaneity of arbitrary processes were developed. In this chapter, spontaneity is discussed in the context of a reactive mixture of gases. Two new state functions are introduced that express spontaneity in terms of the properties of the system only. The Helmholtz energy is the criterion for spontaneity for a reaction in a system whose initial and final states are at the same values of V and T. The Gibbs energy is the criterion for spontaneity for a reaction in a system whose initial and final states are at the same values of P and T. Using the Gibbs energy, a thermodynamic equilibrium constant, K_P, is derived that predicts the equilibrium concentrations of reactants and products in a mixture of reactive ideal gases. ∎

6.1 The Gibbs Energy and the Helmholtz Energy

In Chapter 5, it was shown that the direction of spontaneous change for a process is predicted by $\Delta S + \Delta S_{surroundings} > 0$. In this section, this spontaneity criterion is used to derive two new state functions, the Gibbs and Helmholtz energies. These new state functions provide the basis for all further discussions of equilibrium. The fundamental expression governing spontaneity is the Clausius inequality [Equation (5.33)], written in the form

$$TdS \geq dq \tag{6.1}$$

The equality is satisfied only for a reversible process. Because $dq = dU - dw$,

$$TdS \geq dU - dw \text{ or, equivalently, } -dU + dw + TdS \geq 0 \tag{6.2}$$

As discussed in Section 2.2, a system can do different types of work on the surroundings. It is particularly useful to distinguish between expansion work, in which the work arises from a volume change in the system, and nonexpansion work. We rewrite Equation (6.2) in the form

$$-dU - P_{external}dV + dw_{nonexpansion} + TdS \geq 0 \tag{6.3}$$

This equation expresses the condition of spontaneity for an arbitrary process in terms of the changes in the state functions U, V, S, and T as well as the path-dependent functions $P_{external}dV$ and $w_{nonexpansion}$.

To make a connection with the discussion of spontaneity in Chapter 5, consider a special case of Equation (6.3). For an isolated system, $w = 0$ and $dU = 0$. Therefore, Equation (6.3) reduces to the familiar result derived in Section 5.5:

$$dS \geq 0 \qquad (6.4)$$

Chemists are generally more interested in systems that interact with their environment than in isolated systems. Therefore, the next criteria that are derived define equilibrium and spontaneity for such systems. As was done in Chapters 1 through 5, it is useful to consider transformations at constant temperature and either constant volume or constant pressure. *Note that constant T and P (or V) does not imply that these variables are constant throughout the process, but rather that they are the same for the initial and final states of the process.*

For isothermal processes, $TdS = d(TS)$, and Equation (6.3) can be written in the following form:

$$-dU + TdS \geq -dw_{expansion} - dw_{nonexpansion} \quad \text{or, equivalently,} \qquad (6.5)$$
$$d(U - TS) \leq dw_{expansion} + dw_{nonexpansion}$$

The combination of state functions $U - TS$, which has the units of energy, defines a new state function that we call the **Helmholtz energy,** abbreviated A. Using this definition, the general condition of spontaneity for isothermal processes becomes

$$dA \leq dw_{expansion} + dw_{nonexpansion} \qquad (6.6)$$

Because the equality applies for a reversible transformation, Equation (6.6) provides a way to calculate the maximum work that a system can do on the surroundings in an isothermal process. Example Problem 6.1 illustrates the usefulness of A for calculating the maximum work available through carrying out a chemical reaction.

EXAMPLE PROBLEM 6.1

You wish to construct a fuel cell based on the oxidation of a hydrocarbon fuel. The two choices for a fuel are methane and octane. Calculate the maximum work available through the combustion of these two hydrocarbons, on a per mole and a per gram basis at 298.15 K and 1 bar pressure. The standard enthalpies of combustion are $\Delta H^{\circ}_{combustion}(CH_4, g) = -891 \text{ kJ mol}^{-1}$ and $\Delta H^{\circ}_{combustion}(C_8H_{18}, l) = -5471 \text{ kJ mol}^{-1}$ and $S^{\circ}(C_8H_{18}, l) = 361.1 \text{ J mol}^{-1} \text{k}^{-1}$. Use tabulated values of S° in Appendix A for these calculations. Are there any other factors that should be taken into account in making a decision between these two fuels?

Solution
The combustion reactions are

$$\text{Methane: } CH_4(g) + 2O_2(g) \rightarrow CO_2(g) + 2H_2O(l)$$
$$\text{Octane: } C_8H_{18}(l) + 25/2 \, O_2(g) \rightarrow 8CO_2(g) + 9H_2O(l)$$

$$\Delta A^{\circ}_{methane} = \Delta U^{\circ}_{combustion}(CH_4, g) - T\left(\begin{array}{c} S^{\circ}(CO_2, g) + 2S^{\circ}(H_2O, l) - S^{\circ}(CH_4, g) \\ -2S^{\circ}(O_2, g) \end{array}\right)$$

$$\Delta A^{\circ}_{methane} = \Delta H^{\circ}_{combustion}(CH_4, g)$$
$$- \Delta nRT - T(S^{\circ}(CO_2, g) + 2S^{\circ}(H_2O, l) - S^{\circ}(CH_4, g) - 2S^{\circ}(O_2, g))$$
$$= -891 \times 10^3 \text{J mol}^{-1} + 2 \times 8.314 \text{ J mol}^{-1} \text{K}^{-1} \times 298.15 \text{ K}$$
$$-298.15 \text{ K} \times \left(\begin{array}{c} 213.8 \text{ J mol}^{-1} \text{K}^{-1} + 2 \times 70.0 \text{ J mol}^{-1} \text{K}^{-1} \\ -186.3 \text{ J mol}^{-1} \text{K}^{-1} - 2 \times 205.2 \text{ J mol}^{-1} \text{K}^{-1} \end{array}\right)$$
$$= -814 \text{ kJ mol}^{-1}$$

$$\Delta A_{octane}^\circ = \Delta U_{combustion}^\circ (C_8H_{18}, l)$$
$$-T\left(8S^\circ(CO_2, g) + 9S^\circ(H_2O, l) - S^\circ(C_8H_{18}, l) - \frac{25}{2}S^\circ(O_2, g)\right)$$

$$\Delta A_{octane}^\circ = \Delta H_{combustion}^\circ (C_8H_{18}, l) - \Delta nRT$$
$$-T\left(8S^\circ(CO_2, g) + 9S^\circ(H_2O, l) - S^\circ(C_8H_{18}, l) - \frac{25}{2}S^\circ(O_2, g)\right)$$

$$= -5471 \times 10^3 \, J \, mol^{-1} + \frac{9}{2} \times 8.314 \, J \, mol^{-1} \, K^{-1} \times 298.15 \, K$$

$$-298.15 \, K \times \left(\begin{array}{c} 8 \times 213.8 \, J \, mol^{-1} \, K^{-1} + 9 \times 70.0 \, J \, mol^{-1} \, K^{-1} \\ -361.1 \, J \, mol^{-1} \, K^{-1} - \frac{25}{2} \times 205.2 \, J \, mol^{-1} \, K^{-1} \end{array}\right)$$

$$= -5285 \, kJ \, mol^{-1}$$

On a per mol basis, octane (molecular weight 114.25 g mol^{-1}) is capable of producing a factor of 6.5 more work than methane (molecular weight 16.04 g mol^{-1}). However, on a per gram basis, methane and octane are nearly equal in their ability to produce work (-50.6 kJ g^{-1} versus -46.3 kJ g^{-1}). You might want to choose octane, because it can be stored as a liquid at atmospheric pressure. By contrast, a pressurized tank is needed to store methane as a liquid at 298.15 K.

In discussing the Helmholtz energy, $dT = 0$ was the only constraint applied. We now apply the additional constraint for a constant volume process, $dV = 0$. If only expansion work is possible in the transformation, then $đw_{expansion} = 0$, because $dV = 0$. In this case, the condition that defines spontaneity and equilibrium becomes

$$dA \leq 0 \tag{6.7}$$

The condition for spontaneity at constant T and V takes on a simple form using the Helmholtz energy rather than the entropy, if nonexpansion work is not possible.

Chemical reactions are more commonly studied under constant pressure than constant volume conditions. Therefore, the condition for spontaneity is considered next for an isothermal transformation that takes place at constant pressure, $P = P_{external}$. At constant pressure and temperature, $PdV = d(PV)$ and $TdS = d(TS)$. In this case, Equation (6.3) can be written in the form

$$d(U + PV - TS) = d(H - TS) \leq đw_{nonexpansion} \tag{6.8}$$

The combination of state functions $H - TS$, which has the units of energy, defines a new state function called the **Gibbs energy,** abbreviated G. Using the Gibbs energy, the condition for spontaneity and equilibrium for an isothermal process at constant pressure becomes

$$dG \leq đw_{nonexpansion} \tag{6.9}$$

For a reversible process, the equality holds, and the change in the Gibbs energy is a measure of the maximum nonexpansion work that can be produced in the transformation. If the transformation is carried out in such a way that nonexpansion work is not possible, the condition for spontaneity and equilibrium is

$$dG \leq 0 \tag{6.10}$$

What is the advantage of using the state functions G and A as criteria for spontaneity rather than entropy? The Clausius inequality can be written in the form

$$dS - \frac{dq}{T} \geq 0 \qquad (6.11)$$

Because, as was shown in Section 5.8, $dS_{surroundings} = -dq/T$, the Clausius inequality is equivalent to the spontaneity condition:

$$dS + dS_{surroundings} \geq 0 \qquad (6.12)$$

By introducing G and A, the fundamental conditions for spontaneity have not been changed. However, G and A are expressed only in terms of the macroscopic state variables of the system. By introducing G and A, it is no longer necessary to consider the surroundings explicitly. Knowledge of ΔG and ΔA for the system alone is sufficient to predict the direction of natural change.

Apart from defining the condition of spontaneity, Equation (6.9) is very useful, because it allows one to calculate the **maximum nonexpansion work** that can be produced by a chemical transformation. A particularly important application of this equation is to calculate the electrical work produced by a reaction in an electrochemical cell or fuel cell as shown in Example Problem 6.2. This topic will be discussed in detail in Chapter 11. The redox current that flows between the two half cells is used to do work. By contrast, nonexpansion work is not possible in a conventional combustion process such as that in an automobile engine.

EXAMPLE PROBLEM 6.2

Calculate the maximum nonexpansion work that can be produced by the fuel cell oxidation reactions in Example Problem 6.1.

Solution

$$\Delta G_{methane}^{\circ} = \Delta H_{combustion}^{\circ}(CH_4, g) - T\left(\begin{array}{c} S^{\circ}(CO_2, g) + 2S^{\circ}(H_2O, l) - S^{\circ}(CH_4, g) \\ -2S^{\circ}(O_2, g) \end{array} \right)$$

$$= -891 \times 10^3 \, J \, mol^{-1} - 298.15 \, K$$

$$\times (213.8 \, J \, K^{-1} + 2 \times 70.0 \, J \, K^{-1} - 186.3 \, J \, K^{-1} - 2 \times 205.2 \, J \, K^{-1})$$

$$= -818 \, kJ \, mol^{-1}$$

$$\Delta G_{octane}^{\circ} = \Delta H_{combustion}^{\circ}(C_8H_{18}, l)$$

$$-T\left(8S^{\circ}(CO_2, g) + 9S^{\circ}(H_2O, l) - S^{\circ}(C_8H_{18}, l) - \frac{25}{2}S^{\circ}(O_2, g) \right)$$

$$= -5471 \times 10^3 \, J \, mol^{-1} - 298.15 \, K \times \left(\begin{array}{c} 8 \times 213.8 \, J \, K^{-1} + 9 \times 70.0 \, J \, K^{-1} \\ -361.1 \, J \, K^{-1} - \frac{25}{2} \times 205.2 \, J \, K^{-1} \end{array} \right)$$

$$= -5296 \, kJ \, mol^{-1}$$

Compare this result with that of Example Problem 6.1. What can you conclude about the relative amounts of expansion and nonexpansion work available in these reactions?

It is useful to compare the available work that can be done by a reversible heat engine with the electrical work done by an electrochemical fuel cell using the same chemical reaction. In the reversible heat engine, the maximum available work is the product of the heat withdrawn from the hot reservoir and the efficiency of the heat engine. Consider a heat engine with $T_{hot} = 600 \, K$ and $T_{cold} = 300 \, K$, which has an efficiency of 0.50, and set $q_{hot} = \Delta H_{combustion}^{\circ}$. For these values, the maximum work available from the heat engine is 54% of that available in the electrochemical fuel cell for the combustion of methane. The corresponding value for octane is 52%. Why are these values less than

100%? If the oxidation reactions of Example Problem 6.1 can be carried out as redox reactions using two physically separated half cells, electrical work can be harnessed directly and converted to mechanical work. All forms of work can theoretically (but not practically) be converted to other forms of work with 100% efficiency. However, if chemical energy is converted to heat by burning the hydrocarbon fuel, and the heat is subsequently converted to work using a heat engine, the efficiency is less than 100% as discussed in Section 5.2. It is clear from this comparison why considerable research and development effort is currently being spent on fuel cells.

After this discussion of the usefulness of *G* in predicting the maximum nonexpansion work available through a chemical transformation, the focus turns to the use of *G* to determine the direction of spontaneous change. For macroscopic changes at constant *P* and *T* in which no nonexpansion work is possible, the condition for spontaneity is $\Delta G < 0$ where

$$\Delta G = \Delta H - T\Delta S \tag{6.13}$$

Note that there are two contributions to ΔG that determine if an isothermal chemical transformation is spontaneous. They are the energetic contribution, ΔH, and the entropic contribution, $T\Delta S$. The following conclusions can be drawn based on Equation (6.13):

- The entropic contribution to ΔG is greater for higher temperatures.
- A chemical transformation is always spontaneous if $\Delta H < 0$ (an exothermic reaction) and $\Delta S > 0$.
- A chemical transformation is never spontaneous if $\Delta H > 0$ (an endothermic reaction) and $\Delta S < 0$.
- For all other cases, the relative magnitudes of ΔH and $T\Delta S$ determine if the chemical transformation is spontaneous.

Similarly, for macroscopic changes at constant *V* and *T* in which no nonexpansion work is possible, the condition for spontaneity is $\Delta A < 0$, where

$$\Delta A = \Delta U - T\Delta S \tag{6.14}$$

Again, two contributions determine if an isothermal chemical transformation is spontaneous: ΔU is an energetic contribution, and $T\Delta S$ is an entropic contribution to ΔA. The same conclusions can be drawn from this equation as for those listed earlier, with *U* substituted for *H*.

6.2 The Differential Forms of *U*, *H*, *A*, and *G*

To this point, the state functions *U*, *H*, *A*, and *G* have been defined. In this section, we discuss how these state functions depend on the macroscopic system variables. To do so, the differential forms *dU*, *dH*, *dA*, and *dG* are developed. Starting from the definitions

$$H = U + PV \tag{6.15}$$
$$A = U - TS$$
$$G = H - TS = U + PV - TS$$

the following total differentials can be formed:

$$dU = TdS - PdV \tag{6.16}$$
$$dH = TdS - PdV + PdV + VdP = TdS + VdP \tag{6.17}$$
$$dA = TdS - PdV - TdS - SdT = -SdT - PdV \tag{6.18}$$
$$dG = TdS + VdP - TdS - SdT = -SdT + VdP \tag{6.19}$$

These differential forms express the internal energy as $U(S,V)$, the enthalpy as $H(S,P)$, the Helmholtz energy as $A(T,V)$, and the Gibbs energy as $G(T,P)$. Although other combinations of variables can be used, these **natural variables** are used because the differential expressions are compact.

What information can be obtained from the differential expressions in Equations (6.16) through (6.19)? Because U, H, A, and G are state functions, two different equivalent expressions such as those written for dU here can be formulated:

$$dU = TdS - PdV = \left(\frac{\partial U}{\partial S}\right)_V dS + \left(\frac{\partial U}{\partial V}\right)_S dV \qquad (6.20)$$

For Equation (6.20) to be valid, the coefficients of dS and dV on both sides of the equation must be equal. Applying this reasoning to Equations (6.16) through (6.19), the following expressions are obtained:

$$\left(\frac{\partial U}{\partial S}\right)_V = T \text{ and } \left(\frac{\partial U}{\partial V}\right)_S = -P \qquad (6.21)$$

$$\left(\frac{\partial H}{\partial S}\right)_P = T \text{ and } \left(\frac{\partial H}{\partial P}\right)_S = V \qquad (6.22)$$

$$\left(\frac{\partial A}{\partial T}\right)_V = -S \text{ and } \left(\frac{\partial A}{\partial V}\right)_T = -P \qquad (6.23)$$

$$\left(\frac{\partial G}{\partial T}\right)_P = -S \text{ and } \left(\frac{\partial G}{\partial P}\right)_T = V \qquad (6.24)$$

These expressions state how U, H, A, and G vary with their natural variables. For example, because T and V always have positive values, Equation (6.22) states that H increases as either the entropy or the pressure of the system increases. We discuss how to use these relations for macroscopic changes in the system variables in Section 6.4.

There is also a second way in which the differential expressions in Equations (6.16) through (6.19) can be used. From Section 3.1, we know that because dU is an exact differential:

$$\left(\frac{\partial}{\partial V}\left(\frac{\partial U(S,V)}{\partial S}\right)_V\right)_S = \left(\frac{\partial}{\partial S}\left(\frac{\partial U(S,V)}{\partial V}\right)_S\right)_V$$

Applying the condition that the order of differentiation in the mixed second partial derivative is immaterial to Equations (6.16) through (6.19), we obtain the following four **Maxwell relations:**

$$\left(\frac{\partial T}{\partial V}\right)_S = -\left(\frac{\partial P}{\partial S}\right)_V \qquad (6.25)$$

$$\left(\frac{\partial T}{\partial P}\right)_S = \left(\frac{\partial V}{\partial S}\right)_P \qquad (6.26)$$

$$\left(\frac{\partial S}{\partial V}\right)_T = \left(\frac{\partial P}{\partial T}\right)_V = \frac{\beta}{\kappa} \qquad (6.27)$$

$$-\left(\frac{\partial S}{\partial P}\right)_T = \left(\frac{\partial V}{\partial T}\right)_P = V\beta \qquad (6.28)$$

Equations (6.25) and (6.26) refer to a partial derivative at constant S. What conditions must a transformation at constant entropy satisfy? Because $dS = dq_{reversible}/T$, a transformation at constant entropy refers to a reversible adiabatic process.

The Maxwell relations have been derived using only the property that U, H, A, and G are state functions. These four relations are extremely useful in transforming seemingly obscure partial derivatives in other partial derivatives that can be directly measured. For example, these relations will be used to express U, H, and heat capacities solely in terms of measurable quantities such as κ, β and the state variables P, V, and T in Supplemental Section 6.15.

6.3 The Dependence of the Gibbs and Helmholtz Energies on P, V, and T

State functions A and G are particularly important for chemists because of their roles in determining the direction of spontaneous change in a reaction mixture. Therefore, it is important to know how A changes with T and V, and how G changes with T and P. Because most reactions of interest to chemists are carried out under constant pressure rather than constant volume conditions, more space is devoted to the properties of G than to those of A.

We begin by asking how A changes with T and V. From Section 6.2,

$$\left(\frac{\partial A}{\partial T}\right)_V = -S \text{ and } \left(\frac{\partial A}{\partial V}\right)_T = -P \tag{6.29}$$

where S and P always take on positive values. Therefore, the general statement can be made that the Helmholtz energy of a pure substance decreases as either the temperature or the volume increases.

We proceed in an analogous way for the Gibbs energy. From Section 6.2, Equation (6.24),

$$\left(\frac{\partial G}{\partial T}\right)_P = -S \text{ and } \left(\frac{\partial G}{\partial P}\right)_T = V$$

Whereas the Gibbs energy decreases with increasing temperature, it increases with increasing pressure.

How can Equation (6.24) be used to calculate changes in G with macroscopic changes in the variables T and P? In doing so, each of the variables is considered separately. The total change in G as both T and P are varied is the sum of the separate contributions. This follows because G is a state function. We first discuss the change in G with P.

For a macroscopic change in P at constant T, the second expression in Equation (6.24) is integrated at constant T:

$$\int_{P^\circ}^{P} dG = G(T,P) - G^\circ(T,P^\circ) = \int_{P^\circ}^{P} VdP \tag{6.30}$$

where we have chosen the initial pressure to be the standard state pressure $P^\circ = 1$ bar. This equation takes on different form for liquids and solids, and for gases. For liquids and solids, the volume is to a good approximation independent of P over a limited range in P and

$$G(T,P) = G^\circ(T,P^\circ) + \int_{P^\circ}^{P} VdP \approx G^\circ(T,P^\circ) + V(P - P^\circ) \tag{6.31}$$

By contrast, the volume of a gaseous system changes appreciably with pressure. In calculating the change of G_m with P at constant T, any path connecting the same initial and final states gives the same result. Choosing the reversible path and assuming ideal gas behavior,

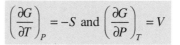

$$G(T,P) = G^\circ(T) + \int_{P^\circ}^{P} VdP = G^\circ(T) + \int_{P^\circ}^{P} \frac{nRT}{P'} dP' = G^\circ(T) + nRT \ln \frac{P}{P^\circ} \tag{6.32}$$

FIGURE 6.1
The molar Gibbs energy of an ideal gas relative to its standard state value is shown as a function of the pressure at 298.15 K.

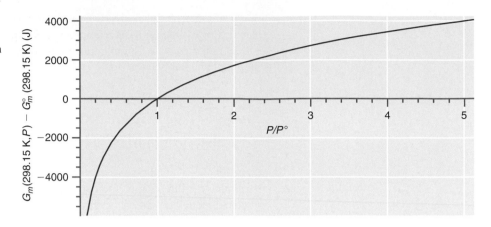

FIGURE 6.1
The molar Gibbs energy of an ideal gas relative to its standard state value is shown as a function of the pressure at 298.15 K.

The functional dependence of the molar Gibbs energy, G_m, on P for an ideal gas is shown in Figure 6.1, where G_m approaches minus infinity as the pressure approaches zero. This is a result of the volume dependence of S that was discussed in Section 5.5, $\Delta S = nR \ln(V_f/V_i)$ at constant T. As $P \to 0$, $V \to \infty$. Because the volume available to a gas molecule is maximized as $V \to \infty$, we find that $S \to \infty$ as $P \to 0$. Therefore, $G = H - TS \to -\infty$ in this limit. A convention for assigning values to the standard state molar Gibbs energies is discussed next.

It is generally not necessary to assign values to the G_m°, because only differences in the Gibbs energy rather than absolute values can be obtained from experiments. However, in order to discuss the chemical potential in Section 6.4, it is useful to introduce a convention that allows an assignment of numerical values to G_m°. With the convention from Section 4.2 that $H_m^\circ = 0$ for an element in its standard state and 298.15 K,

$$G_m^\circ = H_m^\circ - TS_m^\circ = -TS_m^\circ \tag{6.33}$$

for a pure *element* in its standard state at 1 bar and 298.15 K. To obtain an expression for G_m° for a pure *compound* in its standard state, consider ΔG° for the formation reaction. Recall that 1 mol of the product appears on the right side in the formation reaction, and only elements in their standard states appear on the left side:

$$\Delta G_f^\circ = G_{m,product}^\circ - \sum_i v_i G_{m,reactant\,i}^\circ \tag{6.34}$$

$$G_{m,product}^\circ = \Delta G_f^\circ - \sum^{\substack{elements \\ only}} v_i TS_{m,i}^\circ$$

where the v_i are the stoichiometric coefficients of the elemental reactants in the balanced formation reaction, all of which have positive values. Note that this result differs from the corresponding value for the enthalpy, $H_m^\circ = \Delta H_f^\circ$. The values for G_m° obtained in this way are called the **conventional molar Gibbs energies**.[1] This formalism is illustrated in Example Problem 6.3.

EXAMPLE PROBLEM 6.3

Calculate the conventional molar Gibbs energy of (a) Ar(g) and (b) H$_2$O(l) at 1 bar and 298.15 K.

Solution

a. Because Ar(g) is an element in its standard state for these conditions,
$H_m^\circ = 0$ and $G_m^\circ = H_m^\circ - TS_m^\circ = -TS_m^\circ = -298.15\ \text{K} \times 154.8\ \text{J K}^{-1}\ \text{mol}^{-1}$
$= -4.617\ \text{kJ mol}^{-1}$.

---

[1] We note that this convention is not unique. Some authors set $H_m^\circ = G_m^\circ = 0$ for all elements in their standard state at 298.15 K. With this convention, for a compound in its standard state $G_m^\circ = \Delta G_f^\circ$.

b. The formation reaction for $H_2O(l)$ is $H_2(g) + 1/2\ O_2(g) \rightarrow H_2O(l)$

$$G_{m,product}^{\circ} = \Delta G_f^{\circ}(product) - \overset{\substack{elements \\ only}}{\underset{i}{\sum}} v_i TS_{m,i}^{\circ}$$

$$G_m^{\circ}(H_2O, l) = \Delta G_f^{\circ}(H_2O, l) - \left[TS_m^{\circ}(H_2, g) + \frac{1}{2} TS_m^{\circ}(O_2, g) \right]$$

$$= -237.1\ \text{kJ mol}^{-1} - 298.15\ \text{K} \times \left[\begin{array}{l} 130.7\ \text{J K}^{-1}\ \text{mol}^{-1} + \frac{1}{2} \\ \times\ 205.2\ \text{J K}^{-1}\text{mol}^{-1} \end{array} \right]$$

$$= -306.6\ \text{kJ mol}^{-1}$$

These calculations give G_m° at 298.15 K. To calculate the conventional molar Gibbs energy at another temperature, use the Gibbs–Helmholtz equation as discussed next.

We next investigate the dependence of G on T. As will be seen in the next section, the thermodynamic equilibrium constant, K, is related to G/T. Therefore, it is more useful to obtain an expression for the temperature dependence of G/T, than for the temperature dependence of G. The dependence of G/T on temperature is given by

$$\left(\frac{\partial \frac{G}{T}}{\partial T} \right)_P = \frac{1}{T} \left(\frac{\partial G}{\partial T} \right)_P - \frac{G}{T^2} = -\frac{S}{T} - \frac{G}{T^2} = -\frac{G + TS}{T^2} = -\frac{H}{T^2} \tag{6.35}$$

This result is known as the **Gibbs–Helmholtz equation.** Because

$$\frac{d(1/T)}{dT} = -\frac{1}{T^2}$$

the Gibbs–Helmholtz equation can also be written in the form

$$\left(\frac{\partial \frac{G}{T}}{\partial \frac{1}{T}} \right) = \left(\frac{\partial \frac{G}{T}}{\partial T} \frac{dT}{d\left(\frac{1}{T}\right)} \right)_P = -\frac{H}{T^2}(-T^2) = H \tag{6.36}$$

The preceding equation also applies to the change in G and H associated with a process such as a chemical reaction, in which G becomes ΔG. Integrating Equation (6.36) at constant P,

$$\int_{T_1}^{T_2} d\left(\frac{\Delta G}{T} \right) = \int_{T_1}^{T_2} d\left(\frac{\Delta H}{T} \right) \tag{6.37}$$

$$\frac{\Delta G(T_2)}{T_2} = \frac{\Delta G(T_1)}{T_1} + \Delta H(T_1)\left(\frac{1}{T_2} - \frac{1}{T_1} \right)$$

It has been assumed in the second equation that ΔH is independent of T over the temperature interval of interest. If this is not the case, the integral must be evaluated numerically, using tabulated values of ΔH_f° and temperature-dependent expressions of $C_{P,m}$ for reactants and products.

EXAMPLE PROBLEM 6.4

The value of ΔG_f° for Fe(g) is 370.7 kJ mol^{-1} at 298.15 K, and ΔH_f° for Fe(g) is 416.3 kJ mol^{-1} at the same temperature. Assuming that ΔH_f° is constant in the interval 250–400 K, calculate ΔG_f° for Fe(g) at 400 K.

Solution

$$\Delta G_f^\circ(T_2) = T_2\left[\frac{\Delta G_f^\circ(T_1)}{T_1} + \Delta H_f^\circ(T_1)\times\left(\frac{1}{T_2}-\frac{1}{T_1}\right)\right]$$

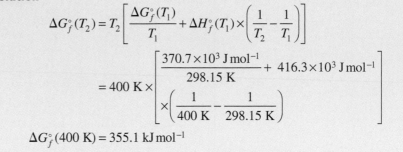

$$\Delta G_f^\circ(400\text{ K}) = 355.1\text{ kJ mol}^{-1}$$

Similar calculations are carried out in Section 6.12, in relating the temperature dependence of the equilibrium constant and the reaction enthalpy.

6.4 The Gibbs Energy of a Reaction Mixture

To this point, the discussion has been limited to systems at a fixed composition. The fact that reactants are consumed and products are generated in chemical reactions requires the expressions derived for state functions such as U, H, S, A, and G to be revised to include these changes in composition. We focus on G in the following discussion. For a reaction mixture containing species 1, 2, 3, ... , G is no longer a function of the variables T and P only. Because it depends on the number of moles of each species, G is written in the form $G = G(T, P, n_1, n_2, n_3, ...)$. The total differential dG is

$$dG = \left(\frac{\partial G}{\partial T}\right)_{P,n_1,n_2...} dT + \left(\frac{\partial G}{\partial P}\right)_{T,n_1,n_2...} dP + \left(\frac{\partial G}{\partial n_1}\right)_{T,P,n_2...} dn_1$$
$$+ \left(\frac{\partial G}{\partial n_2}\right)_{T,P,n_1...} dn_2 + ... \tag{6.38}$$

Note that if the concentration of all species is constant, all of the $dn_i = 0$, and Equation (6.38) reduces to Equation (6.19).

Equation (6.38) can be simplified by defining the **chemical potential, μ_i,** as

$$\mu_i = \left(\frac{\partial G}{\partial n_i}\right)_{P,T,n_j \neq n_i} \qquad j \neq i \tag{6.39}$$

It is important to realize that although μ_i is defined mathematically in terms of an infinitesimal change in the amount dn_i of species i, the chemical potential μ_i is the change in the Gibbs energy per mole of substance i added *at constant concentration*. These two requirements are not contradictory. To keep the concentration constant, one adds a mole of substance i to a huge vat containing many moles of the various species. In this case, the slope of a plot of G versus n_i is the same if the differential $(\partial G/\partial n_i)_{P,T,n_j \neq n_i}$ is formed, where $dn_i \to 0$, or the ratio $(\Delta G/\Delta n_i)_{P,T,n_j \neq n_i}$ is formed, where Δn_i is 1 mol. Using the notation of Equation (6.39), Equation (6.38) can be written as follows:

$$dG = \left(\frac{\partial G}{\partial T}\right)_{P,n_1,n_2\ldots} dT + \left(\frac{\partial G}{\partial P}\right)_{T,n_1,n_2\ldots} dP + \sum_i \mu_i dn_i \qquad (6.40)$$

Now imagine integrating Equation (6.40) at constant composition and at constant T and P from an infinitesimal size of the system where $n_i \to 0$ and therefore $G \to 0$ to a macroscopic size where the Gibbs energy has the value G. Because T and P are constant, the first two terms in Equation (6.40) do not contribute to the integral. Because the composition is constant, μ_i is constant,

$$\int_0^G dG' = \sum_i \mu_i \int_0^{n_i} dn_i' \qquad (6.41)$$

$$G = \sum_i n_i \mu_i$$

Note that because μ_i depends on the number of moles of each species present, it is a function of concentration. If the system consists of a single pure substance A, $G = n_A G_{m,A}$ because G is an extensive quantity. Applying Equation (6.39),

$$\mu_A = \left(\frac{\partial G}{\partial n_A}\right)_{P,T} = \left(\frac{\partial \left[n_A G_{m,A}\right]}{\partial n_A}\right)_{P,T} = G_{m,A}$$

showing that μ_A is equal to the molar Gibbs energy of A *for a pure substance*. As seen later, however, this statement is not true for mixtures.

Why is μ_i called the chemical potential of species i? This can be understood by assuming that the chemical potential for species i has the values μ_i^I in region I, and μ_i^{II} in region II of a given mixture with $\mu_i^I > \mu_i^{II}$. If dn_i moles of species i are transported from region I to region II, at constant T and p, the change in G is given by

$$dG = -\mu_i^I dn_i + \mu_i^{II} dn_i = (\mu_i^{II} - \mu_i^I)dn_i < 0 \qquad (6.42)$$

Because $dG < 0$, this process is spontaneous. *For a given species, transport will occur spontaneously from a region of high chemical potential to one of low chemical potential. The flow of material will continue until the chemical potential has the same value in all regions of the mixture.* Note the analogy between this process and the flow of mass in a gravitational potential or the flow of charge in an electrostatic potential. Therefore, the term *chemical potential* is appropriate.

In this discussion, we have defined an additional criterion for equilibrium in a multicomponent mixture: *At equilibrium, the chemical potential of each individual species is the same throughout a mixture.*

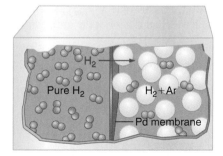

FIGURE 6.2
An isolated system consists of two subsystems. Pure H_2 gas is present on the left of a palladium membrane that is permeable to H_2, but not to argon. The H_2 is contained in a mixture with Ar in the subsystem to the right of the membrane.

6.5 The Gibbs Energy of a Gas in a Mixture

In this and the next section, the conditions for equilibrium in a mixture of ideal gases are derived in terms of the μ_i of the chemical constituents. We will show that the partial pressures of all constituents of the mixture are related by the thermodynamic equilibrium constant, K_P. Consider first the simple system consisting of two volumes separated by a semipermeable membrane, as shown in Figure 6.2. On the left side, the gas consists solely of pure H_2. On the right side, H_2 is present as one constituent of a mixture. The membrane allows only H_2 to pass in both directions.

Once equilibrium has been reached with respect to the concentration of H_2 throughout the system, the hydrogen pressure is the same on both sides of the membrane and

$$\mu_{H_2}^{pure} = \mu_{H_2}^{mixture} \qquad (6.43)$$

Recall from Section 6.3 that the molar Gibbs energy of a pure ideal gas depends on its pressure as $G(T,P) = G°(T) + nRT\ln(P/P°)$. For a pure substance, i, $\mu_i = G_{m,i}$. Therefore, Equation (6.43) can be written in the form

$$\mu_{H_2}^{pure}(T,P_{H_2}) = \mu_{H_2}°(T) + RT\ln\frac{P_{H_2}}{P°} = \mu_{H_2}^{mixture}(T,P_{H_2}) \tag{6.44}$$

The chemical potential of a gas in a mixture depends logarithmically on its partial pressure. Equation (6.44) applies to any mixture, not just to those for which an appropriate semipermeable membrane exists. We, therefore, generalize the discussion by referring to a component of the mixture as A.

The partial pressure of species A in the gas mixture, P_A, can be expressed in terms of x_A, its mole fraction in the mixture, and the total pressure, P:

$$P_A = x_A P \tag{6.45}$$

Using this relationship, Equation (6.44) becomes

$$\mu_A^{mixture}(T,P) = \mu_A°(T) + RT\ln\frac{P}{P°} + RT\ln x_A \tag{6.46}$$

$$= \left(\mu_A°(T) + RT\ln\frac{P}{P°}\right) + RT\ln x_A \quad \text{or}$$

$$\mu_A^{mixture}(T,P) = \mu_A^{pure}(T,P) + RT\ln x_A$$

Because $\ln x_A < 0$, the chemical potential of a gas in a mixture is less than that of the pure gas if P is the same for the pure sample and the mixture. Because matter flows from a region of high to one of low chemical potential, a pure gas initially separated by a barrier from a mixture at the same pressure flows into the mixture as the barrier is removed. This shows that the mixing of gases at constant pressure is a spontaneous process.

6.6 Calculating the Gibbs Energy of Mixing for Ideal Gases

The previous section demonstrated that the mixing of gases is a spontaneous process. We next obtain a quantitative relationship between ΔG_{mixing} and the mole fractions of the individual constituents of the mixture. Consider the system shown in Figure 6.3. The four compartments contain the gases He, Ne, Ar, and Xe at the same temperature and pressure. The volumes of the four compartments differ. To calculate ΔG_{mixing}, we must compare G for the initial state shown in Figure 6.3 and the final state. For the initial state, in which we have four pure separated substances,

$$G_i = G_{He} + G_{Ne} + G_{Ar} + G_{Xe} = n_{He}G_{He}° + n_{Ne}G_{Ne}° + n_{Ar}G_{Ar}° + n_{Xe}G_{Xe}° \tag{6.47}$$

For the final state in which all four components are dispersed in the mixture, from Equation (6.46),

$$G_f = n_{He}(G_{He}° + RT\ln x_{He}) + n_{Ne}(G_{Ne}° + RT\ln x_{Ne}) \tag{6.48}$$
$$+ n_{Ar}(G_{Ar}° + RT\ln x_{Ar}) + n_{Xe}(G_{Xe}° + RT\ln x_{Xe})$$

The Gibbs energy of mixing is $G_f - G_i$ or

$$\Delta G_{mixing} = RTn_{He}\ln x_{He} + RTn_{Ne}\ln x_{Ne} + RTn_{Ar}\ln x_{Ar} + RTn_{Xe}\ln x_{Xe} \tag{6.49}$$
$$= RT\sum_i n_i \ln x_i = nRT\sum_i x_i \ln x_i$$

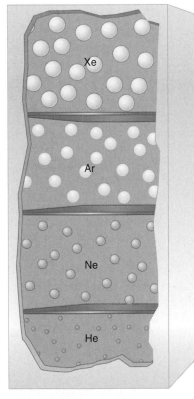

FIGURE 6.3
An isolated system consists of four separate subsystems containing He, Ne, Ar, and Xe, each at a pressure of 1 bar. The barriers separating these subsystems can be removed, leading to mixing.

(handwritten) $\mu = \left(\partial \right.$

(handwritten) always neg for open

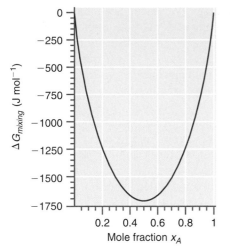

FIGURE 6.4
The Gibbs energy of mixing of the ideal gases A and B as a function of x_A, with $n_A + n_B = 1$ and $T = 298.15$ K.

Note that because each term in the last expression of Equation (6.49) is negative, $\Delta G_{mixing} < 0$. Therefore, mixing is a spontaneous process at constant T and P.

Equation (6.49) allows one to calculate the value of ΔG_{mixing} for any given set of the mole fractions x_i. It is easiest to graphically visualize the results for a binary mixture of species A and B. To simplify the notation, we set $x_A = x$, so that $x_B = 1 - x$. It follows that

$$\Delta G_{mixing} = nRT \left[x \ln x + (1 - x) \ln(1 - x) \right] \tag{6.50}$$

A plot of ΔG_{mixing} versus x is shown in Figure 6.4 for a binary mixture. Note that ΔG_{mixing} is zero for $x_A = 0$ and $x_A = 1$ because only pure substances are present in these limits. Also, ΔG_{mixing} has a minimum for $x_A = 0.5$, because there is a maximal decrease in G that arises from the dilution of both A and B when A and B are present in equal amounts.

The entropy of mixing can be calculated from Equation (6.49):

$$\Delta S_{mixing} = -\left(\frac{\partial \Delta G_{mixing}}{\partial T} \right)_P = -nR \sum_i x_i \ln x_i \tag{6.51}$$

As shown in Figure 6.5, the entropy of mixing is greatest for $x_A = 0.5$. What lies behind the increase in entropy? Each of the two components of the mixture expands from its initial volume to the same final volume. Therefore, ΔS_{mixing} arises purely from the dependence of S on V at constant T,

$$\Delta S = R \left(n_A \ln \frac{V_f}{V_{iA}} + n_B \ln \frac{V_f}{V_{iB}} \right) = R \left(n x_A \ln \frac{1}{x_A} + n x_B \ln \frac{1}{x_B} \right) = -nR(x_A \ln x_A + x_B \ln x_B)$$

if both components and the mixture are at the same pressure.

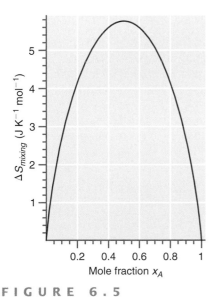

FIGURE 6.5
The entropy of mixing of the ideal gases A and B is shown as a function of the mole fraction of component A at 298.15 K, with $n_A + n_B = 1$.

EXAMPLE PROBLEM 6.5

Consider the system shown in Figure 6.3. Assume that the separate compartments contain 1.0 mol of He, 3.0 mol of Ne, 2.0 mol of Ar, and 2.5 mol of Xe at 298.15 K. The pressure in each compartment is 1 bar.

 a. Calculate ΔG_{mixing}.

 b. Calculate ΔS_{mixing}.

Solution

 a. $\Delta G_{mixing} = RTn_{He} \ln x_{He} + RTn_{Ne} \ln x_{Ne} + RTn_{Ar} \ln x_{Ar} + RTn_{Xe} \ln x_{Xe}$

$$= RT \sum_i n_i \ln x_i = nRT \sum_i x_i \ln x_i$$

$$= 8.5 \text{ mol} \times 8.314 \text{ J K}^{-1} \text{ mol}^{-1} \times 298.15 \text{ K}$$

$$\times \left(\frac{1.0}{8.5} \ln \frac{1.0}{8.5} + \frac{3.0}{8.5} \ln \frac{3.0}{8.5} + \frac{2.0}{8.5} \ln \frac{2.0}{8.5} + \frac{2.5}{8.5} \ln \frac{2.5}{8.5} \right)$$

$$= -2.8 \times 10^4 \text{ J}$$

 b. $\Delta S_{mixing} = -nR \sum_i x_i \ln x_i$

$$= -8.5 \text{ mol} \times 8.314 \text{ J K}^{-1} \text{ mol}^{-1}$$

$$\times \left(\frac{1.0}{8.5} \ln \frac{1.0}{8.5} + \frac{3.0}{8.5} \ln \frac{3.0}{8.5} + \frac{2.0}{8.5} \ln \frac{2.0}{8.5} + \frac{2.5}{8.5} \ln \frac{2.5}{8.5} \right)$$

$$= 93 \text{ J K}^{-1}$$

What is the driving force for the mixing of gases? We saw in Section 6.1 that there are two contributions to ΔG, an enthalpic contribution ΔH and an entropic contribution $T\Delta S$. By calculating ΔH_{mixing} from $\Delta H_{mixing} = \Delta G_{mixing} + T\Delta S_{mixing}$ using Equations (6.50) and (6.51), you will see that for the mixing of ideal gases, $\Delta H_{mixing} = 0$. Because the molecules in an ideal gas do not interact, there is no enthalpy change associated with mixing. We conclude that the mixing of ideal gases is driven entirely by ΔS_{mixing} as shown in Figure 6.5.

Although the mixing of gases is always spontaneous, the same is not true of liquids. Liquids can be either miscible or immiscible. How can this observation be explained? For gases or liquids, $\Delta S_{mixing} > 0$. Therefore, if two liquids are immiscible, $\Delta G_{mixing} > 0$ because $\Delta H_{mixing} > 0$ and $\Delta H_{mixing} > T\Delta S_{mixing}$. If two liquids mix, it is generally energetically favorable for one species to be surrounded by the other species. In this case, $\Delta H_{mixing} < T\Delta S_{mixing}$, and $\Delta G_{mixing} < 0$.

6.7 Expressing Chemical Equilibrium in an Ideal Gas Mixture in Terms of the μ_i

Consider the balanced chemical reaction

$$\alpha A + \beta B + \chi C + \ldots \rightarrow \delta M + \varepsilon N + \gamma O + \ldots \qquad (6.52)$$

in which Greek letters represent stoichiometric coefficients and the uppercase Roman letters represent the reactants and products. We can write an abbreviated expression for this reaction in the form

$$\sum_i v_i X_i = 0 \qquad (6.53)$$

In Equation (6.53), the stoichiometric coefficients of the products are positive, and those of the reactants are negative.

What determines the equilibrium partial pressures of the reactants and products? Imagine that the reaction proceeds in the direction indicated in Equation (6.52) by an infinitesimal amount. The change in the Gibbs energy is given by

$$dG = \sum_i \mu_i dn_i \qquad (6.54)$$

In this equation, the individual dn_i are not independent because they are linked by the stoichiometric equation.

It is convenient at this point to introduce a parameter, ξ, called the **extent of reaction.** If the reaction advances by ξ moles, the number of moles of each species i changes according to

$$n_i = n_i^{initial} + v_i \xi \qquad (6.55)$$

Differentiating this equation leads to $dn_i = v_i d\xi$. By inserting this result in Equation (6.54), we can write dG in terms of ξ. At constant T and P,

$$dG = \left(\sum_i v_i \mu_i \right) d\xi = \Delta G_{reaction} d\xi \qquad (6.56)$$

An advancing reaction can be described in terms of the partial derivative of G with ξ:

$$\left(\frac{\partial G}{\partial \xi} \right)_{T,P} = \sum_i v_i \mu_i = \Delta G_{reaction} \qquad (6.57)$$

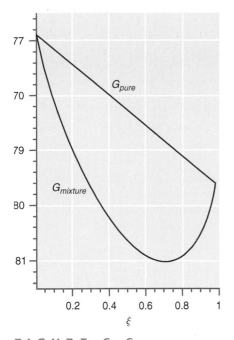

FIGURE 6.6

$G_{mixture}$ and G_{pure} are depicted for the $2\,NO_2(g) \rightleftharpoons N_2O_4(g)$ equilibrium. For this reaction system, the equilibrium position would correspond to $\xi = 1$ in the absence of mixing. The mixing contribution to $G_{mixture}$ shifts ξ_{eq} to values smaller than one.

The direction of spontaneous change is that in which $\Delta G_{reaction}$ is negative. This direction corresponds to $(\partial G/\partial\xi)_{T,P} < 0$. At a given composition of the reaction mixture, the partial pressures of the reactants and products can be determined, and the μ_i can be calculated. Because $\mu_i = \mu_i(T,P,n_A,n_B,....)$, $\sum_i v_i\mu_i = \Delta G_{reaction}$ must be evaluated at specific values of $T,P,n_A,n_B,....$. Based on the value of $\Delta G_{reaction}$, the following conclusions can be drawn:

- If $(\partial G/\partial\xi)_{T,P} < 0$, the reaction proceeds spontaneously as written.
- If $(\partial G/\partial\xi)_{T,P} > 0$, the reaction proceeds spontaneously in the opposite direction.
- If $(\partial G/\partial\xi)_{T,P} = 0$, the reaction system is at equilibrium, and there is no direction of spontaneous change.

The most important value of ξ is ξ_{eq}, corresponding to equilibrium. How can this value be found? To make the following discussion more concrete, we consider the reaction system $2\,NO_2(g) \rightleftharpoons N_2O_4(g)$, with $2 - 2\,\xi$ moles of $NO_2(g)$ and ξ moles of $N_2O_4(g)$ present in a vessel at a constant pressure of 1 bar and 298 K. The parameter ξ could in principle take on all values between zero and one, corresponding to pure $NO_2(g)$ and pure $N_2O_4(g)$, respectively. The Gibbs energy of the pure unmixed reagents and products, G_{pure}, is given by

$$G_{pure} = (2 - 2\xi)G_m^{\circ}(NO_2,g) + \xi G_m^{\circ}(N_2O_4,g) \qquad (6.58)$$

where ΔG_m° is the conventional molar Gibbs energy defined by Equation 6.34. Note that G_{pure} varies linearly with ξ, as shown in Figure 6.6. Because the reactants and products are mixed throughout the range accessible to ξ, G_{pure} is not equal to $G_{mixture}$, which is given by

$$G_{mixture} = G_{pure} + \Delta G_{mixing} \qquad (6.59)$$

Equation (6.59) shows that if G_{pure} alone determined ξ_{eq}, $\xi_{eq} = 0$ or $\xi_{eq} = 1$, depending on whether G_{pure} is lower for reagents or products. For the reaction under consideration, $\Delta G_f^{\circ}(N_2O_4,g) < \Delta G_f^{\circ}(NO_2,g)$ so that $\xi_{eq} = 1$ in this limit. In other words, if G_{pure} alone determined ξ_{eq}, every chemical reaction would go to completion, or the reverse reaction would go to completion. How does mixing influence the equilibrium position for the reaction system? The value of ΔG_{mixing} can be calculated using Equation (6.49). You will show in the end-of-chapter problems that ΔG_{mixing} for this reaction system has a maximum at $\xi_{eq} = 0.55$.

If G_{pure} alone determines ξ_{eq}, then $\xi_{eq} = 1$, and if ΔG_{mixing} alone determines ξ_{eq}, then $\xi_{eq} = 0.55$. However, the minimum in $\Delta G_{reaction}$ is determined by the minimum in $G_{pure} + \Delta G_{mixing}$ rather than in the minimum of the individual components. For our specific case, $\xi_{eq} = 0.72$ at 298 K. The decrease in G on mixing plays a critical role in determining the position of equilibrium in a chemical reaction in that ξ_{eq} is shifted from 1.0 to 0.72.

We summarize the roles of G_{pure} and ΔG_{mixing} in determining ξ_{eq}. For reactions in which $G_{pure}^{reactants}$ and $G_{pure}^{products}$ are very similar, ξ_{eq} will be largely determined by ΔG_{mixing}. For reactions in which $G_{pure}^{reactants}$ and $G_{pure}^{products}$ are very different, ξ_{eq} will not be greatly influenced by ΔG_{mixing} and the equilibrium mixture will essentially consist of pure reactant or pure products.

The direction of spontaneous change is determined by $\Delta G_{reaction}$. How can $\Delta G_{reaction}$ be calculated? We distinguish between calculating $\Delta G_{reaction}^{\circ}$ for the standard state in which $P_i = P^{\circ} = 1$ bar and including the pressure dependence of $\Delta G_{reaction}$. Tabulated values of ΔG_f° for compounds are used to calculate $\Delta G_{reaction}^{\circ}$ at $P^{\circ} = 1$ bar and $T = 298.15$ K as shown in Example Problem 6.6. Just as for enthalpy, ΔG_f° for a pure element in its standard state at the reference temperature is equal to zero because the reactants and products in the formation reaction are identical. Just as for $\Delta H_{reaction}^{\circ}$ discussed in Chapter 4,

$$\Delta G_{reaction}^{\circ} = \sum_i v_i \Delta G_{f,i}^{\circ} \qquad (6.60)$$

EXAMPLE PROBLEM 6.6

Calculate $\Delta G^\circ_{reaction}$ for the reaction $Fe_3O_4(s) + 4H_2(g) \rightarrow 3Fe(s) + 4H_2O(l)$ at 298.15 K.

Solution

$$\Delta G^\circ_{reaction} = 3\Delta G^\circ_f(Fe,s) + 4\Delta G^\circ_f(H_2O,l) - \Delta G^\circ_f(Fe_3O_4,s) - 4\Delta G^\circ_f(H_2,g)$$

$$= 3\times 0\,kJ\,mol^{-1} - 4\times 237.1\,kJ\,mol^{-1} + 1015.4\,kJ\,mol^{-1}$$

$$-4\times 0\,kJ\,mol^{-1}$$

$$= 67.00\,kJ\,mol^{-1}$$

At other temperatures, $\Delta G^\circ_{reaction}$ is calculated as shown in Example Problem 6.7.

EXAMPLE PROBLEM 6.7

Calculate $\Delta G^\circ_{reaction}$ for the reaction in Example Problem 6.6 at 525 K.

Solution

To calculate $\Delta G^\circ_{reaction}$ at the elevated temperature, we use Equation (6.37) assuming that $\Delta H^\circ_{reaction}$ is independent of T:

$$\Delta G^\circ_{reaction}(T_2) = T_2\left[\frac{\Delta G^\circ_{reaction}(298.15\,K)}{298.15\,K} + \Delta H^\circ_{reaction}(298.15\,K)\left(\frac{1}{T_2} - \frac{1}{298.15\,K}\right)\right]$$

$$\Delta H^\circ_{reaction}(298.15\,K) = 3\Delta H^\circ_f(Fe,s) + 4\Delta H^\circ_f(H_2O,l) - \Delta H^\circ_f(Fe_3O_4,s)$$

$$- 4\Delta H^\circ_f(H_2,g)$$

$$= 3\times 0\,kJ\,mol^{-1} - 4\times 285.8\,kJ\,mol^{-1} + 1118.4\,kJ\,mol^{-1}$$

$$- 4\times 0\,kJ\,mol^{-1}$$

$$= -24.80\,kJ\,mol^{-1}$$

$$\Delta G^\circ_{reaction}(525\,K) = 525\,K \times \left[\begin{array}{c}\dfrac{67.00\times 10^3\,J\,mol^{-1}}{298.15\,K} - 24.80 \\ \times 10^3\,J\,mol^{-1}\left(\dfrac{1}{525\,K} - \dfrac{1}{298.15\,K}\right)\end{array}\right]$$

$$= 136.8\,kJ\,mol^{-1}$$

In the next section, the pressure dependence of G is used to calculate $\Delta G_{reaction}$ for arbitrary values of the pressure.

6.8 Calculating $\Delta G_{reaction}$ and Introducing the Equilibrium Constant for a Mixture of Ideal Gases

In this section, we introduce the pressure dependence of μ_i in order to calculate $\Delta G_{reaction}$ for reaction mixtures in which P_i is different from 1 bar. Consider the following reaction that takes place between ideal gas species A, B, C, and D:

$$\alpha A + \beta B \rightleftharpoons \gamma C + \delta D \tag{6.61}$$

Because all four species are present in the reaction mixture, the reaction Gibbs energy is given by

$$\Delta G_{reaction} = \sum_i v_i \Delta G_{f,i} = \gamma \mu_C^\circ + \gamma RT \ln \frac{P_C}{P^\circ} + \delta \mu_D^\circ + \delta RT \ln \frac{P_D}{P^\circ} \qquad (6.62)$$
$$- \alpha \mu_A^\circ - \alpha RT \ln \frac{P_A}{P^\circ} - \beta \mu_B^\circ - \beta RT \ln \frac{P_B}{P^\circ}$$

The terms in the previous equation can be separated into those at the standard condition of $P^\circ = 1$ bar and the remaining terms:

$$\Delta G_{reaction} = \Delta G_{reaction}^\circ + \gamma RT \ln \frac{P_C}{P^\circ} + \delta RT \ln \frac{P_D}{P^\circ} - \alpha RT \ln \frac{P_A}{P^\circ} - \beta RT \ln \frac{P_B}{P^\circ} \qquad (6.63)$$

$$= \Delta G_{reaction}^\circ + RT \ln \frac{\left(\dfrac{P_C}{P^\circ}\right)^\gamma \left(\dfrac{P_D}{P^\circ}\right)^\delta}{\left(\dfrac{P_A}{P^\circ}\right)^\alpha \left(\dfrac{P_B}{P^\circ}\right)^\beta}$$

where

$$\Delta G_{reaction}^\circ = \gamma \mu_C^\circ(T) + \delta \mu_D^\circ(T) - \alpha \mu_A^\circ(T) - \beta \mu_B^\circ(T) = \sum_i v_i \Delta G_{f,i}^\circ \qquad (6.64)$$

Recall from Example Problem 6.6 that standard Gibbs energies of formation rather than chemical potentials are used to calculate $\Delta G_{reaction}^\circ$.

The combination of the partial pressures of reactants and products is called the **reaction quotient of pressures,** which is abbreviated Q_P and defined as follows:

$$Q_P = \frac{\left(\dfrac{P_C}{P^\circ}\right)^\gamma \left(\dfrac{P_D}{P^\circ}\right)^\delta}{\left(\dfrac{P_A}{P^\circ}\right)^\alpha \left(\dfrac{P_B}{P^\circ}\right)^\beta} \qquad (6.65)$$

With these definitions of $\Delta G_{reaction}^\circ$ and Q_P, Equation (6.62) becomes

$$\Delta G_{reaction} = \Delta G_{reaction}^\circ + RT \ln Q_P \qquad (6.66)$$

Note that $\Delta G_{reaction}$ can be separated into two terms, only one of which depends on the partial pressures of reactants and products.

We next show how Equation (6.66) can be used to predict the direction of spontaneous change for given partial pressures of the reactants and products. If the partial pressures of the products C and D are large, and those of the reactants A and B are small compared to their values at equilibrium, Q_P will be large. As a result, $RT \ln Q_P$ will be large and positive, and $\Delta G_{reaction} = \Delta G_{reaction}^\circ + RT \ln Q_P > 0$. In this case, the reaction as written in Equation (6.61) from left to right is not spontaneous, but the reverse of the reaction is spontaneous. Next, consider the opposite extreme. If the partial pressures of the reactants A and B are large, and those of the products C and D are small compared to their values at equilibrium, Q_P will be small. As a result, $RT \ln Q_P$ will be large and negative. In this case, $\Delta G_{reaction} = \Delta G_{reaction}^\circ + RT \ln Q_P < 0$ and the reaction will be spontaneous as written from left to right, as a portion of the reactants combines to form products.

Whereas the two cases that we have considered lead to a change in the partial pressures of the reactants and products, the most interesting case is equilibrium for which $\Delta G_{reaction} = 0$. At equilibrium, $\Delta G_{reaction}^\circ = -RT \ln Q_P$. We denote this special system configuration by adding a superscript *eq* to each partial pressure, and renaming Q_P as K_P. The quantity K_P is called the **thermodynamic equilibrium constant:**

$$0 = \Delta G^{\circ}_{reaction} + RT \ln \frac{\left(\dfrac{P^{eq}_C}{P^{\circ}}\right)^c \left(\dfrac{P^{eq}_D}{P^{\circ}}\right)^d}{\left(\dfrac{P^{eq}_A}{P^{\circ}}\right)^a \left(\dfrac{P^{eq}_B}{P^{\circ}}\right)^b} \quad \text{or, equivalently,} \quad \Delta G^{\circ}_{reaction} = -RT \ln K_P \qquad (6.67)$$

or

$$\ln K_P = -\frac{\Delta G^{\circ}_{reaction}}{RT}$$

Because $\Delta G^{\circ}_{reaction}$ is a function of T only, K_P is also a function of T only. *The thermodynamic equilibrium constant K_P does not depend on the pressure.* Note also that K_P is a dimensionless number.

EXAMPLE PROBLEM 6.8

a. Using data from Table 4.1 (see Appendix A, Data Tables), calculate K_P at 298.15 K for the reaction $CO(g) + H_2O(l) \rightarrow CO_2(g) + H_2(g)$.

b. Based on the value that you obtained for part (a), do you expect the mixture to consist mainly of $CO_2(g)$ and $H_2(g)$ or mainly of $CO(g) + H_2O(l)$ at equilibrium?

Solution

a. $\ln K_P = -\dfrac{1}{RT} \Delta G^{\circ}_{reaction} = -\dfrac{1}{RT} \left(\begin{array}{l} \Delta G^{\circ}_f(CO_2, g) + \Delta G^{\circ}_f(H_2, g) - \Delta G^{\circ}_f(H_2O, l) \\ -\Delta G^{\circ}_f(CO, g) \end{array} \right)$

$= -\dfrac{1}{8.314 \, \text{J mol}^{-1} \, \text{K}^{-1} \times 298.15 \, \text{K}} \times \left(\begin{array}{l} -394.4 \times 10^3 \, \text{J mol}^{-1} + 0 + 237.1 \\ \times 10^3 \, \text{J mol}^{-1} + 137.2 \times 10^3 \, \text{J mol}^{-1} \end{array} \right)$

$= 8.1087$

$K_P = 3.32 \times 10^3$

b. Because $K_P \gg 1$, the mixture will consist mainly of the products $CO_2(g)$ + $H_2(g)$ at equilibrium.

6.9 Calculating the Equilibrium Partial Pressures in a Mixture of Ideal Gases

As shown in the previous section, the partial pressures of the reactants and products in a mixture of gases at equilibrium cannot take on arbitrary values because they are related through K_P. In this section, we show how the equilibrium partial pressures can be calculated for ideal gases. Similar calculations for real gases will be discussed in the next chapter. In Example Problem 6.9, we consider the dissociation of chlorine:

$$Cl_2(g) \rightarrow 2Cl(g) \qquad (6.68)$$

It is useful to set up the calculation in a tabular form as shown in the following example problem.

EXAMPLE PROBLEM 6.9

In this example, n_0 moles of chlorine gas are placed in a reaction vessel, whose temperature can be varied over a wide range, so that molecular chlorine can partially dissociate to atomic chlorine.

a. Derive an expression for K_P in terms of n_0, ξ, and P.

b. Define the degree of dissociation as $\alpha = \xi_{eq}/n_0$, where $2\xi_{eq}$ is the number of moles of $Cl(g)$ present at equilibrium, and n_0 represents the number of moles of $Cl_2(g)$ that would be present in the system if no dissociation occurred. Derive an expression for α as a function of K_P and P.

Solution

a. We set up the following table:

	$Cl_2(g) \rightarrow$	$2Cl(g)$
Initial number of moles	n_0	0
Moles present at equilibrium	$n_0 - \xi_{eq}$	$2\xi_{eq}$
Mole fraction present at equilibrium, x_i	$\dfrac{n_0 - \xi_{eq}}{n_0 + \xi_{eq}}$	$\dfrac{2\xi_{eq}}{n_0 + \xi_{eq}}$
Partial pressure at equilibrium, $P_i = x_i P$	$\left(\dfrac{n_0 - \xi_{eq}}{n_0 + \xi_{eq}}\right)P$	$\left(\dfrac{2\xi_{eq}}{n_0 + \xi_{eq}}\right)P$

We next express K_P in terms of n_0, ξ_{eq}, and P:

$$K_P(T) = \frac{\left(\dfrac{P_{Cl}^{eq}}{P^\circ}\right)^2}{\left(\dfrac{P_{Cl_2}^{eq}}{P^\circ}\right)} = \frac{\left[\left(\dfrac{2\xi_{eq}}{n_0 + \xi_{eq}}\right)\dfrac{P}{P^\circ}\right]^2}{\left(\dfrac{n_0 - \xi_{eq}}{n_0 + \xi_{eq}}\right)\dfrac{P}{P^\circ}} = \frac{4\xi_{eq}^2}{(n_0 + \xi_{eq})(n_0 - \xi_{eq})}\frac{P}{P^\circ} = \frac{4\xi_{eq}^2}{(n_0)^2 - \xi_{eq}^2}\frac{P}{P^\circ}$$

This expression is converted into one in terms of α:

b. $$K_P(T) = \frac{4\xi_{eq}^2}{(n_0)^2 - \xi_{eq}^2}\frac{P}{P^\circ} = \frac{4\alpha^2}{1 - \alpha^2}\frac{P}{P^\circ}$$

$$\left(K_P(T) + 4\frac{P}{P^\circ}\right)\alpha^2 = K_P(T)$$

$$\alpha = \sqrt{\frac{K_P(T)}{K_P(T) + 4\dfrac{P}{P^\circ}}}$$

Because $K_P(T)$ depends strongly on temperature, α will also be a strong function of temperature. Note that α depends on both K_P and P.

Note that whereas $K_P(T)$ is independent of P, α as calculated in Example Problem 6.9 does depend on P. In the particular case considered, α decreases as P increases for constant T. As will be shown in Section 6.12, α depends on P if $\Delta v \neq 0$ for a reaction.

6.10 The Variation of K_P with Temperature

As shown in Example Problem 6.9, the degree of dissociation of Cl_2 is dependent on the temperature. Based on our chemical intuition, we expect the degree of dissociation to be high at high temperatures and low at room temperature. For this to be true, $K_P(T)$ must increase with temperature for this reaction. How can we understand the temperature dependence for this reaction and for reactions in general?

Starting with Equation (6.67), we write

$$\frac{d \ln K_P}{dT} = -\frac{d\left(\Delta G^{\circ}_{reaction}/RT\right)}{dT} = -\frac{1}{R}\frac{d\left(\Delta G^{\circ}_{reaction}/T\right)}{dT} \tag{6.69}$$

Using the Gibbs–Helmholtz equation [Equation (6.35)], the preceding equation reduces to

$$\frac{d \ln K_P}{dT} = -\frac{1}{R}\frac{d(\Delta G^{\circ}_{reaction}/T)}{dT} = \frac{\Delta H^{\circ}_{reaction}}{RT^2} \tag{6.70}$$

Because tabulated values of ΔG°_f are available at 298.15 K, we can calculate $\Delta G^{\circ}_{reaction}$ and K_P at this temperature; K_P can be calculated at the temperature T_f by integrating Equation (6.70) between the appropriate limits:

$$\int_{K_P(298.15\mathrm{K})}^{K_P(T_f)} d \ln K_P = \frac{1}{R} \int_{298.15\mathrm{K}}^{T_f} \frac{\Delta H^{\circ}_{reaction}}{T^2}\,dT \tag{6.71}$$

If the temperature T_f is not much greater than 298.15 K, it can be assumed that $\Delta H^{\circ}_{reaction}$ is constant over the temperature interval. This assumption is better than it might appear at first glance. Although H is strongly dependent on temperature, the temperature dependence of $\Delta H^{\circ}_{reaction}$ is governed by the difference in heat capacities ΔC_p between reactants and products (see Section 3.4). If the heat capacities of reactants and products are nearly the same, $\Delta H^{\circ}_{reaction}$ is nearly independent of temperature. With the assumption that $\Delta H^{\circ}_{reaction}$ is independent of temperature, Equation (6.71) becomes

$$\ln K_P(T_f) = \ln K_P(298.15\ \mathrm{K}) - \frac{\Delta H^{\circ}_{reaction}}{R}\left(\frac{1}{T_f} - \frac{1}{298.15\ \mathrm{K}}\right) \tag{6.72}$$

EXAMPLE PROBLEM 6.10

Using the result of Example Problem 6.9 and the data tables, consider the dissociation reaction $Cl_2(g) \rightarrow 2Cl(g)$.

a. Calculate K_P at 800, 1500, and 2000 K for $P = 0.010$ bar.

b. Calculate the degree of dissociation, α, at 300, 1500, and 2000 K.

Solution

a. $\Delta G^{\circ}_{reaction} = 2\Delta G^{\circ}_f(Cl,g) - \Delta G^{\circ}_f(Cl_2,g) = 2 \times 105.7 \times 10^3\ \mathrm{J\,mol^{-1}} - 0$

$\qquad = 211.4\ \mathrm{kJ\,mol^{-1}}$

$\Delta H^{\circ}_{reaction} = 2\Delta H^{\circ}_f(Cl,g) - \Delta H^{\circ}_f(Cl_2,g) = 2 \times 121.3 \times 10^3\ \mathrm{J\,mol^{-1}} - 0$

$\qquad = 242.6\ \mathrm{kJ\,mol^{-1}}$

$\ln K_P(T_f) = -\dfrac{\Delta G^{\circ}_{reaction}}{RT} - \dfrac{\Delta H^{\circ}_{reaction}}{R}\left(\dfrac{1}{T_f} - \dfrac{1}{298.15\mathrm{K}}\right)$

$\qquad = -\dfrac{211.4 \times 10^3\ \mathrm{J\,mol^{-1}}}{8.314\ \mathrm{J\,K^{-1}\,mol^{-1}} \times 298.15\ \mathrm{K}} - \dfrac{242.6 \times 10^3\ \mathrm{J\,mol^{-1}}}{8.314\ \mathrm{J\,K^{-1}\,mol^{-1}}}$

$\qquad \times \left(\dfrac{1}{T_f} - \dfrac{1}{298.15\ \mathrm{K}}\right)$

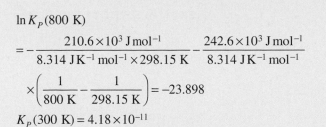

$\ln K_P(800\ \text{K})$

$$= -\frac{210.6 \times 10^3\ \text{J mol}^{-1}}{8.314\ \text{J K}^{-1}\ \text{mol}^{-1} \times 298.15\ \text{K}} - \frac{242.6 \times 10^3\ \text{J mol}^{-1}}{8.314\ \text{J K}^{-1}\ \text{mol}^{-1}}$$

$$\times \left(\frac{1}{800\ \text{K}} - \frac{1}{298.15\ \text{K}} \right) = -23.898$$

$K_P(300\ \text{K}) = 4.18 \times 10^{-11}$

The values for K_P at 1500 and 2000 K are 1.03×10^{-3} and 0.134, respectively.

b. The value of α at 2000 K is given by

$$\alpha = \sqrt{\frac{K_P(T)}{K_P(T) + 4\dfrac{P}{P^\circ}}} = \sqrt{\frac{0.134}{0.134 + 4 \times 0.01}} = 0.878$$

The values of α at 1500 and 800 K are 0.159 and 3.23×10^{-5}, respectively.

The degree of dissociation of Cl_2 increases with temperature as shown in Example Problem 6.10. This is always the case for an endothermic reaction ($\Delta H_{reaction} > 0$), as will be proved in Section 6.13 when Le Chatelier's principle is discussed.

6.11 Equilibria Involving Ideal Gases and Solid or Liquid Phases

In the preceding sections, we discussed chemical equilibrium in a homogeneous system of ideal gases. However, many chemical reactions involve a gas phase in equilibrium with a solid or liquid phase. An example is the thermal decomposition of $CaCO_3(s)$:

$$CaCO_3(s) \rightarrow CaO(s) + CO_2(g) \tag{6.73}$$

In this case, a pure gas is in equilibrium with two solid phases. At equilibrium,

$$\Delta G_{reaction} - \sum_i n_i \mu_i = 0 \tag{6.74}$$

$$0 = \mu_{eq}(\text{CaO},s,P) + \mu_{eq}(CO_2,g,P) - \mu_{eq}(CaCO_3,s,P)$$

Because the equilibrium pressure is $P \neq P^\circ$, the pressure dependence of each species must be taken into account. From Section 6.4, we know that the pressure dependence of G for a solid or liquid is very small:

$$\mu_{eq}(\text{CaO},s,P) \approx \mu^\circ(\text{CaO},s) \text{ and } \mu_{eq}(CaCO_3,s,P) \approx \mu^\circ(CaCO_3,s) \tag{6.75}$$

You will verify the validity of Equation (6.75) in the end-of-chapter problems. Using the dependence of μ on P for an ideal gas, Equation (6.74) becomes

$$0 = \mu^\circ(\text{CaO},s) + \mu^\circ(CO_2,g) - \mu^\circ(CaCO_3,s) + RT \ln \frac{P_{CO_2}}{P^\circ} \text{ or} \tag{6.76}$$

$$\Delta G^\circ_{reaction} = \mu^\circ(\text{CaO},s) + \mu^\circ(CO_2,g) - \mu^\circ(CaCO_3,s) = -RT \ln \frac{P_{CO_2}}{P^\circ}$$

Rewriting this equation in terms of K_P, we obtain

$$\ln K_P = \ln \frac{P_{CO_2}}{P^\circ} = -\frac{\Delta G^\circ_{reaction}}{RT} \tag{6.77}$$

EXAMPLE PROBLEM 6.11

Using the preceding discussion and the tabulated values of ΔG_f° in Appendix A, calculate the $CO_2(g)$ pressure in equilibrium with a mixture of $CaCO_3(s)$ and $CaO(s)$ at 1500, 2000, and 2500 K.

Solution

$$\Delta G_{reaction}^\circ = \Delta G_f^\circ(CaO, s) + \Delta G_f^\circ(CO_2, g) - \Delta G_f^\circ(CaCO_3, s)$$
$$= -603.3 \times 10^3 \, J\,mol^{-1} - 394.4 \times 10^3 \, J\,mol^{-1} + 1081.4 \times 10^3 \, J\,mol^{-1}$$
$$= 83.7\,kJ\,mol^{-1}$$

$$\ln K_P(1500K) = \ln \frac{P_{CO_2}}{P^\circ} = -\frac{\Delta G_{reaction}^\circ}{RT} = -\frac{83.7 \times 10^3 \, J\,mol^{-1}}{8.314 \, J\,K^{-1}\,mol^{-1} \times 1500 \, K} = -6.71157$$

$$P_{CO_2} = e^{-\Delta G_{reaction}^\circ / RT} = 1.22 \times 10^{-3} \, bar$$

$$= 6.51 \times 10^{-3} \, bar \text{ and } 1.78 \times 10^{-2} \, bar \text{ at 2000 and 2500 K, respectively.}$$

If the reaction involves only liquids or solids, the pressure dependence of the chemical potential is generally small and can be neglected. However, it cannot be neglected if $P \gg 1$ bar as shown in Example Problem 6.12.

EXAMPLE PROBLEM 6.12

At 298.15 K, $\Delta G_f^\circ(C, graphite) = 0$, and $\Delta G_f^\circ(C, diamond) = 2.90 \, kJ\,mol^{-1}$. Therefore, graphite is the more stable solid phase at this temperature at $P = P^\circ = 1$ bar. Given that the densities of graphite and diamond are 2.25 and 3.52 kg/L, respectively, at what pressure will graphite and diamond be in equilibrium at 298.15 K?

Solution

At equilibrium $\Delta G = G(C, graphite) - G(C, diamond) = 0$. Using the pressure dependence of G, $(\partial G_m / \partial P)_T = V_m$, we establish the condition for equilibrium:

$$\Delta G = \Delta G_f^\circ(C, graphite) - \Delta G_f^\circ(C, diamond) + (V_m^{graphite} - V_m^{diamond})(\Delta P) = 0$$

$$0 = 0 - 2900 \, J + (V_m^{graphite} - V_m^{diamond})(P - 1 \, bar)$$

$$P = 1 \, bar + \frac{2900 \, J}{M_C \left(\dfrac{1}{\rho_{graphite}} - \dfrac{1}{\rho_{diamond}} \right)}$$

$$= 1 \, bar + \frac{2900 \, J}{12.00 \times 10^{-3} \, kg\,mol^{-1} \times \left(\dfrac{1}{2.25 \times 10^3 \, kg\,m^{-3}} - \dfrac{1}{3.52 \times 10^3\,kg\,m^{-3}} \right)}$$

$$= 10^5 Pa + 1.51 \times 10^9 \, Pa = 1.51 \times 10^4 \, bar$$

Fortunately for all those with diamond rings, although diamond is unstable with respect to graphite at 1 bar and 298 K, the rate of conversion is vanishingly small.

6.12 Expressing the Equilibrium Constant in Terms of Mole Fraction or Molarity

Chemists often find it useful to express the concentrations of reactants and products in units other than partial pressures. Two examples of other units that we will consider in this section are mole fraction and molarity. Note, however, that this discussion is still limited to a mixture of ideal gases. The extension of chemical equilibrium to include neutral and ionic species in aqueous solutions will be made in Chapter 10, after the concept of activity has been introduced.

We first express the equilibrium constant in terms of mole fractions. The mole fraction, x_i, and the partial pressure, P_i, are related by $P_i = x_i P$. Therefore,

$$K_P = \frac{\left(\dfrac{P_C^{eq}}{P^\circ}\right)^c \left(\dfrac{P_D^{eq}}{P^\circ}\right)^d}{\left(\dfrac{P_A^{eq}}{P^\circ}\right)^a \left(\dfrac{P_B^{eq}}{P^\circ}\right)^b} = \frac{\left(\dfrac{x_C^{eq} P}{P^\circ}\right)^c \left(\dfrac{x_D^{eq} P}{P^\circ}\right)^d}{\left(\dfrac{x_A^{eq} P}{P^\circ}\right)^a \left(\dfrac{x_B^{eq} P}{P^\circ}\right)^b} = \frac{\left(x_C^{eq}\right)^c \left(x_D^{eq}\right)^d}{\left(x_A^{eq}\right)^a \left(x_B^{eq}\right)^b} \left(\frac{P}{P^\circ}\right)^{d+c-a-b} \tag{6.78}$$

$$= K_x \left(\frac{P}{P^\circ}\right)^{\Delta v}$$

$$K_x = K_P \left(\frac{P}{P^\circ}\right)^{-\Delta v}$$

Note that just as for K_P, K_x is a dimensionless number.

Because the molarity, c_i, is defined as $c_i = n_i/V = P_i/RT$, we can write $P_i/P^\circ = (RT/P^\circ)c_i$. To work with dimensionless quantities, we introduce the ratio c_i/c°, which is related to P_i/P° by

$$\frac{P_i}{P^\circ} = \frac{c^\circ RT}{P^\circ} \frac{c_i}{c^\circ} \tag{6.79}$$

Using this notation, we can express K_c in terms of K_P:

$$\tag{6.80}$$

$$K_P = \frac{\left(\dfrac{P_C^{eq}}{P^\circ}\right)^c \left(\dfrac{P_D^{eq}}{P^\circ}\right)^d}{\left(\dfrac{P_A^{eq}}{P^\circ}\right)^a \left(\dfrac{P_B^{eq}}{P^\circ}\right)^b} = \frac{\left(\dfrac{c_C^{eq}}{c^\circ}\right)^c \left(\dfrac{c_D^{eq}}{c^\circ}\right)^d}{\left(\dfrac{c_A^{eq}}{c^\circ}\right)^a \left(\dfrac{c_B^{eq}}{c^\circ}\right)^b} \left(\frac{c^\circ RT}{P^\circ}\right)^{d+c-a-b} = K_c \left(\frac{c^\circ RT}{P^\circ}\right)^{\Delta v}$$

$$K_c = K_P \left(\frac{c^\circ RT}{P^\circ}\right)^{-\Delta v}$$

Recall that Δv is the difference in the stoichiometric coefficients of products and reactants. Equations (6.78) and (6.80) show that K_x and K_c are in general different from K_P. They are only equal in the special case that $\Delta v = 0$.

6.13 The Dependence of ξ_{eq} on T and P

Suppose that we have a mixture of reactive gases at equilibrium. Does the equilibrium shift toward reactants or products as T or P is changed? To answer this question we first consider the dependence of K_P on T. Because

$$\frac{d \ln K_P}{dT} = \frac{\Delta H_{reaction}^\circ}{RT^2} \tag{6.81}$$

Note that K_P will change differently with temperature for an exothermic reaction than for an endothermic reaction. For an exothermic reaction, $d \ln K_P/dT < 0$, and ξ_{eq} will shift toward the reactants as T increases. For an endothermic reaction, $d \ln K_P/dT > 0$, and ξ_{eq} will shift toward the products as T increases.

The dependence of ξ_{eq} on pressure can be ascertained from the relationship between K_P and K_x:

$$K_x = K_P \left(\frac{P}{P^\circ} \right)^{-\Delta \nu} \qquad (6.82)$$

Because K_P is independent of pressure, the pressure dependence of K_x arises solely from $(P/P^\circ)^{-\Delta \nu}$. If the number of moles of gaseous products increases as the reaction proceeds, K_x decreases as P increases, and ξ_{eq} shifts back toward the reactants. If the number of moles of gaseous products decreases as the reaction proceeds, K_x increases as P increases, and ξ_{eq} shifts forward toward the products. If $\Delta \nu = 0$, ξ_{eq} is independent of pressure.

A combined change in T and P leads to a superposition of the effects just discussed. According to the French chemist Le Chatelier, reaction systems at chemical equilibrium respond to an outside stress, such as a change in T or P, by countering the stress. Consider the $Cl_2(g) \rightarrow 2Cl(g)$ reaction discussed in Example Problems 6.9 and 6.10 for which $\Delta H_{reaction} > 0$. There, ξ_{eq} responds to an increase in T in such a way that heat is taken up by the system; in other words, more $Cl_2(g)$ dissociates. This counters the stress imposed on the system by an increase in T. Similarly, ξ_{eq} responds to an increase in P in such a way that the volume of the system decreases. Specifically, ξ_{eq} changes in the direction that $\Delta \nu < 0$, and for the reaction under consideration, $Cl(g)$ is converted to $Cl_2(g)$. This shift in the position of equilibrium counters the stress brought about by an increase in P.

SUPPLEMENTAL

6.14 A Case Study: The Synthesis of Ammonia

In this section, we discuss a specific reaction and show both the power and the limitations of thermodynamics in developing a strategy to maximize the rate of a desired reaction. Ammonia synthesis is the primary route to the manufacture of fertilizers in agriculture. The commercial process used today is essentially the same as that invented by the German chemists Robert Bosch and Fritz Haber in 1908. In terms of its impact on human life, the Haber-Bosch synthesis may be the most important chemical process ever invented, because the increased food production possible with fertilizers based on NH_3 has allowed the large increase in the Earth's population in the 20th century to occur. In 2000, more than 2 million tons of ammonia were produced per week using the Haber-Bosch process, and more than 98% of the inorganic nitrogen input to soils used in agriculture worldwide as fertilizer is generated using the Haber-Bosch process.

What useful information can be obtained about the ammonia synthesis reaction from thermodynamics? The overall reaction is

$$1/2 \ N_2(g) + 3/2 \ H_2(g) \rightarrow NH_3(g) \qquad (6.83)$$

Because the reactants are pure elements in their standard states at 298.15 K as well as at all higher temperatures,

$$\Delta H^\circ_{reaction} = \Delta H^\circ_f(NH_3, g) = -45.9 \times 10^3 \ J \, mol^{-1} \ \text{at 298.15 K} \qquad (6.84)$$

$$\Delta G^\circ_{reaction} = \Delta G^\circ_f(NH_3, g) = -16.5 \times 10^3 \ J \, mol^{-1} \ \text{at 298.15 K} \qquad (6.85)$$

Assuming ideal gas behavior, the equilibrium constant K_p is given by

$$K_P = \frac{\left(\dfrac{P_{NH_3}}{P^\circ} \right)}{\left(\dfrac{P_{N_2}}{P^\circ} \right)^{\frac{1}{2}} \left(\dfrac{P_{H_2}}{P^\circ} \right)^{\frac{3}{2}}} = \frac{(x_{NH_3})}{(x_{N_2})^{\frac{1}{2}} (x_{H_2})^{\frac{3}{2}}} \left(\frac{P}{P^\circ} \right)^{-1} \qquad (6.86)$$

We will revisit this equilibrium and include deviations from ideal behavior in Chapter 7. Because the goal is to maximize x_{NH_3}, Equation (6.86) is written in the form

$$x_{NH_3} = (x_{N_2})^{\frac{1}{2}}(x_{H_2})^{\frac{3}{2}}\left(\frac{P}{P^\circ}\right)K_P \tag{6.87}$$

What conditions of temperature and pressure are most appropriate to maximizing the yield of ammonia? Because

$$\frac{d \ln K_P}{dT} = \frac{\Delta H^\circ_{reaction}}{RT^2} < 0 \tag{6.88}$$

K_P and, therefore, x_{NH_3} decrease as T increases. Because $\Delta v = -1$, x_{NH_3} increases with P. The conclusion is that the best conditions to carry out the reaction are high P and low T. In fact, industrial production is carried out for $P > 100$ atm and $T \sim 700$ K.

Our discussion has focused on reaching equilibrium because this is the topic of inquiry in thermodynamics. However, note that many synthetic processes are deliberately carried out far from equilibrium in order to selectively generate a desired product. For example, ethene oxide, which is produced by the partial oxidation of ethene, is an industrially important chemical used to manufacture antifreeze, surfactants, and fungicides. Although the complete oxidation of ethene to CO_2 and H_2O has a more negative value of $\Delta G^\circ_{reaction}$ than the partial oxidation, it is undesirable because it does not produce useful products.

These predictions provide information about the equilibrium state of the system, but lack one important component. How long will it take to reach equilibrium? Unfortunately, thermodynamics gives us no information about the rate at which equilibrium is attained. Consider the following example. Using values of ΔG°_f from the data tables, convince yourself that the oxidation of potassium and coal (essentially carbon) are both spontaneous processes. For this reason, potassium has to be stored in an environment free of oxygen. However, coal is stable in air on the timescale of a human lifetime. Thermodynamics tells us that both processes are spontaneous at 298.15 K, but experience shows us that only one proceeds at a measurable rate. However, by heating coal to an elevated temperature, the oxidation reaction becomes rapid. This result implies that there is an energetic **activation barrier** to the reaction that can be overcome with the aid of thermal energy. As we will see, the same is true of the ammonia synthesis reaction. Although Equations (6.87) and (6.88) predict that x_{NH_3} is maximized as T approaches 0 K, the reaction rate is vanishingly small unless $T > 500$ K.

To this point, we have only considered the reaction system using state functions. However, a chemical reaction consists of a number of individual steps, called a **reaction mechanism.** How might the ammonia synthesis reaction proceed? One possibility is the gas-phase reaction sequence:

$$N_2(g) \rightarrow 2N(g) \tag{6.89}$$

$$H_2(g) \rightarrow 2H(g) \tag{6.90}$$

$$N(g) + H(g) \rightarrow NH(g) \tag{6.91}$$

$$NH(g) + H(g) \rightarrow NH_2(g) \tag{6.92}$$

$$NH_2(g) + H(g) \rightarrow NH_3(g) \tag{6.93}$$

In each of the recombination reactions, other species that are not involved in the reaction can facilitate energy transfer. The enthalpy changes associated with each of the individual steps are known and are indicated in Figure 6.7. This diagram is useful because it gives much more information than the single-value $\Delta H^\circ_{reaction}$. Although the overall process is slightly exothermic, the initial step, which is the dissociation of $N_2(g)$ and $H_2(g)$, is highly endothermic. Heating a mixture of $N_2(g)$ and $H_2(g)$ to high enough temperatures to dissociate a sizable fraction of the $N_2(g)$ would require such a high temperature that all the $NH_3(g)$ formed would dissociate into reactants. This can be shown by carrying out a calculation such as that in Example Problem 6.10. The conclusion is that the gas phase reaction between $N_2(g)$ and $H_2(g)$ is not a practical route to ammonia synthesis.

FIGURE 6.7
An enthalpy diagram is shown for the reaction mechanism in Equations (6.89) through (6.93). The successive steps in the reaction proceed from left to right in the diagram.

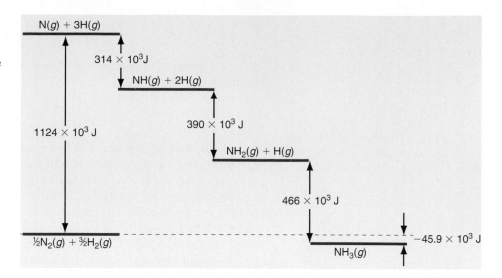

The way around this difficulty is to find another reaction sequence for which the enthalpy changes of the individual step are not prohibitively large. For the ammonia synthesis reaction, such a route is a heterogeneous catalytic reaction, using iron as a catalyst. The mechanism for this path between reactants and products is

$$N_2(g) + \square \rightarrow N_2\,(a) \tag{6.94}$$

$$N_2\,(a) + \square \rightarrow 2N\,(a) \tag{6.95}$$

$$H_2(g) + 2\,\square \rightarrow 2H(a) \tag{6.96}$$

$$N(a) + H(a) \rightarrow NH(a) + \square \tag{6.97}$$

$$NH(a) + H(a) \rightarrow NH_2(a) + \square \tag{6.98}$$

$$NH_2(a) + H(a) \rightarrow NH_3(a) + \square \tag{6.99}$$

$$NH_3(a) \rightarrow NH_3(g) + \square \tag{6.100}$$

The symbol $\square$ denotes an ensemble of neighboring Fe atoms, also called surface sites, which are capable of forming a chemical bond with the indicated entities. The designation (a) indicates that the chemical species is adsorbed (chemically bonded) at a surface site.

The enthalpy change for the overall reaction $N_2(g) + 3/2\,H_2(g) \rightarrow NH_3(g)$ is the same for the mechanisms in Equations (6.89) through (6.93) and (6.94) through (6.100) because H is a state function. This is a characteristic of a catalytic reaction. A catalyst can affect the rate of the forward and back reaction but not the position of equilibrium in a reaction system. Therefore, the catalyst influences the rate with which ξ_{eq} is reached, but it does not affect the value of ξ_{eq}. The enthalpy diagram in Figure 6.8 shows that the enthalpies of the individual steps in the gas phase and surface catalyzed reactions are very different. These enthalpy changes have been determined in experiments, but are not generally listed in thermodynamic tables.

In the catalytic mechanism, the enthalpy for the dissociation of the gas-phase N_2 and H_2 molecules is greatly reduced by the stabilization of the atoms through their bond to the surface sites. However, a small activation barrier to the dissociation of adsorbed N_2 remains. This barrier is the rate-limiting step in the overall reaction. As a result of this barrier, only one in a million gas-phase N_2 molecules incident on the catalyst surface per unit time dissociates and is converted into ammonia. Such a small reaction probability is not uncommon for industrial processes based on heterogeneous catalytic reactions. In industrial reactors, on the order of ~10^7 collisions of the reactants with the catalyst occur in the residence time of the reactants in the reactor. Therefore, even reactions with a reaction probability per collision of 10^{-6} can be carried out with a high yield.

FIGURE 6.8
Enthalpy diagram that compares the homogeneous gas phase and heterogeneous catalytic reactions for the ammonia synthesis reaction. The activation barriers for the individual steps in the surface reaction are shown. The successive steps in the reaction proceed from left to right in the diagram.

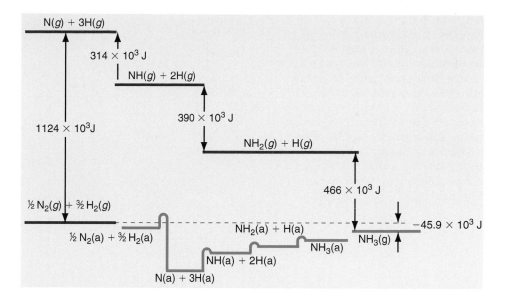

How was Fe chosen as the catalyst for ammonia synthesis? Even today, catalysts are usually optimized using a trial-and-error approach. Haber and Bosch tested hundreds of catalyst materials, and more than 2500 substances have been tried since. On the basis of this extensive screening, it was concluded that a successful catalyst must fulfill two different requirements: It must bind N_2 strongly enough that the molecule can dissociate into N atoms. However, it must not bind N atoms too strongly. If that were the case, the N atoms could not react with H atoms to produce gas-phase NH_3. As was known by Haber and Bosch, both osmium and ruthenium are better catalysts than iron for NH_3 synthesis, but these elements are far too expensive for commercial production. Why are these catalysts better than Fe? Quantum mechanical calculations by Jacobsen *et al.* [*J. American Chemical Society* 123 (2001), 8404−8405] were able to explain the "volcano plot" of the activity versus the N_2 binding energy shown in Figure 6.9.

Jacobsen *et al.* concluded that molybdenum is a poor catalyst because N is bound too strongly, whereas N_2 is bound too weakly. Nickel is a poor catalyst for the opposite reason; N is bound too weakly and N_2 is bound too strongly. Osmium and ruthenium are the best elemental catalysts because they fulfill both requirements well. The calculations suggested that a catalyst that combined the favorable properties of Mo and Co to create

FIGURE 6.9
The calculated turnover frequency (TOF) for NH_3 production (molecules of NH_3 per surface site per second) as a function of the calculated binding energy of N_2 on the surface, relative to that on ruthenium. The reaction conditions are 400°C, 50 bar total pressure, and gas composition $H_2:N_2 = 3:1$ containing 5% NH_3. [Adapted with permission from *Journal of American Chemical Society* and Dieter Kolb. Copyright 2001.]

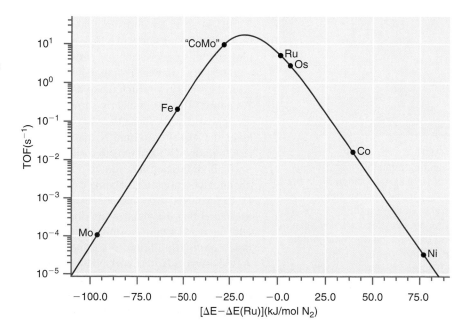

FIGURE 6.10

Experimental data for the TOF for NH_3 production as a function of the NH_3 concentration in the mixture. The Co_3Mo_3N catalyst (shown in the inset) is superior to both Fe and Ru as predicted by the calculations. The reaction conditions are those for Figure 6.9. The number of active surface sites is assumed to be 1% of the total number of surface sites. [Adapted with permission from *Journal of American Chemical Society* and Dieter Kolb. Copyright 2001.]

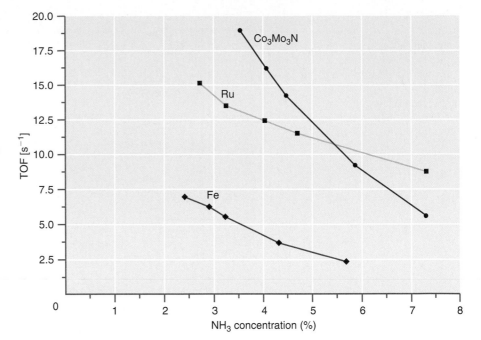

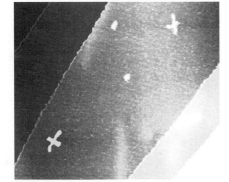

FIGURE 6.11

This scanning tunneling microscope image of a Ru surface shows wide flat terraces separated by monatomic steps. The bright "ribbon" at the step edge is due to Au atoms, which are also seen in the form of four islands on the middle terrace. [Photo courtesy of Professor Dr. Rolf Jurgen Behm/University of Ulm. T. Diemant, T. Hager, H. E. Hoster, H. Rauscher, and R. J. Behm, *Surf. Sci* 141 (2003) 137, fig. 1a.]

Co-Mo surface sites would be a good catalyst, and concluded that the ternary compound Co_3Mo_3N best satisfied these requirements. On the basis of these calculations, the catalyst was synthesized with the results shown in Figure 6.10. The experiments show that Co_3Mo_3N is far more active than Fe, and more active than Ru over most of the concentration range. These experiments show that the turnover frequency for Co_3Mo_3N lies at the point shown on the volcano plot of Figure 6.9. Note that this catalyst is one of the few that have been developed on the basis of theoretical calculations.

An unresolved issue is the nature of the surface sites indicated by □ in Equations (6.94) through (6.100). Are all adsorption sites on a NH_3 synthesis catalyst surface equally reactive? To answer this question, it is important to know that metal surfaces at equilibrium consist of large flat terraces separated by monatomic steps. It was long suspected that the steps rather than the terraces play an important role in ammonia synthesis. This was convincingly shown by Dahl *et al.* [*Physical Review Letters* 83 (1999), 1814−1817] in an elegant experiment. They evaporated a small amount of gold on a Ru crystal. As had been demonstrated by Hwang *et al.* [*Physical Review Letters* 67 (1991), 3279], the inert Au atoms migrate to the step sites on Ru and thereby block these sites, rendering them inactive in NH_3 synthesis. This site blocking is demonstrated using scanning tunneling microscopy as shown in Figure 6.11.

Dahl *et al.* showed that the rate of ammonia synthesis was lowered by a factor of 10^9 after the step sites, which make up only 1% to 2% of all surface sites, were blocked by Au atoms! This result clearly shows that in this reaction only step sites are active. They are more active because they have fewer nearest neighbors than terrace sites. This enables the step sites to bind N_2 more strongly, promoting dissociation into N atoms that react rapidly with H atoms to form NH_3.

This discussion has shown the predictive power of thermodynamics in analyzing chemical reactions. Calculating $\Delta G°_{reaction}$ allowed us to predict that the reaction of N_2 and H_2 to form NH_3 is a spontaneous process at 298.15 K. Calculating $\Delta H°_{reaction}$ allowed us to predict that the yield of ammonia increases with increasing pressure and decreases with increasing temperature. However, thermodynamics cannot be used to determine how rapidly equilibrium is reached in a reaction mixture. If thermodynamic tables of species such as $N_2(a)$, $N(a)$, $NH(a)$, $NH_2(a)$, and $NH_3(a)$ were available, it would be easier for chemists to develop synthetic strategies that are likely to proceed at an appreciable rate. However, whether an activation barrier exists in the conversion of one species to another—and how large the barrier is—are not predictable within thermodynamics.

Therefore, synthetic strategies must rely on chemical intuition, experiments, and quantum mechanical calculations to develop strategies that reach thermodynamic equilibrium at a reasonable rate.

SUPPLEMENTAL

6.15 Expressing *U* and *H* and Heat Capacities Solely in Terms of Measurable Quantities

In this section, we use the Maxwell relations derived in Section 6.2 to express U, H, and heat capacities solely in terms of measurable quantities such as κ, β and the state variables P, V, and T. Often, it is not possible to determine an equation of state for a substance over a wide range of the macroscopic variables. This is the case for most solids and liquids. However, material constants such as heat capacities, the thermal expansion coefficient, and the isothermal compressibility may be known over a limited range of their variables. If this is the case, the enthalpy and internal energy can be expressed in terms of the material constants and the variables V and T or P and T, as shown later. The Maxwell relations play a central role in deriving these formulas.

We begin with the following differential expression for dU that holds for solids, liquids, and gases provided that no nonexpansion work, phase transitions, or chemical reactions occur in the system:

$$dU = TdS - PdV \tag{6.101}$$

Invoking a transformation at constant temperature and dividing both sides of the equation by dV,

$$\left(\frac{\partial U}{\partial V}\right)_T = T\left(\frac{\partial S}{\partial V}\right)_T - P = T\left(\frac{\partial P}{\partial T}\right)_V - P \tag{6.102}$$

To arrive at the second expression in Equation (6.102), we have used the third Maxwell relation Equation (6.27). We next express $(\partial P/\partial T)_V$ in terms of β and κ, where β is the thermal expansion coefficient at constant pressure, and κ is the isothermal compressibility:

$$\left(\frac{\partial P}{\partial T}\right)_V = -\frac{(\partial V/\partial T)_P}{(\partial V/\partial P)_T} = \frac{\beta}{\kappa} \tag{6.103}$$

Equation (6.103) allows us to write Equation (6.102) in the form

$$\left(\frac{\partial U}{\partial V}\right)_T = \frac{\beta T - \kappa P}{\kappa} \tag{6.104}$$

Through this equation, we have linked the internal pressure to easily measured material properties. This allows us to calculate the internal pressure of a liquid as shown in Example Problem 6.13.

EXAMPLE PROBLEM 6.13

At 298 K, the thermal expansion coefficient and the isothermal compressibility of liquid water are $\beta = 2.04 \times 10^{-4} \text{ K}^{-1}$ and $\kappa = 45.9 \times 10^{-6} \text{ bar}^{-1}$.

a. Calculate $(\partial U/\partial V)_T$ for water at 320 K and $P = 1.00$ bar.

b. If an external pressure equal to $(\partial U/\partial V)_T$ were applied to 1.00 m³ of liquid water at 320 K, how much would its volume change? What is the relative change in volume?

c. As shown in Example Problem 3.5, $(\partial U_m/\partial V)_T = a/V_m^2$ for a van der Waals gas. Calculate $(\partial U_m/\partial V)_T$ for $N_2(g)$ at 320 K and $P = 1.00$ atm given that $a = 1.35$ atm L^2 mol^{-2}. Compare the value that you obtain with that of part (a) and discuss the difference.

Solution

a. $$\left(\frac{\partial U}{\partial V}\right)_T = \frac{\beta T - \kappa P}{\kappa} = \frac{2.04 \times 10^{-4} \text{ K}^{-1} \times 320 \text{ K} - 45.9 \times 10^{-6} \text{ bar}^{-1} \times 1.00 \text{ bar}}{45.9 \times 10^{-6} \text{ bar}^{-1}}$$

$$= 1.4 \times 10^3 \text{ bar}$$

b. $$\kappa = -\frac{1}{V}\left(\frac{\partial V}{\partial P}\right)_T$$

$$\Delta V_T = \int_{P_i}^{P_f} V(T)\kappa dP \approx V(T)\kappa(P_f - P_i)$$

$$= 1.00 \text{ m}^3 \times 45.9 \times 10^{-6} \text{ bar} \times (1.4 \times 10^3 \text{ bar} - 1 \text{ bar}) = 6.4 \times 10^{-2} \text{ m}^3$$

$$\frac{\Delta V_T}{V} = \frac{6.4 \times 10^{-2} \text{ m}^3}{1 \text{ m}^3} = 6.4\%$$

c. At atmospheric pressure, we can calculate V_m using the ideal gas law and

$$\left(\frac{\partial U_m}{\partial V}\right)_T = \frac{a}{V_m^2} = \frac{aP^2}{R^2T^2} = \frac{1.35 \text{ atm L}^2 \text{ mol}^{-2} \times (1.00 \text{ atm})^2}{(8.206 \times 10^{-2} \text{ atm L mol}^{-1} \text{ K}^{-1} \times 320 \text{ K})^2} = 1.96 \times 10^{-3} \text{ atm}$$

The internal pressure is much smaller for N_2 gas than for H_2O liquid. This is the case because H_2O molecules in a liquid are relatively strongly attracted to one another, whereas the interaction between gas phase N_2 molecules is very weak.

Using the result in Equation (6.104) for the internal pressure, we can write dU in the form

$$dU_m = \left(\frac{\partial U_m}{\partial T}\right)_V dT + \left(\frac{\partial U_m}{\partial V}\right)_T dV = C_{V,m}dT + \frac{\beta T - \kappa P}{\kappa}dV \qquad (6.105)$$

The usefulness of Equation (6.105) arises from the fact that it contains only the material constants C_V, β, κ, and the macroscopic variables T and V. In Example Problems 6.14 and 6.15 we show that $\Delta U, \Delta H$ and $C_P - C_V$ can be calculated even if the equation of state for the material of interest is not known, as long as the thermal expansion coefficient and the isothermal compressibility are known.

EXAMPLE PROBLEM 6.14

1000 g of liquid Hg undergoes a reversible transformation from an initial state $P_i = 1.00$ bar, $T_i = 300$ K to a final state $P_f = 300$ bar, $T_f = 600$ K. The density of Hg is 13534 kg m^{-3}, $\beta = 18.1 \times 10^{-4}$ K^{-1}, $C_P = 27.98$ J mol^{-1} K^{-1}, and $\kappa = 3.91 \times 10^{-6}$ bar^{-1}, in the range of the variables in this problem. Recall from Section 3.5 that $C_P - C_V = (TV\beta^2/\kappa)$, and that $C_P \approx C_V$ for liquids and solids.

a. Calculate ΔU and ΔH for the transformation. Describe the path assumed for the calculation.

b. Compare the relative contributions to ΔU and ΔH from the change in pressure and the change in temperature. Can the contribution from the change in pressure be neglected?

Solution

a. Because U is a state function, ΔU is independent of the path. We choose the reversible path $P_i = 1.00$ bar, $T_i = 300$ K $\rightarrow P_i = 3.00$ bar, $T_f = 300$ K $\rightarrow P_f = 300$ bar, $T_f = 600$ K. Note that because we have assumed that all the material constants are independent of P and T in the range of interest, the numerical values calculated will depend slightly on the path chosen. However, if the P and T dependence of the material constant were properly accounted for, all paths would give the same value of ΔU.

$$\Delta U = \int_{T_i}^{T_f} C_P dT + \int_{V_i}^{V_f} \frac{\beta T - \kappa P}{\kappa} dV$$

Using the definition $\kappa = -1/V(\partial V/\partial P)_T$, $\ln V_f/V_i = -\kappa(P_f - P_i)$, and $V(P) = V_i e^{-\kappa(P-P_i)}$ we can also express $P(V)$ as $P(V) = P_i - (1/\kappa)\ln(V/V_i)$

$$\Delta U = \int_{T_i}^{T_f} C_P dT + \int_{V_i}^{V_f} \left[\frac{\beta T}{\kappa} - \left(P_i - \frac{1}{\kappa} \ln \frac{V}{V_i} \right) \right] dV$$

$$= \int_{T_i}^{T_f} C_P dT + \left(\frac{\beta T}{\kappa} - P_i \right) \int_{V_i}^{V_f} dV + \frac{1}{\kappa} \int_{V_i}^{V_f} \ln \frac{V}{V_i} dV$$

Using the standard integral $\int \ln (x/a) dx = -x + x \ln (x/a)$

$$\Delta U = \int_{T_i}^{T_f} C_P dT + \left(\frac{\beta T}{\kappa} - P_i \right)(V_f - V_i) + \frac{1}{\kappa} \left[-V + V \ln \frac{V}{V_i} \right]_{V_i}^{V_f}$$

$$= \int_{T_i}^{T_f} C_P dT + \left(\frac{\beta T}{\kappa} - P_i \right)(V_f - V_i) + \frac{1}{\kappa} \left(V_i - V_f + V_f \ln \frac{V_f}{V_i} \right)$$

$$V_i = \frac{m}{\rho} = \frac{1.000 \text{ kg}}{13534 \text{ kg m}^{-3}} = 7.388 \times 10^{-5} \text{ m}^3$$

$$V_f = V_i e^{-\kappa(P_f - P_i)} V_f = 7.388 \times 10^{-5} \text{ m}^3 \exp(-3.91 \times 10^{-6} \text{ bar}^{-1} \times 299 \text{ bar})$$

$$= 7.379 \times 10^{-5} \text{ m}^3$$

$$V_f - V_i = -8.63 \times 10^{-8} \text{ m}^3$$

$$\Delta U = \frac{1000 \text{ g}}{200.59 \text{ g mol}^{-1}} \times 27.98 \text{ J mol}^{-1}\text{K}^{-1} \times (600 \text{ K} - 300 \text{ K})$$

$$- \left(\frac{18.1 \times 10^{-4} \text{ K}^{-1} \times 300 \text{ K}}{3.91 \times 10^{-6} \text{ bar}^{-1}} - 1.00 \text{ bar} \right) \times \frac{10^5 \text{ Pa}}{1 \text{ bar}} \times 8.63 \times 10^{-8} \text{ m}^3$$

$$+ \frac{1}{3.91 \times 10^{-6} \text{ bar}^{-1}} \times \frac{10^5 \text{ Pa}}{1 \text{ bar}} \times \left(\begin{array}{c} 8.63 \times 10^{-8} \text{ m}^3 + 7.379 \times 10^{-5} \text{ m}^3 \\ \times \ln \frac{7.380 \times 10^{-5} \text{ m}^3}{7.388 \times 10^{-5} \text{ m}^3} \end{array} \right)$$

$$= 41.8 \times 10^3 \text{ J} - 1.20 \times 10^3 \text{ J} + 1.1 \text{ J} \approx 40.4 \times 10^3 \text{ J}$$

Using the relationship

$$V(P, 600 \text{ K}) = [1 + \beta(600 \text{ K} - 300 \text{ K})]V(P_i, 300 \text{ K})e^{-\kappa(P-P_i)} = V'e^{-\kappa(P-P_i)}$$

$$\Delta H = \int_{T_i}^{T_f} C_p dT + \int_{P_i}^{P_f} V \, dP = \int_{T_i}^{T_f} C_p dT + V'e^{\kappa P_i} \int_{P_i}^{P_f} e^{-\kappa P} \, dP$$

$$= nC_{P,m}(T_f - T_i) + \frac{V_i'e^{\kappa P_i}}{\kappa}(e^{-\kappa P_i} - e^{-\kappa P_f})$$

$$= \frac{1000 \text{ g}}{200.59 \text{ g mol}^{-1}} \times 27.98 \text{ J mol}^{-1}\text{K}^{-1} \times (600 \text{ K} - 300 \text{ K})$$

$$+ 7.388 \times 10^{-5} \text{m}^3 \times (1 + 300 \text{ K} \times 18.1 \times 10^{-4} \text{ K}^{-1})$$

$$\times \frac{\exp(3.91 \times 10^{-6} \text{bar}^{-1} \times 1 \text{ bar})}{3.91 \times 10^{-6} \text{ bar}^{-1}}$$

$$\times \left[\exp(-3.91 \times 10^{-6} \text{bar}^{-1} \times 1 \text{ bar}) - \exp(-3.91 \times 10^{-6} \text{bar}^{-1} \times 299 \text{ bar}) \right]$$

$$\Delta H = 41.8 \times 10^3 \text{J} + 0.033 \text{ J} = 41.8 \times 10^3 \text{J}$$

b. The temperature dependent contribution to ΔU is 97.1% of the total change in U, and the corresponding value for ΔH is $\approx 100\%$. The contribution from the change in pressure to ΔU is small and to ΔH is negligible.

EXAMPLE PROBLEM 6.15

In the previous Example Problem, it was assumed that $C_P = C_V$.

a. Use $C_{P,m} - C_{V,m} = TV_m\beta^2/\kappa$ and the experimentally determined value $C_{P,m} = 27.98 \text{ J mol}^{-1} \text{ K}^{-1}$ to obtain a value for $C_{V,m}$ for Hg(l) at 300 K.
b. Did the assumption that $C_{P,m} \approx C_{V,m}$ in Example Problem 6.14 introduce an appreciable error in your calculation of ΔU and ΔH?

Solution

a. $C_{P,m} - C_{V,m} = \dfrac{TV_m\beta^2}{\kappa} = \dfrac{300 \text{ K} \times \dfrac{0.20059 \text{ kg mol}^{-1}}{13534 \text{ kg m}^3} \times (18.1 \times 10^{-4}\text{K}^{-1})^2}{3.91 \times 10^{-6}\text{Pa}^{-1}}$

$$= 3.73 \times 10^{-3} \text{ J K}^{-1}\text{mol}^{-1}$$

$$C_{V,m} = C_{P,m} - \frac{TV_m\beta^2}{\kappa}$$

$$= 27.98 \text{ J K}^{-1}\text{mol}^{-1} - 3.73 \times 10^{-3}\text{J K}^{-1}\text{mol}^{-1} \approx 27.98 \text{ J K}^{-1}\text{mol}^{-1}$$

b. ΔU and ΔH are unaffected by assuming that $C_{P,m} \approx C_{V,m}$. It is a very good approximation to set $C_P = C_V$ for a liquid or solid.

Vocabulary

activation barrier	conventional molar Gibbs energies	Gibbs energy
chemical potential	extent of reaction	Gibbs–Helmholtz equation

Helmholtz energy

maximum nonexpansion work

Maxwell relations

natural variables

reaction mechanism

reaction quotient of pressures

thermodynamic equilibrium constant

Questions on Concepts

Q6.1   Under what conditions is $dA \le 0$ a condition that defines the spontaneity of a process?

Q6.2   Under what conditions is $dG \le 0$ a condition that defines the spontaneity of a process?

Q6.3   Which thermodynamic state function gives a measure of the maximum electric work that can be carried out in a fuel cell?

Q6.4   By invoking the pressure dependence of the chemical potential, show that if a valve separating a vessel of pure A from a vessel containing a mixture of A and B is opened, mixing will occur. Both A and B are ideal gases, and the initial pressure in both vessels is 1 bar.

Q6.5   Under what condition is $K_P = K_x$?

Q6.6   It is found that K_P is independent of T for a particular chemical reaction. What does this tell you about the reaction?

Q6.7   The reaction A + B → C + D is at equilibrium for ξ = 0.1. What does this tell you about the variation of G_{pure} with ξ?

Q6.8   The reaction A + B → C + D is at equilibrium for ξ = 0.5. What does this tell you about the variation of G_{pure} with ξ?

Q6.9   Why is it reasonable to set the chemical potential of a pure liquid or solid substance equal to its standard state chemical potential at that temperature independent of the pressure in considering chemical equilibrium?

Q6.10   Is the equation $(\partial U / \partial V)_T = (\beta T - \kappa P) / \kappa$ valid for liquids, solids, and gases?

Q6.11   What is the relationship between the K_P for the two reactions $3/2 H_2 + 1/2 N_2 \rightarrow NH_3$ and $3 H_2 + N_2 \rightarrow 2 NH_3$?

Problems

Problem numbers in **red** indicate that the solution to the problem is given in the *Student's Solutions Manual*.

P6.1   Calculate the maximum nonexpansion work that can be gained from the combustion of benzene(l) and of $H_2(g)$ on a per gram and a per mole basis under standard conditions. Is it apparent from this calculation why fuel cells based on H_2 oxidation are under development for mobile applications?

P6.2   Calculate ΔA for the isothermal compression of 2.00 mol of an ideal gas at 298 K from an initial volume of 35.0 L to a final volume of 12.0 L. Does it matter whether the path is reversible or irreversible?

P6.3   Calculate ΔG for the isothermal expansion of 2.50 mol of an ideal gas at 350 K from an initial pressure of 10.5 bar to a final pressure of 0.500 bar.

P6.4   A sample containing 2.50 mol of an ideal gas at 298 K is expanded from an initial volume of 10.0 L to a final volume of 50.0 L. Calculate ΔG and ΔA for this process for (a) an isothermal reversible path and (b) an isothermal expansion against a constant external pressure of 0.750 bar. Explain why ΔG and ΔA do or do not differ from one another.

P6.5   The pressure dependence of G is quite different for gases and condensed phases. Calculate $G_m(C, solid, graphite,$ 100 bar, 298.15 K) and $G_m(He, g,$ 100 bar, 298.15 K) relative

to their standard state values. By what factor is the change in G_m greater for He than for graphite?

P6.6   Assuming that ΔH_f° is constant in the interval 275 K − 600 K, calculate $\Delta G_f^\circ (H_2O, g,$ 525 K).

P6.7   Calculate $\Delta G_{reaction}^\circ$ for the reaction $CO(g) + 1/2 \, O_2(g) \rightarrow CO_2(g)$ at 298.15 K. Calculate $\Delta G_{reaction}^\circ$ at 650 K assuming that $\Delta H_{reaction}^\circ$ is constant in the temperature interval of interest.

P6.8   Calculate $\Delta A_{reaction}^\circ$ and $\Delta G_{reaction}^\circ$ for the reaction $CH_4(g) + 2 O_2(g) \rightarrow CO_2(g) + 2 H_2O(l)$ at 298 K from the combustion enthalpy of methane and the entropies of the reactants and products.

P6.9   Consider the equilibrium $C_2H_6(g) \rightleftarrows C_2H_4(g) + H_2(g)$. At 1000 K and a constant total pressure of 1 atm, $C_2H_6(g)$ is introduced into a reaction vessel. The total pressure is held constant at 1 atm and at equilibrium the composition of the mixture in mole percent is $H_2(g)$: 26%, $C_2H_4(g)$: 26%, and $C_2H_6(g)$: 48%.

a.   Calculate K_P at 1000 K.

b.   If $\Delta H_{reaction}^\circ$ = 137.0 kJ mol^{-1}, calculate the value of K_P at 298.15K.

c.   Calculate $\Delta G_{reaction}^\circ$ for this reaction at 298.15 K.

P6.10   Consider the equilibrium $NO_2(g) \rightleftarrows NO(g) + 1/2 O_2(g)$. One mole of $NO_2(g)$ is placed in a vessel and allowed to come to equilibrium at a total pressure of 1 atm.

An analysis of the contents of the vessel gives the following results:

T	700 K	800 K
P_{NO}/P_{NO_2}	0.872	2.50

a. Calculate K_P at 700 and 800 K.

b. Calculate $\Delta G^\circ_{reaction}$ for this reaction at 298.15 K, assuming that $\Delta H^\circ_{reaction}$ is independent of temperature.

P6.11 Consider the equilibrium $CO(g) + H_2O(g) \rightleftharpoons CO_2(g) + H_2(g)$. At 1000 K, the composition of the reaction mixture is

Substance	$CO_2(g)$	$H_2(g)$	$CO(g)$	$H_2O(g)$
Mole %	27.1	27.1	22.9	22.9

a. Calculate K_P and $\Delta G^\circ_{reaction}$ at 1000 K.

b. Given the answer to part (a), use the ΔH°_f of the reaction species to calculate $\Delta G^\circ_{reaction}$ at 298.15 K. Assume that $\Delta H^\circ_{reaction}$ is independent of temperature.

P6.12 Consider the reaction $FeO(s) + CO(g) \rightleftharpoons Fe(s) + CO_2(g)$ for which K_P is found to have the following values:

T	600°C	1000°C
K_P	0.900	0.396

a. Calculate $\Delta G^\circ_{reaction}$, $\Delta S^\circ_{reaction}$ and $\Delta H^\circ_{reaction}$ for this reaction at 600°C. Assume that $\Delta H^\circ_{reaction}$ is independent of temperature.

b. Calculate the mole fraction of $CO_2(g)$ present in the gas phase at 600°C.

P6.13 If the reaction $Fe_2N(s) + 3/2H_2(g) \rightleftharpoons 2Fe(s) + NH_3(g)$ comes to equilibrium at a total pressure of 1 atm, analysis of the gas shows that at 700 and 800 K, $P_{NH_3}/P_{H_2} = 2.165$ and 1.083, respectively, if only $H_2(g)$ was initially present in the gas phase and $Fe_2N(s)$ was in excess.

a. Calculate K_P at 700 and 800 K.

b. Calculate $\Delta S^\circ_{reaction}$ at 700 K and 800 K and $\Delta H^\circ_{reaction}$ assuming that it is independent of temperature.

c. Calculate $\Delta G^\circ_{reaction}$ for this reaction at 298.15 K.

P6.14 At 25°C, values for the formation enthalpy and Gibbs energy and $\log_{10} K_P$ for the formation reactions of the various isomers of C_5H_{10} in the gas phase are given by the following table:

Substance	$\Delta H^\circ_f (\text{kJ mol}^{-1})$	$\Delta G^\circ_f (\text{kJ mol}^{-1})$	$\log_{10} K_P$
A = 1-pentene	−20.920	78.605	−13.7704
B = cis-2-pentene	−28.075	71.852	−12.5874
C = trans-2-pentene	−31.757	69.350	−12.1495
D = 2-methyl-1-butene	−36.317	64.890	−11.3680
E = 3-methyl-1-butene	−28.953	74.785	−13.1017
F = 2-methyl-2-butene	−42.551	59.693	−10.4572
G = cyclopentane	−77.24	38.62	−6.7643

Consider the equilibrium $A \rightleftharpoons B \rightleftharpoons C \rightleftharpoons D \rightleftharpoons E \rightleftharpoons F \rightleftharpoons G$, which might be established using a suitable catalyst.

a. Calculate the mole ratios A/G, B/G, C/G, D/G, E/G, F/G present at equilibrium at 25°C.

b. Do these ratios depend on the total pressure?

c. Calculate the mole percentages of the various species in the equilibrium mixture.

P6.15 In this problem, you calculate the error in assuming that $\Delta H^\circ_{reaction}$ is independent of T for a specific reaction. The following data are given at 25° C:

Compound	$CuO(s)$	$Cu(s)$	$O_2(g)$
$\Delta H^\circ_f (\text{kJ mol}^{-1})$	−157		
$\Delta G^\circ_f (\text{kJ mol}^{-1})$	−130		
$C^\circ_{P,m} (\text{J K}^{-1} \text{mol}^{-1})$	42.3	24.4	29.4

a. From equation 6.71,

$$\int_{K_P(T_0)}^{K_P(T_f)} d \ln K_P = \frac{1}{R} \int_{T_0}^{T_f} \frac{\Delta H^\circ_{reaction}(T)}{T^2} dT$$

To a good approximation, we can assume that the heat capacities are independent of temperature over a limited range in temperature, giving $\Delta H^\circ_{reaction}(T) = \Delta H^\circ_{reaction}(T_0) + \Delta C_P(T - T_0)$ where $\Delta C_P = \sum_i v_i C_{P,m}(i)$. By integrating equation 6.71, show that

$$\ln K_P(T) = \ln K_P(T_0) - \frac{\Delta H^\circ_{reaction}(T_0)}{R}\left(\frac{1}{T} - \frac{1}{T_0}\right)$$

$$+ \frac{T_0 \times \Delta C_P}{R}\left(\frac{1}{T} - \frac{1}{T_0}\right) + \frac{\Delta C_P}{R} \ln \frac{T}{T_0}$$

b. Using the result from part (a), calculate the equilibrium pressure of oxygen over copper and $CuO(s)$ at 1200 K. How is this value related to K_P for the reaction $2CuO(s) \rightleftharpoons 2Cu(s) + O_2(g)$?

c. What value would you obtain if you assumed that $\Delta H^\circ_{reaction}$ were constant at its value for 298.15 K up to 1200 K?

P6.16 Show that

$$\left[\frac{\partial(A/T)}{\partial(1/T)}\right]_V = U$$

Write an expression analogous to Equation (6.35) that would allow you to relate ΔA at two temperatures.

P6.17 Calculate $\mu_{O_2}^{mixture}(298.15 \text{ K}, 1 \text{ bar})$ for oxygen in air, assuming that the mole fraction of O_2 in air is 0.200.

P6.18 A sample containing 2.25 moles of He (1 bar, 298 K) is mixed with 3.00 mol of Ne (1 bar, 298 K) and 1.75 mol of Ar (1 bar, 298 K). Calculate ΔG_{mixing} and ΔS_{mixing}.

P6.19 You have containers of pure H_2 and He at 298 K and 1 atm pressure. Calculate ΔG_{mixing} relative to the unmixed gases of

a. a mixture of 10 mol of H_2 and 10 mol of He

b. a mixture of 10 mol of H_2 and 20 mol of He

c. Calculate ΔG_{mixing} if 10 mol of pure He are added to the mixture of 10 mol of H_2 and 10 mole of He.

P6.20 A gas mixture with 4 mol of Ar, x moles of Ne, and y moles of Xe is prepared at a pressure of 1 bar and a temperature of 298 K. The total number of moles in the mixture is three times that of Ar. Write an expression for ΔG_{mixing} in terms of x. At what value of x does the magnitude of ΔG_{mixing} have its maximum value? Calculate ΔG_{mixing} for this value of x.

P6.21 In Example Problem 6.8, K_p for the reaction $CO(g) + H_2O(l) \rightarrow CO_2(g) + H_2(g)$ was calculated to be 3.32×10^3 at 298.15 K. At what temperature does $K_p = 5.00 \times 10^3$? What is the highest value that K_p can have by changing the temperature? Assume that $\Delta H^{\circ}_{reaction}$ is independent of temperature.

P6.22 Calculate K_p at 550 K for the reaction $N_2O_4(l) \rightarrow 2NO_2(g)$ assuming that $\Delta H^{\circ}_{reaction}$ is constant over the interval $298 - 600$ K.

P6.23 Calculate K_p at 475 K for the reaction $NO(g) + 1/2\ O_2(g) \rightarrow NO_2(g)$ assuming that $\Delta H^{\circ}_{reaction}$ is constant over the interval $298 - 600$ K. Do you expect K_p to increase or decrease as the temperature is increased to 550 K?

P6.24 Calculate the degree of dissociation of N_2O_4 in the reaction $N_2O_4(g) \rightarrow 2NO_2(g)$ at 250 K and a total pressure of 0.500 bar. Do you expect the degree of dissociation to increase or decrease as the temperature is increased to 550 K? Assume that $\Delta H^{\circ}_{reaction}$ is independent of temperature.

P6.25 You wish to design an effusion source for Br atoms from $Br_2(g)$. If the source is to operate at a total pressure of 20 Torr, what temperature is required to produce a degree of dissociation of 0.50? What value of the pressure would increase the degree of dissociation to 0.65 at this temperature?

P6.26 A sample containing 2.00 moles of N_2 and 6.00 mol of H_2 are placed in a reaction vessel and brought to equilibrium at 20.0 bar and 750 K in the reaction $1/2\ N_2(g) + 3/2\ H_2(g) \rightarrow NH_3(g)$.

a. Calculate K_p at this temperature.

b. Set up an equation relating K_p and the extent of reaction as in Example Problem 6.9.

c. Using a numerical equation solver, calculate the number of moles of each species present at equilibrium.

P6.27 Consider the equilibrium in the reaction $3O_2(g) \rightleftharpoons 2O_3(g)$, with $\Delta H^{\circ}_{reaction} = 285.4 \times 10^3$ J mol^{-1} at 298 K. Assume that $\Delta H^{\circ}_{reaction}$ is independent of temperature.

a. Without doing a calculation, predict whether the equilibrium position will shift toward reactants or products as the pressure is increased.

b. Without doing a calculation, predict whether the equilibrium position will shift toward reactants or products as the temperature is increased.

c. Calculate K_p at 550 K.

d. Calculate K_x at 550 K and 0.500 bar.

P6.28 You place 2.00 mol of $NOCl(g)$ in a reaction vessel. Equilibrium is established with respect to the decomposition reaction $NOCl(g) \rightleftharpoons NO(g) + 1/2\ Cl_2(g)$.

a. Derive an expression for K_p in terms of the extent of reaction ξ.

b. Simplify your expression for part (a) in the limit that ξ is very small.

c. Calculate ξ and the degree of dissociation of NOCl in the limit that ξ is very small at 375 K and a pressure of 0.500 bar.

d. Solve the expression derived in part (a) using a numerical equation solver for the conditions stated in the previous part. What is the relative error in ξ made using the approximation of part (b)?

P6.29 $Ca(HCO_3)_2(s)$ decomposes at elevated temperatures according to the stoichiometric equation $Ca(HCO_3)_2(s) \rightarrow CaCO_3(s) + H_2O(g) + CO_2(g)$.

a. If pure $Ca(HCO_3)_2(s)$ is put into a sealed vessel, the air is pumped out, and the vessel and its contents are heated, the total pressure is 0.115 bar. Determine K_p under these conditions.

b. If the vessel initially also contains 0.225 bar $H_2O(g)$, what is the partial pressure of $CO_2(g)$ at equilibrium?

P6.30 Assume that a sealed vessel at constant pressure of 1 bar initially contains 2.00 mol of $NO_2(g)$. The system is allowed to equilibrate with respect to the reaction $2\ NO_2(g) \rightleftharpoons N_2O_4(g)$. The number of moles of $NO_2(g)$ and $N_2O_4(g)$ at equilibrium is $2.00 - 2\xi$ and ξ, respectively, where ξ is the extent of reaction.

a. Derive an expression for the entropy of mixing as a function of ξ.

b. Graphically determine the value of ξ for which ΔS_{mixing} has its maximum value.

c. Write an expression for G_{pure} as a function of ξ. Use Equation 6.34 to obtain values of G_m° for NO_2 and N_2O_4.

d. Plot $G_{mixture} = G_{pure} + \Delta G_{mixing}$ as a function of ξ for $T = 298$ K and graphically determine the value of ξ for which $G_{mixture}$ has its minimum value. Is this value the same as for part (b)?

CHAPTER 7

The Properties of Real Gases

The ideal gas law is only accurate for gases at low values of the density. To design production plants that use real gases at high pressures, real gas equations of state valid for gases at higher densities are needed. Such equations must take the finite volume of a molecule and the intermolecular potential into account and accurately describe the $P–V$ relationship of a given gas at a fixed value of T within their range of validity. However, they cannot be used to compare different gases, because they contain parameters that have unique values for a given gas. Therefore, it is convenient to introduce dimensionless reduced variables in place of P, V, and T. Equations of state expressed in terms of reduced variables are useful in estimating the volume of a real gas at given values of P and T. A further consequence of nonideality is that the chemical potential of a real gas must be expressed in terms of its fugacity rather than its partial pressure. Fugacities rather than pressures must also be used in calculating the thermodynamic equilibrium constant K_p for a real gas. ∎

7.1 Real Gases and Ideal Gases

To this point, the ideal gas equation of state has been assumed to be sufficiently accurate to describe the $P–V–T$ relationship for a real gas. This assumption has allowed calculations of expansion work and of the equilibrium constant, K_P, in terms of partial pressures using the ideal gas law. In fact, the ideal gas law provides an accurate description of the $P–V–T$ relationship for many gases, such as He, for a wide range of P, V, and T values. However, it describes the $P–V$ relationship for water within ±10% only for $T > 1300$ K, as shown in Section 7.4. What is the explanation for this different behavior of He and H_2O? Is it possible to derive a "universal" equation of state that can be used to describe the $P–V$ relationship for gases as different as He and H_2O?

In Section 1.5, the two main deficiencies in the microscopic model on which the ideal gas law is based were discussed. The first assumption is that gas molecules are point masses. However, molecules occupy a finite volume; therefore, a real gas cannot be compressed to a volume that is less than the total molecular volume. The second assumption is that the molecules in the gas do not interact, but molecules in a real gas do interact with one another through a potential as depicted in Figure 1.6. Because the potential has a finite range, its effect is negligible at low densities, which correspond to large distances between molecules. Additionally, at low densities, the molecular volume is negligible compared with the volume that the gas occupies. Therefore, the $P–V–T$

relationship of a real gas is the same as that for an ideal gas at sufficiently low densities and high temperatures. At higher densities and low temperatures, molecular interactions cannot be neglected. Because of these interactions, the pressure of a real gas can be higher or lower than that for an ideal gas at the same density and temperature. What determines which of these two cases applies? The questions raised in this section are the major themes of this chapter.

7.2 Equations of State for Real Gases and Their Range of Applicability

7.1 van der Waals Equation of State

In this section, several equations of state for real gases are discussed. The range of the variables P, V, and T over which they accurately describe a real gas are also discussed. Such equations of state must exhibit a limiting P–V–T behavior identical to that for an ideal gas at low density. They must also correctly model the deviations for ideal gas behavior that real gases exhibit at moderate and high densities. The first two equations of state considered here include two parameters, a and b, that must be experimentally determined for a given gas. The parameter a is a measure of the strength of the attractive part of the intermolecular potential, and b is a measure of the minimum volume that a mole of molecules can occupy. Real gas equations of state are best viewed as empirical equations whose functional form has been chosen to fit experimentally determined P–V–T data. The most widely used is the **van der Waals equation of state:**

$$P = \frac{RT}{V_m - b} - \frac{a}{V_m^2} = \frac{nRT}{V - nb} - \frac{n^2 a}{V^2} \tag{7.1}$$

A second useful equation of state is the **Redlich-Kwong equation of state,** which is given by

$$P = \frac{RT}{V_m - b} - \frac{a}{\sqrt{T}} \frac{1}{V_m(V_m + b)} = \frac{nRT}{V - nb} - \frac{n^2 a}{\sqrt{T}} \frac{1}{V(V + nb)} \tag{7.2}$$

Although the same symbols are used for parameters a and b in both equations of state, they have different values for a given gas.

Figure 7.1 shows that the degree to which the ideal gas, van der Waals, and Redlich-Kwong equations of state correctly predict the P–V behavior of CO_2 depends on P, V, and T. At 426 K, all three equations of state reproduce the correct P–V behavior reasonably well over the range shown, with the ideal gas law having the largest error. By contrast, the three equations of state give significantly different results at 310 K. The ideal gas law gives unacceptably large errors, and the Redlich-Kwong equation of state is more accurate than is the van der Waals equation. We will have more to say about the range over which the ideal gas law is reasonably accurate when discussing the compression factor in Section 7.3.

A third widely used equation of state for real gases is the **Beattie-Bridgeman equation of state.** This equation uses five experimentally determined parameters to fit P–V–T data. Because of its complexity, it is not used in this chapter.

$$P = \frac{RT}{V_m^2}\left(1 - \frac{c}{V_m T^3}\right)(V_m + B) - \frac{A}{V_m^2} \quad \text{with} \tag{7.3}$$

$$A = A_0\left(1 - \frac{a}{V_m}\right) \quad \text{and} \quad B = B_0\left(1 - \frac{b}{V_m}\right)$$

FIGURE 7.1
Isotherms for CO_2 are shown at **(a)** 426 K and **(b)** 310 K using the van der Waals equation of state (yellow curve), the Redlich-Kwong equation of state (blue curve), and the ideal gas equation of state (red curve). The black dots are accurate values taken from the *NIST Chemistry Webbook*.

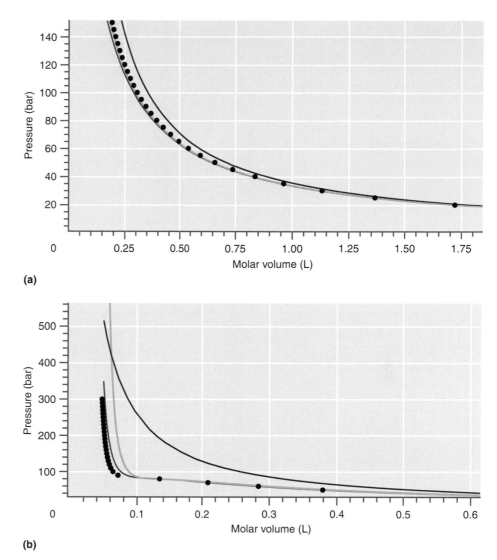

(a)

(b)

A further important equation of state for real gases has a different form than any of the previous equations. The **virial equation of state** is written in the form of a power series in $1/V_m$

$$P = RT \left[\frac{1}{V_m} + \frac{B(T)}{V_m^2} + ... \right] \qquad (7.4)$$

The power series does not converge at high pressures where V_m becomes small. The $B(T), C(T)$, and so on are called the second, third, and so on virial coefficients. This equation is more firmly grounded in theory than the previously discussed three equations because a series expansion is always valid in its convergence range. In practical use, the series is usually terminated after the second virial coefficient because values for the higher coefficients are not easily obtained. Table 7.1 (see Appendix A, Data Tables) lists values for the second virial coefficient for selected gases for different temperatures. If $B(T)$ is negative (positive), the attractive (repulsive) part of the potential dominates at that value of T. Statistical thermodynamics can be used to relate the virial coefficients with the intermolecular potential function. As you will show in the end-of-chapter problems, $B(T)$ for a van der Waals gas is given by $B(T) = b - (a/RT)$.

FIGURE 7.2

Calculated isotherms are shown for CO_2, modeled as a van der Waals gas. The gas and liquid (blue) regions, and the gas–liquid (yellow) coexistence region are shown. The dashed curve was calculated using the ideal gas law. The isotherm at $T = 304.12$ K is at the critical temperature and is called the critical isotherm.

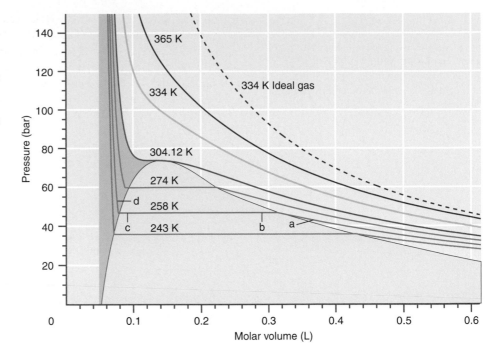

The principal limitation of the ideal gas law is that it does not predict that a gas can be liquefied under appropriate conditions. Consider the approximate P–V diagram for CO_2 shown in Figure 7.2. (It is approximate because it is based on the van der Waals equation of state, rather than experimental data.) Each of the curves is an isotherm corresponding to a fixed temperature. The behavior predicted by the ideal gas law is shown for $T = 334$ K. Consider the isotherm for $T = 258$ K. Starting at large values of V, the pressure rises as V decreases, and then becomes constant over a range of values of V. The value of V at which P becomes constant depends on T. As the volume of the system is decreased further, the pressure suddenly increases rapidly as V decreases.

The reason for this unusual dependence of P on V_m becomes clear when the experiment is carried out in the transparent piston and cylinder assembly shown in Figure 7.3. This system consists of either a single phase or two phases separated by a sharp interface, depending on the values of T and V_m. For points a, b, c, and d on the 258 K isotherm of Figure 7.2, the system has the following composition: At point a, the system consists entirely of $CO_2(g)$. However, at points b and c, a sharp interface separates $CO_2(g)$ and $CO_2(l)$. Along the line linking points b and c, the system contains $CO_2(g)$ and $CO_2(l)$ in equilibrium with one another. The proportion of liquid to gas changes, but the pressure remains constant. The temperature-dependent equilibrium pressure is called the **vapor**

FIGURE 7.3

The volume and the composition of a system containing CO_2 at 258 K is shown at the points a, b, c, and d indicated in Figure 7.2. The liquid and gas volumes are not shown to scale.

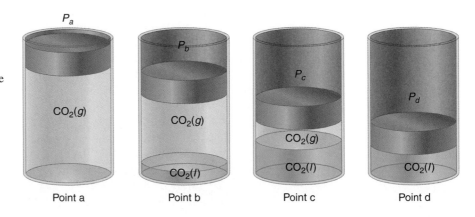

pressure of the liquid. As the volume is decreased further, the system becomes a single-phase system again, consisting of $CO_2(l)$ only. The pressure changes slowly with increasing V if $V > 0.33$ L because $CO_2(g)$ is quite compressible. However, the pressure changes rapidly with decreasing V if $V < 0.6$ L because $CO_2(l)$ is nearly incompressible. The single-phase regions and the two-phase gas–liquid coexistence region are shown in Figure 7.2.

If the same experiment is carried out at successively higher temperatures, it is found that the range of V_m in which two phases are present becomes smaller, as seen in the 243, 258, and 274 K isotherms of Figure 7.2. The temperature at which the range of V_m has shrunk to a single value is called the **critical temperature,** T_c. For CO_2, $T_c = 304.12$ K. At $T = T_c$, the isotherm exhibits an inflection point so that

$$\left(\frac{\partial P}{\partial V_m}\right)_{T=T_c} = 0 \quad \text{and} \quad \left(\frac{\partial^2 P}{\partial V_m^2}\right)_{T=T_c} = 0 \tag{7.5}$$

What is the significance of the critical temperature? Critical behavior will be discussed in Chapter 8. At this point, it is sufficient to know that as the critical point is approached, the density of $CO_2(l)$ decreases and the density of $CO_2(g)$ increases, and at $T = T_c$, the densities are equal. Above T_c, no interface is observed in the experiment depicted in Figure 7.3, and liquid and gas phases can no longer be distinguished. The term *supercritical fluid* is used instead. As will be discussed in Chapter 8, T_c and the corresponding values P_c and V_c, which together are called the **critical constants,** take on particular significance in describing the phase diagram of a pure substance. The critical constants for a number of different substances are listed in Table 7.2 (see Appendix A, Data Tables).

The parameters a and b for the van der Waals and Redlich-Kwong equations of state are chosen so that the equation of state best represents real gas data. This can be done by using the values of P, V, and T at the critical point, T_c, P_c, and V_c as shown in Example Problem 7.1. Parameters a and b calculated from critical constants in this way are listed in Table 7.4 (see Appendix A, Data Tables).

EXAMPLE PROBLEM 7.1

At $T = T_c$, $(\partial P / \partial V_m)_{T=T_c} = 0$ and $(\partial^2 P / \partial V_m^2)_{T=T_c} = 0$. Use this information to determine a and b in the van der Waals equation of state in terms of the experimentally determined values T_c and P_c.

Solution

$$\left(\frac{\partial\left[\dfrac{RT}{V_m-b} - \dfrac{a}{V_m^2}\right]}{\partial V_m}\right)_{T=T_c} = -\frac{RT_c}{(V_{mc}-b)^2} + \frac{2a}{V_{mc}^3} = 0$$

$$\left(\frac{\partial^2\left[\dfrac{RT}{V_m-b} - \dfrac{a}{V_m^2}\right]}{\partial V_m^2}\right)_{T=T_c} = \left(\frac{\partial\left[-\dfrac{RT}{(V_m-b)^2} + \dfrac{2a}{V_m^3}\right]}{\partial V_m}\right)_{T=T_c} = \frac{2RT_c}{(V_{mc}-b)^3} - \frac{6a}{V_{mc}^4} = 0$$

Equating RT_c from these two equations gives

$$RT_c = \frac{2a}{V_{mc}^3}(V_{mc}-b)^2 = \frac{3a}{V_{mc}^4}(V_{mc}-b)^3, \text{ which simplifies to} \frac{3}{2V_{mc}}(V_{mc}-b) = 1$$

The solution to this equation is $V_{mc} = 3b$. Substituting this result into $(\partial P / \partial V_m)_{T=T_c} = 0$ gives

$$-\frac{RT_c}{(V_{mc}-b)^2} + \frac{2a}{V_{mc}^3} = -\frac{RT_c}{(2b)^2} + \frac{2a}{(3b)^3} = 0 \quad \text{or} \quad T_c = \frac{2a}{(3b)^3}\frac{(2b)^2}{R} = \frac{8a}{27Rb}$$

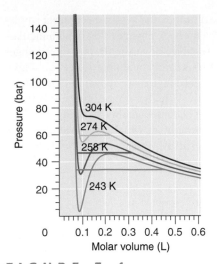

FIGURE 7.4

Van der Waals isotherms show the relationship between pressure and molar volume for CO_2 at the indicated temperatures. The Maxwell construction is shown. No oscillations in the calculated isotherms occur for $T \geq T_c$.

7.2 Comparison between Ideal Gas Law and van der Waals Equation

Substituting these results for T_c and V_{mc} in terms of a and b into the van der Waals equation gives the result $P_c = a/27b^2$. We only need two of the critical constants T_c, P_c, and V_{mc} to determine a and b. Because the measurements for P_c and T_c are more accurate than for V_{mc}, we use these constants to obtain expressions for a and b. These results for P_c and T_c can be used to express a and b in terms of P_c and T_c. The results are

$$b = \frac{RT_c}{8P_c} \text{ and } a = \frac{27R^2T_c^2}{64P_c}$$

A similar analysis for the Redlich-Kwong equation gives

$$a = \frac{R^2T_c^{5/2}}{9P_c(2^{1/3}-1)} \text{ and } b = \frac{(2^{1/3}-1)RT_c}{3P_c}$$

What is the range of validity of the van der Waals and Redlich-Kwong equations? No two-parameter equation of state is able to reproduce the isotherms shown in Figure 7.2 for $T < T_c$ because it cannot reproduce either the range in which P is constant or the discontinuity in $(\partial P/\partial V)_T$ at the ends of this range. This failure is illustrated in Figure 7.4, in which isotherms calculated using the van der Waals equation of state are plotted for some of the values of T as in Figure 7.2. Below T_c, all calculated isotherms have an oscillating region that is unphysical because V increases as P increases. In the **Maxwell construction,** the oscillating region is replaced by the horizontal line for which the areas above and below the line are equal, as indicated in Figure 7.4. The Maxwell construction is used in generating the isotherms shown in Figure 7.2.

The Maxwell construction can be justified on theoretical grounds, but the equilibrium vapor pressure determined in this way for a given value of T is only in qualitative agreement with experiment. The van der Waals and Redlich-Kwong equations of state do a good job of reproducing P–V isotherms for real gases only in the single-phase gas region $T > T_c$ and for densities well below the critical density, $\rho_c = M/V_{mc}$, where V_{mc} is the molar volume at the critical point. The Beattie-Bridgeman equation, which has three more adjustable parameters, is accurate above T_c for higher densities.

7.3 The Compression Factor

How large is the error in P–V curves if the ideal gas law is used rather than the van der Waals or Redlich-Kwong equations of state? To address this question, it is useful to introduce the **compression factor,** z, defined by

$$z = \frac{V_m}{V_m^{ideal}} = \frac{PV_m}{RT} \tag{7.6}$$

For the ideal gas, $z = 1$ for all values of P and V_m. If $z > 1$, the real gas exerts a greater pressure than the ideal gas, and if $z < 1$, the real gas exerts a smaller pressure than the ideal gas for the same values of T and V_m.

The compression factor for a given gas is a function of temperature, as shown in Figure 7.5. In this figure, z has been calculated for N_2 at two values of T using the van der Waals and Redlich-Kwong equations of state. The dots are calculated from accurate values of V_m taken from the *NIST Chemistry Workbook*. Although the results calculated using these equations of state are not in quantitative agreement with accurate results for all P and T, the trends in the functional dependence of z on P for different T values are correct. Because it is inconvenient to rely on tabulated data for individual gases, we focus on $z(P)$ curves calculated from real gas equations of state. Both the van der Waals

FIGURE 7.5
The calculated compression factor for N_2 is shown as a function of pressure for $T = 200$ and 400 K and compared with accurate values. The blue and red curves have been calculated using the van der Waals and Redlich-Kwong equations of state, respectively. The Redlich-Kwong equation does not give physically meaningful solutions above $P = 160$ bar for 200 K, where $\rho = 2.5\rho_c$. The dots are calculated from accurate values for V_m taken from the NIST Chemistry Webbook.

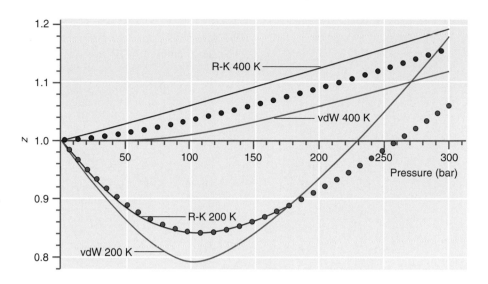

and Redlich-Kwong equations predict that for $T = 200$ K, z initially decreases with pressure. The compression factor only becomes greater than the ideal gas value of one for pressures in excess of 200 bar. For $T = 425$ K, z increases linearly with T. This functional dependence is also predicted by the Redlich-Kwong equation. The van der Waals equation predicts that the initial slope is zero, and z increases slowly with P. For both temperatures, $z \rightarrow 1$ as $P \rightarrow 0$. This result shows that the ideal gas law is obeyed if P is sufficiently small.

To understand why the low pressure value of the compression factor varies with temperature for a given gas, we use the van der Waals equation of state. Consider the variation of the compression factor with P at constant T,

$$\left(\frac{\partial z}{\partial P}\right)_T = \left(\frac{\partial z}{\partial \left[RT/V_m\right]}\right)_T = \frac{1}{RT}\left(\frac{\partial z}{\partial \left[1/V_m\right]}\right)_T$$

for a van der Waals gas in the ideal gas limit as $1/V_m \rightarrow 0$. As shown in Example Problem 7.2, the result $(\partial z/\partial P)_T = b - a/RT$ is obtained.

EXAMPLE PROBLEM 7.2

Show that the slope of z as a function of P as $P \rightarrow 0$ is related to the van der Waals parameters by

$$\lim_{p \rightarrow 0}\left(\frac{\partial z}{\partial P}\right)_T = \frac{1}{RT}\left(b - \frac{a}{RT}\right)$$

Solution
Rather than differentiating z with respect to P, we transform the partial derivative to one involving V_m:

$$z = \frac{V_m}{V_{m,ideal}} = \frac{PV_m}{RT} = \frac{\left(\dfrac{RT}{V_m - b} - \dfrac{a}{V_m^2}\right)V_m}{RT} = \frac{V_m}{V_m - b} - \frac{a}{RTV_m}$$

$$\left(\frac{\partial z}{\partial P}\right)_T = \left(\frac{\partial z}{\partial \left[\dfrac{RT}{V_m}\right]}\right)_T = \frac{1}{RT}\left(\frac{\partial \left[\dfrac{V_m}{V_m - b} - \dfrac{a}{RTV_m}\right]}{\partial \dfrac{1}{V_m}}\right)_T = \frac{1}{RT}\left(\frac{\partial \left[\dfrac{V_m}{V_m - b} - \dfrac{a}{RTV_m}\right]}{\partial V_m}\right)_T \frac{dV_m}{d\dfrac{1}{V_m}}$$

We have transformed the differentiation with respect to $1/V_m$ to one involving V_m. The substitution of RT/V_m for P is only valid in the low density limit. Because

$$\frac{dV_m}{d\frac{1}{V_m}} = \left(\frac{d\frac{1}{V_m}}{dV_m}\right)^{-1} = -V_m^2$$

$$\left(\frac{\partial z}{\partial P}\right)_T = \frac{-V_m^2}{RT}\left(\frac{1}{V_m - b} - \frac{V_m}{(V_m - b)^2} + \frac{a}{RTV_m^2}\right)$$

$$= \frac{V_m^2}{RT}\left(-\frac{(V_m - b)}{(V_m - b)^2} + \frac{V_m}{(V_m - b)^2} - \frac{a}{V_m^2 RT}\right)$$

$$= \frac{1}{RT}\left(\frac{bV_m^2}{(V_m - b)^2} - \frac{a}{RT}\right)$$

As $P \to 0$, $V_m \to \infty$ and $bV_m^2/(V_m - b)^2 \to b$. Therefore,

$$\lim_{P \to 0}\left(\frac{\partial z}{\partial P}\right)_T = \frac{1}{RT}\left(b - \frac{a}{RT}\right)$$

From Example Problem 7.2, the van der Waals equation predicts that the initial slope of the z versus P curve is zero if $b = a/RT$. The corresponding temperature is known as the **Boyle temperature** T_B

$$T_B = \frac{a}{Rb} \tag{7.7}$$

WWW

7.3 Compression Factor and Molar Volume

Values for the Boyle temperature of several gases are shown in Table 7.3.

Because the parameters a and b are substance dependent, T_B is different for each gas. Using the van der Waals parameters for N_2, $T_B = 425$ K, whereas the experimentally determined value is 327 K. The agreement is qualitative rather than quantitative. At the Boyle temperature both $z \to 1$ and $(\partial z/\partial P)_T \to 0$ as $P \to 0$, which is the behavior exhibited by an ideal gas. It is only at $T = T_B$ that a real gas exhibits ideal behavior as $P \to 0$ with respect to $\lim_{P \to 0}(\partial z/\partial P)_T$. Above the Boyle temperature, $(\partial z/\partial P)_T > 0$ as $P \to 0$, and below the Boyle temperature, $(\partial z/\partial P)_T < 0$ as $P \to 0$. These inequalities provide a criterion to predict whether z increases or decreases with pressure at low pressures for a given value of T.

Note that the initial slope of a z versus P plot is determined by the relative magnitudes of b and a/RT. Recall that the repulsive interaction is represented through b, and the attractive part of the potential is represented through a. From the discussion above, $\lim_{P \to 0}(\partial z/\partial P)_T$ is always positive at high temperatures, because $b - (a/RT) \to b > 0$ as $T \to \infty$. This means that molecules primarily feel the repulsive part of the potential for $T \gg T_B$. Conversely, for $T \ll T_B$, $\lim_{P \to 0}(\partial z/\partial P)_T$ will always be negative, because the molecules primarily feel the attractive part of the potential. For high enough values of P, $z > 1$ for all gases, showing that the repulsive part of the potential dominates at high gas densities, regardless of the value of T.

Next consider the functional dependence of z on P for different gases at a single temperature. Calculated values for oxygen, hydrogen, and ethene obtained using the van der Waals equation at $T = 400$ K are shown in Figure 7.6 together with accurate results for these gases. It is seen that $\lim_{P \to 0}(\partial z/\partial P)_T$ at 400 K is positive for H_2, negative for ethane, and approximately zero for O_2. These trends are correctly predicted by the van der Waals equation. How can this behavior be explained? It is useful to compare the shape of the curves for H_2 and ethene curves with those in Figure 7.5 for N_2 at different temperatures. This comparison suggests that at 400 K the gas temperature is well above T_B for H_2 so that the repulsive part of the potential dominates. By contrast, it suggests that 400 K is well be-

FIGURE 7.6

The compression factor is shown as a function of pressure for $T = 400$ K for three different gases. The solid lines have been calculated using the van der Waals equation of state. The dots are calculated from accurate values for V_m taken from the *NIST Chemistry Webbook.*

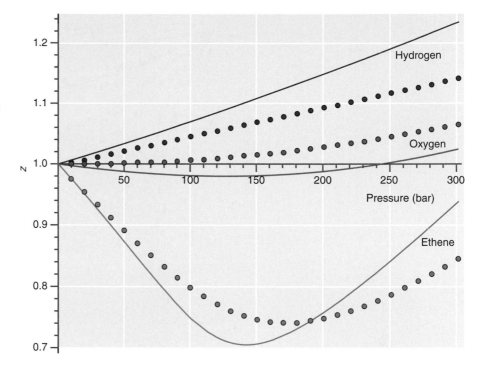

TABLE 7.3

Boyle Temperatures of Selected Gases

Gas	T_B (K)	Gas	T_B (K)
He	23	O_2	400
H_2	110	CH_4	510
Ne	122	Kr	575
N_2	327	Ethene	735
CO	352	H_2O	1250

Source: Calculated from data in Lide, D. R., Ed., *CRC Handbook of Thermophysical and Thermochemical Data.* CRC Press, Boca Raton, FL, 1994.

low T_B for ethene, and the attractive part of the potential dominates. For O_2, $T_B = 400$ K, and as is seen in Table 7.3, $T_B = 735$ K for ethene, and 110 K for H_2.

This result can be generalized. If $\lim_{P \to 0} (\partial z/\partial P)_T < 0$ for a particular gas, $T < T_B$, and the attractive part of the potential dominates. If $\lim_{P \to 0} (\partial z/\partial P)_T > 0$ for a particular gas, $T > T_B$, and the repulsive part of the potential dominates.

7.4 The Law of Corresponding States

As shown in Section 7.3, the compression factor is a convenient way to quantify deviations from the ideal gas law. In calculating z using the van der Waals and Redlich-Kwong equations of state, different parameters must be used for each gas. Is it possible to find an equation of state for real gases that does not explicitly contain material-dependent parameters? Real gases differ from one another primarily in the value of the molecular volume and in the depth of the attractive potential. Because molecules that have a stronger attractive interaction exist as liquids to higher temperatures, one might think that the critical temperature is a measure of the depth of the attractive potential. Similarly, one might think that the critical volume is a measure of the molecular volume. If this is the case, different gases should behave similarly if T, V, and P are measured relative to their critical values.

These considerations suggest the following hypothesis. Different gases have the same equation of state if each gas is described by the dimensionless reduced variables $T_r = T/T_c$, $P_r = P/P_c$, and $V_{mr} = V_m/V_{mc}$, rather than by T, P, and V_m. The preceding statement is known as the **law of corresponding states.** If two gases have the same values of T_r, P_r, and V_{mr}, they are in corresponding states. The values of P, V, and T can be very different for two gases that are in corresponding states. For example, H_2 at 12.93 bar and 32.98 K in a volume of 64.2×10^{-3} L and Br_2 at 103 bar and 588 K in a volume of 127×10^{-3} L are in the same corresponding state.

In the following, we justify the law of corresponding states and show that it is obeyed by many gases. The parameters a and b can be eliminated from the van der Waals equation of state by expressing the equation in terms of the reduced variables T_r, P_r, and V_{mr}. This can be seen by writing the van der Waals equation in the form

$$P_r P_c = \frac{RT_r T_c}{V_{mr} V_{mc} - b} - \frac{a}{V_{mr}^2 V_{mc}^2} \tag{7.8}$$

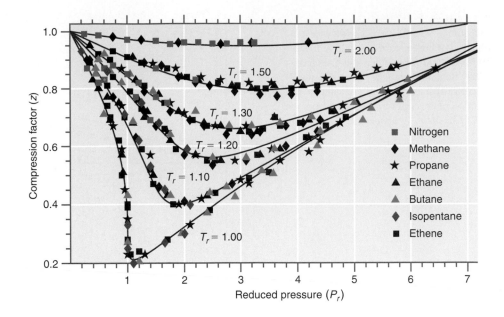

Next replace T_c, V_{mc}, and P_c by the relations derived in Example Problem 7.1:

$$P_c = \frac{a}{27b^2}, \quad V_{mc} = 3b, \quad \text{and} \quad T_c = \frac{8a}{27Rb} \qquad (7.9)$$

Equation (7.8) becomes

$$\frac{aP_r}{27b^2} = \frac{8aT_r}{27b(3bV_{mr} - b)} - \frac{a}{9b^2V_{mr}^2} \quad \text{or} \qquad (7.10)$$

$$P_r = \frac{8T_r}{3V_{mr} - 1} - \frac{3}{V_{mr}^2}$$

Equation (7.10) relates T_r, P_r, and V_{mr} without reference to the parameters a and b. Therefore, it has the character of a universal equation, like the ideal gas equation of state. To a good approximation, the law of corresponding states is obeyed for a large number of different gases as shown in Figure 7.7, as long as $T_r > 1$. However, Equation (7.10) is not really universal because the material-dependent quantities enter through values of P_c, T_c, and V_c rather than through a and b.

The law of corresponding states implicitly assumes that two parameters are sufficient to describe an intermolecular potential. This assumption is best for molecules that are nearly spherical, because for such molecules the potential is independent of the molecular orientation. It is not nearly as good for dipolar molecules such as HF, for which the potential is orientation dependent.

The results shown in Figure 7.7 demonstrate the validity of the law of corresponding states. How can this result be generalized and applied to a specific gas? The goal is to calculate z for specific values of P_r and T_r, and to use these values to estimate the error in using the ideal gas law. A convenient way to display these results is in the form of a graph. Calculated results for z, using the van der Waals equation of state as a function of P_r for different values of T_r, are shown in Figure 7.8. For a given gas and specific P and T values, P_r and T_r can be calculated. A value of z can then be read from the curves in Figure 7.8. What general trends can be predicted using the calculated curves?

From Figure 7.8, we see that for $P_r < 5.5$, $z < 1$ as long as $T_r < 2$. This means that the real gas exerts a smaller pressure than an ideal gas in this range of T_r and P_r. We

FIGURE 7.8
Compression factor, z, as a function of P_r for the T_r values indicated. The curves were calculated using the van der Waals equation of state.

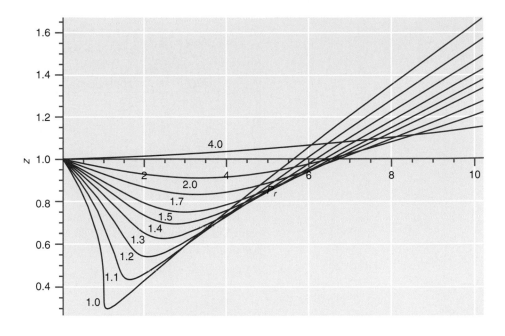

conclude that for these values of T_r and P_r, the molecules are more influenced by the attractive part of the potential than the repulsive part that arises from the finite molecular volume. However, $z > 1$ for $P_r > 7$ for all values of T_r as well as for all values of P_r if $T_r > 4$. Under these conditions, the real gas exerts a larger pressure than an ideal gas. The molecules are more influenced by the repulsive part of the potential than the attractive part.

Using the compression factor, the error in assuming that the pressure can be calculated using the ideal gas law can be defined by

$$\text{Error} = 100\frac{z-1}{z} \qquad (7.11)$$

Figure 7.8 shows that the ideal gas law is in error by less than 30% in the range of T_r and P_r where the repulsive part of the potential dominates if $P_r < 8$. However, the error can be as great as -300% in the range of T_r and P_r where the attractive part of the potential dominates. The error is greatest near $T_r = 1$, because at this value of the reduced temperature, the liquid and gaseous phases can coexist. At or slightly above the critical temperature, a real gas is much more compressible than an ideal gas. These curves can be used to estimate the temperature range over which the molar volume for $H_2O(g)$ predicted by the ideal gas law is within 10% of the result predicted by the van der Waals equation of state over a wide range of pressure. From Figure 7.8, this is true only if $T_r > 2.0$, or if $T > 1300$ K.

EXAMPLE PROBLEM 7.3

Using Figure 7.8 and the data in Table 7.2 (see Appendix A, Data Tables), calculate the volume occupied by 1.00 kg of CH_4 gas at $T = 230$ K and $P = 68.0$ bar. Calculate $V - V_{ideal}$ and the relative error in V if V were calculated from the ideal gas equation of state.

Solution
From Table 7.2, T_r and P_r can be calculated:

$$T_r = \frac{230\ K}{190.56\ K} = 1.21 \quad \text{and} \quad P_r = \frac{68.0\ bar}{45.99\ bar} = 1.48$$

From Figure 7.8, $z = 0.63$.

$$V = \frac{znRT}{P} = \frac{0.70 \times \dfrac{1000 \text{ g}}{16.04 \text{ g mol}^{-1}} \times 0.08314 \text{ L bar K}^{-1} \text{ mol}^{-1} \times 230 \text{ K}}{68.0 \text{ bar}} = 11.0 \text{ L}$$

$$V - V_{ideal} = 11.0 \text{ L} - \frac{11.0 \text{ L}}{0.70} = -6.5 \text{ L}$$

$$\frac{V - V_{ideal}}{V} = -\frac{6.5 \text{ L}}{11.0 \text{ L}} = -58\%$$

Because the critical variables can be expressed in terms of the parameters a and b as shown in Example Problem 7.1, the compression factor at the critical point can also be calculated. For the van der Waals equation of state,

$$z_c = \frac{P_c V_c}{RT_c} = \frac{1}{R} \times \frac{a}{27b^2} \times 3b \times \frac{27Rb}{8a} = \frac{3}{8} \tag{7.12}$$

Equation (7.12) predicts that the critical compressibility is independent of the parameters a and b and should, therefore, have the same value for all gases. A comparison of this prediction with the experimentally determined value of z_c in Table 7.2 shows qualitative, but not quantitative agreement. A similar analysis using the critical parameters obtained from the Redlich-Kwong equation also predicts that the critical compression factor should be independent of a and b. In this case, $z_c = 0.333$. This value is in better agreement with the values listed in Table 7.2 than that calculated using the van der Waals equation.

7.5 Fugacity and the Equilibrium Constant for Real Gases

As shown in the previous section, the pressure exerted by a real gas can be greater or less than that for an ideal gas. We next discuss how this result affects the value of the equilibrium constant for a mixture of reactive gases. For a pure ideal gas, the chemical potential as a function of the pressure has the form (see Section 6.3)

$$\mu(T,P) = \mu°(T) + RT \ln \frac{P}{P°} \tag{7.13}$$

To construct an analogous expression for a real gas, we write

$$\mu(T,P) = \mu°(T) + RT \ln \frac{f}{f°} \tag{7.14}$$

where the quantity f is called the **fugacity** of the gas. The fugacity can be viewed as the effective pressure that a real gas exerts. For densities corresponding to the attractive range of the intermolecular potential, $G_m^{real} < G_m^{ideal}$ and $f < P$. For densities corresponding to the repulsive range of the intermolecular potential, $G_m^{real} > G_m^{ideal}$ and $f > P$. These relationships are depicted in Figure 7.9.

The fugacity has the limiting behavior that $f \to P$ as $P \to 0$. The standard state of fugacity, denoted $f°$, is defined as the value that the fugacity would have if the gas behaved ideally at 1 bar pressure. This is equivalent to saying that $f° = P°$. This standard state is a hypothetical standard state, because a real gas will not exhibit ideal behavior at a pressure of one bar. However, the standard state defined in this way makes Equation (7.14) become identical to Equation (7.13) in the ideal gas limit $f \to P$ as $P \to 0$.

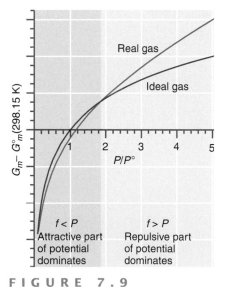

FIGURE 7.9
For densities corresponding to the attractive range of the potential, $P < P_{ideal}$. Therefore, $G_m^{real} < G_m^{ideal}$ and $f < P$. The inequalities are reversed for densities corresponding to the repulsive range of the potential.

The preceding discussion provides a method to calculate the fugacity for a given gas. How are the fugacity and the pressure related? For any gas, real or ideal, at constant T,

$$dG_m = V_m dP \tag{7.15}$$

Therefore,

$$d\mu_{ideal} = V_m^{ideal} dP \tag{7.16}$$

$$d\mu_{real} = V_m^{real} dP$$

Equation (7.16) shows that because $V_m^{ideal} \neq V_m^{real}$, the chemical potential of a real gas will change differently with pressure than the chemical potential of an ideal gas. We form the difference

$$d\mu_{real} - d\mu_{ideal} = (V_m^{real} - V_m^{ideal})dP \tag{7.17}$$

and integrate Equation (7.17) from an initial pressure P_i to a final pressure P:

$$\int_{P_i}^{P} (d\mu_{real} - d\mu_{ideal}) = [\mu_{real}(P) - \mu_{real}(P_i)] - [\mu_{ideal}(P) - \mu_{ideal}(P_i)] \tag{7.18}$$

$$= \int_{P_i}^{P} (V_m^{real} - V_m^{ideal})dP'$$

The previous equations allows us to calculate the difference in chemical potential between a real and an ideal gas at pressure P. Now let $P_i \rightarrow 0$. In this limit, $\mu_{real}(P_i) = \mu_{ideal}(P_i)$ because all real gases approach ideal gas behavior at a sufficiently low pressure. Equation (7.18) becomes

$$\mu_{real}(P) - \mu_{ideal}(P) = \int_{0}^{P} (V_m^{real} - V_m^{ideal})dP' \tag{7.19}$$

Equation (7.19) provides a way to calculate the fugacity of a real gas. Using Equations (7.13) and (7.14) for $\mu_{real}(P)$ and $\mu_{ideal}(P)$, P and f can be related by

$$\ln f = \ln P + \frac{1}{RT}\int_{0}^{P}(V_m^{real} - V_m^{ideal})dP' \tag{7.20}$$

Because tabulated values of the compression factor z of real gases are widely available, it is useful to rewrite Equation (7.20) in terms of z by substituting $z = V_m^{real}/V_m^{ideal}$. The result is

$$\ln f = \ln P + \int_{0}^{P}\frac{z-1}{P'}dP' \quad \text{or} \quad f = P\exp\left[\int_{0}^{P}\left(\frac{z-1}{P'}\right)dP\right] \quad \text{or} \quad f = \gamma(P,T)P \tag{7.21}$$

Equation (7.21) provides a way to calculate the fugacity if z is known as a function of pressure. It is seen that f and P are related by the proportionality factor, γ, which is called the **fugacity coefficient.** However, γ is not a constant; it depends on both P and T.

7.4 Fugacity and the Fugacity Coefficient

EXAMPLE PROBLEM 7.4

For $T > T_B$, the equation of state $P(V_m - b) = RT$ is an improvement over the ideal gas law because it takes the finite volume of the molecules into account. Derive an expression for the fugacity coefficient for a gas that obeys this equation of state.

Solution

Because $z = PV_m/RT = 1 + Pb/RT$, the fugacity coefficient is given by

$$\ln \gamma = \ln \frac{f}{P} = \int_0^P \frac{z-1}{P'} dP' = \int_0^P \frac{b}{RT} dP' = \frac{bP}{RT}$$

Equivalently, $\gamma = e^{bP/RT}$. Note that $\gamma > 1$ for all values of P and T because the equation of state does not take the attractive part of the intermolecular potential into account.

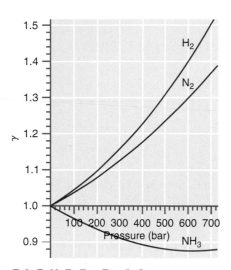

FIGURE 7.10

The fugacity coefficients for H_2, N_2, and NH_3 are plotted as a function of the partial pressure of the gases for $T = 700$ K. The calculations were carried out using the Beattie-Bridgeman equation of state.

Calculated values for γ (P,T) for H_2, N_2, and NH_3 are shown in Figure 7.10 for $T = 700$ K. It can be seen that $\gamma \to 1$ as $P \to 0$. This must be the case because the fugacity and pressure are equal in the ideal gas limit. The following general statements can be made about the relationship between γ and z: γ $(P,T) > 1$ if the integrand of Equation (7.21) satisfies the condition $(z-1)/P > 0$ for all pressures up to P. Similarly, γ $(P,T) < 1$ if $(z-1)/P < 0$ for all pressures up to P.

These predictions can be related to the Boyle temperature. If $T > T_B$, then γ $(P,T) > 1$ for all pressures, the fugacity is greater than the pressure, and $\gamma > 1$. However, if $T < T_B$, then γ $(P,T) < 1$, and the fugacity is smaller than the pressure. The last statement does not hold for very high values of P_r, because as shown in Figure 7.8, $z > 1$ for all values of T at such high relative pressures. Are these conclusions in accord with the curves in Figure 7.10? The Boyle temperatures for H_2, N_2, and NH_3 are 110, 327, and 995 K, respectively. Because at 700 K, $T > T_B$ for H_2 and N_2, but $T < T_B$ for NH_3, we conclude that $\gamma > 1$ at all pressures for H_2 and N_2, and $\gamma < 1$ at all but very high pressures for NH_3. These conclusions are consistent with the results shown in Figure 7.10.

The fugacity coefficient can also be graphed as a function of T_r and P_r. This is convenient because it allows γ for any gas to be estimated once T and P have been expressed as reduced variables. Graphs of γ as functions of P_r and T_r are shown in Figure 7.11. The curves have been calculated using Beattie-Bridgeman parameters for N_2, and are known to be accurate for N_2 over the indicated range of reduced temperature and pressure. Their applicability to other gases assumes that the law of corresponding states holds. As Figure 7.7 shows, this is generally a good assumption.

What are the consequences of the fact that except in the dilute gas limit $f \neq P$? Because the chemical potential of a gas in a reaction mixture is given by Equation (7.14),

FIGURE 7.11

The fugacity coefficient is plotted as a function of the reduced pressure for the indicated values of the reduced temperature.

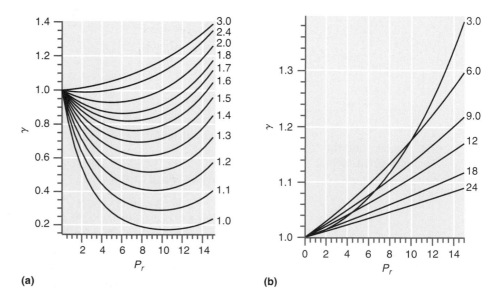

(a)   (b)

the thermodynamic equilibrium constant for a real gas, K_f, must be expressed in terms of the fugacities. Therefore K_f for the reaction $3/2H_2 + 1/2N_2 \rightarrow NH_3$ is given by

$$K_f = \frac{\left(\dfrac{f_{NH_3}}{f^\circ}\right)}{\left(\dfrac{f_{N_2}}{f^\circ}\right)^{1/2}\left(\dfrac{f_{H_2}}{f^\circ}\right)^{3/2}} = \frac{\left(\dfrac{\gamma_{NH_3}P_{NH_3}}{P^\circ}\right)}{\left(\dfrac{\gamma_{N_2}P_{N_2}}{P^\circ}\right)^{1/2}\left(\dfrac{\gamma_{H_2}P_{H_2}}{P^\circ}\right)^{3/2}} = K_P\frac{\gamma_{NH_3}}{(\gamma_{N_2})^{1/2}(\gamma_{H_2})^{3/2}} \qquad (7.22)$$

We next calculate the error in using K_P rather than K_f to calculate an equilibrium constant, using the ammonia synthesis reaction at 700 K and a total pressure of 400 bar as an example. In the industrial synthesis of ammonia, the partial pressures of H_2, N_2, and NH_3 are typically 270, 90, and 40 bar, respectively. The calculated fugacity coefficients under these conditions are $\gamma_{H_2} = 1.11$, $\gamma_{N_2} = 1.04$, and $\gamma_{NH_3} = 0.968$ using the Beattie-Bridgeman equation of state. Therefore,

$$K_f = K_P\frac{(0.968)}{(1.04)^{1/2}(1.11)^{3/2}} = 0.917\, K_P \qquad (7.23)$$

Note that K_f is smaller than K_p for the conditions of interest. The NH_3 concentration at equilibrium is proportional to the equilibrium constant, as shown in Section 6.14.

$$x_{NH_3} = (x_{N_2})^{1/2}(x_{H_2})^{3/2}\left(\frac{P}{P^\circ}\right)K_f \qquad (7.24)$$

What conclusions can be drawn from this calculation? If K_P were used rather than K_f, the calculated mole fraction of ammonia would be in error by -9%, which is a significant error in calculating the economic viability of a production plant. However, for pressures near 1 bar, activity coefficients for most gases are very close to unity as can be seen in Figure 7.11. Therefore, under typical laboratory conditions, the fugacity of a gas can be set equal to its partial pressure if P, V, and T are not close to their critical values.

Vocabulary

Beattie-Bridgeman equation of state	fugacity	van der Waals equation of state
Boyle temperature	fugacity coefficient	vapor pressure
compression factor	law of corresponding states	virial equation of state
critical constants	Maxwell construction	
critical temperature	Redlich-Kwong equation of state	

Questions on Concepts

Q7.1 Explain why the oscillations in the two-phase coexistence region using the Redlich-Kwong and van der Waals equations of state (see Figure 7.4) do not correspond to reality.

Q7.2 Explain the significance of the Boyle temperature.

Q7.3 The value of the Boyle temperature increases with the strength of the attractive interactions between molecules. Arrange the Boyle temperatures of the gases Ar, CH_4, and C_6H_6 in increasing order.

Q7.4 Will the fugacity coefficient of a gas above the Boyle temperature be less than one at low pressures?

Q7.5 Using the concept of the intermolecular potential, explain why two gases in corresponding states can be expected to have the same value for z.

Q7.6 By looking at the a and b values for the van der Waals equation of state, decide whether 1 mole of O_2 or H_2O has the higher pressure at the same value of T and V.

Q7.7 Consider the comparison made between accurate results and those based on calculations using the van der Waals and Redlich-Kwong equations of state in Figures 7.1 and 7.5. Is it clear that one of these equations of state is better than the other under all conditions?

Q7.8 Why is the standard state of fugacity, $f°$, equal to the standard state of pressure, $P°$?

Q7.9 For a given set of conditions, the fugacity of a gas is greater than the pressure. What does this tell you about the interaction between the molecules of the gas?

Q7.10 A system containing argon gas is at pressure P_1 and temperature T_1. How would you go about estimating the fugacity coefficient of the gas?

Problems

Problem numbers in **red** indicate that the solution to the problem is given in the *Student's Solutions Manual*.

P7.1 A sample containing 35.0 g of Ar is enclosed in a container of volume 0.165 L at 390 K. Calculate P using the ideal gas, van der Waals, and Redlich-Kwong equations of state. Based on your results, does the attractive or repulsive contribution to the interaction potential dominate under these conditions?

P7.2 Calculate the density of $O_2(g)$ at 375 K and 385 bar using the ideal gas and the van der Waals equations of state. Use a numerical equation solver to solve the van der Waals equation for V_m or use an iterative approach starting with V_m equal to the ideal gas result. Based on your result, does the attractive or repulsive contribution to the interaction potential dominate under these conditions?

P7.3 At 500 K and 400 bar, the experimentally determined density of N_2 is 7.90 mol L^{-1}. Compare this with values calculated from the ideal and Redlich-Kwong equations of state. Use a numerical equation solver to solve the Redlich-Kwong equation for V_m or use an iterative approach starting with V_m equal to the ideal gas result. Discuss your results.

P7.4 The observed Boyle temperatures of H_2, N_2, and CH_4 are 110, 327, and 510 K, respectively. Compare these values with those calculated for a van der Waals gas with the appropriate parameters.

P7.5 Calculate the van der Waals parameters of methane from the values of the critical constants.

P7.6 Calculate the Redlich-Kwong parameters of methane from the values of the critical constants.

P7.7 Use the law of corresponding states and Figure 7.8 to estimate the molar volume of methane at $T = 285$ K and $P = 180$ bar.

P7.8 Calculate the P and T values for which $H_2(g)$ is in a corresponding state to $Xe(g)$ at 450 K and 85.0 bar.

P7.9 Assume that the equation of state for a gas can be written in the form $P(V_m - b(T)) = RT$. Derive an expression for $\beta = 1/V (\partial V/\partial T)_P$ and $\kappa = -1/V (\partial V/\partial P)_T$ for such a gas in terms of $b(T)$, $db(T)/dT$, P, and V_m.

P7.10 One mole of Ar initially at 298 K undergoes an adiabatic expansion against a pressure $P_{external} = 0$ from a volume of 20.0 L to a volume of 65.0 L. Calculate the final temperature using the ideal gas and van der Waals equations of state.

P7.11 One mole of Ar undergoes an isothermal reversible expansion from an initial volume of 1.00 L to a final volume of 65.0 L at 298 K. Calculate the work done in this process using the ideal gas and van der Waals equations of state. What percentage of the work done by the van der Waals gas arises from the attractive potential?

P7.12 For a van der Waals gas, $z = (V_m/V_m - b) - a/RTV_m$. Expand the first term of this expression in a Taylor series in the limit $V_m \gg b$ to obtain $z \approx 1 + (b - a/RT)(1/V_m)$.

P7.13 Show that $T\beta = 1 + T(\partial \ln z/\partial T)_P$ and that $P\kappa = 1 - P(\partial \ln z/\partial P)_T$.

P7.14 A van der Waals gas has a value of $z = 1.00084$ at 298 K and 1 bar and the Boyle temperature of the gas is 125 K. Because the density is low, you can calculate V_m from the ideal gas law. Use this information and the result of Problem 7.12 to estimate a and b.

P7.15 The experimental critical constants of H_2O are $T_c = 647.14$ K, $P_c = 220.64$ bar, and $V_c = 55.95 \times 10^{-3}$ L. Use the values of P_c and T_c to calculate V_c. Assume that H_2O behaves as (a) an ideal gas, (b) a van der Waals gas, and (c) a Redlich-Kwong gas at the critical point. For parts (b) and (c), use the formulas for the critical compression factor. Compare your answers with the experimental value.

P7.16 Another equation of state is the Bertholet equation, $V_m = (RT/P) + b - (a/RT^2)$. Derive expressions for $\beta = 1/V(\partial V/\partial T)_P$ and $\kappa = -1/V (\partial V/\partial P)_T$ from the Bertholet equation in terms of V, T, and P.

P7.17 For the Bertholet equation, $V_m = (RT/P) + b - (a/RT^2)$, find an expression for the Boyle temperature in terms of a, b, and R.

P7.18 The experimentally determined density of H_2O at 1200 bar and 800 K is 537 g L^{-1}. Calculate z and V_m from this information. Compare this result with what you would have estimated from Figure 7.8. What is the relative error in using Figure 7.8 for this case?

P7.19 The volume of a spherical molecule can be estimated as $V = b/4N_A$ where b is the van der Waals parameter and N_A is Avogadro's number. Justify this relationship by considering a spherical molecule of radius r, with volume $V = 4/3\,\pi r^3$. What is the volume centered at the molecule that is excluded for the center of mass of a second molecule in terms of V? Multiply this volume by N_A and set it equal to b. Apportion this volume equally among the molecules to arrive at $V = b/4N_A$. Calculate the radius of a methane molecule from the value of its van der Waals parameter b.

P7.20 At what temperature does the slope of the z versus P curve as $P \rightarrow 0$ have its maximum value for a van der Waals gas? What is the value of the maximum slope?

P7.21 Show that the van der Waals and Redlich-Kwong equations of state reduce to the ideal gas equation of state in the limit of low density.

P7.22 Show that the second virial coefficient for a van der Waals gas is given by

$$B(T) = \frac{1}{RT}\left(\frac{\partial z}{\partial \frac{1}{V_m}}\right)_T = b - \frac{a}{RT}$$

P7.23 For a gas at a given temperature, the compressibility is described by the empirical equation

$$z = 1 - 9.00 \times 10^{-3}\frac{P}{P^\circ} + 4.00 \times 10^{-5}\left(\frac{P}{P^\circ}\right)^2$$

where $P^\circ = 1$ bar. Calculate the activity coefficient for $P = 100, 200, 300, 400$, and 500 bar. For which of these values is the activity coefficient greater than one?

P7.24 For values of z near one, it is a good approximation to write $z(P) = 1 + (\partial z/\partial P)_T P$. If $z = 1.00054$ at 0°C and 1 bar, and the Boyle temperature of the gas is 220 K, calculate the values of V_m, a and b for the van der Waals gas.

P7.25 Calculate the critical volume for ethane using the data for T_c and P_c in Table 7.2 (see Appendix A, Data Tables) assuming (a) the ideal gas equation of state and (b) the van der Waals equation of state. Use an iterative approach to obtain V_c from the van der Waals equation, starting with the ideal gas result. How well do the calculations agree with the tabulated values for V_c?

Web-Based Simulations, Animations, and Problems

W7.1 In this problem, the student gains facility in using the van der Waals equation of state. A set of isotherms will be generated by varying the temperature and initial volume using sliders for a given substance. Buttons allow a choice among more than 20 gases.

a. The student generates a number of P–V curves at and above the critical temperature for the particular gas, and explains trends in the ratio P_{vdW}/P_{ideal} as a function of V for a given T, and as a function of T for a given V.

b. The compression factor z is calculated for two gases at the same value of T_r and is graphed versus P and P_r. The degree to which the law of corresponding states is valid is assessed.

W7.2 A quantitative comparison is made between the ideal gas law and the van der Waals equation of state for one of more than 20 different gases. The temperature is varied using sliders and P_{ideal}, P_{vdW}, the relative error $P_{vdW}/P_{vdW} - P_{ideal}$, and the density of gas relative to that at the critical point are calculated. The student is asked to determine the range of

pressures and temperatures in which the ideal gas law gives reasonably accurate results.

W7.3 The compression factor and molar volume are calculated for an ideal and a van der Waals gas as a function of pressure and temperature. These variables can be varied using sliders. Buttons allow a choice among more than 20 gases. The relative error $V_{vdW}/V_{vdW} - V_{ideal}$ and the density of gas relative to that at the critical point are calculated. The student is asked to determine the range of pressures and temperatures in which the ideal gas law gives reasonably accurate results for the molar volume.

W7.4 The fugacity and fugacity coefficient are determined as a function of pressure and temperature for a model gas. These variables can be varied using sliders. The student is asked to determine the Boyle temperature and also the pressure range in which the fugacity is either more or less than the ideal gas pressure for the temperature selected.

CHAPTER 8

Phase Diagrams and the Relative Stability of Solids, Liquids, and Gases

It is our experience that the solid form of matter is favored at low temperatures, and that most substances can exist in liquid and gaseous phases at higher temperatures. In this chapter, criteria are developed that allow one to determine which of these phases is favored at a given temperature and pressure. The conditions under which two or three phases of a pure substance can coexist in equilibrium are also discussed, as well as the unusual properties of supercritical fluids. P–T, P–V, and P–V–T phase diagrams summarize all of this information in a form that is very useful to chemists. ∎

8.1 What Determines the Relative Stability of the Solid, Liquid, and Gas Phases?

Substances are found in solid, liquid, and gaseous phases. **Phase** refers to a form of matter that is uniform with respect to chemical composition and the state of aggregation on both microscopic and macroscopic length scales. For example, liquid water in a beaker is a single-phase system, but a mixture of ice in liquid water consists of two distinct phases, each of which is uniform on microscopic and macroscopic length scales. Although a substance may exist in several different solid phases, it can only exist in a single gaseous state. Most substances have a single liquid state, although there are exceptions such as helium, which can be a normal liquid or a superfluid. In this section, the conditions under which a pure substance spontaneously forms a solid, liquid, or gas are discussed.

Experience demonstrates that as T is lowered from 300 to 250 K at atmospheric pressure, liquid water is converted to the solid phase. Similarly, as liquid water is heated to 400 K at atmospheric pressure, it vaporizes to form a gas. Experience also shows that if a solid block of carbon dioxide is placed in an open container at 1 bar, it sublimes over time without passing through a liquid phase. Because of this property, solid CO_2 is known as dry ice. These observations can be generalized to state that the solid phase is the most stable state of a substance at sufficiently low temperatures, and that the gas phase is the most stable state of a substance at sufficiently high temperatures. The liquid state is stable at intermediate temperatures if it exists at the pressure of interest. What determines which of the solid, liquid, or gas phases is most stable at a given temperature and pressure?

As discussed in Chapter 6, the criterion for stability at constant temperature and pressure is that the Gibbs energy, $G(T,P, n)$, be minimized. Because for a pure substance,

$$\mu = \left(\frac{\partial G}{\partial n}\right)_{T,P} = \left(\frac{\partial [nG_m]}{\partial n}\right)_{T,P} = G_m$$

where n designates the number of moles of substance in the system, $d\mu = dG_m$, and we can express the differential $d\mu$ as

$$d\mu = -S_m dT + V_m dP \tag{8.1}$$

which is identical in content to Equation (6.19). From this equation, how μ varies with changes in P and T can be determined:

$$\left(\frac{\partial \mu}{\partial T}\right)_P = -S_m \text{ and } \left(\frac{\partial \mu}{\partial P}\right)_T = V_m \tag{8.2}$$

Because S_m and V_m are always positive, μ decreases as the temperature increases, and it increases as the pressure increases. Section 5.4 demonstrated that S varies slowly with T (as $\ln T$). Therefore, over a limited range in T, a plot of μ versus T at constant P is a straight line with a negative slope.

It is also known from experience that heat is absorbed as a solid melts to form a liquid, and as a liquid vaporizes to form a gas. Both processes are endothermic. This observation shows that $\Delta S = \Delta H / T$ is positive for both of these reversible constant temperature phase changes. Because the heat capacity is always positive for a solid, liquid, or gas, the entropy of the three phases follows this order:

$$S_m^{gas} > S_m^{liquid} > S_m^{solid} \tag{8.3}$$

The functional relation between μ and T for the solid, liquid, and gas phases is graphed at a given value of P in Figure 8.1. The entropy of a phase is the magnitude of the slope of the μ versus T line, and the relative entropies of the three phases are given by Equation (8.3). The stable state of the system at any given temperature is that phase which has the lowest μ.

Assume that the initial state of the system is described by the dot in Figure 8.1. It can be seen that the most stable phase is the solid phase, because μ for the liquid and gas phases is much larger than that for the solid. As the temperature is increased, the chemical potential falls as μ remains on the solid line. However, because the slope of the liquid and gas lines is greater than that for the liquid, each of these μ versus T lines will intersect the solid line for some value of T. In Figure 8.1, the liquid line intersects the solid line at T_m, which is called the melting temperature. At this temperature, the solid and liquid phases coexist and are in thermodynamic equilibrium. However, if the temperature is raised by an infinitesimal amount dT, the solid will melt completely because the liquid phase has the lower chemical potential at $T_m + dT$. Similarly, the liquid and gas phases are in thermodynamic equilibrium at T_b. For $T > T_b$, the system is entirely in the gas phase. Note that the progression of solid $\to$ liquid $\to$ gas as T increases at this value of P can be explained with no other information than that $(\partial \mu / \partial T)_P = -S_m$ and that $S_m^{gas} > S_m^{liquid} > S_m^{solid}$.

If the temperature is changed too quickly, the equilibrium state of the system may not be reached. For example, it is possible to form a superheated liquid, in which the liquid phase is metastable above T_b. Superheated liquids are dangerous, because of the large volume increase that occurs if the system suddenly converts to the stable vapor phase. Boiling chips are often used in chemical laboratories to avoid the formation of superheated liquids. Similarly, it is possible to form a supercooled liquid, in which case the liquid is metastable below T_m. Glasses are made by cooling a viscous liquid fast enough to avoid crystallization. These disordered materials lack the periodicity of crystals but behave mechanically like solids. Seed crystals can be used to avoid supercooling if the viscosity of the liquid is not too high and the cooling rate is sufficiently slow. Liquid crystals, which can be viewed as a state of matter intermediate between a solid and a liquid, are discussed in Section 8.8.

In Figure 8.1, we consider changes with T at constant P. How is the relative stability of the three phases affected if P is changed at constant T? From Equation (8.2), $(\partial \mu / \partial P)_T = V_m$ and $V_m^{gas} \gg V_m^{liquid}$. For most substances, $V_m^{liquid} > V_m^{solid}$. Therefore, the μ versus T line for

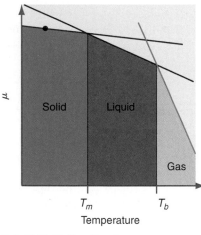

FIGURE 8.1

The chemical potential of a substance in the solid (blue line), liquid (red line), and gaseous (orange line) states is plotted as a function of the temperature for a given value of pressure. The substance melts at the temperature T_m, corresponding to the intersection of the solid and liquid lines. It boils at the temperature T_b, corresponding to the intersection of the liquid and gas lines. The temperature ranges in which the different phases are the most stable are indicated by shaded areas.

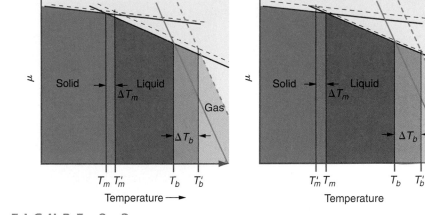

FIGURE 8.2

The solid lines show μ as a function of temperature for all three phases at $P = P_1$. The dashed lines show the same information for $P = P_2$, where $P_2 > P_1$. The unprimed temperatures refer to $P = P_1$ and the primed temperatures refer to $P = P_2$. The left diagram applies if $V_m^{liquid} > V_m^{solid}$. The right diagram applies if $V_m^{liquid} < V_m^{solid}$. The shifts in the solid and liquid lines are greatly exaggerated. The colored areas correspond to the temperature range in which the phases are most stable. The shaded area between T_m and T_m' is either solid or liquid, depending on P. The shaded area between T_b and T_b' is either liquid or gas, depending on P.

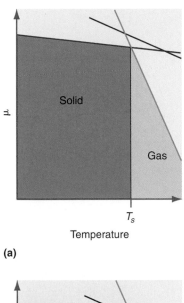

(a)

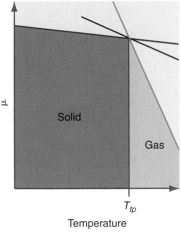

(b)

FIGURE 8.3

The chemical potential of a substance in the solid (blue line), liquid (brown line), and gaseous (orange line) states is plotted as a function of the temperature for a fixed value of pressure. **(a)** The pressure lies below the triple point pressure, and the solid sublimes. **(b)** The pressure corresponds to the triple point pressure. At T_{tp}, all three phases coexist in equilibrium. The colored areas correspond to the temperature range in which the phases are the most stable. The liquid phase is not stable in part (a), and is only stable at the single temperature T_{tp}, in part (b).

the gas changes much more rapidly (by a factor of ~1000) with pressure than the liquid and solid lines. This behavior is illustrated in Figure 8.2, where it can be seen that the point at which the solid and the liquid lines intersect shifts as the pressure is increased. Because $V_m^{gas} \gg V_m^{liquid} > 0$, an increase in the pressure always leads to a **boiling point elevation.** An increase in the pressure leads to a **freezing point elevation** if $V_m^{liquid} > V_m^{solid}$ and to a **freezing point depression** if $V_m^{liquid} < V_m^{solid}$, as is the case for water. Few substances obey the relation $V_m^{liquid} < V_m^{solid}$ and the consequences of this unusual behavior for water will be discussed in Section 8.2.

The μ versus T line for a gas changes much more rapidly with P than the liquid and solid lines. As a consequence, changes in P can change the way in which a system progresses through the phases with increasing T from the "normal" order solid $\rightarrow$ liquid $\rightarrow$ gas shown in Figure 8.1. For example, the sublimation of dry ice at 298 K and 1 bar can be explained using Figure 8.3a. For CO_2 at the given pressure, the μ versus T line for the liquid intersects the corresponding line for the solid at a higher temperature than the gaseous line. Therefore, the solid $\rightarrow$ liquid transition is energetically unfavorable with respect to the solid $\rightarrow$ gas transition at this pressure. Under these conditions, the solid sublimes and the transition temperature T_s is called the **sublimation temperature.** There is also a pressure at which the μ versus T lines for all three phases intersect. The P, V_m, T values for this point specify the **triple point,** so named because all three phases coexist in equilibrium at this point. This case is shown in Figure 8.3b. Triple point temperatures for a number of substances are listed in Table 8.1 (see Appendix A, Data Tables).

8.2 The Pressure–Temperature Phase Diagram

As shown in the previous section, at a given value of pressure and temperature, a system containing a pure substance may consist of a single phase, two phases in equilibrium, or three phases in equilibrium. This usefulness of a **phase diagram** is that it displays this information graphically. Although any two of the macroscopic system variables P, V, and T can be used to construct a phase diagram, the P–T diagram is particularly useful. In this

A *P–T* phase diagram displays single-phase regions, coexistence curves for which two phases coexist at equilibrium, and a triple point. The processes corresponding to paths *a*, *b*, *c*, and *d* are described in the text. Two solid–liquid coexistence curves are shown. For most substances, the solid line, which has a positive slope, is observed. For water, the red dashed line corresponding to a negative slope is observed.

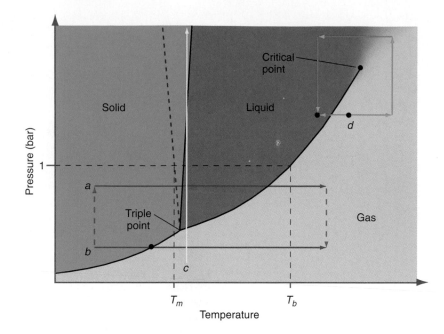

section, the features of a *P–T* phase diagram are discussed that are common to pure substances. Phase diagrams are generally determined experimentally, because material-specific forces between atoms determine the temperatures and pressures at which different phases are stable. Increasingly, large-scale calculations have become sufficiently accurate that major features of phase diagrams can be obtained using microscopic theoretical models. However, as shown in Section 8.3, thermodynamics can say a great deal about the phase diagram without considering the microscopic properties of the system.

The ***P–T* phase diagram,** a sample of which is shown in Figure 8.4, displays stability regions for a pure substance as a function of pressure and temperature. Most *P,T* points correspond to a single solid, liquid, or gas phase. At the triple point, all three phases coexist. The triple point of water is 273.16 K and 611 Pa. All *P,T* points for which the same two phases coexist at equilibrium fall on a curve. Such a curve is called a **co-existence curve.** Three separate coexistence curves are shown in Figure 8.4, corresponding to solid–gas, solid–liquid, and gas–solid coexistence. As shown in Section 8.4, the slopes of the solid–gas and liquid–gas curves are always positive. The slope of the solid–liquid curve can be either positive or negative.

The boiling point of a substance is defined as the temperature at which the vapor pressure of the substance is equal to the external pressure. The **standard boiling temperature** is the temperature at which the vapor pressure of the substance is 1 bar. The **normal boiling temperature** is the temperature at which the vapor pressure of the substance is 1 atm. Values of the normal boiling and freezing temperatures for a number of substances are shown in Table 8.2. Because 1 bar is slightly less than 1 atm, the standard boiling temperature is slightly less than the normal boiling temperature. Along two-phase curves in which one of the coexisting phases is the gas, *P* refers to the **vapor pressure** of the substance. In all regions, *P* refers to the hydrostatic pressure that would be exerted on the pure substance if it were confined in a piston and cylinder assembly.

The solid–liquid coexistence curve traces out the melting point as a function of pressure. The magnitude of the slope of this curve is large, as proven in Section 8.4. Therefore, T_m is a weak function of the pressure. The slope of this curve is positive, and the melting temperature increases with pressure if the solid is more dense than the liquid. This is the case for most substances. The slope is negative and the melting temperature decreases with pressure if the solid is less dense than the liquid. Water is one of the few substances that exhibits this behavior. Imagine the fate of aquatic plants and animals in climate zones where the temperature routinely falls below 0°C in the winter if water be-

TABLE 8.2

Melting and Boiling Temperatures and Enthalpies of Transition at 1 atm Pressure

Substance	Name	mp (K)	ΔH_{fusion} (kJ mol^{-1}) at T_m	bp (K)	$\Delta H_{vaporization}$ (kJ mol^{-1})at T_b
Ar	Argon	83.8	1.12	87.3	6.43
Cl_2	Chlorine	171.6	6.41	239.18	20.41
Fe	Iron	1811	13.81	3023	349.5
H_2	Hydrogen	13.81	0.12	20.4	0.90
H_2O	Water	273.15	6.010	373.15	40.65
He	Helium	0.95	0.021	4.22	0.083
I_2	Iodine	386.8	14.73	457.5	41.57
N_2	Nitrogen	63.5	0.71	77.5	5.57
Na	Sodium	370.87	2.60	1156	98.0
NO	Nitric oxide	109.5	2.3	121.41	13.83
O_2	Oxygen	54.36	0.44	90.7	6.82
SO_2	Sulfur dioxide	197.6		263.1	24.94
Si	Silicon	1687	50.21	2628	359
W	Tungsten	3695	52.31	5933	422.6
Xe	Xenon	161.4	1.81	165.11	12.62
CCl_4	Carbon tetrachloride	250	3.28	349.8	29.82
CH_4	Methane	90.68	0.94	111.65	8.19
CH_3OH	Methanol	175.47	3.18	337.7	35.21
CO	Carbon monoxide	68	0.83	81.6	6.04
C_2H_4	Ethene			169.38	13.53
C_2H_6	Ethane	90.3	2.86	184.5	14.69
C_2H_5OH	Ethanol	159.0	5.02	351.44	38.56
C_3H_8	Propane	85.46	3.53	231.08	19.04
C_5H_5N	Pyridine			388.38	35.09
C_6H_6	Benzene	278.68	9.95	353.24	30.72
C_6H_5OH	Phenol	314.0	11.3	455.02	45.69
$C_6H_5CH_3$	Toluene	178.16	6.85	383.78	33.18
$C_{10}H_8$	Naphthalene	353.3	17.87	491.14	43.18

Sources: Data from Lide, D. R., Ed., *Handbook of Chemistry and Physics,* 83rd ed. CRC Press, Boca Raton, FL, 2002; Lide, D. R., Ed., *CRC Handbook of Thermophysical and Thermochemical Data.* CRC Press, Boca Raton, FL, 1994; and Blachnik, R., Ed., *D'Ans Lax Taschenbuch für Chemiker und Physiker,* 4th ed. Springer, Berlin, 1998.

haved "normally." Lakes would begin to freeze over at the water–air interface, and the ice formed would fall to the bottom of the lake. This would lead to more ice formation until the whole lake was full of ice. In cool climate zones, it is likely that ice at the bottom of the lakes would remain throughout the summer. The aquatic life that we are familiar with would not survive under such a scenario.

The slope of the liquid–gas coexistence curve is much smaller than that of the solid–liquid coexistence curve, as proven in Section 8.4. Therefore, the boiling point is a much stronger function of the pressure than the freezing point. The boiling point always increases with pressure. This property is utilized in a pressure cooker. Increasing the pressure in a pressure cooker by 1 bar increases the boiling temperature of water by approximately 20°C. The rate of the chemical processes involved in cooking increase

FIGURE 8.5
The temperature versus heat curve is shown for the process corresponding to path *a* in Figure 8.4. The temperature rises linearly with q_P in single-phase regions, and remains constant along the two-phase curves as the relative amounts of the two phases in equilibrium change (not to scale).

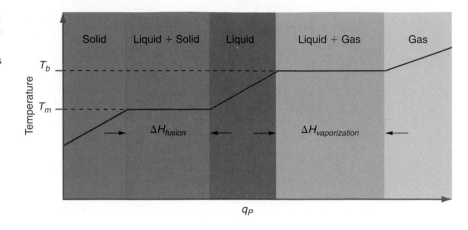

exponentially with T. Therefore, a pressure cooker operating at $P = 2$ bar can cook food in 20% to 40% of the time required for cooking at atmospheric pressure. By contrast, a mountain climber in the Himalayas would find that the boiling temperature of water is reduced by approximately 10°C. Cooking takes significantly longer under these conditions.

Whereas the solid–gas and liquid–solid coexistence curves extend indefinitely, the liquid–gas line ends at the **critical point,** characterized by $T = T_c$ and $P = P_c$. For $T > T_c$ and $P > P_c$, the liquid and gas phases have the same density, so that it is not meaningful to refer to distinct phases. Substances for which $T > T_c$ and $P > P_c$ are called **supercritical fluids**. As discussed in Section 8.8, supercritical fluids have unusual properties that make them useful in chemical technologies.

Each of the paths labeled *a*, *b*, *c*, and *d* in Figure 8.4 corresponds to a process that demonstrates the usefulness of the *P–T* phase diagram. In the following, each process is considered individually. Process *a* follows a constant pressure (isobaric) path. An example of this path is heating ice. The system is initially in the solid single-phase region. Assume that heat is added to the system at a constant rate using current flow through a resistive heater. Because the pressure is constant, $q_P = \Delta H$. Furthermore, as discussed in Section 2.4, $\Delta H \approx C_P \Delta T$ in a single-phase region. Combining these equations, $\Delta T = q_P / C_P^{solid}$. Along path *a*, the temperature increases linearly with q_P in the single-phase solid region as shown in Figure 8.5. At the melting temperature T_m, heat continues to be absorbed by the system as the solid is transformed into a liquid. This system now consists of two distinct phases, solid and liquid. As heat continues to flow into the system, the temperature will not increase until the system consists entirely of liquid. The heat taken up per mol of the system at the constant temperature T_m is ΔH_{fusion}.

The temperature again increases linearly with q_P and $\Delta T = q_P / C_P^{liquid}$ until the boiling point is reached. At this temperature, the system consists of two phases, liquid and gas. The temperature remains constant until all the liquid has been converted into gas. The heat taken up per mol of the system at the constant temperature T_b is $\Delta H_{vaporization}$. Finally, the system enters the single-phase gas region. Along path *b*, the pressure is less than the triple point pressure. Therefore, the liquid phase is not stable, and the solid is converted directly into the gaseous form. As Figure 8.4 shows, there is only one two-phase interval along this path. A diagram indicating the relationship between temperature and heat flow is shown in Figure 8.6.

Note that the initial and final states in process *b* can be reached by an alternative route described by the vertical dashed arrows in Figure 8.4. The pressure of the system in process *b* can be increased to the value of the initial state of process *a* at constant temperature. The process follows that described for process *a*, after which the pressure is returned to the final pressure of process *b*. Invoking Hess's law, the enthalpy change for this pathway and for pathway *b* are equal. Now imagine that the constant pressure for processes *a* and *b* differs only by an infinitesimal amount, although that for *a* is higher than the triple point pressure, and that for *b* is lower than the triple

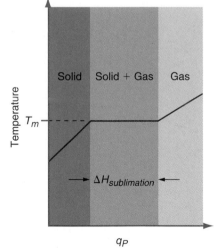

FIGURE 8.6
The temperature versus heat curve is shown for the process corresponding to path *b* in Figure 8.4. The temperature rises linearly with q_P in single-phase regions, and remains constant along the two-phase curves as the relative amounts of the two phases in equilibrium change (not to scale).

point pressure. We can express this mathematically by setting the pressure for process a equal to $P_{tp} + dP$, and that for process b equal to $P_{tp} - dP$. We examine the limit $dP \rightarrow 0$. In this limit, $\Delta H \rightarrow 0$ for the two steps in the process indicated by the dashed arrows because $dP \rightarrow 0$. Therefore, ΔH for the transformation solid $\rightarrow$ liquid $\rightarrow$ gas in process a and for the transformation solid $\rightarrow$ gas in process b must be identical. We conclude that

$$\Delta H = \Delta H_{sublimation} = \Delta H_{fusion} + \Delta H_{vaporization} \qquad (8.4)$$

Path c indicates an isothermal process in which the pressure is increased. The initial state of the system is the single-phase gas region. As the gas is compressed, it is liquefied as it crosses the solid–liquid coexistence curve. As the pressure is increased further, the sample freezes as it crosses the liquid–solid coexistence curve. Crystallization is exothermic, and heat must flow to the surroundings as the liquid solidifies. If the process is reversed, heat must flow into the system to keep T constant as the solid melts.

If T is below the triple point temperature, the liquid exists at equilibrium only if the slope of the liquid–solid coexistence curve is negative, as is the case for water. Liquid water below the triple point temperature can freeze at constant T if the pressure is lowered sufficiently to cross the liquid–solid coexistence curve. In an example of such a process, a thin wire to which a heavy weight is attached on each end is stretched over a block of ice. With time, it is observed that the wire lies within the ice block and eventually passes through the block. There is no visible evidence of the passage of the wire in the form of a narrow trench. What happens in this process? Because the wire is thin, the force on the wire results in a high pressure in the area of the ice block immediately below the wire. This high pressure causes local melting of the ice below the wire. Melting allows the wire to displace the liquid water, which flows to occupy the volume immediately above the wire. Because in this region water no longer experiences a high pressure, it freezes again and hides the passage of the wire.

The consequences of having a critical point in the gas–liquid coexistence curve are illustrated in Figure 8.4. Process d, indicated by the double-headed arrow, is a constant pressure heating or cooling such that the gas–liquid coexistence curve is crossed. In a reversible process, a clearly visible interface will be observed along the two-phase gas–liquid coexistence curve. However, the same overall process can be carried out in four steps indicated by the single-headed arrows. In this case, two-phase coexistence is not observed, because the gas–liquid coexistence curve is not crossed. The overall transition is the same along both paths, namely, gas is transformed into liquid. However, no interface will be observed in this process.

EXAMPLE PROBLEM 8.1

Draw a generic P–T phase diagram like that shown in Figure 8.4. Draw pathways in the diagram that correspond to the processes described here:

a. You hang wash out to dry at a temperature below the triple point. Initially, the water in the wet clothing has frozen. However, after a few hours in the sun, the clothing is warmer, dry, and soft.

b. A small amount of ethanol is contained in a thermos bottle. A test tube is inserted into the neck of the thermos bottle through a rubber stopper. A few minutes after filling the test tube with liquid nitrogen, the ethanol is no longer visible at the bottom of the bottle.

c. A transparent cylinder and piston assembly contains only a pure liquid in equilibrium with its vapor pressure. An interface is clearly visible between the two phases. When you increase the temperature by a small amount, the interface disappears.

Solution

The phase diagram with the paths is shown here:

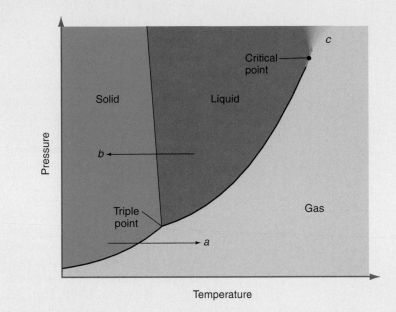

Paths a and b are not unique. Path a must occur at a pressure lower than the triple point pressure and process b must occur at a pressure greater than the triple point pressure. Path c will lie on the liquid–gas coexistence line up to the critical point, but can deviate once $T > T_c$ and $P > P_c$.

A P–T phase diagram for water at high P values is shown in Figure 8.7. Water has several solid phases that are stable in different pressure ranges because they have different densities. Eleven different crystalline forms of ice have been identified up to a pressure of 10^{12} atm. For a comprehensive collection of material on the phase diagram

FIGURE 8.7

The P–T phase diagram for H_2O is shown pressures up to 3.5×10^{10} bar. (Reprinted with permission from D. R. Lide, Ed., *CRC Handbook of Chemistry and Physics*, 83rd ed., Figure 3, page 12-202, CRC Press, Boca Raton, FL, 2002.)

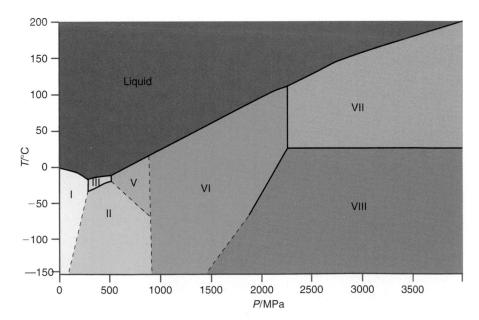

FIGURE 8.8
Two different crystal structures are shown. Hexagonal ice (left) is the stable form of ice under atmospheric conditions. Ice VI (right) is only stable at elevated pressures as shown in the phase diagram of Figure 8.7. The dashed lines indicate hydrogen bonds.

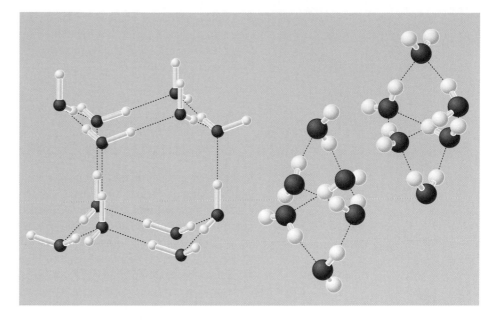

of water, see *http://www.lsbu.ac.uk/water/phase.html*. Note in Figure 8.7 that ice VI does not melt until the temperature is raised to ~ 100°C for $P \approx 2000$ MPa.

Hexagonal ice (ice I) is the normal form of ice and snow. The structure shown in Figure 8.8 may be thought of as consisting of a set of parallel sheets, connected to one another through hydrogen bonding. Hexagonal ice has a fairly open structure with a density of 0.931 g cm^{-3} near the triple point. Figure 8.8 also shows the crystal structure of ice VI. All water molecules in this structure are hydrogen bonded to four other molecules. Ice VI is much more closely packed than hexagonal ice, and has a density of 1.31 g cm^{-3} at 1.6 GPa, at which pressure the density of liquid water is 1.18 g cm^{-3}. Note that ice VI will not float on liquid water.

As shown in Figure 8.7, phase diagrams can be quite complex for simple substances because a number of solid phases can exist as P and T are varied. A further example is sulfur, which can also exist in several different solid phases. A portion of the phase diagram for sulfur is shown in Figure 8.9, and the solid phases are described by the symmetry of their unit cells. Note that several points correspond to three-phase equilibria.

FIGURE 8.9
The $P–T$ phase diagram for sulfur (not to scale).

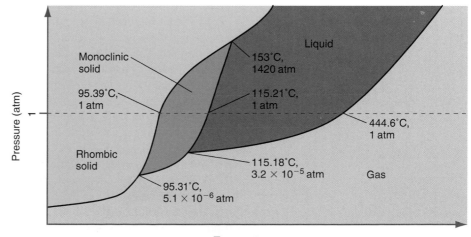

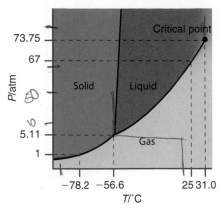

FIGURE 8.10

The $P-T$ phase diagram for CO_2 (not to scale).

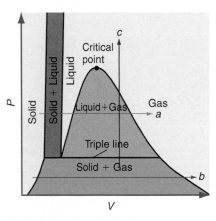

FIGURE 8.11

A $P-V$ phase diagram displays single- and two-phase coexistence regions, a critical point, and a triple line. The two-phase coexistence areas are colored.

By contrast, the CO_2 phase diagram shown in Figure 8.10 is simpler. It is similar in structure to that of water, but the solid–liquid coexistence curve has a positive slope. Several of the end-of-chapter problems and questions refer to the phase diagrams in Figures 8.7, 8.9, and 8.10.

8.3 The Pressure–Volume and Pressure–Volume–Temperature Phase Diagrams

In the previous section, the regions of stability and equilibrium among the solid, liquid, and gas phases were described using a $P-T$ phase diagram. Any phase diagram that includes only two of the three state variables diagram is limited, because it does not contain information on the third variable. We first complement the information contained in the $P-T$ phase diagram with a **$P-V$ phase diagram,** and then combine these two representations into a $P-V-T$ phase diagram. Figure 8.11 shows a $P-V$ phase diagram for a substance for which $V_m^{liquid} > V_m^{solid}$.

Significant differences are seen in the way that two- and three-phase coexistence are represented in the two-phase diagrams. The two-phase coexistence curves of the $P-T$ phase diagram become two-phase regions in the $P-V$ phase diagram, because the volume of a system in which two phases coexist varies with the relative amounts of the material in each phase. For pressures well below the critical point, the range in V over which the gas and liquid coexist is large compared to the range in V over which the solid and liquid coexist because $V_m^{solid} < V_m^{liquid} \ll V_m^{gas}$. Therefore, the gas–liquid coexistence region is broader than the solid–liquid coexistence region. Note that the triple point in the $P-T$ phase diagram becomes a triple line in the $P-V$ diagram. Although P and T have unique values at the triple point, V can range between a maximum value for which the system consists almost entirely of gas with traces of the liquid and solid phases, and a minimum value for which the system consists almost entirely of solid with traces of the liquid and gas phases.

The usefulness of the $P-V$ diagram is illustrated by tracing several processes in Figure 8.11. In process a, a solid is converted to a gas by increasing the temperature in an isobaric process for which P is greater than the triple point pressure. This same process was depicted in Figure 8.4. In the $P-V$ phase diagram, it is clear that this process involves large changes in volume in the two-phase coexistence region over which the temperature remains constant, which was not obvious in Figure 8.4. Process b shows an isobaric transition from solid to gas for P below the triple point pressure, for which the system has only one two-phase coexistence region. Process c shows a constant volume transition from a system consisting of solid and vapor in equilibrium to a supercritical fluid. How does the temperature change along this path?

Process b in Figure 8.11 is known as **freeze drying.** Assume that the food to be freeze dried is placed in a vessel at $-10°C$, and the system is allowed to equilibrate at this temperature. The partial pressure of $H_2O(g)$ in equilibrium with the ice crystals in the food under these conditions is 260 Pa. A vacuum pump is now started, and the gas phase is pumped away. The temperature reaches a steady-state value determined by heat conduction into the sample and heat loss through sublimation. The solid has an equilibrium vapor pressure determined by the steady-state temperature. As the pump removes water from the gas phase, ice spontaneously sublimes in order to keep the gas-phase water partial pressure at its equilibrium value. If the capacity of the vacuum pump is sufficient to bring the partial pressure of water below its equilibrium value at $-10°C$, all the ice in the food sample will sublime. After the sublimation is complete, the food has been freeze dried.

All the information on the values of P, V, and T corresponding to single-phase regions, two-phase regions, and the triple point displayed in the $P-T$ and $P-V$ phase diagrams is best displayed in a three-dimensional **$P-V-T$ phase diagram.** Such a diagram is shown in Figure 8.12 for an ideal gas, which does not exist in the form of condensed phases. It

FIGURE 8.12
A *P–V–T* diagram for an ideal gas.
Constant pressure, constant volume, and
constant temperature paths are shown as
black, red, and blue curves, respectively.

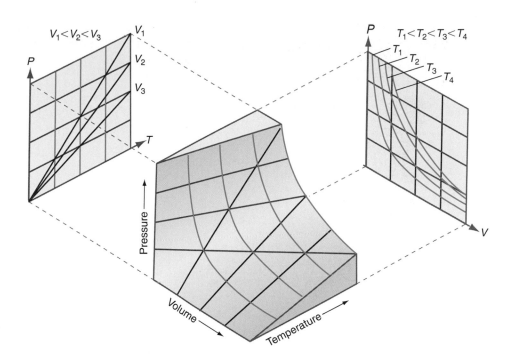

FIGURE 8.12
A *P–V–T* diagram for an ideal gas.
Constant pressure, constant volume, and
constant temperature paths are shown as
black, red, and blue curves, respectively.

is easy to obtain the *P–T* and *P–V* phase diagrams from the *P–V–T* phase diagram. The *P–T* phase diagram is a projection of the three-dimensional surface on the *P–T* plane, and the *P–V* phase diagram is a projection of the three-dimensional surface on the *P–V* plane.

Figure 8.13 shows a *P–V–T* diagram for a substance that expands upon melting. The usefulness of the *P–V–T* phase diagram can be illustrated by revisiting the isobaric conversion of a solid to a gas at a temperature above the triple point shown as process *a* in Figure 8.4. This process is shown as the path $a \rightarrow b \rightarrow c \rightarrow d \rightarrow e \rightarrow f$ in Figure 8.13. We can see now that the temperature increases along the segments $a \rightarrow b$, $c \rightarrow d$, and $e \rightarrow f$, all of which lie with single-phase regions, and remains constant along the segments $b \rightarrow c$ and $d \rightarrow e$, which lie within two-phase regions. Similarly, process *c* in Figure 8.4 is shown as the path $g \rightarrow h \rightarrow i \rightarrow k \rightarrow l \rightarrow m$ in Figure 8.13.

FIGURE 8.13
A *P–V–T* phase diagram is shown for a substance that contracts on freezing. The indicated processes are discussed in the text.

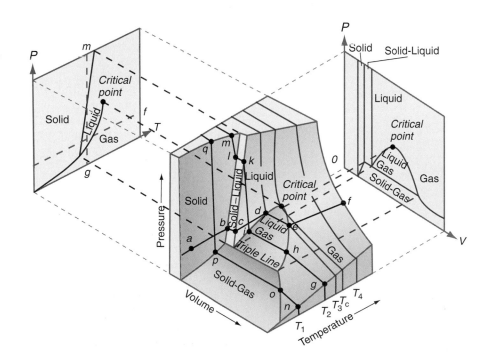

8.4 Providing a Theoretical Basis for the P–T Phase Diagram

In this section, a theoretical basis is provided for the coexistence curves that separate different single-phase regions in the P–T phase diagram. Along the coexistence curves, two phases are in equilibrium. From Section 5.6, we know that if two phases, α and β, are in equilibrium at a pressure P and temperature T, their chemical potentials must be equal:

$$\mu_\alpha(P,T) = \mu_\beta(P,T) \tag{8.5}$$

If the macroscopic variables are changed by a small amount, $P, T \rightarrow P + dP, T + dT$ such that the system pressure and temperature still lie on the coexistence curve, then

$$\mu_\alpha(P,T) + d\mu_\alpha = \mu_\beta(P,T) + d\mu_\beta \tag{8.6}$$

In order for the two phases to remain in equilibrium,

$$d\mu_\alpha = d\mu_\beta \tag{8.7}$$

Because $d\mu$ can be expressed in terms of dT and dP,

$$d\mu_\alpha = -S_{m\alpha}dT + V_{m\alpha}dP \quad \text{and} \quad d\mu_\beta = -S_{m\beta}dT + V_{m\beta}dP \tag{8.8}$$

The expressions for $d\mu$ can be equated, giving

$$-S_{m\alpha}dT + V_{m\alpha}dP = -S_{m\beta}dT + V_{m\beta}dP \quad \text{or} \tag{8.9}$$

$$(S_{m\beta} - S_{m\alpha})dT = (V_{m\beta} - V_{m\alpha})dP$$

Assume that as $P, T \rightarrow P + dP, T + dT$, an incremental amount of phase α is transformed to phase β. In this case, $\Delta S_m = S_{m\beta} - S_{m\alpha}$ and $\Delta V_m = V_{m\beta} - V_{m\alpha}$. Rearranging Equation (8.9) gives the **Clapeyron equation:**

$$\frac{dP}{dT} = \frac{\Delta S_m}{\Delta V_m} \tag{8.10}$$

The importance of the Clapeyron equation is that it allows one to calculate the slope of the coexistence curves in a P–T phase diagram if ΔS_m and ΔV_m for the transition are known. One can use the Clapeyron equation to estimate the slope of the solid–liquid coexistence curve. At the melting temperature,

$$\Delta G_m^{fusion} = \Delta H_m^{fusion} - T\Delta S_m^{fusion} = 0 \tag{8.11}$$

Therefore, the ΔS_m values for the fusion transition can be calculated from the enthalpy of fusion and the fusion temperature. Values of the normal fusion and vaporization temperatures, as well as ΔH_m for fusion and vaporization, are shown in Table 8.2 for a number of different elements and compounds. Although there is a significant variation in these values, for our purposes, it is sufficient to use the average value of $\Delta S_m^{fusion} = 22$ J mol^{-1} K^{-1} calculated from the data in Table 8.2 in order to estimate the slope of the solid–liquid coexistence curve.

For the fusion transition, ΔV is small because the densities of the solid and liquid states are quite similar. The average ΔV_m^{fusion} for Ag, AgCl, Ca, CaCl$_2$, K, KCl, Na, NaCl, and H$_2$O is $+ 4 \times 10^{-6}$ m^3. Of these substances, only H$_2$O has a negative value for ΔV_m^{fusion}. We next use the average values of ΔS_m^{fusion} and ΔV_m^{fusion} to estimate the slope of the solid–liquid coexistence curve:

$$\left(\frac{dP}{dT}\right)_{fusion} = \frac{\Delta S_m^{fusion}}{\Delta V_m^{fusion}} \approx \frac{22 \text{ J mol}^{-1} \text{ K}^{-1}}{\pm 4 \times 10^{-6} \text{ m}^3 \text{ mol}^{-1}} \tag{8.12}$$

$$= \pm 5.5 \times 10^6 \text{ Pa K}^{-1} = \pm 55 \text{ bar K}^{-1}$$

Inverting this result, $(dT/dP)_{fusion} \approx \pm 0.02 \, \text{K bar}^{-1}$. An increase of P by ~50 bar is required to change the melting temperature by one degree. This result explains the very steep solid–liquid coexistence curve shown in Figure 8.4.

The same analysis applies to the liquid–gas coexistence curve. Because $\Delta H_m^{vaporization}$ and $\Delta V_m^{vaporization} = V_m^{gas} - V_m^{liquid}$ are always positive, $(dP/dT)_{vaporization}$ is always positive. The average of $\Delta S_m^{vaporization}$ for the substances shown in Table 8.2 is 95 J mol^{-1} K^{-1}. This value is in accord with **Trouton's rule,** which states that $\Delta S_m^{vaporization} \approx 90 \, \text{J mol}^{-1} \, \text{K}^{-1}$ for liquids. The rule fails for liquids in which there are strong interactions between molecules such as $-OH$ or $-NH_2$ groups capable of forming hydrogen bonds.

The molar volume of an ideal gas is approximately 20 L mol^{-1} in the temperature range in which many liquids boil. Because $V_m^{gas} \gg V_m^{liquid}$, $\Delta V_m^{vaporization} \approx 20 \times 10^{-3} \, \text{m}^3 \, \text{mol}^{-1}$. The slope of the liquid–gas coexistence line is given by

$$\left(\frac{dP}{dT} \right)_{vaporization} = \frac{\Delta S_m^{vaporization}}{\Delta V_m^{vaporization}} \approx \frac{95 \, \text{J mol}^{-1} \, \text{K}^{-1}}{2 \times 10^{-2} \, \text{m}^3 \, \text{mol}^{-1}} \approx 5 \times 10^3 \, \text{Pa K}^{-1} \tag{8.13}$$
$$= 5 \times 10^{-2} \, \text{bar K}^{-1}$$

This slope is a factor of 10^3 smaller than the slope for the liquid–solid coexistence curve. Inverting this result, $(dT/dP)_{vaporization} \approx 20 \, \text{K bar}^{-1}$. This result shows that it takes only a modest increase in the pressure to increase the boiling point of a liquid by a significant amount. For this reason, a pressure cooker does not need to be able to withstand high pressures. Note that the slope of the liquid–gas coexistence curve in Figure 8.4 is much less than that of the solid–liquid coexistence curve. The slope of both curves increases with T because ΔS increases with T.

The solid–gas coexistence curve can also be analyzed using the Clapeyron equation. Because entropy is a state function, the entropy change for the processes solid $(P,T) \rightarrow$ gas (P,T) and solid $(P,T) \rightarrow$ liquid $(P,T) \rightarrow$ gas (P,T) must be the same. Therefore, $\Delta S_m^{sublimation} = \Delta S_m^{fusion} + \Delta S_m^{vaporization} > \Delta S_m^{vaporization}$. Because the molar volume of the gas is much larger than that of the solid or liquid, $\Delta V_m^{sublimation} \approx \Delta V_m^{vaporization}$. We conclude that $(dP/dT)_{sublimation} > (dP/dT)_{vaporization}$. Therefore, the slope of the solid–gas coexistence curve will be greater than that of the liquid–gas coexistence curve. Because this comparison applies to a common value of the temperature, it is best made for temperatures just above and just below the triple point temperature. This difference in slope of these two coexistence curves is exaggerated in Figure 8.4.

8.5 Using the Clapeyron Equation to Calculate Vapor Pressure as a Function of T

From watching a pot of water as it is heated on a stove, it is clear that the vapor pressure of a liquid increases rapidly with increasing temperature. The same conclusion holds for a solid below the triple point. To calculate the vapor pressure at different temperatures, the Clapeyron equation must be integrated. Consider the solid–liquid coexistence curve:

$$\int_{P_i}^{P_f} dP = \int_{T_i}^{T_f} \frac{\Delta S_m^{fusion}}{\Delta V_m^{fusion}} dT = \int_{T_i}^{T_f} \frac{\Delta H_m^{fusion}}{\Delta V_m^{fusion}} \frac{dT}{T} \approx \frac{\Delta H_m^{fusion}}{\Delta V_m^{fusion}} \int_{T_i}^{T_f} \frac{dT}{T} \tag{8.14}$$

where the integration is along the solid–liquid coexistence curve. In the last step, it has been assumed that ΔH_m^{fusion} and ΔV_m^{fusion} are independent of T over the temperature range of interest. Assuming that $(T_f - T_i)/T_i$ is small, the previous equation can be simplified to give

$$P_f - P_i = \frac{\Delta H_m^{fusion}}{\Delta V_m^{fusion}} \ln \frac{T_f}{T_i} = \frac{\Delta H_m^{fusion}}{\Delta V_m^{fusion}} \ln \frac{T_i + \Delta T}{T_i} \approx \frac{\Delta H_m^{fusion}}{\Delta V_m^{fusion}} \frac{\Delta T}{T_i} \tag{8.15}$$

The last step uses the result $\ln(1 + x) = x$ for $x \ll 1$, obtained by expanding $\ln(1 + x)$ in a Taylor series about $x = 0$. We see that ΔP varies linearly with ΔT in this limit. The value of the slope dP/dT was discussed in the previous section.

For the liquid–gas coexistence curve, we have a different result, because $\Delta V \approx V^{gas}$. Assuming that the ideal gas law holds, then

$$\frac{dP}{dT} = \frac{\Delta S_m^{vaporization}}{\Delta V_m^{vaporization}} \approx \frac{\Delta H_m^{vaporization}}{TV^{gas}} = \frac{P\Delta H_m^{vaporization}}{RT^2} \qquad (8.16)$$

$$\frac{dP}{P} = \frac{\Delta H_m^{vaporization}}{R} \frac{dT}{T^2}$$

Assuming that $\Delta H_m^{vaporization}$ remains constant over the range of temperature of interest,

$$\int_{P_i}^{P_f} \frac{dP}{P} = \frac{\Delta H_m^{vaporization}}{R} \times \int_{T_i}^{T_f} \frac{dT}{T^2} \qquad (8.17)$$

$$\ln\frac{P_f}{P_i} = -\frac{\Delta H_m^{vaporization}}{R} \times \left(\frac{1}{T_f} - \frac{1}{T_i}\right)$$

The same procedure is followed for the solid–gas coexistence curve. The result is the same as Equation (8.17) with $\Delta H_m^{sublimation}$ substituted for $\Delta H_m^{vaporization}$. Equation (8.17) provides a way to determine the enthalpy of vaporization for a liquid by measuring its vapor pressure as a function of temperature, as shown in Example Problem 8.2. In this discussion, it has been assumed that $\Delta H_m^{vaporization}$ is independent of temperature. More accurate values of the vapor pressure as a function of temperature can be obtained by fitting experimental data. This leads to an expression for the vapor pressure as a function of temperature. These functions for selected liquids and solids are listed in Tables 8.3 and 8.4 (see Appendix A, Data Tables).

EXAMPLE PROBLEM 8.2

The normal boiling temperature of benzene is 353.24 K, and the vapor pressure of liquid benzene is 1.00×10^4 Pa at 20.0°C. The enthalpy of fusion is 9.95 kJ mol^{-1}, and the vapor pressure of solid benzene is 88.0 Pa at −44.3°C. Calculate the following:

a. $\Delta H_m^{vaporization}$

b. $\Delta S_m^{vaporization}$

c. Triple point temperature and pressure

Solution

a. We can calculate $\Delta H_m^{vaporization}$ using the Clapeyron equation because we know the vapor pressure at two different temperatures:

$$\ln\frac{P_f}{P_i} = -\frac{\Delta H_m^{vaporization}}{R}\left(\frac{1}{T_f} - \frac{1}{T_i}\right)$$

$$\Delta H_m^{vaporization} = -\frac{R\ln\dfrac{P_f}{P_i}}{\left(\dfrac{1}{T_f} - \dfrac{1}{T_i}\right)} = -\frac{8.314\ \text{J mol}^{-1}\,\text{K}^{-1} \times \ln\dfrac{101{,}325\ \text{Pa}}{1.00 \times 10^4\ \text{Pa}}}{\left(\dfrac{1}{353.24\ \text{K}} - \dfrac{1}{273.15 + 20.0\ \text{K}}\right)}$$

$$= 33.2\ \text{kJ mol}^{-1}$$

b. $\Delta S_m^{vaporization} = \dfrac{\Delta H_m^{vaporization}}{T_b} = \dfrac{33.2 \times 10^3\,\text{J mol}^{-1}}{353.24\,\text{K}} = 93.9\,\text{J mol}^{-1}\,\text{K}^{-1}$

c. At the triple point, the vapor pressures of the solid and liquid are equal:

$$\ln \frac{P_{tp}^{liquid}}{P^\circ} = \ln \frac{P_i^{liquid}}{P^\circ} - \frac{\Delta H_m^{vaporization}}{R}\left(\frac{1}{T_{tp}} - \frac{1}{T_i^{liquid}}\right)$$

$$\ln \frac{P_{tp}^{solid}}{P^\circ} = \ln \frac{P_i^{solid}}{P^\circ} - \frac{\Delta H_m^{sublimation}}{R}\left(\frac{1}{T_{tp}} - \frac{1}{T_i^{solid}}\right)$$

$$\ln \frac{P_i^{liquid}}{P^\circ} - \ln \frac{P_i^{solid}}{P^\circ} - \frac{\Delta H_m^{sublimation}}{RT_i^{solid}} + \frac{\Delta H_m^{vaporization}}{RT_i^{liquid}} = \frac{(\Delta H_m^{vaporization} - \Delta H_m^{sublimation})}{RT_{tp}}$$

$$T_{tp} = \frac{(\Delta H_m^{vaporization} - \Delta H_m^{sublimation})}{R\left(\ln \dfrac{P_i^{liquid}}{P^\circ} - \ln \dfrac{P_i^{solid}}{P^\circ} - \dfrac{\Delta H_m^{sublimation}}{RT_i^{solid}} + \dfrac{\Delta H_m^{vaporization}}{RT_i^{liquid}}\right)}$$

$$= \frac{9.95 \times 10^3\,\text{J mol}^{-1}}{8.314\,\text{J K}^{-1}\text{mol}^{-1} \times \left(\begin{array}{c}\ln \dfrac{10{,}000\,\text{Pa}}{1\,\text{Pa}} - \ln \dfrac{88.0\,\text{Pa}}{1\,\text{Pa}} - \dfrac{(33.2 \times 10^3 + 9.95 \times 10^3)\text{J mol}^{-1}}{8.314\,\text{J K}^{-1}\,\text{mol}^{-1} \times 228.9\,\text{K}} \\ + \dfrac{33.2 \times 10^3\,\text{J mol}^{-1}}{8.314\,\text{J K}^{-1}\,\text{mol}^{-1} \times 293.15\,\text{K}}\end{array}\right)}$$

$= 277\,\text{K}$

We calculate the triple point pressure using the Clapeyron equation:

$$\ln \frac{P_f}{P_i} = -\frac{\Delta H_m^{vaporization}}{R}\left(\frac{1}{T_f} - \frac{1}{T_i}\right)$$

$$\ln \frac{P_{tp}}{101325} = -\frac{33.2 \times 10^3\,\text{J mol}^{-1}}{8.314\,\text{J mol}^{-1}\,\text{K}^{-1}} \times \left(\frac{1}{277\,\text{K}} - \frac{1}{353.24\,\text{K}}\right)$$

$$\ln \frac{P_{tp}}{P^\circ} = 8.41465$$

$$P_{tp} = 4.51 \times 10^3\,\text{Pa}$$

8.6 The Vapor Pressure of a Pure Substance Depends on the Applied Pressure

Consider the piston and cylinder assembly containing water at 25°C shown in Figure 8.14. The equilibrium vapor pressure of water at this temperature is $P^* = 3.16 \times 10^3$ Pa, or 0.0316 bar. Therefore, if the weightless piston is loaded with a mass sufficient to generate a pressure of 1 bar, the system is in the single-phase liquid region of the phase diagram in Figure 8.4. This state of the system is shown in Figure 8.14a. The size of the mass is reduced so that the pressure is exactly equal to the vapor pressure of water. The system now lies in the two-phase liquid–gas region of the phase diagram described by the liquid–gas coexistence curve. The piston can be pulled outward or pushed inward while maintaining this pressure. This action leads to a larger or smaller volume for the gas phase, but the pressure will remain constant at 3.16×10^3 Pa as long as the temperature of the system remains constant. This state of the system is shown in Figure 8.14b.

Keeping the temperature constant, enough argon gas is introduced into the cylinder such that the sum of the argon and H_2O partial pressures is 1 bar. This state of the

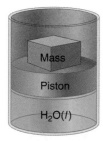

$P_{external} = 1.00$ bar

(a)

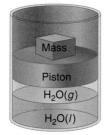

$P_{external} = 0.0316$ bar

(b)

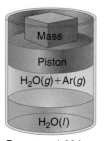

$P_{external} = 1.00$ bar

(c)

FIGURE 8.14

A piston and cylinder assembly at 298 K is shown with the contents being pure water (a) at a pressure greater than the vapor pressure, (b) at a pressure equal to the vapor pressure, and (c) at 1 bar for a mixture of argon and water.

system is shown in Figure 8.14c. What is the vapor pressure of water in this case, and does it differ from that for the system shown in Figure 8.14b? The vapor pressure P is used to denote the partial pressure of water in the gas phase, and $\mathbf{P}$ to denote the sum of the argon and water partial pressures.

To calculate the partial pressure of water in the argon–water mixture, the following equilibrium condition holds:

$$\mu_{liquid}(T,\mathbf{P}) = \mu_{gas}(T,P) \tag{8.18}$$

Differentiating this expression with respect to $\mathbf{P}$, we obtain

$$\left(\frac{\partial \mu_{liquid}(T,\mathbf{P})}{\partial \mathbf{P}}\right)_T = \left(\frac{\partial \mu_{gas}(T,P)}{\partial P}\right)_T \left(\frac{\partial P}{\partial \mathbf{P}}\right)_T \tag{8.19}$$

Because $d\mu = -S_m dT + V_m dP$, $(d\mu/dP)_T = V_m$, and the previous equation becomes

$$V_m^{liquid} = V_m^{gas}\left(\frac{\partial P}{\partial \mathbf{P}}\right)_T \quad \text{or} \quad \left(\frac{\partial P}{\partial \mathbf{P}}\right)_T = \frac{V_m^{liquid}}{V_m^{gas}} \tag{8.20}$$

This equation shows that the vapor pressure P increases if the total pressure $\mathbf{P}$ increases. However, the rate of increase is small because the ratio $V_m^{liquid}/V_m^{gas} \ll 1$. It is reasonable to replace V_m^{gas} with the ideal gas value RT/P in Equation (8.20). This leads to the equation

$$\frac{RT}{P}dP = V_m^{liquid}d\mathbf{P} \quad \text{or} \quad RT\int_{P_0}^{P}\frac{dP'}{P'} = V_m^{liquid}\int_{P_0}^{\mathbf{P}}d\mathbf{P}' \tag{8.21}$$

Integrating Equation (8.21) gives

$$RT\ln\left(\frac{P}{P_0}\right) = V_m^{liquid}(\mathbf{P} - P_0) \tag{8.22}$$

For the specific case under consideration, $P_0 = 0.0316$ bar, $\mathbf{P} = 1$ bar, and $V_m^{liquid} = 1.8 \times 10^{-5}$ m³:

$$\ln\left(\frac{P}{P_0}\right) = \frac{V_m^{liquid}(\mathbf{P}-P_0)}{RT} = \frac{1.8\times10^{-5}\text{ m}^3 \times (1-0.0316)\times10^5\text{ Pa}}{8.314\text{ J mol}^{-1}\text{ K}^{-1}\times298\text{ K}} = 7.04\times10^{-4}$$

$$P = 1.0007\ P_0 \approx 0.0316\text{ bar}$$

For an external pressure of 1 bar, the effect is negligible. However, for $\mathbf{P} = 100$ bar, $P = 0.0339$ bar, amounting to an increase in the vapor pressure of 7%.

8.7 Surface Tension

In discussing the liquid phase, the effect of the boundary surface on the properties of the liquid has been neglected. In the absence of a gravitational field, a liquid droplet will assume a spherical shape, because in this geometry the maximum number of molecules is surrounded by neighboring molecules. Because the interaction between molecules in a liquid is attractive, minimizing the surface-to-volume ratio minimizes the energy. How does the energy of the droplet depend on its surface area? Starting with the equilibrium spherical shape, assume that the droplet is distorted to create more area while keeping the volume constant. The work associated with the creation of additional surface area at constant V and T is

$$dA = \gamma d\sigma \tag{8.23}$$

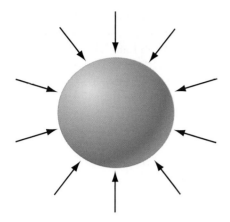

FIGURE 8.15
The forces acting on a spherical droplet that arise from surface tension.

where A is the Helmholtz energy, γ is the surface tension, and σ is the unit element of area. The **surface tension** has the units of energy/area or J m^{-2}, which is equivalent to N m^{-1} (Newtons per meter). Because $dA < 0$ for a spontaneous process at constant V and T, Equation (8.23) predicts that a liquid, or a bubble, or a liquid film suspended in a wire frame will tend to minimize its surface area.

Consider the spherical droplet depicted in Figure 8.15. There must be a force acting on the droplet in the radially inward direction for the liquid to assume a spherical shape. An expression for the force can be generated as follows. If the radius of the droplet is increased from r to $r + dr$, the area increases by

$$d\sigma = 4\pi(r+dr)^2 - 4\pi r^2 = 4\pi(r^2 + 2rdr + (dr)^2) - 4\pi r^2 \approx 8\pi rdr \qquad (8.24)$$

From Equation (8.23), the work done in the expansion of the droplet is $8\pi\gamma rdr$. The force, which is normal to the surface of the droplet, is the work divided by the distance or

$$F = 8\pi\gamma r \qquad (8.25)$$

The net effect of this force is to generate a pressure differential across the droplet surface. At equilibrium, there is a balance between the inward and outward acting forces. The inward acting force is the sum of the force exerted by the external pressure and the force arising from the surface tension, whereas the outward acting force arises solely from the pressure in the liquid:

$$4\pi r^2 P_{outer} + 8\pi\gamma r = 4\pi r^2 P_{inner} \quad \text{or} \qquad (8.26)$$

$$P_{inner} = P_{outer} + \frac{2\gamma}{r}$$

Note that $P_{inner} - P_{outer} \rightarrow 0$ as $r \rightarrow \infty$. Therefore, the pressure differential exists only for a curved surface. From the geometry in Figure 8.15, it is apparent that the higher pressure is always on the concave side of the interface. Values for the surface tension for a number of liquids are listed in Table 8.5.

One effect of this pressure differential is that the vapor pressure of a droplet depends on its radius. By substituting numbers in Equation (8.26) and using Equation (8.22) to calculate the vapor pressure, we find that the vapor pressure of a 10^{-7} m water droplet is increased by 1%, that of a 10^{-8} m droplet is increased by 11%, and that of a 10^{-9} m droplet is increased by 270%. [At such a small diameter, the application of Equation (8.26) is questionable because the size of an individual water molecule is comparable to the droplet diameter. Therefore, a microscopic theory is needed to describe the forces within the droplet.] This effect plays a role in the formation of liquid droplets in a condensing gas such as fog. Small droplets evaporate more rapidly than large droplets, and the vapor condenses on the larger droplets, allowing them to grow at the expense of small droplets.

TABLE 8.5

Surface Tension of Selected Liquids at 298 K

Formula	Name	γ (mN m^{-1})	Formula	Name	(mN m^{-1})
Br$_2$	Bromine	40.95	CS$_2$	Carbon disulfide	31.58
H$_2$O	Water	71.99	C$_2$H$_5$OH	Ethanol	21.97
Hg	Mercury	485.5	C$_6$H$_5$N	Pyridine	36.56
CCl$_4$	Carbon tetrachloride	26.43	C$_6$H$_6$	Benzene	28.22
CH$_3$OH	Methanol	22.07	C$_8$H$_{18}$	Octane	21.14

Source: Data from Lide, D. R., Ed., *Handbook of Chemistry and Physics,* 83rd ed. CRC Press, Boca Raton, FL, 2002.

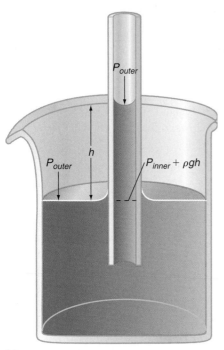

(a)

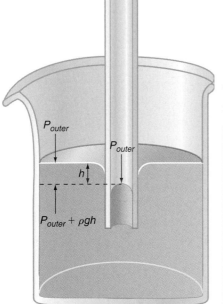

(b)

FIGURE 8.16
(a) If the liquid wets the interior wall of the capillary, a capillary rise is observed. The combination Pyrex–water exhibits this behavior. **(b)** If the liquid does not wet the capillary surface, a capillary depression is observed. The combination Pyrex–mercury exhibits this behavior.

Capillary rise and **capillary depression** are other consequences of the pressure differential across a curved surface. Assume that a capillary of radius r is partially immersed in a liquid. When the liquid comes in contact with a solid surface, there is a natural tendency to minimize the energy of the system. If the surface tension of the liquid is lower than that of the solid, the liquid will wet the surface, as shown in Figure 8.16a. However, if the surface tension of the liquid is higher than that of the solid, the liquid will avoid the surface, as shown in Figure 8.16b. In either case, there is a pressure differential in the capillary across the gas–liquid interface, because the interface is curved. If we assume that the liquid–gas interface is tangent to the interior wall of the capillary at the solid–liquid interface, the radius of curvature of the interface is equal to the capillary radius.

The difference in the pressure across the curved interface, $2\gamma/r$, is balanced by the weight of the column in the gravitational field, $\rho g h$. Therefore, the capillary rise or depression is given by

$$h = \frac{2\gamma}{\rho g r} \tag{8.27}$$

In the preceding discussion, it was assumed that either (1) the liquid completely wets the interior surface of the capillary, in which case the liquid coats the capillary walls, but does not fill the core, or (2) the liquid is completely nonwetting, in which case the liquid does not coat the capillary walls, but fills the core. In a more realistic model, the interaction is intermediate between these two extremes. In this case, the liquid–surface is characterized by the **contact angle** θ, as shown in Figure 8.17.

Complete **wetting** corresponds to $\theta = 0°$ and complete **nonwetting** corresponds to $\theta = 180°$. For intermediate cases,

$$P_{inner} = P_{outer} + \frac{2\gamma}{r\cos\theta} \text{ and } h = \frac{2\gamma}{\rho g r \cos\theta} \tag{8.28}$$

The measurement of the contact angle is one of the main experimental methods used to measure the difference in surface tension at the solid–liquid interface.

EXAMPLE PROBLEM 8.3

The six-legged water strider supports itself on the surface of a pond on four of its legs. Each of these legs causes a depression to be formed in the pond surface. Assume that each depression can be approximated as a hemisphere of radius 1.2×10^{-4} m and that θ (as in Figure 8.17) is 0°. Calculate the force that one of the insect's legs exerts on the pond.

Solution

$$\Delta P = \frac{2\gamma}{r\cos\theta} = \frac{2 \times 71.99 \times 10^{-3} \text{ N m}^{-1}}{1.2 \times 10^{-4} \text{ m} \times 1} = 1.20 \times 10^3 \text{ Pa}$$

$$F = PA = P \times \pi r^2 = 1.20 \times 10^3 \text{ Pa} \times \pi (1.2 \times 10^{-4} \text{ m})^2 = 5.4 \times 10^{-5} \text{ N}$$

EXAMPLE PROBLEM 8.4

Water is transported upward in trees through channels in the trunk called xylem. Although the diameter of the xylem channels varies from species to species, a typical value is 2.0×10^{-5} m. Is capillary rise sufficient to transport water to the top of a redwood tree that is 100 m high? Assume complete wetting of the xylem channels.

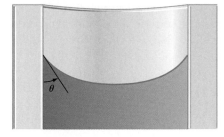

FIGURE 8.17
For cases intermediate between wetting and nonwetting, the contact angle θ lies in the range $0° < \theta < 180°$.

Solution
From Equation (8.28),

$$h = \frac{2\gamma}{\rho g r \cos\theta} = \frac{2 \times 71.99 \times 10^{-3}\,\text{N m}^{-1}}{997\,\text{kg m}^{-3} \times 9.81\,\text{m s}^{-2} \times 2.0 \times 10^{-5}\,\text{m} \times 1} = 0.74\,\text{m}$$

No, capillary rise is not sufficient to account for water supply to the top of a redwood tree.

As Example Problem 8.4 shows, capillary rise is insufficient to account for water transport to the leaves in all but the smallest plants. The property of water that accounts for water supply to the top of a redwood is its high **tensile strength.** Imagine pulling on a piston and cylinder containing only liquid water to create a negative pressure. How hard can you pull on the water without "breaking" the water column? The answer to this question depends on whether bubbles are nucleated in the liquid. This phenomenon is called cavitation. If cavitation occurs, the bubbles will grow rapidly as the piston is pulled outward, and the bubble pressure is given by Equation (8.28), where $P_{external}$ is the vapor pressure of water. The height of the water column in this case is limited to about 9.7 m. However, bubble nucleation is a kinetic phenomenon initiated at certain sites at the wall surrounding the water, and under the conditions present in xylem tubes, it is largely suppressed. In the absence of bubble nucleation, theoretical calculations predict that the tensile strength of water is sufficient that negative pressure in excess of 1000 atm can be generated. The pressure is negative because the water is under tension rather than compression. Experiments on water inclusions in very small cracks in natural rocks have verified these estimates. However, bubble nucleation occurs at much lower negative pressures in capillaries similar in diameter to xylem tubes. Even in these capillaries, negative pressures of more than 50 atm have been observed.

How does the high tensile strength of water explain the transport of water to the top of a redwood? If one cuts into a tree near its base, the sap oozes rather than spurts out, showing that the pressure in the xylem tubes is ~1 atm at the base of a tree. Imagine the redwood in its infancy as a seedling. Capillary rise is sufficient to fill the xylem tubes to the top of the plant. As the tree grows, the water can be pulled upward because of its high tensile strength. As the height of the tree increases, the pressure at the top becomes increasingly negative. As long as cavitation does not occur, the water column remains intact. As water evaporates from the leaves, it is resupplied from the roots through the pressure gradient in the xylem tubes that arises from the weight of the column. If the tree (and each xylem tube) grows to a height of ~100 m, and $P = 1$ atm at the base, the pressure at the top must be ~ -9 atm, from $\Delta P = \rho g h$. We again encounter a negative pressure because the water is under tension. If water did not have a sufficiently high tensile strength, gas bubbles would form in the xylem tubes. This would disrupt the flow of sap, and tall trees could not exist.

8.8 Chemistry in Supercritical Fluids

Chemical reactions can take place in the gas phase or in solution (homogeneous reactions) or on the surfaces of solids (heterogeneous reactions). Solvents with suitable properties can influence both the yield and the selectivity of reactions in solution. It has been found that the use of supercritical fluids as solvents increases the number of parameters that chemists have at their disposal to tune a reaction system to meet their needs. Why are supercritical fluids useful as solvents in chemical reactions?

Supercritical fluids (SCFs) near the critical point with $T_r \sim 1.0$–1.1 and $P_r \sim 1$–2 have a density that is an appreciable fraction of the liquid-phase density. They are unique in that they exhibit favorable properties of liquids and gases. Because the density is high, the solubility of solid substances is quite high, but the diffusion of solutes in the fluid is higher than in the normal liquid. This is the case because the density of the SCF is lower than in the normal liquid. For similar reasons, the viscosity of SCFs is lower than that in normal liquids. As a result, mass transfer is faster, and the overall reaction rate can be increased. Because a SCF is more gas-like than a liquid, the solubility of gases can be much higher in the SCF. This property is of particular usefulness in enhancing the reactivity of reactions in which one of the reactants is a gas, such as oxidation or hydrogenation. For example, hydrogen generally has a low solubility in organic solvents, but is quite soluble in SCFs.

Carbon dioxide and water exhibit the unusual properties of SCFs. Because the values for the critical constants of CO_2 (P_c = 73.74 bar and T_c = 304 K) are easily attainable, it is possible to use chemically inert supercritical CO_2 to replace toxic organic solvents in the dry cleaning industry. Although supercritical CO_2 is a good solvent for nonpolar molecules, it must be mixed with other substances to achieve a required minimum solubility for polar substances. This can be done without increasing the toxicity of the process greatly. The critical constants of H_2O are considerably higher (P_c = 220.64 bar and T_c = 647 K), so that the demands on the vessel used to contain the fluid are higher than for CO_2. However, as supercritical H_2O is formed, many of the hydrogen bonds in the normal liquid are broken. As a result, the dielectric constant can be varied between the normal value for the liquid of 80, and 5. At the lower end of this range, supercritical water acts like a nonpolar solvent, and is effective in dissolving organic materials. Supercritical water has considerable potential for use at the high temperatures at which organic solvents begin to decompose. For example, it can be used to destroy toxic substances through oxidation in the decontamination of contaminated groundwater. One challenge in using supercritical water is that it is highly corrosive, placing demands on the materials used to construct a usable facility.

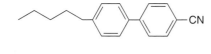

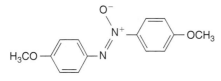

FIGURE 8.18

Liquid crystals are generally formed from polar organic molecules with a rod-like shape.

8.9 Liquid Crystals and LCD Displays

Liquid crystals are an exception to the general statement (superfluid He is another) that there are three equilibrium states of matter, namely, solids, liquids, and gases. Note that **glasses,** which are liquids of such high viscosity that they cannot achieve equilibrium on the timescale of a human lifetime, are a commonly encountered nonequilibrium state of matter. An example is SiO_2 in the form of window glass. The properties of **liquid crystals** are intermediate between liquids and solids. Molecules that form liquid crystals are typically rod shaped and about 2.5 nm in length. Several such molecules are shown in Figure 8.18.

How do liquid crystals differ from the other states of matter? The ordering in solid, liquid, and liquid crystal phases of such molecules is shown schematically in Figure 8.19. Whereas the solid phase is perfectly ordered and the liquid phase has no

FIGURE 8.19

Solid **(a)** liquid **(b)** and liquid crystal **(c)** phases are shown.

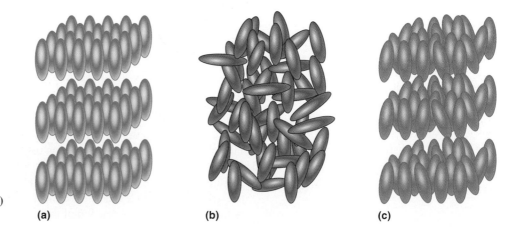

(a)   (b)   (c)

residual order, the liquid crystal phase retains some order. The axes of all molecules deviate only by a small amount from their value in the liquid crystal structure shown. A phase with this kind oriental ordering is called a **nematic phase.**

The **twisted nematic phase** shown in Figure 8.20 is of particular importance, because liquid crystals with this structure are the basis of the multibillion-dollar liquid crystal display (LCD) industry and also the basis for sensor strips that change color with temperature. The twisted nematic phase consists of parallel planes in which the angle of the preferred orientation direction increases in a well-defined manner as the number of layers increases. If light is incident on such a crystal, it is partially reflected and partially transmitted from a number of layers. A maximum in the reflection occurs if the angle of incidence and the spacing between layers is such that constructive interference occurs in the light reflected from successive layers. Because this occurs only in a narrow range of wavelengths, the film appears colored, with the color determined by the wavelength. As the temperature increases or decreases, the layer spacing and, therefore, the color of the reflected light changes. For this reason, the color of the liquid crystal film changes with temperature.

The way in which an **LCD display** functions is illustrated in Figure 8.21. A twisted nematic liquid crystal is sandwiched between two transparent conducting electrodes. The thickness of the film is such that the orientational direction rotates by a total of 90° through the film. Ambient light is incident on an upper polarizing filter that allows only one direction of polarization to pass. The rod-like molecules act as a waveguide and rotate the plane of polarization as the light passes through the liquid crystal film. Therefore, the light passes through the lower polarizer, which is rotated by 90° with respect to the first.

If an electric field is applied to the electrodes, the rotation of the orientational direction from plane to plane no longer occurs because the polar molecules align along the electric field. Therefore, the plane of polarization of the light transmitted by the upper polarizer is not rotated as it passes through the film. As a result, the light is not able to pass through the second polarizer. Now imagine the lower electrode to be in the form of a mirror. In the absence of the electric field (Figure 8.21a), the display will appear bright because the plane of polarization of the light reflected from the mirror is again rotated 90° as it passes back through the film. Consequently, it passes through the upper polarizer. By contrast, no light is reflected if the field is on (Figure 8.21b) and the display appears dark. Now imagine each of the electrodes to be patterned as shown in Figure 8.21c, and it will be clear why in a liquid crystal watch display dark numbers are observed on a light background.

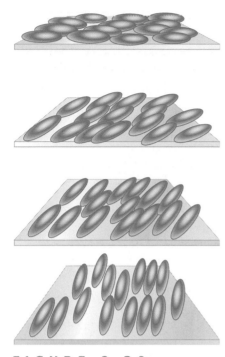

FIGURE 8.20
In a twisted nematic phase, the direction of preferential orientation of the rod-like molecules increases in a well-defined manner from one plane to the next.

FIGURE 8.21
A LCD consists of a twisted nematic liquid film enclosed between parallel transparent conducting electrodes. Polarizers whose transmission direction is rotated 90 degrees with respect to one another are mounted on the electrodes. **(a)** Light passing through the first polarizer is transmitted by the second polarizer because the plane of polarization of the light is rotated by the crystal. **(b)** The orientational ordering of the twisted nematic phase is destroyed by application of the electric field. No light is passed. **(c)** Arrangement of electrodes in a LCD alphanumeric display.

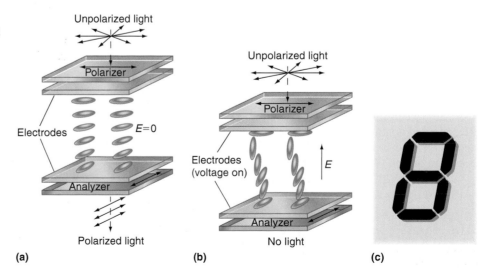

(a)   (b)   (c)

Vocabulary

boiling point elevation

capillary depression

capillary rise

Clapeyron equation

coexistence curve

contact angle

critical point

freeze drying

freezing point depression

freezing point elevation

glass

LCD display

liquid crystals

nematic phase

nonwetting

normal boiling temperature

phase

phase diagram

P–T phase diagram

P–V phase diagram

P–V–T phase diagram

standard boiling temperature

sublimation temperature

supercritical fluids

surface tension

tensile strength

triple point

Trouton's rule

twisted nematic phase

vapor pressure

wetting

Questions on Concepts

Q8.1 At a given temperature, a liquid can coexist with its gas at a single value of the pressure. However, you can sense the presence of $H_2O(g)$ above the surface of a lake by the humidity, and it is still there if the barometric pressure rises or falls at constant temperature. How is this possible?

Q8.2 Why is it reasonable to show the μ versus T segments for the three phases as straight lines as is done in Figure 8.1? More realistic curves would have some curvature. Is the curvature upward or downward on a μ versus T plot?

Q8.3 Show the paths $n \rightarrow o \rightarrow p \rightarrow q$ and $a \rightarrow b \rightarrow c \rightarrow d \rightarrow e \rightarrow f$ of the P–V–T phase diagram of Figure 8.13 in the P–T phase diagram of Figure 8.4.

Q8.4 Figure 8.5 is not drawn to scale. What would be the relative lengths on the q_P axis of the liquid + solid, liquid, and liquid + gas segments for water if the drawing were to scale and the system consisted of H_2O?

Q8.5 Why is $\Delta H_{sublimation} = \Delta H_{fusion} + \Delta H_{vaporization}$?

Q8.6 A triple point refers to a point in a P–T phase diagram for which three phases are in equilibrium. Do all triple points correspond to a gas–liquid–solid equilibrium?

Q8.7 Why are the triple point temperature and the normal freezing point very close in temperature for most substances?

Q8.8 As the pressure is increased at –45°C, ice I is converted to ice II. Which of these phases has the lower density?

Q8.9 What is the physical origin of the pressure difference across a curved liquid–gas interface?

Q8.10 Why does the triple point in a P–T diagram become a triple line in a P–V diagram?

Q8.11 Give a molecular level explanation as to why the surface tension of $Hg(l)$ is not zero.

Problems

Problem numbers in **red** indicate that the solution to the problem is given in the *Student's Solutions Manual*.

P8.1 In this problem, you will calculate the differences in the chemical potentials of ice and supercooled water, and of steam and superheated water all at 1 atm pressure shown schematically in Figure 8.1. For this problem, $S^{\circ}_{H_2O,s}$ = 48.0 J mol^{-1} K^{-1}, $S^{\circ}_{H_2O,l}$ = 70.0 J mol^{-1} K^{-1}, and $S^{\circ}_{H_2O,g}$ = 188.8 J mol^{-1} K^{-1}.

a. By what amount does the chemical potential of water exceed that of ice at $-5.00°C$?

b. By what amount does the chemical potential of water exceed that of steam at 105.00°C?

P8.2 The phase diagram of NH_3 can be characterized by the following information. The normal melting and boiling temperatures are 195.2 and 239.82 K, respectively, the triple point pressure and temperature are 6077 Pa and 195.41 K, respectively. The critical point parameters are 112.8×10^5 Pa and 405.5 K. Make a sketch of the P-T phase diagram (not necessarily to scale) for NH_3. Place a point in the phase diagram for the following conditions. State which and how many phases are present.

a. 195.41 K, 1050 Pa

b. 195.41 K, 6077 Pa

c. 237.51 K, 101325 Pa

d. 420 K, 130×10^5 Pa

e. 190 K, 6077 Pa

P8.3 Within what range can you restrict the values of P and T if the following information is known about CO_2? Use Figure 8.10 to answer this question.

a. As the temperature is increased, the solid is first converted to the liquid and subsequently to the gaseous state.

b. As the pressure on a cylinder containing pure CO_2 is increased from 65 to 80 atm, no interface delineating liquid and gaseous phases is observed.

c. Solid, liquid, and gas phases coexist at equilibrium.

d. An increase in pressure from 10 to 50 atm converts the liquid to the solid.

e. An increase in temperature from $-80°$ to $20°C$ converts a solid to a gas with no intermediate liquid phase.

P8.4 Within what range can you restrict the values of P and/or T if the following information is known about sulfur? Use Figure 8.9 to answer this problem.

a. Only the monoclinic solid phase is observed for $P = 1$ atm.

b. When the pressure on the vapor is increased, the liquid phase is formed.

c. Solid, liquid, and gas phases coexist at equilibrium.

d. As the temperature is increased, the rhombic solid phase is converted to the liquid directly.

e. As the temperature is increased at 1 atm, the monoclinic solid phase is converted to the liquid directly.

P8.5 The vapor pressure of liquid SO_2 is 2232 Pa at 201 K, and $\Delta H_{vaporization} = 24.94$ kJ mol^{-1}. Calculate the normal and standard boiling points. Does your result for the normal boiling point agree with that in Table 8.1? If not, suggest a possible cause.

P8.6 For water, $\Delta H_{vaporization}$ is 40.65 kJ mol^{-1}, and the normal boiling point is 373.15 K. Calculate the boiling point for water on the top of a mountain of height 5500 m, where the normal barometric pressure is 380 Torr.

P8.7 Use the values for ΔG_f(ethanol, l) and ΔG_f (ethanol, g) from Appendix A to calculate the vapor pressure of ethanol at 298.15 K.

P8.8 Use the vapor pressures of ClF_3 given in the following table to calculate the enthalpy of vaporization using a graphical method or a least squares fitting routine.

T (°C)	P (Torr)	T (°C)	P (Torr)
−246.97	29.06	−233.14	74.31
−41.51	42.81	−30.75	86.43
−35.59	63.59	−27.17	107.66

P8.9 Use the following vapor pressures of 1-butene given here to calculate the enthalpy of vaporization using a graphical method or a least squares fitting routine.

T (K)	P (atm)
273.15	1.268
275.21	1.367
277.60	1.490
280.11	1.628
283.15	1.810

P8.10 Use the vapor pressures of Cl_2 given in the following table to calculate the enthalpy of vaporization using a graphical method or a least squares fitting routine.

T (K)	P (atm)	T (K)	P (atm)
227.6	0.585	283.15	4.934
238.7	0.982	294.3	6.807
249.8	1.566	305.4	9.173
260.9	2.388	316.5	12.105
272.0	3.483	327.6	15.676

P8.11 Use the vapor pressures of n-butane given in the following table to calculate the enthalpy of vaporization using a graphical method or a least squares fitting routine.

T (K)	P (Torr)	T (K)	P (Torr)
187.45	5.00	220.35	60.00
195.35	10.00	228.95	100.00
204.25	20.00	241.95	200.0
214.05	40.00	256.85	400.0

P8.12 Use the vapor pressures of ice given here to calculate the enthalpy of sublimation using a graphical method or a least squares fitting routine.

T (°C)	P(Torr)
−28.00	0.3510
−29.00	0.3169
−30.00	0.2859
−31.00	0.2575
−32.00	0.2318

P8.13 Carbon tetrachloride melts at 250 K. The vapor pressure of the liquid is 10,539 Pa at 290 K and 74,518 Pa at 340 K. The vapor pressure of the solid is 270 Pa at 232 K and 1092 Pa at 250 K.

a. Calculate $\Delta H_{vaporization}$ and $\Delta H_{sublimation}$.

b. Calculate ΔH_{fusion}.

c. Calculate the normal boiling point and $\Delta S_{vaporization}$ at the boiling point.

d. Calculate the triple point pressure and temperature.

P8.14 It has been suggested that the surface melting of ice plays a role in enabling speed skaters to achieve peak performance. Carry out the following calculation to test this hypothesis. At 1 atm pressure, ice melts at 273.15 K, $\Delta H_{fusion} = 6010$ J mol^{-1}, the density of ice is 920 kg m^{-3}, and the density of liquid water is 997 kg m^{-3}.

a. What pressure is required to lower the melting temperature by 5.0°C?

b. Assume that the width of the skate in contact with the ice has been reduced by sharpening to 25×10^{-3} cm, and that the length of the contact area is 15 cm. If a skater of mass 85 kg is balanced on one skate, what pressure is exerted at the interface of the skate and the ice?

c. What is the melting point of ice under this pressure?

d. If the temperature of the ice is $-5.0°C$, do you expect melting of the ice at the ice–skate interface to occur?

P8.15 Solid iodine, $I_2(s)$, at 25°C has an enthalpy of sublimation of 56.30 kJ mol^{-1}. The $C_{P,m}$ of the vapor and solid phases at that temperature are 36.9 and 54.4 J K^{-1} mol^{-1}, respectively. The sublimation pressure at 25°C is 0.30844 Torr. Calculate the sublimation pressure of the solid at the melting point (113.6°C) assuming

a. that the enthalpy of sublimation and the heat capacities do not change with temperature.

b. that the enthalpy of sublimation at temperature T can be calculated from the equation $\Delta H°_{sublimation}(T)$ $= \Delta H°_{sublimation}(T_0) + \Delta C_P(T - T_0)$

P8.16 Carbon disulfide, $CS_2(l)$, at 25°C has a vapor pressure of 0.4741 bar and an enthalpy of vaporization of 27.66 kJ mol^{-1}. The $C_{P,m}$ of the vapor and liquid phases at that temperature are 45.4 and 75.7 J K^{-1} mol^{-1}, respectively. Calculate the vapor pressure of $CS_2(l)$ at 100.0°C assuming

a. that the enthalpy of sublimation and the heat capacities do not change with temperature.

b. that the enthalpy of sublimation at temperature T can be calculated from the equation $\Delta H°_{sublimation}(T)$ $= \Delta H°_{sublimation}(T_0) + \Delta C_P(T - T_0)$

P8.17 Consider the transition between two forms of solid tin, $Sn(s, gray) \rightleftarrows Sn(s, white)$. The two phases are in equilibrium at 1 bar and 18°C. The densities for gray and white tin are 5750 and 7280 kg m^{-3}, respectively, and $\Delta S_{transition} = 8.8$ J K^{-1} mol^{-1}. Calculate the temperature at which the two phases are in equilibrium at 200 bar.

P8.18 You have collected a tissue specimen that you would like to preserve by freeze drying. To ensure the integrity of the specimen, the temperature should not exceed -10.5°C. The vapor pressure of ice at 273.16 K is 611 Pa; $\Delta H°_{fusion} = 6.01$ kJ mol^{-1} and $\Delta H°_{vaporization} = 40.65$ kJ mol^{-1}. What is the maximum pressure at which the freeze drying can be carried out?

P8.19 The vapor pressure of methanol (l) is given by

$$\ln\left(\frac{P}{Pa}\right) = 23.593 - \frac{3.6791 \times 10^3}{\frac{T}{K} - 31.317}$$

a. Calculate the standard boiling temperature.

b. Calculate $\Delta H_{vaporization}$ at 298 K and at the standard boiling temperature.

P8.20 The vapor pressure of a liquid can be written in the empirical form known as the Antoine equation, where $A(1)$, $A(2)$, and $A(3)$ are constants determined from measurements:

$$\ln\frac{P(T)}{Pa} = A(1) - \frac{A(2)}{\frac{T}{K} + A(3)}$$

Starting with this equation, derive an equation giving $\Delta H_{vaporization}$ as a function of temperature.

P8.21 The vapor pressure of an unknown solid is approximately given by $\ln(P/Torr) = 22.413 - 2035(K/T)$, and

the vapor pressure of the liquid phase of the same substance is approximately given by $\ln(P/Torr) = 18.352 - 1736(K/T)$.

a. Calculate $\Delta H_{vaporization}$ and $\Delta H_{sublimation}$.

b. Calculate ΔH_{fusion}.

c. Calculate the triple point temperature and pressure.

P8.22 The densities of a given solid and liquid of molecular weight 122.5 at its normal melting temperature of 427.15 K are 1075 and 1012 kg m^{-3}, respectively. If the pressure is increased to 120 bar, the melting temperature increases to 429.35 K. Calculate $\Delta H°_{fusion}$ and $\Delta S°_{fusion}$ for this substance.

P8.23 In Equation (8.13), $(dP/dT)_{vaporization}$ was calculated by assuming that $V_m^{gas} \gg V_m^{liquid}$. In this problem, you will test the validity of this approximation. For water at its normal boiling point of 373.13 K, $\Delta H_{vaporization} = 40.65 \times 10^3$ J mol^{-1}, $\rho_{liquid} = 958.66$ kg m^{-3}, and $\rho_{gas} = 0.58958$ kg m^{-3}. Compare the calculated values for $(dP/dT)_{vaporization}$ with and without the approximation of Equation (8.13). What is the relative error in making the approximation?

P8.24 The variation of the vapor pressure of the liquid and solid forms of a pure substance near the triple point are given by $\ln(P_{solid}/Pa) = -8750(K/T) + 31.143$ and $\ln(P_{liquid}/Pa) = -4053(K/T) + 21.10$. Calculate the temperature and pressure at the triple point.

P8.25 Calculate the vapor pressure of CS_2 at 298 K if He is added to the gas phase at a partial pressure of 200 bar. The vapor pressure of CS_2 is given by the empirical equation

$$\ln\frac{P(T)}{Pa} = 20.801 - \frac{2.6524 \times 10^3}{\frac{T}{K} - 33.402}$$

The density of CS_2 at this temperature is 1255.5 kg m^{-3}. By what factor does the vapor pressure change?

P8.26 Use the vapor pressures for PbS given in the following table to estimate the temperature and pressure of the triple point and also the enthalpies of fusion, vaporization, and sublimation.

Phase	T (°C)	P (Torr)
Solid	1048	40.0
Solid	1108	100
Liquid	1221	400
Liquid	1281	760

P8.27 Use the vapor pressures for C_2N_2 given in the following table to estimate the temperature and pressure of the triple point and also the enthalpies of fusion, vaporization, and sublimation.

Phase	T (°C)	P (Torr)
Solid	−62.7	40.0
Solid	−51.8	100
Liquid	−33.0	400
Liquid	−21.0	760

P8.28   A reasonable approximation to the vapor pressure of krypton is given by $\log_{10}(P/\text{Torr}) = b - 0.05223(a/T)$

For solid krypton, $a = 10065$ and $b = 7.1770$. For liquid krypton, $a = 9377.0$ and $b = 6.92387$. Use these formulas to estimate the triple point temperature and pressure and also the enthalpies of vaporization, fusion, and sublimation of krypton.

P8.29   The normal melting point of H_2O is 273.15 K, and $\Delta H_{fusion} = 6010$ J mol^{-1}. Calculate the decrease in the normal freezing point at 100 and 500 bar assuming that the density of the liquid and solid phases remains constant at 997 and 917 kg m^{-3}, respectively.

P8.30   Autoclaves that are used to sterilize surgical tools require a temperature of 120°C to kill bacteria. If water is used for this purpose, at what pressure must the autoclave operate? The normal boiling point of water is 373.15 K, and $\Delta H_{vaporization}^{\circ} = 40.656 \times 10^3$ J mol^{-1} at the normal boiling point.

P8.31   The vapor pressure of $H_2O(l)$ is 23.766 Torr at 298.15 K. Use this value to calculate $\Delta G_f^{\circ}(H_2O, g) - \Delta G_f^{\circ}(H_2O, l)$. Compare your result with those in Table 4.1.

P8.32   Calculate the difference in pressure across the liquid–air interface for a water droplet of radius 150 nm.

P8.33   Calculate the factor by which the vapor pressure of a droplet of methanol of radius 1.00×10^{-4} m at 45.0°C in equilibrium with its vapor is increased with respect to a very large droplet. Use the tabulated value of the density and the surface tension at 298 K from Appendix A for this problem. (*Hint:* You need to calculate the vapor pressure of methanol at this temperature.)

P8.34   Calculate the vapor pressure of water droplets of radius 1.25×10^{-8} m at 360 K in equilibrium with water vapor. Use the tabulated value of the density and the surface tension at 298 K from Appendix A for this problem. (*Hint:* You need to calculate the vapor pressure of water at this temperature.)

P8.35   In Section 8.7, it is stated that the maximum height of a water column in which cavitation does not occur is ~9.7 m. Show that this is the case at 298 K.

CHAPTER 9

Ideal and Real Solutions

In an ideal solution of A and B, the A–B interactions are the same as the A–A and B–B interactions. In this case, the equilibrium between the solution concentration and gas-phase partial pressure is described by Raoult's law for each component. In an ideal solution, the vapor over a solution is enriched in the most volatile component, allowing a separation into its components through fractional distillation. Nonvolatile solutes lead to a decrease of the vapor pressure above a solution. Such solutions exhibit a freezing point depression and a boiling point elevation. These properties depend only on the concentration and not on the identity of the nonvolatile solutes for an ideal solution. Real solutions are described by a modification of the ideal dilute solution model. In an ideal dilute solution, the solvent obeys Raoult's law, and the solute obeys Henry's law. This model is limited in its applicability. To quantify deviations from the ideal dilute solution model, we introduce the concept of the *activity*. The activity of a component of the solution is defined with respect to a standard state. Knowledge of the activities of the various components of a reactive mixture is essential in modeling chemical equilibrium. The thermodynamic equilibrium constant for real solutions is calculated by expressing the reaction quotient Q in terms of activities rather than concentrations. ∎

9.1 Defining the Ideal Solution

In an ideal gas, the atoms or molecules do not interact with one another. Clearly, this is not a good model for a liquid, because without attractive interactions, a gas will not condense. The attractive interaction in liquids varies greatly, as shown by the large variation in boiling points among the elements; for instance, helium has a normal boiling point of 4.2 K, whereas hafnium has a boiling point of 5400 K.

In developing a model for solutions, the vapor phase that is in equilibrium with the solution must be taken into account. Consider pure liquid benzene in a beaker placed in a closed room. Because the liquid is in equilibrium with the vapor phase above it, there is a nonzero partial pressure of benzene in the air surrounding the beaker. This pressure is called the vapor pressure of benzene at the temperature of the liquid. What happens when toluene is added to the beaker? The partial pressure of benzene is reduced, and the vapor phase now contains both benzene and toluene. For this particular mixture, the partial pressure of each component (i) above the liquid is given by

$$P_i = x_i P_i^* \quad i = 1,2 \tag{9.1}$$

where x_i is the mole fraction of that component in the liquid. This equation states that the partial pressure of each of the two components is directly proportional to the vapor pressure of the corresponding pure substance, P_i^*, and that the proportionality constant is x_i. Equation (9.1) is known as **Raoult's law** and is the definition of an **ideal solution.** Raoult's law holds for each substance in an ideal solution over the range from $0 \le x_i \le 1$. In binary solutions, one refers to the component that has the higher value of x_i as the **solvent** and the component that has the lower value of x_i as the **solute.**

Few solutions satisfy Raoult's law. However, it is useful to study the thermodynamics of ideal solutions and to introduce departures from ideal behavior later. Why is Raoult's law not generally obeyed over the whole concentration range of a binary solution consisting of molecules A and B? Equation (9.1) is only obeyed if the A–A, B–B, and A–B interactions are all equally strong. This criterion is satisfied for a mixture of benzene and toluene because the two molecules are very similar in size, shape, and chemical properties. However, it is not satisfied for arbitrary molecules. Raoult's law is an example of a **limiting law;** the solvent in a real solution obeys Raoult's law as the solution becomes highly dilute.

Raoult's law is derived in Example Problem 9.1 and can be rationalized using the model depicted in Figure 9.1. In the solution, molecules of solute are distributed in the solvent. The solution is in equilibrium with the gas phase, and the gas-phase composition is determined by a dynamic balance between evaporation from the solution and condensation from the gas phase, as indicated for one solvent and one solute molecule in Figure 9.1.

FIGURE 9.1
Schematic model of a solution. The white and black spheres represent solvent and solute molecules, respectively.

EXAMPLE PROBLEM 9.1

Assume that the rates of evaporation, R_{evap}, and condensation, R_{cond}, of the solvent from the surface of pure liquid solvent are given by the expressions

$$R_{evap} = A k_{evap}$$

$$R_{cond} = A k_{cond} P_{solvent}^*$$

where A is the surface area of the liquid and k_{evap} and k_{cond} are the rate constants for evaporation and condensation, respectively. Derive a relationship between the vapor pressure of the solvent above a solution and above the pure solvent.

Solution
For the pure solvent, the equilibrium vapor pressure is found by setting the rates of evaporation and condensation equal:

$$R_{evap} = R_{cond}$$

$$A k_{evap} = A k_{cond} P_{solvent}^*$$

$$P_{solvent}^* = \frac{k_{evap}}{k_{cond}}$$

Next, consider the ideal solution. In this case, the rate of evaporation is reduced by the factor $x_{solvent}$.

$$R_{evap} = A k_{evap} x_{solvent}$$

$$R_{cond} = A k_{cond} P_{solvent}$$

and at equilibrium

$$R_{evap} = R_{cond}$$

$$Ak_{evap}x_{solvent} = Ak_{cond}P_{solvent}$$

$$P_{solvent} = \frac{k_{evap}}{k_{cond}}x_{solvent} = P^*_{solvent}x_{solvent}$$

The derived relationship is Raoult's law.

9.2 The Chemical Potential of a Component in the Gas and Solution Phases

If the liquid and vapor phases are in equilibrium, the following equation holds for each component of the solution, where μ_i is the chemical potential of species i:

$$\mu_i^{solution} = \mu_i^{vapor} \tag{9.2}$$

Recall from Section 6.3 that the chemical potential of a substance in the gas phase is related to its partial pressure, P_i, by

$$\mu_i^{vapor} = \mu_i^{\circ} + RT\ln\frac{P_i}{P^{\circ}} \tag{9.3}$$

where μ_i° is the chemical potential of pure component i in the gas phase at the standard state pressure $P^{\circ} = 1$ bar. Because at equilibrium $\mu_i^{solution} = \mu_i^{vapor}$, Equation (9.3) can be written in the form

$$\mu_i^{solution} = \mu_i^{\circ} + RT\ln\frac{P_i}{P^{\circ}} \tag{9.4}$$

For pure liquid i in equilibrium with its vapor, $\mu_i^*(\text{liquid}) = \mu_i^*(\text{vapor}) = \mu_i^*$. Therefore, the chemical potential of the pure liquid is given by

$$\mu_i^* = \mu_i^{\circ} + RT\ln\frac{P_i^*}{P^{\circ}} \tag{9.5}$$

Subtracting Equation (9.5) from (9.4) gives

$$\mu_i^{solution} = \mu_i^* + RT\ln\frac{P_i}{P_i^*} \tag{9.6}$$

For an ideal solution, $P_i = x_iP_i^*$. Combining Equations (9.6) and (9.1), the central equation describing ideal solutions is obtained:

$$\mu_i^{solution} = \mu_i^* + RT\ln x_i \tag{9.7}$$

This equation relates the chemical potential of a component in an ideal solution to the chemical potential of the pure liquid form of component i and the mole fraction of that component in the solution. This equation is most useful in describing the thermodynamics of solutions in which all components are volatile and miscible in all proportions.

Keeping in mind that $\mu_i = G_{i,m}$, the form of Equation (9.7) is identical to that derived for the Gibbs energy of a mixture of gases in Section 6.5. Therefore, one can derive relations for the thermodynamics of mixing to form ideal solutions that are identical to those developed in Section 6.6. We repeat them here as Equation (9.8). Note in particular that ΔH_{mixing} and ΔV_{mixing} are zero for an ideal solution:

$$\Delta G_{mixing} = nRT \sum_i x_i \ln x_i \qquad (9.8)$$

$$\Delta S_{mixing} = -\left(\frac{\partial \Delta G_{mixing}}{\partial T}\right)_P = -nR \sum_i x_i \ln x_i$$

$$\Delta V_{mixing} = \left(\frac{\partial \Delta G_{mixing}}{\partial P}\right)_{T,n_1,n_2} = 0 \quad \text{and}$$

$$\Delta H_{mixing} = \Delta G_{mixing} + T \Delta S_{mixing} = nRT \sum_i x_i \ln x_i - T\left(nR \sum_i x_i \ln x_i\right) = 0$$

EXAMPLE PROBLEM 9.2

An ideal solution is made from 5.00 mol of benzene and 3.25 mol of toluene. Calculate ΔG_{mixing} and ΔS_{mixing} at 298 K and 1 bar pressure. Is mixing a spontaneous process?

Solution

The mole fractions of the components in the solutions are $x_{benzene} = 0.606$ and $x_{toluene} = 0.394$.

$$\Delta G_{mixing} = nRT \sum_i x_i \ln x_i$$

$$= 8.25 \text{ moles} \times 8.314 \text{ J mol}^{-1} \text{ K}^{-1} \times 298\text{K} \times (0.606 \ln 0.606 + 0.394 \ln 0.394)$$

$$= -13.7 \times 10^3 \text{ J}$$

$$\Delta S_{mixing} = -nR \sum_i x_i \ln x_i$$

$$= -8.25 \text{ moles} \times 8.314 \text{ J mol}^{-1} \text{ K}^{-1} \times (0.606 \ln 0.606 + 0.394 \ln 0.394)$$

$$= 46.0 \text{ J K}^{-1}$$

Mixing is spontaneous because $\Delta G_{mixing} < 0$. If two liquids are miscible, it is always true that $\Delta G_{mixing} < 0$.

9.3 Applying the Ideal Solution Model to Binary Solutions

Although the ideal solution model can be applied to any number of components, the focus in this chapter is on simplifying the mathematics, so binary solutions, which consist of only two components, will be used. Because Raoult's law holds for both components of the mixture, $P_1 = x_1 P_1^*$ and $P_2 = x_2 P_2^* = (1 - x_1)P_2^*$. The total pressure in the gas phase varies linearly with the mole fraction of each of its components in the liquid:

$$P_{total} = P_1 + P_2 = x_1 P_1^* + (1 - x_1)P_2^* = P_2^* + (P_1^* - P_2^*)x_1 \qquad (9.9)$$

The individual partial pressures as well as P_{total} above a benzene–ethene chloride (CH_2Cl_2) solution are shown in Figure 9.2. Small deviations from Raoult's law are seen. Such deviations are typical, because few solutions obey Raoult's law exactly. Nonideal solutions, which generally exhibit large deviations form Raoult's law, are discussed in Section 9.9.

The concentration unit used in Equation (9.9) is the mole fraction of each component in the liquid phase. The mole fraction of each component in the gas phase can also

FIGURE 9.2
The vapor pressure of benzene (yellow), ethene chloride (red), and the total vapor pressure (blue) above the solution is shown as a function of $x_{ethene\ chloride}$. The symbols are data points [J. v. Zawidski, *Zeitschrift für Physikalische Chemie*, 35 (1900) 129]. The solid lines are polynomial fits to the data. The dashed lines are calculated using Raoult's law.

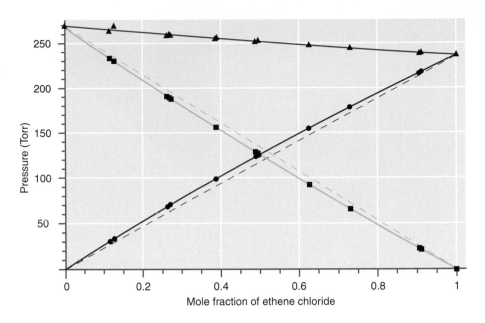

be calculated. Using the symbols y_1 and y_2 to denote the gas-phase mole fractions and the definition of the partial pressure, we can write

$$y_1 = \frac{P_1}{P_{total}} = \frac{x_1 P_1^*}{P_2^* + (P_1^* - P_2^*)x_1} \tag{9.10}$$

To obtain the pressure in the vapor phase as a function of y_1, we first solve Equation (9.10) for x_1:

$$x_1 = \frac{y_1 P_2^*}{P_1^* + (P_2^* - P_1^*)y_1} \tag{9.11}$$

and obtain P_{total} from $P_{total} = P_2^* + (P_1^* - P_2^*)x_1$

$$P_{total} = \frac{P_1^* P_2^*}{P_1^* + (P_2^* - P_1^*)y_1} \tag{9.12}$$

Equation (9.12) can be rearranged to give an equation for y_1 in terms of the vapor pressures of the pure components and the total pressure:

$$y_1 = \frac{P_1^* P_{total} - P_1^* P_2^*}{P_{total}(P_1^* - P_2^*)} \tag{9.13}$$

The variation of the total pressure with x_1 and y_1 is not the same, as is seen in Figure 9.3. In Figure 9.3a, the system consists of a single-phase liquid for pressures above the curve and of a two-phase vapor–liquid mixture for $P–x_1$ points lying on the curve. Only points lying above the curve are meaningful, because points lying below the curve do not correspond to equilibrium states at which the liquid is present. If the system were placed in such an unstable state, liquid would evaporate to bring the pressure up to the value on the liquid–vapor coexistence curve. If there is not enough liquid to bring the pressure up to the curve, all the liquid will evaporate and x_1 is no longer a meaningful variable. In Figure 9.3b, the system consists of a single-phase vapor for pressures below the curve and of a two-phase vapor–liquid mixture for $P–y_1$ points lying on the curve. For the reason discussed earlier, points lying above the curve do not correspond to equilibrium states. The excess vapor would condense to form liquid.

FIGURE 9.3

The total pressure above a benzene–toluene ideal solution is shown for different values of **(a)** the mole fraction of benzene in the solution and **(b)** as a function of the mole fraction of benzene in the vapor. Points on the curves correspond to vapor–liquid coexistence. Only the curves and the shaded areas are of physical significance as explained in text.

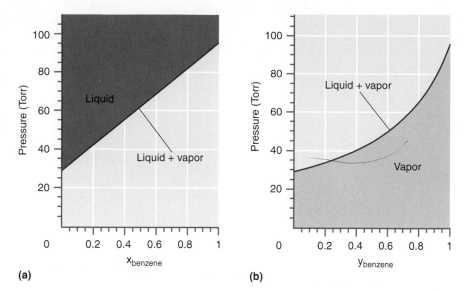

(a)   (b)

Note that the pressure is plotted as a function of different variables in the two parts of Figure 9.3. To compare the gas phase and liquid composition at a given total pressure, both are graphed as a function of $Z_{benzene}$, which is called the **average composition** of benzene in the whole system in Figure 9.4. The average composition Z is defined by

$$Z_{benzene} = \frac{n_{benzene}^{liquid} + n_{benzene}^{vapor}}{n_{toluene}^{liquid} + n_{toluene}^{vapor} + n_{benzene}^{liquid} + n_{benzene}^{vapor}} = \frac{n_{benzene}}{n_{total}}$$

In the region labeled "Liquid" in Figure 9.4, the system consists entirely of a liquid phase, and $Z_{benzene} = x_{benzene}$. In the region labeled "Vapor," the system consists entirely of a gaseous phase and $Z_{benzene} = y_{benzene}$. The area separating the single-phase liquid and vapor regions corresponds to the two-phase liquid–vapor coexistence region.

To demonstrate the usefulness of this **pressure–average composition (P–Z) diagram,** consider a constant temperature process in which the pressure of the system is decreased from the value corresponding to point a in Figure 9.4. Because $P > P_{total}$, the system is initially entirely in the liquid phase. As the pressure is decreased at constant composition, the system remains entirely in the liquid phase until the constant composi-

FIGURE 9.4

A P–Z phase diagram is shown for a benzene–toluene ideal solution. The upper curve shows the vapor pressure as a function of $x_{benzene}$. The lower curve shows the vapor pressure as a function of $y_{benzene}$. Above the two curves, the system is totally in the liquid phase, and below the two curves, the system is totally in the vapor phase. The area intermediate between the two curves shows the liquid–vapor coexistence region. The horizontal lines connecting the curves are called tie lines.

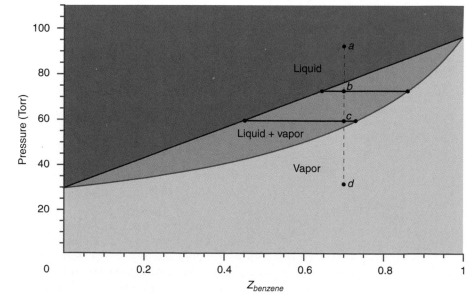

tion line intersects the P versus $x_{benzene}$ curve. At this point, the system enters the two-phase vapor–liquid coexistence region. As point b is reached, what are the values for $x_{benzene}$ and $y_{benzene}$, the mole fractions of the vapor and liquid phases? These values can be determined by constructing a tie line in the two-phase coexistence region.

A **tie line** (see Figure 9.4) is a horizontal line at the pressure of interest that connects the P versus $x_{benzene}$ and P versus $y_{benzene}$ curves. Note that for all values of the pressure, $y_{benzene}$ is greater than $x_{benzene}$, showing that the vapor phase is always enriched in the more volatile or higher vapor pressure component in comparison with the liquid phase.

EXAMPLE PROBLEM 9.3

An ideal solution is made from 5.00 mol of benzene and 3.25 mol of toluene. At 298 K, the vapor pressure of the pure substances are $P^*_{benzene} = 96.4\,\text{Tor}$ and $P^*_{toluene} = 28.9\,\text{Torr}$.

  a. The pressure above this solution is reduced from 760 Torr. At what pressure does the vapor phase first appear?

  b. What is the composition of the vapor under these conditions?

Solution

  a. The mole fractions of the components in the solution are $x_{benzene} = 0.606$ and $x_{toluene} = 0.394$. The vapor pressure above this solution is

$$P_{total} = x_{benzene}P^*_{benzene} + x_{toluene}P^*_{toluene} = 0.606 \times 96.4\,\text{Torr} + 0.394 \times 28.9\,\text{Torr}$$
$$= 69.8\,\text{Torr}$$

No vapor will be formed until the pressure has been reduced to this value.

  b. The composition of the vapor at a total pressure of 69.8 Torr is given by

$$y_{benzene} = \frac{P^*_{benzene}P_{total} - P^*_{benzene}P^*_{toluene}}{P_{total}(P^*_{benzene} - P^*_{toluene})}$$
$$= \frac{96.4\,\text{Torr} \times 69.8\,\text{Torr} - 96.4\,\text{Torr} \times 28.9\,\text{Torr}}{69.8\,\text{Torr} \times (96.4\,\text{Torr} - 28.9\,\text{Torr})} = 0.837$$
$$y_{toluene} = 1 - y_{benzene} = 0.163$$

Note that the vapor is enriched relative to the liquid in the more volatile component, which has the lower boiling temperature.

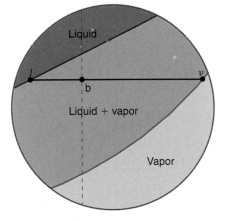

FIGURE 9.5
An enlarged region of the two-phase coexistence region of Figure 9.4 is shown. The vertical line through point b indicates a transition at constant system composition. The lever rule (see text) is used to determine what fraction of the system is in the liquid and vapor phases.

To calculate the relative amount of material in each of the two phases in a coexistence region, we derive the lever rule for a binary solution of the components A and B. Figure 9.5 shows a magnified portion of Figure 9.4 centered at the tie line that passes through point b. We derive the lever rule using the following geometrical argument. The lengths of the line segments lb and bv are given by

$$lb = Z_B - x_B = \frac{n^{tot}_B}{n^{tot}} - \frac{n^{liq}_B}{n^{tot}_{liq}} \tag{9.14}$$

$$bv = y_B - Z_B = \frac{n^{vapor}_B}{n^{tot}_{vapor}} - \frac{n^{tot}_B}{n^{tot}} \tag{9.15}$$

The subscripts and superscripts on n indicate the number of moles of component B in the vapor and liquid phases and the total number of moles of B in the system. If Equation (9.14) is multiplied by n^{tot}_{liq}, Equation (9.15) is multiplied by n^{tot}_{vapor}, and the two equations are subtracted, we find that

$$lb\ n^{tot}_{liq} - bv\ n^{tot}_{vapor} = \frac{n^{tot}_B}{n^{tot}}(n^{tot}_{liq} + n^{tot}_{vapor}) - (n^{liq}_B + n^{vapor}_B) = n^{tot}_B - n^{tot}_B = 0 \tag{9.16}$$

We conclude that $\qquad \dfrac{n^{tot}_{liq}}{n^{tot}_{vap}} = \dfrac{bv}{lb}$

It is convenient to restate this result as

$$n_{liq}^{tot}(Z_B - x_B) = n_{vapor}^{tot}(y_B - Z_B) \tag{9.17}$$

Equation (9.17) is called the **lever rule** in analogy with the torques acting on a lever of length $lb + bv$ with the fulcrum positioned at point b. For the specific case shown in Figure 9.5, $n_{liq}^{tot}/n_{vapor}^{tot} = 2.34$. Therefore, 70% of the total number of moles in the system is in the liquid phase, and 30% is in the vapor phase.

EXAMPLE PROBLEM 9.4

For the benzene–toluene solution of Example Problem 9.3, calculate

a. the total pressure

b. the liquid composition

c. the vapor composition when 1.50 mol of the solution has been converted to vapor.

Solution

The lever rule relates the average composition, $Z_{benzene} = 0.606$, and the liquid and vapor compositions:

$$n_{vapor}(y_{benzene} - Z_{benzene}) = n_{liq}(Z_{benzene} - x_{benzene})$$

Entering the parameters of the problem, this equation simplifies to

$$6.75 x_{benzene} + 1.50 y_{benzene} = 5.00$$

The total pressure is given by

$$P_{total} = x_{benzene}P_{benzene}^* + (1 - x_{benzene})P_{toluene}^* = [96.4 x_{benzene} + 28.9(1 - x_{benzene})]\text{Torr}$$

and the vapor composition is given by

$$y_{benzene} = \frac{P_{benzene}^* P_{total} - P_{benzene}^* P_{toluene}^*}{P_{total}(P_{benzene}^* - P_{toluene}^*)} = \left[\frac{96.4\dfrac{P_{total}}{\text{Torr}} - 2786}{67.5\dfrac{P_{total}}{\text{Torr}}}\right]$$

These three equations in three unknowns can be solved by using an equation solver or by eliminating the variables by combining equations. For example, the first equation can be used to express $y_{benzene}$ in terms of $x_{benzene}$. This result can be substituted into the second and third equations to give two equations in terms of $x_{benzene}$ and p_{total}. The solution for $x_{benzene}$ obtained from these two equations can be substituted in the first equation to give $y_{benzene}$. The answers are $x_{benzene} = 0.561$, $y_{benzene} = 0.810$, and $P_{total} = 66.8$ Torr.

9.4 The Temperature–Composition Diagram and Fractional Distillation

The enrichment of the vapor phase above a solution in the more volatile component is the basis for fractional distillation, an important separation technique in chemistry and in the chemical industry. It is more convenient to discuss fractional distillation using a **temperature–composition diagram** than using the pressure–average composition diagram discussed in the previous section. The temperature–composition diagram gives the temperature of the solution as a function of the average system composition for a predetermined total vapor pressure, P_{total}. It is convenient to use the value $P_{total} = 1$ atm, so that the vertical axis is the normal boiling point of the solution. Figure 9.6 shows a

FIGURE 9.6

FIGURE 9.6
The boiling temperature of an ideal solution of the components benzene and toluene is plotted versus the average system composition, $Z_{benzene}$. The upper red curve shows the boiling temperature as a function of $y_{benzene}$, and the lower red curve shows the boiling temperature as a function of $x_{benzene}$. The area intermediate between the two curves shows the vapor–liquid coexistence region.

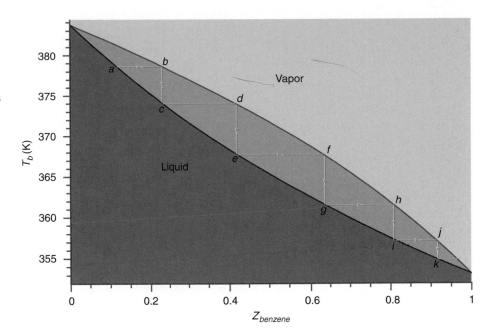

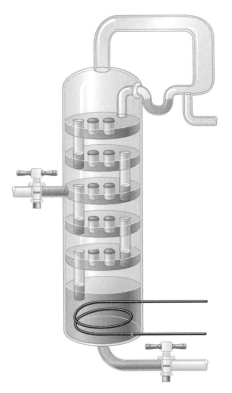

FIGURE 9.7
Schematic of a fractional distillation column. The solution to be separated into its components is introduced at the bottom of the column. The resistive heater provides the energy needed to vaporize the liquid. It can be assumed that the liquid and vapor are at equilibrium at each level of the column. The equilibrium temperature decreases from the bottom to the top of the column.

boiling temperature–composition diagram for a benzene–toluene solution. Neither the $T_b - x_{benzene}$ nor the $T_b - y_{benzene}$ curves are linear in a temperature–composition diagram. Note that because the more volatile component has the lower boiling point, the vapor and liquid regions are inverted when compared with the pressure–average composition diagram.

The principle of **fractional distillation** can be illustrated using the sequence of lines labeled *a* through *k* in Figure 9.6. The vapor above the solution at point *a* is enriched in benzene. If this vapor is separated from the original solution and condensed by lowering the temperature, the resulting liquid will have a higher mole fraction of benzene than the original solution. As for the original solution, the vapor above this separately collected liquid is enriched in benzene. As this process is repeated, the successively collected vapor samples become more enriched in benzene. In the limit of a very large number of steps, the last condensed samples are essentially pure benzene. The multistep procedure described earlier is very cumbersome because it requires the collection and evaporation of many different samples. In practice, the separation into pure benzene and toluene is accomplished using a distillation column, shown schematically in Figure 9.7.

Rather than have the whole system at a uniform temperature, a distillation column operates with a temperature gradient so that the top part of the column is at a lower temperature than the solution being boiled off. The horizontal segments *ab*, *cd*, *ef*, and so on in Figure 9.6 correspond to successively higher (cooler) portions of the column. Each of these segments corresponds to one of the distillation stages in Figure 9.7. Because the vapor is moving upward through the condensing liquid, heat and mass exchange facilitates equilibration between the two phases.

Distillation plays a major role in the separation of crude oil into its various components. The lowest boiling liquid is gasoline, followed by kerosene, diesel fuel and heating oil, and gas oil, which is further broken down into lower molecular weight components by catalytic cracking. It is important in distillation to keep the temperature of the boiling liquid as low as possible to avoid the occurrence of thermally induced reactions. A useful way to reduce the boiling temperature is to distill the liquid under a partial vacuum by pumping on the gas phase. The boiling point of a typical liquid mixture can be reduced by approximately 100°C if the pressure is reduced from 760 to 20 Torr.

Although the principle of fractional distillation is the same for real solutions, it is not possible to separate a binary solution into its pure components if the nonideality is strong enough. If the A–B interactions are more attractive than the A–A and B–B interactions, the boiling point of the solution will go through a maximum at a concentration intermediate between $x_A = 0$ and $x_A = 1$. An example of such a case is an acetone–chloroform

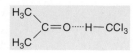

FIGURE 9.8
Hydrogen bond formation between acetone and chloroform leads to the formation of a maximum boiling azeotrope.

TABLE 9.1

Composition and Boiling Temperatures of Selected Azeotropes

Azeotropic Mixture	Boiling Temperature of Components (°C)	Mole Fraction of First Component	Azeotrope Boiling Point (°C)
Water–ethanol	100/78	0.096	78.2
Water–trichloromethane	100/61	0.160	56.1
Water–benzene	100/80	0.295	69.3
Water–toluene	100/111	0.444	84.1
Ethanol–hexane	78/69	0.332	58.7
Ethanol–benzene	78/80	0.440	67.9
Ethyl acetate–hexane	78/69	0.394	65.2
Carbon disulfide–acetone	46/56	0.608	39.3
Toluene–acetic acid	111/118	0.625	100.7

Source: Lide, D. R., Ed., *Handbook of Chemistry and Physics,* 83rd ed. CRC Press, Boca Raton, FL, 2002.

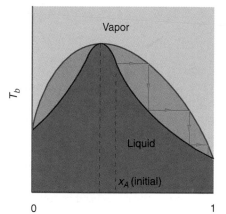

(a)

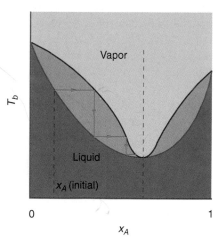

(b)

FIGURE 9.9
A boiling point diagram is shown for **(a)** maximum and **(b)** minimum boiling point azeotropes. The dashed lines indicate the initial composition of the solution and the composition of the azeotrope. The sequence of horizontal segments with arrows corresponds to successively higher (cooler) portions of the column.

mixture. The hydrogen on the chloroform forms a hydrogen bond with the oxygen in acetone as shown in Figure 9.8, leading to stronger A–B than A–A and B–B interactions.

A boiling point diagram for this case is shown in Figure 9.9a. At the maximum boiling temperature, the liquid and vapor composition lines are tangent to one another. Fractional distillation of such a solution beginning with an initial value of x_A greater than that corresponding to the maximum boiling point is shown schematically in the figure. The component with the lowest boiling point will initially emerge at the top of the distillation column. In this case, it will be pure component A. However, the liquid left in the heated flask will not be pure component B, but rather the solution corresponding to the concentration at which the maximum boiling point is reached. Continued boiling of the solution at this composition will lead to evaporation at constant composition. Such a mixture is called an azeotrope, and because $T_{b,azeotrope} > T_{b,A}$, $T_{b,B}$, it is called a maximum boiling azeotrope. An example for a maximum boiling azeotrope is a mixture of H_2O ($T_b = 100°C$) and HCOOH ($T_b = 100°C$) at $x_{H_2O} = 0.427$, which boils at 107.2°C. Other commonly occurring azeotropic mixtures are listed in Table 9.1.

If the A–B interactions are less attractive than the A–A and B–B interactions, a minimum boiling azeotrope can be formed. A schematic boiling point diagram for such an azeotrope is also shown in Figure 9.9b. Fractional distillation of such a solution beginning with an initial value of x_A less than that corresponding to the minimum boiling point leads to a liquid with the azeotropic composition initially emerging at the top of the distillation column. An example for a minimum boiling azeotrope is a mixture of CS_2 ($T_b = 46.3°C$) and acetone ($T_b = 56.2°C$) at $x_{CS_2} = 0.608$, which boils at 39.3°C.

It is still possible to collect one component of an azeotropic mixture using the property that the azeotropic composition is a function of the total pressure as shown in Figure 9.10. The mixture is first distilled at atmospheric pressure and the volatile distillate is collected. This vapor is condensed and subsequently distilled at a reduced pressure for which the azeotrope contains less A (more B). If this mixture is distilled, the azeotrope evaporates, leaving some pure B behind.

9.5 The Gibbs–Duhem Equation

In this section, we show that the chemical potentials of the two components in a binary solution are not independent. This is an important result, because it allows the chemical potential of a nonvolatile solute such as sucrose in a volatile solvent such as

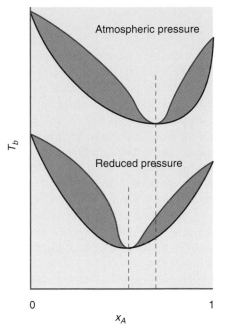

FIGURE 9.10
Because the azeotropic composition depends on the total pressure, pure B can be recovered from the A–B mixture by first distilling the mixture under atmospheric pressure and, subsequently, under a reduced pressure. Note that the boiling temperature is lowered as the pressure is reduced.

water to be determined. Such a solute has no measurable vapor pressure; therefore, its chemical potential cannot be measured directly. As shown later, its chemical potential can be determined knowing only the chemical potential of the solvent as a function of concentration.

From Chapter 6, the differential form of the Gibbs energy is given by

$$dG = -SdT + VdP + \sum_i \mu_i dn_i \tag{9.18}$$

For a binary solution at constant T and P, this equation reduces to

$$dG = \mu_1 dn_1 + \mu_2 dn_2 \tag{9.19}$$

Imagine starting with an infinitesimally small amount of a solution at constant T and P. The amount is gradually increased at constant composition. Because of this restriction, the chemical potentials are unchanged as the size of the system is changed. Therefore, the μ_i can be taken out of the integral:

$$\int_0^G dG' = \mu_1 \int_0^{n_1} dn_1' + \mu_2 \int_0^{n_2} dn_2' \quad \text{or} \tag{9.20}$$

$$G = \mu_1 n_1 + \mu_2 n_2$$

The primes have been introduced to avoid using the same symbol for the integration variable and the upper limit. The total differential of the last equation is

$$dG = \mu_1 dn_1 + \mu_2 dn_2 + n_1 d\mu_1 + n_2 d\mu_2 \tag{9.21}$$

The previous equation differs from Equation (9.19) because, in general, we have to take changes of the composition of the solution into account. Therefore, μ_1 and μ_2 must be regarded as variables. Setting Equations (9.19) and (9.21) equal, one obtains the **Gibbs–Duhem equation** for a binary solution, which can be written in either of two forms:

$$n_1 d\mu_1 + n_2 d\mu_2 = 0 \quad \text{or} \quad x_1 d\mu_1 + x_2 d\mu_2 = 0 \tag{9.22}$$

This equation states that the chemical potentials of the components in a binary solution are not independent. If the change in the chemical potential of the first component is $d\mu_1$, the change of the chemical potential of the second component is given by

$$d\mu_2 = -\frac{n_1 d\mu_1}{n_2} \tag{9.23}$$

The use of the Gibbs–Duhem equation is illustrated in Example Problem 9.5.

EXAMPLE PROBLEM 9.5

One component in a solution follows Raoult's law, $\mu_1^{solution} = \mu_1^* + RT \ln x_1$ over the entire range $0 \le x_1 \le 1$. Using the Gibbs–Duhem equation, show that the second component must also follow Raoult's law.

Solution
From Equation (9.20),

$$d\mu_2 = -\frac{x_1 d\mu_1}{x_2} = -\frac{n_1}{n_2} d(\mu_1^* + RT \ln x_1) = -RT \frac{x_1}{x_2} \frac{dx_1}{x_1}$$

Because $x_1 + x_2 = 1$, then $dx_2 = -dx_1$ and $d\mu_2 = RT \, dx_2/x_2$. Integrating this equation, one obtains $\mu_2 = RT \ln x_2 + C$, where C is a constant of integration. This constant can be evaluated by examining the limit $x_2 \to 1$. This limit corresponds to the pure substance 2 for which $\mu_2 = \mu_2^* = RT \ln 1 + C$. We conclude that $C = \mu_2^*$ and, therefore, $\mu_2^{solution} = \mu_2^* + RT \ln x_2$.

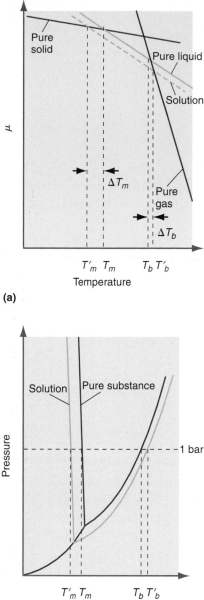

(a)

(b)

FIGURE 9.11

Illustration of the boiling point elevation and freezing point depression in two different ways: **(a)** These effects arise because the chemical potential of the solution is lowered through addition of the nonvolatile solute, while the chemical potential of the vapor and liquid is unaffected at constant pressure. **(b)** The same information from part (a) is shown using a *P–T* phase diagram.

9.6 Colligative Properties

Many solutions consist of nonvolatile solutes that have limited solubility in a volatile solvent. Examples are solutions of sucrose or sodium chloride in water. Important properties of these solutions, including boiling point elevation, freezing point depression, and osmotic pressure are found to depend only on the solute concentration—not on the nature of the solute. These properties are called **colligative properties.** In this section, colligative properties are discussed using the model of the ideal solution. Corrections for nonideality require that activities be used in place of concentrations, as shown in Section 9.13.

As discussed in connection with Figure 9.1, the vapor pressure above a solution containing a solute is lowered with respect to the pure solvent. As will be shown in this and in the next three sections, the solution exhibits a freezing point depression, a boiling point elevation, and an osmotic pressure. Generally, the solute does not crystallize out with the solvent during freezing because the solute cannot be easily integrated into the solvent crystal structure. For this case, the change in chemical potential on dissolution of a nonvolatile solute can be understood using Figure 9.11a.

Only the liquid chemical potential is affected through formation of the solution. Although the gas pressure is lowered by the addition of the solute, the chemical potential of the gas is unaffected because it remains pure and because the comparison in Figure 9.11a is made at constant pressure. The chemical potential of the solid is unaffected because of the assumption that the solute does not crystallize out with the solvent. As shown in the figure, the melting temperature T_m, defined as the intersection of the solid and liquid μ versus T curves, is lowered by dissolution of a nonvolatile solute in the solvent. Similarly, the boiling temperature T_b is raised by dissolution of a nonvolatile solute in the solvent.

The same information is shown in a *P–T* phase diagram in Figure 9.11b. Because the vapor pressure above the solution is lowered by the addition of the solute, the liquid–gas coexistence curve intersects the solid–gas coexistence curve at a lower temperature than for the pure solvent. This intersection defines the triple point, and it must also be the origin of the solid–liquid coexistence curve. Therefore, the solid–liquid coexistence curve is shifted to lower temperatures through the dissolution of the nonvolatile solute. The overall effect of the shifts in the solid–gas and solid–liquid coexistence curves is a **freezing point depression** and a boiling point elevation, both of which depend only on the concentration, and not on the identity, of the solute. The preceding discussion is qualitative in nature. In the next two sections, quantitative relationships are developed between the colligative properties and the concentration of the nonvolatile solute.

9.7 The Freezing Point Depression and Boiling Point Elevation

If the solution is in equilibrium with the pure solid solvent, the following relation must be satisfied:

$$\mu_{solution} = \mu^*_{solid} \tag{9.24}$$

Recall that $\mu_{solution}$ refers to the chemical potential of the solvent in the solution, and μ^*_{solid} refers to the chemical potential of the pure solvent in solid form. From Equation (9.7), we can express $\mu_{solution}$ in terms of the chemical potential of the pure solvent and its concentration and rewrite Equation (9.24) in the form

$$\mu^*_{solvent} + RT \ln x_{solvent} = \mu^*_{solid} \tag{9.25}$$

This equation can be solved for $\ln x_{solvent}$:

$$\ln x_{solvent} = \frac{\mu^*_{solid} - \mu^*_{solvent}}{RT} \tag{9.26}$$

The difference in chemical potentials $\mu^*_{solid} - \mu^*_{solvent} = -\Delta G_{fusion,m}$ so that

$$\ln x_{solvent} = \frac{-\Delta G_{fusion,m}}{RT} \tag{9.27}$$

Because we are interested in how the freezing temperature is related to $x_{solvent}$ at constant pressure, the partial derivative $(\partial T / \partial x_{solvent})_P$ is needed. This quantity can be obtained by differentiating Equation (9.27) with respect to $x_{solvent}$:

$$\left(\frac{\partial \ln x_{solvent}}{\partial x_{solvent}}\right)_P = \frac{1}{x_{solvent}} = -\frac{1}{R}\left(\frac{\partial \frac{\Delta G_{fusion,m}}{T}}{\partial T}\right)_P \left(\frac{\partial T}{\partial x_{solvent}}\right)_I \tag{9.28}$$

The first partial derivative on the right side of Equation (9.28) can be simplified using the Gibbs–Helmholtz equation (see Section 6.3), giving

$$\frac{1}{x_{solvent}} = \frac{\Delta H_{fusion,m}}{RT^2}\left(\frac{\partial T}{\partial x_{solvent}}\right)_P \quad \text{or} \tag{9.29}$$

$$\frac{dx_{solvent}}{x_{solvent}} = d\ln x_{solvent} = \frac{\Delta H_{fusion,m}}{R}\frac{dT}{T^2} \quad (\text{constant } P)$$

This equation can be integrated between the limits given by the pure solvent ($x_{solvent} = 1$) for which the fusion temperature is T_{fusion}, and an arbitrary small concentration of solute for which the fusion temperature is T:

$$\int_1^{x_{solvent}} \frac{dx}{x} = \int_{T_{fusion}}^{T} \frac{\Delta H_{fusion,m}}{R}\frac{dT'}{T'^2} \tag{9.30}$$

For $x_{solvent}$ not very different from one, $\Delta H_{fusion,m}$ is assumed to be independent of T, and Equation (9.30) simplifies to

$$\frac{1}{T} = \frac{1}{T_{fusion}} - \frac{R\ln x_{solvent}}{\Delta H_{fusion,m}} \tag{9.31}$$

It is more convenient in discussing solutions to use the molality of the solute rather than the mole fraction of the solvent as the concentration unit. For dilute solutions $\ln x_{solvent} = \ln(n_{solvent}/n_{solvent} + m_{solute}M_{solvent}n_{solvent}) = -\ln(1 + M_{solvent}n_{solute})$ and Equation (9.31) can be rewritten in terms of the molality (m) rather than the mole fraction. The result is

$$\Delta T_f = -\frac{RM_{solvent}T_{fusion}^2}{\Delta H_{fusion,m}}m_{solute} = -K_f m_{solute} \tag{9.32}$$

$m = \frac{\Delta T_f}{-K_f}$

where $M_{solvent}$ is the molecular mass of the solvent. In going from Equation 9.31 to Equation 9.32 we have made the approximation $\ln(1 + M_{solvent}m_{solute}) \approx M_{solvent}m_{solute}$ and $1/T - 1/T_{fusion} \approx -\Delta T_f/T_{fusion}^2$ Note that K_f depends only on the properties of the solvent and is primarily determined by the molecular mass and the enthalpy of fusion. For most solvents, the magnitude of K_f lies between 1.5 and 10 as shown in Table 9.2. However, K_f can reach unusually high values, for example 30 (K kg)/mole for carbon tetrachloride and 40 (K kg)/mole for camphor.

The **boiling point elevation** can be calculated using the same argument with $\Delta G_{vaporization}$ and $\Delta H_{vaporization}$ substituted for ΔG_{fusion} and ΔH_{fusion}. The resulting equations are

$$\left(\frac{\partial T}{\partial m_{solute}}\right)_{p,m\to 0} = \frac{RM_{solvent}T_{vaporization}^2}{\Delta H_{vaporization,m}} \quad \text{and} \tag{9.33}$$

$$\Delta T_b = \frac{RM_{solvent}T^2}{\Delta H_{vaporization,m}}m_{solute} = K_b m_{solute} \tag{9.34}$$

$m = \frac{\Delta T_b}{K_b}$

$\frac{\Delta T_f}{-K_f} = \frac{\Delta T_b}{K_b}$

Because $\Delta H_{vaporization,m} > \Delta H_{fusion,m}$, it follows that $K_f > K_b$. Typically, K_b values range between 0.5 and 5.0 K/(mole kg^{-1}) as shown in Table 9.2. Note also that, by

TABLE 9.2

Freezing Point Depression and Boiling Point Elevation Constants

Substance	Standard Freezing Point (K)	K_f (K kg mol^{-1})	Standard Boiling Point (K)	K_b (K kg mol^{-1})
Acetic acid	289.6	3.59	391.2	3.08
Benzene	278.6	5.12	353.3	2.53
Camphor	449	40	482.3	5.95
Carbon disulfide	161	3.8	319.2	2.40
Carbon tetrachloride	250.3	30	349.8	4.95
Cyclohexane	279.6	20.0	353.9	2.79
Ethanol	158.8	2.0	351.5	1.07
Phenol	314	7.27	455.0	3.04
Water	273.15	1.86	373.15	0.51

Source: Lide, D. R., Ed., *Handbook of Chemistry and Physics,* 83rd ed. CRC Press, Boca Raton, FL, 2002.

convention, both K_f and K_b are positive; hence the negative sign in Equation (9.32) is replaced by a positive sign in Equation (9.34).

EXAMPLE PROBLEM 9.6

In this example, 4.50 g of a substance dissolved in 125 g of CCl_4 leads to an elevation of the boiling point of 0.65 K. Calculate the freezing point depression, the molecular mass of the substance, and the factor by which the vapor pressure of CCl_4 is lowered.

Solution

$$\Delta T_m = \left(\frac{K_f}{K_b} \right) \Delta T_b = \frac{30 \text{ K/(mole kg}^{-1})}{4.95 \text{ K/(mole kg}^{-1})} \times 0.65 \text{ K} = 3.9 \text{ K}$$

To avoid confusion, we use the symbol m for molality and **m** for mass. We solve for the molecular mass m_{solute} using Equation (9.34):

$$\Delta T_b = K_b m_{solute} = K_b \times \left(\frac{\mathbf{m}_{solute} / M_{solute}}{\mathbf{m}_{solvent} \text{(kg)}} \right)$$

$$M_{solute} = \frac{K_b \mathbf{m}_{solute}}{\mathbf{m}_{solvent} \Delta T_b}$$

$$M_{solute} = \frac{4.95 \text{ K kg mol}^{-1} \times 4.50 \text{ g}}{0.125 \text{ kg} \times 0.65 \text{ K}} = 274 \text{ g mol}^{-1}$$

We solve for the factor by which the vapor pressure of the solvent is reduced by using Raoult's law:

$$\frac{P_{solvent}}{P_{solvent}^*} = x_{solvent} = 1 - x_{solute} = 1 - \frac{n_{solute}}{n_{solute} + n_{solvent}}$$

$$= 1 - \frac{\dfrac{4.50 \text{ g}}{274 \text{ g mol}^{-1}}}{\left(\dfrac{4.50 \text{ g}}{274 \text{ g mol}^{-1}} \right) + \left(\dfrac{125 \text{ g}}{153.8 \text{ g mol}^{-1}} \right)} = 0.98$$

FIGURE 9.12

An osmotic pressure arises if a solution containing a solute that cannot pass through the membrane boundary is immersed in the pure solvent.

9.8 The Osmotic Pressure

Some membranes allow the passage of small molecules like water, yet do not allow larger molecules like sucrose to pass through them. Such a **semipermeable membrane** is an essential component in medical technologies such as kidney dialysis, which is described later. If a sac of such a membrane containing a solute that cannot pass through the membrane is immersed in a beaker containing the pure solvent, then initially the solvent diffuses into the sac. Diffusion ceases when equilibrium is attained, and at equilibrium, the pressure is higher in the sac than in the surrounding solvent. This result is shown schematically in Figure 9.12. The process in which the solvent diffuses through a membrane and dilutes a solution is known as **osmosis.** The amount by which the pressure in the solution is raised is known as the **osmotic pressure.**

To understand the origin of the osmotic pressure, denoted by π, the equilibrium condition is applied to the contents of the sac and the surrounding solvent:

$$\mu_{solvent}^{solution}(T, P + \pi, x_{solvent}) = \mu_{solvent}^{*}(T, P) \tag{9.35}$$

Using Raoult's law to express the concentration dependence of $\mu_{solution}$,

$$\mu_{solvent}^{solution}(T, P + \pi, x_{solvent}) = \mu_{solvent}^{*}(T, P + \pi) + RT \ln x_{solvent} \tag{9.36}$$

Because μ for the solvent is lower in the solution than in the pure solvent, only an increased pressure in the solution can raise its μ sufficiently to achieve equilibrium with the pure solvent.

The dependence of μ on pressure and temperature is given by $d\mu = dG_m = V_m dP - S_m dT$. At constant T we can write

$$\mu_{solvent}^{*}(T, P + \pi, x_{solvent}) - \mu_{solvent}^{*}(T, P) = \int_{P}^{P+\pi} V_m^{*} dP' \tag{9.37}$$

where V_m^{*} is the molar volume of the pure solvent and P is the pressure in the solvent outside the sac. Because a liquid is nearly incompressible, it is reasonable to assume that V_m^{*} is independent of P to evaluate the integral in the previous equation. Therefore, $\mu_{solvent}^{*}(T, P + \pi, x_{solvent}) - \mu_{solvent}^{*}(T, P) = V_m^{*}\pi$, and Equation (9.36) reduces to

$$\pi V_m^{*} + RT \ln x_{solvent} = 0 \tag{9.38}$$

For a dilute solution, $n_{solvent} \gg n_{solute}$, and

$$\ln x_{solvent} = \ln(1 - x_{solute}) \approx -x_{solute} = -\frac{n_{solute}}{n_{solute} + n_{solvent}} \approx -\frac{n_{solute}}{n_{solvent}} \tag{9.39}$$

Equation (9.39) can be simplified further by recognizing that for a dilute solution, $V \approx n_{solvent} V_m^{*}$. With this substitution, Equation (9.38) becomes

$$\pi = \frac{n_{solute} RT}{V} \tag{9.40}$$

which is known as the **van't Hoff equation.** Note the similarity in form between this equation and the ideal gas law.

An important application of the selective diffusion of the components of a solution through a membrane is dialysis. In healthy individuals, the kidneys remove waste products from the bloodstream, whereas individuals with damaged kidneys use a dialysis machine for this purpose. Blood from the patient is shunted through tubes made of a selectively porous membrane surrounded by a flowing sterile solution made up of water, sugars, and other components. Blood cells and other vital components of blood are too large to fit through the pores in the membranes, but urea and salt flow out of the bloodstream through membranes into the sterile solution and are removed as waste.

EXAMPLE PROBLEM 9.7

Calculate the osmotic pressure generated at 298 K if a cell with a total solute concentration of 0.500 mol L^{-1} is immersed in pure water. The cell wall is permeable to water molecules, but not to the solute molecules.

Solution

$$\pi = \frac{n_{solute}RT}{V} = 0.500 \text{ mol L}^{-1} \times 8.206 \times 10^{-2} \text{ L atm K}^{-1} \text{ mol}^{-1} \times 298 \text{ K} = 12.2 \text{ atm}$$

As this calculation shows, the osmotic pressure generated for moderate solute concentrations can be quite high. Hospital patients have died after pure water has accidentally been injected into their blood vessels, because the osmotic pressure is sufficient to burst the walls of blood cells.

Plants use osmotic pressure to achieve mechanical stability in the following way. A plant cell is bounded by a cellulose cell wall that is permeable to most components of the aqueous solutions that they encounter. A semipermeable cell membrane through which water, but not solute molecules, can pass is located just inside the cell wall. When the plant has sufficient water, the cell membrane expands, pushing against the cell wall, which gives the plant stalk a high rigidity. However, if there is a drought, the cell membrane is not totally filled with water and it moves away from the cell wall. As a result, only the cell wall contributes to the rigidity, and plants droop under these conditions.

Another important application involving osmotic pressure is the desalination of seawater using reverse osmosis. Seawater is typically 1.1 molar in NaCl. If such a solution is separated from pure water with a semipermeable membrane that allows the passage of water, but not of the solvated Na^+ and Cl^- ions, Equation (9.40) shows that an osmotic pressure of 27 bar is built up. If the side of the membrane on which the seawater is found is subjected to a pressure greater than 27 bar, H_2O from the seawater will flow through the membrane so that separation of pure water from the seawater occurs. This process is called reverse osmosis. The challenge in carrying out reverse osmosis on the industrial scale needed to provide a coastal city with potable water is to produce robust membranes that accommodate the necessary flow rates without getting fouled by algae and also effectively separate the ions from the water. The mechanism that leads to rejection of the ions is not fully understood. However, it is not based on the size of pores within the membrane alone. The major mechanism involves polymeric membranes that carry a surface charge within the pores in their hydrated state. These membrane-anchored ions repel the mobile Na^+ and Cl^- ions, while allowing the passage of the neutral water molecule.

9.9 Real Solutions Exhibit Deviations from Raoult's Law

In Sections 9.1 through 9.8, the discussion has been limited to ideal solutions. However, in general, if two volatile and miscible liquids are combined to form a solution, Raoult's law is not obeyed. This is the case because the A–A, B–B, and A–B interactions in a binary solution of A and B are unequal. If the A–B interactions are less (more) attractive than the A–A and A–A interactions, positive (negative) deviations from Raoult's law will be observed. An example of a binary solution with positive deviations from Raoult's law is CS_2–acetone. Experimental data for this system are shown in Table 9.3 and the data are plotted in Figure 9.13. How can a thermodynamic framework analogous to that pre-

TABLE 9.3

Partial and Total Pressures above a CS_2–Acetone Solution

x_{CS_2}	P_{CS_2} (Torr)	$P_{acetone}$ (Torr)	P_{total} (Torr)	x_{CS_2}	P_{CS_2} (Torr)	$P_{acetone}$ (Torr)	P_{total} (Torr)
0	0	343.8	343.8	0.4974	404.1	242.1	646.2
0.0624	110.7	331.0	441.7	0.5702	419.4	232.6	652.0
0.0670	119.7	327.8	447.5	0.5730	420.3	232.2	652.5
0.0711	123.1	328.8	451.9	0.6124	426.9	227.0	653.9
0.1212	191.7	313.5	505.2	0.6146	427.7	225.9	653.6
0.1330	206.5	308.3	514.8	0.6161	428.1	225.5	653.6
0.1857	258.4	295.4	553.8	0.6713	438.0	217.0	655.0
0.1991	271.9	290.6	562.5	0.6713	437.3	217.6	654.9
0.2085	283.9	283.4	567.3	0.7220	446.9	207.7	654.6
0.2761	323.3	275.2	598.5	0.7197	447.5	207.1	654.6
0.2869	328.7	274.2	602.9	0.8280	464.9	180.2	645.1
0.3502	358.3	263.9	622.2	0.9191	490.7	123.4	614.1
0.3551	361.3	262.1	623.4	0.9242	490.0	120.3	610.3
0.4058	379.6	254.5	634.1	0.9350	491.9	109.4	601.3
0.4141	382.1	253.0	635.1	0.9407	492.0	103.5	595.5
0.4474	390.4	250.2	640.6	0.9549	496.2	85.9	582.1
0.4530	394.2	247.6	641.8	0.9620	500.8	73.4	574.2
0.4933	403.2	242.8	646.0	0.9692	502.0	62.0	564.0
				1	512.3	0	512.3

Source: J. v. Zawidski, *Zeitschrift für Physikalische Chemie* 35 (1900) 129.

sented for the ideal solution in Section 9.2 be developed for real solutions? This issue is addressed throughout the rest of this chapter.

Figure 9.13 shows that the partial and total pressures above a real solution can differ substantially from the behavior predicted by Raoult's law. Another way that ideal and real solutions differ is that the set of equations denoted Equation (9.8), which describes the change in volume, entropy, enthalpy, and Gibbs energy that results from mixing, are not applicable to real solutions. For real solutions, these equations can only be written in a much less explicit form. Assuming that A and B are miscible,

$$\Delta G_{mixing} < 0 \qquad (9.41)$$
$$\Delta S_{mixing} > 0$$
$$\Delta V_{mixing} \neq 0$$
$$\Delta H_{mixing} \neq 0$$

Whereas $\Delta G_{mixing} < 0$ and $\Delta S_{mixing} > 0$ always hold for miscible liquids, ΔV_{mixing} and ΔH_{mixing} can be positive or negative, depending on the nature of the A–B interaction in the solution.

As indicated in Equation (9.41), the volume change upon mixing is not generally zero. Therefore, the volume of a solution will not be given by

$$V_m^{ideal} = x_A V_{m,A}^* + (1 - x_A)V_{m,B}^* \qquad (9.42)$$

FIGURE 9.13
The data in Table 9.3 are plotted versus x_{CS_2}. The dashed lines show the expected behavior if Raoult's law were obeyed.

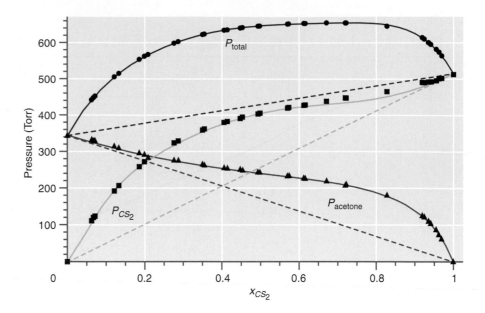

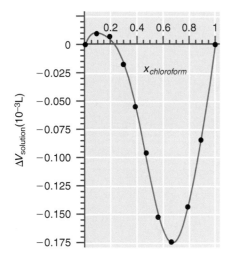

FIGURE 9.14

Deviations in the volume from the behavior expected for 1 mol of an ideal solution [Equation (9.42)] are shown for the acetone–chloroform system as a function of the mole fraction of chloroform.

as expected for 1 mol of an ideal solution, where $V_{m,A}^*$ and $V_{m,B}^*$ are the molar volumes of the pure substances A and B. Figure 9.14 shows $\Delta V_m = V_m^{real} - V_m^{ideal}$ for an acetone–chloroform solution as a function of $x_{chloroform}$. Note that ΔV_m can be positive or negative for this solution, depending on the value of $x_{chloroform}$. The deviations from ideality are small, but are clearly evident.

The deviation of the volume from ideal behavior can best be understood by defining the concept of **partial molar quantities.** This concept is illustrated by discussing the **partial molar volume.** The volume of 1 mol of pure water at 25°C is 18.1 cm³. However, if 1 mol of water is added to a large volume of an ethanol–water solution with $x_{H_2O} = 0.75$, the volume of the solution increases by only 16 cm³. This is the case because the local structure around a water molecule in the solution is more compact than in pure water. The partial molar volume of a component in a solution is defined as the volume by which the solution changes if 1 mol of the component is added to such a large volume that the solution composition can be assumed constant. This statement is expressed mathematically in the following form:

$$\overline{V}_1(P,T,n_1,n_2) = \left(\frac{\partial V}{\partial n_1}\right)_{P,T,n_2} \tag{9.43}$$

With this definition, the volume of a binary solution is given

$$V = n_1\overline{V}_1(P,T,n_1,n_2) + n_2\overline{V}_2(P,T,n_1,n_2) \tag{9.44}$$

Note that because the partial molar volumes depend on the concentration of all components, the same is true of the total volume.

One can form partial molar quantities for any extensive property of a system (for example U, H, G, A, and S). Partial molar quantities (other than the chemical potential, which is the partial molar Gibbs energy) are usually denoted by the appropriate symbol topped by a horizontal bar. The partial molar volume is a function of P, T, n_1, and n_2, and $\overline{V}_i$ can be greater than or less than the molar volume of the pure component. Therefore, the volume of a solution of two miscible liquids can be greater than or less than the sum of the volumes of the pure components of the solution. Figure 9.15 shows data for the partial volumes of acetone and chloroform in an

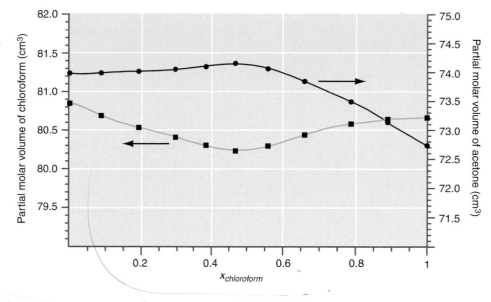

FIGURE 9.15
The partial molar volumes of chloroform (yellow curve) and acetone (red curve) in a chloroform–acetone binary solution are shown as a function of $x_{chloroform}$.

acetone–chloroform binary solution at 298 K. Note that the changes in the partial molar volumes with concentration are small, but not negligible.

In Figure 9.15, we can see that $\bar{V}_1$ increases if $\bar{V}_2$ decreases and vice versa. This is the case because partial molar volumes are related in the same way as the chemical potentials are related in the Gibbs–Duhem equation [Equation (9.22)]. In terms of the partial molar volumes, the Gibbs–Duhem equation takes the form

$$x_1 d\bar{V}_1 + x_2 d\bar{V}_2 = 0 \quad \text{or} \quad d\bar{V}_1 = -\frac{x_2}{x_1} d\bar{V}_2 \qquad (9.45)$$

Therefore, as seen in Figure 9.15, if $\bar{V}_2$ changes by an amount $d\bar{V}_2$ over a small concentration interval, $\bar{V}_1$ will change in the opposite direction. The Gibbs–Duhem equation is applicable to both ideal and real solutions.

9.10 The Ideal Dilute Solution

Although no simple model exists that describes all real solutions, there is a limiting law that merits further discussion. In this section, we describe the model of the ideal dilute solution, which provides a useful basis for describing the properties of real solutions if they are dilute. Just as for an ideal solution, at equilibrium the chemical potentials of a component in the gas and solution phases of a real solution are equal. As for an ideal solution, the chemical potential of a component in a real solution can be separated into two terms, a standard state chemical potential and a term that depends on the partial pressure.

$$\mu_i^{solution} = \mu_i^* + RT \ln \frac{P_i}{P_i^*} \qquad (9.46)$$

Recall that for a pure substance $\mu_i^*(\text{vapor}) = \mu_i^*(\text{liquid}) = \mu_i^*$. Because the solution is not ideal, $P_i \neq x_i P_i^*$.

First consider only the solvent in a dilute binary solution. To arrive at an equation for $\mu_i^{solution}$ that is similar to Equation (9.7) for an ideal solution, we define the dimensionless **activity**, $a_{solvent}$, of the solvent by

$$a_{solvent} = \frac{P_{solvent}}{P_{solvent}^*} \qquad (9.47)$$

Note that $a_{solvent} = x_{solvent}$ for an ideal solution. For a nonideal solution, the activity and the mole fraction are related through the **activity coefficient** $\gamma_{solvent}$, defined by

$$\gamma_{solvent} = \frac{a_{solvent}}{x_{solvent}} \tag{9.48}$$

The activity coefficient quantifies the degree to which the solution is nonideal. The activity plays the same role for a component of a solution that the fugacity plays for a real gas in expressing deviations from ideal behavior. In both cases, ideal behavior is observed in the appropriate limit, namely, $P \rightarrow 0$ for the gas, and $x_{solute} \rightarrow 0$ for the solution. To the extent that there is no atomic-scale model that tells us how to calculate γ, it should be regarded as a correction factor that exposes the inadequacy of our model, rather than as a fundamental quantity. As will be discussed in Chapter 10, there is such an atomic-scale model for dilute electrolyte solutions.

How is the chemical potential of a component related to its activity? Combining Equations (9.46) and (9.47), one obtains a relation that holds for all components of a real solution:

$$\mu_i^{solution} = \mu_i^* + RT \ln a_i \tag{9.49}$$

Equation (9.49) is a central equation in describing real solutions. It is the starting point for the discussion in the rest of this chapter.

The preceding discussion focused on the solvent in a dilute solution. However, the ideal dilute solution is defined by the conditions, $x_{solute} \rightarrow 0$ and $x_{solvent} \rightarrow 1$. Because the solvent and solute are considered in different limits, we use different expressions to relate the mole fraction of a component and the partial pressure of the component above the solution.

Consider the partial pressure of acetone as a function of x_{CS_2} shown in Figure 9.16. Although Raoult's law is not obeyed over the whole concentration range, it is obeyed in the limit that $x_{acetone} \rightarrow 1$ and $x_{CS_2} \rightarrow 0$. In this limit, the average acetone molecule at the surface of the solution is surrounded by acetone molecules. Therefore, to a good approximation, $P_{acetone} = x_{acetone} P_{acetone}^*$ as $x_{acetone} \rightarrow 1$. Because the majority species is defined to be the solvent, we see that Raoult's law is obeyed for the solvent in a dilute solution. This limiting behavior is also observed for CS_2 *in the limit in which it is the solvent*, as seen in Figure 9.13.

Consider the opposite limit in which $x_{acetone} \rightarrow 0$. In this case, the average acetone molecule at the surface of the solution is surrounded by CS_2 molecules. Therefore, the molecule experiences very different interactions with its neighbors than if it were surrounded by acetone molecules. For this reason, $P_{acetone} \neq x_{acetone} P_{acetone}^*$ as $x_{acetone} \rightarrow 0$. However, it is apparent from Figure 9.16 that $P_{acetone}$ also varies linearly with $x_{acetone}$ in this limit. This behavior is described by the following equation:

FIGURE 9.16
The partial pressure of acetone from Table 9.3 is plotted as a function of x_{CS_2}. Note that the $P_{acetone}$ follows Raoult's law as $x_{CS_2} \rightarrow 0$, and Henry's law as $x_{CS_2} \rightarrow 1$.

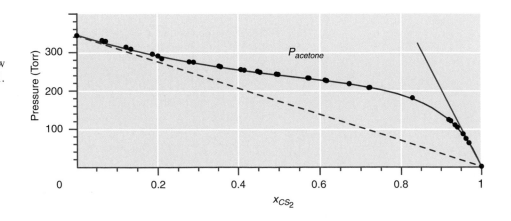

TABLE 9.4

Henry's Law Constants for Aqueous Solutions Near 298 K

Substance	k_H (Torr)	k_H (bar)
Ar	2.80×10^7	3.72×10^4
C_2H_6	2.30×10^7	3.06×10^4
CH_4	3.07×10^7	4.08×10^4
CO	4.40×10^6	5.84×10^3
CO_2	1.24×10^6	1.65×10^3
H_2S	4.27×10^5	5.68×10^2
He	1.12×10^8	1.49×10^6
N_2	6.80×10^7	9.04×10^4
O_2	3.27×10^7	4.95×10^4

Source: Alberty, R. A., and Silbey, R. S., *Physical Chemistry,* John Wiley & Sons, New York, 1992.

$$P_{acetone} = x_{acetone} k_H^{acetone} \quad \text{as } x_{acetone} \to 0 \tag{9.50}$$

This relationship is known as **Henry's law,** and the constant k_H is known as the **Henry's law constant.** The value of the constant depends on the nature of the solute and solvent and quantifies the degree to which deviations from Raoult's law occur. As the solution approaches ideal behavior, $k_H^i \to P_i^*$. For the data shown in Figure 9.13, $k_H^{CS_2} = 2010$ Torr and $k_H^{acetone} = 1950$ Torr. Note that these values are substantially greater than the vapor pressures of the pure substances, which are 512.3 and 343.8 Torr, respectively. The Henry's law constants are less than the vapor pressures of the pure substances if the system exhibits negative deviations from Raoult's law. Henry's law constants for aqueous solutions are listed for a number of solutes in Table 9.4.

Based on these results, the **ideal dilute solution** is defined. *An ideal dilute solution is a solution in which the solvent is described using Raoult's law and the solute is described using Henry's law.* As shown by the data in Figure 9.13, the partial pressures above the CS_2–acetone mixture are consistent with this model in either of two limits, $x_{acetone} \to 1$ or $x_{CS_2} \to 1$. In the first of these limits, we consider acetone to be the solvent and CS_2 to be the solute. Acetone is the solute and CS_2 is the solvent in the second limit.

9.11 Activities Are Defined with Respect to Standard States

The ideal dilute solution model's predictions that Raoult's law is obeyed for the solvent and Henry's law is obeyed for the solute are not obeyed over a wide range of concentration. The concept of the activity coefficient introduced in Section 9.10 is used to quantify these deviations. In doing so, it is useful to define the activities in such a way that the solution approaches ideal behavior in the limit of interest, which is generally $x_A \to 0$, or $x_A \to 1$. With this choice, the activity approaches the concentration, and it is reasonable to set the activity coefficient equal to one. Specifically, $a_i \to x_i$ as $x_i \to 1$ for the solvent, and $a_i \to x_i$ as $x_i \to 0$ for the solute. The reason for this choice is that numerical values for activity coefficients are generally not known. Choosing the standard state as described earlier ensures that the concentration (divided by the unit concentration to make it dimensionless), which is easily measured, is a good approximation to the activity.

In Section 9.10 the activity and activity coefficient for the solvent in a dilute solution ($x_{solvent} \to 1$) were defined by the relations

$$a_i = \frac{P_i}{P_i^*} \quad \text{and} \quad \gamma_i = \frac{a_i}{x_i} \tag{9.51}$$

As shown in Figure 9.13, the activity approaches unity as $x_{solvent} \to 1$. We refer to an activity calculated using Equation (9.51) as being based on a **Raoult's law standard state.** The standard state chemical potential based on Raoult's law is $\mu_{solvent}^*$, which is the chemical potential of the pure solvent.

However, this definition of the activity and choice of a standard state is not optimal for the solute at a low concentration, because the solute obeys Henry's law rather than Raoult's law and, therefore, the activity coefficient will differ appreciably from one. In this case,

$$\mu_{solute}^{solution} = \mu_{solute}^* + RT \ln \frac{k_H^{solute} x_{solute}}{P_{solute}^*} = \mu_{solute}^{*H} + RT \ln x_{solute} \quad \text{as } x_{solute} \to 0 \tag{9.52}$$

The standard state chemical potential is the value of the chemical potential when $x_i = 1$. We see that the Henry's law standard state chemical potential is given by

$$\mu_{solute}^{*H} = \mu_{solute}^* + RT \ln \frac{k_H^{solute}}{P_{solute}^*} \tag{9.53}$$

The activity and activity coefficient based on Henry's law are defined, respectively, by

$$a_i = \frac{P_i}{k_i^H} \quad \text{and} \quad \gamma_i = \frac{a_i}{x_i} \tag{9.54}$$

We note that Henry's law is still obeyed for a solute that has such a small vapor pressure that we refer to it as a nonvolatile solute.

The **Henry's law standard state** is a state in which the pure solute has a vapor pressure $k_{H,solute}$ rather than its actual value P^*_{solute}. It is a hypothetical state that does not exist. Recall that the value $k_{H,solute}$ is obtained by extrapolation from the low coverage range in which Henry's law is obeyed. Although this definition may seem peculiar, only in this way can we ensure that $a_{solute} \to x_{solute}$ and $\gamma_{solute} \to 1$ as $x_{solute} \to 0$. We reiterate the reason for this choice. If the preceding conditions are satisfied, the concentration (divided by the unit concentration to make it dimensionless) is, to a good approximation, equal to the activity. Therefore, the equilibrium constant can be calculated without having numerical values for activity coefficients. We emphasize the difference in the Raoult's law and Henry's law standard states because it is the standard state chemical potentials μ^*_{solute} and μ^{*H}_{solute} that are used to calculate $\Delta G°$ and the thermodynamic equilibrium constant, K. Different choices for standard states will result in different numerical values for K. The standard chemical potential μ^{*H}_{solute} refers to the hypothetical standard state in which $x_{solute} = 1$, and each solute species is in an environment characteristic of the infinitely dilute solution.

We now consider a less well-defined situation. For solutions in which the components are miscible in all proportions, such as the CS_2–acetone system, either a Raoult's law or a Henry's law standard state can be defined, as we show with sample calculations in Example Problems 9.8 and 9.9. This is the case because there is no unique choice for the standard state over the entire concentration range in such a system. Numerical values for the activities and activity coefficients will differ, depending on whether the Raoult's law or the Henry's law standard state is used.

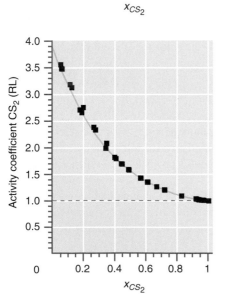

FIGURE 9.17

The activity and activity coefficient for CS_2 in a CS_2–acetone solution based on a Raoult's law standard state are shown as a function of x_{CS_2}.

EXAMPLE PROBLEM 9.8

Calculate the activity and activity coefficient for CS_2 at $x_{CS_2} = 0.3502$ using data from Table 9.3. Assume a Raoult's law standard state.

Solution

$$a_{CS_2}^R = \frac{P_{CS_2}}{P^*_{CS_2}} = \frac{358.3\,\text{Torr}}{512.3\,\text{Torr}} = 0.6994$$

$$\gamma_{CS_2}^R = \frac{a_{CS_2}^R}{x_{CS_2}} = \frac{0.6994}{0.3502} = 1.997$$

The activity and activity coefficients for CS_2 are concentration dependent. Results calculated as in Example Problem 9.2 using a Raoult's law standard state are shown in Figure 9.17 as a function of x_{CS_2}. For this solution, $\gamma_{CS_2}^R > 1$ for all values of the concentration for which $x_{CS_2} < 1$. Note that $\gamma_{CS_2}^R \to 1$ as $x_{CS_2} \to 1$ as the model requires. The activity and activity coefficients for CS_2 using a Henry's law standard state are shown in Figure 9.18 as a function of x_{CS_2}. For this solution, $\gamma_{CS_2}^H < 1$ for all values of the concentration for which $x_{CS_2} > 0$. Note that $\gamma_{CS_2}^H \to 1$ as $x_{CS_2} \to 0$ as the model requires. Which of these two possible standard states should be chosen? There is a good answer to this question only in the limits $x_{CS_2} \to 0$ or $x_{CS_2} \to 1$. For intermediate concentrations, either standard state can be used.

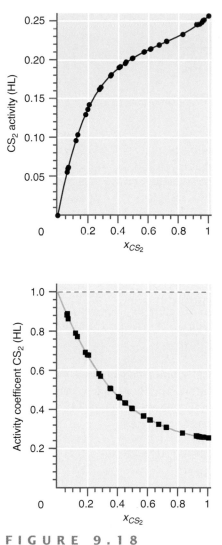

FIGURE 9.18
The activity and activity coefficient for CS_2 in a CS_2–acetone solution based on a Henry's law standard state are shown as a function of x_{CS_2}.

EXAMPLE PROBLEM 9.9

Calculate the activity and activity coefficient for CS_2 at $x_{CS_2} = 0.3502$ using data from Table 9.3. Assume a Henry's law standard state.

Solution

$$a_{CS_2}^H = \frac{P_{CS_2}}{k_{H,CS_2}} = \frac{358.3\,\text{Torr}}{2010\,\text{Torr}} = 0.1783$$

$$\gamma_{CS_2}^H = \frac{a_{CS_2}^H}{x_{CS_2}} = \frac{0.1783}{0.3502} = 0.5090$$

The Henry's law standard state just discussed is defined with respect to concentration measured in units of the mole fraction. This is not a particularly convenient scale, and either the molarity or molality concentration scales are generally used in laboratory experiments. The mole fraction of the solute can be converted to the molality scale by dividing the first expression in Equation 9.55 by $n_{solvent}M_{solvent}$:

$$x_{solute} = \frac{n_{solute}}{n_{solvent} + n_{solute}} = \frac{m_{solute}}{\dfrac{1}{M_{solvent}} + m_{solute}} \tag{9.55}$$

where m_{solute} is the molality of the solute, and $M_{solvent}$ is the molecular mass of the solvent in kg mol^{-1}. We see that $m_{solute} \to x/M_{solvent}$ as $x \to 0$. Using molality as the concentration unit, the activity and activity coefficient of the solute are defined by

$$a_{solute}^{molality} = \frac{P_{solute}}{k_H^{molality}} \quad \text{with} \quad a_{solute}^{molality} \to m_{solute} \quad \text{as} \quad m_{solute} \to 0 \quad \text{and} \tag{9.56}$$

$$\gamma_{solute}^{molality} = \frac{a_{solute}^{molality}}{m_{solute}} \quad \text{with} \quad \gamma_{solute}^{molality} \to 1 \quad \text{as} \quad m_{solute} \to 0 \tag{9.57}$$

The Henry's law constants and activity coefficients determined on the mole fraction scale must be recalculated to describe a solution characterized by its molality or molarity, as is shown in Example Problem 9.10. Similar conversions can be made if the concentration of the solution is expressed in terms of molality.

What is the standard state if concentrations rather than mole fractions are used? The standard state in this case is the hypothetical state in which Henry's law is obeyed by a solution that is 1.0 molar (or 1.0 molal) in the solute concentration. It is a hypothetical state because at this concentration, substantial deviations from Henry's law will be observed. Although this definition may seem peculiar at first reading, only in this way can we ensure that the activity becomes equal to the molarity (or molality), and the activity coefficient approaches one, as the solute concentration approaches zero.

EXAMPLE PROBLEM 9.10

a. Derive a general equation, valid for dilute solutions, relating the Henry's law constants for the solute on the mole fraction and molarity scales.

b. Determine the Henry's law constants for acetone on the molarity scale. Use the results for the Henry's law constants on the mole fraction scale cited in Section 9.10. The density of acetone is 789.9 g L^{-1}.

Solution

a. We use the symbol c_{solute} to designate the solute molarity, and $c°$ to indicate a 1 molar concentration.

$$\frac{dP}{d\left(\dfrac{c_{solute}}{c^\circ}\right)} = \frac{dP}{dx_{solute}}\frac{dx_{solute}}{d\left(\dfrac{c_{solute}}{c^\circ}\right)}$$

To evaluate this equation, we must determine

$$\frac{dx_{solute}}{d\left(\dfrac{c_{solute}}{c^\circ}\right)}$$

$$x_{solute} = \frac{n_{solute}}{n_{solute}+n_{solvent}} \approx \frac{n_{solute}}{n_{solvent}} = \frac{n_{solute}M_{solvent}}{V_{solution}\rho_{solution}} = c_{solute}\frac{M_{solvent}}{\rho_{solution}}$$

Therefore,

$$\frac{dP}{d\left(\dfrac{c_{solute}}{c^\circ}\right)} = \frac{c^\circ M_{solvent}}{\rho_{solution}}\frac{dP}{dx_{solute}}$$

b. $$\frac{dP}{d\left(\dfrac{c_{solute}}{c^\circ}\right)} = \frac{1\ \text{mol L}^{-1}\times 58.08\ \text{g mol}^{-1}}{789.9\ \text{g L}^{-1}}\times 1950\ \text{Torr} = 143.4\ \text{Torr}$$

The colligative properties discussed for an ideal solution in Sections 9.7 and 9.8 refer to the properties of the solvent in a dilute solution. The Raoult's law standard state applies to this case and in an ideal dilute solution, Equations (9.32), (9.34), and (9.40) can be used with activities replacing concentrations:

$$\Delta T_f = -K_f \gamma m_{solute} \tag{9.58}$$

$$\Delta T_b = K_b \gamma m_{solute}$$

$$\pi = \gamma c_{solute} RT$$

The activity coefficients are defined with respect to molality for the boiling point elevation ΔT_b and freezing point depression ΔT_f, and with respect to molarity for the osmotic pressure π. Equations (9.58) provide a useful way to determine activity coefficients as shown in Example Problem 9.11.

EXAMPLE PROBLEM 9.11

In 500 g of water, 24.0 g of a nonvolatile solute of molecular weight 241 g mol^{-1} is dissolved. The observed freezing point depression is 0.359°C. Calculate the activity coefficient of the solute.

Solution

$$\Delta T_f = -K_f \gamma m_{solute}; \quad \gamma = -\frac{\Delta T_f}{K_f m_{solute}}$$

$$\gamma = \frac{0.359\ \text{K}}{1.86\ \text{K kg mol}^{-1}\times \dfrac{24.0}{241\times 0.500}\ \text{mol kg}^{-1}} = 0.969$$

9.12 Henry's Law and the Solubility of Gases in a Solvent

The ideal dilute solution model can be applied to the solubility of gases in liquid solvents. An example for this type of solution equilibrium is the amount of N_2 absorbed by water at sea level, which is considered in Example Problem 9.12. In this case, one of the components of the solution is a liquid and the other is a gas. The equilibrium of interest between the solution and the vapor phase is

$$N_2(\text{aqueous}) \rightleftharpoons N_2(\text{vapor}) \tag{9.59}$$

The chemical potential of the dissolved N_2 is given by

$$\mu_{N_2}^{solution} = \mu_{N_2}^{*H}(\text{vapor}) + RT \ln a_{solute} \tag{9.60}$$

In this case, a Henry's law standard state is the appropriate choice, because the nitrogen is sparingly soluble in water. The mole fraction of N_2 in solution, x_{N_2}, is given by

$$x_{N_2} = \frac{n_{N_2}}{n_{N_2} + n_{H_2O}} \approx \frac{n_{N_2}}{n_{H_2O}} \tag{9.61}$$

The amount of dissolved gas is given by

$$n_{N_2} = n_{H_2O} x_{N_2} = n_{H_2O} \frac{P_{N_2}}{k_H^{N_2}} \tag{9.62}$$

Example Problem 9.12 shows how Equation (9.62) is used to model the dissolution of a gas in a liquid.

EXAMPLE PROBLEM 9.12

The average human with a body weight of 70 kg has a blood volume of 5.00 L. The Henry's law constant for the solubility of N_2 in H_2O is 9.04×10^4 bar at 298 K. Assume that this is also the value of the Henry's law constant for blood and that the density of blood is 1.00 kg L^{-1}.

a. Calculate the number of moles of nitrogen absorbed in this amount of blood in air of composition 80% N_2 at sea level, where the pressure is 1 bar, and at a pressure of 50 bar.

b. Assume that a diver accustomed to breathing compressed air at a pressure of 50 bar is suddenly brought to sea level. What volume of N_2 gas is released as bubbles in the diver's bloodstream?

Solution

a. $n_{N_2} = n_{H_2O} \dfrac{P_{N_2}}{k_H^{N_2}}$

$$= \frac{5.0 \times 10^3 \text{g}}{18.02 \text{ g mol}^{-1}} \times \frac{0.80 \text{ bar}}{9.04 \times 10^4 \text{ bar}}$$

$$= 2.5 \times 10^{-3} \text{ mol at 1 bar total pressure}$$

At 50 bar, $n_{N_2} = 50 \times 2.5 \times 10^{-3} \text{ mol} = 0.125 \text{ mol}$.

b. $V = \dfrac{nRT}{P}$

$$= \frac{(0.125 \text{ mol} - 2.5 \times 10^{-3} \text{ mol}) \times 8.314 \times 10^{-2} \text{ L bar mol}^{-1}\text{K}^{-1} \times 300 \text{ K}}{1.00 \text{ bar}}$$

$$= 3.1 \text{ L}$$

The symptoms induced by the release of air into the bloodstream are known to divers as the bends. The volume of N_2 just calculated is far more than is needed to cause the formation of arterial blocks due to gas-bubble embolisms.

9.13 Chemical Equilibrium in Solutions

The concept of activity can be used to express the thermodynamic equilibrium constant in terms of activities for real solutions. Consider a reaction between solutes in a solution. At equilibrium, the following relation must hold:

$$\left(\sum_j v_j \mu_j (\text{solution})\right)_{equilibrium} = 0 \tag{9.63}$$

where the subscript states that the individual chemical potentials must be evaluated under equilibrium conditions. Each of the chemical potentials in Equation (9.63) can be expressed in terms of a standard state chemical potential and a concentration-dependent term. Assume a Henry's law standard state for each solute. Equation (9.63) then takes the form

$$\sum_j v_j \mu_j^{*H}(\text{solution}) + RT \sum_j \ln(a_i^{eq})^{v_j} = 0 \tag{9.64}$$

Using the relation between the Gibbs energy and the chemical potential, the previous equation can be written in the form

$$\Delta G_{reaction}^\circ = -RT \sum_j \ln(a_i^{eq})^{v_j} = -RT \ln K \tag{9.65}$$

The equilibrium constant in terms of activities is given by

$$K = \prod_i (a_i^{eq})^{v_j} = \prod_i (\gamma_i^{eq})^{v_j} \left(\frac{c_i^{eq}}{c^\circ}\right)^{v_j} \tag{9.66}$$

where the symbol Π indicates that the terms following the symbol are multiplied with one another. This equilibrium constant is the fundamental thermodynamic equilibrium for all systems. It can be viewed as a generalization of K_P, defined in Equation (9.67). The equilibrium constant K defined by Equation (9.66) can be applied to equilibria involving gases, liquids, dissolved species, and solids. For gases, the fugacities divided by f° (see Section 7.5) are the activities.

To obtain a numerical value for K, the standard state Gibbs reaction energy ΔG_H° must be known. As for gas-phase reactions, the $\Delta G_{reaction}^\circ$ must be determined experimentally. This can be done by measuring the individual activities of the species in solution and calculating K from these results. After a series of $\Delta G_{reaction}^\circ$ for different reactions has been determined, they can be combined to calculate the $\Delta G_{reaction}^\circ$ for other reactions, as discussed for reaction enthalpies in Chapter 4. Because of the significant interactions between the solutes and the solvent, K values depend on the nature of the solvent, and for electrolyte solutions to be discussed in Chapter 10, they additionally depend on the ionic strength.

An equilibrium constant in terms of molarities or molalities can also be defined starting from Equation (9.66) and setting all activity coefficients equal to one. This is most appropriate for a dilute solution of a nonelectrolyte, using a Henry's law standard state:

$$K = \prod_i (\gamma_i^{eq})^{v_j} \left(\frac{c_i^{eq}}{c^\circ}\right)^{v_j} \approx \prod_i \left(\frac{c_i^{eq}}{c^\circ}\right)^{v_j} \tag{9.67}$$

EXAMPLE PROBLEM 9.13

a. Write the equilibrium constant for the reaction $N_2(aq, m) \rightleftharpoons N_2(g, P)$ in terms of activities at 25°C, where m is the molarity of $N_2(aq)$.

b. By making suitable approximations, convert the equilibrium constant of part (a) into one in terms of pressure and molarity only.

Solution

a. $K = \prod_i (a_i^{eq})^{\nu_j} = \prod_i (\gamma_i^{eq})^{\nu_j} \left(\dfrac{c_i^{eq}}{c^\circ} \right)^{\nu_j} = \dfrac{\left(\dfrac{\gamma_{N_2,g} P}{P^\circ} \right)}{\left(\dfrac{\gamma_{N_2,aq} m}{m^\circ} \right)}$

$= \dfrac{\gamma_{N_2,g}}{\gamma_{N_2,aq}} \dfrac{\left(\dfrac{P}{P^\circ} \right)}{\left(\dfrac{m}{m^\circ} \right)}$

b. Using a Henry's law standard state for dissolved N_2, $\gamma_{N_2,aq} \approx 1$, because the concentration is very low. Similarly, because N_2 behaves like an ideal gas up to quite high pressures at 25°C, $\gamma_{N_2,g} \approx 1$. Therefore,

$$K \approx \dfrac{\left(\dfrac{P}{P^\circ} \right)}{\left(\dfrac{m}{m^\circ} \right)}$$

Note that in this case, the equilibrium constant is simply the Henry's law constant in terms of molarity.

The numerical values for the dimensionless thermodynamic equilibrium constant depend on the choice of the standard states for the components involved in the reaction. The same is true for ΔG°. Therefore, it is essential to know which standard state has been assumed before an equilibrium constant is used. The activity coefficients of most neutral solutes are close to one with the appropriate choice of standard state. Therefore, example calculations of chemical equilibrium using activities will be deferred until electrolyte solutions are discussed in Chapter 10. For such solutions, γ_{solute} differs substantially from one, even for dilute solutions.

Vocabulary

activity

activity coefficient

average composition

boiling point elevation

colligative properties

fractional distillation

freezing point depression

Gibbs–Duhem equation

Henry's law

Henry's law constant

Henry's law standard state

ideal dilute solution

ideal solution

lever rule

limiting law

osmosis

osmotic pressure

partial molar quantities

partial molar volume

pressure–average composition diagram

Raoult's law

Raoult's law standard state

semipermeable membrane

solute

solvent

temperature–composition diagram

tie line

van't Hoff equation

Questions on Concepts

Q9.1 Using the differential form of G, $dG = VdP - SdT$, show that if $\Delta G_{mixing} = nRT \sum_i x_i \ln x_i$, then $\Delta H_{mixing} = \Delta V_{mixing} = 0$.

Q9.2 For a pure substance, the liquid and gaseous phases can only coexist for a single value of the pressure at a given temperature. Is this also the case for an ideal solution of two volatile liquids?

Q9.3 Fractional distillation of a particular binary liquid mixture leaves behind a liquid consisting of both components in which the composition does not change as the liquid is boiled off. Is this behavior characteristic of a maximum or a minimum boiling point azeotrope?

Q9.4 Why is the magnitude of the boiling point elevation less than that of the freezing point depression?

Q9.5 Why is the preferred standard state for the solvent in an ideal dilute solution the Raoult's law standard state? Why is the preferred standard state for the solute in an ideal dilute solution the Henry's law standard state? Is there a preferred standard state for the solution in which $x_{solvent} = x_{solute} = 0.5$?

Q9.6 Is a whale likely to get the bends when it dives deep into the ocean and resurfaces? Answer this question by

considering the likelihood of a diver getting the bends if he or she dives and resurfaces on one lung full of air as opposed to breathing air for a long time at the deepest point of the dive.

Q9.7 The statement "The boiling point of a typical liquid mixture can be reduced by approximately 100°C if the pressure is reduced from 760 to 20 Torr" is found in Section 9.4. What figure(s) in Chapter 8 can you identify to support this statement in a qualitative sense?

Q9.8 Explain why chemists doing quantitative work using liquid solutions prefer to express concentration in terms of molality rather than molarity.

Q9.9 Explain the usefulness of a tie line on a P-Z phase diagram such as that of Figure 9.4.

Q9.10 Explain why colligative properties depend only on the concentration, and not on the identity of the molecule.

Problems

Problem numbers in **red** indicate that the solution to the problem is given in the *Student's Solutions Manual*.

P9.1 At 303 K, the vapor pressure of benzene is 118 Torr and that of cyclohexane is 122 Torr. Calculate the vapor pressure of a solution for which $x_{benzene} = 0.25$ assuming ideal behavior.

P9.2 A volume of 5.50 L of air is bubbled through liquid toluene at 298 K, thus reducing the mass of toluene in the beaker by 2.38 g. Assuming that the air emerging from the beaker is saturated with toluene, determine the vapor pressure of toluene at this temperature.

P9.3 An ideal solution is formed by mixing liquids A and B at 298 K. The vapor pressure of pure A is 180 Torr and that of pure B is 82.1 Torr. If the mole fraction of A in the vapor is 0.450, what is the mole fraction of A in the solution?

P9.4 A and B form an ideal solution. At a total pressure of 0.900 bar, $y_A = 0.450$ and $x_A = 0.650$. Using this information, calculate the vapor pressure of pure A and of pure B.

P9.5 A and B form an ideal solution at 298K, with $x_A = 0.600$, $P_A^* = 105$ Torr and $P_B^* = 63.5$ Torr.

a. Calculate the partial pressures of A and B in the gas phase.

b. A portion of the gas phase is removed and condensed in a separate container.

Calculate the partial pressures of A and B in equilibrium with this liquid sample at 298 K.

P9.6 The vapor pressures of 1-bromobutane and 1-chlorobutane can be expressed in the form

$$\ln \frac{P_{bromo}}{Pa} = 17.076 - \frac{1584.8}{\frac{T}{K} - 111.88}$$

and

$$\ln \frac{P_{chloro}}{Pa} = 20.612 - \frac{2688.1}{\frac{T}{K} - 55.725}$$

Assuming ideal solution behavior, calculate x_{bromo} and y_{bromo} at 300.0 K and a total pressure of 8741 Pa.

P9.7 Assume that 1-bromobutane and 1-chlorobutane form an ideal solution. At 273 K, $P_{chloro}^* = 3790$ Pa and $P_{bromo}^* = 1394$ Pa. When only a trace of liquid is present at 273 K, $y_{chloro} = 0.75$.

a. Calculate the total pressure above the solution.

b. Calculate the mole fraction of 1-chlorobutane in the solution.

c. What value would z_{chloro} have in order for there to be 4.86 mol of liquid and 3.21 mol of gas at a total pressure equal to that in part (a)? [*Note:* This composition is different from that of part (a).]

P9.8 An ideal solution at 298 K is made up of the volatile liquids A and B, for which $P_A^* = 125$ Torr and $P_B^* = 46.3$ Torr. As the pressure is reduced from 450 Torr, the first vapor is observed at a pressure of 70.0 Torr. Calculate x_A.

P9.9 At −47°C, the vapor pressure of ethyl bromide is 10.0 Torr and that of ethyl chloride is 40.0 Torr. Assume that the solution is ideal. Assume there is only a trace of liquid present and the mole fraction of ethyl chloride in the vapor is 0.80 and answer these questions:

a. What is the total pressure and the mole fraction of ethyl chloride in the liquid?

b. If there are 5.00 mol of liquid and 3.00 mol of vapor present at the same pressure as in part (a), what is the overall composition of the system?

P9.10 At $-31.2°C$, pure propane and n-butane have vapor pressures of 1200 and 200 Torr, respectively.

a. Calculate the mole fraction of propane in the liquid mixture that boils at $-31.2°C$ at a pressure of 760 Torr.

b. Calculate the mole fraction of propane in the vapor that is in equilibrium with the liquid of part (a).

P9.11 In an ideal solution of A and B, 3.50 mol are in the liquid phase and 4.75 mol are in the gaseous phase. The overall composition of the system is $Z_A = 0.300$ and $x_A = 0.250$. Calculate y_A.

P9.12 Given the vapor pressures of the pure liquids and the overall composition of the system, what are the upper and lower limits of pressure between which liquid and vapor coexist in an ideal solution?

P9.13 At $39.9°C$, a solution of ethanol ($x_1 = 0.9006$, $P_1^* = 130.4$ Torr) and isooctane ($P_2^* = 43.9$ Torr) forms a vapor phase with $y_1 = 0.6667$ at a total pressure of 185.9 Torr.

a. Calculate the activity and activity coefficient of each component.

b. Calculate the total pressure that the solution would have if it were ideal.

P9.14 Ratcliffe and Chao [*Canadian Journal of Chemical Engineering* 47, (1969), 148] obtained the following tabulated results for the variation of the total pressure above a solution of isopropanol ($P_1^* = 1008$ Torr) and n-decane ($P_2^* = 48.3$ Torr) as a function of the mole fraction of the n-decane in the solution and vapor phases. Using these data, calculate the activity coefficients for both components using a Raoult's law standard state.

P (Torr)	x_2	y_2
942.6	0.1312	0.0243
909.6	0.2040	0.0300
883.3	0.2714	0.0342
868.4	0.3360	0.0362
830.2	0.4425	0.0411
786.8	0.5578	0.0451
758.7	0.6036	0.0489

P9.15 At $39.9°C$, the vapor pressure of water is 55.03 Torr (component A) and that of methanol (component B) is 255.6 Torr. Using data from the following table, calculate the activity coefficients for both components using a Raoult's law standard state.

x_A	y_A	P (Torr)
0.0490	0.0175	257.9
0.3120	0.1090	211.3
0.4750	0.1710	184.4
0.6535	0.2550	156.0
0.7905	0.3565	125.7

P9.16 The partial pressures of Br_2 above a solution containing CCl_4 as the solvent at $25°C$ are found to have the values listed in the following table as a function of the mole

fraction of Br_2 in the solution [G. N. Lewis and H. Storch, *J. American Chemical Society* 39 (1917), 2544]. Use these data and a graphical method to determine the Henry's law constant for Br_2 in CCl_4 at $25°C$.

x_{Br_2}	P (Torr)	x_{Br_2}	P (Torr)
0.00394	1.52	0.0130	5.43
0.00420	1.60	0.0236	9.57
0.00599	2.39	0.0238	9.83
0.0102	4.27	0.0250	10.27

P9.17 The data from Problem P9.16 can be expressed in terms of the molality rather than the mole fraction of Br_2. Use the data from the following table and a graphical method to determine the Henry's law constant for Br_2 in CCl_4 at $25°C$ in terms of molality.

m_{Br_2}	P (Torr)	m_{Br_2}	P (Torr)
0.026	1.52	0.086	5.43
0.028	1.60	0.157	9.57
0.039	2.39	0.158	9.83
0.067	4.27	0.167	10.27

P9.18 The partial molar volumes of ethanol in a solution with $x_{H_2O} = 0.60$ at $25°C$ are 17 and 57 cm^3 mol^{-1}, respectively. Calculate the volume change upon mixing sufficient ethanol with 2 mol of water to give this concentration. The densities of water and ethanol are 0.997 and 0.7893 g cm^{-3}, respectively, at this temperature.

P9.19 A solution is prepared by dissolving 32.5 g of a nonvolatile solute in 200 g of water. The vapor pressure above the solution is 21.85 Torr and the vapor pressure of pure water is 23.76 Torr at this temperature. What is the molecular weight of the solute?

P9.20 The heat of fusion of water is 6.008×10^3 J mol^{-1} at its normal melting point of 273.15 K. Calculate the freezing point depression constant K_f.

P9.21 The dissolution of 5.25 g of a substance in 565 g of benzene at 298 K raises the boiling point by $0.625°C$. Note that $K_f = 5.12$ K kg mol^{-1}, $K_b = 2.53$ K kg mol^{-1}, and the density of benzene is 876.6 kg m^{-3}. Calculate the freezing point depression, the ratio of the vapor pressure above the solution to that of the pure solvent, the osmotic pressure, and the molecular weight of the solute. $P_{benzene}^* = 103$ Torr at 298 K.

P9.22 A sample of glucose ($C_6H_{12}O_6$) of mass 1.25 g is placed in a test tube of radius 1.00 cm. The bottom of the test tube is a membrane that is semipermeable to water. The tube is partially immersed in a beaker of water at 298 K so that the bottom of the test tube is only slightly below the level of the water in the beaker. The density of water at this temperature is 997 kg m^{-3}. After equilibrium is reached, how high is the water level of the water in the tube above that in the beaker? What is the value of the osmotic pressure? You may find the approximation $\ln(1/1 + x) \approx -x$ useful.

P9.23 The osmotic pressure of an unknown substance is measured at 298 K. Determine the molecular weight if the concentration of this substance is 25.5 kg m^{-3} and the osmotic pressure is 4.50×10^4 Pa. The density of the solution is 997 kg m^{-3}.

P9.24 An ideal dilute solution is formed by dissolving the solute A in the solvent B. Write expressions equivalent to Equations (9.9) through (9.13) for this case.

P9.25 A solution is made up of 184.2 g of ethanol and 108.1 g of H$_2$O. If the volume of the solution is 333.4 cm^3 and the partial molar volume of H$_2$O is 17.0 cm^3, what is the partial molar volume of ethanol under these conditions?

P9.26 Calculate the solubility of H$_2$S in 1 L of water if its pressure above the solution is 3.25 bar. The density of water at this temperature is 997 kg m^{-3}.

P9.27 The densities of pure water and ethanol are 997 and 789 kg m^{-3}, respectively. The partial molar volumes of ethanol and water in a solution with $x_{ethanol} = 0.20$ are 55.2 and 17.8×10^{-3} L mol^{-1}, respectively. Calculate the change in volume relative to the pure components when 1.00 L of a solution with $x_{ethanol} = 0.20$ is prepared.

P9.28 At a given temperature, a nonideal solution of the volatile components A and B has a vapor pressure of 832 Torr. For this solution, $y_A = 0.404$. In addition, $x_A = 0.285$, $P_A^* = 591$ Torr, and $P_B^* = 503$ Torr. Calculate the activity and activity coefficient of A and B.

P9.29 Calculate the activity and activity coefficient for CS$_2$ at $x_{CS_2} = 0.7220$ using the data in Table 9.3 for both a Raoult's law and a Henry's law standard state.

P9.30 At high altitudes, mountain climbers are unable to absorb a sufficient amount of O$_2$ into their bloodstreams to maintain a high activity level. At a pressure of 1 bar, blood is typically 95% saturated with O$_2$, but near 18,000 feet where the pressure is 0.50 bar, the corresponding degree of saturation is 71%. Assuming that the Henry's law constant for blood is the same as for water, calculate the amount of O$_2$ dissolved in 1.00 L of blood for pressures of 1 bar and 0.500 bar. Air contains 20.99% O$_2$ by volume. Assume that the density of blood is 998 kg m^{-3}.

Electrolyte Solutions

Electrolyte solutions are quite different from the ideal and real solutions of neutral solutes discussed in Chapter 9. The fundamental reason for this difference is that solutes in electrolyte solutions exist as solvated positive and negative ions. In Chapter 4, the formation enthalpy and Gibbs energy of a pure element in its standard state at 1 bar of pressure were set equal to zero. These assumptions allow the formation enthalpies and Gibbs energies of compounds to be calculated using the results of thermochemical experiments. In electrolyte solutions, an additional assumption, $\Delta G_f^\circ(H^+, aq) = 0$, is made to allow the formation enthalpies, Gibbs energies, and entropies of individual ions to be determined. Why are electrolyte and nonelectrolyte solutions so different? The Coulomb interactions between ions in an electrolyte solution are of much longer range than the interactions between neutral solutes. For this reason, electrolyte solutions deviate from ideal behavior at much lower concentrations than do solutions of neutral solutes. Although a formula unit of an electrolyte dissociates into positive and negative ions, only the mean activity and activity coefficient of these ions is accessible through experiments. The Debye–Hückel limiting law provides a useful way to calculate activity coefficients for dilute electrolyte solutions. ■

10.1 The Enthalpy, Entropy, and Gibbs Energy of Ion Formation in Solutions

In this chapter, materials, called **electrolytes,** are discussed that dissociate into positively and negatively charged mobile solvated ions when dissolved in an appropriate solvent. Consider the following overall reaction in water:

$$1/2 H_2(g) + 1/2 Cl_2(g) \rightarrow H^+(aq) + Cl^-(aq) \tag{10.1}$$

in which $H^+(aq)$ and $Cl^-(aq)$ represent mobile solvated ions. Although similar in structure, Equation (10.1) represents a reaction that is quite different than the gas-phase dissociation of an HCl molecule to give $H^+(g)$ and $Cl^-(g)$. For this reaction, $\Delta H_{reaction}$ is -167.2 kJ mol^{-1} when measured with constant pressure calorimetry (see Chapter 4). The shorthand notation $H^+(aq)$ and $Cl^-(aq)$ refers to positive and negative ions as well as their associated hydration shell. The hydration shell is essential in lowering the energy of the ions, thereby making the previous reaction spontaneous. Although energy flow into the system is required to dissociate and ionize hydrogen and chlorine, even more energy is gained in the orientation of the dipolar water molecule around the ions in the **solvation shell.** Therefore, the reaction is exothermic.

The standard state enthalpy for this reaction can be written in terms of formation enthalpies:

$$\Delta H^{\circ}_{reaction} = \Delta H^{\circ}_{reaction}(H^+, aq) + \Delta H^{\circ}_{reaction}(Cl^-, aq) \tag{10.2}$$

There is no contribution of $H_2(g)$ and $Cl_2(g)$ to $\Delta H^{\circ}_{reaction}$ in Equation (10.2) because ΔH°_f for a pure element in its standard state is zero.

Unfortunately, no direct calorimetric experiment can measure only the heat of formation of the solvated anion or cation. This is the case because the solution must remain electrically neutral; therefore, any dissociation reaction of a neutral solute must produce both anions and cations. As we have seen in Chapter 4, tabulated values of formation enthalpies, entropies, and Gibbs energies for various chemical species are very useful. How can this information be obtained for individual solvated cations and anions?

The discussion in the rest of this chapter is restricted to aqueous solutions, for which water is the solvent. Values of thermodynamic functions for anions and cations in aqueous solutions can be obtained by making an appropriate choice for the zero of $\Delta H^{\circ}_f, \Delta G^{\circ}_f$, and S°. By convention, the formation Gibbs energy for $H^+(aq)$ at unit activity is set equal to zero at all temperatures:

$$\Delta G^{\circ}_f(H^+, aq) = 0 \quad \text{for all } T \tag{10.3}$$

With this choice,

$$S^{\circ}_f(H^+, aq) = \left(\frac{\partial \Delta G^{\circ}_f(H^+, aq)}{\partial T}\right)_P = 0 \quad \text{and} \tag{10.4}$$

$$\Delta H^{\circ}_f(H^+, aq) = \Delta G^{\circ}_f(H^+, aq) + T S^{\circ}_f(H^+, aq) = 0$$

Using the convention of Equation (10.3), which has the consequences shown in Equation (10.4), the values of $\Delta H^{\circ}_f, \Delta G^{\circ}_f$, and S° for an individual ion can be assigned numerical values, as shown next.

As discussed earlier, $\Delta H^{\circ}_{reaction}$ for the reaction $1/2\ H_2(g) + 1/2\ Cl_2(g) \rightarrow H^+(aq) + Cl^-(aq)$ can be directly measured. The value of $\Delta G^{\circ}_{reaction}$ can be determined from $\Delta G^{\circ}_{reaction} = -RT \ln K$ by measuring the degree of dissociation in the reaction, using the solution conductivity, and $\Delta S^{\circ}_{reaction}$ can be determined from the relation

$$\Delta S^{\circ}_{reaction} = \frac{\Delta H^{\circ}_{reaction} - \Delta G^{\circ}_{reaction}}{T}$$

Using the conventions stated in Equations (10.3) and (10.4) together with the conventions regarding ΔH°_f and ΔG°_f for pure elements, $\Delta H^{\circ}_{reaction} = \Delta H^{\circ}_f(Cl^-, aq)$, $\Delta G^{\circ}_{reaction} = \Delta G^{\circ}_f(Cl^-, aq)$ and

$$\Delta S^{\circ}_{reaction} = S^{\circ}_f(Cl^-, aq) - \frac{1}{2} S^{\circ}_f(H_2, g) - \frac{1}{2} S^{\circ}_f(Cl_2, g)$$

for the reaction under discussion. In this way, the numerical values $\Delta H_f(Cl^-, aq) = -167.2$ kJ mol^{-1}, $S^{\circ}_f(Cl^-, aq) = 56.5$ J K^{-1} mol^{-1}, and $\Delta G^{\circ}_f(Cl^-, aq) = -131.2$ kJ mol^{-1} can be obtained.

These values can be used to determine the formation functions of other ions. To illustrate how this is done, consider the following reaction:

$$NaCl(s) \rightarrow Na^+(aq) + Cl^-(aq) \tag{10.5}$$

for which the standard reaction enthalpy is found to be $+3.90$ kJ mol^{-1}. For this reaction,

$$\Delta H_{reaction} = \Delta H_f(Cl^-, aq) + \Delta H_f(Na^+, aq) - \Delta H_f(NaCl, s) \tag{10.6}$$

We use the tabulated value of $\Delta H_f^\circ(NaCl, s) = -411.2$ kJ mol^{-1} and the value for $\Delta H_f^\circ(Cl^-, aq)$ just determined to obtain a value for $\Delta H_f^\circ(Na^+, aq) = -240.1$ kJ mol^{-1}. Proceeding to other reactions that involve either $Na^+(aq)$ or $Cl^-(aq)$, the enthalpies of formation of the counter ions can be determined. This procedure can be extended to include other ions. Values for ΔG_f° and S_f° can be determined in a similar fashion. Values for ΔH_f°, ΔG_f°, and S_f° for aqueous ionic species are tabulated in Table 10.1. These thermodynamic quantities are called **conventional formation enthalpies, conventional Gibbs energies of formation,** and **conventional formation entropies** because of the convention described earlier.

Note that ΔH_f°, ΔG_f°, and S_f° for ions are defined relative to $H^+(aq)$. Negative values for ΔH_f° indicate that the formation of the solvated ion is more exothermic than the formation of $H^+(aq)$. A similar statement can be made for ΔG_f°. Generally speaking, ΔH_f° for multiply charged ions is more negative than that of singly charged ions, and ΔH_f° for a given charge is more negative for smaller ions because of the stronger electrostatic attraction between the multiply charged or smaller ion and the water in the solvation shell.

Recall from Section 5.8 that the entropy of an atom or molecule was shown to be always positive. This is not the case for solvated ions because the entropy is measured relative to $H^+(aq)$. The entropy decreases as the hydration shell is formed because liquid water molecules are converted to relatively immobile molecules. Ions with a negative value for the conventional standard entropy such as $Mg^{2+}(aq)$, $Zn^{2+}(aq)$, and $PO_4^{3-}(aq)$ have a larger charge-to-size ratio than $H^+(aq)$. For this reason, the solvation shell is more tightly bound. Conversely, ions with a positive value for the standard entropy such as $Na^+(aq)$, $Cs^+(aq)$, and $NO_3^-(aq)$ have a smaller charge-to-size ratio than $H^+(aq)$.

TABLE 10.1
Conventional Formation Enthalpies, Gibbs Energies, and Entropies of Selected Aqueous Anions and Cations

Ion	ΔH_f° (kJ mol^{-1})	ΔG_f° (kJ mol^{-1})	S_f° (J K^{-1} mol^{-1})
$Ag^+(aq)$	105.6	77.1	72.7
$Br^-(aq)$	-121.6	-104.0	82.4
$Ca^{2+}(aq)$	-542.8	-553.6	-53.1
$Cl^-(aq)$	-167.2	-131.2	56.5
$Cs^+(aq)$	-258.3	-292.0	133.1
$Cu^+(aq)$	71.7	50.0	40.6
$Cu^{2+}(aq)$	64.8	65.5	-99.6
$F^-(aq)$	-332.6	-278.8	-13.8
$H^+(aq)$	0	0	0
$I^-(aq)$	-55.2	-51.6	111.3
$K^+(aq)$	-252.4	-283.3	102.5
$Li^+(aq)$	-278.5	-293.3	13.4
$Mg^{2+}(aq)$	-466.9	-454.8	-138.1
$NO_3^-(aq)$	-207.4	-111.3	146.4
$Na^+(aq)$	-240.1	-261.9	59.0
$OH^-(aq)$	-230.0	-157.2	-10.9
$PO_4^{3-}(aq)$	-1277.4	-1018.7	-220.5
$SO_4^{2-}(aq)$	-909.3	-744.5	20.1
$Zn^{2+}(aq)$	-153.9	-147.1	-112.1

Source: Lide, D. R., Ed., *Handbook of Chemistry and Physics,* 83rd ed., CRC Press, Boca Raton, FL, 2002.

10.2 Understanding the Thermodynamics of Ion Formation and Solvation

As discussed in the preceding section, $\Delta H_f^\circ, \Delta G_f^\circ$, and S_f° cannot be determined for an individual ion in a calorimetric experiment. However, as seen next, values for thermodynamic functions associated with individual ions can be calculated with a reasonable level of confidence using a thermodynamic model. This result allows the conventional values of $\Delta H_f^\circ, \Delta G_f^\circ$, and S_f° to be converted to absolute values for individual ions. In the following discussion, the focus is on ΔG_f°.

We first discuss the individual contributions to ΔG_f°, and do so by analyzing the following sequence of steps that describe the formation of $H^+(aq)$ and $Cl^-(aq)$:

$1/2\ H_2(g) \rightarrow H(g)$	$\Delta G^\circ = 203.3\ \text{kJ mol}^{-1}$
$1/2\ Cl_2(g) \rightarrow Cl(g)$	$\Delta G^\circ = 105.7\ \text{kJ mol}^{-1}$
$H(g) \rightarrow H^+(g) + e^-$	$\Delta G^\circ = 1312\ \text{kJ mol}^{-1}$
$Cl(g) + e^- \rightarrow Cl^-(g)$	$\Delta G^\circ = -349\ \text{kJ mol}^{-1}$
$Cl^-(g) \rightarrow Cl^-(aq)$	$\Delta G^\circ = \Delta G^\circ_{solvation}(Cl^-, aq)$
$H^+(g) \rightarrow H^+(aq)$	$\Delta G^\circ = \Delta G^\circ_{solvation}(H^+, aq)$

$$1/2\ H_2(g) + 1/2\ Cl_2(g) \rightarrow H^+(aq) + Cl^-(aq) \qquad \Delta G^\circ = -131.2\ \text{kJ mol}^{-1}$$

This information is shown pictorially in Figure 10.1. Because G is a state function, both the green and yellow shaded paths must have the same ΔG value. The first two reactions in this sequence are the dissociation of the molecules in the gas phase. The second two reactions are the formation of ions, and ΔG° is determined from the known ionization energy and electron affinity. The change in the Gibbs energy for the overall process is

$$\Delta G^\circ = \Delta G^\circ_{solvation}(Cl^-, aq) + \Delta G^\circ_{solvation}(H^+, aq) + 1272\ \text{kJ mol}^{-1} \qquad (10.7)$$

Equation 10.7 allows us to relate the $\Delta G_{solution}$ of the solvated H^+ and Cl^- ions with ΔG for the overall reaction.

As Equation (10.7) shows, $\Delta G^\circ_{solvation}$ plays a critical role in the determination of the Gibbs energies of ion formation. Although $\Delta G^\circ_{solvation}$ of an individual cation or anion cannot be determined experimentally, it can be estimated using a model developed by Max Born. In this model, the solvent is treated as a uniform fluid with the appropriate dielectric constant, and the ion is treated as a charged sphere. How can $\Delta G^\circ_{solvation}$ be calculated with these assumptions? At constant T and P, the nonexpansion work for a reversible process equals ΔG for the process. Therefore, if the reversible work associated with solvation can be calculated, ΔG for the process is known. Imagine a process in which a neutral atom A gains the charge q, first in a vacuum and secondly in a uniform dielectric medium. The value of $\Delta G^\circ_{solvation}$ of an ion with a charge q is the reversible work for the process $(A \rightarrow A^q)_{solution}$ minus that for the reversible process $(A \rightarrow A^q)_{vacuum}$.

FIGURE 10.1
ΔG° is shown pictorially for two different paths starting with $1/2\ H_2(g)$ and $1/2\ Cl_2(g)$ and ending with $H^+(aq) + Cl^-(aq)$. The units for the numbers are kJ mol^{-1}. Because ΔG is the same for both paths, $\Delta G^\circ_{solvation}(H^+, aq)$ can be expressed in terms of gas-phase dissociation and ionization energies.

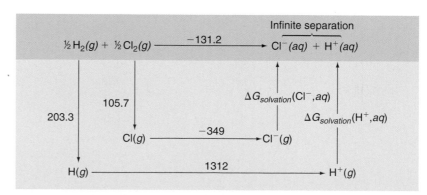

The electrical potential around a sphere of radius r with the charge q' is given by $\phi = q'/4\pi\varepsilon r$. From electrostatics, the work in charging the sphere by the additional amount dq is ϕdq. Therefore, the work in charging a neutral sphere in vacuum to the charge q is

$$w = \int_0^q \frac{q'dq'}{4\pi\varepsilon_0 r} = \frac{1}{4\pi\varepsilon_0 r}\int_0^q q'dq' = \frac{q^2}{8\pi\varepsilon_0 r} \tag{10.8}$$

where ε_0 is the permittivity of free space. The work of the same process in a solvent is $q^2/8\pi\varepsilon_0\varepsilon_r r$, where ε_r is the relative permittivity (dielectric constant) of the solvent. Consequently, the electrostatic component to the molar solvation Gibbs energy for an ion of charge $q = ze$ is given by

$$\Delta G^\circ_{solvation} = \frac{z^2 e^2 N_A}{8\pi\varepsilon_0 r}\left(\frac{1}{\varepsilon_r} - 1\right) \tag{10.9}$$

Because $\varepsilon_r > 1$, $\Delta G^\circ_{solvation} < 0$, showing that solvation is a spontaneous process. Values for ε_r for a number of solvents are listed in Table 10.2 (see Appendix A, Data Tables).

To test this model, one compares measured values of the absolute values of $\Delta G^\circ_{solvation}$ with the functional form proposed in Equation (10.9). This requires knowledge of $\Delta G^\circ_{solvation}(H^+, aq)$ and experimentally determined values of $\Delta G^\circ_{solvation}$ referenced to $H^+(aq)$. It turns out that $\Delta G^\circ_{solvation}(H^+, aq)$ can be calculated by comparing values of ΔG°_f for gaseous positive and negative ions of the negatively charged halogen ions and the positively charged alkali metal ions. A similar analysis can be used to obtain $\Delta H^\circ_{solvation}(H^+, aq)$ and $S^\circ_{solvation}(H^+, aq)$. Because the justification is involved, the results are simply stated here. Because of the models used, these results exhibit greater uncertainty than do the conventional values referenced to $H^+(aq)$:

$$\Delta H^\circ_{solvation}(H^+, aq) \approx -1090 \, \text{kJ mol}^{-1} \tag{10.10}$$
$$\Delta G^\circ_{solvation}(H^+, aq) \approx -1050 \, \text{kJ mol}^{-1}$$
$$S^\circ_{solvation}(H^+, aq) \approx -130 \, \text{J mol}^{-1}\,\text{K}^{-1}$$

The values listed in Equation (10.10) can be used to calculate absolute values of $H^\circ_{solvation}$, $\Delta G^\circ_{solvation}$, and $S^\circ_{solvation}$ for other ions from the conventional values referenced to $H^+(aq)$. These calculated absolute values can be used to test the validity of the Born model. If the model is valid, a plot of $\Delta G^\circ_{solvation}$ versus z^2/r should give a straight line as shown by Equation (10.9) and the data points for individual ions should lie on the line. The results are shown in Figure 10.2, where r is the ionic radius obtained from crystal structure determinations.

The first and second clusters of data points in Figure 10.2 are for singly and doubly charged ions, respectively. The data are compared with the result predicted by Equation (10.9) in Figure 10.2a. As can be seen from the figure, the trends are reproduced, but there is no quantitative agreement. The agreement can be considerably improved by using an effective radius for the ion rather than the ionic radius from crystal structure determinations. The effective radius is defined as the distance from the center of the ion to the center of charge in the dipolar water molecule. Latimer, Pitzer, and Slansky [*J. Chemical Physics*, 7 (1939) 109] found the best agreement with the Born equation by adding 0.085 nm to the crystal radius of positive ions, and 0.100 nm to the crystal radius for negative ions to account for the fact that the H_2O molecule is not a point dipole. This difference is explained by the fact that the center of charge in the water molecule is closer to positive ions than to negative ions. Figure 10.2b shows that the agreement obtained between the predictions of Equation (10.9) and experimental values is very good if this correction to the ionic radii is made.

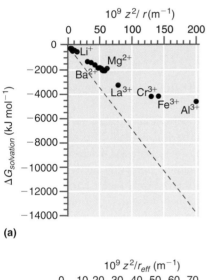

(a)

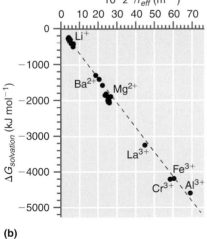

(b)

FIGURE 10.2

(a) The solvation energy calculated using the Born model is shown as a function of z^2/r. (b) The same results are shown as a function of z^2/r_{eff}. (See text.) The dashed line shows the behavior predicted by Equation (10.9).

Figure 10.2 shows good agreement between the predictions of the Born model and calculated values for $\Delta G°_{solvation}$. However, because of uncertainties about the numerical values of the ionic radii and for the dielectric constant of the immobilized water in the solvation shell of an ion, values obtained from different models differ by about ±50 kJ mol^{-1} for the solvation enthalpy and Gibbs energy and ±10 J K^{-1} mol^{-1} for the solvation entropy. Because this absolute uncertainty is large compared to the relative uncertainty of the thermodynamic functions using the convention described in Equations (10.3) and (10.4), absolute values of $\Delta H°_f$, $\Delta G°_f$, and $S°_f$ are not generally used by chemists for solvated ions in aqueous solutions.

10.3 Activities and Activity Coefficients for Electrolyte Solutions

The ideal dilute solution model presented for the activity and activity coefficient of components of real solutions in Chapter 9 is not valid for electrolyte solutions. This is the case because solute–solute interactions are dominated by the long-range electrostatic forces present between ions in electrolyte solutions. For example, the reaction that occurs when NaCl is dissolved in water is

$$NaCl(s) + H_2O(l) \rightarrow Na^+(aq) + Cl^-(aq) \tag{10.11}$$

In dilute solutions, NaCl is completely dissociated. Therefore, the solute–solute interactions are electrostatic in nature.

Although the concepts of activity and activity coefficients introduced in Chapter 9 are applicable to electrolytes, these concepts must be formulated differently for electrolytes to include the Coulomb interactions among ions. We next discuss a model for electrolyte solutions that is stated in terms of the chemical potentials, activities, and activity coefficients of the individual ionic species. However, only mean activities and activity coefficients (to be defined later) can be determined through experiments.

Consider the Gibbs energy of the solution, which can be written as

$$G = n_{solvent}\mu_{solvent} + n_{solute}\mu_{solute} \tag{10.12}$$

For the general electrolyte $A_{v_+}B_{v_-}$ that dissociates completely, one can also write an equivalent expression for G:

$$G = n_{solvent}\mu_{solvent} + n_+\mu_+ + n_-\mu_- = n_{solvent}\mu_{solvent} + n_{solute}(v_+\mu_+ + v_-\mu_-) \tag{10.13}$$

where v_+ and v_- are the stoichiometric coefficients of the cations and anions, respectively, produced upon dissociation of the electrolyte. In shorthand notation, an electrolyte is called a 1–1 electrolyte if $v_+ = 1$ and $v_- = 1$. Similarly, for a 2–3 electrolyte, $v_+ = 2$ and $v_- = 3$. Because Equations (10.12) and (10.13) describe the same solution, they are equivalent. Therefore,

$$\mu_{solute} = v_+\mu_+ + v_-\mu_- \tag{10.14}$$

Although this equation is formally correct for a strong electrolyte, one can never make a solution of either cations or anions alone, because any solution is electrically neutral. Therefore, it is useful to define a **mean ionic chemical potential** $\mu_\pm$ for the solute

$$\mu_\pm = \frac{\mu_{solute}}{v} = \frac{v_+\mu_+ + v_-\mu_-}{v} \tag{10.15}$$

where $v = v_+ + v_-$. The reason for doing so is that $\mu_\pm$ can be determined experimentally, whereas μ_+ and μ_- are not accessible through experiments.

The next task is to relate the chemical potentials of the solute and its individual ions to the activities of these species. For the individual ions,

$$\mu_+ = \mu_+^\circ + RT \ln a_+ \quad \text{and} \quad \mu_- = \mu_-^\circ + RT \ln a_- \tag{10.16}$$

where the standard chemical potentials of the ions, μ_+° and μ_-°, are based on a Henry's law standard state. Substituting Equation (10.16) in (10.15), an equation for the mean ionic chemical potential is obtained that is similar in structure to the expressions that we derived for the ideal dilute solution:

$$\mu_\pm = \mu_\pm^\circ + RT \ln a_\pm \tag{10.17}$$

The **mean ionic activity** $a_\pm$ is related to the individual ion activities by

$$a_\pm^\nu = a_+^{\nu_+} a_-^{\nu_-} \quad \text{or} \quad a_\pm = (a_+^{\nu_+} a_-^{\nu_-})^{1/\nu} \tag{10.18}$$

EXAMPLE PROBLEM 10.1

Write the mean ionic activities of NaCl, K_2SO_4, and H_3PO_4 in terms of the ionic activities of the individual anions and cations. Assume complete dissociation.

Solution

$$a_{NaCl}^2 = a_{Na^+} a_{Cl^-} \quad \text{or} \quad a_{NaCl} = \sqrt{a_{Na^+} a_{Cl^-}}$$

$$a_{K_2SO_4}^3 = a_{K^+}^2 a_{SO_4^{2-}} \quad \text{or} \quad a_{K_2SO_4} = (a_{K^+}^2 a_{SO_4^{2-}})^{1/3}$$

$$a_{H_3PO_4}^4 = a_{H^+}^3 a_{PO_4^{3-}} \quad \text{or} \quad a_{H_3PO_4} = (a_{H^+}^3 a_{PO_4^{3-}})^{1/4}$$

If the ionic activities are referenced to the concentration units of molality, then

$$a_+ = \frac{m_+}{m^\circ} \gamma_+ \quad \text{and} \quad a_- = \frac{m_-}{m^\circ} \gamma_- \tag{10.19}$$

where $m_+ = \nu_+ m$ and $m_- = \nu_- m$. Because the activity is unitless, the molality must be referenced to a standard state concentration chosen to be $m^\circ = 1$ mol kg^{-1}. As in Chapter 9, a hypothetical standard state based on molality is defined. In this standard state, Henry's law, which is valid in the limit $m \to 0$, is obeyed up to a concentration of $m = 1$ molal. Substitution of Equation (10.19) in Equation (10.18) shows that

$$a_\pm^\nu = \left(\frac{m_+}{m^\circ}\right)^{\nu_+} \left(\frac{m_-}{m^\circ}\right)^{\nu_-} \gamma_+^{\nu_+} \gamma_-^{\nu_-} \tag{10.20}$$

To simplify this notation, we define the **mean ionic molality** $m_\pm$ and **mean ionic activity coefficient** $\gamma_\pm$ by

$$m_\pm^\nu = m_+^{\nu_+} m_-^{\nu_-} \tag{10.21}$$

$$m_\pm = (\nu_+^{\nu_+} \nu_-^{\nu_-})^{1/\nu} m \quad \text{and}$$

$$\gamma_\pm^\nu = \gamma_+^{\nu_+} \gamma_-^{\nu_-}$$

$$\gamma_\pm = (\gamma_+^{\nu_+} \gamma_-^{\nu_-})^{1/\nu}$$

With these definitions, the mean ionic activity is related to the mean ionic activity coefficient and mean ionic molality as follows:

$$a_\pm^\nu = \left(\frac{m_\pm}{m^\circ}\right)^\nu \gamma_\pm^\nu \quad \text{or} \quad a_\pm = \left(\frac{m_\pm}{m^\circ}\right) \gamma_\pm \tag{10.22}$$

Equations (10.19) through (10.22) relate the activities, activity coefficients, and molalities of the individual ionic species to mean ionic quantities and measurable properties of the system such as the molality and activity of the solute. With these definitions, Equation (10.17) defines the chemical potential of the electrolyte solute in terms of its activity:

$$\mu_{solute} = \mu_{solute}^\circ + RT \ln a_\pm^\nu \tag{10.23}$$

Equations (10.20) and (10.21) can be used to express the chemical potential of the solute in terms of measurable or easily accessible quantities:

$$\mu_{solute} = \mu_\pm = \left[\mu_\pm^\circ + RT \ln(\nu_+^{\nu_+} \nu_-^{\nu_-}) \right] + \nu RT \ln\left(\frac{m}{m^\circ}\right) + \nu RT \ln \gamma_\pm \tag{10.24}$$

The first term in the square bracket is defined by the "normal" standard state, which is usually taken to be a Henry's law standard state. The second term is obtained from the chemical formula for the solute. These two terms can be combined to create a new standard state $\mu_\pm^{\circ\circ}$ defined by the terms in the square brackets in Equation (10.24):

$$\mu_{solute} = \mu_\pm = \mu_\pm^{\circ\circ} + \nu RT \ln\left(\frac{m}{m^\circ}\right) + \nu RT \ln \gamma_\pm \tag{10.25}$$

The first two terms in Equation (10.25) correspond to the "ideal" ionic solution, which is associated with $\gamma_\pm = 1$.

The last term in Equation (10.25), which is the most important term in this discussion, contains the deviations from ideal behavior. The mean activity coefficient $\gamma_\pm$ can be obtained through experiment. For example, the activity coefficient of the solvent can be determined by measuring the boiling point elevation, the freezing point depression, or the lowering of the vapor pressure above the solution upon solution formation. The activity of the solute is obtained from that of the solvent using the Gibbs–Duhem equation. As shown in Section 11.8, $\gamma_\pm$ can also be determined through measurements on electrochemical cells. In addition, a very useful theoretical model allows $\gamma_\pm$ to be calculated for dilute electrolytic solutions. This model is discussed in the next section.

10.4 Calculating $\gamma_\pm$ Using the Debye–Hückel Theory

There is no model that adequately explains the deviations from ideality for the solutions of neutral solutes discussed in Sections 9.1 through 9.4. This is the case because the deviations arise through A–A, B–B, and A–B interactions that are specific to components A and B. This precludes a general model that holds for arbitrary A and B. However, the situation for solutions of electrolytes is different.

Deviations from ideal solution behavior occur at a much lower concentration for electrolytes than for nonelectrolytes, because the dominant interaction between the ions in an electrolyte is a long-range electrostatic Coulomb interaction rather than a short-range van der Waals or chemical interaction. Because of its long range, the Coulomb interaction among the ions cannot be neglected even for very dilute solutions of electrolytes. The Coulomb interaction allows a model of electrolyte solutions to be formulated for the following reason. The attractive or repulsive interaction of two ions depends only on their charge and separation, and not on their chemical identity. Therefore, the solute–solute interactions can be modeled knowing only the charge on the ions, and the model becomes independent of the identity of the solute species.

Measurements of activity coefficients in electrolyte solutions show that $\gamma_\pm < 1$ for dilute solutions in the limit $m \to 0$. Because $\gamma_\pm < 1$, the chemical potential of the solute in a dilute solution is lower than that for a solution of uncharged solute species. Why is this the case? The lowering of μ_{solute} arises because the net electrostatic interaction among the ions surrounding an arbitrarily chosen central ion is attractive rather than repulsive. The model that describes the lowering of the energy of electrolytic solutions is due to Peter Debye and Erich Hückel. Rather than derive their results, the essential features of their model are described next.

The solute ions in the solvent give rise to a spatially dependent electrostatic potential, ϕ, which can be calculated if the spatial distribution of ions is known. In dilute electrolyte solutions, the energy increase or decrease experienced by an ion of charge $\pm ze$ if the potential ϕ could be turned on suddenly is small compared to the thermal energy. This condition can be expressed in the form

$$\left|\pm ze\phi\right| \ll kT \tag{10.26}$$

In Equation (10.26), e is the charge on a proton, and k is Boltzmann's constant. In this limit, the dependence of ϕ on the spatial coordinates and the spatial distribution of the ions around an arbitrary central ion can be calculated. In contrast to the potential around an isolated ion in a dielectric medium, which is described by

$$\phi_{isolated\ ion}(r) = \frac{\pm ze}{4\pi\varepsilon_r\varepsilon_0 r} \tag{10.27}$$

the potential in the dilute electrolyte solution has the form

$$\phi_{solution}(r) = \frac{\pm ze}{4\pi\varepsilon_r\varepsilon_0 r}\exp(-\kappa r) \tag{10.28}$$

In Equations (10.27) and (10.28), ε_0 and ε_r are the permittivity of free space and the relative permittivity (dielectric constant) of the dielectric medium or solvent, respectively.

The Debye–Hückel theory shows that κ is related to the individual charges on the ions and to the solute molality m by

$$\kappa^2 = e^2 N_A m\left(\frac{v_+ z_+^2 + v_- z_-^2}{\varepsilon_0 \varepsilon_r kT}\right) \tag{10.29}$$

From this formula, it can be seen that the screening becomes more effective as the concentration of the ionic species increases. Screening is also more effective for multiply charged ions and for larger values of v_+ and v_-.

The ratio

$$\frac{\phi_{solution}(r)}{\phi_{isolated\ ion}(r)} = e^{-\kappa r}$$

is plotted in Figure 10.3 for different values of m for an aqueous solution of a 1–1 electrolyte. Note that the potential falls off much more rapidly with the radial distance, r, in the electrolyte solution than in the uniform dielectric medium. Note also

FIGURE 10.3
The ratio of the falloff in the electrostatic potential in the electrolyte solution to that for an isolated ion is shown as a function of the radial distance for three different molarities of a 1–1 electrolyte such as NaCl.

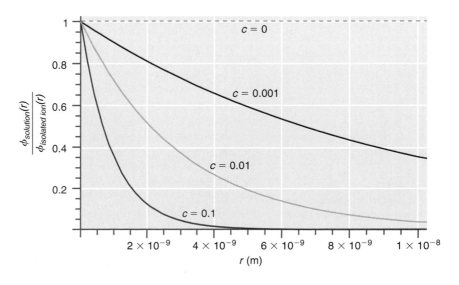

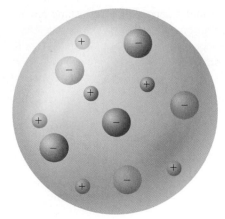

FIGURE 10.4
Pictorial rendering of the arrangement of ions about an arbitrary ion in an electrolyte solution. The central ion is more likely to have oppositely charged ions as neighbors. The large circle represents a sphere of radius $r \sim 8/\kappa$. From a point outside of this sphere, the charge on the central ion is essentially totally screened.

that the potential falls off more rapidly with increasing concentration of the electrolyte. The origin of this effect is that ions of sign opposite to the central ion are more likely to be found close to the central ion. These surrounding ions form a diffuse ion cloud around the central ion, as shown pictorially in Figure 10.4. If a spherical surface is drawn centered at the central ion, the net charge within the surface can be calculated. Calculations show that the net charge has the same sign as the central charge, falls off rapidly with distance, and is close to zero for $\kappa r \sim 8$. For larger values of κr, the central ion is completely screened by the diffuse ion cloud, meaning that the net charge is zero. Because of the lowering of energy that arises from the electrostatic interactions within the diffuse charge cloud, $\gamma_\pm < 1$ in dilute electrolyte solutions. The net effect of the diffuse ion cloud is to screen the central ion from the rest of the solution, and the quantity $1/\kappa$ is known as the **Debye–Hückel screening length.** Larger values of κ correspond to a smaller diffuse cloud, and a more effective screening.

It is convenient to combine the concentration-dependent terms that contribute to the screening length in the **ionic strength** I, which is defined by

$$I = \frac{m}{2}\sum_i (v_{i+}z_{i+}^2 + v_{i-}z_{i-}^2) = \frac{1}{2}\sum_i (m_{i+}z_{i+}^2 + m_{i-}z_{i-}^2) \qquad (10.30)$$

EXAMPLE PROBLEM 10.2

Calculate I for (a) a 0.050 molal solution of NaCl and for (b) a Na_2SO_4 solution of the same molality.

Solution

a. $I_{NaCl} = \dfrac{m}{2}(v_+z_+^2 + v_-z_-^2) = \dfrac{0.050\,\text{mol kg}^{-1}}{2} \times (1+1) = 0.050\,\text{mol kg}^{-1}$

b. $I_{Na_2SO_4} = \dfrac{m}{2}(v_+z_+^2 + v_-z_-^2) = \dfrac{0.050\,\text{mol kg}^{-1}}{2} \times (2+4) = 0.15\,\text{mol kg}^{-1}$

Using the definition of the ionic strength, Equation (10.29) can be written in the form

$$\kappa = \left(\sqrt{\frac{2e^2 N_A}{\varepsilon_0 kT}}\right)\sqrt{\frac{I}{\varepsilon_r}} = 2.91\times10^{10}\sqrt{\frac{I/\text{mol kg}^{-1}}{\varepsilon_r}}\ m^{-1} \text{ at } 298\text{ K} \quad (10.31)$$

The first term in this equation contains only fundamental constants that are independent of the solvent and solute as well as the temperature. The second term contains the ionic strength of the solution and the unitless relative permittivity of the solvent. For the more conventional units of mol L^{-1}, and for water, for which $\varepsilon_r = 78.5$, $\kappa = 3.29\times10^9\sqrt{I}\ m^{-1}$ at 298 K.

By calculating the charge distribution of the ions around the central ion and the work needed to charge these ions up to their charges z_+ and z_- from an initially neutral state, Debye and Hückel were able to obtain an expression for the mean ionic activity coefficient. It is given by

$$\ln\gamma_\pm = -|z_+z_-|\frac{e^2\kappa}{8\pi\varepsilon_0\varepsilon_r kT} \qquad (10.32)$$

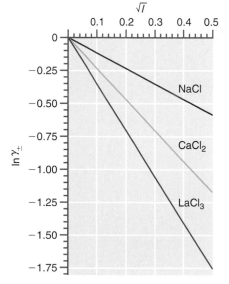

FIGURE 10.5
The decrease in the Debye–Hückel mean activity coefficient with the square root of the ionic strength is shown for a 1–1, a 1–2, and a 1–3 electrolyte, all of the same molality in the solute.

This equation is known as the **Debye–Hückel limiting law.** It is called a limiting law because Equation (10.32) is only obeyed for small values of the ionic strength. Note that because of the negative sign in Equation (10.32), $\gamma_\pm < 1$. From the concentration dependence of κ shown in Equation (10.31), the model predicts that $\ln\gamma_\pm$ decreases with the ionic strength as $\sqrt{I}$. This dependence is shown in Figure 10.5. Although all three solutions have the same solute concentration, they have different values for z^+ and z^-. For this reason, the three lines have a different slope.

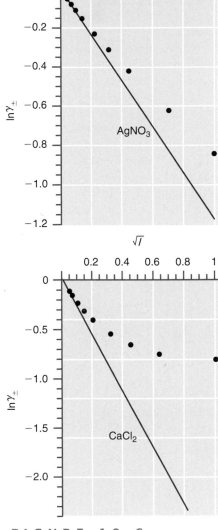

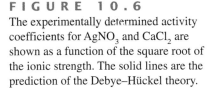

Equation (10.32) can be simplified for a particular choice of solvent and temperature. For aqueous solutions at 298.15 K, the result is

$$\log \gamma_\pm = -0.5092\left|z_+ z_-\right|\sqrt{I} \quad \text{or} \quad \ln \gamma_\pm = -1.173\left|z_+ z_-\right|\sqrt{I} \qquad (10.33)$$

How well does the Debye–Hückel limiting law agree with experimental data? Figure 10.6 shows a comparison of the model with data on the aqueous solutions of $AgNO_3$ and $CaCl_2$. In each case, $\ln \gamma_\pm$ is plotted versus $\sqrt{I}$. The Debye–Hückel limiting law predicts that the data will fall on the line indicated in each figure. The data points deviate from the predicted behavior above $\sqrt{I} = 0.1$ for $AgNO_3$ ($m = 0.01$), and above $\sqrt{I} = 0.006$ for $CaCl_2$ ($m = 0.002$). In the limit that $I \rightarrow 0$, the limiting law is obeyed. However, the deviations are noticeable at a concentration for which a neutral solute would exhibit ideal behavior.

The deviations continue to increase with increasing ionic strength. Figure 10.7 shows experimental data for $ZnBr_2$ out to $\sqrt{I} = 5.5$, corresponding to $m = 10$. Note that, although the Debye–Hückel limiting law is obeyed as $I \rightarrow 0$, $\ln \gamma_\pm$ goes through a minimum and begins to increase with increasing ionic strength. At the highest value of the ionic strength plotted, $\gamma_\pm = 2.32$, which is greater than one. Although the deviations from ideal behavior are less pronounced in Figure 10.6, the trend is the same for all the solutes. The mean ionic activity coefficient $\gamma_\pm$ falls off more slowly with the ionic strength than predicted by the Debye–Hückel limiting law. The behavior shown in Figure 10.7 is typical for most electrolytes; after passing through a minimum, $\gamma_\pm$ rises with increasing ionic strength, and for high values of I, $\gamma_\pm > 1$. Experimental values for $\gamma_\pm$ for a number of solutes at different concentrations in aqueous solution are listed in Table 10.3 (see Appendix A, Data Tables).

There are a number of reasons why the experimental values of $\gamma_\pm$ differ at high ionic strength from those calculated from the Debye–Hückel limiting law. They mainly involve the assumptions made in the model. It has been assumed that the ions can be treated as point charges with zero volume, whereas ions and their associated solvation shells occupy a finite volume. As a result, there is an increase in the repulsive interaction among ions in an electrolyte over that predicted for point charges; this increase becomes more important as the concentration increases. Repulsive interactions raise the energy of the solution and, therefore, increase $\gamma_\pm$. The Debye–Hückel model also assumes that the solvent can be treated as a structureless dielectric medium. However, the ion is surrounded by a relatively ordered primary solvation shell, as well as by more loosely bound water molecules. The atomic level structure of the solvation shell is not adequately represented by using the dielectric strength of bulk solvent. Another factor that has not been taken into account is that as the concentration increases, some ion pairing occurs such that the concentration of ionic species is less than would be calculated assuming complete dissociation.

FIGURE 10.6

The experimentally determined activity coefficients for $AgNO_3$ and $CaCl_2$ are shown as a function of the square root of the ionic strength. The solid lines are the prediction of the Debye–Hückel theory.

FIGURE 10.7

Experimentally determined values for the mean activity coefficient for $ZnBr_2$ are shown as a function of the square root of the ionic strength. The solid line is the prediction of the Debye–Hückel theory.

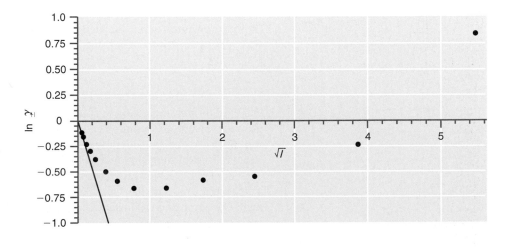

Additionally, consider the fact that the water molecules in the solvation shell have effectively been removed from the solvent. For example, in an aqueous solution of H_2SO_4, approximately nine H_2O molecules are tightly bound per dissolved H_2SO_4 formula unit. Therefore, the number of moles of H_2O as solvent in 1 L of a one molar H_2SO_4 solution is reduced from 55 for pure H_2O to 46 in the solution. Consequently, the actual solute molarity is larger than that calculated by assuming that all the H_2O is in the form of solvent. Because the activity increases linearly with the actual molarity, $\gamma_\pm$ increases as the solute concentration increases. If there were no change in the enthalpy of solvation with concentration, all the H_2O molecules would be removed from the solvent at a concentration of six molar H_2SO_4. Clearly, this assumption is unreasonable. What actually happens is that solvation becomes energetically less favorable as the H_2SO_4 concentration increases. This corresponds to a less negative value of $\ln\gamma_\pm$, or equivalently to an increase in $\gamma_\pm$. Summing up, many factors explain why the Debye–Hückel limiting law is only valid for small concentrations. Because of the complexity of these different factors, there is no simple formula based on theory that can replace the Debye–Hückel limiting law. However, the main trends exhibited in Figures 10.6 and 10.7 are reproduced in more sophisticated theories of electrolyte solutions.

Because none of the usual models are valid at high concentrations, empirical models that "improve" on the Debye–Hückel model by predicting an increase in $\gamma_\pm$ for high concentrations are in widespread use. An empirical modification of the Debye–Hückel limiting law that has the form

$$\log_{10}\gamma_\pm = -0.51\left|z_+ z_-\right|\left[\frac{\left(\dfrac{I}{m^\circ}\right)^{1/2}}{1+\left(\dfrac{I}{m^\circ}\right)^{1/2}} - 0.20\left(\frac{I}{m^\circ}\right)\right] \tag{10.34}$$

is known as the **Davies equation.** As seen in Figure 10.8, this equation for $\gamma_\pm$ shows the correct limiting behavior for low I values, and the trend at higher values of I is in better agreement with the experimental results shown in Figures 10.6 and 10.7.

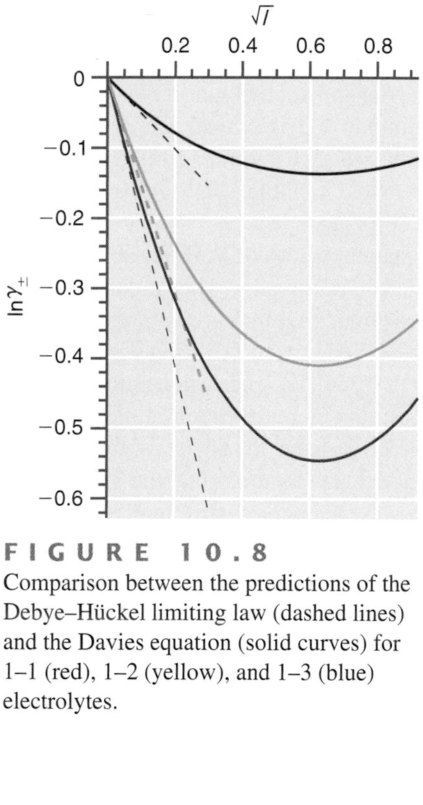

FIGURE 10.8
Comparison between the predictions of the Debye–Hückel limiting law (dashed lines) and the Davies equation (solid curves) for 1–1 (red), 1–2 (yellow), and 1–3 (blue) electrolytes.

10.5 Chemical Equilibrium in Electrolyte Solutions

As discussed in Section 9.13, the equilibrium constant in terms of activities is given by Equation (9.66):

$$K = \prod_i (a_i^{eq})^{\nu_j} \tag{10.35}$$

It is convenient to define the activity of a species relative to its molarity. In this case,

$$a_i = \gamma_i \frac{c_i}{c^\circ} \tag{10.36}$$

where γ_i is the activity coefficient of species i. We next specifically consider chemical equilibrium in electrolyte solutions, illustrating that activities rather than concentrations must be taken into account to accurately model equilibrium concentrations. We restrict our considerations to the range of ionic strengths for which the Debye–Hückel limiting law is valid. As an example, we calculate the degree of dissociation of MgF_2 in water. The equilibrium constant in terms of molarities for ionic salts is usually given the symbol K_{sp}, where the subscript refers to the solubility product. The equilibrium constant K_{sp} is unitless and has the value of 6.4×10^{-9} for the

TABLE 10.4

Solubility Product Constants (Molarity Based) for Selected Salts

Salt	K_{sp}	Salt	K_{sp}
AgBr	4.9×10^{-13}	CaSO$_4$	4.9×10^{-6}
AgCl	1.8×10^{-10}	Mg(OH)$_2$	5.6×10^{-11}
AgI	8.5×10^{-17}	Mn(OH)$_2$	1.9×10^{-13}
Ba(OH)$_2$	5.0×10^{-3}	PbCl$_2$	1.6×10^{-5}
BaSO$_4$	1.1×10^{-10}	Pb SO$_4$	1.8×10^{-8}
CaCO$_3$	3.4×10^{-9}	ZnS	1.6×10^{-23}

Source: Lide, D. R., Ed., *Handbook of Chemistry and Physics,* 83rd ed., CRC Press, Boca Raton, FL, 2002.

reaction shown in Equation (10.37). Values for K_{sp} are generally tabulated for reduced concentration units of molarity (c/c°) rather than molality (m/m°), and values for selected substances are listed in Table 10.4. For highly dilute solutions, the numerical value of the concentration is the same on both scales.

We next consider dissociation of MgF$_2$ in an aqueous solution:

$$MgF_2(s) \rightarrow Mg^{2+}(aq) + 2F^-(aq) \tag{10.37}$$

Because the activity of the pure solid can be set equal to one,

$$K_{sp} = a_{Mg^{2+}} a_{F^-}^2 = \left(\frac{c_{Mg^{2+}}}{c^\circ}\right)\left(\frac{c_{F^-}}{c^\circ}\right)^2 \gamma_\pm^3 = 6.4 \times 10^{-9} \tag{10.38}$$

From the stoichiometry of the overall equation, we know that $c_{F^-} = 2c_{Mg^{2+}}$, but Equation (10.38) still contains two unknowns, $\gamma_\pm$ and c_{F^-}, that we solve for iteratively.

First assume that $\gamma_\pm = 1$ and solve Equation (10.38) for c_{F^-}, giving $c_{Mg^{2+}} = 1.17 \times 10^{-3}$ mol L^{-1}. Next calculate the ionic strength from

$$I = \frac{1}{2}(z_+^2 m_+ + z_-^2 m_-) = \frac{1}{2}(4 \times 1.17 \times 10^{-3} + 2.34 \times 10^{-3}) = 3.51 \times 10^{-3} \text{ mol L}^{-1} \tag{10.39}$$

and recalculate $\gamma_\pm$ from the Debye–Hückel limiting law of Equation (10.29), giving $\gamma_\pm = 0.870$. Substitute this improved value of $\gamma_\pm$ in Equation (10.38), giving $c_{Mg^{2+}} = 1.34 \times 10^{-3}$ mol L^{-1}. A second iteration gives $\gamma_\pm = 0.862$ and $c_{Mg^{2+}} = 1.36 \times 10^{-3}$ mol L^{-1}, and a third iteration gives $\gamma_\pm = 0.861$ and $c_{Mg^{2+}} = 1.36 \times 10^{-3}$ mol L^{-1}. This result is sufficiently accurate. These results show that assuming that $\gamma_\pm = 1$ leads to an unacceptably large error of 14% in $c_{Mg^{2+}}$.

Another effect of the ionic strength on solubility is described by the terms *salting in* and *salting out*. The behavior shown in Figure 10.7, in which the activity coefficient first decreases and subsequently increases with concentration, affects the solubility of a salt in the following way. For a salt such as MgF$_2$, the product $[Mg^{2+}][F^-]^2 \gamma_\pm^3 = K$ is constant as the concentration varies for constant T, because the thermodynamic equilibrium constant K depends only on T. Therefore, the concentrations $[Mg^{2+}]$ and $[F^-]$ change in an opposite way to $\gamma_\pm$. At small values of the ionic strength, $\gamma_\pm < 1$, and the solubility increases as $\gamma_\pm$ decreases with concentration until the minimum in a plot of $\gamma_\pm$ versus I is reached. This effect is known as **salting in.** For high values of the ionic strength, $\gamma_\pm > 1$ and the solubility is less than at low values of I. This effect is known as **salting out.** Salting in and salting out are frequently encountered in studies of the solubility of proteins in aqueous electrolyte solutions.

Vocabulary

conventional formation enthalpies

conventional formation entropies

conventional Gibbs energies of formation

Davies equation

Debye–Hückel limiting law

Debye–Hückel screening length

electrolyte

ionic strength

mean ionic activity

mean ionic activity coefficient

mean ionic chemical potential

mean ionic molality

salting in

salting out

solvation shell

Questions on Concepts

Q10.1 Tabulated values of standard entropies of some aqueous ionic species are negative. Why is this statement consistent with the third law of thermodynamics?

Q10.2 Why is the value for the dielectric constant for water in the solvation shell around ions less than that for bulk water?

Q10.3 Why is it possible to formulate a general theory for the activity coefficient for electrolyte solutions, but not for nonelectrolyte solutions?

Q10.4 Why are activity coefficients calculated using the Debye–Hückel limiting law always less than one?

Q10.5 Why does an increase in the ionic strength in the range where the Debye–Hückel law is valid lead to an increase in the solubility of a weakly soluble salt?

Q10.6 Discuss how the Debye–Hückel screening length changes as the (a) temperature, (b) dielectric constant, and (c) ionic strength of an electrolyte solution are increased.

Q10.7 What is the correct order of the following inert electrolytes in their ability to increase the degree of dissociation of acetic acid?

a. $0.001m$ NaCl

b. $0.001m$ KBr

c. $0.10m$ CuCl$_2$

Q10.8 Why is it not appropriate to use ionic radii from crystal structures to calculate $\Delta G^\circ_{solvation}$ of ions using the Born model?

Q10.9 Why is it not possible to measure the activity coefficient of Na$^+(aq)$?

Q10.10 What can you conclude about the interaction between ions in an electrolyte solution if the mean ionic activity coefficient is greater than one?

Problems

Problem numbers in **red** indicate that the solution to the problem is given in the *Student's Solutions Manual*.

P10.1 Calculate $\Delta H^\circ_{reaction}$ and $\Delta G^\circ_{reaction}$ for the reaction AgNO$_3(aq)$ + KCl$(aq) \rightarrow$ AgCl(s) + KNO$_3(aq)$.

P10.2 Calculate $\Delta H^\circ_{reaction}$ and $\Delta G^\circ_{reaction}$ for the reaction Ba(NO$_3)_2(aq)$ + 2 KCl$(aq) \rightarrow$ BaCl$_2(s)$ +2 KNO$_3(aq)$.

P10.3 Calculate $\Delta S^\circ_{reaction}$ for the reaction AgNO$_3(aq)$ + KCl$(aq) \rightarrow$ AgCl(s) + KNO$_3(aq)$.

P10.4 Calculate $\Delta S^\circ_{reaction}$ for the reaction Ba(NO$_3)_2(aq)$ + 2 KCl$(aq) \rightarrow$ BaCl$_2(s)$ + 2 KNO$_3(aq)$.

P10.5 Calculate $\Delta G^\circ_{solvation}$ in an aqueous solution for Cl$^-(aq)$ using the Born model. The radius of the Cl$^-$ ion is 1.81×10^{-10}m.

P10.6 Calculate the value of $m_\pm$ in 5.0×10^{-4} molal solutions of (a) KCl, (b) Ca(NO$_3)_2$, and (c) ZnSO$_4$. Assume complete dissociation.

P10.7 Express $\mu_\pm$ in terms of μ_+ and μ_- for (a) NaCl, (b) MgBr$_2$, (c) Li$_3$PO$_4$, and (d) Ca(NO$_3)_2$. Assume complete dissociation.

P10.8 Express $a_\pm$ in terms of a_+ and a_- for (a) Li$_2$CO$_3$, (b) CaCl$_2$, (c) Na$_3$PO$_4$, and (d) K$_4$Fe(CN)$_6$. Assume complete dissociation.

P10.9 Express $\gamma_\pm$ in terms of γ_+ and γ_- for (a) SrSO$_4$, (b) MgBr$_2$, (c) K$_3$PO$_4$, and (d) Ca(NO$_3)_2$. Assume complete dissociation.

P10.10 Calculate the ionic strength in a solution that is $0.0050m$ in K$_2$SO$_4$, $0.0010m$ in Na$_3$PO$_4$, and $0.0025m$ in MgCl$_2$.

P10.11 Calculate the mean ionic activity of a $0.0150m$ K$_2$SO$_4$ solution for which the mean activity coefficient is 0.465.

P10.12 Calculate the mean ionic molality and mean ionic activity of a $0.150m$ Ca(NO$_3)_2$ solution for which the mean ionic activity coefficient is 0.165.

P10.13 In the Debye–Hückel theory, the counter charge in a spherical shell of radius r and thickness dr around the central ion of charge $+q$ is given by $-q\kappa^2 re^{-\kappa r}dr$. Calculate the most probable value of r, r_{mp}, from this expression. Evaluate r_{mp} for a $0.050m$ solution of NaCl at 298 K.

P10.14 Calculate the Debye–Hückel screening length $1/\kappa$ at 298 K in a $0.00100m$ solution of NaCl.

P10.15 Calculate the probability of finding an ion at a distance greater than $1/\kappa$ from the central ion.

P10.16 Using the Debye–Hückel limiting law, calculate the value of $\gamma_\pm$ in $5.0 \times 10^{-3}\,m$ solutions of (a) KCl, (b) $Ca(NO_3)_2$, and (c) $ZnSO_4$. Assume complete dissociation.

P10.17 Calculate I, $\gamma_\pm$, and $a_\pm$ for a $0.0250m$ solution of $AlCl_3$ at 298 K. Assume complete dissociation.

P10.18 Calculate I, $\gamma_\pm$, and $a_\pm$ for a $0.0250m$ solution of K_2SO_4 at 298 K. Assume complete dissociation. How confident are you that your calculated results will agree with experimental results?

P10.19 Calculate I, $\gamma_\pm$, and $a_\pm$ for a $0.0325m$ solution of $K_4Fe(CN)_6$ at 298 K.

P10.20 Calculate the solubility of $BaSO_4$ ($K_{sp} = 1.08 \times 10^{-10}$) (a) in pure H_2O and (b) in an aqueous solution with $I = 0.0010$ mol kg^{-1}.

P10.21 Dichloroacetic acid has a dissociation constant of $K_a = 3.32 \times 10^{-2}$. Calculate the degree of dissociation for a $0.125m$ solution of this acid (a) using the Debye–Hückel limiting law and (b) assuming that the mean ionic activity coefficient is one.

P10.22 Chloroacetic acid has a dissociation constant of $K_a = 1.38 \times 10^{-3}$. (a) Calculate the degree of dissociation for a $0.0825m$ solution of this acid using the Debye-Hückel

limiting law. (b) Calculate the degree of dissociation for a $0.0825\,m$ solution of this acid that is also $0.022m$ in KCl using the Debye–Hückel limiting law.

P10.23 The equilibrium constant for the hydrolysis of dimethylamine,

$$(CH_3)_2NH(aq) + H_2O(aq) \rightarrow CH_3NH_3^+(aq) + OH^-(aq)$$

is 5.12×10^{-4}. Calculate the extent of hydrolysis for (a) a $0.125m$ solution of $(CH_3)_2NH$ in water and (b) a solution that is also $0.045m$ in $NaNO_3$.

P10.24 From the data in Table 10.3 (see Appendix A, Data Tables), calculate the activity of the electrolyte in $0.100m$ solutions of

a. KCl   b. H_2SO_4 c. $MgCl_2$

P10.25 Calculate the mean ionic molality, $m_\pm$, in $0.0500m$ solutions of (a) $Ca(NO_3)_2$, (b) NaOH, (c) $MgSO_4$, and (d) $AlCl_3$.

P10.26 Calculate the ionic strength of each of the solutions in Problem P10.26.

P10.27 At 25°C, the equilibrium constant for the dissociation of acetic acid, K_a, is 1.75×10^{-5}. Using the Debye–Hückel limiting law, calculate the degree of dissociation in $0.100m$ and $1.00m$ solutions. Compare these values with what you would obtain if the ionic interactions had been ignored.

P10.28 Estimate the degree of dissociation of a $0.100m$ solution of acetic acid ($K_a = 1.75 \times 10^{-5}$) that is also $0.500m$ in the strong electrolyte given in parts (a)–(c). Use the data tables to obtain $\gamma_\pm$, as the electrolyte concentration is too high to use the Debye–Hückel limiting law.

a. $Ca(Cl)_2$ b. KCl   c. $MgSO_4$

CHAPTER 11

Electrochemical Cells, Batteries, and Fuel Cells

If a metal electrode is immersed in an aqueous solution containing a metal cation, an equilibrium is established. This equilibrium leads to a negative charge formation on the electrode. This configuration of electrode and solution is called a half-cell. Two half-cells can be combined to form an electrochemical cell. The equilibrium condition in an electrochemical cell is that the electrochemical potential, rather than the chemical potential, of a species is the same in all parts of the cell. The electrochemical potential can be changed through the application of an electrical potential external to the cell. This allows the direction of spontaneous change in the cell reaction to be reversed. Electrochemical cells can be used to determine the equilibrium constant in the cell reaction and to determine the mean activity coefficient of a solute. Electrochemical cells can also be used to provide power, in which case they are called batteries. Electrochemical cells in which the reactants can be supplied continuously are called fuel cells. ■

11.1 The Effect of an Electrical Potential on the Chemical Potential of Charged Species

If a Zn electrode is partially immersed in an aqueous solution of $ZnSO_4$, a slight negative charge builds up on the Zn electrode, and an equally large positive charge builds up in the surrounding solution. As a result of this charging, there is a difference in the electrical potential of the electrode and the solution, as depicted in Figure 11.1. The charge separation in the system arises through the dissociation equilibrium

$$Zn(s) \rightleftarrows Zn^{2+}(aq) + 2e^- \tag{11.1}$$

Whereas the Zn^{2+} ions go into the solution, the electrons remain on the Zn electrode. The equilibrium position in this reaction lies far toward $Zn(s)$. At equilibrium, fewer than 10^{-14} mol of the $Zn(s)$ dissolve for an electrode similar in dimensions to a ballpoint pen in a small beaker of water. However, this minuscule amount of charge transfer between the electrode and the solution is sufficient to create a difference of approximately 1 V in the electrical potential between the Zn electrode and the electrolyte solution. Because the value of the equilibrium constant depends on the identity of the

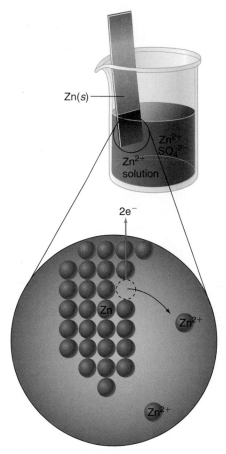

FIGURE 11.1

When a Zn electrode is immersed in water, a very small amount of the Zn goes into solution as Zn^{2+} (aq), leaving two electrons behind on the Zn electrode per ion formed. Although the charge buildup on the electrode and in the solution is very small, it leads to a difference on the order of 1 V in the electrical potential of the solution and the electrode.

metal electrode, the difference in the electrical potential between the electrode and the solution also depends on the identity of the metal. It is important to realize that this electrical potential affects the energy of all charged particles in the solution.

The chemical potential of a **neutral** atom or molecule is not affected if a small electrical potential is applied to the environment containing the species. However, this is not the case for a **charged** species such as a Na^+ ion in an electrolyte solution. The work needed to transfer dn moles of charge reversibly from a chemically uniform phase at an electrical potential ϕ_1 to a second, otherwise identical phase at an electrical potential ϕ_2 can be calculated. From electrostatics, the work is equal to the product of the charge and the difference in the electrical potential between the two locations:

$$dw_{rev} = (\phi_2 - \phi_1)dQ \tag{11.2}$$

In this equation, $dQ = -zFdn$ is the charge transferred through the potential, z is the charge in units of the electron charge ($+1, -1, +2, -2, \dots$), and the **Faraday constant** F is the absolute magnitude of the charge associated with 1 mol of a singly charged species. The Faraday constant has the numerical value $F = 96,485$ Coulombs mole^{-1} (C mol^{-1}).

Because the work being carried out in this reversible process is nonexpansion work, $dw_{rev} = dG$, which is the difference in the **electrochemical potential,** $\tilde{\mu}$, of the charged particle in the two phases:

$$dG = \tilde{\mu}_2 dn - \tilde{\mu}_1 dn \tag{11.3}$$

The electrochemical potential is a generalization of the chemical potential to include the effect of an electrical potential on a charged particle. It is the sum of the normal chemical potential, μ, and a term that results from the nonzero value of the electrical potential:

$$\tilde{\mu} = \mu + z\phi F \tag{11.4}$$

Note that with this definition $\tilde{\mu} \to \mu$ as $\phi \to 0$.

Combining Equations (11.2) and (11.4) gives

$$\tilde{\mu}_2 - \tilde{\mu}_1 = +z(\phi_2 - \phi_1)F \quad \text{or} \quad \tilde{\mu}_2 = \tilde{\mu}_1 + z(\phi_2 - \phi_1)F \tag{11.5}$$

Because only the difference in the electrical potential between two points can be measured, one can set $\phi_1 = 0$ in Equation (11.5) to obtain the result

$$\tilde{\mu}_2 = \tilde{\mu}_1 + z\phi_2 F \tag{11.6}$$

This result shows that charged particles in two otherwise identical phases have different values for the electrochemical potential if the phases are at different electrical potentials. Because the particles will flow in a direction that decreases their electrochemical potential, the flow of negatively charged particles in a conducting phase is toward a region of more positive electric potential. The opposite is true for positively charged particles.

Equation (11.6) is the basis for understanding all electrochemical reactions. In an electrochemical environment, the equilibrium condition is

$$\Delta G = \sum_i v_i \tilde{\mu}_i = 0, \quad \text{rather than} \quad \Delta G = \sum_i v_i \mu_i = 0 \tag{11.7}$$

Chemists have a limited ability to change the chemical potential μ_i of a neutral species in solution by varying P, T, and concentration. However, because the electrochemical potential can be varied through the application of an electrical potential, it can be changed easily. It is possible to vary $\tilde{\mu}_i$ to a far greater extent than μ_i, because $z_i \phi F$ can be larger in magnitude than μ_i and can have either the same or the opposite sign. As Equation (11.7) shows, this also applies to ΔG. Because a change in the electrical potential can lead to a change in the sign of ΔG, the direction of spontaneous change in a reaction system can be changed simply by applying suitable electrical potentials within the system.

11.2 Conventions and Standard States in Electrochemistry

To understand the conventions and standard states in electrochemistry, it is useful to consider an **electrochemical cell** such as the one shown in Figure 11.2. This particular cell is known as the **Daniell cell,** after its inventor. On the left, a Zn electrode is immersed in a solution of $ZnSO_4$. The solute is completely dissociated to form $Zn^{2+}(aq)$ and $SO_4^{2-}(aq)$. On the right, a Cu electrode is immersed in a solution of $CuSO_4$, which is completely dissociated to form $Cu^{2+}(aq)$ and $SO_4^{2-}(aq)$. The two half-cells are connected by an ionic conductor known as a salt bridge. The **salt bridge** consists of an electrolyte such as KCl suspended in a gel. A salt bridge allows current to flow between the half-cells while preventing the mixing of the solutions. A metal wire fastened to each electrode allows the electron current to flow through the external part of the circuit.

FIGURE 11.2
Schematic diagram of the Daniell cell. Zn^{2+}/Zn and Cu^{2+}/Cu half-cells are connected through a salt bridge in the internal circuit. A voltmeter is shown in the external circuit. The inset shows the atomic level processes that occur at each electrode.

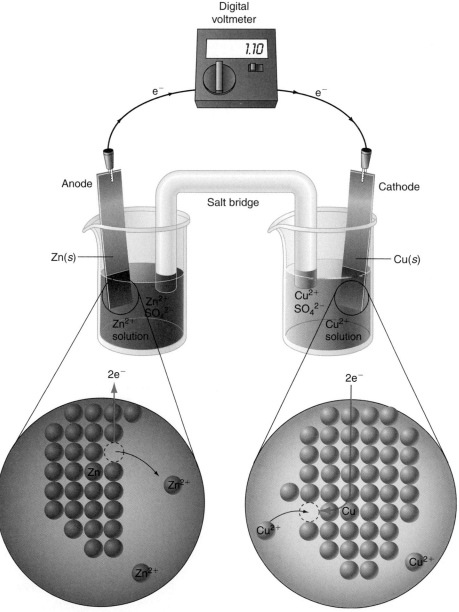

The measurement of electrical potentials is discussed using this cell. Because of the equilibrium established between $Zn(s)$ and Zn^{2+} (aq) and between $Cu(s)$ and Cu^{2+} (aq) [see Equations (11.20) and (11.21)], there is a difference in the electrical potential between the metal electrode and the solution in each of the two half-cells. Can this electrical potential be measured directly? Assume that the probes are two chemically inert Pt wires. One Pt wire is placed on the Zn electrode, and the second Pt wire is placed in the $ZnSO_4$ solution. In this measurement, the measured voltage is the difference in electrical potential between a Pt wire connected to a Zn electrode in a $ZnSO_4$ solution, and a Pt electrode in a $ZnSO_4$ solution. A difference in electrical potential can only be measured between one phase and a second phase *of identical composition*. For example, the difference in electrical potential across a resistor is measured by contacting the metal wire at each end of the resistor with metal probes connected to the terminals of a voltmeter.

On the basis of this discussion, it is clear that the electrical potential in the solution cannot be directly measured. Therefore, the convenient standard state $\phi = 0$ is chosen for the electrical potential of all ions in the solution. With this choice,

$$\tilde{\mu}_i = \mu_i \text{ (ions in solution)} \tag{11.8}$$

Adopting this standard state simplifies calculations, because μ_i can be calculated from the solute concentration at low concentrations using the Debye–Hückel limiting law discussed in Section 10.4 or at higher concentrations if the activity coefficients are known.

Next, consider appropriate standard states for electrons and ions in a metal electrode. As shown in Equation (11.6), the electrochemical potential consists of two parts, a chemical component and a component that depends on the electrical potential. For an electron in a metal, there is no way to determine the relative magnitude of the two components. Therefore, the convention is established in which the standard state chemical potential of an electron in a metal electrode is zero so that

$$\tilde{\mu}_{e^-} = -\phi F \text{ (electrons in metal electrode)} \tag{11.9}$$

For a metal electrode at equilibrium,

$$M^{z+} + ze^- \rightleftarrows M \tag{11.10}$$

Because M is a neutral species, $\tilde{\mu}_M = \mu_M$. Therefore, at equilibrium, the electrochemical potentials of the metal, the metal cation, and the electron are related by

$$\mu_M = \tilde{\mu}_{M^{z+}} + z\tilde{\mu}_{e^-} \tag{11.11}$$

Using the relations $\tilde{\mu}_{e^-} = -\phi F$ and $\tilde{\mu}_{M^{z+}} = \mu_{M^{z+}} + zF\phi$ in Equation (11.11),

$$\mu_M = \mu_{M^{z+}} + zF\phi - zF\phi = \mu_{M^{z+}} \quad \text{and for a pure metal at 1 bar} \tag{11.12}$$

$$\mu_M^\circ = \mu_{M^{z+}}^\circ = 0$$

Recall that $G_{M,m}^\circ = \mu_M^\circ = 0$ because the electrode is a pure element in its standard state. Note that this convention refers to the chemical potential rather than the electrochemical potential. Combining Equations (11.11) and (11.12),

$$\tilde{\mu}_{M^{z+}} = \mu_{M^{z+}}^\circ + zF\phi = zF\phi \tag{11.13}$$

Equations (11.9), (11.12), and (11.13) define the standard states for a neutral metal atom in the electrode, the metal cation, and the electron in the electrode.

Even with these conventions, a further problem arises when trying to obtain numerical values for the electrical potentials of different half-cells. Using the experimental setup of Figure 11.2, only the electrical potential difference between two half-cells can be measured, as opposed to the absolute electrical potential of each half-cell. Therefore, it is convenient to choose one half-cell as a reference and arbitrarily assign a fixed electrical potential to this half-cell. Once this is done, the electrical potential associated with any other half-cell can be determined by combining it with the reference half-cell. The measured potential difference across the cell is associated with the half-cell of interest. It is next shown that the standard hydrogen electrode (SHE) fulfills the role of a reference half-cell of zero potential without further assumptions.

The equilibrium reaction in this half-cell is

$$H^+(aq) + e^- \rightleftharpoons \frac{1}{2}H_2(g) \tag{11.14}$$

and the equilibrium in the half-cell is described by

$$\mu_{H^+}(aq) + \tilde{\mu}_{e^-} = \frac{1}{2}\mu_{H_2(g)} \tag{11.15}$$

It is useful to separate $\mu_{H_2(g)}$ and $\mu_{H^+}(aq)$ into a standard state portion and one that depends on the activity and use Equation (11.9) for the electrochemical potential of the electron. The preceding equation then takes the form

$$\mu_{H^+}^\circ + RT \ln a_{H^+} - F\phi_{H^+/H_2} = \frac{1}{2}\mu_{H_2}^\circ + \frac{1}{2}RT \ln f_{H_2} \tag{11.16}$$

where f is the fugacity of the hydrogen gas. Solving Equation (11.16) for ϕ_{H^+/H_2}

$$\phi_{H^+/H_2} = \frac{\mu_{H^+}^\circ - \frac{1}{2}\mu_{H_2}^\circ}{F} - \frac{RT}{F}\ln\frac{f_{H_2}^{1/2}}{a_{H^+}} \tag{11.17}$$

For unit activities of all species, the cell has its standard state potential, designated ϕ_{H^+/H_2}°. Because $\mu_{H_2}^\circ = 0$,

$$\phi_{H^+/H_2}^\circ = \frac{\mu_{H^+}^\circ - \frac{1}{2}\mu_{H_2}^\circ}{F} = \frac{\mu_{H^+}^\circ}{F} \tag{11.18}$$

Recall from Section 10.1 the convention that $\Delta G_f^\circ(H^+, aq) = \mu_{H^+}^\circ = 0$ was established. As a result of this convention,

$$\phi_{H^+/H_2}^\circ = 0 \tag{11.19}$$

The standard hydrogen electrode is a convenient **reference electrode** against which the potentials of all other half-cells can be measured. A schematic drawing of this electrode is shown in Figure 11.3. To achieve equilibrium on a short timescale, this reaction $H^+(aq) + e^- \rightleftharpoons 1/2\,H_2(g)$ is carried out over a Pt catalyst. It is also necessary to establish a standard state for the activity of $H^+(aq)$. It is customary to use a Henry's law standard state based on molarity. Therefore, $a_i \rightarrow c_i$ and $\gamma_i \rightarrow 1$ as $c_i \rightarrow 0$. The standard state is a (hypothetical) aqueous solution of $H^+(aq)$ that shows ideal solution behavior at a concentration of $c^\circ = 1$ mol L^{-1}.

The usefulness of the result $\phi_{H^+/H_2}^\circ = 0$ is that values for the electrical potential can be assigned to individual half-cells by measuring their potential relative to the H^+/H_2 half-cell. For example, the cell potential of the electrochemical cell in Figure 11.4 is assigned to the Zn/Zn^{2+} half-cell if the $H^+(aq)$ and $H_2(g)$ activities both have the value one. Although it is convenient to reference values of half-cell potentials to that of the standard hydrogen electrode, this does not imply that half-cell potentials cannot be measured. The determination of absolute half-cell potentials is discussed in Supplemental Section 11.16.

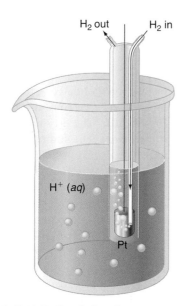

H₂ out   H₂ in

H⁺ (aq)

Pt

FIGURE 11.3

The standard hydrogen electrode (SHE) consists of a solution of an acid such as HCl, H$_2$ gas, and a catalyst that allows the equilibrium in the half-cell reaction to be established rapidly. The activities of H$_2$ and H$^+$ are equal to one.

In a cell consisting of an arbitrary half-cell and the standard hydrogen electrode, the entire cell voltage is assigned to the arbitrary half-cell.

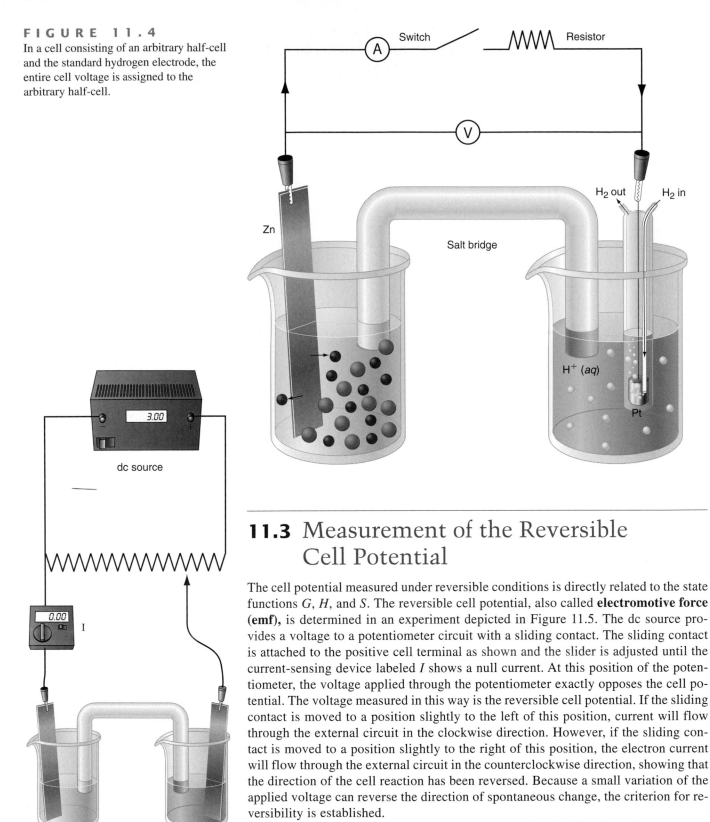

F I G U R E   1 1 . 5

Schematic diagram showing how the reversible cell potential is measured.

11.3 Measurement of the Reversible Cell Potential

The cell potential measured under reversible conditions is directly related to the state functions G, H, and S. The reversible cell potential, also called **electromotive force (emf),** is determined in an experiment depicted in Figure 11.5. The dc source provides a voltage to a potentiometer circuit with a sliding contact. The sliding contact is attached to the positive cell terminal as shown and the slider is adjusted until the current-sensing device labeled I shows a null current. At this position of the potentiometer, the voltage applied through the potentiometer exactly opposes the cell potential. The voltage measured in this way is the reversible cell potential. If the sliding contact is moved to a position slightly to the left of this position, current will flow through the external circuit in the clockwise direction. However, if the sliding contact is moved to a position slightly to the right of this position, the electron current will flow through the external circuit in the counterclockwise direction, showing that the direction of the cell reaction has been reversed. Because a small variation of the applied voltage can reverse the direction of spontaneous change, the criterion for reversibility is established.

This measurement also demonstrates how easily the direction of spontaneous change can be reversed. This reversal is accomplished by changing the electrochemical potential of the electrons in one of the electrodes relative to that in the other electrode.

11.4 Chemical Reactions in Electrochemical Cells and the Nernst Equation

What reactions occur in the Daniell cell shown in Figure 11.2? If the half-cells are connected through the external circuit, Zn atoms leave the Zn electrode to form Zn^{2+} in solution, and Cu^{2+} ions are deposited as Cu atoms on the Cu electrode. In the external circuit, it is observed that electrons flow through the wires and the resistor in the direction from the Zn electrode to the Cu electrode. These observations are consistent with the following electrochemical reactions:

Left half-cell: $Zn(s) \rightleftarrows Zn^{2+}(aq) + 2e^-$ (11.20)

Right half-cell: $Cu^{2+}(aq) + 2e^- \rightleftarrows Cu(s)$ (11.21)

Overall: $Zn(s) + Cu^{2+}(aq) \rightleftarrows Zn^{2+}(aq) + Cu(s)$ (11.22)

In the left half-cell, Zn is being oxidized to Zn^{2+}, and in the right half-cell, Cu^{2+} is being reduced to Cu. By convention, the electrode at which oxidation occurs is called the **anode,** and the electrode at which reduction occurs is called the **cathode.** Each half-cell in an electrochemical cell must contain a species that can exist in an oxidized and a reduced form. For a general redox reaction, the reactions at the anode and cathode and the overall reaction can be written as follows:

Anode: $Red_1 \rightarrow Ox_1 + v_1 e^-$ (11.23)

Cathode: $Ox_2 + v_2 e^- \rightarrow Red_2$ (11.24)

Overall: $v_2 Red_1 + v_1 Ox_2 \rightarrow v_2 Ox_1 + v_1 Red_2$ (11.25)

Note that electrons do not appear in the overall reaction, because the electrons produced at the anode are consumed at the cathode.

How are the cell voltage and the ΔG for the overall reaction related? This important relationship can be determined from the electrochemical potentials of the species involved in the overall reaction of the Daniell cell:

$$\Delta G = \tilde{\mu}_{Zn^{2+}} + \tilde{\mu}_{Cu} - \tilde{\mu}_{Cu^{2+}} - \tilde{\mu}_{Zn} = \tilde{\mu}^{\circ}_{Zn^{2+}} - \tilde{\mu}^{\circ}_{Cu^{2+}} + RT \ln \frac{a_{Zn^{2+}}}{a_{Cu^{2+}}} \quad (11.26)$$

$$= \Delta G^{\circ} + RT \ln \frac{a_{Zn^{2+}}}{a_{Cu^{2+}}}$$

In this equation, $\tilde{\mu}_{Cu} = \tilde{\mu}^{\circ}_{Cu} = 0$, and the same is true for Zn because the cell is at a pressure of 1 bar. If this reaction is carried out reversibly, the electrical work done is equal to the product of the charge and the potential difference through which the charge is moved. However, the reversible work at constant pressure is also equal to ΔG. Therefore, we can write the following equation:

$$\Delta G = -nF\Delta\phi \quad (11.27)$$

In Equation (11.27), $\Delta\phi$ is the measured potential difference generated by the spontaneous chemical reaction for particular values of $a_{Zn^{2+}}$ and $a_{Cu^{2+}}$, and n is the number of moles of electrons involved in the redox reaction. It is seen that the measured cell voltage is directly proportional to ΔG. For a reversible reaction, the symbol E is used in place of $\Delta\phi$, and E is referred to as the electromotive force. Using this definition, we rewrite Equation (11.27) as follows:

$$-2FE = \Delta G^{\circ} + RT \ln \frac{a_{Zn^{2+}}}{a_{Cu^{2+}}} \quad (11.28)$$

FIGURE   11.6

The cell potential E varies linearly with log Q. The slope of a plot of $(E - E^\circ)/(RT/F)$ is inversely proportional to the number of electrons transferred in the redox reaction.

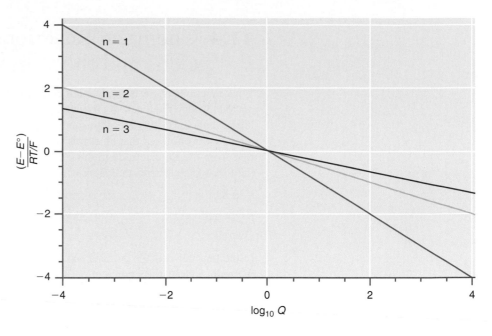

For standard state conditions, $a_{Zn^{2+}} = a_{Cu^{2+}} = 1$, and Equation (11.28) takes the form $\Delta G^\circ = -2FE^\circ$. This definition of E° allows Equation (11.28) to be rewritten as

$$E = E^\circ - \frac{RT}{2F} \ln \frac{a_{Zn^{2+}}}{a_{Cu^{2+}}} \tag{11.29}$$

For a general overall electrochemical reaction involving the transfer of n moles of electrons,

$$E = E^\circ - \frac{RT}{nF} \ln Q \tag{11.30}$$

where Q is the familiar reaction quotient. The preceding equation is known as the **Nernst equation.** At 298.15 K, the Nernst equation can be written in the form

$$E = E^\circ - \frac{0.05916\ \text{V}}{n} \log_{10} Q \tag{11.31}$$

This function is graphed in Figure 11.6. The Nernst equation allows the emf for an electrochemical cell to be calculated if the activity is known for each species and if E° is known.

The Nernst equation has been derived on the basis of the overall cell reaction. For a half-cell, an equation of a similar form can be derived. The equilibrium condition for the half-cell reaction

$$\text{Ox} + \nu e^- \rightarrow \text{Red} \tag{11.32}$$

is given by

$$\mu_{Ox}^{n+} + n\tilde{\mu}_{e^-} \rightleftarrows \mu_{Red} \tag{11.33}$$

Using the convention for the electrochemical potential of an electron in a metal electrode [Equation (11.9)], Equation (11.33) can be written in the form

$$\mu_{Ox^{n+}}^\circ + RT \ln a_{Ox^{n+}} - nF\phi_{Ox/Red} = \mu_{Red}^\circ + RT \ln a_{Red} \tag{11.34}$$

$$\phi_{Ox/Red} = -\frac{\mu_{Red}^\circ - \mu_{Ox^{n+}}^\circ}{nF} - \frac{RT}{nF} \ln \frac{a_{Red}}{a_{Ox^{n+}}}$$

$$E_{Ox/Red} = E_{Ox/Red}^\circ - \frac{RT}{nF} \ln \frac{a_{Red}}{a_{Ox^{n+}}}$$

The last line in Equation (11.34) has the same form as the Nernst equation, but the activity of the electrons does not appear in Q. An example of the application of Equation (11.34) to a half-cell reaction is shown in Example Problem 11.1.

EXAMPLE PROBLEM 11.1

Calculate the potential of the H^+/H_2 half-cell when $a_{H^+} = 0.770$ and $f_{H_2} = 1.13$.

Solution

$$E = E° - \frac{0.05916 \text{ V}}{n} \log_{10} \frac{a_{H^+}}{\sqrt{f_{H_2}}} = 0 - \frac{0.05916 \text{ V}}{1} \log_{10} \frac{0.770}{\sqrt{1.13}} = 0.0083 \text{ V}$$

11.5 Combining Standard Electrode Potentials to Determine the Cell Potential

A representative set of standard potentials is listed in Tables 11.1 and 11.2 (see Appendix A, Data Tables). By convention, half-cell emfs are always tabulated as reduction potentials. However, whether the reduction or the oxidation reaction is spontaneous in a half-cell is determined by the relative emfs of the two half-cells that make up the electrochemical cell. Because $\Delta G° = -nFE°$, and ΔG for the oxidation and reduction reactions are equal in magnitude and opposite in sign,

$$E°_{reduction} = -E°_{oxidation} \tag{11.35}$$

How is the cell potential related to the potentials of the half-cells? The standard potential of a half-cell is given by $E° = -\Delta G°/nF$ and is an intensive property, because although both $\Delta G°$ and n are extensive quantities, the ratio $\Delta G°/n$ is an intensive quantity. In particular, $E°$ is not changed if all the stoichiometric coefficients are multiplied by any integer, because both $\Delta G°$ and n are changed by the same factor. Therefore,

$$E°_{cell} = E°_{reduction} + E°_{oxidation} \tag{11.36}$$

even if the balanced reaction for the overall cell is multiplied by an arbitrary number. In Equation (11.36), the standard potentials on the right refer to the half-cells.

EXAMPLE PROBLEM 11.2

An electrochemical cell is constructed using a half-cell for which the reduction reaction is given by

$$Fe(OH)_2(s) + 2e^- \rightarrow Fe(s) + 2OH^-(aq) \qquad E° = -0.877 \text{ V}$$

It is combined with the half-cells for which the reduction reaction is given by

  a. $Al^{3+}(aq) + 3e^- \rightarrow Al(s)$ $E° = -1.66 \text{ V}$

  b. $AgBr(s) + e^- \rightarrow Ag(s) + Br^-(aq)$ $E° = +0.071 \text{ V}$

The activity of all species is one for both reactions. Write the overall reaction for the cells in the direction of spontaneous change. Is the Fe reduced or oxidized in the spontaneous reaction?

Solution

The emf for the cell is the sum of the emfs for the half-cells, one of which is written as an oxidation reaction, and the other of which is written as a reduction reaction. For the direction of spontaneous change, $E° = E°_{reduction} + E°_{oxidation} > 0$. Note that the two half-cell reactions are combined with the appropriate stoichiometry so that electrons do not appear in the overall equation. Note also that the half-cell emfs are not changed in multiplying the overall reactions by the integers necessary to eliminate the electrons in the overall equation, because $E°$ is an intensive rather than an extensive quantity.

a. $3Fe(OH)_2(s) + 2Al(s) \rightarrow 3Fe(s) + 6OH^-(aq) + 2Al^{3+}(aq)$

$E° = -0.877\ V + 1.66\ V = +0.783\ V$

The Fe is reduced.

b. $Fe(s) + 2OH^-(aq) + 2AgBr(s) \rightarrow Fe(OH)_2(s) + 2Ag(s) + 2Br^-(aq)$

$E° = 0.071\ V + 0.877\ V = +0.95\ V$

The Fe is oxidized.

Because ΔG is a state function, the cell potential for a third half-cell can be obtained from the cell potentials of two half-cells if they have an oxidation or reduction reaction in common. The procedure is analogous to the use of Hess's law in Chapter 4 and is illustrated in Example Problem 11.3.

EXAMPLE PROBLEM 11.3

You are given the following reduction reactions and $E°$ values:

$Fe^{3+}(aq) + e^- \rightarrow Fe^{2+}(aq)$ $\qquad\qquad$ $E° = +0.771\ V$

$Fe^{2+}(aq) + 2e^- \rightarrow Fe(s)$ $\qquad\qquad$ $E° = -0.447\ V$

Calculate $E°$ for the half-cell reaction $Fe^{3+}(aq) + 3e^- \rightarrow Fe(s)$.

Solution

We calculate the desired value of $E°$ by converting the given $E°$ values to $\Delta G°$ values, and combining these reduction reactions to obtain the desired equation.

$$Fe^{3+}(aq) + e^- \rightarrow Fe^{2+}(aq)$$

$$\Delta G° = -nFE° = -1 \times 96485\ C\ mol^{-1} \times 0.771\ V = -74.39\ kJ\ mol^{-1}$$

$$Fe^{2+}(aq) + 2e^- \rightarrow Fe(s)$$

$$\Delta G° = -nFE° = -2 \times 96485\ C\ mol^{-1} \times (-0.447\ V) = 86.26\ kJ\ mol^{-1}$$

We next add the two equations as well as their $\Delta G°$ to obtain

$$Fe^{3+}(aq) + 3e^- \rightarrow Fe(s)$$

$$\Delta G° = -74.39\ kJ\ mol^{-1} + 86.26\ kJ\ mol^{-1} = 11.87\ kJ\ mol^{-1}$$

$$E°_{Fe^{3+}/Fe} = -\frac{\Delta G°}{nF} = \frac{-11.87 \times 10^3\ J\ mol^{-1}}{3 \times 96485\ C\ mol^{-1}} = -0.041\ V$$

The $E°$ values cannot be combined directly, because they are intensive rather than extensive quantities.

The preceding calculation can be generalized as follows. Assume that n_1 electrons are transferred in the reaction with the potential $E°_{A/B}$, and n_2 electrons are transferred in the reaction with the potential $E°_{B/C}$. If n_3 electrons are transferred in the reaction with the potential $E°_{A/C}$, then $n_3 E°_{A/C} = n_1 E°_{A/B} + n_2 E°_{B/C}$.

11.6 Obtaining Reaction Gibbs Energies and Reaction Entropies from Cell Potentials

Section 11.4 demonstrated that $\Delta G = -nF\Delta\phi$. Therefore, if the cell potential is measured under standard conditions,

$$\Delta G^{\circ} = -nFE^{\circ} \tag{11.37}$$

If E° is known, $\Delta G^{\circ}_{reaction}$ can be determined using Equation (11.37). For example, E° for the Daniell cell is $+1.10$ V. Therefore, $\Delta G^{\circ}_{reaction}$ for the reaction $Zn(s) + Cu^{2+}(aq) \rightleftarrows Zn^{2+}(aq) + Cu(s)$ is

$$\Delta G^{\circ}_{reaction} = -nFE^{\circ} = -2 \times 96{,}485\,C\,mol^{-1} \times 1.10\,V = -212\,kJ\,mol^{-1} \tag{11.38}$$

The reaction entropy is related to $\Delta G^{\circ}_{reaction}$ by

$$\Delta S^{\circ}_{reaction} = -\left(\frac{\partial \Delta G^{\circ}_{reaction}}{\partial T}\right)_{P} = nF\left(\frac{\partial E^{\circ}}{\partial T}\right)_{P} \tag{11.39}$$

Therefore, a measurement of the temperature dependence of E° can be used to determine $\Delta S^{\circ}_{reaction}$, as shown in Example Problem 11.4.

EXAMPLE PROBLEM 11.4

The standard potential of the cell formed by combining the $Cl_2/Cl^-(aq)$ half-cell with the standard hydrogen electrode is $+1.36$ V, and $(\partial E^{\circ}/\partial T)_P = -1.20 \times 10^{-3}\,V\,K^{-1}$. Calculate $\Delta S^{\circ}_{reaction}$ for the reaction $H_2(g) + Cl_2(g) \rightarrow 2H^+(aq) + 2Cl^-(aq)$. Compare your result with the value that you obtain by using the values of S° in the data tables from Appendix A.

Solution

From Equation (11.39),

$$\Delta S^{\circ}_{reaction} = -\left(\frac{\partial \Delta G^{\circ}_{reaction}}{\partial T}\right)_{P} = nF\left(\frac{\partial E^{\circ}}{\partial T}\right)_{P}$$

$$= 2 \times 96480\,C\,mol^{-1} \times (-1.2 \times 10^{-3}\,V\,K^{-1}) = -2.3 \times 10^2\,J\,K^{-1}$$

From Table 10.1 (see Appendix A, Data Tables),

$$\Delta S^{\circ}_{reaction} = 2S^{\circ}(H^+, aq) + 2S^{\circ}(Cl^-, aq) - S^{\circ}(H_2, g) - S^{\circ}(Cl_2, g)$$

$$= 2 \times 0 + 2 \times 56.5\,J\,K^{-1}mol^{-1} - 130.7\,J\,K^{-1}mol^{-1} - 223.1\,J\,K^{-1}mol^{-1}$$

$$= -240.8\,J\,K^{-1}mol^{-1}$$

The limited precision of the temperature dependence of E° limits the precision in the determination of $\Delta S^{\circ}_{reaction}$.

11.7 The Relationship between the Cell EMF and the Equilibrium Constant

If the redox reaction is allowed to proceed until equilibrium is reached, $\Delta G = 0$, so that $E = 0$. For the equilibrium state, the reaction quotient $Q = K$. Therefore,

$$E^{\circ} = \frac{RT}{nF} \ln K \tag{11.40}$$

Equation (11.40) shows that a measurement of the standard state cell potential, for which $a_i = 1$ for all species in the redox reaction, allows K for the overall reaction to be determined. Although this statement is true, it is not practical to adjust all activities to the value one. The determination of $E°$ is discussed in the next section.

Electrochemical measurements provide a powerful way to determine the equilibrium constant in an electrochemical cell. To determine K, the overall reaction must be separated into the oxidation and reduction half-reactions, and the number of electrons transferred must be determined. For example, consider the reaction

$$2MnO_4^-(aq) + 6H^+(aq) + 5\ HOOCCOOH(aq) \rightleftharpoons 2Mn^{2+}(aq) + 8H_2O(l) \qquad (11.41)$$
$$+ 10CO_2(g)$$

The oxidation and reduction half-reactions are

$$MnO_4^-(aq) + 8H^+(aq) + 5e^- \rightleftharpoons Mn^{2+}(aq) + 4H_2O(l) \qquad E° = +1.51\ V \quad (11.42)$$
$$HOOCCOOH(aq) \rightleftharpoons 2H^+(aq) + 2CO_2(g) + 2e^- \qquad E° = +0.49\ V \quad (11.43)$$

To eliminate the electrons in the overall equation, the first equation must be multiplied by 2, and the second equation must be multiplied by 5; $E°$ is unchanged by doing so. However, n is affected by the multipliers. In this case $n = 10$. Therefore,

$$\Delta G° = -nFE° = -10 \times 96485\ C\,mol^{-1} \times (1.51\ V + 0.49\ V) = -1.93 \times 10^2\ kJ\,mol^{-1} \quad (11.44)$$
$$\ln K = -\frac{\Delta G°}{RT} = \frac{1.93 \times 10^2\ kJ\,mol^{-1}}{8.314\ J\,K^{-1}mol^{-1} \times 298.15\ K} = 778$$

As this result shows, the equilibrium corresponds to essentially complete conversion of reactants to products. Example Problem 11.5 shows the same calculation for the Daniell cell.

EXAMPLE PROBLEM 11.5

For the Daniell cell $E° = 1.10$ V. Calculate K for the reaction

$$Zn(s) + Cu^{2+}(aq) \rightarrow Zn^{2+}(aq) + Cu(s).$$

Solution

$$\ln K = \frac{nF}{RT}E° = \frac{2 \times 96485\ C\,mol^{-1} \times 1.10\ V}{8.314\ J\,K^{-1}mol^{-1} \times 298.15\ K}$$
$$= 85.63$$
$$K = 1.55 \times 10^{37}$$

Note that the equilibrium constant calculated in Example Problem 11.5 is so large that it could not have been measured by determining the activities of $a_{Zn^{2+}}$ and $a_{Cu^{2+}}$ by spectroscopic methods. This would require a measurement technique that is accurate over more than 30 orders of magnitude in the activity. By contrast, the equilibrium constant in an electrochemical cell can be determined with high accuracy using only a voltmeter.

A further example of the use of electrochemical measurements to determine equilibrium constants is the solubility constant for a weakly soluble salt. If the overall reaction corresponding to dissolution can be generated by combining half-cell potentials, then the solubility constant can be calculated from the potentials. For example, the following half-cell reactions can be combined to calculate the solubility product of AgBr.

$$AgBr(s) + e^- \rightarrow Ag(s) + Br^-(aq) \qquad E° = 0.07133\ V \qquad \text{and}$$
$$\underline{Ag(s) \rightarrow Ag^+(aq) + e^- \qquad\qquad E° = -0.7996\ V}$$
$$AgBr(s) \rightarrow Ag^+(aq) + Br^-(aq) \qquad E° = -0.7283\ V$$

$$\ln K_{sp} = \frac{nF}{RT}E° = \frac{1 \times 96485\ C\,mol^{-1} \times (-0.7283\ V)}{8.314\ J\,K^{-1}mol^{-1} \times 298.15\ K} = -28.35$$

The value of the solubility constant is $K_{sp} = 4.88 \times 10^{-13}$.

EXAMPLE PROBLEM 11.6

A concentration cell consists of two half-cells that are identical except for the activities of the redox components. Consider two half-cells based on the Ag^+ $(aq) + e^- \rightarrow Ag(s)$ reaction. The left half-cell contains $AgNO_3$ at unit activity, and the right half-cell initially had the same concentration of $AgNO_3$, but just enough $NaCl(aq)$ has been added to completely precipitate the $Ag^+(aq)$ as $AgCl$. Write an equation for the overall cell reaction. If the emf of this cell is 0.29 V, what is K_{sp} for $AgCl$?

Solution

The overall reaction is $Ag^+(aq, a = 1) \rightarrow Ag^+(aq, a)$. In the right half-cell we have the equilibrium $AgCl \rightleftharpoons Ag^+(aq, a_+) + Cl^-(aq, a_+)$, so that $a_{Ag^+} a_{Cl^-} = a_\pm^2 = K_{sp}$. Because the half-cell reactions are the same, $E° = 0$ and

$$E = -\frac{0.05916 \text{ V}}{n} \log_{10} \frac{a_\pm}{1} = -\frac{0.05916}{1} \log_{10} \sqrt{K_{sp}} = -\frac{0.05916 \text{ V}}{2} \log_{10} K_{sp}$$

$$\log_{10} K_{sp} = -\frac{2E}{0.05916} = -\frac{2 \times (0.29 \text{ V})}{0.05916 \text{ V}} = -9.804$$

$$K_{sp} = 1.6 \times 10^{-10}$$

11.8 Determination of $E°$ and Activity Coefficients Using an Electrochemical Cell

The main problem in determining standard potentials lies in knowing the value of the activity coefficient $\gamma_\pm$ for a given solute concentration. The best strategy is to carry out measurements of the cell potential at low concentrations, where $\gamma_\pm \rightarrow 1$, rather than near unit activity, where $\gamma_\pm$ differs appreciably from one. Consider an electrochemical cell consisting of the Ag^+/Ag and SHE half-cells at 298 K. The cell reaction is $Ag^+(aq) + 1/2H_2(g) \rightarrow Ag(s) + H^+(aq)$ and $Q = (a_{Ag^+})$. Assume that the Ag^+ arises from the dissociation of $AgNO_3$. Recall that the activity of an individual ion cannot be measured directly. It must be calculated from the measured activity $a_\pm \gamma$ and the definition $a_\pm^\nu = a_+^{\nu+} a_-^{\nu-}$. In this case, $a_\pm^2 = a_{Ag^+} a_{NO_3^-}$ and $a_\pm = a_{Ag^+} = a_{NO_3^-}$. Similarly, $\gamma_\pm = \gamma_{Ag^+} = \gamma_{NO_3^-}$ and $m_{Ag^+} = m_{NO_3^-} = m_\pm = m$, and E is given by

$$E = E°_{Ag^+/Ag} + \frac{RT}{F} \ln a_{Ag^+} = E°_{Ag^+/Ag} + \frac{RT}{F} \ln(m/m°) + \frac{RT}{F} \ln \gamma_\pm \qquad (11.45)$$

At low enough concentrations, the Debye–Huckel limiting law is valid and $\log \gamma_\pm = -0.5090 \sqrt{m_\pm}$ at 298 K. Using this relation, Equation (11.45) can be rewritten in the form

$$E - 0.05916 \text{ V} \log_{10}(m/m°) = E°_{Ag^+/Ag} - (0.05916 \text{ V})(0.5090)\sqrt{(m/m°)} \quad (11.46)$$

$$= E°_{Ag^+/Ag} - 0.03011\sqrt{(m/m°)}$$

The left-hand side of this equation can be calculated from measurements and plotted as a function of $\sqrt{(m/m°)}$. The results will resemble the graph shown in Figure 11.7. An extrapolation of the line that best fits the data to $m = 0$ gives $E°$ as the intercept with the vertical axis. Once $E°$ has been determined, Equation (11.45) can be used to calculate $\gamma_\pm$.

Electrochemical cells provide a powerful method of determining activity coefficients, because cell potentials can be measured more accurately and more easily than colligative properties such as freezing point depression or boiling point elevation. Note that although the Debye–Hückel limiting law was used to determine $E°$, it is not necessary to use the limiting law to calculate activity coefficients once $E°$ is known.

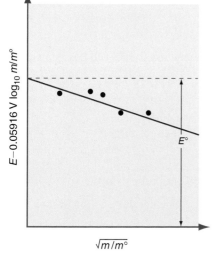

FIGURE 11.7

The value of $E°$ and the activity coefficient can be measured by plotting the left-hand side of Equation (11.46) against the square root of the molality.

11.9 Cell Nomenclature and Types of Electrochemical Cells

It is useful to use an abbreviated notation to describe an electrochemical cell. This notation includes all species involved in the cell reaction, and phase boundaries within the cell, which are represented by a vertical line. As will be seen later in this section, the metal electrodes appear at the ends of this notation, and the half-cell in which oxidation occurs is written on the left, and the half-cell in which reduction occurs is written on the right.

We briefly discuss an additional small contribution to the cell potential that arises from the differing diffusion rates of large and small ions in an electrical field. As an electrochemical reaction proceeds, ions that diffuse rapidly across a liquid–liquid junction, such as H^+, will travel farther than ions that diffuse slowly, such as Cl^-, in a given time. At steady state, a dipole layer is built up across this junction, and the rates of ion transfer through this dipole layer become equal. This kinetic effect will give rise to a small **junction potential** between two liquids of different composition or concentration. Such a junction potential is largely eliminated by a salt bridge. An interface for which the junction potential has been eliminated is indicated by a pair of vertical lines.

For example, the abbreviated notation for the Daniell cell containing a salt bridge is

$$Zn(s)|ZnSO_4(aq)\|CuSO_4(aq)|Cu(s) \tag{11.47}$$

and a cell made up of the Zn/Zn^{2+} half-cell and the standard hydrogen electrode is described by

$$Zn(s)|ZnSO_4(aq)\|H^+(aq)|H_2(g)|Pt(s) \tag{11.48}$$

The overall reaction in this cell is $Zn(s) + 2H^+(aq) \rightarrow Zn^{2+}(aq) + H_2(g)$. By convention, a single dashed line is used to indicate a liquid–liquid interface across which charge transfer can occur. In general, in such a cell the solutions are physically separated by a porous membrane to prevent mixing of the solutions. In this case, the junction potential has not been eliminated.

The half-cell and overall reactions can be determined from the abbreviated notation in the following way. An electron is transferred from the electrode at the far left of the abbreviated notation to the electrode on the far right through the external circuit. The number of electrons is then adjusted to fit the half-cell reaction. This procedure is illustrated in Example Problem 11.7.

EXAMPLE PROBLEM 11.7

Determine the half-cell reactions and the overall reaction for the cell designated

$$Ag(s)|AgCl(s)|Cl^-(aq, a_\pm = 0.0010)\|Fe^{2+}(aq, a_\pm = 0.50)\,Fe^{3+}(aq, a_\pm = 0.10)|Pt(s)$$

Solution

$$Ag(s) + Cl^-(aq) \rightarrow AgCl(s) + e^-$$
$$Fe^{3+}(aq) + e^- \rightarrow Fe^{2+}(aq)$$

The overall reaction is

$$Ag(s) + Cl^-(aq) + Fe^{3+}(aq) \rightarrow AgCl(s) + Fe^{2+}(aq)$$

Only after the cell potential is calculated is it clear whether the reaction or the reverse reaction is spontaneous.

We have already discussed several types of half-cells. The hydrogen standard electrode involves the equilibrium between a gas and a dissolved species. A second such electrode is the Cl_2/Cl^- electrode, for which the reduction reaction is described by

$$Cl_2(g) + 2e^- \rightleftarrows 2Cl^-(aq) \qquad E° = +1.36 \text{ V} \qquad (11.49)$$

Another type of half-cell that is frequently encountered involves a metal and a metal ion in solution. Both half-cells in the Daniell cell fall into this category.

$$Zn^{2+}(aq) + 2e^- \rightleftarrows Zn(s) \qquad E° = -0.76 \text{ V} \qquad (11.50)$$

A number of half-cells consist of a metal, an insoluble salt containing the metal, and an aqueous solution containing the anion of the salt. Two examples of this type of half-cell are the Ag–AgCl half-cell for which the reduction reaction is

$$AgCl(s) + e^- \rightarrow Ag(s) + Cl^-(aq) \qquad E° = +0.22 \text{ V} \qquad (11.51)$$

and the calomel (mercurous chloride) electrode, which is frequently used as a reference electrode in electrochemical cells:

$$Hg_2Cl_2(s) + 2e^- \rightleftarrows 2Hg(l) + 2Cl^-(aq) \qquad E° = +0.27 \text{ V} \qquad (11.52)$$

In an oxidation–reduction half-cell, both species are present in solution, and the electrode is an inert conductor such as Pt, which allows an electrical connection to be made to the solution. In the Fe^{3+}/Fe^{2+} half-cell, the reduction reaction is

$$Fe^{3+}(aq) + e^- \rightleftarrows Fe^{2+}(aq) \qquad E° = 0.771 \text{ V} \qquad (11.53)$$

11.10 The Electrochemical Series

Tables 11.1 and 11.2 (see Appendix A, Data Tables) list the reduction potentials of commonly encountered half-cells. The emf of a cell constructed from two of these half-cells with standard reduction potentials $E_1°$ and $E_2°$ is given by

$$E_{cell}° = E_1° - E_2° \qquad (11.54)$$

The potential $E_{cell}°$ will be positive and, therefore, $\Delta G < 0$, if the reduction potential for reaction 1 is more positive than that of reaction 2. Therefore, the relative strength of a species as an oxidizing agent follows the order of the numerical value of its reduction potential in Table 11.2. The **electrochemical series** shown in Table 11.3 is obtained if the oxidation of neutral metals to their most common oxidation state is considered. For example, the entry for gold in Table 11.3 refers to the reduction reaction

$$Au^{3+}(aq) + 3e^- \rightleftarrows Au(s) \qquad E° = 1.498 \text{ V}$$

In a redox couple formed from two entries in the list shown in Table 11.3, the species lying higher in the table will be reduced, and the species lying lower in the list will be oxidized in the spontaneous reaction. For example, this list predicts that the spontaneous reaction in the copper–zinc couple is $Zn(s) + Cu^{2+}(aq) \rightleftarrows Zn^{2+}(aq) + Cu(s)$, and not the reverse reaction.

EXAMPLE PROBLEM 11.8

For the reduction of the permanganate ion MnO_4^- to Mn^{2+} in an acidic solution, $E°$ is +1.51 V. The reduction reactions and standard potentials for Zn^{2+}, Ag^+, and Au^+ are given here:

$$Zn^{2+}(aq) + 2e^- \rightleftarrows Zn(s) \qquad E° = -0.7618 \text{ V}$$
$$Ag^+(aq) + e^- \rightleftarrows Ag(s) \qquad E° = 0.7996 \text{ V}$$
$$Au^+(aq) + e^- \rightleftarrows Au(s) \qquad E° = 1.692 \text{ V}$$

Which of these metals will be oxidized by the MnO_4^- ion?

TABLE 11.3

The Electrochemical Series

Most Strongly Reducing (The metal is least easily oxidized.)
Gold (most positive reduction potential)
Platinum
Palladium
Silver
Rhodium
Copper
Mercury
(Hydrogen; zero reduction potential by convention)
Lead
Tin
Nickel
Iron
Zinc
Chromium
Vanadium
Manganese
Magnesium
Sodium
Calcium
Potassium
Rubidium
Cesium
Lithium (most negative reduction potential)
Least Strongly Reducing (The metal is most easily oxidized.)

Solution

The cell potentials assuming the reduction of the permanganate ion and oxidation of the metal are

$$\text{Zn:} \quad 1.51\text{ V} + 0.761\text{ V} = 2.27\text{ V} > 0$$
$$\text{Ag:} \quad 1.51\text{ V} - 0.7996\text{ V} = 0.710\text{ V} > 0$$
$$\text{Au:} \quad 1.51\text{ V} - 1.692\text{ V} = -0.18\text{ V} < 0$$

If $E° > 0$, $\Delta G < 0$. On the basis of the sign of the cell potential, we conclude that only Zn and Ag will be oxidized by the MnO_4^- ion.

11.11 Thermodynamics of Batteries and Fuel Cells

Batteries and fuel cells are electrochemical cells that are designed to maximize the ratio of output power to the cell weight or volume. **Batteries** contain the reactants needed to support the overall electrochemical reaction, whereas **fuel cells** are designed to accept a continuous flow of reactants from the surroundings. Batteries that cannot be recharged are called primary batteries, whereas rechargeable batteries are called secondary batteries.

It is useful to compare the relative amount of work that can be produced through an electrochemical reaction with the work that a heat engine could produce using the same overall reaction. The maximum electrical work is given by

$$w_{electrical} = -\Delta G = -\Delta H \left(1 - \frac{T\Delta S}{\Delta H} \right) \tag{11.55}$$

whereas the maximum work available through a reversible heat engine operating between T_1 and T_2 is

$$w_{thermal} = q_{hot}\varepsilon = -\Delta H \left(\frac{T_h - T_c}{T_h} \right) \tag{11.56}$$

where ε is the efficiency of a reversible heat engine (see Section 5.2). To compare the maximum thermal and electrical work, we use the overall reaction for the familiar lead-acid battery used in cars. For this reaction, $\Delta G° = -376.97\text{ kJ mol}^{-1}$, $\Delta H° = -227.58\text{ kJ mol}^{-1}$, and $\Delta S° = 501.1\text{ J K}^{-1}\text{mol}^{-1}$. Assuming $T_1 = 600$ K and $T_2 = 300$ K and that the battery operates at 300 K, then

$$\frac{w_{electrical}}{w_{thermal}} = 3.31 \tag{11.57}$$

This calculation shows that much more work can be produced in the electrochemical reaction than in the thermal reaction. This comparison does not even take into account that the lead-acid battery can be recharged, whereas the thermal reaction can only be run once.

11.12 The Electrochemistry of Commonly Used Batteries

The lead-acid battery was invented in 1859 and is still widely used in automobiles. Because the power required to start an automobile engine is on the order of a kilowatt, the current capacity of such a battery must be on the order of a hundred amperes. Additionally, such a battery must be capable of 500 to 1500 recharging cycles from a deep discharge. In recharging batteries, the reaction product in the form of a solid must be converted back to the reactant, also in the form of a solid. Because the solids in general have a different crystal structure and density, the conversion induces mechanical stress in the anode and cathode, which ultimately leads to a partial disintegration of these electrodes. This is the main factor that limits the number of charge–discharge cycles that a battery can tolerate.

The electrodes in the lead-acid battery consist of Pb powder and finely divided PbO and $PbSO_4$ supported on a Pb frame. The electrodes are supported in a container containing concentrated H_2SO_4. The cell reactions at the cathode and anode are

$$PbO_2(s) + 4H^+(aq) + SO_4^{2-}(aq) + 2e^- \rightleftarrows PbSO_4(s) + 2H_2O(l) \tag{11.58}$$

$$E° = 1.685 \text{ V}$$

$$Pb(s) + SO_4^{2-}(aq) \rightleftarrows PbSO_4(s) + 2e^- \qquad E° = -0.356 \text{ V} \tag{11.59}$$

and the overall reaction is

$$PbO_2(s) + Pb(s) + 2H_2SO_4 \rightleftarrows 2PbSO_4(s) + 2H_2O(l) \qquad E° = 2.04 \text{ V} \tag{11.60}$$

In the preceding reactions, the arrow pointing to the right indicates the discharge direction, and the arrow pointing in the opposite direction indicates the charging direction.

Six such cells connected in series are required for a battery that provides a nominal potential of 12 V. The lead-acid battery is very efficient in that more than 90% of the electrical charge used to charge the battery is available in the discharge part of the cycle. This means that side reactions such as the electrolysis of water play a minimal role in charging the battery. However, only about 50% of the lead in the battery is converted between PbO_2 and $PbSO_4$. Because Pb has a large atomic mass, this limited convertibility decreases the power per unit weight figure of merit for the battery. Parasitic side reactions also lead to a self-discharge of the cell without current flowing in the external circuit. For the lead-acid battery, the capacity is diminished by approximately 0.5% per day through self-discharge.

An equally frequently encountered battery is the alkaline cell, for which a schematic drawing is shown in Figure 11.8. The individual elements are indicated. The anode in this cell is powdered zinc, and the cathode is in the form of a MnO_2 paste mixed with powdered carbon to impart conductivity. KOH is used as the electrolyte. The anode and cathode reactions are

Anode: $Zn(s) + 2OH^-(aq) \rightarrow ZnO(s) + H_2O(l) + 2e^- \qquad E° = 1.1 \text{ V} \tag{11.61}$
Cathode: $2MnO_2(s) + H_2O(l) + 2e^- \rightarrow Mn_2O_3(s) + 2OH^-(aq) \tag{11.62}$
$$E° = -0.76 \text{ V}$$

The nickel metal hydride battery is finding use in hybrid vehicles that rely on dc motors to drive the vehicle in city traffic and use an engine for freeway driving. One manufacturer uses 38 modules of 6 cells, each with a nominal voltage of 1.2 V and a total voltage of 274 V to power the vehicle. The capacity of the battery pack is ~1800 W-h. The anode and cathode reactions are

Anode: $MH(s) + OH^-(aq) \rightleftarrows M + H_2O(l) + e^- \qquad E° = 0.83 \text{ V} \tag{11.63}$
Cathode: $NiOOH(s) + H_2O(l) + 2e^- \rightleftarrows Ni(OH)_2(s) + OH^-(aq) \tag{11.64}$
$$E° = 0.52 \text{ V}$$

The electrolyte is $KOH(aq)$, and the overall reaction is

$$MH(s) + NiOOH(s) \rightleftarrows M + Ni(OH)_2(s) \qquad E° = 1.35 \text{ V} \tag{11.65}$$

where M designates an alloy that can contain V, Ti, Zr, Ni, Cr, Co, and Fe.

Lithium ion batteries are of particular interest for two reasons. Because the cell potential is high (~3.7 V), a single cell produces nearly the same voltage as three alkaline batteries in series. Additionally, because Li has a low atomic mass and a large

FIGURE 11.8
Schematic diagram of an alkaline cell.

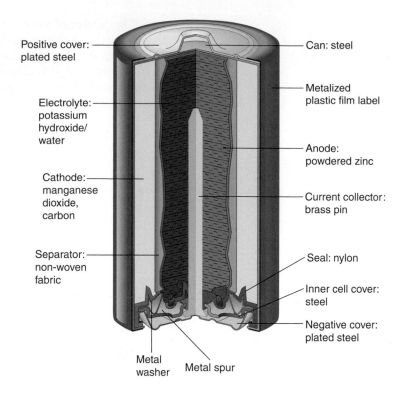

Positive cover: plated steel

Can: steel

Metalized plastic film label

Electrolyte: potassium hydroxide/ water

Anode: powdered zinc

Cathode: manganese dioxide, carbon

Current collector: brass pin

Seal: nylon

Separator: non-woven fabric

Inner cell cover: steel

Negative cover: plated steel

Metal washer   Metal spur

cell potential, the energy density (A h kg⁻¹) of a lithium cell is larger than that of a lead storage battery by a factor of ~15. Rechargeable lithium batteries have the following half-cell reactions:

Anode: $$LiCoO_2(s) \rightleftarrows Li_{1-n}CoO_2(s) + ne^- \qquad (11.66)$$

Cathode: $$C(s) + nLi^+ + ne^- \rightleftarrows CLi_x \qquad (11.67)$$

The right and left arrows indicate the charge and discharge directions, respectively. The overall cell reaction is

$$LiCoO_2(s) + C(s) \rightleftarrows Li_{1-n}CoO_2(s) + CLi_x \qquad E° \sim 3.7 \text{ V} \qquad (11.68)$$

and the fully charged battery has a cell potential of ~3.7 V. The structure of $LiCoO_2(s)$ and CLi_x are shown schematically in Figure 11.9. CLi_x designates Li atoms intercalated between sheets of graphite; it is not a stoichiometric compound.

11.13 Fuel Cells

The primary advantage of fuel cells over batteries is that they can be continually refueled and do not require a downtime for recharging. A number of different technologies are used in fuel cells. Most are still in the research and development stage, and only phosphoric acid fuel cells are available as off-the-shelf technology. All fuel cells currently use O_2 or air as the oxidant and either H_2, methane, or H_2 produced by reforming hydrocarbons or alcohols as the fuel. The goal of much current research is the development of direct methanol oxidation fuel cells. Because methanol is a liquid with an energy density that is only a factor of two below that of gasoline, storing the fuel is much less of an engineering challenge than storing gaseous H_2. An early application of this technology is likely to be micro-fuel cells that displace batteries in computers and cell phones. With an energy density that is an order of mag-

FIGURE 11.9

The cell voltage in a lithium battery is generated by moving the lithium between a lattice site in LiCoO$_2$ and an intercalation position between sheets of graphite.

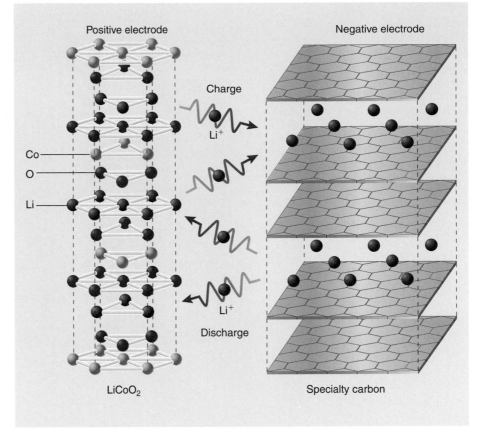

nitude greater than currently available batteries of the same volume, fuel cells are likely to have a major impact in providing power to handheld devices. The following discussion is restricted to the most widely understood technology, the **proton exchange membrane fuel cell (PEMFC).**

Proton exchange membrane fuel cells using H$_2$ and O$_2$ as the reactants were originally used in the NASA *Gemini* space flights of the 1960s. The principles underlying this technology have not changed in the intervening years. However, advances in technology have significantly increased the power generated per unit weight as well as the power generated per unit area of the electrodes. A schematic drawing of a single PEMFC and the relevant half-cell reactions is shown in Figure 11.10. The anode and cathode have channels that transport H$_2$ and O$_2$ to the catalyst and membrane. An intermediate diffusion layer ensures uniform delivery of reactants and separation of gases from the water formed in the reaction. The membrane is coated with a catalyst (usually Pt or Pt alloys) that facilitates the H$^+$ formation and the reaction between O$_2$ and H$^+$. Individual units can be arranged back to back to form a stack so that the potential across the stack is a multiple of the individual cell potential.

The heart of this fuel cell is the proton exchange membrane, which functions as a solid electrolyte. This membrane facilitates the passage of H$^+$ from the anode to the cathode in the internal circuit, and it also prevents electrons and negative ions from moving in the opposite direction. The membrane must be thin (~10–100 μm) to allow rapid charge transport at reasonably high current densities and must be unreactive under the potentials present in the cell. The most widely used membranes are polymeric forms of perfluorosulfonic acids. These membranes are quite conductive if fully hydrated. The internal structure of these membranes is depicted in Figure 11.11. A cartoon that illustrates how the membrane facilitates H$^+$ transport is shown in Figure 11.11a. It depicts the membranes as being made up of spherical cavities, or

FIGURE 11.10
Schematic diagram of a proton exchange membrane fuel cell. The half-cell reactions are shown. The channels in the anode and cathode facilitate the supply of O_2 and H_2 to the cell and carry away the H_2O reaction product. The gas diffusion layer ensures that the reactants are uniformly distributed over the membrane surface.

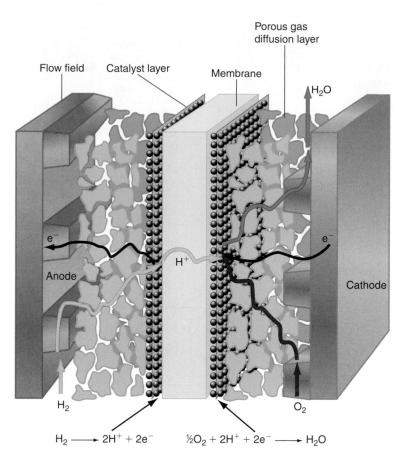

$$H_2 \longrightarrow 2H^+ + 2e^- \qquad \tfrac{1}{2}O_2 + 2H^+ + 2e^- \longrightarrow H_2O$$

inverted micelles ~4 nm in diameter, connected by cylindrical channels ~1 nm in diameter. The interior of the cavities and channels are lined with SO_3^- groups. The $-CF_2-$ backbone of the polymer provides rigidity, makes the membrane unreactive, and increases the acidity of the terminal SO_3^- groups. The positively charged H^+ can migrate through this network under the influence of the electrical potential gradient across the membrane that arises through the half-cell reactions. By contrast, negatively charged species in a cavity cannot migrate through the network, because they are repelled by the negatively charged SO_3^- groups in the narrow channels that connect adjacent cavities. Figure 11.11b gives a more realistic picture of the interconnected network of cavities and channels.

The electrochemistry of the two half-cell reactions shown in Figure 11.10 is well understood. The main challenges in manufacturing a fuel cell based on the overall reaction $H_2 + 1/2 \, O_2 \rightarrow H_2O$ lie in maintaining a high cell potential at the necessary current output. A number of strategies have been pursued to improve the performance of proton exchange membrane fuel cells. Optimizing the interface between the gas diffusion element and the membrane has been the most important area of progress. The use of porous and finely divided materials allows the electrode surface area to be maximized per unit volume of the fuel cell. Careful integration of the catalyst and membrane surface also allows the amount of the expensive Pt-based catalysts to be minimized. A future goal is to develop operating conditions that do not require highly pure H_2. Currently, high purity is required because trace amounts (~50 ppm) of CO poison the Pt catalyst surface and render it inactive for H_2 dissociation. Fuel cells that operate at temperatures above 100°C are much less sensitive to poisoning by CO. However, operation at these temperatures requires proton exchange membranes that use ionic liquids other than water to maintain a high conductivity.

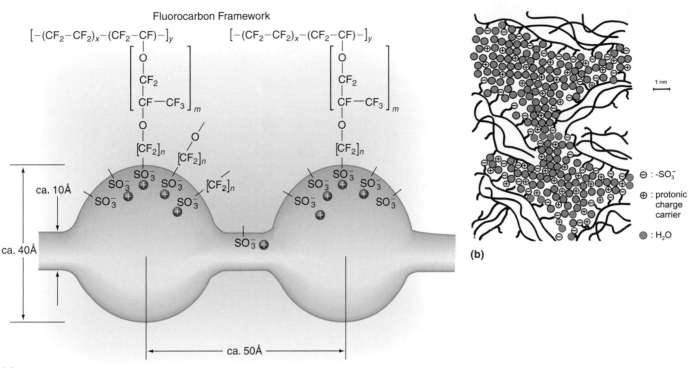

Fluorocarbon Framework

$$[-(CF_2-CF_2)_x-(CF_2-CF)-]_y$$

ca. 10Å

ca. 40Å

ca. 50Å

(a)

1 nm

⊖ : -SO₃⁻

⊕ : protonic charge carrier

● : H₂O

(b)

FIGURE 11.11

Diagram of the proton exchange membrane, indicating the functional groups present in cavities and in the connecting channels. [Part **(a)** reprinted with permission from C. H. Hamann *et al., Electrochemistry,* Figure 7.6, page 297, Wiley-VCH, New York, 1998; part **(b)** "A proton exchange membrane." K. D. Kreuer, Hydrocarbon Membranes, chapter 33 in W. Vielstich et al., Eds., *Handbook of Fuel Cells,* Volume 3, Fuel Cell Technology and Applications, Part 1, Figure 1, page 423, John Wiley and Sons, Ltd, 2003. Courtesy of Dr. Klaus-Dieter Kreuer, Maier Department, Max-Planck-Institut für Festkörperforschung.]

SUPPLEMENTAL

11.14 Electrochemistry at the Atomic Scale

The half-cell reaction $Cu^{2+}(aq) + 2e^- \rightleftarrows Cu(s)$ only describes an overall process. What is known about this reaction at an atomic scale? In addressing this question, we draw extensively on a review article by Dieter Kolb [*Surface Science* 500 (2002) 722–740].

We first describe the structure of the interface between the solution and a metal electrode in an electrochemical cell, which is known as the electrical double layer. To a good approximation, the interface can be modeled as a parallel plate capacitor as depicted in Figure 11.12a. In this figure, the positively charged plate of the capacitor is the metal electrode, and the negative plate is made of negative ions that are surrounded by their solvations shells. The two types of negative ions at the interface are distinguished by the forces that hold them in this region. Specifically bound ions are those that form a chemical bond with one or more of the metal atoms at the surface of the electrode. Examples of specifically bound ions are Cl^- and Br^-. It is advantageous for these ions to form bonds with the surface, because the water molecules in their solvation shell are not as strongly bound as for positive ions such as Na^+ and K^+. The part of the solvation shell directed toward the electrode atoms to which the ion is bonded is missing. The plane that goes through the center of the specifically adsorbed ions is known as the inner Helmholtz plane. The second type of ion that is found close to the positively charged electrode consists of fully solvated ions, which are called nonspecifically adsorbed ions. The plane that goes through the center of the nonspecifically adsorbed ions is known as the outer Helmholtz plane. Fully solvated positive ions are found outside the electrical double layer.

As shown in Figure 11.12b, the falloff in the electrical potential between the metal electrode and the electrolyte solution occurs in a very small distance of approximately 3 nm, resulting in an electrical field in the electrical double layer as large as 3×10^7 V cm^{-1}. Because the charge that can be accommodated at the surface of the electrode by the specifically and nonspecifically adsorbed ions is on the order of 0.1–0.2 electron per atom of metal at the surface, the electrical double layer has a very large capacitance of 20–50 μC cm^{-2}. For a typical difference in potential between the electrode and the solution of ~1 V, the energy stored in the interfacial capacitance can be as large as 150 kJ kg^{-1}. Because this density is so high, electrochemical capacitors can be used to

FIGURE 11.12

(a) The electrical double layer, showing specifically and nonspecifically adsorbed ions and water molecules, and the inner and outer Helmholtz planes. **(b)** Variation of the electrical potential in the electrical double layer, which has a thickness ~ 3 nm. The solid line passes through the center of a nonspecifically adsorbed ion, and the dashed line passes through the center of a specifically adsorbed ion. **(c)** The electrochemical cell. The region that is shown greatly magnified in parts (a) and (b) is indicated. [Courtesy of Dr. D. M. Kolb, Department of Electrochemistry, University of Ulm.]

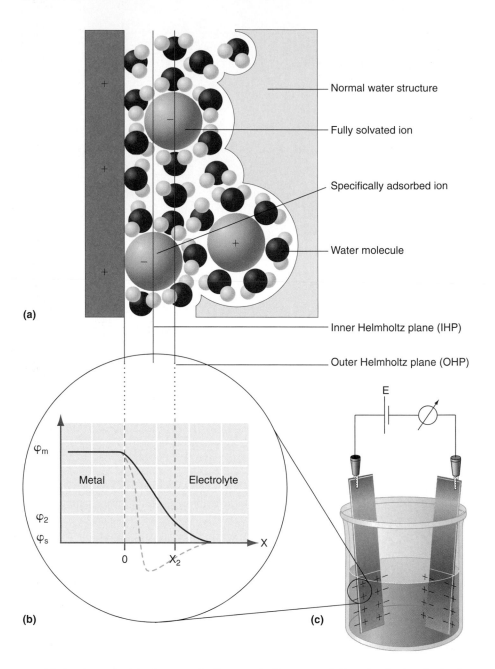

Normal water structure

Fully solvated ion

Specifically adsorbed ion

Water molecule

(a)

Inner Helmholtz plane (IHP)

Outer Helmholtz plane (OHP)

φ_m

Metal

Electrolyte

φ_2

φ_s

0   X_2 X

(b)

(c)

provide backup electrical energy in the event of a power failure in electronic devices. Note in Figure 11.12b that the electrical potential has fallen to the value in the middle of the solution at a distance of approximately 5 nm. All important aspects of an electrochemical reaction occur in this region immediately adjacent to the electrode surface.

Having described the structure of the solution adjacent to the electrode surface, what can be said about the structure of the electrode surface at an atomic scale? Much of what is known about the structure of the electrode surface at an atomic scale has been obtained using the scanning tunneling microscope (STM) discussed in Chapter 5 of Engel's *Quantum Chemistry and Spectroscopy,* and the reader is encouraged to review this material before proceeding. In an STM, a metal tip is positioned within 0.5–2 nm of a conducting surface, and at these distances, electrons can tunnel between the tip and surface. The tunneling current falls off exponentially with the tip–surface distance. Using piezoelectric elements, the tip height is varied as the tip is scanned over the surface to keep the tunneling current at a constant value, typically 1 nA. The voltage applied to the piezoelectric element normal to the surface is a direct measure of the height of the surface at a point parallel to the surface. These voltage values are used to construct a topographical map of the surface. A schematic picture of the essential elements of an STM is shown in Figure 11.13.

FIGURE 11.13
Schematic picture of an STM. The piezo elements labeled x and y are used to scan the tip parallel to the surface, and the z piezo element is used to vary the tip–surface distance. The inset shows a greatly magnified image of the region between the end of the tip and the surface. The dashed line is a cut through a contour map of the surface resulting from the measurement. [Courtesy of Dr. D. M. Kolb, Department of Electrochemistry, University of Ulm.]

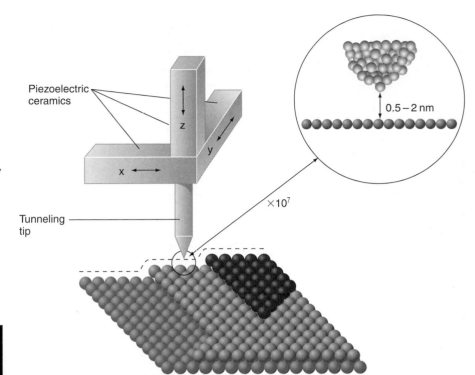

FIGURE 11.14
Very sharp STM tips can be prepared using electrochemical etching. In an electrochemical STM, the shank of the tip is coated with an insulating material to suppress ohmic conduction between the tip and surface through the electrolyte solution. [Courtesy of Dr. D. M. Kolb, Department of Electrochemistry, University of Ulm.]

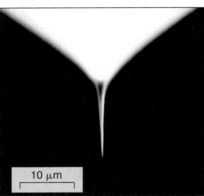

In an electrochemical STM, the tip is coated with an insulating material except in the immediate vicinity of the tunneling region, as illustrated in Figure 11.14, to avoid current passing from the tip to the surface through the conducting electrolyte solution.

The structure of an electrode surface in an electrochemical environment depends on the possible reactions that the electrode can undergo in the solution, and the applied potential. For a material such as platinum, which is not easily oxidized, it is possible by careful preparation to obtain an electrode that is virtually identical to what is expected from terminating the crystal structure of the electrode material. An example of this ideal case is shown in Figure 11.15a for a platinum electrode surface. The great

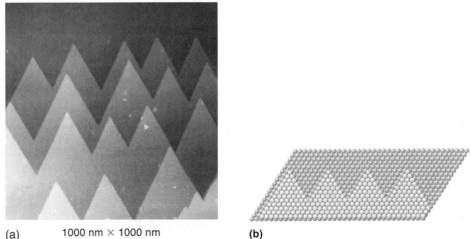

(a)   1000 nm × 1000 nm   (b)

FIGURE 11.15
(a) A 1000-nm × 1000-nm STM image of a well-prepared platinum electrode exposing a close-packed layer of Pt atoms in the surface. The image shows several stacked terraces, which differ in height by one atomic layer. **(b)** Ball model showing the atomic layer structure underlying the image of part (a). Only step edges in which the step edge of a terrace consists of a close-packed row of atoms is observed in part (a). Note the lateral offset in successively higher terraces, which are characteristic of the stacking in planes of a face-centered cubic lattice. [Part (a) Courtesy of Dr. D. M. Kolb, Department of Electrochemistry, University of Ulm.]

10nm × 10nm

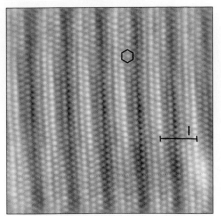

$E_{SCE} = -0.2V$

FIGURE 11.16

A 10-nm × 10-nm image of a well-prepared gold electrode. Each dot corresponds to an individual Au atom. The topmost layer undergoes a reconstruction to form a close-packed layer of hexagonal symmetry as indicated by the hexagon. Because this hexagonal layer is not in registry with the underlying square lattice, the surface plane is buckled with the repeat unit labeled l. At the bottom, E_{SCE} indicates that the potential is measured relative to a standard calomel electrode, for which the half-cell reaction is $Hg_2Cl_2(s) + 2e^- \rightarrow 2Hg(l) + 2Cl^-(aq)$. [Courtesy of Dr. D. M. Kolb, Department of Electrochemistry, University of Ulm.]

majority of Pt atoms are in their ideal lattice positions. A much smaller number are located at the steps that form the edges of large terraces that make up the long-range structure of the surface. A detail showing the arrangement of the Pt atoms at the edge of a terrace is shown in Figure 11.15b. Although the step atoms are a small minority of the total number of surface atoms, they play a significant role in electrochemical processes, as shown later.

For the case of the platinum electrode, the surface corresponds to the most densely packed plane of the face-centered cubic lattice, which has the lowest surface energy. What is the atomic level structure of an electrode surface if the surface plane is not that of lowest surface energy? An example for this case is shown in Figure 11.16. The gold surface consisted of a square array of Au atoms rather than the lowest energy close-packed layer, which has hexagonal symmetry. To minimize its energy, the topmost surface layer undergoes a reconstruction to a close-packed layer with the same structure as shown in Figure 11.15b. Because the hexagonal surface layer and the underlying square lattice do not have the same symmetry, the topmost layer of the electrode is periodically buckled.

The process in which the surface of the electrode undergoes the reconstruction described earlier can be imaged with the electrochemical STM, as shown in Figure 11.17. The light areas in this image are gold islands, around which the reconstruction must detour. Because of the square symmetry of the underlying layers, the reconstruction proceeds in two domains along two directions oriented at 90° with respect to one another. Note the defects in the reconstructed layer that arise from a meandering of a reconstructed part of a domain, defects that arise through the intersection of two domains of different orientation, as well as defects due to the Au islands present on the surface.

We now return to the influence of surface defects in electrochemical reactions. If a Au electrode is immersed in a $CuSO_4$ solution, and the potential is adjusted appropriately, Cu will be deposited on the electrode, as described by the reaction $Cu^{2+}(aq) + 2e^- \rightarrow Cu(s)$. How does the Cu layer grow on the Au surface? The answer is provided by the STM image shown in Figure 11.18. The initial stage of the Cu film is the formation of small Cu islands that are exclusively located at step edges of the underlying Au surface. This result can be understood by realizing that the coordination number of a Cu atom at a gold site at the step edge is larger than on a flat portion of the surface. As more copper is deposited, the initially formed islands grow laterally and eventually merge to form a uniform layer.

Several important conclusions can be made concerning the growth of an electrochemically deposited metal film from these studies. The fact that island growth is observed rather than random deposition of Cu atoms indicates that the initially adsorbed Cu

FIGURE 11.17

STM images of a gold electrode in which the reconstruction of the topmost layer is not complete. The light areas are higher lying Au islands. Note that the reconstruction proceeds along two perpendicular directions. This occurs because the underlying lattice structure has square symmetry. [Courtesy of Dr. D. M. Kolb, Department of Electrochemistry, University of Ulm.]

80nm × 80nm

14nm × 14nm

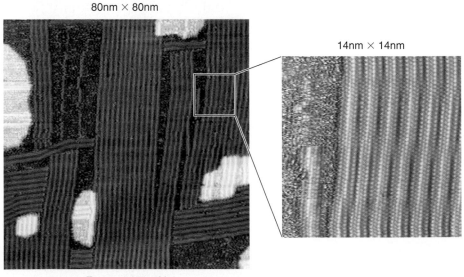

$E_{SCE} = -250$ mV

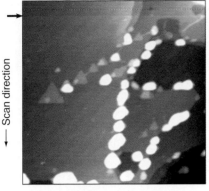

Scan direction

170 nm × 170 nm

FIGURE 11.18

Image of a gold electrode in which a small amount of copper has been electrochemically deposited. Note that the Cu is deposited in the form of small islands that are anchored at step edges of the underlying gold electrode. The onset of the deposition is initiated by the horizontal arrow in the figure. At that point, the electrode potential was increased to a value that allows the reaction $Cu^{2+}(aq) + 2e^- \rightarrow Cu(s)$ to proceed. [Courtesy of Dr. D. M. Kolb, Department of Electrochemistry, University of Ulm.]

atoms have a high mobility parallel to the surface. They diffuse across a limited region of the surface until they encounter the edge of an island. Because these edge sites have a higher coordination number than a site on the terrace, they are strongly bound and diffuse no further. The other conclusion that can be drawn is that the metal film grows one layer at a time. A second layer is not nucleated until the underlying layer is nearly completed. Layer-by-layer growth ensures that a compact and crystalline film is formed.

As the deposition of Cu on a gold electrode shows, step edges play a central role in determining the formation of electrochemically deposited layers. Although small in number, surface defects such as step edges or vacancies play a major role in the chemistry of a wide range of chemical processes including catalytic cracking of crude oil and the corrosion of metals.

Electrochemical scanning tunneling microscopy has also provided new insight into the atomic level processes underlying the electrochemical dissolution of a metal electrode. To carry out these studies, an STM was constructed that allowed 25 images to be obtained per second. A comparison of images obtained at closely spaced times allows researchers to follow the ongoing processes [O. Magnussen *et al.*, *Electrochimica Acta*, 46 (2001), 3725–3733]. A mode of growth was observed in which individual atoms were added to a single row at the edge of a terrace as shown in Figure 11.19.

However, a second and unexpected mode of growth and dissolution is observed in which entire segments of a row can grow or dissolve collectively. By adjusting the electrode potential slightly, either Cu growth or dissolution can be initiated. Figure 11.20a shows a sequence of images for electrode dissolution, and Figure 11.20b shows a sequence

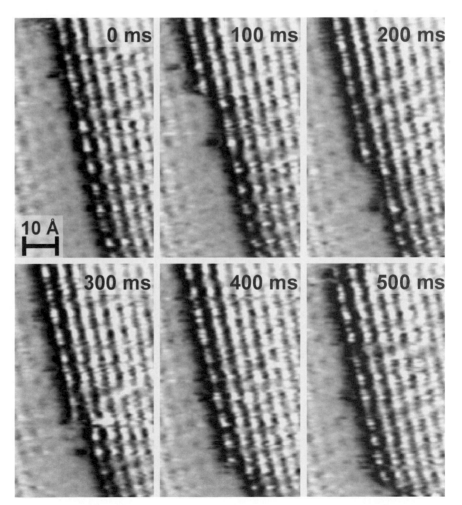

FIGURE 11.19

A series of images taken at the indicated time intervals is shown as Cu is deposited electrochemically on a copper electrode. Note that an individual row at the edge of the upper terrace grows from the top to the bottom of the image. [Photo courtesy of Professor Dr. Rolf Jurgen Behm/University of Ulm. Magnussen *et al.*, "In-Situ Atomic-Scale Studies of the Mechanisms and Dynamics of Metal STM," Figure 1, *Electrochemica Acta* 46 (2001), 3725–3733, p. 3727.]

FIGURE 11.20
Examples of a concerted removal of whole
or multiple steps for a copper electrode:
(a) dissolution and **(b)** growth. The times
at which the images were acquired relative
to the first image are shown. [Photo
courtesy of Professor Dr. Rolf Jurgen
Behm/University of Ulm. Magnussen *et
al.,* "In-Situ Atomic-Scale Studies of the
Mechanisms and Dynamics of Metal
STM," Figure 2, *Electrochemica Acta* 46
(2001), 3725–3733.]

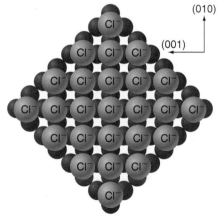

FIGURE 11.21
Because they are larger, the Cl^- ions
cannot form the same structure as the
underlying Cu surface. Instead, they form
a square lattice that is rotated by 45° with
respect to the copper unit cell.

for Cu deposition on the electrode. In both cases, entire row or row segments are deposited at once rather than in the atom-by-atom fashion depicted in Figure 11.19.

The Cu electrode has a square symmetry in this case, and one might assume that growth and dissolution occur with equal probability in either of two directions oriented 90° to one another. However, the images in Figure 11.20 indicate that this is not the case. Growth or dissolution occurs only in the direction labeled (001) and not along the direction labeled (010). This behavior is a consequence of the atomic level structure of the Cl^- specifically adsorbed ions of the HCl electrolyte. This structure is shown in Figure 11.21, and consists of a square lattice that has a larger unit cell length than the underlying Cu lattice, and is rotated 45° with respect to the Cu lattice.

The adsorbed Cl^- layer removes the equivalence of the (010) and (001) directions as is shown in Figure 11.22. The kinks formed along step edges along these two directions have a different structure. Therefore, they are expected to have a different reactivity in the dissolution/process. As these experiments show, specifically adsorbed ions play a central role in an electrochemical reaction.

The ability to verify that an electrode is both crystalline and has a low defect density using STM has made the interpretation of experimental results obtained through classical electrochemical techniques significantly easier. An example is cyclic voltammetry, in which the electrode potential is varied linearly with time and then the potential is changed in the opposite direction and returned to the initial value. As the potential is varied, the electrode current is measured. Because the area in a plot of current versus time is the electrical charge, cyclic voltammetry provides a way to identify regions of potential in which significant amounts of charge are transferred to or from the electrode. These are the regions in which electrochemical reactions proceed. An example of a cyclic voltammogram for the deposition of less than one monolayer of Cu on a gold electrode is shown in Figure 11.23.

STM studies show that Cu can form a low-density ordered structure on the surface that has a saturated coverage of 0.67 monolayer, and a monolayer structure that has the same structure as the underlying Au electrode. Metal deposition begins to occur at a potential of 0.4 V and the surface is fully covered as the potential is lowered to 0 V. Deposition corresponds to a positive current in Figure 11.23 (the scan direction is from right to left in the figure), and dissolution corresponds to a negative current (left to right in the figure). Two well-defined peaks are observed in both the growth and dissolution directions. The emf associated with these peaks can be used to calculate the ΔG_f° associated with the formation of a chemically bonded Cu atom at the site on the Au surface characteristic of the different geometries of the two ordered Cu phases. Both of these peaks occur at a potential less than that needed to deposit copper on a copper electrode, be-

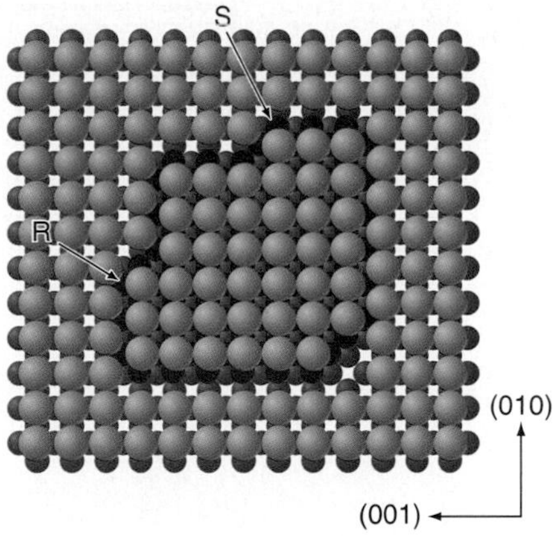

FIGURE 11.22
The kinks formed in step edges along the (010) and (001) directions have a different structure. This explains the different reactivity observed along these directions as shown in Figure 11.20. The small light and dark circles denote Cu atoms, and the large gray circles denote specifically adsorbed Cl⁻ ions. [Courtesy of Professor Dr. Rolf Jurgen Behm/University of Ulm.]

cause Cu atoms are more strongly bound to surface sites on the Au surface than on the copper surface. One refers to underpotential deposition to emphasize that the potential is less than that expected for a material deposited on its own lattice. The combination of a structural determination at the atomic level provided by STM with the I–V relation provided by cyclic voltammetry is well suited to understanding electrochemical processes at surfaces.

FIGURE 11.23
(a) A cyclic voltamogram is shown for the deposition/dissolution of Cu on a well-ordered single-crystal gold electrode. By measuring the area under the peaks, the total amount of Cu deposited or dissolved can be determined. This result is plotted in part **(b)**. The amount by which peaks A and B are displaced in voltage in the dissolution process from their values in the deposition process is a measure of the rate of electron transfer in the processes. [Courtesy of Dr. D. M. Kolb, Department of Electrochemistry, University of Ulm.]

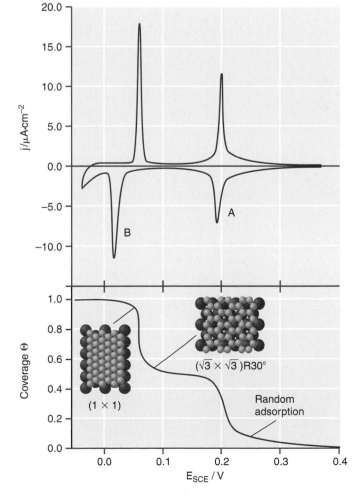

FIGURE 11.24

Possible current paths between a tool and an electrode immersed in an electrolyte solution. Before a reaction can proceed, the capacitance of the electrical double layer must be built up to its full value. [Reprinted with permission from A. L. Trimmer *et al.,* "Single-Step Electrochemical Machining of Complex Nanostructures with Ultrashort Voltage Pulses," Figure 1, *Applied Physics Letters* 82(19) (2003), 3327, © 2003 American Institute of Physics.]

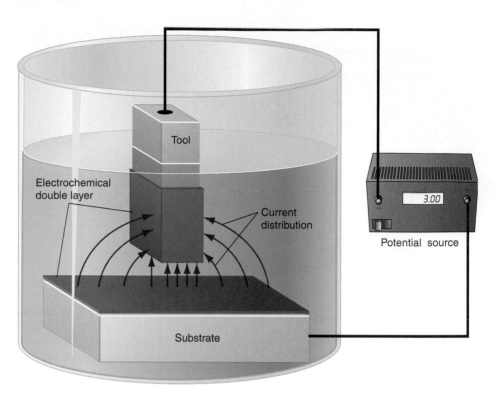

SUPPLEMENTAL

11.15 Using Electrochemistry for Nanoscale Machining

Many current technologies require fabrication of miniature devices with dimensional control in the micron to nanometer scale. For example, lithographic methods are used to make the masks employed in the fabrication of integrated circuits. Until recently, electrochemical machining techniques have not been able to provide dimensional control better than 10 μm for a reason related to the capacitance of the electrical double layer. If the external voltage is changed in an electrochemical cell, how long does it take for the cell to reach equilibrium with respect to the potential distribution? As discussed in the previous section, the electrical double region can be modeled as a large capacitor that must be charged up to the value of the new potential as the external voltage is changed. The current must flow through the electrolyte solution, which has an ohmic resistance, and the cell can be modeled as a series R-C circuit with R equal to the electrical resistance of the solution, and C equal to the capacitance of the electrical double layer.

If a metal tool is placed very close to an electrode, the resistance associated with a current path between the tool and the electrode depends on the path, as shown in Figure 11.24. Those paths immediately below the tool correspond to those of smallest electrical resistance. It can be shown that the time-dependent voltage across the capacitor in an R-C series circuit is given by

$$V(t) = V_0(1 - e^{-t/RC}) \qquad (11.69)$$

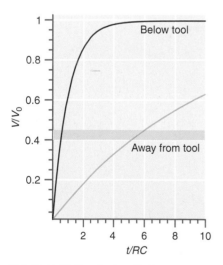

FIGURE 11.25

The ratio V/V_0 is shown as a function of t/RC for a point immediately below the tool, and away from the tool at a point where the resistance along the path is higher by a factor of 10.

The rate at which V/V_0 increases for a point on the surface directly below the tool and a point well to one side of the tool is shown in Figure 11.25. Below the tool, the potential is built up to its full value at much shorter times than outside of this region. Until the potential is built up to the value required for an electrochemical reaction of interest, the reaction will not proceed. Rolf Schuster and coworkers [*Electrochimica Acta* 48 (2003), 20] have used this fact to supply the external voltage in the form of a

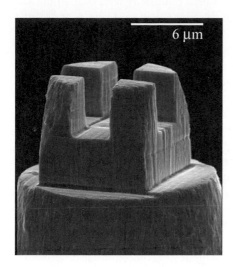

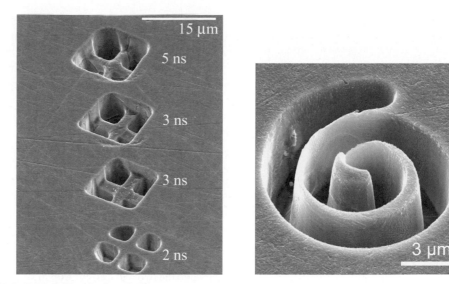

FIGURE 11.26

Results obtained using electrochemical machining of nickel in a HCl solution using the tool shown in the left image. Note the dependence of the resolution on pulse length in the middle image. [Left and middle images reprinted with permission from A. L. Trimmer *et al., Applied Physics Letters* (2003), 82, 3327–3329; right image reprinted with permission from M. Kock *et al., Electrochemica Acta* Volume 48, Issues 20–22, 30 September 2003, pages 3213–3219, Fig 7, p. 3218.]

short pulse, rather than as a dc voltage. As a consequence, the reaction of interest occurs directly under the tool, and only to a negligible extent outside of this region. For example, if the dissolution reaction in which material is removed from the work piece only takes place at potentials higher than that indicated by the horizontal bar in Figure 11.25, it will only take place under the tool for a short enough pulse. Note that if a dc potential is applied instead of a pulse, the reaction will not be localized.

Examples of the patterns that have been generated in this way by metal dissolution are shown in Figure 11.26. A resolution of ~20 nm has been obtained by using pulses of 200-ps duration.

SUPPLEMENTAL

11.16 Absolute Half-Cell Potentials

The cell potentials listed in Tables 11.1 and 11.2 (see Appendix A, Data Tables) are all measured relative to the standard hydrogen electrode. Because an electrochemical cell consists of two half-cells, any common reference for the half-cell potentials cancels when calculating the cell potential. For this reason, equilibrium constants, activity coefficients, and the direction of spontaneous change can all be calculated without absolute half-cell potentials. However, there is no reason why a half-cell potential cannot be determined in principle. The difficulty is in devising an appropriate experiment to measure this quantity. Recall that putting one lead of a voltmeter on the electrode and the second lead in the solution will not give the half-cell potential as discussed in Section 11.2. Although not necessary for discussing electrochemical reactions, absolute half-cell potentials are useful in formulating a microscopic model of electrochemical processes. For example, absolute values for half-cell potentials allow the solvation Gibbs energy of the ion involved in the redox reaction to be calculated directly, rather than to rely on a calculation as outlined in Section 10.2. We next describe a method for the determination of half-cell potentials that was formulated by R. Gomer and G. Tryson [*J. Chem. Phys.* 66

(1977), 4413–4424]. The following discussion assumes familiarity with the particle in a box model of the conduction electrons in a metal (see Chapters 4 and 5 of Engel's *Quantum Chemistry and Spectroscopy*).

Consider the electrochemical cell shown schematically in Figure 11.27. The conduction electrons in each metal electrode are described using the particle in a box model, and all energy levels up the highest occupied level, called the Fermi level, are filled. The vertical distance corresponds to energy for a negative test charge, and the horizontal axis corresponds to distance. The double layer acts as a capacitor, and shifts the potential in the solution relative to the Fermi level by the amount qV_{MS1} or qV_{MS2} as shown in Figure 11.27. The change in potential within the double layer is depicted as being linear rather than as shown correctly in Figure 11.12. To remove an electron from the metal, it is necessary to give an electron at the Fermi level an amount of energy equal to the work function, ϕ. The **work function** of a metal is analogous to the ionization energy of an atom, and different metals have different values of the work function. Because of the presence of the double layer at each electrode, a negatively charged ion or an electron has a lower energy in the bulk of the solution than at the surface of an electrode.

How do the two half-cells line up in an energy diagram? This question can be answered using the fact that the electrochemical potential of an electron in a metal is the Fermi energy. If two metals are placed in contact, electrons flow from one metal to the other until the electrochemical potential of an electron is the same in each metal. This is equivalent to saying that the Fermi levels of the two electrodes lie at the same energy. The equilibrium

$$M^{n+}(aq) + ne^{-} \rightarrow M(s) \tag{11.70}$$

that is established separately in each half-cell determines the drop in electrical potential across the double layer and thereby shifts the Fermi level of each metal electrode relative to the energy of the bulk solution. Therefore, the Fermi levels of the two electrodes are exactly offset by the Gibbs energy:

$$\Delta G = -eE_{cell} \tag{11.71}$$

How can an absolute half-cell potential be defined that is consistent with the energy diagram in Figure 11.27? To answer this question, the reversible work and, therefore, the Gibbs energy associated with the half-cell reaction $M^{n+}(aq) + ne^{-} \rightarrow M(s)$ is calculated. For simplicity, it is assumed that $n = 1$. Because G is a state function, any convenient path between the initial states of the metal atom in the solid and the final state of the ion in the bulk solution and the electron at the Fermi level of the electrode can be chosen. Breaking this overall process down into the individual steps shown in the following equations simplifies the analysis. The ΔG associated with each step is indicated.

FIGURE 11.27

An energy diagram for a cell. The energies associated with differences in electrical potential are shown as a product of the test charge (negative in this case) and the potential difference. The vertical bar in the center indicates a salt bridge junction. The decrease in energy on the solution side of each electrode is due to the electrical double layer.

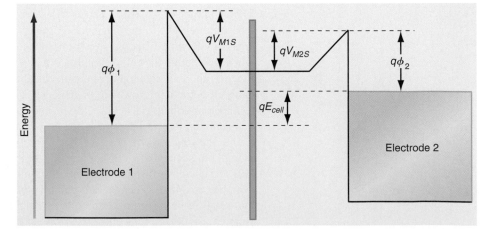

$$M_1(g) \to M_1(g) \qquad\qquad \Delta G = \Delta G_{vaporization,1} \qquad (11.72)$$

$$M_1(g) \to M_1^+(g) + e^-(M_1) \qquad \Delta G = q(I_1 - \phi_1) \qquad (11.73)$$

$$M_1^+(g) \to M_1^+(\text{solution, at } M_1) \qquad \Delta G = \Delta G_{solvation,1} \qquad (11.74)$$

$$M_1^+(aq, \text{ at } M_1) \to M_1^+(aq, \text{ bulk solution}) \qquad \Delta G = qV_{M_1S} \qquad (11.75)$$

Note that in the final step, the solvated ion at the electrode is transferred through the double layer into the solution. The overall process is

$$M_1(s) \to M_1^+ (aq, \text{ bulk solution}) + e^-(M_1),$$ and the ΔG for the overall process is

$$\Delta G_1 = \Delta G_{vaporization,1} + q(I_1 - \phi_1) + \Delta G_{solvation,1} + qV_{M_1S} \qquad (11.76)$$

The analogous equation for electrode 2 is

$$\Delta G_2 = \Delta G_{vaporization,2} + q(I_2 - \phi_2) + \Delta G_{solvation,2} + qV_{M_2S} \qquad (11.77)$$

If the two half-cells are in equilibrium, $\Delta G_1 = \Delta G_2$ and

$$(\Delta G_1^{vap} - \Delta G_2^{vap}) + (\Delta G_1^{solv} - \Delta G_2^{solv}) + q(I_1 - I_2) = q(V_{M_2S} - \phi_2) - q(V_{M_1S} - \phi_1) \qquad (11.78)$$

$$= qE_{cell}$$

Equating $q(V_{M_1S} - \phi_1) - q(V_{M_2S} - \phi_2)$ with qE_{cell} follows from the energy diagram shown in Figure 11.27.

Equation (11.78) shows that a plausible definition of each of the half-cell potentials in terms of the physical parameters of the problem is given by

$$E_1 = V_{M_1S} - \phi_1 \quad \text{and} \quad E_2 = V_{M_2S} - \phi_2 \qquad (11.79)$$

Note that the half-cell potential is not the difference in the electrical potential between the electrode and the bulk solution. Only if the potential is defined in this way will $E = E_1 + E_2$, as can be seen in Figure 11.27.

Although Equation (11.79) defines an absolute half-cell potential, numerical values for the half-cell can only be determined if V_{MS} and ϕ are known. Note that the work function in Equation (11.79) is not the work function in air or in vacuum, but the work function in solution. The value of ϕ will vary for these different environments because of molecular layers absorbed on the metal surface. As discussed in Section 11.2, V_{MS} cannot be measured directly using a voltmeter and a probe inserted in the solution. However, V_{MS} can be measured without contacting the solution using the experimental setup shown schematically in Figure 11.28.

FIGURE 11.28
Apparatus used to determine absolute half-cell potentials.

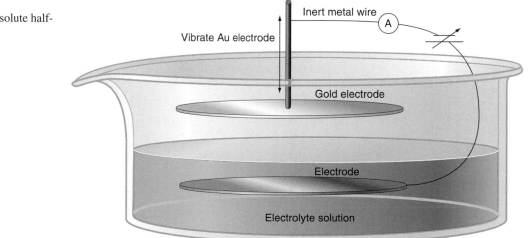

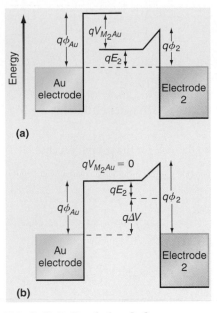

FIGURE 11.29

These energy diagrams are used to explain how the absolute half-cell potential can be determined using the experiment illustrated in Figure 11.28. **(a)** The variable voltage in series with the wire is set equal to zero. **(b)** The variable voltage is adjusted so that the current flow in the external circuit is equal to zero.

The electrode of interest is immersed in the electrolyte solution. A reference electrode (in this case gold) in the form of a flat plate is oriented parallel to the surface of the electrolyte solution. Gold is chosen because the work function in air is known. The Au and metal electrodes are connected by a metal wire, and a variable dc voltage source is inserted in the wire. The gold electrode is mounted in such a way that its distance to the electrolyte surface can be varied in a periodic fashion. The energy diagram shown in Figure 11.29 is used to understand how the experimental setup can be used to measure V_{MS}. By connecting the electrode in the solution with the gold electrode in air, their Fermi levels are equalized as shown in Figure 11.29a. Note that there is no double-layer potential on the electrode in air. The difference in the electrical potential of the gold electrode and the surface of the electrolyte surface is designated as V_{M_2Au}.

Because of the connecting wire shown in Figure 11.28, the Au electrode and the electrochemical double layer of the immersed electrode form a capacitor that charges up because of the potential difference between the two plates that arises through equalizations of the Fermi levels in the electrodes. First consider the case in which the adjustable voltage in Figure 11.28 is set equal to zero. The potential difference between the surface of the electrolyte and the parallel gold electrode is shown in Figure 11.29a. As the Au plate is vibrated, the capacitance changes and, therefore, an alternating current flows in the external circuit as the capacitor is alternately charged and discharged. Next carefully adjust the variable voltage in the wire as the Au electrode vibrates until no current can be detected in the external circuit. If no current flows, the voltage difference between the Au electrode and the electrolyte surface must be equal to zero. This case corresponds to the energy diagram of Figure 11.29b. The null current must correspond to the condition

$$\Delta V = V_{M_2Au} \tag{11.80}$$

Note that in this experiment, V_{MS} can be determined experimentally without inserting a probe into the liquid. Because ϕ_{Au} is known, $E_2 = V_{M_2Au} - \phi_{Au}$ can be calculated, and therefore, the absolute half-cell potential can be determined.

Using this method, Gomer and Tryson [*J. Chem. Phys.* 66 (1977), 4413–4424] determined the absolute half-cell potential of the standard hydrogen electrode to be $E^{\circ}_{SHE} = -4.73 \pm 0.05$ V. Once this value is known, the absolute half-cell potentials of all other half-cell potentials can be determined by adding –4.73 V to the values in Tables 11.1 and 11.2 (see Appendix A, Data Tables). The preceding discussion has omitted the mention of a small dipole layer that exists at the surface of an aqueous solution because of the preferential orientation of water dipoles at the air–water interface. Although the magnitude of this dipole cannot be measured directly, calculations indicate that its magnitude is ~50 mV. This value has been taken into account in obtaining the result $E^{\circ}_{SHE} = -4.73 \pm 0.05$ V.

Once the absolute half-cell potential is known, the important (and otherwise experimentally inaccessible) quantity $\Delta G_{solvation}$ can be calculated. From Equations (11.72) through (11.75), the relation between the half-cell potential and the energetic parameters in these equations is

$$E_{half-cell} = -\frac{1}{q}(\Delta G_{vaporization} + \Delta G_{solvation} + Iq) \tag{11.81}$$

This equation allows $\Delta G_{solvation}$ of an individual ion to be determined from the appropriate half-cell potential. The solvation Gibbs energy calculated in this way is in good agreement with the values calculated from the Born model if the ionic radii obtained from crystal structures are corrected for the water of solvation as discussed in Section 10.2.

Vocabulary

anode

battery

cathode

charged

Daniell cell

electrochemical cell

electrochemical potential

electrochemical series

electromotive force (emf)

Faraday constant

fuel cell

junction potential

Nernst equation

neutral

proton exchange membrane fuel cell (PEMFC)

reference electrode

salt bridge

work function

Questions on Concepts

Q11.1 What is the difference in the chemical potential and the electrochemical potential for an ion and for a neutral species in solution? Under what conditions is the electrochemical potential equal to the chemical potential for an ion?

Q11.2 Show that if $\Delta G_f^{\circ}(\text{H}^+, aq) = 0$ for all T, the potential of the standard hydrogen electrode is zero.

Q11.3 To determine standard cell potentials, measurements are carried out in very dilute solutions rather than at unit activity. Why is this the case?

Q11.4 Explain why the magnitude of the maximum work available from a battery can be greater than the magnitude of the reaction enthalpy of the overall cell reaction.

Q11.5 The temperature dependence of the potential of a cell is vanishingly small. What does this tell you about the thermodynamics of the cell reaction?

Q11.6 Why is the capacitance of an electrolytic capacitor so high compared with conventional capacitors?

Q11.7 Can specifically adsorbed ions in the electrochemical double layer influence electrode reactions?

Q11.8 How can one conclude from Figure 11.18 that Cu atoms can diffuse rapidly over a well-ordered Au electrode in an electrochemical cell?

Q11.9 How is it possible to deposit Cu on a Au electrode at a potential lower than that corresponding to the reaction $\text{Cu}^{2+}(aq) + 2\text{e}^- \rightleftarrows \text{Cu}(s)$?

Q11.10 Why is it possible to achieve high-resolution electrochemical machining by applying a voltage pulse rather than a dc voltage to the electrode being machined?

Problems

Problem numbers in **red** indicate that the solution to the problem is given in the *Student's Solutions Manual*.

P11.1 Calculate $\Delta G_{reaction}^{\circ}$ and the equilibrium constant at 298.15 K for the reaction $\text{Hg}_2\text{Cl}_2(s) \rightarrow 2\text{Hg}(l) + \text{Cl}_2(g)$.

P11.2 Calculate $\Delta G_{reaction}^{\circ}$ and the equilibrium constant at 298.15 K for the reaction

$$\text{Cr}_2\text{O}_7^{2-}(aq) + 3\,\text{H}_2(g) + 8\,\text{H}^+(aq) \rightarrow 2\,\text{Cr}^{3+}(aq) + 7\,\text{H}_2\text{O}(l).$$

P11.3 Using half-cell potentials, calculate the equilibrium constant at 298.15 K for the reaction $2\,\text{H}_2\text{O}(l) \rightarrow 2\,\text{H}_2(g) + \text{O}_2(g)$. Compare your answer with that calculated using ΔG_f° values from Table 11.1 (see Appendix A). What is the value of E° for the overall reaction that makes the two methods agree exactly?

P11.4 For the half-cell reaction $\text{AgBr}(s) + \text{e}^- \rightarrow \text{Ag}(s) + \text{Br}^-(aq)$, $E^{\circ} = +0.0713$ V. Using this result and $\Delta G_f^{\circ}(\text{AgBr}, s) = -96.9$ kJ mol^{-1}, determine $\Delta G_f^{\circ}(\text{Br}^-, aq)$.

P11.5 For the half-cell reaction $\text{Hg}_2\text{Cl}_2(s) + 2\text{e}^- \rightarrow 2\text{Hg}(l) + 2\text{Cl}^-(aq)$, $E^{\circ} = +0.27$ V. Using this result and $\Delta G_f^{\circ}(\text{AgBr}, s) = -96.9$ kJ mol^{-1}, determine $\Delta G_f^{\circ}(\text{Cl}^-, aq)$.

P11.6 By finding appropriate half-cell reactions, calculate the equilibrium constant at 298.15 K for the following reactions:

a. $4\,\text{NiOOH} + 2\,\text{H}_2\text{O} \rightarrow 4\,\text{Ni(OH)}_2 + \text{O}_2$

b. $4\,\text{NO}_3^- + 4\,\text{H}^+ \rightarrow 4\,\text{NO} + 2\,\text{H}_2\text{O} + 3\,\text{O}_2$

P11.7 By finding appropriate half-cell reactions, calculate the equilibrium constant at 298.15 K for the following reactions:

a. $2 Cd(OH)_2 \rightarrow 2 Cd + O_2 + 2H_2O$

b. $2MnO_4^{2-} + 2H_2O \rightarrow 2MnO_2 + 4OH^- + O_2$

P11.8 Consider the cell
$Hg(l)|Hg_2SO_4(s)|FeSO_4(aq, a = 0.0100)|Fe(s)$.

a. Write the cell reaction.

b. Calculate the cell potential, the equilibrium constant for the cell reaction, and $\Delta G°$ at 25°C.

P11.9 Determine the half-cell reactions and the overall cell reaction, calculate the cell potential, and determine the equilibrium constant at 298.15 K for the cell

$$Cu(s)|Cu^{2+}(aq, a_{Cu^{2+}} = 0.0150)|H^+(aq, a_{H^+} = 0.100)|H_2(g)|Pt(s)$$

Is the cell reaction spontaneous as written?

P11.10 Determine the half-cell reactions and the overall cell reaction, calculate the cell potential and determine the equilibrium constant at 298.15 K for the cell

$$Ag(s)|AgCl(s)|Cl^-(aq, a_{Cl^-} = 0.00500)|Cd^{2+}(aq, a_{Cd^{2+}} = 0.100)|Cd(s)$$

Is the cell reaction spontaneous as written?

P11.11 Determine the half-cell reactions and the overall cell reaction, calculate the cell potential, and determine the equilibrium constant at 298.15 K for the cell

$$Pt(s)|Mn^{2+}(aq, a_{\pm} = 0.0150),$$
$$Mn^{3+}(aq, a_{\pm} = 0.200)|Zn^{2+}(aq, a_{\pm} = 0.100)|Zn(s)$$

Is the cell reaction spontaneous as written?

P11.12 The half-cell potential for the reaction $O_2(g) + 4H^+ (aq) + 4e^- \rightarrow 2H_2O$ is +1.03 V at 298.15 K when $a_{O_2} = 1$. Determine a_{H^+}.

P11.13 You are given the following half-cell reactions:

$Pd^{2+}(aq) + 2e^- \rightleftarrows Pd(s)$ $\qquad E° = 0.83$ V

$PdCl_4^{2-}(aq) + 2e^- \rightleftarrows Pd(s) + 4Cl^-(aq)$ $E° = 0.64$ V

a. Calculate the equilibrium constant for the reaction

$$Pd^{2+}(aq) + 4Cl^-(aq) \rightleftarrows PdCl_4^{2-}(aq)$$

b. Calculate $\Delta G°$ for this reaction.

P11.14 Determine $E°$ for the reaction $Cr^{2+} + 2e^- \rightarrow Cr$ from the one-electron reduction potential for Cr^{3+} and the three-electron reduction potential for Cr^{3+} given in Table 11.1 (see Appendix A).

P11.15 The Edison storage cell is described by

$$Fe(s)|FeO(s)|KOH(aq, a)|Ni_2O_3(s)|NiO(s)|Ni(s)$$

and the half-cell reactions are as follows:

$Ni_2O_3(s) + H_2O(l) + 2e^- \rightleftarrows 2NiO(s) + 2OH^-(aq)$
$$E° = 0.40 \text{ V}$$
$FeO(s) + H_2O(l) + 2e^- \rightleftarrows Fe(s) + 2OH^-(aq)$
$$E° = -0.87 \text{ V}$$

a. What is the overall cell reaction?

b. How does the cell potential depend on the activity of the KOH?

c. How much electrical work can be obtained per kilogram of the active materials in the cell?

P11.16 Determine K_{sp} for AgBr at 298.15 K using the electrochemical cell described by

$$Ag(s)|AgBr(s)|Br^-(aq, a_{Br^-})||Ag^+(aq, a_{Ag^+})|Ag(s)$$

P11.17 The standard potential $E°$ for a given cell is 1.100 V at 298.15 K and $(\partial E°/\partial T)_P = -6.50 \times 10^{-5}$ V K^{-1}. Calculate $\Delta G°_{reaction}, \Delta S°_{reaction}$, and $\Delta H°_{reaction}$. Assume that $n = 2$.

P11.18 For a given overall cell reaction, $\Delta S°_R = 17.5$ J mol^{-1} K^{-1} and $\Delta H°_R = -225.0$ kJ mol^{-1}. Calculate $E°$ and $(\partial E°/\partial T)_P$. Assume that $n = 1$.

P11.19 Consider the Daniell cell for the indicated molalities: $Zn(s)|ZnSO_4(aq, 0.300m)||CuSO_4(aq, 0.200m)|Cu(s)$. The activity coefficient $\gamma_{\pm}$ has the value 0.1040 for $CuSO_4$ and 0.08350 for $ZnSO_4$ at the indicated concentrations. Calculate E by setting the activity equal to the molality and by using the above values of the $\gamma_{\pm}$. How large is the relative error if the concentrations, rather than the activities, are used?

P11.20 The standard half-cell potential for the reaction $O_2(g) + 4H^+(aq) + 4e^- \rightarrow 2H_2O$ is +1.03 V at 298.15 K. Calculate E for a 0.500-molal solution of HCl for $a_{O_2} = 1$ (a) assuming that the a_{H^+} is equal to the molality and (b) using the measured mean ionic activity coefficient for this concentration, $\gamma_{\pm} = 0.757$. How large is the relative error if the concentrations, rather than the activities, are used?

P11.21 Consider the half-cell reaction $O_2(g) + 4H^+ (aq) + 4e^- \rightarrow 2H_2O$. By what factor are n, Q, E, and $E°$ changed if all the stoichiometric coefficients are multiplied by the factor two? Justify your answers.

P11.22 The cell potential E for the cell $Pt(s)|H_2(g, a_{H_2} = 1)|H^+(aq, a_{H^+} = 1)||NaCl(aq, m = 0.300)|AgCl(s)|Ag(s)$ is + 0.260 V. Determine γ_{Cl^-} assuming that $\gamma_{\pm} = \gamma_{Na^+} = \gamma_{Cl^-}$.

P11.23 Consider the Daniell cell, for which the overall cell reaction is $Zn(s) + Cu^{2+}(aq) \rightarrow Zn^{2+}(aq) + Cu(s)$. The concentrations of $CuSO_4$ and $ZnSO_4$ are 2.500 and 1.100×10^{-3} m, respectively.

a. Calculate E setting the activities of the ionic species equal to their molalities.

b. Calculate $\gamma_{\pm}$ for each of the half-cell solutions using the Debye–Huckel limiting law.

c. Calculate E using the mean ionic activity coefficients determined in part (b).

P11.24 Consider the cell $Pt(s)|H_2(g, 1 atm)|H^+(aq, a = 1)||Fe^{3+}(aq), Fe^{2+}(aq)|Pt(s)$ given that $Fe^{3+} + e^- \rightleftarrows Fe^{2+}$ and $E° = 0.771$ V.

a. If the cell potential is 0.712 V, what is the ratio of $Fe^{2+}(aq)$ to $Fe^{3+}(aq)$?

b. What is the ratio of these concentrations if the cell potential is 0.830 V?

c. Calculate the fraction of the total iron present as $Fe^{3+}(aq)$ at cell potentials of 0.650, 0.700, 0.750, 0.771, 0.800, and 0.900 V. Graph the result as a function of the cell potential.

P11.25 Consider the couple $Ox + e^- \rightleftarrows Red$ with the oxidized and reduced species at unit activity. What must be the value of $E°$ for this half-cell if the reductant R is to liberate hydrogen at 1 atm from

a. an acid solution with $a_{H^+} = 1$?

b. water at pH = 7?

c. Is hydrogen a better reducing agent in acid or basic solution?

P11.26 Consider the half-cell reaction $AgCl(s) + e^- \rightleftarrows Ag(s) + Cl^-(aq)$. If $\mu°(AgCl, s) = -109.71$ kJ mol^{-1}, and if $E° = +0.222$ V for this half-cell, calculate the standard Gibbs energy of formation of $Cl^-(aq)$.

P11.27 The data in the following table have been obtained for the potential of the cell $Pt(s)|H_2(g, f = 1 \text{ atm})|HCl(aq, m)|AgCl(s)|Ag(s)$ as a function of m at 25°C.

m (mol kg^{-1})	E (V)	m(mol kg^{-1})	E (V)	m (mol kg^{-1})	E (V)
0.00100	0.57915	0.0200	0.43024	0.500	0.27231
0.00200	0.54425	0.0500	0.38588	1.000	0.23328
0.00500	0.49846	0.100	0.35241	1.500	0.20719
0.0100	0.46417	0.200	0.31874	2.000	0.18631

a. Determine $E°$ using a graphical method.

b. Calculate $\gamma_\pm$ for HCl at $m = 0.00100, 0.0100$, and 0.100 mol kg^{-1}.

P11.28 Harnet and Hamer [*J. American Chemical Society* 57 (1935), 33] report values for the potential of the cell $Pt(s)|PbSO_4(s)|H_2SO_4(aq, a)|PbSO_4(s)|PbO_2(s)|Pt(s)$ over a wide range of temperature and H_2SO_4 concentrations. In $1m$ H_2SO_4, their results were described by $E(V) = 1.91737 + 56.1 \times 10^{-6}t + 108 \times 10^{-8}t^2$, where t is the temperature on the Celsius scale. Calculate ΔG, ΔH, and ΔS for the cell reaction at 0° and 25°C.

P11.29 Between 0° and 90°C, the potential of the cell $Pt(s)|H_2(g, f = 1 \text{ atm})|HCl(aq, m = 0.100)|AgCl(s)|Ag(s)$ is described by the equation $E(V) = 0.35510 - 0.3422 \times 10^{-4}t - 3.2347 \times 10^{-6}t^2 + 6.314 \times 10^{-9}t^3$, where t is the temperature on the Celsius scale. Write the cell reaction and calculate ΔG, ΔH, and ΔS for the cell reaction at 50°C.

P11.30 Consider the reaction $Sn(s) + Sn^{4+}(aq) \rightleftarrows 2Sn^{2+}(aq)$. If metallic tin is in equilibrium with a solution of Sn^{2+} in which $a_{Sn^{2+}} = 0.100$, what is the activity of Sn^{4+} at equilibrium?

Probability

The concept of probability is central to many areas of chemistry. The characterization of large assemblies of molecules, from experimental observations to theoretical descriptions, relies on the concepts of statistics and probability. Given the utility of these concepts in chemistry, the central ideas of probability theory are presented in this chapter, including permutations, configurations, and probability distribution functions. ∎

12.1 Why Probability?

At this point in our exploration of physical chemistry, two limiting perspectives of matter have been introduced. One perspective is the microscopic viewpoint of quantum mechanics, in which matter is described through a detailed analysis of its atomic and molecular components. This approach is elegant in its detail and was triumphant in describing many experimental observations that escaped classical descriptions of matter. For example, the observation of discrete emission from the hydrogen atom can only be explained using quantum theory. So successful is this approach that it stands as one of the greatest human accomplishments of the 20th century, and the ramifications of the quantum perspective continue to be explored.

Given the success of quantum theory in describing aspects of nature that are beyond the reach of classical mechanics, some might be tempted to dismiss classical descriptions of matter such as thermodynamics as irrelevant; however, this is not the case. The macroscopic perspective inherent to thermodynamics is extremely powerful in its ability to predict the outcome of chemical events. Thermodynamics involves the numerous relationships between macroscopic observables and relates experimental measurements of macroscopic properties to predictions of chemical behavior. Perhaps the most impressive aspect of thermodynamics is its ability to predict reaction spontaneity. By simply considering the Gibbs or Helmholtz energy difference between reactants and products, it is possible to state with certainty if a reaction will occur spontaneously. Even though it has impressive predictive utility, thermodynamics offers little help when one wants to know not only if a reaction will occur, but why the reaction occurs in the first place. What are the molecular details that give rise to Gibbs energy, and why should this quantity vary from one species to the next? Unfortunately, answers to these questions are beyond the descriptive bounds of thermodynamics. Can the detailed molecular descriptions available from quantum mechanics be used to formulate an answer to these questions? This type of approach demands that the quantum perspective converge with that of thermodynamics, and the link between these perspectives is developed in the following four chapters.

Statistical mechanics provides a methodology that allows for the translation of the microscopic properties of matter into macroscopic behavior. In this approach, a thermodynamic system is described as a collection of smaller units, a reduction in scale that can be taken to the atomic or molecular level. Starting from the microscopic perspective, statistical mechanics allows one to take detailed quantum descriptions of atoms and molecules and determine the corresponding thermodynamic properties. For example, consider a system consisting of 1 mol of gaseous HCl. We can take our knowledge of the quantum energetics of HCl and use this information in combination with statistical mechanics to determine thermodynamic properties of the system such as internal energy, heat capacity, entropy, and other properties described earlier in this book.

But the question remains as to how this statistical bridge will be built. The task at hand is to consider a single atom or molecule and scale this perspective up to assemblies on the order of 10^{23}! Such an approach necessitates a quantitative description of chemistry as a collection of events or observables, a task that is readily met by probability theory. Therefore, the mathematical tools of probability theory are required before proceeding with a statistical development. Probability is a concept of central utility in discussing chemical systems; and the mathematical tools developed in this section will find wide application in subsequent chapters.

12.2 Basic Probability Theory

Probability theory was developed beginning in the late 1600s as a mathematical formalism for describing games of chance. Consistent with the origins of this field, the majority of illustrative examples employed in this chapter involve games of chance. Central to probability theory are **variables,** or quantities that can change in value throughout the course of an experiment or series of events. A simple example is the outcome of a coin toss, with the variable being the side of the coin observed after tossing the coin. The variable can assume one of two values—heads or tails—and the value the variable assumes will change from one coin toss to the next. Variables can be partitioned into two categories: discrete variables and continuous variables.

Discrete variables assume only a limited number of specific values. The outcome of a coin toss is an excellent example of a discrete variable in that the outcome of the toss can assume only one of two values: heads or tails. For another example, imagine a classroom with 100 desks in it, each of which is numbered. If we define the variable *chair number* as being the number on the chair, then this variable can assume values ranging from 1 to 100 in integer values. The possible values a variable can assume are collectively called the **sample space** of the variable. In the chair example, the sample space is equal to the collection of integers from 1 to 100, that is, $\{1, 2, 3, \ldots, 100\}$.

Continuous variables can assume any value within a set of limits. For example, the variable X that can have any value in the range of $1 \leq X \leq 100$. Thermodynamics provides another well-known example of a continuous variable: temperature. The absolute temperature scale ranges from 0 K to infinity, with the variable, temperature, able to assume any value between these two limits. For continuous variables, the sample space is defined by the limiting values of the variable.

The treatment of probability differs depending on whether the variable of interest is discrete or continuous. Probability for discrete variables is mathematically simpler to describe; therefore, we focus on the discrete case first and then generalize to the continuous case later in Section 12.5.

Once a variable and its corresponding sample space have been defined, the question becomes to what extent the variable will assume a given value from the sample space. In other words, we are interested in the **probability** that the variable will assume a certain value. Imagine a lottery where balls numbered 1 to 50 are randomly mixed inside a machine, and a single ball is selected. What is the probability that the ball chosen will have the value one (①)? If the chance of selecting any ball is equal, then the probability of selecting ① is simply 1/50. Does this mean that we will select ① only once every 50 selections? Consider each ball selection as an individual experiment, and after each experiment the selected ball is thrown back into the machine and another experiment is performed. The probability of selecting ① in any experiment is 1/50, but if the outcome of each experiment is independent of other outcomes, then it is not inconceivable that ① will not be selected after 50 trials, *or* that ① will be retrieved more than once. However, if the experiment is performed many times, the end result will be the retrieval of ① 1/50th of the total number of trials. This simple example illustrates a very important point: probabilities dictate the likelihood that the variable will assume a given value as determined from an infinite number of experiments. As scientists, we are not able to perform an infinite number of experiments; therefore, the extension of probabilities to situations involving a limited number of experiments is done with the understanding that the probabilities provide an approximate expectation of an experimental outcome.

In the lottery example, the probability of selecting any single ball is 1/50. Because there are 50 balls total, the sum of the probabilities for selecting each individual ball must be equal to 1. Consider a variable, X, for which the sample space consists of M values denoted as $\{x_1, x_2, \dots, x_M\}$. The probability that the variable X will assume one of these values (p_i) is:

$$0 \le p_i \le 1 \tag{12.1}$$

where the subscript indicates one of the values contained in the set space ($i = 1, 2, \dots, M$). Furthermore, X must assume some value from the sample set in a given experiment dictating that the sum of all probabilities will equal unity:

$$p_1 + p_2 + \dots + p_M = \sum_{i=1}^{M} p_i = 1 \tag{12.2}$$

In Equation (12.2), the sum of probabilities has been indicated by the summation sign, with the limits of summation indicating that the sum is taken over the entire sample space, from $i = 1$ to M. The combination of the sample space, $S = \{x_1, x_2, \dots, x_M\}$, and corresponding probabilities, $P = \{p_1, p_2, \dots, p_M\}$, is known as the **probability model** for the experiment.

EXAMPLE PROBLEM 12.1

What is the probability model for the lottery experiment described in the preceding text?

Solution

The value of the ball that is retrieved in an individual experiment is the variable of interest and it can take on integer values from 1 to 50. Therefore, the sample space is

$$S = \{1, 2, 3, \dots, 50\}.$$

If the probability of retrieving any individual ball is equal, and there are 50 balls total, then the probabilities are given by

$$P = \{p_1, p_2, \dots, p_{50}\} \text{ with all } p_i = 1/50$$

Finally, we note that the sum of all probabilities is equal to 1:

$$P_{total} = \sum_{i=1}^{50} p_i = \left(\frac{1}{50}\right)_1 + \left(\frac{1}{50}\right)_2 + \dots + \left(\frac{1}{50}\right)_{50} = 1$$

The preceding discussion described the probability associated with a single experiment; however, there are times when one is more interested in the probability associated with a given outcome for a series of experiments, that is, the **event probability.** For example, imagine tossing a coin four times. What is the probability that at least two heads are observed after four tosses? All of the potential outcomes for this series of experiments are given in Figure 12.1. Of the 16 potential outcomes, 11 have at least two heads. Therefore, the probability of obtaining at least two heads after tossing a coin four times is 11/16, or the number of outcomes of interest divided by the total number of outcomes.

Let the sample space associated with a particular variable be S where $S = \{s_1, s_2, \ldots, s_N\}$, and let the probability that the outcome or event of interest, E, occurs be equal to P_E. Finally, there are j values in S corresponding to the outcome of interest. If the probability of observing an individual value is equivalent, then P_E is given by

$$P_E = \frac{1}{N} + \frac{1}{N} + \ldots + \frac{1}{N} = \frac{j}{N} \qquad (12.3)$$

This expression states that the probability of the event outcome of interest occurring is equal to the sum of the probabilities for each individual value in sample space corresponding to the desired outcome. Alternatively, if there are N values in sample space, and E of these values correspond to the event of interest, then

$$P_E = \frac{E}{N} \qquad (12.4)$$

EXAMPLE PROBLEM 12.2

What is the probability of selecting a heart from a standard deck of 52 cards?

Solution

In a standard deck of cards, each suit has 13 cards and there are four suits total (hearts, spades, clubs, and diamonds). The sample space consists of the 52 cards, of which 13 correspond to the event of interest (selecting a heart). Therefore,

$$P_E = \frac{E}{N} = \frac{13}{52} = \frac{1}{4}$$

12.2.1 The Fundamental Counting Principle

In the preceding examples, the number of ways a given event could be accomplished was determined by counting. This approach is reasonable when dealing with just a few experiments, but what if we toss a coin 50 times and are interested in the probability of the coin landing heads 20 times out of the 50 tosses? Clearly, writing down every possible outcome and counting would be a long and tedious process.

A more efficient method with which to determine the number of arrangements is illustrated by the following example. Imagine you are the instructor of a class consisting

FIGURE 12.1
Potential outcomes after tossing a coin four times. Red signifies heads and blue signifies tails.

of 30 students and need to assemble these students into a line. How many arrangements of students are possible? There are 30 possibilities for the selection of the first student in line, 29 possibilities for the second, and so on until the last student is placed in line. If the probabilities for picking any given student are equal, then the total number of ways to arrange the students (W) is

$$W = (30)(29)(28)...(2)(1) = 30! = 2.65 \times 10^{32}$$

The exclamation point symbol (!) in this expression is referred to as factorial, with $n!$ indicating the product of all values from 1 to n. The preceding result in its most general form is known as the **fundamental counting principle.**

FUNDAMENTAL COUNTING PRINCIPLE: For a series of manipulations $\{M_1, M_2 ..., M_j\}$ having n_i ways to accomplish each manipulation $\{n_1, n_2, ..., n_j\}$, the total number of ways to perform the entire series of manipulations ($Total_M$) is the product of the number of ways to perform each manipulation under the assumption that the ways are independent:

$$Total_M = (n_1)(n_2)...(n_j) \tag{12.5}$$

EXAMPLE PROBLEM 12.3

How many five-card arrangements are possible from a standard deck of 52 cards?

Solution

Employing the fundamental counting principle, each manipulation is a card that you have in your hand. Therefore, five manipulations are possible. There are 52 possible cards we could receive as our first card, or 52 ways to accomplish the first manipulation ($n_1 = 52$). Next, there are 51 possible cards we could receive as the second card in our hand, or 51 ways to accomplish the second manipulation ($n_2 = 51$). Following this logic:

$$Total_M = (n_1)(n_2)(n_3)(n_4)(n_5)$$
$$= (52)(51)(50)(49)(48) = 311,875,200$$

EXAMPLE PROBLEM 12.4

In Section 10.8 of Engel's *Quantum Chemistry and Spectroscopy*, the possible spin states for the first excited state of He with the electron configuration $1s^1 2s^1$ were discussed. Using the fundamental counting principle, how many possible spin states are expected for this excited state?

Solution

Because the electrons are in different orbitals, they do not have to be spin paired. Therefore, there are two choices for the spin state of the first electron, and two choices for the spin state of the second electron such that

$$Total_M = (n_1)(n_2) = (2)(2) = 4$$

12.2.2 Permutations

In the example of a classroom with 30 students, we found that there are 30! unique ways of arranging the students into a line, or 30! **permutations.** The total number of objects that are arranged is known as the order of the permutation, denoted as n, such that there are $n!$ total permutations of n objects. We have assumed that the entire set of n objects is used, but how many permutations are possible if only a subset of objects is employed in constructing the permutation? Let $P(n,j)$ represent the number of permutations possible using a subset of j objects from the total group of n; $P(n,j)$ is equal to

$$P(n, j) = n(n-1)(n-2)...(n-j+1) \tag{12.6}$$

Equation (12.6) can be rewritten by noting that:

$$n(n-1)...(n-j+1) = \frac{n(n-1)...(1)}{(n-j)(n-j-1)...(1)} = \frac{n!}{(n-j)!} \tag{12.7}$$

Therefore, $P(n,j)$ is given by the following relationship:

$$P(n, j) = \frac{n!}{(n-j)!} \tag{12.8}$$

EXAMPLE PROBLEM 12.5

The coach of a basketball team has 12 players on the roster, but can only play 5 players at one time. How many 5-player arrangements are possible using the 12-player roster?

Solution

For this problem the permutation order (n) is 12, and the subset (j) is 5 such that

$$P(n, j) = P(12,5) = \frac{12!}{(12-5)!} = 95,040$$

12.2.3 Configurations

The previous section discussed the number of ordered arrangements or permutations possible using a given number of objects. However, many times one is instead interested in the number of *unordered* arrangements that are possible. The basketball team example of Example Problem 12.5 is an excellent illustration of this point. In a basketball game, the coach is generally more concerned with which five players are in the game at any one time rather than the order in which they entered the game. An unordered arrangement of objects is referred to as a **configuration.** Similar to permutations, configurations can also be constructed using all objects in the set being manipulated (n), or just subset of objects with size j. We will refer to configurations using the nomenclature $C(n,j)$.

For a conceptual example of configurations, consider the four colored balls shown in Figure 12.2. How many three-ball configurations and associated permutations can be made using these balls? The possibilities are illustrated in the figure. Notice that a configuration is simply a collection of three colored balls, and that a permutation corresponds to an ordered arrangement of these same balls such that each configuration has six associated permutations. This observation can be used to develop the following relationship between configurations and permutations:

$$C(n, j) = \frac{P(n, j)}{j!}$$

where $C(n,j)$ is the number of configurations that are possible using a subset of j objects from a total number of n objects. Substituting the definition of $P(n,j)$ from Equation (12.8) into the preceding expression results in the following relationship:

$$C(n, j) = \frac{P(n, j)}{j!} = \frac{n!}{j!(n-j)!} \tag{12.9}$$

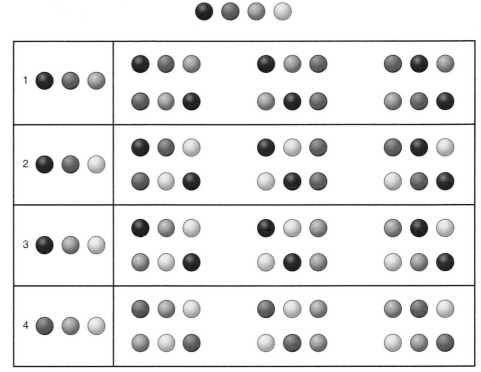

FIGURE 12.2
Illustration of configurations and permutations using four colored balls. The left-hand column presents the four possible three-color configurations, and the right-hand column presents the six permutations corresponding to each configuration.

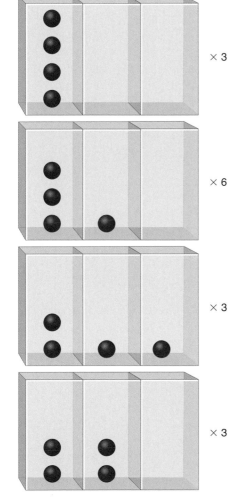

FIGURE 12.3
Configurations associated with arranging four identical particles (red balls) in three states (rectangles). The number of permutations associated with each configuration is given by the numerals to the right of each configuration.

EXAMPLE PROBLEM 12.6

If you are playing cards with a standard 52-card deck, how many possible 5-card combinations or "hands" are there?

Solution
Each 5-card hand is a configuration or subset taken from the 52-card deck. Therefore, $n = 52, j = 5$, and

$$C(n, j) = C(52,5) = \frac{52!}{5!(52-5)!} = \frac{52!}{(5!)(47!)} = 2,598,960$$

This result should be contrasted with the 311,875,200 permutations obtained in Example Problem 12.3. Generally, card players are interested in what five cards are in their possession, and not the order in which they arrived.

12.2.4 A Counting Example: Bosons and Fermions (Advanced)

The concepts of permutation and configuration will be exceedingly important in the next chapter, but we can introduce a simple counting problem here to demonstrate the connection between probability theory and chemistry. Consider the following question: how many ways can a set of n indistinguishable particles be arranged into x equally accessible states, with each state capable of holding any number of particles? This counting problem is encountered when describing particles known as **bosons,** in which multiple particles can occupy the same state. Photons and particles of integer spin such as ^{4}He are bosons, and these particles follow **Bose–Einstein statistics.** To describe the arrangement of such particles over a collection of states, we first consider a manageable example consisting of four particles and three states, as illustrated in Figure 12.3. In the figure, each particle is shown as a red circle, and the three states are shown by rectangles. The four possible configurations have a collection of associated permutations given by the numerals on the right side of each configuration. For example, the second configuration with three particles in one state, one in a second

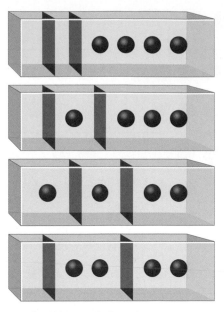

FIGURE 12.4
Second model for the four-particle/three-state arrangement depicted in Figure 12.3. In this model, states correspond to regions inside a single rectangle created by movable walls (black rectangles).

state, and no particles in the third state has six associated permutations. The total number of possible arrangements is equal to the total number of permutations, or 15 in this example.

Another way to envision the possible configurations in this example is shown in Figure 12.4. Here, the particles are confined to a single box with two movable walls allowing for three separate partitions. The figure demonstrates that there are again four possible configurations, identical to the result shown in Figure 12.3. The advantage of this depiction is that we can envision this problem as counting the number of permutations associated with a collection of six indistinguishable objects: four particles and two movable walls. For a given number of states, x, there will be $x - 1$ movable walls. Combining the walls with the n particles, the total number of permutations is

$$P_{BE} = \frac{(n + x - 1)!}{n!(x - 1)!} \tag{12.10}$$

The subscript BE refers to Bose–Einstein, reflecting the fact that we are discussing bosons. Employing Equation (12.10) to our illustrative example with $n = 4$ and $x = 3$, $P_{BE} = 15$ consistent with the result from simply counting the number of configurations. This result can be understood using the probability concepts introduced to this point. Specifically, the total number of objects being manipulated is $n + x - 1$ corresponding to $(n + x - 1)!$ permutations if the particles and walls were distinguishable. To account for the fact that the particles and walls are indistinguishable, we divided by $n!$ and $(x - 1)!$ resulting in the final expression for P_{BE}.

The second type of particles are **fermions,** which are species of noninteger spin such as electrons or ^{3}He. In a collection of fermions, each particle has a unique set of quantum numbers such that no two particles will occupy the same state. Fermions follow **Fermi–Dirac statistics.** Suppose we have n fermions to distribute over x states. How many possible arrangement permutations are there? There will be x choices for placement of the first particle, $x - 1$ choices for the second, and so forth so that the total number of state arrangements is given by

$$x(x - 1)(x - 2) \cdots (x - n + 1) = \frac{x!}{(x - n)!}$$

However, as in the boson example, the particles are indistinguishable, requiring that we divide the above expression by $n!$ resulting in

$$P_{FD} = \frac{x!}{n!(x - n)!} \tag{12.11}$$

You have already encountered this counting problem in Chapter 10 of Engel's *Quantum Chemistry and Spectroscopy* when discussing term symbols, as Example Problem 12.7 illustrates.

EXAMPLE PROBLEM 12.7

How many quantum states are possible for a carbon atom with the configuration $1s^2 2s^2 2p^2$?

Solution
Only the two electrons in the p orbitals contribute to defining the number of states such that $n = 2$. Next, there are three p orbitals and two possible spin orientations in each orbital such that $x = 6$. Therefore, the total number of quantum states, or arrangement permutations, is

$$P_{FD} = \frac{x!}{n!(x - n)!} = \frac{6!}{2!(4!)} = 15$$

12.2.5 Binomial Probabilities

The probability of an event occurring, E, out of a total number of possible outcomes, N, was denoted as P_E in Equation (12.4). We can also define the **complement** of P_E as the probability of an outcome other than that associated with the event of interest, and denote this quantity as P_{EC}. With this definition, the sum of P_E and P_{EC} are related as follows:

$$P_E + P_{EC} = 1 \tag{12.12}$$

Equation (12.12) states that the probability of the event of interest occurring combined with the event not occurring must be equal to unity, and provides a definition for experiments known as **Bernoulli trials.** In such a trial, the outcome of a given experiment will be a success (i.e., the outcome of interest) or a failure (i.e., not the outcome of interest). The tossing of a coin is a Bernoulli trial in which the outcome "heads" can be considered a success and "tails" a failure (or vice versa). A collection of Bernoulli trials is known as a **binomial experiment,** and these simple experiments provide a suitable framework in which to explore probability distributions. It is critical to note that the outcome of each trial is independent of the outcome of any other trial in the experiment. Consider a binomial experiment in which a coin is tossed four times and this question is asked: "What is the probability of observing heads (or a successful outcome) every time?" Because the probability of success for each trial is 1/2, the total probability is the product of the success probabilities for each trial:

$$P_E = \left(\frac{1}{2}\right)\left(\frac{1}{2}\right)\left(\frac{1}{2}\right)\left(\frac{1}{2}\right) = \frac{1}{16}$$

Note that this is the answer one would reach by considering all of the possible permutations encountered when flipping a coin four times as illustrated in Figure 12.1.

For a series of Bernoulli trials in which the probability of success for a single trial is P_E, the probability of obtaining j successes in a trial consisting of n trials is given by

$$P(j) = C(n, j)(P_E)^j(1 - P_E)^{n-j} \tag{12.13}$$

The $(P_E)^j$ term in the preceding expression is the product of probabilities for the j successful trials. Because n total trials were performed, $(n - j)$ trials must have failed, and the probability of these trials occurring is given by $(1 - P_E)^{n-j}$. But why does the total number of configurations appear in Equation (12.13)? The answer to this question lies in the difference between permutations and configurations. Again, consider a series of four coin tosses where the outcome of interest is the exact permutation {H, T, T, H}. If H is a successful trial occurring with probability P_E, then the probability of observing this permutation is

$$P = (P_E)(1 - P_E)(1 - P_E)(P_E) = (P_E)^2(1 - P_E)^2 = \left(\frac{1}{2}\right)^2\left(\frac{1}{2}\right)^2 = \frac{1}{16} \tag{12.14}$$

but this is also the probability for observing {H, H, H, H}. That is, the probability of observing a specific order of trial outcomes, or a single permutation, is equivalent. If the outcome of interest is two successful trials that can occur in any order, then the probabilities for all possible permutations corresponding to two successful trials must be added, and this is accomplished by the inclusion of $C(n,j)$ in Equation (12.13).

EXAMPLE PROBLEM 12.8

Imagine tossing a coin 50 times. What are the probabilities of having the coin land heads up 10 times (i.e., 10 successful experiments) and 25 times?

Solution

The trial of interest consists of 50 separate experiments; therefore, $n = 50$. We will first consider the case of 25 successful experiments where $j = 25$. The probability (P_{25}) is

$$P_{25} = C(n, j)(P_E)^j(1 - P_E)^{n-j}$$
$$= C(50, 25)(P_E)^{25}(1 - P_E)^{25}$$
$$= \left(\frac{50!}{(25!)(25!)}\right)\left(\frac{1}{2}\right)^{25}\left(\frac{1}{2}\right)^{25} = (1.26 \times 10^{14})(8.88 \times 10^{-16}) = 0.11$$

Performing the same steps for the case where $j = 10$, we find that

$$P_{10} = C(n, j)(P_E)^j(1 - P_E)^{n-j}$$
$$= C(50, 10)(P_E)^{10}(1 - P_E)^{40}$$
$$= \left(\frac{50!}{(10!)(40!)}\right)\left(\frac{1}{2}\right)^{10}\left(\frac{1}{2}\right)^{40} = (1.03 \times 10^{10})(8.88 \times 10^{-16}) = 9.1 \times 10^{-6}$$

12.3 Stirling's Approximation

When calculating $P(n,j)$ and $C(n,j)$, it is necessary to evaluate factorial quantities. In the examples encountered so far, n and j were sufficiently small such that these quantities could be evaluated on a calculator. However, this approach to evaluating factorial quantities is limited to relatively small numbers. For example, 100! is equal to 9.3×10^{157}, which is an extremely large number and beyond the range of many calculators. Furthermore, we are interested in extending the probability concepts we have developed up to chemical systems for which $n \sim 10^{23}$! The factorial of such a large number is simply beyond the computational ability of most calculators.

Fortunately, approximation methods are available that will allow us to calculate the factorial of large numbers. The most famous of these methods is known as **Stirling's approximation,** which provides a simple method by which to calculate the natural log of $N!$. A simplified version of this approximation is

$$\ln N! = N \ln N - N \tag{12.15}$$

Equation (12.15) is readily derived as follows:

$$\ln(N!) = \ln[(N)(N - 1)(N - 2)...(2)(1)] \tag{12.16}$$
$$= \ln(N) + \ln(N - 1) + \ln(N - 2) + ... + \ln(2) + \ln(1)$$
$$= \sum_{n=1}^{N} \ln(n) \approx \int_1^N \ln(n)dn$$
$$= N \ln N - N - (1 \ln 1 - 1) \approx N \ln N - N$$

In this derivation, the summation over n is replaced by integration, an acceptable approximation when N is large. The final result is obtained by evaluating the integral over the limits indicated. Note that the main assumption inherent in this approximation is that N is a large number. The central concern in applying Stirling's approximation is whether N is sufficiently large to justify its application. Example Problem 12.9 illustrates this point.

EXAMPLE PROBLEM 12.9

Evaluate $\ln(N!)$ for $N = 10$, 50, and 100 using a calculator, and compare the result to that obtained using Stirling's approximation.

Solution

For $N = 10$ using a calculator we can determine that $N! = 3.63 \times 10^6$ and $\ln(N!) = 15.1$. Using Stirling's approximation:

$$\ln(N!) = N \ln N - N = 10 \ln(10) - 10 = 13.0$$

This value represents a 13.9% error relative to the exact result, a substantial difference. The same procedure for $N = 50$ and 100 results in the following:

N	$\ln(N!)$ Calculated	$\ln(N!)$ Stirling	Error (%)
50	148.5	145.6	2.0
100	363.7	360.5	0.9

Example Problem 12.9 demonstrates that there are significant differences between the exact and approximate results even for $N = 100$. The example also demonstrates that the magnitude of this error decreases as N increases. For the chemical systems encountered in subsequent chapters, N will be $\sim 10^{23}$, many orders of magnitude larger than the values studied in this example. Therefore, for our purposes Stirling's approximation represents an elegant and sufficiently accurate method by which to evaluate the factorial of large quantities.

12.4 Probability Distribution Functions

Returning to the coin-tossing experiment, we now ask what the probability of obtaining a given outcome (i.e., number of heads) will be after tossing a coin 50 times. Using Equation (12.13), a table of probability as a function of the number of heads can be constructed with $n = 50$ (total number of tosses) and $j =$ the number of heads (i.e., the number of successful tosses):

Number of Heads	Probability	Number of Heads	Probability
0	8.88×10^{-16}	30	0.042
1	4.44×10^{-14}	35	2.00×10^{-3}
2	1.09×10^{-12}	40	9.12×10^{-6}
5	1.88×10^{-9}	45	1.88×10^{-9}
10	9.12×10^{-6}	48	1.09×10^{-12}
15	2.00×10^{-3}	49	4.44×10^{-14}
20	0.042	50	8.88×10^{-16}
25	0.112		

Rather than reading all of the probability values from a table, this same information can be presented graphically by plotting the probability as a function of outcome. This plot for the case where $P_E = 0.5$ (i.e., the coin landing heads or tails is equally likely) is shown as the red line in Figure 12.5. Notice that the maximum probability is predicted to be 25 heads such that this outcome is the most probable (as intuition suggests). A second feature of this distribution of probabilities is that the actual value of the probability corresponding to 25 successful trials is not unity, but 0.112. However, summing all of the outcome probabilities reveals that

$$P_0 + P_1 + \ldots + P_{50} = \sum_{j=0}^{50} P_j = 1 \tag{12.17}$$

Figure 12.5 presents the variation in probability of event outcome as a function of the number of heads observed after flipping a coin 50 times. This plot also demonstrates that

FIGURE 12.5
Plot of the probability of the number of heads being observed after flipping a coin 50 times. The red curve represents the distribution of probabilities for $P_E = 0.5$, the blue curve for $P_E = 0.3$, and the yellow curve for $P_E = 0.7$.

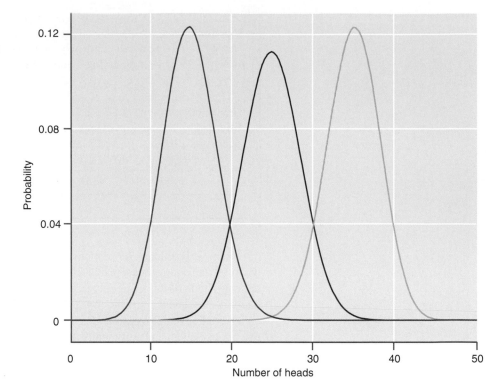

the probability formula for Bernoulli trials can be used to describe the variation or distribution of probability versus event outcome. Therefore, the probability expression we have employed can also be thought of as the distribution function for a binomial experiment. From Equation (12.13), the probability of observing j successful trials following n total trials is given by

$$P(j) = \frac{n!}{j!(n-j)!}(P_E)^j(1-P_E)^{n-j} \quad \text{for} \quad j = 0, 1, 2, ..., n \qquad (12.18)$$

In the coin-flipping experiment just discussed, the assumption was made that the coin can land either heads or tails with equal probability such that $P_E = 0.5$. What would happen if a coin were used where the probability of observing heads is 0.7? First, we would expect the probability distribution of trial outcomes to vary with respect to the case where $P_E = 0.5$. For the experiment in which $P_E = 0.7$ and $(1 - P_E) = 0.3$, the distribution function $P(j)$ can be calculated and compared to the previous case, as illustrated in Figure 12.5. Comparison of the probability distributions demonstrates that the most probable outcome has shifted to a greater number of heads, H = 35 in the case of $P_E = 0.7$. The outcome H = 15 is most probable when $P_E = 0.3$. A final aspect to note is that in addition to the maximum probability changing from H = 25 as P_E is changed from 0.5 to some other value, the probability for the most likely outcome also increases. For example, for $P_E = 0.5$, the maximum probability is 0.112 for H = 25. If $P_E = 0.7$, the maximum probability now becomes 0.128 for H = 35. That is, not only does the most probable outcome depend on P_E, but the probability of observing the most probable outcome also changes with P_E.

The coin-tossing experiment provides an excellent example of a probability distribution function and motivates a more formal definition of such functions. A **probability distribution function** represents the probability of a variable (X) having a given value, with the probability described by a function

$$P(X_i) \propto f_i \qquad (12.19)$$

In Equation (12.19), $P(X_i)$ is the probability that the variable X will have some value X_i in the sample space. This equation states that the set of probabilities $\{P(X_1), P(X_2), \ldots, P(X_M)\}$ will be proportional to the value of the distribution function evaluated for the corresponding value of the variable given by $\{f_1, f_2, \ldots, f_M\}$. In the coin toss example, the variable *number of heads* could assume a value between 0 and 50, and Equation (12.13) was employed to determine the corresponding probability, $P(X_i)$, for each possible value of the variable. For a binomial experiment, Equation (12.13) represents the function f in Equation (12.19). We can express Equation (12.19) as an equality by introducing a proportionality constant (C), and

$$P(X_i) = Cf_i \tag{12.20}$$

Imposing the requirement that the total probability be equal to unity results in

$$\sum_{i=1}^{M} P(X_i) = 1 \tag{12.21}$$

Equation (12.21) is readily evaluated to yield the proportionality constant:

$$1 = \sum_{i=1}^{M} Cf_i = Cf_1 + Cf_2 + \ldots + Cf_M$$

$$1 = C(f_1 + f_2 + \ldots + f_M) = C\sum_{i=1}^{M} f_i$$

$$C = \frac{1}{\displaystyle\sum_{i=1}^{M} f_i} \tag{12.22}$$

Substitution into our original expression for probability provides the final result of interest:

$$P(X_i) = \frac{f_i}{\displaystyle\sum_{i=1}^{M} f_i} \tag{12.23}$$

Equation (12.23) states that the probability of a variable having a given value from the sample space is given by the value of the probability function for this outcome divided by the sum of the probabilities for all possible outcomes. Equation (12.23) is a general expression for probability, and we will use this construct in the upcoming chapter when defining the Boltzmann distribution, one of the central results of statistical thermodynamics.

EXAMPLE PROBLEM 12.10

What is the normalized probability of receiving any one card from a standard deck of 52 cards?

Solution
The variable of interest is the card received, which can be any one of 52 cards (that is, the sample space for the variable consists of the 52 cards in the deck). The probability of receiving any card is equal so that $f_i = 1$ for $i = 1$ to 52. With these definitions, the probability becomes:

$$P(X_i) = \frac{f_i}{\displaystyle\sum_{i=1}^{52} f_i} = \frac{1}{\displaystyle\sum_{i=1}^{52} f_i} = \frac{1}{52}$$

12.5 Probability Distributions Involving Discrete and Continuous Variables

To this point we have assumed that the variable of interest is discrete. As such, probability distributions can be constructed by calculating the corresponding probability when X assumes each value from the sample set. However, what if the variable X is continuous? In this case the probability is with respect to the variable having a value within a portion of the domain of X, denoted as dX. Therefore, $P(X)$ is the **probability density** and $P(X)dX$ is the probability that the variable X has a value in the range of dX. By analogy with the development for discrete variables, the probability is given by

$$P(X)dX = Cf(X)dX \tag{12.24}$$

which states that the probability $P(X)dX$ is proportional to some function $f(X)dX$ as yet undefined. The normalization condition is applied to ensure that the total probability is unity over the domain of the variable ($X_1 \leq X \leq X_2$):

$$\int_{X_1}^{X_2} P(X)dX = C \int_{X_1}^{X_2} f(X)dX = 1 \tag{12.25}$$

The second equality in Equation (12.25) dictates that

$$C = \frac{1}{\int_{X_1}^{X_2} f(X)dX} \tag{12.26}$$

Equation (12.26) is identical to the discrete variable result with the exception that summation has been replaced by integration since the domain consists of a continuous possibility of variable values. With the preceding definition of the proportionality constant, the probability is defined as

$$P(X)dX = \frac{f(X)dX}{\int_{X_1}^{X_2} f(X)dX} \tag{12.27}$$

The similarity of this expression to the corresponding expression for discrete variables [Equation (12.23)] illustrates an important point. When working with probability distributions corresponding to continuous variables, integration over the domain of the variable is performed. In contrast, summation is performed when the variable is discrete. Although a change in the mathematical approach occurs between discrete and continuous variables, it is important to realize that the conceptual description of probability is unchanged. We have already encountered an example of a continuous probability distribution in Chapter 3 of Engel's *Quantum Chemistry and Spectroscopy*. Specifically, the normalization condition was applied to the spatial domain of the wave function through the following relationship [Equation (3.2) of Engel's *Quantum Chemistry and Spectroscopy*]:

$$\int_{-\infty}^{\infty} \psi^*(x,t)\psi(x,t)dx = 1$$

The product of the wave function with its complex conjugate represents the probability density for the spatial location of the particle, and it is equivalent to $P(X)$ in Equation (12.24). By integrating this product over all space, all possible locations of the particle are included such that the probability of the particle being somewhere must be one.

FIGURE 12.6
Probability distributions for the
translational kinetic energy of an ideal gas.

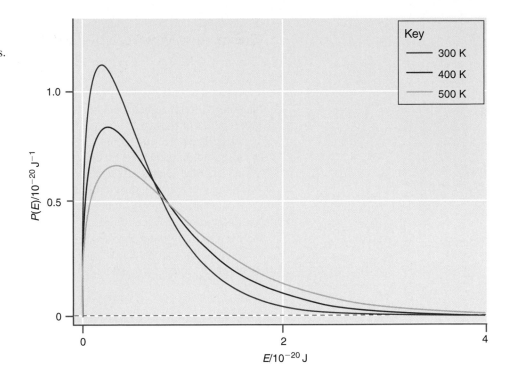

FIGURE 12.6
Probability distributions for the
translational kinetic energy of an ideal gas.

For another example of probability distribution functions involving continuous variables, consider the following probability distribution function associated with the translational kinetic energy of ideal gas particles (E), a topic that will be discussed in detail in Chapter 16:

$$P(E)dE = 2\pi \left(\frac{1}{\pi RT}\right)^{3/2} E^{1/2} e^{-E/RT} dE \tag{12.28}$$

In Equation (12.28), T is temperature and R is the ideal gas constant. The variable of interest, E, is continuous in the domain from 0 to infinity. Figure 12.6 presents an illustration of this distribution function for three temperatures: 300, 400, and 500 K. Notice that energy corresponding to maximum probability changes as a function of temperature, and that the probability of finding a particle at larger kinetic energies also increases with temperature. This example demonstrates that a substantial amount of information regarding the behavior of a chemical system can be succinctly presented using probability distributions.

12.5.1 Continuous Representation of Discrete Variables

It is theoretically possible to evaluate probabilities involving discrete variables by summation; however, we will encounter situations in later chapters where summation involves a prohibitively large number of terms. In such cases, treating a discrete variable as continuous simplifies matters greatly. However, one concern with this approach is the error introduced by including values for the discrete variable that are not contained in the sample set. Under what conditions are such errors acceptable? Consider the following unnormalized probability distribution function:

$$P(X) = e^{-0.3X} \tag{12.29}$$

We are interested in normalizing this distribution by determining the total probability P_{total}. First, X is treated as a discrete variable with the sample space consisting of integer values ranging from 0 to 100, and P_{total} is

$$P_{total} = \sum_{X=0}^{100} e^{-0.3X} = 3.86 \tag{12.30}$$

Next, X is treated as continuous with the domain of the sample set equal to $0 \le X \le 100$. The corresponding expression for the total probability is

$$P_{total} = \int_0^{100} e^{-0.3X} dX = 3.33 \tag{12.31}$$

The only difference between these last two equations is that the summation over discrete values of the variable X has been replaced by integration over the range of the variable. The preceding comparison demonstrates that the continuous approximation is close to the exact result given by summation. In general, if the differences between values the function can assume are small relative to the domain of interest, then treating a discrete variable as continuous is appropriate. This issue will become critical when the various energy levels of an atom or molecule are discussed. Specifically, the approximation of Equation (12.31) will be used to treat translational and rotational states from a continuous perspective where direct summation is impractical. In the remainder of this text, situations in which the continuous approximation is not valid will be carefully noted.

EXAMPLE PROBLEM 12.11

Determine the total probability for the following distribution function:

$$P(X) = e^{-0.05X}$$

First, treat the variable X as discrete with the sample space consisting of integer values ranging from 0 to 100, and then redo the calculation treating X as a continuous variable with a range of $0 \le X \le 100$.

Solution

This is the same analysis performed in the previous experiment, but now the differences between values the probability can assume are smaller in comparison to the previous distribution. Therefore, we would expect the discrete and continuous results to be closer in value. First, treating X as a discrete variable, we get

$$P_{total} = \sum_{X=0}^{100} e^{-0.05X} = 20.4$$

Next, treating X as continuous,

$$P_{total} = \int_0^{100} e^{-0.05X} dX = 19.9$$

Comparison of these results demonstrates that the difference between the discrete and continuous treatments is smaller for this more "fine-grain" distribution.

12.6 Characterizing Distribution Functions

Distribution functions provide all the probability information available for a given system; however, there will be times when a full description of the distribution function is not required. Imagine that you have performed a series of experiments in which only certain aspects of a molecular distribution are studied. The quantities of the distribution function of interest are components of the distribution addressed by experiment. Consider the translational kinetic energy distribution depicted in Figure 12.6. What if all we were interested in was the kinetic energy where the probability distribution function was at a maximum? In this case, full knowledge of the probability distribution is not needed. Instead, a single quantity for comparison to experiment, or a "benchmark" value, is all that is needed. In this section, benchmark values of substantial utility when characterizing distribution functions are presented.

12.6.1 Average Values

The **average value** of a quantity is perhaps the most useful—and obvious—way to characterize a distribution function. Consider a function, $g(X)$, whose value is dependent on variable X. The average value of this function is dependent on the probability distribution associated with the variable X. If the probability distribution describing the likelihood of X assuming some value is known, this distribution can be employed to determine the average value for the function as follows:

$$\langle g(X) \rangle = \sum_{i=1}^{M} g(X_i) P(X_i) = \frac{\displaystyle\sum_{i=1}^{M} g(X_i) f_i}{\displaystyle\sum_{i=1}^{M} f_i} \tag{12.32}$$

Equation (12.32) states that in order to determine the average value of the function $g(X)$, one simply sums the values of this function determined for each value of the sample set, $\{X_1, X_2, \ldots, X_M\}$, multiplied by the probability of the variable assuming this value. The summation in the denominator provides for normalization of the probability distribution. The angle brackets around $g(X)$ (i.e., $\langle \ldots \rangle$) denotes the average value of the function, a quantity that is referred to as an **expectation value.**

EXAMPLE PROBLEM 12.12

Imagine that you are at a carnival and observe a dart game in which you throw a single dart at the following target:

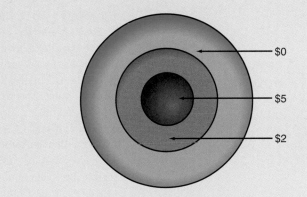

The dollar amounts in the figure indicate the amount of money you will win if you hit the corresponding part of the target with your dart. The geometry of the target is such that the radii of the circles increase linearly between successive areas. Finally, you are a sufficiently accomplished dart thrower that you will not miss the target altogether. If it costs you $1.50 each time you throw a dart, is this carnival game something you should play?

Solution
We will assume that the probability of hitting each section of the target is directly proportional to the area. If the radius of the inner circle is r, then the radii of the second and third circles are $2r$ and $3r$, respectively. The area of the inner curve, A_1, is πr^2 and the areas of the outer two circles are

$$A_2 = \pi(2r)^2 - \pi r^2 = 3A_1$$
$$A_3 = \pi(3r)^2 - \pi(2r)^2 = 5A_1$$

Finally, total area is given by the sum of the individual areas, or $9A_1$, such that the probability of hitting each part of the target is

$$f_1 = \frac{A_1}{A_1 + A_2 + A_3} = \frac{A_1}{9A_1} = \frac{1}{9}$$

$$f_2 = \frac{A_2}{A_1 + A_2 + A_3} = \frac{3A_1}{9A_1} = \frac{3}{9}$$

$$f_3 = \frac{A_3}{A_1 + A_2 + A_3} = \frac{5A_1}{9A_1} = \frac{5}{9}$$

The sum of probabilities equals 1 so that the probability distribution is normalized. The quantity of interest is the average payout, or $<n_\$>$, given by

$$\langle n_\$ \rangle = \frac{\sum_{i=1}^{3} n_\$ f_i}{\sum_{i=1}^{3} f_i} = \frac{(\$5)\dfrac{1}{9} + (\$2)\dfrac{3}{9} + (\$0)\dfrac{5}{9}}{1} = \$1.22$$

Comparison of the average payout to the amount you pay to throw a dart (\$1.50) suggests that this game is a money-losing activity (from the dart-thrower's perspective). Notice in this example that the "experimental" quantities you are comparing are the amount of money you have to pay and the amount of money you can expect to receive on average, which is the benchmark of the distribution of primary interest when you make a decision to play.

12.6.2 Distribution Moments

Some of the most widely used benchmark values for distributions involve functions of the form x^n where n is an integer. If $n = 1$, the corresponding function $<x>$ is referred to as the **first moment** of the distribution function, and it is equal to the average value of the distribution as discussed earlier. If $n = 2$, $<x^2>$ is referred to as the **second moment** of the distribution. Finally, the square root of $<x^2>$ is referred to as the root-mean-squared or rms value. The first and second moments, as well as the rms value, are extremely useful quantities for characterizing distribution functions. In the discussion of molecular motion in Chapters 16 and 17, we will find that a collection of molecules will have a distribution of translational velocities and speeds. Rather than discussing the entire velocity or speed distribution, the moments of the distribution are utilized to describe the system of interest. **Distribution moments** are readily calculated for both discrete and continuous distributions as Example Problem 12.13 illustrates.

EXAMPLE PROBLEM 12.13

Consider the following distribution function:

$$P(x) = Cx^2 e^{-ax^2}$$

Probability distributions of this form are encountered when describing speed distributions of ideal gas particles. In this probability distribution function, C is a normalization constant that ensures the total probability is equal to one, and a is a constant. The domain of interest is given by $\{0 \leq x \leq \infty\}$. Are the mean and rms values for this distribution the same?

Solution

First, the normalization condition is applied to the probability:

$$1 = \int_0^\infty Cx^2 e^{-ax^2}\,dx = C\int_0^\infty x^2 e^{-ax^2}\,dx$$

The integral is easily evaluated using the integral tables provided in Appendix B, Math Supplement:

$$\int_0^\infty x^2 e^{-ax^2}\,dx = \frac{1}{4a}\sqrt{\frac{\pi}{a}}$$

resulting in the following definition for the normalization constant:

$$C = \frac{1}{\dfrac{1}{4a}\sqrt{\dfrac{\pi}{a}}} = \frac{4a^{3/2}}{\sqrt{\pi}}$$

With this normalization constant, the mean and rms values of the distribution function can be determined. The mean is equal to $<x>$, which is determined as follows:

$$\langle x \rangle = \int_0^\infty x P(x)\,dx = \int_0^\infty x\left(\frac{4a^{3/2}}{\sqrt{\pi}}x^2 e^{-ax^2}\right)dx$$

$$= \frac{2}{\sqrt{\pi a}}$$

Proceeding in a similar fashion for the second moment ($<x^2>$),

$$\langle x^2 \rangle = \int_0^\infty x^2 P(x)\,dx = \int_0^\infty x^2\left(\frac{4a^{3/2}}{\sqrt{\pi}}x^2 e^{-ax^2}\right)dx$$

$$= \frac{3}{2a}$$

The resulting rms value is:

$$\text{rms} = \sqrt{\langle x^2 \rangle} = \sqrt{\frac{3}{2a}}$$

The mean and rms values for the distribution are not equivalent. The normalized distribution and the location of the mean and rms values for the case where $a = 0.3$ are shown here:

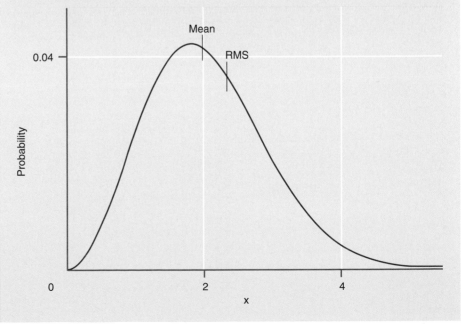

12.6.3 Variance

The **variance** (denoted as σ^2) provides a measure of the width of a distribution. The variance is defined as the average deviation squared from the mean of the distribution:

$$\sigma^2 = \langle (x - \langle x \rangle)^2 \rangle = \langle x^2 - 2x\langle x \rangle + \langle x \rangle^2 \rangle \tag{12.33}$$

It is important to remember that the value for the mean (or first moment) is a constant value and should not be confused with the variable. Equation (12.33) can be simplified by using the following two general properties for the averages involving two functions, $b(x)$ and $d(x)$:

$$\langle b(x) + d(x) \rangle = \langle b(x) \rangle + \langle d(x) \rangle \tag{12.34}$$

$$\langle cb(x) \rangle = c\langle b(x) \rangle \tag{12.35}$$

In the second property, c is simply a constant. Using these properties the expression for variance becomes:

$$\sigma^2 = \langle x^2 - 2x\langle x \rangle + \langle x \rangle^2 \rangle = \langle x^2 \rangle - \langle 2x\langle x \rangle \rangle + \langle x \rangle^2$$

$$= \langle x^2 \rangle - 2\langle x \rangle \langle x \rangle + \langle x \rangle^2$$

$$\boxed{\sigma^2 = \langle x^2 \rangle - \langle x \rangle^2} \tag{12.36}$$

In other words, the variance of a distribution is equal to the difference between the second moment and the square of the first moment.

To illustrate the use of variance as a benchmark, consider the following distribution function, which is referred to as a **Gaussian distribution:**

$$P(X)dX = \frac{1}{(2\pi\sigma^2)^{1/2}} e^{-(X-\delta)^2/2\sigma^2} \tag{12.37}$$

The Gaussian distribution is the "bell-shaped curve" of renown in the social sciences, employed widely through chemistry and physics, and well known to any college students concerned about their grades. It is the primary distribution function utilized to describe error in experimental measurements. The variable X is continuous in the domain from negative infinity and infinity, $\{-\infty \leq X \leq \infty\}$). The Gaussian probability distribution has a maximum at $X = \delta$. The width of the distribution is determined by the variance σ^2 with an increase in the variance corresponding to increased width of the distribution. The dependence of the Gaussian distribution on variance is illustrated in Figure 12.7 using Equation (12.37) with $\delta = 0$. As the variance of the distribution increases from $\sigma^2 = 0.4$ to 2.0, the width of the probability distribution increases.

EXAMPLE PROBLEM 12.14

Determine the variance for the distribution function $P(x) = Cx^2e^{-ax^2}$.

Solution
This is the distribution function employed in Example Problem 12.13, and we have for the first and second moments:

$$\langle x \rangle = \frac{2}{\sqrt{\pi a}}$$

$$\langle x^2 \rangle = \frac{3}{2a}$$

Using these values in the expression for variance, we find

$$\sigma^2 = \langle x^2 \rangle - \langle x \rangle^2 = \frac{3}{2a} - \left(\frac{2}{\sqrt{\pi a}} \right)^2$$

$$= \frac{3}{2a} - \frac{4}{\pi a} = \frac{0.23}{a}$$

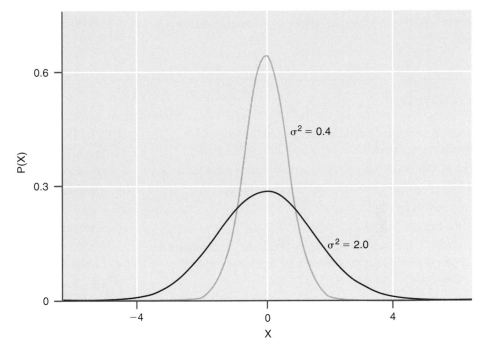

FIGURE 12.7
The influence of variance on Gaussian probability distribution functions. The figure presents the evolution in probability, $P(X)$, as a function of the variable X. The distributions for two values of the variance of the distribution, σ^2, are presented: 2.0 (red line) and 0.4 (yellow line). Notice that an increase in the variance corresponds to an increase in the width of the distribution.

For Further Reading

Bevington, P., and D. Robinson, *Reduction and Error Analysis for the Physical Sciences.* McGraw-Hill, New York, 1992.

Dill, K., and S. Bromberg, *Molecular Driving Forces.* Garland Science, New York, 2003.

McQuarrie, D., *Mathematical Methods for Scientists and Engineers.* University Science Books, Sausalito, CA, 2003.

Nash, L. K., *Elements of Statistical Thermodynamics.* Addison-Wesley, San Francisco, 1972.

Ross, S., *A First Course in Probability Theory*, 3rd ed. Macmillan, New York, 1988.

Taylor, J. R., *An Introduction to Error Analysis.* University Science Books, Mill Valley, CA, 1982.

Vocabulary

average values	distribution moments	probability
Bernoulli trial	event probability	probability density
binomial experiment	expectation value	probability distribution function
Bose–Einstein statistics	Fermi–Dirac statistics	probability model
bosons	fermions	sample space
complement	first moment	second moment
configuration	fundamental counting principle	Stirling's approximation
continuous variable	Gaussian distribution	variable
discrete variable	permutation	variance

Questions on Concepts

Q12.1 What is the difference between a configuration and a permutation?

Q12.2 What are the elements of a probability model, and how do they differ for continuous and discrete variables?

Q12.3 How does Figure 12.2 change if one is concerned with two versus three colored-ball configurations and permutations?

Q12.4 What must the outcome of a binomial experiment be if $P_E = 1$?

Q12.5 Why is normalization of a probability distribution important? What would one have to consider when working with a probability distribution that was not normalized?

Q12.6 What properties of atomic and molecular systems could you imagine describing using probability distributions?

Q12.7 When is the higher moment of a probability distribution more useful as a benchmark value as opposed to simply using the mean of the distribution?

Problems

Problem numbers in **red** indicate that the solution to the problem is given in the *Student's Solutions Manual*.

P12.1 Suppose that you draw a card from a standard deck of 52 cards. What is the probability of drawing:

a. an ace of any suit?

b. the ace of spades?

c. How would your answers to parts (a) and (b) change if you were allowed to draw three times, replacing the card drawn back into the deck after each draw?

P12.2 You are dealt a hand consisting of 5 cards from a standard deck of 52 cards. Determine the probability of obtaining the following hands:

a. a flush (five cards of the same suit)

b. a king, queen, jack, ten, and ace of the same suit (a "royal flush")

P12.3 A pair of standard dice are rolled. What is the probability of observing the following:

a. The sum of the dice is equal to 7.

b. The sum of the dice is equal to 9.

c. The sum of the dice is less than or equal to 7.

P12.4 Answer Problem P12.3 assuming that "shaved" dice are used so that the number 6 appears twice as often as any other number.

P12.5 Evaluate the following:

a. The number of permutations employing all objects in a six-object set

b. The number of permutations employing four objects from a six-object set

c. The number of permutations employing no objects from a six-object set

d. $P(50,10)$

P12.6 Determine the number of permutations of size 3 that can be made from the set $\{1, 2, 3, 4, 5, 6\}$. Write down all of the permutations.

P12.7 Determine the numerical values for the following:

a. The number of configurations employing all objects in a six-object set

b. The number of configurations employing four objects from a six-object set

c. The number of configurations employing no objects from a six-object set

d. $C(50,10)$

P12.8 Radio station call letters consist of four letters (for example, KUOW).

a. How many different station call letters are possible using the 26 letters in the English alphabet?

b. Stations west of the Mississippi River must use the letter K as the first call letter. Given this requirement, how many different station call letters are possible if repetition is allowed for any of the remaining letters?

c. How many different station call letters are possible if repetition is not allowed for any of the letters?

P12.9 Four bases (A, C, T, and G) appear in DNA. Assume that the appearance of each base in a DNA sequence is random.

a. What is the probability of observing the sequence AAGACATGCA?

b. What is the probability of finding the sequence GGGGGAAAAA?

c. How do your answers to parts (a) and (b) change if the probability of observing A is twice that of the probabilities used in parts (a) and (b) of this question when the preceding base is G?

P12.10 In the neck of the flask depicted in the following figure, five white balls rest on five black balls. Suppose the balls are tipped back into the flask, shaken, and the flask is reinverted. What is the probability that the order depicted in the figure will be seen?

a. How many 9-player batting orders are possible given that the order of batting is important?

b. How many 9-player batting orders are possible given that the all-star designated hitter must be in the order batting in the fourth spot?

c. How many 9-player fielding teams are possible, under the assumption that the location of the players on the field is not important?

P12.14 Imagine an experiment in which you flip a coin four times. Furthermore, the coin is balanced fairly such that the probability of landing heads or tails is equivalent. After tossing the coin 10 times, what is the probability of observing

a. no heads?   c. five heads?

b. two heads?   d. eight heads?

P12.15 Imagine performing the coin-flip experiment of Problem P12.14, but instead of using a fair coin, a weighted coin is employed for which the probability of landing heads is twofold greater than landing tails. After tossing the coin 10 times, what is the probability of observing

a. no heads?   c. five heads?

b. two heads?   d. eight heads?

P12.16 In Chapter 17 we will model particle diffusion as a random walk in one dimension. In such processes, the probability of moving an individual step in the $+x$ or $-x$ direction is equal to zero. Imagine starting at $x = 0$ and performing a random walk in which 20 steps are taken.

a. What is the farthest distance the particle can possibly move in the $+x$ direction? What is the probability of this occurring?

b. What is the probability the particle will not move at all?

c. What is the probability of the particle moving half the maximum distance in the x direction?

d. Plot the probability of the particle moving a given distance versus distance. What does the probability distribution look like? Is the probability normalized?

P12.17 Simplify the following expressions:

a. $\dfrac{n!}{(n-2)!}$ b. $\dfrac{n!}{\left(\dfrac{n}{2}!\right)^2}$

P12.18 You are at a carnival and are considering playing the dart game described in Example Problem 12.12; however, you are confident of your dart-throwing skills such that the probability of hitting the center area of the target is three times greater than the probability determined by area. Assuming the confidence in your skills is warranted, is it a good idea to play?

P12.19 Radioactive decay can be thought of as an exercise in probability theory. Imagine that you have a collection of

P12.11 The Washington State Lottery consists of drawing five balls numbered 1 to 43, and a single ball numbered 1 to 23 from a separate machine.

a. What is the probability of hitting the jackpot in which the values for all six balls are correctly predicted?

b. What is the probability of predicting just the first five balls correctly?

c. What is the probability of predicting the first five balls in the exact order they are picked?

P12.12 Fermions and bosons demonstrate different distribution statistics over a set of quantum states. However, in Chapter 13 we will encounter the Boltzmann distribution in which we essentially ignore the differentiation between fermions and bosons. This is appropriate only in the "dilute limit" where the number of available states far outnumbers the number of particles. To illustrate this convergence:

a. Determine the number of arrangement permutations possible for 3 bosons and 10 states, and repeat this calculation for fermions.

b. Repeat the calculations from part (a) for 3 particles, but now 100 states. What do you notice about the difference between the two results?

P12.13 Consider the 25 players on a professional baseball team. At any point, 9 players are on the field.

radioactive nuclei at some initial time (N_0) and are interested in how many nuclei will still remain at a later time (N). For first-order radioactive decay, $N/N_0 = e^{-kt}$. In this expression, k is known as the decay constant and t is time.

a. What is the variable of interest in describing the probability distribution?

b. At what time will the probability of nuclei undergoing radioactive decay be 0.50?

P12.20 In Chapter 13, we will encounter the energy distribution $P(\varepsilon) = Ae^{-\varepsilon/kT}$, where $P(\varepsilon)$ is the probability of a molecule occupying a given energy state, ε is the energy of the state, k is a constant equal to 1.38×10^{-23} J K^{-1}, and T is temperature. Imagine that there are three energy states at 0, 100, and 500 J mol^{-1}.

a. Determine the normalization constant for this distribution.

b. What is the probability of occupying the highest energy state at 298 K?

c. What is the average energy at 298 K?

d. Which state makes the largest contribution to the average energy?

P12.21 Assume that the probability of occupying a given energy state is given by the relationship provided in Problem P12.20.

a. Consider a collection of three total states with the first state located at $\varepsilon = 0$ and others at kT and $2kT$, respectively, relative to this first state. What is the normalization constant for the probability distribution?

b. How would your answer change if there are five states with $\varepsilon = kT$ in addition to the single states at $\varepsilon = 0$ and $\varepsilon = 2kT$?

c. Determine the probability of occupying the energy level $\varepsilon = kT$ for the cases in which one and five states exist at this energy.

P12.22 Consider the following probability distribution corresponding to a particle located between point $x = 0$ and $x = a$:

$$P(x)dx = C \sin^2\left[\frac{\pi x}{a}\right] dx$$

a. Determine the normalization constant, C.

b. Determine $<x>$.

c. Determine $<x^2>$.

d. Determine the variance.

P12.23 Consider the probability distribution for molecular velocities in one dimension (v_x) given by $P(v_x)dv_x = Ce^{-mv_x^2/2kT}dv_x$.

a. Determine the normalization constant, C.

b. Determine $\langle v_x \rangle$.

c. Determine $\langle v_x^2 \rangle$.

d. Determine the variance.

P12.24 In nonlinear optical switching devices based on dye-doped polymer systems, the spatial orientation of the dye molecules in the polymer is an important parameter. These devices are generally constructed by orienting dye molecules with a large dipole moment using an electric field. Imagine placing a vector along the molecular dipole moment such that the molecular orientation can be described by the orientation of this vector in space relative to the applied field (z direction) as illustrated here:

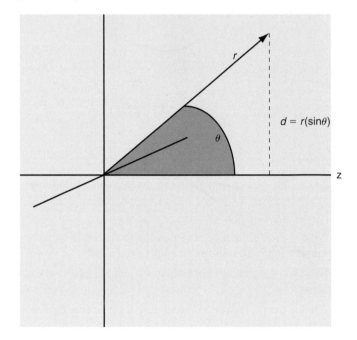

For random molecular orientation about the z axis, the probability distribution describing molecular orientation along the z axis is given by $P(\theta) = \sin\theta d\theta / \int_0^{\pi} \sin\theta d\theta$. Orientation is quantified using moments of $\cos\theta$.

a. Determine $\langle \cos\theta \rangle$ for this probability distribution.

b. Determine $\langle \cos^2\theta \rangle$ for this probability distribution.

CHAPTER 13

The Boltzmann Distribution

Employing statistical concepts, one can determine the most probable distribution of energy in a molecular system. This distribution, referred to as the Boltzmann distribution, represents the most probable configuration of energy for a molecular system at equilibrium and also gives rise to the thermodynamic properties of the system. In this chapter, the Boltzmann distribution is derived starting with the probability concepts introduced in the previous chapter. This distribution is then applied to some elementary examples to demonstrate how the distribution of energy in a molecular system depends on both the available energy and the energy-level spacings that characterize the system. The concepts outlined here provide the conceptual framework required to apply statistical mechanics to molecular systems. ∎

13.1 Microstates and Configurations

We begin by extending the concepts of probability theory introduced in the previous chapter to chemical systems. Although such an extension of probability theory may appear difficult, the sheer size of chemical systems makes the application of these statistical concepts straightforward. Recall the discussion of permutations and configurations illustrated by tossing a coin four times. The possible outcomes for this experiment are presented in Figure 13.1.

Figure 13.1 illustrates that there are five possible outcomes for the trial: from no heads to all heads. Which of these trial outcomes is most likely? Using probability theory, we found that the most probable outcome will be the one with the most possible ways to achieve that outcome. In the coin toss example, the configuration "2 Head" has the greatest number of ways to achieve this configuration; therefore, the "2 Head" configuration represents the most likely outcome. In the language of probability theory, the configuration with the largest number of corresponding permutations will be the most likely trial outcome. As discussed in the previous chapter, the probability, P_E, of this configuration representing the trial outcome is given by

$$P_E = \frac{E}{N} \tag{13.1}$$

where E is the number of permutations associated with the event of interest, and N is the total number of possible permutations. This relatively simple equation provides the key idea for this entire chapter: *the most likely configurational outcome for a trial is the configuration with the greatest number of associated permutations*. The application of this idea to macroscopic molecular systems is aided by the fact that for systems containing

FIGURE 13.1
Possible configurations and permutations for a Bernoulli trial consisting of flipping a coin four times. Blue indicates tails and red indicates heads.

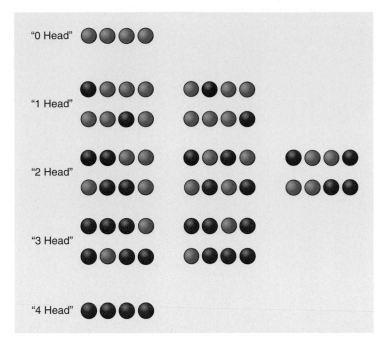

a large number of units, one configuration will have vastly more associated permutations than any other configuration. As such, this configuration will be the only one that is observed to an appreciable extent.

The application of probability theory to chemical systems is achieved by considering system energy. Specifically, we are interested in developing a formalism that is capable of identifying the most likely configuration or distribution of energy in a chemical system. To begin, consider a simple "molecular" system consisting of three quantum harmonic oscillators that share a total of three quanta of energy. The energy levels for the oscillators are given by:

$$E_n = h\nu\left(n + \frac{1}{2}\right) \qquad \text{for } n = 0, 1, 2, \ldots, \text{infinity} \qquad (13.2)$$

In this equation, h is Planck's constant, ν is the oscillator frequency, and n is the quantum number associated with a given energy level of the oscillator. This quantum number can assume integer values starting from 0 and increasing to infinity; therefore, the energy levels of the quantum harmonic oscillator consist of a ladder or "manifold" of equally spaced levels. The lowest level ($n = 0$) has an energy of $h\nu/2$ referred to as the zero point energy. A modified version of the harmonic oscillator is employed here in which the ground-state energy ($n = 0$) is zero so that the level energies are given by

$$E_n = h\nu n \qquad \text{for } n = 0, 1, 2, \ldots, \text{infinity} \qquad (13.3)$$

When considering atomic and molecular systems in the upcoming chapters, the relative difference in energy levels will prove to be the relevant quantity of interest, and a similar modification will be employed. The final important point to note is that the oscillators are distinguishable. That is, each oscillator in the three-oscillator collection can be readily identified. This assumption must be relaxed when describing molecular systems such as an ideal gas because atomic or molecular motion prohibits identification of each gaseous particle. The extension of this approach to indistinguishable particles is readily accomplished; therefore, starting with the distinguishable case will not prove problematic.

In this example system, the three oscillators are of equal frequency such that the energy levels for all three are identical. Three quanta of energy are placed into this system, and the question becomes "What is the most probable distribution of energy." From the preceding discussion, probability theory dictates that the most likely distribution of energy corresponds to the configuration with the largest number of associated permutations. How can the concepts of configuration and permutations be connected to

FIGURE 13.2

Configurations and associated permutations involving the distribution of three quanta of energy over three distinguishable oscillators.

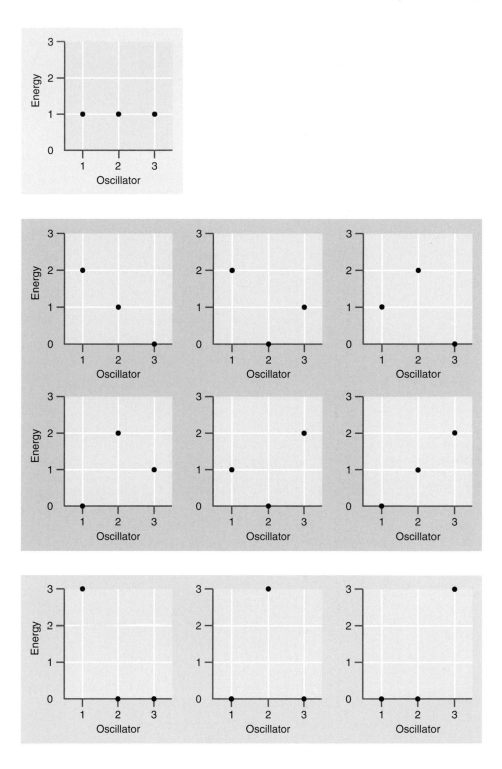

oscillators and energy? Two central definitions are introduced here that will prove useful in making this connection: a **configuration** is a general arrangement of total energy available to the system, and a **microstate** is a specific arrangement of energy that describes the energy contained by each individual oscillator. This definition of a configuration is equivalent to the definition of configuration from Chapter 12, and microstates are equivalent to permutations as previously defined. To determine the most likely configuration of energy in the example system, we will simply count all of the possible microstates and arrange them with respect to their corresponding configurations.

In the first configuration depicted in Figure 13.2, each oscillator has one quantum of energy such that all oscillators populate the $n = 1$ energy level. Only one permutation is associated with this configuration. In terms of the nomenclature just introduced, there is one microstate corresponding to this energy configuration. In the next configuration illustrated in the figure, one oscillator contains two quanta of energy, a second contains one quantum of energy, and a third contains no energy. Six potential arrangements correspond to this general distribution of energy; that is, six microstates correspond to this configuration. The last configuration depicted is one in which all three quanta of energy reside on a single oscillator. Because there are three choices for which oscillator will have all three quanta of energy, there are three corresponding microstates for this configuration. It is important to note that the total energy of all of the arrangements just mentioned is the same and that the only difference is the distribution of the energy over the oscillators.

Which configuration of energy would we expect to observe? Just like the coin tossing example, we expect to see the energy configuration that has the largest number of microstates. In this example, that configuration is the second one discussed, or the "2, 1, 0" configuration. If all microstates depicted have an equal probability of being observed, the probability of observing the 2,1,0 configuration is simply the number of microstates associated with this configuration divided by the total number of microstates available, or

$$P_E = \frac{E}{N} = \frac{6}{6+3+1} = \frac{6}{10} = 0.6$$

Note that although this example involves a "molecular" system, the concepts encountered can be generalized to probability theory. Whether tossing a coin or distributing energy among distinguishable oscillators, the ideas are the same.

13.1.1 Counting Microstates and Weight

The three-oscillator example provides an approach for finding the most probable configuration of energy for a chemical system: determine all of the possible configurations of energy and corresponding microstates and identify the configuration with the greatest number of microstates. Clearly, this would be an extremely laborious task for a chemical system of interesting size. Fortunately, there are ways to obtain a quantitative count of all of the microstates associated with a given configuration without actually "counting" them. First, recall from Chapter 12 that the total number of possible permutations given N objects is $N!$. For the most probable 2,1,0 configuration described earlier, there are three objects of interest (i.e., three oscillators) such that $N! = 3! = 6$. This is exactly the same number of microstates associated with this configuration. But what of the other configurations? Consider the 3, 0, 0 configuration in which one oscillator has all three quanta of energy. In assigning quanta of energy to this system to construct each microstate, there are three choices of where to place the three quanta of energy, and two remaining choices for zero quanta. However, this latter choice is redundant in that it does not matter which oscillator receives zero quanta first. The two conceptually different arrangements correspond to exactly the same microstate and, thus, are indistinguishable. To determine the number of microstates associated with such distributions of energy, the total number of possible permutations is divided by a factor that corrects for overcounting which for the '3, 0, 0' configuration is accomplished as follows:

$$\text{Number of microstates} = \frac{3!}{2!} = 3$$

This expression is simply the probability expression for the number of permutations available using a subgroup from an overall group of size N. Therefore, if no two oscillators reside in the same energy level, then the total number of microstates available is given by $N!$, where N is the number of oscillators. However, if two or more oscillators occupy the same energy state (including the zero-energy state), then we need to divide by a term that corrects for overcounting of identical permutations. The total number of microstates associated with a given configuration of energy is referred to as the **weight** of the configuration, W, which is given by

$$W = \frac{N!}{a_0!\,a_1!\,a_2!...a_n!} = \frac{N!}{\prod_n a_n!} \tag{13.4}$$

In Equation (13.4), W is the weight of the configuration of interest, N is the number of units over which energy is distributed, and the a_n terms represent the number of units occupying the nth energy level. The a_n quantities are referred to as **occupation numbers** because they describe how many units occupy a given energy level. For example, in the 3, 0, 0 configuration presented in Figure 13.2, $a_0 = 2$, $a_3 = 1$, and all other $a_n = 0$ (with $0! = 1$). The denominator in our expression for weight is evaluated by taking the product of the factorial of the occupation numbers ($a_n!$), with this product denoted by the Π symbol (which is analogous to the Σ symbol denoting summation). Equation (13.4) is not limited to our specific example, but is a general relationship that applies to any collection of distinguishable units for which only one state is available at a given energy level. The situation in which multiple states exist at a given energy level is discussed later in this chapter.

EXAMPLE PROBLEM 13.1

What is the weight associated with the configuration corresponding to observing 40 heads after flipping a coin 100 times? How does this weight compare to that of the most probable outcome?

Solution

Using the expression for weight of Equation (13.4), the coin flip can be envisioned as a system in which two states can be populated: heads or tails. In addition, the number of distinguishable units is 100, the number of coin tosses. Using these definitions,

$$W = \frac{N!}{a_H!\,a_T!} = \frac{100!}{40!\,60!} = 1.37 \times 10^{28}$$

The most probable outcome corresponds to the configuration where 50 heads are observed such that

$$W = \frac{N!}{a_H!\,a_T!} = \frac{100!}{50!\,50!} = 1.01 \times 10^{29}$$

13.1.2 The Dominant Configuration

The three-oscillator example from the preceding section illustrates a few key ideas that will guide us toward our development of the Boltzmann distribution. Specifically, weight is the total number of permutations corresponding to a given configuration. The probability of observing a configuration is given by the weight of that configuration divided by the total weight:

$$P_i = \frac{W_i}{W_1 + W_2 + ... + W_N} = \frac{W_i}{\displaystyle\sum_{j=1}^{N} W_j} \tag{13.5}$$

where P_i is the probability of observing configuration i, W_i is the weight associated with this configuration, and the denominator represents the sum of weights for all possible configurations. Equation (13.5) predicts that the configuration with the largest weight will have the greatest probability of being observed. The configuration with the largest weight is referred to as the **dominant configuration.**

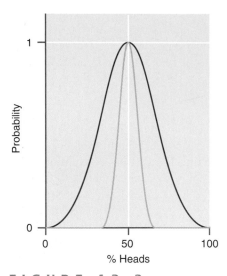

FIGURE 13.3

Comparison of relative probability (probability/maximum probability) for outcomes of a coin-flip trial in which the number of tosses is 10 (red line) and 100 (yellow line). The x axis is the percentage of tosses that are heads. Notice that all trials have a maximum probability at 50% heads; however, as the number of tosses increases, the probability distribution becomes more centered about this value as evidenced by the decrease in distribution width.

Given the definition of the dominant configuration, the question arises as to how dominant this configuration is relative to other configurations. A conceptual answer to this question is provided by an experiment in which a coin is tossed 10 times. How probable is the outcome of four heads relative to five heads? Although the outcome of $n_H = 5$ has the largest weight, the weight of $n_H = 4$ is of comparable value. Therefore, observing the $n_H = 4$ configuration would not be at all surprising. But what if the coin were flipped 100 times? How likely would the outcome of $n_H = 40$ be relative to $n_H = 50$? This question was answered in Example Problem 13.1, and the weight of $n_H = 50$ was significantly greater than $n_H = 40$. As the number of tosses increases, the probability of observing heads 50% of the time becomes greater relative to any other outcome. In other words, the configuration associated with observing 50% heads should become more dominant configuration as the number of coin tosses increases. An illustration of this expectation is presented in Figure 13.3 where the relative weights associated with observing heads a certain percentage of the time after tossing a coin 10 and 100 times are presented. The figure demonstrates that after 10 tosses, the relative probability of observing something other than five heads is still appreciable. However, as the number of coin tosses increases, the probability distribution narrows such that the final result of 50% heads becomes dominant. Taking this argument to sizes associated with molecular assemblies, imagine performing Avogadro's number of coin tosses! Given the trend illustrated in Figure 13.3, one would expect the probability to be sharply peaked at 50% heads. In other words, as the number of experiments in the trial increases, the 50% heads configuration will not only become the most probable, but the weight associated with this configuration will become so large that the probability of observing another outcome is minuscule. Indeed, the most probable configuration evolves into the dominant configuration as the size of the system increases.

EXAMPLE PROBLEM 13.2

Consider a collection of 10,000 particles with each particle capable of populating one of three energy levels having energies, 0, ε, and 2ε with a total available energy of 5000ε. Under the constraint that the total number of particles and total energy be constant, determine the dominant configuration.

Solution

With constant total energy, only one of the energy-level populations is independent. Treating the number of particles in the highest energy level (N_3) as the independent variable, the number of particles in the intermediate (N_2) and lowest (N_1) energy levels is given by

$$N_2 = 5000 - 2N_3$$
$$N_1 = 10,000 - N_2 - N_3$$

Because the number of particles in a given energy level must be greater than or equal to 0, the preceding equations demonstrate that N_3 can range from 0 to 2500. Given the size of the state populations, it is more convenient to calculate the natural log of the weight associated with each configuration of energy as a function of N_3 by using Equation (13.4) in the following form:

$$\ln W = \ln\left(\frac{N!}{N_1!N_2!N_3!}\right) = \ln N! - \ln N_1! - \ln N_2! - \ln N_3!$$

Each term here can be readily evaluated using Stirling's approximation. The results of this calculation are presented in Figure 13.4. This figure demonstrates that $\ln(W)$ has a maximum value at $N_3 \approx 1200$ (or 1162 to be precise). The dominance of this configuration is also illustrated in Figure 13.4, where the weight of the configurations corresponding to the allowed values of N_3 are compared with that of the dominant configuration. Even for this relatively simple system having only 10,000 particles, the weight is sharply peaked at a configuration corresponding to the dominant configuration.

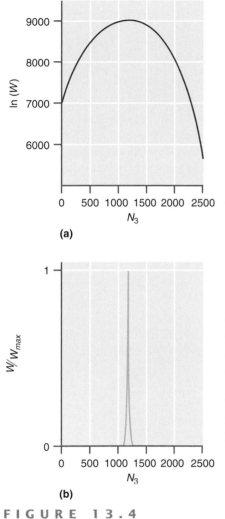

FIGURE 13.4
Illustration of the dominant configuration for a system consisting of 10,000 particles with each particle having three energy levels at energies of 0, ε, and 2ε as discussed in Example Problem 13.2. The number of particles populating the higher energy level is N_3, and the energy configurations are characterized by the population in this level. **(a)** Variation in the natural log of the weight, $\ln(W)$, for energy configurations as a function of N_3, demonstrating that $\ln(W)$ has a maximum at $N_3 \approx 1200$. **(b)** Variation in the weight associated with a given configuration to that of the dominant configuration. The weight is sharply peaked around $N_3 \approx 1200$ corresponding to the dominant configuration of energy.

13.2 Derivation of the Boltzmann Distribution

As the size of the system increases, a single configurational outcome will have such a large relative weight that only this configuration will be observed. In this limit, it becomes pointless to define all possible configurations, and only the outcome associated with the dominant configuration is of interest. A method is needed by which to identify this configuration directly. Inspection of Figures 13.3 and 13.4 reveals that the dominant configuration can be determined as follows. Because the dominant configuration has the largest associated weight, any change in outcome corresponding to a different configuration will be reflected by a reduction in weight. Therefore, the dominant configuration can be identified by locating the peak of the curve corresponding to weight as a function of **configurational index,** denoted by χ. Because W, and will be large for molecular systems, it is more convenient to work with $\ln W$, and the search criterion for the dominant configuration becomes

$$\frac{d \ln W}{d\chi} = 0 \tag{13.6}$$

This expression is a mathematical definition of the dominant configuration, and it states that if a configuration space is searched by monitoring the change in $\ln W$ as a function of configurational index, a maximum will be observed that corresponds to the dominant configuration. A graphical description of the search criterion is presented in Figure 13.5.

The distribution of energy associated with the dominant configuration is known as the **Boltzmann distribution.** We begin our derivation of this distribution by taking the natural log of the weight using the expression for weight developed previously and applying Stirling's approximation:

$$\ln W = \ln N! - \ln \prod_n a_n! \tag{13.7}$$

$$= N \ln N - \sum_n a_n \ln a_n$$

The criterion for the dominant configuration requires differentiation of $\ln W$ by some relevant configurational index, but what is this index? We are interested in the distribution of energy among a collection of molecules, or the number of molecules that resides in a given energy level. Because the number of molecules residing in a given energy level is the occupation number, a_n, the occupation number provides a relevant configurational index. Recognizing this, differentiation of $\ln W$ with respect to a_n yields the following:

$$\frac{d \ln W}{da_n} = \frac{dN}{da_n} \ln N + N \frac{d \ln N}{da_n} - \sum_n \frac{d(a_n \ln a_n)}{da_n} \tag{13.8}$$

To evaluate the partial differentials on the right-hand side of Equation (13.8), the following relationships are used. (A discussion of partial derivatives is contained in Appendix B, Math Supplement.) First, the sum over occupation numbers is equal to the number of objects denoted by N:

$$N = \sum_n a_n \tag{13.9}$$

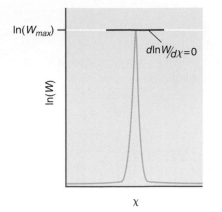

FIGURE 13.5
Mathematical definition of the dominant configuration. The change in the natural log of weight, $\ln(W)$, as a function of configuration index X is presented. If we determine the change in $\ln(W)$ as a function of configuration index, the change will equal zero at the maximum of the curve corresponding to the location of the dominant configuration.

This last equation makes intuitive sense. The objects in our collection must be in one of the available energy levels; therefore, summation over the occupation numbers is equivalent to counting all objects. With this definition, the partial derivative of N with respect to a_n is simply one. Second, the following mathematical relationship is employed:

$$\frac{d \ln x}{dx} = \frac{1}{x}$$

such that

$$\frac{d \ln W}{da_n} = \ln N + N\left(\frac{1}{N}\right) - \left(\ln a_n + 1\right) = -\ln\left(\frac{a_n}{N}\right) \tag{13.10}$$

Comparison of Equation (13.10) and the search criteria for the dominant configuration suggests that if $\ln(a_n/N) = 0$ our search is complete. However, this assumes that the occupation numbers are independent, yet the sum of the occupation numbers must equal N dictating that reduction of one occupation number must be balanced by an increase in another. In other words, one object in the collection is free to gain or lose energy, but a corresponding amount of energy must be lost or gained elsewhere in the system. We can ensure that both the number of objects and total system energy are constant by requiring that

$$\sum_n da_n = 0 \quad \text{and} \quad \sum_n \varepsilon_n da_n = 0 \tag{13.11}$$

The da_n terms denote change in occupation number a_n. In the expression on the right, ε_n is the energy associated with a specific energy level. Because the conservation of N and energy has not been required to this point, these conditions are now included in the derivation by introducing Lagrange multipliers, α and β, as weights for the corresponding constraints into the differential as follows:

$$d \ln W = 0 = \sum_n -\ln\left(\frac{a_n}{N}\right) da_n + \alpha \sum_n da_n - \beta \sum_n \varepsilon_n da_n \tag{13.12}$$

$$= \sum_n \left(-\ln\left(\frac{a_n}{N}\right) + \alpha - \beta \varepsilon_n\right) da_n$$

The Lagrange method of undetermined multipliers is described in the Math Supplement. Formally, this technique allows for maximization of a function that is dependent on many variables that are constrained among themselves. The key step in this approach is to determine the identity of α and β by noting that in Equation (13.12) the equality is only satisfied when the collection of terms in parentheses is equal to zero:

$$0 = -\ln\left(\frac{a_n}{N}\right) + \alpha - \beta \varepsilon_n \tag{13.13}$$

$$\ln\left(\frac{a_n}{N}\right) = \alpha - \beta \varepsilon_n$$

$$\frac{a_n}{N} = e^\alpha e^{-\beta \varepsilon_n}$$

$$a_n = N e^\alpha e^{-\beta \varepsilon_n}$$

At this juncture the Lagrange multipliers can be defined. First, α is defined by summing both sides of the preceding equality over all energy levels. Recognizing that $\Sigma a_n = N$,

$$N = \sum_n a_n = N e^\alpha \sum_n e^{-\beta \varepsilon_n}$$

$$1 = e^\alpha \sum_n e^{-\beta \varepsilon_n}$$

$$e^\alpha = \frac{1}{\displaystyle\sum_n e^{-\beta \varepsilon_n}} \tag{13.14}$$

This last equality is a central result in statistical mechanics. The denominator in Equation (13.14) is referred to as the **partition function,** q, and is defined as follows:

$$q = \sum_n e^{-\beta\varepsilon_n} \tag{13.15}$$

The partition function represents the sum over all terms that describes the probability associated with the variable of interest, in this case ε_n, or the energy of level n. Using the partition function with Equation (13.13), the probability of occupying a given energy level, p_n, becomes

$$p_n = \frac{a_n}{N} = e^{\alpha}e^{-\beta\varepsilon_n} = \frac{e^{-\beta\varepsilon_n}}{q} \tag{13.16}$$

Equation (13.16) is the final result of interest. It quantitatively describes the probability of occupying a given energy for the dominant configuration of energy. This well-known and important result is referred to as the Boltzmann distribution. We can compare Equation (13.16) to the expression from Chapter 12 for probability involving discrete variables, where probability was defined as follows:

$$P(X_i) = \frac{f_i}{\sum_{i=1}^{M} f_i} \tag{13.17}$$

This comparison demonstrates that the Boltzmann distribution is nothing more than a statement of probability, with the partition function serving to normalize the probability distribution.

Quantitative application of Equation (13.17) to a molecular system is still problematic since we have yet to evaluate β, the second of our Lagrange multipliers. However, a conceptual discussion can provide some insight into how this distribution describes molecular systems. Imagine a collection of harmonic oscillators as in the previous example, but instead of writing down all microstates and identifying the dominant configuration, the dominant configuration is instead given by the Boltzmann distribution law of Equation (13.16). This law establishes that the probability of observing an oscillator in a given energy level is dependent on level energy (ε_n) as $\exp(-\beta\varepsilon_n)$. Because the exponent must be unitless, β must have units of inverse energy.[1] Recall that the energy levels of the harmonic oscillator (neglecting zero point energy) are $\varepsilon_n = nh\nu$ for $n = 0, 1, 2, \ldots, \infty$. Therefore, our conceptual example will employ oscillators where $h\nu = \beta^{-1}$ as illustrated in Figure 13.6.

FIGURE 13.6
Two example oscillators. In case 1, the energy spacing is β^{-1}, and the energy spacing is $\beta^{-1}/2$ in case 2.

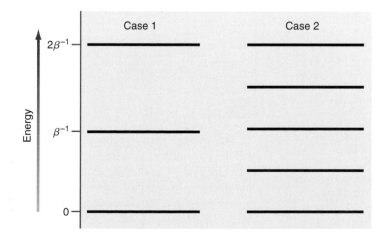

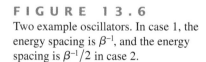

---

[1] This simple unit analysis points to an important result: that β is related to energy. As will be shown, $1/\beta$ provides a measure of the energy available to the system.

Using this value for the energy spacings, the exponential terms in the Boltzmann distribution are easily evaluated:

$$e^{-\beta\varepsilon_n} = e^{-\beta(n/\beta)} = e^{-n} \tag{13.18}$$

The partition function is evaluated by performing the summation over the energy levels:

$$q = \sum_{n=0}^{\infty} e^{-n} = 1 + e^{-1} + e^{-2} + \ldots \tag{13.19}$$

$$= \frac{1}{1 - e^{-1}} = 1.58$$

In the last step of this example, we have used the following series expression (where $x < 1$):

$$\frac{1}{1-x} = 1 + x + x^2 + \ldots$$

With the partition function, the probability of an oscillator occupying the first three levels ($n = 0$, 1, and 2) is

$$p_0 = \frac{e^{-\beta\varepsilon_0}}{q} = \frac{e^{-0}}{1.58} = 0.633 \tag{13.20}$$

$$p_1 = \frac{e^{-\beta\varepsilon_1}}{q} = \frac{e^{-1}}{1.58} = 0.233$$

$$p_2 = \frac{e^{-\beta\varepsilon_2}}{q} = \frac{e^{-2}}{1.58} = 0.086$$

EXAMPLE PROBLEM 13.3

For the example just discussed, what is the probability of finding an oscillator in energy levels $n \geq 3$?

Solution

The Boltzmann distribution is a normalized probability distribution. As such, the sum of all probabilities equals unity:

$$p_{total} = 1 = \sum_{n=0}^{\infty} p_n$$

$$1 = p_0 + p_1 + p_2 + \sum_{n=3}^{\infty} p_n$$

$$1 - (p_0 + p_1 + p_2) = 0.048 = \sum_{n=3}^{\infty} p_n$$

In other words, only 4.8% of the oscillators in our collection will be found in levels $n \geq 3$.

13.1 Behavior of the Partition Function

We continue with our conceptual example by asking the following question: "How will the probability of occupying a given level vary with a change in energy separation between levels?" In the first example, the energy spacings were equal to β^{-1}. A reduction in energy-level spacings to half this value requires that $h\nu = \beta^{-1}/2$. It is important to note that β has not changed relative to the previous example; only the separation in energy levels has changed. With this new energy separation, the exponential terms in the Boltzmann distribution become

$$e^{-\beta\varepsilon_n} = e^{-\beta(n/2\beta)} = e^{-n/2} \tag{13.21}$$

Substituting this equation into the expression for the partition function,

$$q = \sum_{n=0}^{\infty} e^{-n/2} = 1 + e^{-1/2} + e^{-1} + \ldots + e^{-\infty} \tag{13.22}$$

$$= \frac{1}{1 - e^{-1/2}} = 2.54$$

Using this value for the partition function, the probability of occupying the first three levels ($n = 0$, 1, and 2) corresponding to this new spacing is

$$p_0 = \frac{e^{-\beta\varepsilon_0}}{q} = \frac{e^{-0}}{2.54} = 0.394 \tag{13.23}$$

$$p_1 = \frac{e^{-\beta\varepsilon_1}}{q} = \frac{e^{-1/2}}{2.54} = 0.239$$

$$p_2 = \frac{e^{-\beta\varepsilon_2}}{q} = \frac{e^{-1}}{2.54} = 0.145$$

Comparison with the previous system probabilities in Equation (13.20) illustrates some interesting results. First, with a decrease in energy-level spacings, the probability of occupying the lowest energy level ($n = 0$) decreases, whereas the probability of occupying the other energy levels increases. Reflecting this change in probabilities, the value of the partition function has also increased. Since the partition function represents the sum of the probability terms over all energy levels, an increase in the magnitude of the partition function reflects an increase in the probability of occupying higher energy levels. That is, the partition function provides a measure of the number of energy levels that are occupied for a given value of β.

EXAMPLE PROBLEM 13.4

For the preceding example with decreased energy-level spacings presented, what is the probability of finding an oscillator in energy states $n \geq 3$?

Solution

The calculation from the previous example is used to find that

$$\sum_{n=3}^{\infty} p_n = 1 - (p_0 + p_1 + p_2) = 0.222$$

Consistent with the discussion, the probability of occupying higher energy levels has increased substantially with a reduction in level spacings.

13.2.1 Degeneracy

To this point, the assumption has been that only one state is present at a given energy level; however, when discussing atomic and molecular systems, in some situations more than a single state will be present. Multiple states at a given energy level is referred to as **degeneracy,** and degeneracy is incorporated into the expression for the partition function as follows:

$$q = \sum_n g_n e^{-\beta\varepsilon_n} \tag{13.24}$$

Equation (13.24) is identical to the previous definition of q from Equation (13.15) with the exception that the term g_n has been included. This term represents the number of

FIGURE 13.7
Illustration of degeneracy. In system 1, one state is present at energies 0 and β^{-1}. In system 2, the energy spacing is the same, but at energy β^{-1} two states are present such that the degeneracy at this energy is two.

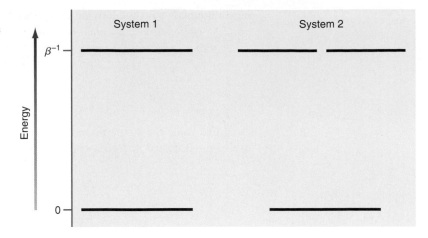

states present at a given energy level, or the degeneracy of the level. The corresponding expression for the probability of occupying energy level ε_i is

$$p_i = \frac{g_i e^{-\beta \varepsilon_i}}{q} \tag{13.25}$$

In Equation (13.25), g_i is the degeneracy of the level with energy-level ε_i, and q is as defined in Equation (13.24).

How does degeneracy influence probability? Consider Figure 13.7 in which a system with single states at energy 0 and β^{-1} is shown with a similar system in which two states are present at energy β^{-1}. The partition function for the first system is

$$q_{system1} = \sum_n g_n e^{-\beta \varepsilon_n} = 1 + e^{-1} = 1.37 \tag{13.26}$$

For the second system, q is

$$q_{system2} = \sum_n g_n e^{-\beta \varepsilon_n} = 1 + 2e^{-1} = 1.74 \tag{13.27}$$

The corresponding probability of occupying a state at energy β^{-1} for the two systems is given by

$$p_{system1} = \frac{g_i e^{-\beta \varepsilon_i}}{q} = \frac{e^{-1}}{1.34} = 0.27 \tag{13.28}$$

$$p_{system2} = \frac{g_i e^{-\beta \varepsilon_i}}{q} = \frac{2e^{-1}}{1.74} = 0.42$$

Notice that the probability of occupying a state at energy β^{-1} is greater for system 2, the system with degeneracy. This increase reflects the fact that two states are now available for population at this energy. However, this increase is not simply twice that of the non-degenerate case (system 1) since the value of the partition function also changes due to degeneracy.

13.3 Dominance of the Boltzmann Distribution

The search criterion for the Boltzmann distribution as illustrated in Figure 13.5 dictates that there will be an energy distribution for which the number of corresponding microstates is greatest. However, the question remains as to just how dominant the Boltzmann distribution is. Another way to approach this question is to ask "What exactly is the width of the curve presented in Figure 13.5?" Is there sufficient breadth in this distribution that a slight change in configuration results in another energy distribution, different from that of the dominant configuration, but with a reasonable probability of being observed?

Consider an isolated macroscopic assembly of distinguishable units.[2] The dominant configuration will be that having the largest number of microstates corresponding to weight $= W_{max}$. In addition, consider a slightly different configuration having weight $= W$, and $W < W_{max}$. Let α_n be the fractional change in the number of units present in the nth energy state:

$$\alpha_n = \frac{a'_n - a_n}{a_n} = \frac{\delta_n}{a_n} \tag{13.29}$$

In this expression, δ_n is simply the change in occupation number for level a_n, to some new configuration (denoted as a'_n) relative to the occupation numbers corresponding to the dominant configuration. The system is isolated dictating that both the number of particles and the total amount of energy are conserved, resulting in $\Delta N = 0$ and $\Delta E = 0$. With these definitions, the ratio of W_{max}/W is given by the following equation:

$$\frac{W_{max}}{W} = \frac{\dfrac{N!}{\prod_n a_n!}}{\dfrac{N!}{\prod_n (a_n + \delta_n)!}} = \prod_n \left(a_n + \frac{\delta_n}{2}\right)^{\delta_n} \tag{13.30}$$

The last equality in Equation (13.30) is satisfied when $|\delta_n| \ll a_n$. The ratio of weights provides a measure of the number of microstates associated with the modified configuration versus the Boltzmann distribution. Because W will be quite large for an assembly of molecules, the natural log of this ratio is used:

$$\ln\left(\frac{W_{max}}{W}\right) = \sum_n \ln\left(a_n + \frac{\delta_n}{2}\right)^{\delta_n} \tag{13.31}$$

$$= \sum_n \delta_n \ln\left(a_n + \frac{\delta_n}{2}\right)$$

$$= \sum_n \delta_n \ln a_n \left(1 + \frac{\delta_n}{2a_n}\right)$$

$$= \sum_n \delta_n \ln a_n + \sum_n \delta_n \ln\left(1 + \frac{\delta_n}{2a_n}\right)$$

The last equality in Equation (13.31) can be simplified by noting that $\ln(1 \pm z) = \pm z$ for small z. Because δ_n represents a fractional change in occupation number corresponding to the dominant configuration, $\delta_n/2a_n \ll 1$. Therefore,

$$\ln\left(\frac{W_{max}}{W}\right) = \sum_n \delta_n \ln a_n + \sum_n \frac{\delta_n^2}{2a_n} \tag{13.32}$$

Recall the following definition, which related the fractional change in occupation number to the occupation numbers themselves:

$$\alpha_n = \frac{\delta_n}{a_n} \tag{13.33}$$

---

[2] A full presentation of the following derivation appears in L. K. Nash, "On the Boltzmann Distribution Law," *J. Chemical Education* 59 (1982), 824.

Substitution of the preceding relationship for δ_n results in

$$\ln\left(\frac{W_{max}}{W}\right) = \sum_n a_n\alpha_n \ln a_n + \sum_n \frac{a_n}{2}(\alpha_n^2) \tag{13.34}$$

$$= \sum_n a_n\alpha_n(\ln a_0 - \beta\varepsilon_n) + \sum_n \frac{a_n}{2}(\alpha_n^2)$$

$$= \sum_n a_n\alpha_n(\ln a_0) - \sum_n \beta\varepsilon_n a_n\alpha_n + \sum_n \frac{a_n}{2}(\alpha_n^2)$$

$$= \ln a_0 \sum_n a_n\alpha_n - \beta\sum_n \varepsilon_n a_n\alpha_n + \sum_n \frac{a_n}{2}(\alpha_n^2)$$

$$= \sum_n \frac{a_n}{2}(\alpha_n^2)$$

The last step in this derivation was performed by realizing that $\Delta N = 0$ (and the first summation is equal to zero) and $\Delta E = 0$ (so that the second summation is also equal to zero). The root-mean-squared deviation in occupation number relative to those of the dominant configuration is defined as follows:

$$\alpha_{rms} = \left[\frac{\sum_n a_n\alpha_n^2}{N}\right]^{1/2} \tag{13.35}$$

Given this equation, the ratio of weights corresponding to the Boltzmann distribution, W_{max}, and the slightly modified distribution, W, reduces to

$$\frac{W_{max}}{W} = e^{N\alpha_{rms}^2/2} \tag{13.36}$$

To apply this relationship to a chemical system, imagine that the system of interest is a mole of molecules such that $N = 6.022 \times 10^{23}$. In addition, the fractional change in occupation number will be exceedingly small corresponding to $\alpha_{rms} = 10^{-10}$ (that is, one part in 10^{10}). Using these values,

$$\frac{W_{max}}{W} = e^{\frac{(6.022\times10^{23})(10^{-20})}{2}} \approx e^{3000}$$

Evaluation of the resulting exponential term will result in an overflow error on your calculator (unless you have an exceptionally good calculator). The ratio of weights is an extremely large number, and it demonstrates that a minute change in configuration will result in a significant reduction in weight. Clearly, the width of the curve illustrated in Figure 13.5 is extremely small in a system where N is on the order of Avogadro's number, and the most probable distribution is virtually the only distribution that will be observed for a macroscopic assembly of units. Of the total number of microstates available to a large assembly of units, the vast majority of these microstates correspond to the dominant configuration and a subset of configurations that differ from the dominant configuration by exceedingly small amounts such that the macroscopic properties of the assembly will be identical to those of the dominant configuration. In short, the macroscopic properties of the assembly are defined by the dominant configuration.

13.4 Physical Meaning of the Boltzmann Distribution Law

How does one know other configurations exist if the dominant configuration is all one expects to see for a system at equilibrium? Furthermore, are the other nondominant configurations of the system of no importance? Modern experiments are capable of displacing systems from equilibrium and monitoring the system as it relaxes back toward equilibrium. Therefore, the capability exists to experimentally prepare a nondominant configuration so that these configurations can be studied. An illuminating, conceptual answer to this question is provided by the following logical arguments. First, consider the central postulate of statistical mechanics:

> Every possible microstate of an isolated assembly of units occurs with equal probability.

How does one know that this postulate is true? For example, imagine a collection of 100 oscillators having 100 quanta of energy. The total number of microstates available to this system is on the order of 10^{200}, which is an extremely large number. Now, imagine performing an experiment in which the energy content of each oscillator is measured such that the corresponding microstate can be established. Also assume that a measurement can be performed every 10^{-9} s (1 nanosecond) such that microstates can be measured at a rate of 10^9 microstates per second. Even with such a rapid determination of microstates, it would take us 10^{191} s to count every possible microstate, a period of time that is much larger than the age of the universe! In other words, the central postulate cannot be verified experimentally. However, we will operate under the assumption that the central postulate is true because statistical mechanical descriptions of chemical systems have provided successful and accurate descriptions of macroscopic systems.

Even if the validity of the central postulate is assumed, the question of its meaning remains. To gain insight into this question, consider a large or macroscopic collection of distinguishable and identical oscillators. Furthermore, the collection is isolated resulting in both the total energy and the number of oscillators being constant. Finally, the oscillators are free to exchange energy such that any configuration of energy (and, therefore, any microstate) can be achieved. The system is set free to evolve, and the following features are observed:

1. All microstates are equally probable; however, one has the greatest probability of observing a microstate associated with the dominant configuration.

2. As demonstrated in the previous section, configurations having a significant number of microstates will be only infinitesimally different from the dominant configuration. The macroscopic properties of the system will be identical to that of the dominant configuration. Therefore, with overwhelming probability, one will observe a macroscopic state of the system characterized by the dominant configuration.

3. Continued monitoring of the system will result in the observation of macroscopic properties of the systems that appear unchanging, although energy is still being exchanged between the oscillators in our assembly. This macroscopic state of the system is called the **equilibrium state.**

4. Given items 1 through 3, the equilibrium state of the system is characterized by the dominant configuration.

This logical progression brings us to an important conclusion: *the Boltzmann distribution law describes the energy distribution associated with a chemical system at equilibrium.* In terms of probability, the fact that all microstates have equal probability of being observed does not translate into an equal probability of observing all configurations. As illustrated in Section 13.3, the vast majority of microstates correspond to the Boltzmann distribution, thereby dictating that the most probable configuration that will be observed is the one characterized by the Boltzmann distribution.

13.5 The Definition of β

Use of the Boltzmann distribution requires an operative definition for β, preferably one in which this quantity is defined in terms of measurable system variables. Such a definition can be derived by considering the variation in weight, W, as a function of total energy contained by an assembly of units, E. To begin, imagine an assembly of 10 oscillators having only three quanta of total energy. In this situation, the majority of the oscillators occupy the lowest energy states, and the weight corresponding to the dominant configuration should be small. However, as energy is deposited into the system, the oscillators will occupy higher energy states and the denominator in Equation (13.4) will be reduced, resulting in an increase in W. Therefore, one would expect E and W to be correlated.

The relationship between E and W can be determined by taking the natural log of Equation (13.4):

$$\ln W = \ln N! - \ln \prod_n a_n!$$ (13.37)

$$= \ln N! - \sum_n \ln a_n!$$

Interest revolves around the change in W with respect to E, a relationship that requires the total differential of W:

$$d \ln W = -\sum_n d \ln a_n!$$ (13.38)

$$= -\sum_n \ln a_n da_n$$

The result provided by Equation (13.38) was derived using Stirling's approximation to evaluate $\ln(a_n!)$ and recognizing that $\sum_n da_n = 0$. Simplification of Equation (13.38) is accomplished using the Boltzmann relationship to define the ratio between the occupation number for an arbitrary energy level, ε_n, versus the lowest or ground energy level ($\varepsilon_0 = 0$):

$$\frac{a_n}{a_0} = \frac{\dfrac{Ne^{-\beta\varepsilon_n}}{q}}{\dfrac{Ne^{-\beta\varepsilon_o}}{q}} = e^{-\beta\varepsilon_n}$$ (13.39)

$$\ln a_n = \ln a_0 - \beta\varepsilon_n$$ (13.40)

In the preceding steps, the partition function, q, and N are the same and simply cancel. Taking this expression for $\ln(a_n)$ and substituting into Equation (13.38) yields

$$d \ln W = -\sum_n (\ln a_0 - \beta\varepsilon_n) da_n$$ (13.41)

$$= -\ln a_0 \sum_n da_n + \beta \sum_n \varepsilon_n da_n$$

The first summation in Equation (13.41) represents the total change in occupation numbers, and it is equal to the change in the total number of oscillators in the system. Because the system is closed with respect to the number of oscillators, $dN = 0$ and the first summation is also equal to zero. The second term represents the change in total energy of the system (dE) accompanying the deposition of energy into the system:

$$\sum_n \varepsilon_n da_n = dE$$ (13.42)

With this last equality, the relationship between β, weight, and total energy is finally derived:

$$d \ln W = \beta dE$$ (13.43)

This last equality is quite remarkable and provides significant insight into the physical meaning of β. We began by recognizing that weight increases in proportion with the energy available to the system, and β is simply the proportionality constant in this relationship. Unit

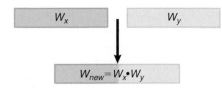

FIGURE 13.8

Two assemblies of distinguishable units, denoted x and y, are brought into thermal contact.

analysis of Equation (13.43) also demonstrates that β must have units of inverse energy as inferred previously.

Associating β with measurable system variables is the last step in deriving a full definition for the Boltzmann distribution. This step can be accomplished through the following conceptual experiment. Imagine two separate systems of distinguishable units at equilibrium having associated weights W_x and W_y. Next, these assemblies are brought into thermal contact, and the composite system is allowed to evolve toward equilibrium, as illustrated in Figure 13.8. Furthermore, the composite system is isolated from the surroundings such that the total energy available to the composite system is the sum of energy contained in the individual assemblies. The total weight of the combined system immediately after establishing thermal contact is the product of W_x and W_y. If the two systems are initially at different equilibrium conditions, the instantaneous composite system weight will be less than the weight of the composite system at equilibrium. Since the composite weight will increase as equilibrium is approached

$$d(W_x \cdot W_y) \geq 0 \qquad (13.44)$$

This inequality can be simplified by applying the chain rule for differentiation (see the Math Supplement):

$$W_y dW_x + W_x dW_y \geq 0 \qquad (13.45)$$

$$\frac{dW_x}{W_x} + \frac{dW_y}{W_y} \geq 0$$

$$d \ln W_x + d \ln W_y \geq 0$$

Substitution of Equation (13.43) into the last expression of Equation (13.45) results in

$$\beta_x dE_x + \beta_y dE_y \geq 0 \qquad (13.46)$$

where β_x and β_y are the corresponding β values associated with the initial assemblies x and y. Correspondingly, dE_x and dE_y refer to the change in total energy for the individual assemblies. Because the composite system is isolated from the surroundings, any change in energy for assembly x must be offset by a corresponding change in assembly y:

$$dE_x + dE_y = 0 \qquad (13.47)$$

$$dE_x = -dE_y$$

Now, if dE_x is positive, then by Equation (13.46),

$$\beta_x \geq \beta_y \qquad (13.48)$$

Can the preceding result be interpreted in terms of system variables? This question can be answered by considering the following. If dE_x is positive, energy flows into assembly x from assembly y. Thermodynamics dictates that because temperature is a measure of internal kinetic energy, an increase in the energy will be accompanied by an increase in the temperature of assembly x. A corresponding decrease in the temperature of assembly y will also occur. Therefore, before equilibrium is established, thermodynamic considerations dictate that

$$T_y \geq T_x \qquad (13.49)$$

In order for Equations (13.48) and (13.49) to be true, β must be inversely related to T. Furthermore, from unit analysis of Equation (13.43) we know that β must have units of inverse energy. This requirement is met by including a proportionality constant in the relationship between β and T, resulting in the final expression for β:

$$\beta = \frac{1}{kT} \qquad (13.50)$$

The constant in Equation (13.50), k, is referred to as **Boltzmann's constant** and has a numerical value of 1.381×10^{-23} J K^{-1}. The product of k and Avogadro's number is equal to R, the ideal gas constant (8.314 J mol^{-1} K^{-1}). Although the joule is the SI unit for energy, much of the information regarding molecular energy levels is derived from spectroscopic measurements. These spectroscopic quantities are generally expressed in units of wavenumbers (cm^{-1}). The **wavenumber** is simply the number of waves in an electromagnetic field per centimeter. Conversion from wavenumbers to joules is performed by multiplying the quantity in wavenumbers by Planck's constant, h, and the speed of light, c. In Example Problem 13.5, the vibrational energy levels for I$_2$ are given by the vibrational frequency of the oscillator, $\tilde{\nu}$ = 208 cm^{-1}. Using this spectroscopic information, the vibrational level energies in joules are

$$E_n = nhc\tilde{\nu} = n(6.626 \times 10^{-34}\,\text{J s})(3 \times 10^{10}\,\text{cm s}^{-1})(208\,\text{cm}^{-1}) \qquad (13.51)$$
$$= n(4.13 \times 10^{-21}\,\text{J})$$

At times the conversion from wavenumbers to joules will prove inconvenient. In such cases, Boltzmann's constant can be expressed in units of wavenumbers instead of joules where $k = 0.695$ cm^{-1} K^{-1}. In this case, the spectroscopic quantities in wavenumbers can be used directly when evaluating partition functions and other statistical-mechanical expressions.

EXAMPLE PROBLEM 13.5

The vibrational frequency of I$_2$ is 208 cm^{-1}. What is the probability of I$_2$ populating the $n = 2$ vibrational level if the molecular temperature is 298 K?

Solution

Molecular vibrational energy levels can be modeled as harmonic oscillators; therefore, this problem can be solved employing a strategy identical to the one just presented. To evaluate the partition function q, the "trick" used earlier was to write the partition function as a series and use the equivalent series expression:

$$q = \sum_n e^{-\beta \varepsilon_n} = 1 + e^{-\beta hc\tilde{\nu}} + e^{-2\beta hc\tilde{\nu}} + e^{-3\beta hc\tilde{\nu}} + \dots$$

$$= \frac{1}{1 - e^{-\beta hc\tilde{\nu}}}$$

Since $\tilde{\nu} = 208$ cm^{-1} and $T = 298$ K, the partition function is

$$q = \frac{1}{1 - e^{-\beta hc\tilde{\nu}}} = \frac{1}{1 - \exp\left[-\left(\dfrac{(6.626 \times 10^{-34}\,\text{J s})(3.00 \times 10^{10}\,\text{cm s}^{-1})(208\,\text{cm}^{-1})}{(1.38 \times 10^{-23}\,\text{J K}^{-1})(298\,\text{K})}\right)\right]}$$

$$= \frac{1}{1 - e^{-1}} = 1.58$$

This result is then used to evaluate the probability of occupying the second vibrational state ($n = 2$) as follows:

$$p_2 = \frac{e^{-2\beta hc\tilde{\nu}}}{q} = \frac{\exp\left[-2\left(\dfrac{(6.626 \times 10^{-34}\,\text{J s})(3.00 \times 10^{10}\,\text{cm s}^{-1})(208\,\text{cm}^{-1})}{(1.38 \times 10^{-23}\,\text{J K}^{-1})(298\,\text{K})}\right)\right]}{1.58} = 0.086$$

The last result in Example Problem 13.5 should look familiar. An identical example was worked earlier in this chapter where the energy-level spacings were equal to β^{-1} (case 1 in Figure 13.6) and the probability of populating states for which $n = 0, 1,$ and 2 was determined. This previous example in combination with the molecular example just presented illustrates that the exponential term $\beta\varepsilon_n$ in the Boltzmann distribution and partition function can be thought of as a comparative term that describes the ratio of the energy needed to populate a given energy level versus the thermal energy available to the system, as quantified by kT. Energy levels that are significantly higher in energy than kT are not likely to be populated, whereas the opposite is true for energy levels that are small relative to kT.

Example Problem 13.5 is reminiscent of the development presented in Chapter 8 of Engel's *Quantum Chemistry and Spectroscopy* in which the use of the Boltzmann distribution to predict the relative population in vibrational and rotational states and the effect of these populations on vibrational and rotational transition intensities was presented. In addition, the role of the Boltzmann distribution in nuclear magnetic resonance spectroscopy (NMR) was outlined in Chapter 18 of the *Quantum Chemistry and Spectroscopy* and is further explored in the following example problem.

EXAMPLE PROBLEM 13.6

In NMR spectroscopy, energy separation between spin states is created by placing nuclei in a magnetic field. Protons have two possible spin states: $+1/2$ and $-1/2$. The energy separation between these two states, ΔE, is dependent on the strength of the magnetic field, and is given by

$$\Delta E = g_N \beta_N B = (2.82 \times 10^{-26}\,\text{J T}^{-1})B$$

where B is the magnetic field strength in teslas (T). Also, g_N and β_N are the nuclear factor and nuclear magneton for a proton, respectively (see Section 18.1 of *Quantum Chemistry and Spectroscopy*). Early NMR spectrometers employed magnetic field strengths of approximately 1.45 T. What is the ratio of the population between the two spin states given this magnetic field strength and $T = 298$ K?

Solution

Using the Boltzmann distribution, the occupation number for energy levels is given by

$$a_n = \frac{Ne^{-\beta\varepsilon_n}}{q}$$

where N is the number of particles, ε_n is the energy associated with the level of interest, and q is the partition function. Using the preceding equation, the ratio of occupation numbers is given by

$$\frac{a_{1/2}}{a_{-1/2}} = \frac{\dfrac{Ne^{-\beta\varepsilon_{1/2}}}{q}}{\dfrac{Ne^{-\beta\varepsilon_{-1/2}}}{q}} = e^{-\beta\left(\varepsilon_{1/2} - \varepsilon_{-1/2}\right)} = e^{-\beta\Delta E}$$

Substituting for ΔE and β (and taking care that units cancel), the ratio of occupation numbers is given by

$$\frac{a_{1/2}}{a_{-1/2}} = e^{-\beta\Delta E} = \exp\left[\frac{-(2.82 \times 10^{-26}\,\text{J T}^{-1})(1.45\,\text{T})}{(1.38 \times 10^{-23}\,\text{J K}^{-1})(298\,\text{K})}\right]$$

$$= e^{-(9.94 \times 10^{6})} = 0.999990$$

In other words, in this system the energy spacing is significantly smaller than the energy available (kT) such that the higher energy spin state is populated to a significant extent, and is nearly equal in population to that of the lower energy state.

For Further Reading

Chandler, D., *Introduction to Modern Statistical Mechanics*. Oxford, New York, 1987.

Hill, T., *Statistical Mechanics. Principles and Selected Applications*. Dover, New York, 1956.

McQuarrie, D., *Statistical Mechanics*. Harper & Row, New York, 1973.

Nash, L. K., "On the Boltzmann Distribution Law." *J. Chemical Education* 59 (1982), 824.

Nash, L. K., *Elements of Statistical Thermodynamics*. Addison-Wesley, San Francisco, 1972.

Noggle, J. H., *Physical Chemistry*. HarperCollins, New York, 1996.

Vocabulary

Boltzmann distribution	degeneracy	occupation number
Boltzmann's constant	dominant configuration	partition function
configuration	equilibrium state	wavenumber
configurational index	microstate	weight

Questions on Concepts

Q13.1 What is the difference between a configuration and a microstate?

Q13.2 How does one calculate the number of microstates associated with a given configuration?

Q13.3 What is an occupation number? How is this number used to describe energy distributions?

Q13.4 Explain the significance of the Boltzmann distribution. What does this distribution describe?

Q13.5 What is degeneracy? Can you conceptually relate the expression for the partition function without degeneracy to that with degeneracy?

Q13.6 How is β related to temperature? What are the units of kT?

Problems

Problem numbers in **red** indicate that the solution to the problem is given in the *Student's Solutions Manual*.

P13.1

a. What is the possible number of microstates associated with tossing a coin N times and having it come up H times heads and T times tails?

b. For a series of 1000 tosses, what is the total number of microstates associated with 50% heads and 50% tails?

c. How less probable is the outcome that the coin will land 40% heads and 60% tails?

P13.2

a. Realizing that the most probable outcome from a series of N coin tosses is $N/2$ heads and $N/2$ tails, what is the expression for W_{max} corresponding to this outcome?

b. Given your answer for part (a), derive the following relationship between the weight for an outcome other than the most probable and W_{max}:

$$\log\left(\frac{W}{W_{max}}\right) = -H\log\left(\frac{H}{N/2}\right) - T\log\left(\frac{T}{N/2}\right)$$

c. We can define the deviation of a given outcome from the most probable outcome using a "deviation index," $\alpha = (H - T)/N$. Show that the number of heads or tails can be expressed as $H = (N/2)(1 + \alpha)$ and $T = (N/2)(1 - \alpha)$.

d. Finally, demonstrate that $W/W_{max} = e^{-N\alpha^2}$.

P13.3 Consider the case of 10 oscillators and eight quanta of energy. Determine the dominant configuration of energy for this system by identifying energy configurations and calculating the corresponding weights. What is the probability of observing the dominant configuration?

P13.4 Determine the weight associated with the following card hands:

a. Having any five cards

b. Having five cards of the same suit (known as a "flush")

P13.5 For a two-level system, the weight of a given energy distribution can be expressed in terms of the number of systems, N, and the number of systems occupying the excited state, n_1. What is the expression for weight in terms of these quantities?

P13.6 The probability of occupying a given excited state, p_i, is given by $p_i = n_i/N = e^{-\beta\varepsilon_i}/q$, where n_i is the occupation number for the state of interest, N is the number of particles, and ε_i is the energy of the level of interest. Demonstrate that the preceding expression is independent of the definition of energy for the lowest state.

P13.7 Barometric pressure can be understood using the Boltzmann distribution. The potential energy associated with being a given height above the Earth's surface is mgh, where m is the mass of the particle of interest, g is the acceleration due to gravity, and h is height. Using this definition of the potential energy, derive the following expression for pressure: $P = P_o e^{-mgh/kT}$. Assuming that the temperature remains at 298 K, what would you expect the relative pressures of N_2 and O_2 to be at the tropopause, the boundary between the troposphere and stratosphere roughly 11 km above the Earth's surface? At the Earth's surface, the composition of air is roughly 78% N_2, 21% O_2, and the remaining 1% is other gases.

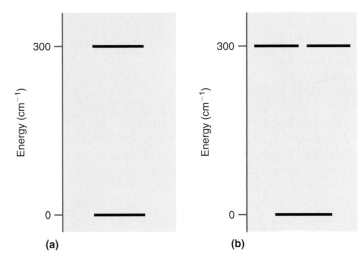

(a)　　　　　　　　　　**(b)**

P13.8 Consider the following energy-level diagrams:

a. At what temperature will the probability of occupying the second energy level be 0.15 for the states depicted in part (a) of the figure?

b. Perform the corresponding calculation for the states depicted in part (b) of the figure. Before beginning the calculation, do you expect the temperature to be higher or lower than that determined in part (a) of this problem? Why?

P13.9 Consider the following energy-level diagrams, modified from Problem P13.8 by the addition of another excited state with energy of 600 cm^{-1}:

a. At what temperature will the probability of occupying the second energy level be 0.15 for the states depicted in part (a) of the figure?

b. Perform the corresponding calculation for the states depicted in part (b) of the figure.

(*Hint:* You may find this problem easier to solve numerically using a spreadsheet program such as Excel.)

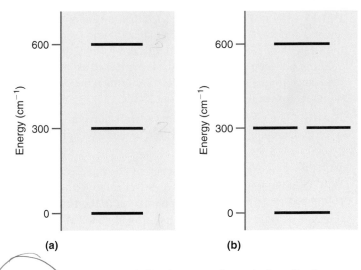

(a)　　　　　　　　　　**(b)**

P13.10 Consider the following sets of populations for four equally spaced energy levels:

ε/k (K)	Set A	Set B	Set C
300	5	3	4
200	7	9	8
100	15	17	16
0	33	31	32

a. Demonstrate that the sets have the same energy.

b. Determine which of the sets is the most probable.

c. For the most probable set, is the distribution of energy consistent with a Boltzmann distribution?

P13.11 A set of 13 particles occupies states with energies of 0, 100, and 200 cm^{-1}. Calculate the total energy and number of microstates for the following energy configurations:

a. $a_0 = 8$, $a_1 = 5$, and $a_2 = 0$

b. $a_0 = 9$, $a_1 = 3$, and $a_2 = 1$

c. $a_0 = 10$, $a_1 = 1$, and $a_2 = 2$

Do any of these configurations correspond to the Boltzmann distribution?

P13.12 For a set of nondegenerate levels with energy $\varepsilon/k = 0$, 100, and 200 K, calculate the probability of occupying each state when $T = 50$, 500, and 5000 K. As the temperature continues to increase, the probabilities will reach a limiting value. What is this limiting value?

P13.13 Consider a collection of molecules where each molecule has two nondegenerate energy levels that are separated by 6000 cm^{-1}. Measurement of the level populations demonstrates that there are 8 times more molecules in the ground state than in the upper state. What is the temperature of the collection?

P13.14 The ^{13}C nucleus is a spin 1/2 particle as is a proton. However, the energy splitting for a given field strength is roughly 1/4 of that for a proton. Using a 1.45-T magnet as in Example Problem 13.6, what is the ratio of populations in the excited and ground spin states for ^{13}C at 298 K?

P13.15 ^{14}N is a spin 1 particle such that the energy levels are at 0 and $\pm \gamma B \hbar$, where γ is the magnetogyric ratio and B is the strength of the magnetic field. In a 4.8-T field, the energy splitting between any two spin states expressed as the resonance frequency is 14.45 MHz. Determine the occupation numbers for the three spin states at 298 K.

P13.16 The vibrational frequency of I_2 is 208 cm^{-1}. At what temperature will the population in the first excited state be half that of the ground state?

P13.17 The vibrational frequency of Cl_2 is 525 cm^{-1}. Will the temperature be higher or lower relative to I_2 (see Problem P13.16) at which the population in the first excited vibrational state is half that of the ground state? What is this temperature?

P13.18 Determine the partition function for the vibrational degrees of freedom of Cl_2 ($\tilde{\nu} = 525$ cm^{-1}) and calculate the probability of occupying the first excited vibrational level at 300 and 1000 K. Determine the temperature at which identical probabilities will be observed for F_2 ($\tilde{\nu} = 917$ cm^{-1}).

P13.19 A two-level system is characterized by an energy separation of 1.3×10^{-18} J. At what temperature will the population of the ground state be 5 times greater than that of the excited state?

P13.20 The lowest two electronic energy levels of the molecule NO are illustrated here:

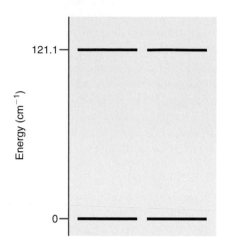

Determine the probability of occupying one of the higher energy states at 100, 500, and 2000 K.

Web-Based Simulations, Animations, and Problems

W13.1 In this simulation the behavior of the partition function for a harmonic oscillator with temperature and oscillator frequency is explored. The variation in q with temperature for an oscillator where $\tilde{\nu} = 1000$ cm^{-1} is studied, and variation in the individual level contributions to the partition function is studied. In addition, the variation in level contributions at fixed temperature, but as the oscillator frequency is varied is depicted. This simulation provides insight into the elements of the partition function and the variation of this function with temperature and energy-level spacings.

CHAPTER 14

Ensemble and Molecular Partition Functions

The relationship between the microscopic description of individual molecules and the macroscopic properties of a collection of molecules is a central concept in statistical mechanics. In this chapter, the relationship between the partition function that describes a collection of noninteracting molecules and the partition function describing an individual molecule is developed. We demonstrate that the molecular partition function can be decomposed into the product of partition functions for each energetic degree of freedom, and the functional form of these partition functions is derived. The concepts outlined in this chapter provide the basic foundation on which statistical thermodynamics resides. ∎

14.1 The Canonical Ensemble

An **ensemble** is defined as a collection of identical units or replicas of a system. For example, a mole of water can be envisioned as an ensemble with Avogadro's number of identical units of water molecules. The ensemble provides a theoretical concept by which the microscopic properties of matter can be related to the corresponding thermodynamic system properties as expressed in the following postulate:

> The average value for a property of the ensemble corresponds to the time-averaged value for the corresponding macroscopic property of the system.

What does this postulate mean? Imagine the individual units of the ensemble sampling the available energy space; the energy content of each unit is measured at a single time, and the measured unit energies are used to determine the average energy for the ensemble. According to the postulate, this energy will be equivalent to the average energy of the ensemble as measured over time. This idea, first formulated by J. W. Gibbs in the late 1800s, lies at the heart of statistical thermodynamics and is explored in this chapter.

To connect ensemble average values and thermodynamic properties, we begin by imagining a collection of identical copies of the system as illustrated in Figure 14.1. These copies of the system are held fixed in space such that they are distinguishable.

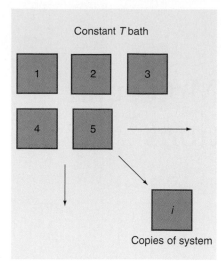

Constant *T* bath

Copies of system

FIGURE 14.1

The canonical ensemble is comprised of a collection of identical systems having fixed temperature, volume, and number of particles. The units are embedded in a constant *T* bath. The arrows indicate that an infinite number of copies of the system comprises the ensemble.

The volume, V, temperature, T, and number of particles in each system, N, are constant. An ensemble in which V, T, and N are constant is referred to as a **canonical ensemble.** The term *canonical* means "by common practice" because this is the ensemble employed unless the problem of interest dictates that other variables be kept constant. Note that other quantities can be constant to construct other types of ensembles. For example, if N, V, and energy are held fixed, the corresponding ensemble is referred to as *microcanonical*. However, for the purposes of this text, the canonical ensemble will prove sufficient.

In the canonical ensemble, each ensemble member is embedded in a temperature bath such that the total ensemble energy is constant. Furthermore, the walls that define the volume of the units can conduct heat, allowing for energy exchange with the surroundings. The challenge is to link the statistical development presented in the previous chapter to a similar statistical description for this ensemble. We begin by considering the total energy of the ensemble, E_c, which is given by

$$E_c = \sum_i a_{(c)i} E_i \tag{14.1}$$

In Equation (14.1), the terms $a_{(c)i}$ are the occupation numbers corresponding to the number of ensemble members having energy E_i. Proceeding exactly as in the previous chapter, the weight, W_c, associated with a specific configuration of energy among the N_c members of the ensemble is given by

$$W_c = \frac{N_c!}{\prod_i a_{(c)i}!} \tag{14.2}$$

This relationship can be used to derive the probability of finding an ensemble unit at energy E_i:

$$p(E_i) = \frac{W_i e^{-\beta E_i}}{Q} \tag{14.3}$$

Equation (14.3) looks very similar to the probability expression derived previously. In this equation, W_i can be thought of as the number of states present at a given energy E_i. The quantity Q in Equation (14.3) is referred to as the **canonical partition function** and is defined as follows:

$$Q = \sum_i e^{-\beta E_i} \tag{14.4}$$

In Equation (14.4), the summation is over all energy levels. The probability defined in Equation (14.3) is dependent on two factors: W_i, or the number of states present at a given energy that will increase with energy, and a Boltzmann term $e^{-\beta E_i}/Q$ that describes the probability of an ensemble unit having energy E_i that decreases exponentially with energy. The generic behavior of each term with energy is depicted in Figure 14.2. The product of these terms will reach a maximum corresponding to the average ensemble energy. The figure illustrates that an individual unit of the ensemble will have an energy that is equal to or extremely close to the average energy, and that units having energy far from this value will be exceedingly rare. We know this to be the case from experience. Imagine a swimming pool filled with water divided up into one-liter units. If the thermometer at the side of the pool indicates that the water temperature is 18°C, someone diving into the pool will not be worried that the liter of water immediately under their head will spontaneously freeze. That is, the temperature measured in one part of the pool is sufficient to characterize the temperature of the water in any part of the pool. Figure 14.2 provides an illustration of the statistical aspects underlying this expectation.

The vast majority of systems in the ensemble will have energy <*E*>; therefore, the thermodynamic properties of the unit are representative of the thermodynamic properties of the ensemble, demonstrating the link between the microscopic unit and the

FIGURE 14.2

For the canonical ensemble, the probability of a member of the ensemble having a given energy is dependent on the product of W_i, the number of states present at a given energy, and the Boltzmann distribution function for the ensemble. The product of these two factors results in a probability distribution that is peaked about the average energy, $<E>$.

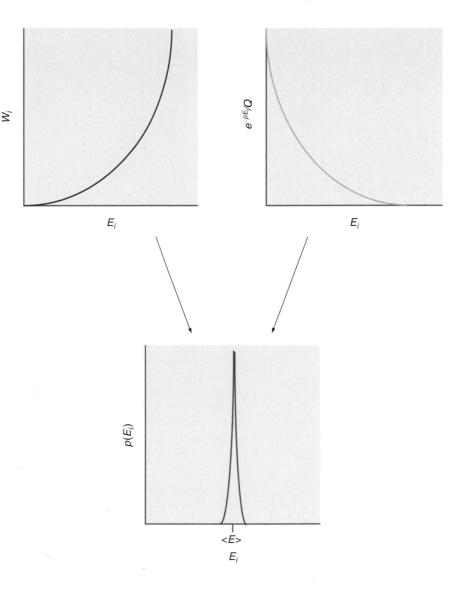

macroscopic ensemble. To make this connection mathematically exact the canonical partition function, Q, must be related to the partition function describing the individual members of the ensemble, q.

14.2 Relating Q to q for an Ideal Gas

In relating the canonical partition function, Q, to the partition function describing the members of the ensemble, q, our discussion is limited to systems consisting of independent "ideal" particles in which the interactions between particles is negligible (for example, an ideal gas). The relationship between Q and q is derived by considering an ensemble made up of two distinguishable units, A and B, as illustrated in Figure 14.3. For this simple ensemble, the partition function is

$$Q = \sum_n e^{-\beta E_n} = \sum_n e^{-\beta(\varepsilon_{A_n} + \varepsilon_{B_n})} \tag{14.5}$$

In this expression, ε_{A_n} and ε_{B_n} refer to the energy levels associated with unit A and B, respectively.

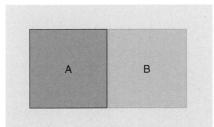

FIGURE 14.3

A two-unit ensemble. In this ensemble, the two units, A and B, are distinguishable.

Assuming that the energy levels are quantized such that they can be indexed as 0, 1, 2, and so forth. Employing this idea, Equation (14.5) becomes

$$Q = \sum_n e^{-\beta(\varepsilon_A + \varepsilon_B)} = e^{-\beta(\varepsilon_{A_0} + \varepsilon_{B_0})} + e^{-\beta(\varepsilon_{A_0} + \varepsilon_{B_1})} + e^{-\beta(\varepsilon_{A_0} + \varepsilon_{B_2})} + \dots$$

$$+ e^{-\beta(\varepsilon_{A_1} + \varepsilon_{B_0})} + e^{-\beta(\varepsilon_{A_1} + \varepsilon_{B_1})} + e^{-\beta(\varepsilon_{A_1} + \varepsilon_{B_2})} + \dots$$

$$+ e^{-\beta(\varepsilon_{A_2} + \varepsilon_{B_0})} + e^{-\beta(\varepsilon_{A_2} + \varepsilon_{B_1})} + e^{-\beta(\varepsilon_{A_2} + \varepsilon_{B_2})} + \dots$$

$$= (e^{-\beta\varepsilon_{A_0}} + e^{-\beta\varepsilon_{A_1}} + e^{-\beta\varepsilon_{A_2}} + \dots)(e^{-\beta\varepsilon_{B_0}} + e^{-\beta\varepsilon_{B_1}} + e^{-\beta\varepsilon_{B_2}} + \dots)$$

$$= (q_A)(q_B)$$

$$= q^2$$

The last step in the derivation is accomplished by recognizing that the ensemble units are identical such that the partition functions are also identical. Extending the preceding result to a system with N distinguishable units, the canonical partition function is found to be simply the product of unit partition functions

$$Q = q^N \quad \text{for } N \text{ distinguishable units} \qquad (14.6)$$

Thus far no mention has been made of the size of the identical systems comprising the ensemble. The systems can be as small as desired, including just a single molecule. Taking the single-molecule limit, a remarkable conclusion is reached: the canonical ensemble is nothing more than the product of the molecular partition functions as discussed in Chapter 13. This is the direct connection between the microscopic and macroscopic perspectives that we have been searching for. The quantized energy levels of the molecular (or atomic) system are embedded in the **molecular partition function,** q, and this partition function can be used to define the partition function for the ensemble, Q. Finally, Q can be directly related to the thermodynamic properties of the ensemble.

The preceding derivation assumed that the ensemble members were distinguishable. This might be the case for a collection of molecules coupled to a surface where they cannot move, but what happens to this derivation when the ensemble is in the gaseous state? Clearly, the translational motion of the gas molecules will make identification of each individual molecule impossible. Therefore, how does Equation (14.6) change if the units are indistinguishable? A simple counting example will help to answer to this question. Consider three distinguishable oscillators (A, B, and C) with three total quanta of energy as described in the previous chapter. The dominant configuration of energy was with the oscillators in three separate energy states, denoted "2, 1, 0." The energy states relative to the oscillators can be arranged in six different ways:

A	B	C
2	1	0
2	0	1
1	2	0
0	2	1
1	0	2
0	1	2

However, if the three oscillators are indistinguishable, there is no difference among the arrangements listed. In effect, there is only one arrangement of energy that should be counted. This problem was encountered in Chapter 12 when discussing indistinguishable particles, and in such cases the total number of permutations was divided by $N!$ where N was the number of units in the collection. Extending this logic to a molecular ensemble dictates that the canonical partition function for indistinguishable particles have the following form:

$$Q = \frac{q^N}{N!} \quad \text{for } N \text{ indistinguishable units} \qquad (14.7)$$

Equation (14.7) is correct in the limit for which the number of energy levels available is significantly greater than the number of particles. This discussion of statistical mechanics is limited to systems for which this is true, and the validity of this statement is demonstrated later in this chapter. It is also important to keep in mind that Equation (14.7) is limited to ideal systems of noninteracting particles such as an ideal gas.

14.3 Molecular Energy Levels

The relationship between the canonical and molecular partition functions provides the link between the microscopic and macroscopic descriptions of the system. The molecular partition function can be evaluated by considering molecular energy levels. For polyatomic molecules, there are four **energetic degrees of freedom** to consider in constructing the molecular partition function:

1. Translation
2. Rotation
3. Vibration
4. Electronic

Assuming the energetic degrees of freedom are not coupled, the total molecular partition function that includes all of these degrees of freedom can be decomposed into a product of partition functions corresponding to each degree of freedom. An equivalent approach was taken when separating the molecular Hamiltonian into translational, rotational, and vibrational components as was done in Section 7.2 of Engel's *Quantum Chemistry and Spectroscopy*. Let ε_{Total} represent the energy associated with a given molecular energy level. This energy will depend on the translational, rotational, vibrational, and electronic level energies as follows:

$$\varepsilon_{Total} = \varepsilon_T + \varepsilon_R + \varepsilon_V + \varepsilon_E \qquad (14.8)$$

Recall that the molecular partition function is obtained by summing over molecular energy levels. Using the expression for the total energy and substituting into the expression for the partition function, the following expression is obtained:

$$\begin{aligned}
q_{Total} &= \sum g_{Total} e^{-\beta \varepsilon_{Total}} \qquad (14.9) \\
&= \sum (g_T g_R g_V g_E) e^{-\beta(\varepsilon_T + \varepsilon_R + \varepsilon_V + \varepsilon_E)} \\
&= \sum (g_T e^{-\beta \varepsilon_T})(g_R e^{-\beta \varepsilon_R})(g_V e^{-\beta \varepsilon_V})(g_E e^{-\beta \varepsilon_E}) \\
&= q_T q_R q_V q_E
\end{aligned}$$

This relationship demonstrates that the total molecular partition function is simply the product of partition functions for each molecular energetic degree of freedom. Using this definition for the molecular partition function, the final relationships of interest are

$$Q_{Total} = q_{Total}^N \qquad \text{(distinguishable)} \qquad (14.10)$$

$$Q_{Total} = \frac{1}{N!} q_{Total}^N \qquad \text{(indistinguishable)} \qquad (14.11)$$

All that remains to derive are partition functions for each energetic degree of freedom, a task that is accomplished in the remainder of this chapter.

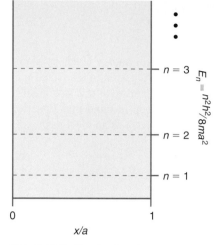

FIGURE 14.4
Particle-in-a-box model for translational energy levels.

14.4 Translational Partition Function

Translational energy levels correspond to the translational motion of atoms or molecules in a container of volume V. Rather than work directly in three-dimensions, a one-dimensional model is first employed and then later extended to three dimensions. From quantum mechanics, the energy levels of a molecule confined to a box were described by the "particle-in-a-box" model as illustrated in Figure 14.4. In this figure, a particle with mass m is free to move in the domain $0 \leq x \leq a$, where a is the length of the box. Using the expression for the energy levels provided in Figure 14.4, the partition function for translational energy in one dimension becomes

$$q_{T,1D} = \sum_{n=1}^{\infty} e^{\frac{-\beta n^2 h^2}{8ma^2}} \tag{14.12}$$

Notice that the summation consists of an infinite number of terms. Furthermore, a closed-form expression for this series does not exist such that it appears one must evaluate the sum directly. However, a way around this apparently impossible task becomes evident when the spacing between energy translational energy states is considered, as illustrated in the following example.

EXAMPLE PROBLEM 14.1

What is the difference in energy between the $n = 2$ and $n = 1$ states for molecular oxygen constrained by a one-dimensional box having a length of 1 cm?

Solution

The energy difference is obtained by using the expression for the one-dimensional particle-in-a-box model as follows:

$$\Delta E = E_2 - E_1 = 3E_1 = \frac{3h^2}{8ma^2}$$

The mass of an O_2 molecule is 5.31×10^{-25} kg such that

$$\Delta E = \frac{3(6.626 \times 10^{-34} \text{ J s})^2}{8(5.31 \times 10^{-25} \text{ kg})(0.01 \text{ m})^2}$$

$$= 3.10 \times 10^{-38} \text{ J}$$

Converting to units of cm^{-1}:

$$\Delta E = \frac{3.10 \times 10^{-38} \text{ J}}{hc} = 1.57 \times 10^{-15} \text{ cm}^{-1}$$

At 298 K, the amount of thermal energy available as given by the product of Boltzmann's constant and temperature, kT, is 208 cm^{-1}. Clearly, the spacings between translational energy levels are extremely small relative to kT at room temperature.

Because numerous translational energy levels are accessible at room temperature, the summation in Equation (14.12) can be replaced by integration with negligible error:

$$q_T = \sum e^{-\beta \alpha n^2} \approx \int_0^{\infty} e^{-\beta \alpha n^2} dn \tag{14.13}$$

In this expression, the following substitution was made to keep the collection of constant terms compact:

$$\alpha = \frac{h^2}{8ma^2} \tag{14.14}$$

The integral in Equation (14.13) is readily evaluated (see Appendix B, Math Supplement):

$$q_T \approx \int_0^\infty e^{-\beta\alpha n^2}\, dn = \frac{1}{2}\sqrt{\frac{\pi}{\beta\alpha}} \tag{14.15}$$

Substituting in for α, the **translational partition function** in one dimension becomes

$$q_{T,1D} = \left(\frac{2\pi m}{h^2\beta}\right)^{1/2} a \tag{14.16}$$

This expression can be simplified by defining the **thermal de Broglie wavelength,** or simply the **thermal wavelength,** as follows:

$$\Lambda = \left(\frac{h^2\beta}{2\pi m}\right)^{1/2} \tag{14.17}$$

such that

$$q_{T,1D} = \frac{a}{\Lambda} = (2\pi mkT)^{1/2}\frac{a}{h} \tag{14.18}$$

length of box

Referring to Λ as the thermal wavelength reflects the fact that the average momentum of a gas particle, p, is equal to $(mkT)^{1/2}$. Therefore, Λ is essentially h/p, or the de Broglie wavelength of the particle as defined in Section 1.5 of *Quantum Chemistry and Spectroscopy*. Extension of the one-dimensional result to three dimensions is straightforward. The translational degrees of freedom are considered separable; therefore, the three-dimensional translational partition function is the product of one-dimensional partition functions for each dimension:

$$q_{T,3D} = q_{T_x} q_{T_y} q_{T_z}$$

$$= \left(\frac{a_x}{\Lambda}\right)\left(\frac{a_y}{\Lambda}\right)\left(\frac{a_z}{\Lambda}\right)$$

$$= \left(\frac{1}{\Lambda}\right)^3 a_x a_y a_z$$

$$= \left(\frac{1}{\Lambda}\right)^3 V$$

$$q_{T,3D} = \frac{V}{\Lambda^3} = (2\pi mkT)^{3/2}\frac{V}{h^3} \tag{14.19}$$

where V is volume and Λ is the thermal wavelength [Equation (14.17)]. Notice that the translational partition is a function of both V and T. Recall the discussion from the previous chapter in which the partition function was described conceptually as providing a measure of the number of energy states available to the system at a given temperature. The increase in q_T with volume reflects the fact that as volume is increased, the translational energy-level spacings decrease such that more states are available for population at a given T. Given the small energy spacings between translational energy levels relative to kT at room temperature, we might expect that at room temperature a significant number of translational energy states are accessible. The following example provides a test of this expectation.

EXAMPLE PROBLEM 14.2

What is the translational partition function for Ar confined to a volume of 1 L at 298 K?

Solution

Evaluation of the translational partition function is dependent on determining the thermal wavelength [Equation (14.17)]:

$$\Lambda = \left(\frac{h^2\beta}{2\pi m}\right)^{1/2} = \frac{h}{(2\pi mkT)^{1/2}}$$

The mass of Ar is 6.63×10^{-26} kg. Using this value for m, the thermal wavelength becomes

$$\Lambda = \frac{6.626 \times 10^{-34}\,\text{J s}}{(2\pi(6.63\times10^{-26}\,\text{kg})(1.38\times10^{-23}\,\text{J K}^{-1})(298\,\text{K}))^{1/2}}$$
$$= 1.60 \times 10^{-11}\,\text{m}$$

The units of volume must be such that the partition function is unitless. Therefore, conversion of volume to units of cubic meters (m^3) is performed as follows:

$$V = 1\,\text{L} = 1000\,\text{mL} = 1000\,\text{cm}^3\left(\frac{1\,\text{m}}{100\,\text{cm}}\right)^3 = 0.001\,\text{m}^3$$

The partition function is simply the volume divided by the thermal wavelength cubed:

$$q_{T,3D} = \frac{V}{\Lambda^3} = \frac{0.001\,\text{m}^3}{(1.60\times10^{-11}\,\text{m})^3} = 2.44 \times 10^{29}$$

The magnitude of the translational partition function determined in Example Problem 14.2 illustrates that a vast number of translation energy states are available at room temperature. In fact, the number of accessible states is roughly 10^6 times larger than Avogadro's number, illustrating that the assumption that many more states are available relative to units in the ensemble (Section 14.2) is reasonable.

14.5 Rotational Partition Function: Diatomics

A **diatomic** molecule consists of two atoms joined by a chemical bond as illustrated in Figure 14.5. In treating rotational motion of diatomic molecules, the rigid rotor approximation is employed in which the bond length is assumed to remain constant during rotational motion and effects such as centrifugal distortion are neglected.

In deriving the rotational partition function, an approach similar to that used in deriving the translational partition function is employed. Within the rigid-rotor approximation, the quantum mechanical description of rotational energy levels for diatomic molecules dictates that the energy of a given rotational state, E_J, is dependent on the rotational quantum number, J, as explained in Chapter 8 of *Quantum Chemistry and Spectroscopy*:

$$E_J = BJ(J+1) \quad \text{for } J = 0, 1, 2, \cdots \tag{14.20}$$

where J is the quantum number corresponding to rotational energy level and can take on integer values beginning with zero. The quantity B is the **rotational constant** and is given by

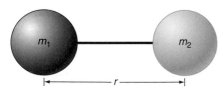

FIGURE 14.5

Schematic representation of a diatomic molecule, consisting of two masses (m_1 and m_2) joined by a chemical bond with the separation of atomic centers equal to the bond length, r.

$$B = \frac{h}{8\pi^2 cI} \tag{14.21}$$

where I is the moment of inertia, which is equal to

$$I = \mu r^2 \tag{14.22}$$

In the expression for the moment of inertia, r is the distance separating the two atomic centers and μ is the reduced mass, which for a diatomic consisting of atoms having masses m_1 and m_2 is equal to

$$\mu = \frac{m_1 m_2}{m_1 + m_2} \tag{14.23}$$

Because diatomic molecules differ depending on the masses of atoms in the molecule and the bond length, the value of the rotational constant is molecule dependent. Using the preceding expression for the rotational energy, the rotational partition function can be constructed by simply substituting into the general form of the molecular partition function:

$$q_R = \sum_J g_J e^{-\beta hcBJ(J+1)} \tag{14.24}$$

In this expression, the energies of the levels included in the summation are given by $hcBJ(J + 1)$. However, notice that the expression for the rotational partition function contains an addition term, g_J, that represents the number of rotational states present at a given energy level, or the degeneracy of the rotational energy level. To determine the degeneracy, refer to Chapter 7 in *Quantum Chemistry and Spectroscopy* and the discussion of the rigid rotor and the time-independent Schrödinger equation:

$$H\psi = E\psi \tag{14.25}$$

For the rigid rotor, the Hamiltonian (H) is proportional to the square of the total angular momentum given by the operator $\hat{l}^2$. The eigenstates of this operator are the spherical harmonics with the following eigenvalues:

$$\hat{l}^2\psi = \hat{l}^2 Y_{l,m}(\theta,\phi) = \frac{h^2}{4\pi^2} l(l+1) Y_{l,m}(\theta,\phi) \tag{14.26}$$

In this expression, l is a quantum number corresponding to total angular momentum, and it ranges from 0, 1, 2, ... , to infinity. The spherical harmonics are also eigenfunctions of the $\hat{l}_z$ operator corresponding to the z component of the angular momentum. The corresponding eigenvalues employing the $\hat{l}_z$ operator are given by

$$\hat{l}_z Y_{l,m}(\theta,\phi) = \frac{h}{2\pi} m Y_{l,m}(\theta,\phi) \tag{14.27}$$

Possible values for the quantum number m in Equation (14.27) are dictated by the quantum number l:

$$m = -l ... 0 ... l \quad \text{or} \quad (2l+1) \tag{14.28}$$

Thus, the degeneracy of the rotational energy levels originates from the quantum number m because all values of m corresponding to a given quantum number l will have the same total angular momentum and, therefore, the same energy. Using the value of $(2l + 1)$ for the degeneracy, the rotational partition function is

$$q_R = \sum_J (2J+1) e^{-\beta hcBJ(J+1)} \tag{14.29}$$

As written, evaluation of Equation (14.29) involves summation over all rotational states. A similar issue was encountered when the expression for the translational partition function was evaluated. The spacings between translational levels were very small relative to kT such that the partition function could be evaluated by integration rather than discrete summation. Are the rotational energy-level spacings also small relative to kT such that integration can be performed instead of summation?

To answer this question, consider the energy-level spacings for the rigid rotor presented in Figure 14.6. The energy of a given rotational state (in units of the rotational constant B) are presented as a function of the rotational quantum number, J. The energy-level spacings are multiples of B. The value of B will vary depending on the molecule of interest, with representative values provided in Table 14.1. Inspection of the table reveals a few interesting trends. First, the rotational constant depends on the atomic mass, with an increase in atomic mass resulting in a reduction in the rotational constant. Second, the values for B are quite different; therefore, any comparison of rotational state energies to kT will depend on the diatomic of interest. For example, at 298 K, $kT = 208$ cm^{-1}, which is roughly equal to the energy of the $J = 75$ level of I_2. For this species, the energy-level spacings are clearly much smaller than kT and integration of the partition function is appropriate. However, for H_2 the $J = 2$ energy level is greater than kT so that integration would be inappropriate, and evaluation of the partition function by direct summation must be performed. In the remainder of this chapter, we assume that integration of the rotational partition function is appropriate unless stated otherwise.

With the assumption that the rotational energy-level spacings are small relative to kT, evaluation of the rotational partition function is performed with integration over the rotational states:

$$q_R = \int_0^\infty (2J+1)e^{-\beta hcBJ(J+1)}dJ \tag{14.30}$$

Evaluation of the preceding integral is simplified by recognizing the following:

$$\frac{d}{dJ}e^{-\beta hcBJ(J+1)} = -\beta hcB(2J+1)e^{-\beta hcBJ(J+1)}$$

Using this relationship, the expression for the **rotational partition function** can be rewritten and the result evaluated as follows:

$$q_R = \int_0^\infty (2J+1)e^{-\beta hcBJ(J+1)}dJ = \int_0^\infty \frac{-1}{\beta hcB}\frac{d}{dJ}e^{-\beta hcBJ(J+1)}dJ$$

$$= \frac{-1}{\beta hcB}e^{-\beta hcBJ(J+1)}\Big|_0^\infty = \frac{1}{\beta hcB}$$

$$q_R = \frac{1}{\beta hcB} = \frac{kT}{hcB} \tag{14.31}$$

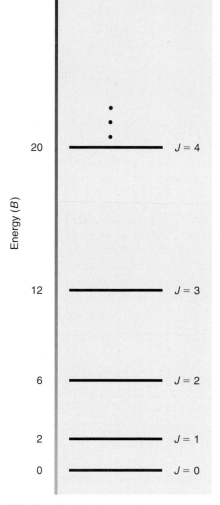

FIGURE 14.6
Rotational energy levels as a function of the rotational quantum number J. The energy of a given rotational state is equal to $BJ(J + 1)$.

TABLE 14.1

Rotational Constants for Some Representative Diatomic Molecules

Molecule	$B(\text{cm}^{-1})$	Molecule	$B(\text{cm}^{-1})$
H^{35}Cl	10.595	H$_2$	60.853
H^{37}Cl	10.578	^{14}N^{16}O	1.7046
D^{35}Cl	5.447	^{127}I^{127}I	0.03735

Source: Herzberg, G., *Molecular Spectra and Molecular Structure, Volume 1: Spectra of Diatomic Molecules.* Krieger Publishing, Melbourne, FL, 1989.

14.5.1 The Symmetry Number

The expression for the rotational partition function of a diatomic molecule provided in the previous section is correct for heterodiatomic species in which the two atoms comprising the diatomic are not equivalent. HCl is a heterodiatomic species because the two atoms in the diatomic, H and Cl, are not equivalent. However, the expression for the rotational partition function must be modified when applied to homodiatomic molecules such as N_2. A simple illustration of why such a modification is necessary is presented in Figure 14.7. In the figure, rotation of the heterodiatomic results in a species that is distinguishable from the molecule before rotation. However, the same 180° rotation applied to a homodiatomic results in a configuration that is equivalent to the prerotation form. This difference in behavior is similar to the differences between canonical partition functions for distinguishable and indistinguishable units. In the partition function case, the result for the distinguishable case was divided by $N!$ to take into account the "overcounting" of nonunique microstates encountered when the units are indistinguishable. In a similar spirit, for homodiatomic species the number of classical rotational states (i.e., distinguishable rotational configurations) is overcounted by a factor of 2.

To correct our rotational partition function for overcounting, we can simply divide the expression for the rotational partition function by the number of equivalent rotational configurations. This factor is known as the **symmetry number,** σ, and is incorporated into the partition function as follows:

$$q_R = \frac{1}{\sigma \beta hcB} = \frac{kT}{\sigma hcB} \tag{14.32}$$

The concept of a symmetry number can be extended to molecules other than diatomics. For example, consider a trigonal pyramidal molecule such as NH_3 as illustrated in Figure 14.8. Imagine that performing a 120° rotation about an axis through the nitrogen atom and the center of the triangle made the three hydrogens. The resulting configuration would be exactly equivalent to the previous configuration before rotation. Furthermore, a second 120° rotation would produce a third configuration. A final 120° would result in the initial prerotation configuration. Therefore, NH_3 has three equivalent rotational configurations; therefore, $\sigma = 3$.

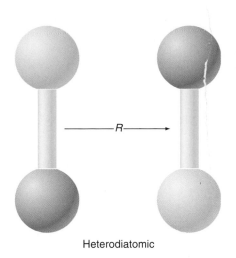

Heterodiatomic

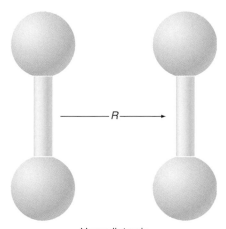

Homodiatomic

FIGURE 14.7
A 180° rotation of heterodiatomic and homodiatomic molecules.

FIGURE 14.8
Rotational configurations for NH_3.

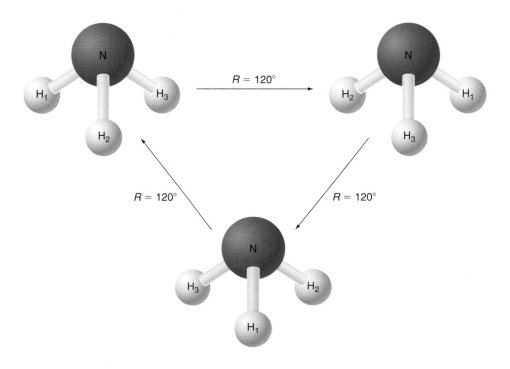

EXAMPLE PROBLEM 14.3

What is the symmetry number for methane (CH_4)?

Solution

To determine the number of equivalent rotational configurations, we will proceed in a fashion similar to that employed for NH_3. The tetrahedral structure of methane is shown in the following figure:

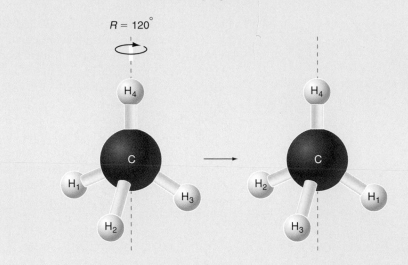

Similar to NH_3, three equivalent configurations can be generated by 120° rotation about the axis depicted by the dashed line in the figure. Furthermore, we can draw four such axes of rotation aligned with each of the four C—H bonds. Therefore, there are 12 total rotational configurations for CH_4 corresponding to $\sigma = 12$.

14.5.2 Rotational Level Populations and Spectroscopy

In Chapter 8 of *Quantum Chemistry and Spectroscopy,* the relationship between the populations in various rotational energy levels and the rotational-vibrational infrared absorption intensities were described. With the rotational partition function, we are in a position to explore this relationship in detail. The probability of occupying a given rotational energy level, p_J, is given by

$$p_J = \frac{g_J e^{-\beta hcBJ(J+1)}}{q_R} = \frac{(2J+1)e^{-\beta hcBJ(J+1)}}{q_R} \tag{14.33}$$

$H^{35}Cl$ where $B = 10.595$ cm^{-1} was previously employed to illustrate the relationship between p_J and absorption intensity. At 300 K the rotational partition function for $H^{35}Cl$ is

$$q_R = \frac{1}{\sigma\beta hcB} = \frac{kT}{\sigma hcB} = \frac{(1.38\times10^{-23}\,\text{J K}^{-1})(300\,\text{K})}{(1)(6.626\times10^{-34}\,\text{J s})(3.00\times10^{10}\,\text{cm s}^{-1})(10.595\,\text{cm}^{-1})} \tag{14.34}$$

$$= 19.7$$

With q_R, the level probabilities can be readily determined using Equation (14.33), and the results of this calculation are presented in Figure 14.9. The intensity of the P and R branch transitions illustrated in Figure 8.16 of *Quantum Chemistry and Spectroscopy* are proportional to the probability of occupying a given J level. This dependence is reflected by the evolution in rotational-vibrational transitions intensity as a function of J. The transition moment demonstrates modest J dependence as well such that the correspondence between transition intensity and rotational level population is not exact.

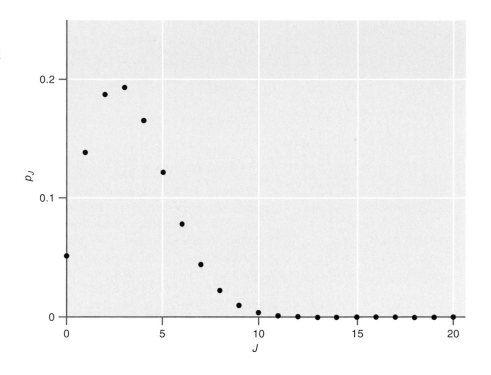

FIGURE 14.9
Probability of occupying a rotational
energy level, p_J, as a function of rotational
quantum number J for H³⁵Cl at 300 K.

EXAMPLE PROBLEM 14.4

In a rotational spectrum of HBr ($B = 8.46$ cm^{-1}), the maximum intensity is
observed for the $J = 4$ to 5 transition. At what temperature was the spectrum
obtained?

Solution

The information provided for this problem dictates that the $J = 4$ rotational
energy level was the most populated at the temperature at which the spectrum
was taken. To determine the temperature, we first determine the change in
occupation number for the rotational energy level a_J versus J as follows:

$$a_J = \frac{N(2J+1)e^{-\beta hcBJ(J+1)}}{q_R} = \frac{N(2J+1)e^{-\beta hcBJ(J+1)}}{\left(\dfrac{1}{\beta hcB}\right)}$$

$$= N\beta hcB(2J+1)e^{-\beta hcBJ(J+1)}$$

Next, we take the derivative of a_J with respect to J and set the derivative equal
to zero to find the maximum of the function:

$$\frac{da_J}{dJ} = 0 = \frac{d}{dJ} N\beta hcB(2J+1)e^{-\beta hcBJ(J+1)}$$

$$0 = \frac{d}{dJ}(2J+1)e^{-\beta hcBJ(J+1)}$$

$$0 = 2e^{-\beta hcBJ(J+1)} - \beta hcB(2J+1)^2 e^{-\beta hcBJ(J+1)}$$

$$0 = 2 - \beta hcB(2J+1)^2$$

$$2 = \beta hcB(2J+1)^2 = \frac{hcB}{kT}(2J+1)^2$$

$$T = \frac{(2J+1)^2 hcB}{2k}$$

Substitution of $J = 4$ into the preceding expression results in the following temperature at which the spectrum was obtained:

$$T = \frac{(2J+1)^2 hcB}{2k} = \frac{(2(4)+1)^2 (6.626 \times 10^{-34} \text{ J s})(3.00 \times 10^{10} \text{ cm s}^{-1})(8.46 \text{ cm}^{-1})}{2(1.38 \times 10^{-23} \text{ J K}^{-1})}$$

$$= 494 \text{ K}$$

14.5.3 An Advanced Topic: The Rotational States of H_2

The rotational-level distribution of H_2 provides an elegant example of the influence of molecular symmetry on the partition function. Molecular hydrogen exists in two forms, one in which the nuclear spins are paired (***para*-hydrogen**) and one in which the spins are aligned (***ortho*-hydrogen**). Because the hydrogen nuclei are spin 1/2 particles, they are fermions. The Pauli exclusion principle dictates that when two identical fermions interchange position, the overall wave function describing the system must by antisymmetric (or change sign) with interchange. The wave function in this case can be separated into a product of spin and rotational components. The spin component of the wave function is considered first. For *para*-hydrogen, rotation results in the interchange of two nuclei (A and B) having opposite spin (α and β) such that the spin component of the wave function should be antisymmetric with nuclei interchange due to rotation. This requirement is accomplished using the following linear combination of nuclear spin states:

$$\psi_{spin,para} = \alpha(A)\beta(B) - \alpha(B)\beta(A)$$

Interchange of the nuclear labels A and B corresponding to rotation will result in the preceding wave function changing sign such that the spin component of the *para*-hydrogen wave function is antisymmetric with respect to interchange.

For *ortho*-hydrogen, rotation results in the interchange of two nuclei with the same spin; therefore, the spin wave function should be symmetric. Three combinations of nuclear spin states meet this requirement:

$$\psi_{spin,ortho} = \{\alpha(A)\alpha(B), \beta(A)\beta(B), \alpha(A)\beta(B) + \beta(A)\alpha(B)\}$$

In summary, the spin component of the wave function is antisymmetric with respect to nuclei interchange for *para*-hydrogen, but symmetric for *ortho*-hydrogen.

Next, consider the symmetry of the rotational component of the wave function. It can be shown that the symmetry of rotational states is dependent on the rotational quantum number, J. If J is an even integer ($J = 0, 2, 4, 6, \ldots$) the corresponding rotational wave function is symmetric with respect to interchange, and if J is odd ($J = 1, 3, 5, 7, \ldots$) the wave function is antisymmetric. Because the wave function is the product of spin and rotational components, this product must be antisymmetric. Therefore, the rotational wave function for *para*-hydrogen is restricted to even-J levels, and for *ortho*-hydrogen is restricted to odd-J levels. Finally, the nuclear-spin-state degeneracy for *ortho*- and *para*-hydrogen is three and one, respectively. Therefore, the rotational energy levels for *ortho*-hydrogen have an additional threefold degeneracy.

A collection of molecular hydrogen will contain both *ortho*- and *para*-hydrogen so that the rotational partition function is

$$q_R = \frac{1}{4}\left[1 \sum_{J=0,2,4,6,\ldots} (2J+1)e^{-\beta hcBJ(J+1)} + 3 \sum_{J=1,3,5,\ldots} (2J+1)e^{-\beta hcBJ(J+1)}\right] \quad (14.35)$$

In Equation (14.35), the first term in brackets corresponds to *para*-hydrogen, and the second term to *ortho*-hydrogen. In essence, this expression for q_R represents average H_2 consisting of one part *para*-hydrogen and three parts *ortho*-hydrogen. Notice that the symmetry number is omitted in Equation (14.35) because overcounting of the allowed rotational levels has already been taken into account by restricting the summations to even or odd J. At high temperatures, the value of q_R determined using Equation (14.35) will, to good approximation, equal that obtained using Equation (14.32) with $\sigma = 2$, as the following example illustrates.

EXAMPLE PROBLEM 14.5

What is the rotational partition function for H_2 at 1000 K?

Solution

The rotational partition function for H_2 assuming that the high-temperature limit is valid is given by

$$q_R = \frac{1}{\sigma \beta hcB} = \frac{1}{2\beta hcB}$$

With $B = 60.589 \text{ cm}^{-1}$ (Table 14.1):

$$q_R = \frac{1}{2\beta hcB} = \frac{kT}{2hcB} = \frac{(1.38 \times 10^{-23} \text{ J K}^{-1})(1000 \text{ K})}{2(6.626 \times 10^{-34} \text{ J s})(3.00 \times 10^{10} \text{ cm s}^{-1})(60.589 \text{ cm}^{-1})} = 5.74$$

Evaluation of the rotational partition function by direct summation is performed as follows:

$$q_R = \frac{1}{4}\left[1 \sum_{J=0,2,4,6,\dots} (2J+1)e^{-\beta hcBJ(J+1)} + 3 \sum_{J=1,3,5,\dots} (2J+1)e^{-\beta hcBJ(J+1)} \right] = 5.91$$

Comparison of these two expressions demonstrates that the high-T expression for q_R with $\sigma = 2$ provides a good estimate for the value of the rotational partition function of H_2.

14.5.4 The Rotational Temperature

Whether the rotational partition function should be evaluated by direct summation or integration is entirely dependent on the size of the rotational energy spacings relative to the amount of thermal energy available (kT). This comparison is facilitated through the introduction of the **rotational temperature,** Θ_R, defined as rotational constant divided by Boltzmann's constant:

$$\Theta_R = \frac{hcB}{k} \tag{14.36}$$

Unit analysis of Equation (14.36) dictates that Θ_R has units of temperature. We can rewrite the expression for the rotational partition function in terms of the rotational temperature as follows:

$$q_R = \frac{1}{\sigma \beta hcB} = \frac{kT}{\sigma hcB} = \frac{T}{\sigma \Theta_R}$$

A second application of the rotational temperature is as a comparative metric to the temperature at which the partition function is being evaluated. Figure 14.10 presents a comparison between q_R for $H^{35}Cl$ ($\Theta_R = 15.24 \text{ K}$) determined by summation [Equation (14.29)] and using the integrated form of the partition function [Equation (14.36)]. At low temperatures, significant differences between the summation and integrated results are evident. At higher temperatures, the summation result remains larger than the integrated result; however, both results predict that q_R will increase linearly with temperature. At high temperatures, the error in using the integrated result decreases such that for temperature where $T/\Theta_R \geq 10$, use of the integrated form of the

FIGURE 14.10
Comparison of q_R for $H^{35}Cl$ ($\Theta_R = 15.24$ cm^{-1}) determined by summation and by integration. Although the summation result remains greater than the integrated result at all temperatures, the fractional difference decreases with elevated temperatures such that the integrated form provides a sufficiently accurate measure of q_R when $T/\Theta_R > 10$.

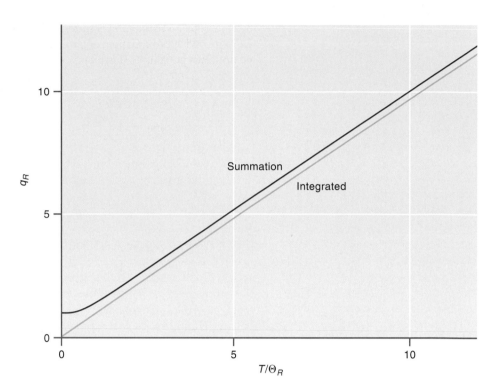

rotational partition function is reasonable. The integrated form of the partition function is referred to as the **high-temperature** or **high-T limit** because it is applicable when kT is significantly greater than the rotational energy spacings. The following example illustrates the use of the rotational temperature in deciding which functional form of the rotational partition function to use.

EXAMPLE PROBLEM 14.6

Evaluate the rotational partition functions for I_2 at $T = 100$ K.

Solution

Because $T = 100$ K, it is important to ask how kT compares to the rotational energy-level spacings. Using Table 14.1, $B(I_2) = 0.0374$ cm^{-1} corresponding to rotational temperatures of

$$\Theta_R(I_2) = \frac{hcB}{k} = \frac{(6.626 \times 10^{-34}\,\text{J s})(3.00 \times 10^{10}\,\text{cm s}^{-1})(0.0374\,\text{cm}^{-1})}{1.38 \times 10^{-23}\,\text{J K}^{-1}} = 0.054\,\text{K}$$

Comparison of these rotational temperatures to 100 K indicates that the high-temperature expression for the rotational partition function is valid for I_2:

$$q_R(I_2) = \frac{T}{\sigma\Theta_R} = \frac{100\,\text{K}}{(2)(0.054\,\text{K})} = 926$$

14.6 Rotational Partition Function: Polyatomics

In the diatomic systems described in the preceding section, there are two nonvanishing moments of inertia as illustrated in Figure 14.11. For **polyatomic** molecules (more than two atoms) the situation can become more complex.

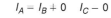

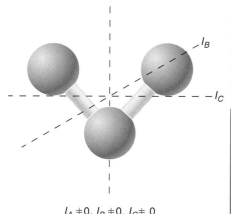

$I_A = I_B + 0 \quad I_C - 0$

$I_A \neq 0, I_B \neq 0, I_C \neq 0$

FIGURE 14.11
Moments of inertia for diatomic and nonlinear polyatomic molecules. Note that in the case of the diatomic, $I_C = 0$ in the limit that the atomic masses are considered to be point masses that reside along the axis connecting the two atomic centers. Each moment of inertia will have a corresponding rotational constant.

If the polyatomic system is linear, there are again only two nonvanishing moments of inertia such that a linear polyatomic molecule can be treated using the same formalism as diatomic molecules. However, if the polyatomic molecule is not linear, then there are three nonvanishing moments of inertia. Therefore, the partition function that describes the rotational energy levels must take into account rotation about all three axes. Derivation of this partition function is not trivial; therefore, the result without derivation is stated here:

$$q_R = \frac{\sqrt{\pi}}{\sigma} \left(\frac{1}{\beta hc B_A} \right)^{1/2} \left(\frac{1}{\beta hc B_B} \right)^{1/2} \left(\frac{1}{\beta hc B_C} \right)^{1/2} \tag{14.37}$$

The subscript on B in Equation (14.37) indicates the corresponding moment of inertia as illustrated in Figure 14.11, and σ is the symmetry number as discussed earlier. In addition, the assumption is made that the polyatomic is "rigid" during rotational motion. The development of the rotational partition function for diatomic systems provides some intuition into the origin of this partition function. One can envision each moment of inertia contributing $(\beta hc B)^{-1/2}$ to the overall partition function. In the case of diatomics or linear polyatomics, the two nonvanishing moments of inertia are equivalent such that the product of the contribution from each moment results in the expression for the diatomic derived earlier. For the nonlinear polyatomic system, the partition function is the product of the contribution from each of the moments of inertia, which may or may not be equivalent as indicated by the subscripts on the corresponding rotational constants in the partition function presented earlier.

EXAMPLE PROBLEM 14.7

Evaluate the rotational partition functions for the following species at 298 K. You can assume that the high-temperature expression is valid.

a. OCS ($B = 1.48$ cm^{-1})

b. ONCl ($B_A = 2.84$ cm^{-1}, $B_B = 0.191$ cm^{-1}, $B_C = 0.179$ cm^{-1})

c. CH$_2$O ($B_A = 9.40$ cm^{-1}, $B_B = 1.29$ cm^{-1}, $B_C = 1.13$ cm^{-1})

Solution

a. OCS is a linear molecule as indicated by the single rotational constant. In addition, the molecule is asymmetric such that $\sigma = 1$. Using the rotational constant, the rotational partition function is

$$q_R = \frac{1}{\sigma \beta hc B} = \frac{kT}{hcB} = \frac{(1.38 \times 10^{-23} \, \text{J K}^{-1})(298 \, \text{K})}{(6.626 \times 10^{-34} \, \text{J s})(3.00 \times 10^{10} \, \text{cm s}^{-1})(1.48 \, \text{cm}^{-1})} = 140$$

b. ONCl is a nonlinear polyatomic. It is asymmetric such that $\sigma = 1$, and the partition function becomes:

$$q_R = \frac{\sqrt{\pi}}{\sigma} \left(\frac{1}{\beta hc B_A} \right)^{1/2} \left(\frac{1}{\beta hc B_B} \right)^{1/2} \left(\frac{1}{\beta hc B_C} \right)^{1/2}$$

$$= \sqrt{\pi} \left(\frac{kT}{hc} \right)^{3/2} \left(\frac{1}{B_A} \right)^{1/2} \left(\frac{1}{B_B} \right)^{1/2} \left(\frac{1}{B_C} \right)^{1/2}$$

$$= \sqrt{\pi} \left(\frac{(1.38 \times 10^{-23} \, \text{J K}^{-1})(298 \, \text{K})}{(6.626 \times 10^{-34} \, \text{J s})(3.00 \times 10^{10} \, \text{cm s}^{-1})} \right)^{3/2} \left(\frac{1}{2.84 \, \text{cm}^{-1}} \right)^{1/2} \left(\frac{1}{0.191 \, \text{cm}^{-1}} \right)^{1/2}$$

$$\times \left(\frac{1}{0.179 \, \text{cm}^{-1}} \right)^{1/2}$$

$$= 16{,}940$$

c. CH_2O is a nonlinear polyatomic. However, the symmetry of this molecule is such that $\sigma = 2$. With this value for the symmetry number, the rotational partition function becomes:

$$q_R = \frac{\sqrt{\pi}}{\sigma}\left(\frac{1}{\beta hcB_A}\right)^{1/2}\left(\frac{1}{\beta hcB_B}\right)^{1/2}\left(\frac{1}{\beta hcB_C}\right)^{1/2}$$

$$= \frac{\sqrt{\pi}}{2}\left(\frac{kT}{hc}\right)^{3/2}\left(\frac{1}{B_A}\right)^{1/2}\left(\frac{1}{B_B}\right)^{1/2}\left(\frac{1}{B_C}\right)^{1/2}$$

$$= \frac{\sqrt{\pi}}{2}\left(\frac{(1.38\times10^{-23}\,\mathrm{J\,K^{-1}})(298\ \mathrm{K})}{(6.626\times10^{-34}\,\mathrm{J\,s})(3.00\times10^{10}\,\mathrm{cm\,s^{-1}})}\right)^{3/2}\left(\frac{1}{9.40\ \mathrm{cm^{-1}}}\right)^{1/2}\left(\frac{1}{1.29\ \mathrm{cm^{-1}}}\right)^{1/2}$$

$$\times\left(\frac{1}{1.13\ \mathrm{cm^{-1}}}\right)^{1/2}$$

$$= 711$$

Note that the values for all three partition functions indicate that a substantial number of rotational states are populated at room temperature.

14.7 Vibrational Partition Function

The quantum mechanical model for vibrational degrees of freedom is the harmonic oscillator. In this model, each vibrational degree of freedom is characterized by a quadratic potential as illustrated in Figure 14.12. The energy levels of the harmonic oscillator are as follows:

$$E_n = hc\tilde{\nu}\left(n + \frac{1}{2}\right) \tag{14.38}$$

This equation demonstrates that the energy of a given level, E_n, is dependent on the quantum number n, which can take on integer values beginning with zero ($n = 0, 1, 2, \ldots$). The frequency of the oscillator, or vibrational frequency, is given by $\tilde{\nu}$ in units of $\mathrm{cm^{-1}}$. Note that the energy of the $n = 0$ level is not zero, but $hc\tilde{\nu}/2$. This residual energy is known as the zero point energy and was discussed in detail during the quantum mechanical development of the harmonic oscillator. The expression for E_n provided in Equation 14.38 can be used to construct the vibrational partition function as follows:

$$q_V = \sum_{n=0}^{\infty} e^{-\beta E_n} \tag{14.39}$$

$$= \sum_{n=0}^{\infty} e^{-\beta hc\tilde{\nu}\left(n+\frac{1}{2}\right)}$$

$$= e^{-\beta hc\tilde{\nu}/2}\sum_{n=0}^{\infty} e^{-\beta hc\tilde{\nu}n}$$

The sum can be rewritten using the series identity:

$$\frac{1}{1 - e^{-\alpha x}} = \sum_{n=0}^{\infty} e^{-n\alpha x} \tag{14.40}$$

With this substitution, we arrive at the following expression for the **vibrational partition function:**

$$q_V = \frac{e^{-\beta hc\tilde{\nu}/2}}{1 - e^{-\beta hc\tilde{\nu}}} \quad \text{(with zero point energy)} \tag{14.41}$$

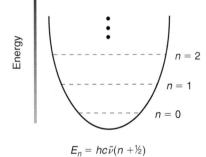

$$E_n = hc\tilde{\nu}(n + \tfrac{1}{2})$$

FIGURE 14.12

The harmonic oscillator model. Each vibrational degree of freedom is characterized by a quadratic potential. The energy levels corresponding to this potential are evenly spaced.

Although this expression is correct as written, at times it is advantageous to redefine the vibrational energy levels such that $E_0 = 0$. In other words, the energy of all levels is decreased by an amount equal to the zero point energy. Why would this be an advantageous thing to do? Consider the calculation of the probability of occupying a given vibrational energy level, p_n, as follows:

$$p_n = \frac{e^{-\beta E_n}}{q_V} = \frac{e^{-\beta hc\tilde{v}\left(n+\frac{1}{2}\right)}}{\frac{e^{-\beta hc\tilde{v}/2}}{1-e^{-\beta hc\tilde{v}}}} = \frac{e^{-\beta hc\tilde{v}/2}e^{-\beta hc\tilde{v}n}}{\frac{e^{-\beta hc\tilde{v}/2}}{1-e^{-\beta hc\tilde{v}}}} = e^{-\beta hc\tilde{v}n}\left(1-e^{-\beta hc\tilde{v}}\right) \qquad (14.42)$$

Notice that in Equation (14.42) the zero point energy contributions for both the energy level and the partition function cancel. Therefore, the relevant energy for determining p_n is not the absolute energy of a given level, but the *relative* energy of the level. Given this, one can simply eliminate the zero point energy, resulting in the following expression for the vibrational partition function:

$$q_V = \frac{1}{1-e^{-\beta hc\tilde{v}}} \quad \text{(without zero point energy)} \qquad (14.43)$$

It is important to be consistent in including or not including zero point energy. For example, what if one were to perform the probability calculation presented earlier including zero point energy for the vibrational state of interest, but not including zero point energy in the expression for the vibrational partition function? Proceeding as before, we arrive at the following *incorrect* result:

$$p_n = \frac{e^{-\beta E_n}}{q_V} = \frac{e^{-\beta hc\tilde{v}\left(n+\frac{1}{2}\right)}}{\frac{1}{1-e^{-\beta hc\tilde{v}}}} = \frac{e^{-\beta hc\tilde{v}/2}e^{-\beta hc\tilde{v}n}}{\frac{1}{1-e^{-\beta hc\tilde{v}}}} = e^{-\beta hc\tilde{v}/2}e^{-\beta hc\tilde{v}n}(1-e^{-\beta hc\tilde{v}})$$

Note that the zero point energy terms did not cancel as in the previous case, reflecting the fact that the energy of state n is defined differently relative to the expression for the partition function. In summary, once a decision has been made to include or ignore zero point energy, the approach taken must be consistently applied.

EXAMPLE PROBLEM 14.8

At what temperature will the vibrational partition function for I_2 ($\tilde{v} = 208$ cm^{-1}) be greatest: 298 or 1000 K?

Solution

Because the partition function is a measure of the number of states that are accessible given the amount of energy available (kT), we would expect the partition function to be greater for $T = 1000$ K relative to $T = 298$ K. We can confirm this expectation by numerically evaluating the vibrational partition function at these two temperatures:

$$(q_V)_{298K} = \frac{1}{1-e^{-\beta hc\tilde{v}}} = \frac{1}{1-e^{-hc\tilde{v}/kT}}$$

$$= \frac{1}{1-\exp\left[-\dfrac{(6.626\times10^{-34}\,\text{J s})(3.00\times10^{10}\,\text{cm s}^{-1})(208\,\text{cm}^{-1})}{(1.38\times10^{-23}\,\text{J K}^{-1})(298\,\text{K})}\right]} = 1.58$$

$$(q_V)_{1000K} = \frac{1}{1-e^{-\beta hc\tilde{v}}}$$

$$= \frac{1}{1-\exp\left[-\dfrac{(6.626\times10^{-34}\,\text{J s})(3.00\times10^{10}\,\text{cm s}^{-1})(208\,\text{cm}^{-1})}{(1.38\times10^{-23}\,\text{J K}^{-1})(1000\,\text{K})}\right]} = 3.86$$

Consistent with our expectation, the partition function increases with temperature, indicating that more states are accessible at elevated temperatures. The variation of q_V with temperature for I_2 is shown here:

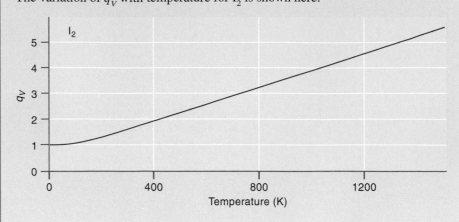

14.7.1 Beyond Diatomics: Multidimensional q_V

The expression for the vibrational partition function derived in the preceding subsection is for a single vibrational degree of freedom and is sufficient for diatomic molecules. However, triatomics and larger molecules (collectively referred to as polyatomics) require a different form for the partition function that takes into account all vibrational degrees of freedom. To define the vibrational partition function for polyatomics, we first need to know how many vibrational degrees of freedom there will be. A polyatomic molecule consisting of N atoms has $3N$ total degrees of freedom corresponding to three cartesian degrees of freedom for each atom. The atoms are connected by chemical bonds; therefore, the atoms are not free to move independently of each other. First, the entire molecule can translate through space; therefore, three of the $3N$ total degrees of freedom correspond to translational motion of the entire molecule. Next, a rotational degree of freedom will exist for each nonvanishing moment of inertia. As discussed in the section on rotational motion, linear polyatomics have two rotational degrees of freedom because there are two nonvanishing moments of inertia, and nonlinear polyatomic molecules have three rotational degrees of freedom. The remaining degrees of freedom are vibrational such that the number of vibrational degrees of freedom are

Linear polyatomics	$3N - 5$	(14.44)
Nonlinear polyatomics	$3N - 6$	(14.45)

Note that a diatomic molecule can be viewed as linear polyatomic with $N = 2$, and the preceding expressions dictate that there is only one vibrational degree of freedom $[3(2) - 5 = 1]$ as stated earlier.

The final step in deriving the partition function for a polyatomic system is to recognize that within the harmonic approximation, the vibrational degrees of freedom are separable, and each vibration can be treated as a separate energetic degree of freedom. In Section 14.3, various forms of molecular energy were shown to be separable so that the total molecular partition function is simply the sum of the partition functions for each energetic degree of freedom. Similar logic applies to vibrational degrees of freedom where the total vibrational partition function is simply the product of vibrational partition functions for each vibrational degree of freedom:

$$(q_V)_{Total} = \sum_{i=1}^{3N-5 \text{ or } 3N-6} (q_V)_i \qquad (14.46)$$

In Equation (14.46), the total vibrational partition function is equal to the product of vibrational partition functions for each vibrational mode (denoted by the subscript i). There will be $3N - 5$ or $3N - 6$ mode-specific partition functions depending on the geometry of the molecule.

EXAMPLE PROBLEM 14.9

The triatomic chlorine dioxide (OClO) has three vibrational modes of frequency: 450, 945, and 1100 cm^{-1}. What is the value of the vibrational partition function for $T = 298$ K?

Solution

The total vibrational partition function is simply the product of partition functions for each vibrational degree of freedom. Setting the zero point energy equal to zero, we find that

$$q_{450} = \frac{1}{1 - e^{-\beta hc(450 \text{ cm}^{-1})}} = \frac{1}{1 - \exp\left[-\dfrac{(6.626 \times 10^{-34} \text{ J s})(3.00 \times 10^{10} \text{ cm s}^{-1})(450 \text{ cm}^{-1})}{(1.38 \times 10^{-23} \text{ J s})(298 \text{ K})}\right]}$$

$$= 1.12$$

$$q_{945} = \frac{1}{1 - e^{-\beta hc(945 \text{ cm}^{-1})}} = \frac{1}{1 - \exp\left[-\dfrac{(6.626 \times 10^{-34} \text{ J s})(3.00 \times 10^{10} \text{ cm s}^{-1})(945 \text{ cm}^{-1})}{(1.38 \times 10^{-23} \text{ J s})(298 \text{ K})}\right]}$$

$$= 1.01$$

$$q_{1100} = \frac{1}{1 - e^{-\beta hc(1100 \text{ cm}^{-1})}}$$

$$= \frac{1}{1 - \exp\left[-\dfrac{(6.626 \times 10^{-34} \text{ J s})(3.00 \times 10^{10} \text{ cm s}^{-1})(1100 \text{ cm}^{-1})}{(1.38 \times 10^{-23} \text{ J s})(298 \text{ K})}\right]} = 1.00$$

$$(q_V)_{Total} = \prod_{i=1}^{3N-6} (q_V)_i = (q_{450})(q_{950})(q_{1100}) = (1.12)(1.01)(1.00) = 1.13$$

Note that the total vibrational partition function is close to unity. This is consistent with the fact that the vibrational energy spacings for all modes are significantly greater than kT such that few states other than $n = 0$ are populated.

14.7.2 High Temperature Approximation to q_V

Similar to the development of rotations, the **vibrational temperature** (Θ_V) is defined as the frequency of a given vibrational degree of freedom divided by k:

$$\Theta_V = \frac{hc\tilde{\nu}}{k} \tag{14.47}$$

Unit analysis of Equation (14.47) dictates that Θ_V will have units of temperature (K). We can incorporate this term into our expression for the vibrational partition function as follows:

$$q_V = \frac{1}{1 - e^{-\beta hc\tilde{\nu}}} = \frac{1}{1 - e^{-hc\tilde{\nu}/kT}} = \frac{1}{1 - e^{-\Theta_V/T}} \tag{14.48}$$

The utility of this form of the partition function is that the relationship between vibrational energy and temperature becomes transparent. Specifically, as T becomes large relative to

Θ_V, the exponent becomes smaller and the exponential term approaches one. The denominator in Equation (14.48) will decrease such that the vibrational partition function will increase. If the temperature becomes sufficiently large relative to Θ_V, q_V can be reduced to a simpler form. The Math Supplement (Appendix B) provides the following series expression for $\exp(-x)$:

$$e^{-x} = 1 - x + \frac{x^2}{2} - \dots$$

For the vibrational partition function in Equation (14.48), $x = -\Theta_V/T$. When $T \gg \Theta_V$, x becomes sufficiently small that only the first two terms can be included in the series expression for $\exp(-x)$ since higher order terms are negligible. Substituting into the expression for the vibrational partition function:

$$q_V = \frac{1}{1 - e^{-\Theta_V/T}} = \frac{1}{1 - \left(1 - \frac{\Theta_V}{T}\right)} = \frac{T}{\Theta_V} \tag{14.49}$$

This result is the high-temperature (or high-T) limit for the vibrational partition function.

$$q_V = \frac{T}{\Theta_V} \quad \text{(high-T limit)} \tag{14.50}$$

When is Equation (14.50) appropriate for evaluating q_V as opposed to the exact expression? The answer to this question depends on both the vibrational frequency of interest and the temperature. Figure 14.13 provides a comparison for I_2 ($\Theta_V = 299$ K) between the exact expression without zero point energy [Equation (14.48)] and high-T expression [Equation (14.50)] for q_V. Similar to the case for rotations, the two results are significantly different at low temperature, but predict the same linear dependence on T at elevated temperatures. When $T \geq 10\,\Theta_V$, the fractional difference between the high-T and exact results is sufficiently small that the high-T result for q_V can be used. For the majority of molecules, this temperature will be extremely high as shown in Example Problem 14.10.

EXAMPLE PROBLEM 14.10

At what temperature is the high-T limit for q_V appropriate for F_2 ($\tilde{\nu} = 917$ cm^{-1})?

Solution
The high-T limit is applicable when $T = 10\,\Theta_V$. The vibrational temperature for F_2 is

$$\Theta_V = \frac{hc\tilde{\nu}}{k} = \frac{(6.626 \times 10^{-34}\,\text{J s})(3.00 \times 10^{10}\,\text{cm s}^{-1})(917\,\text{cm}^{-1})}{1.38 \times 10^{-23}\,\text{J K}^{-1}} = 1319\,\text{K}$$

Therefore, the high-T limit is applicable when T = ~13,000 K. To make sure this is indeed the case, we can compare the value for q_V determined by both the full expression for the partition function and the high-T approximation:

$$q_v = \frac{1}{1 - e^{-\Theta_V/T}} = \frac{1}{1 - e^{-1319\,\text{K}/13,000\,\text{K}}} = 10.4$$

$$= \frac{T}{\Theta_V} = \frac{13,000\,\text{K}}{1319\,\text{K}} = 9.9$$

Comparison of the two methods for evaluating the partition function demonstrates that the high-T limit expression provides a legitimate estimate of the partition function at this temperature. However, the temperature at which this is true is exceedingly high.

FIGURE 14.13
Comparison between the exact
[Equation (14.48)] and high-T [Equation
(14.50)] results for q_V. Parameters
employed in the calculation correspond to
$I_2\,(\Theta_V = 299\text{ K})$. The fractional difference
between the two results is small for
temperatures where $T/\Theta_V \geq 10$.

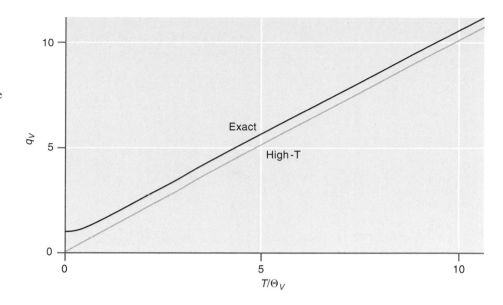

14.7.3 Degeneracy and q_V

The total vibrational partition function for a polyatomic molecule is the product of the
partition functions for each vibrational degree of freedom. What if two or more of these
vibrational degrees of freedom have the same frequency such that the energy levels are
degenerate? It is important to keep in mind that there will always be $3N - 5$ or $3N - 6$
vibrational degrees of freedom depending on geometry. In the case of degeneracy, two
or more of these degrees of freedom will have the same vibrational energy spacings, or
the same vibrational temperature. The total partition function is still the product of partition functions for each vibrational degree of freedom; however, the degenerate vibrational modes have identical partition functions. There are two ways to incorporate the
effects of vibrational degeneracy into the expression for the vibrational partition function. First, one can simply use the existing form of the partition function and keep track
of all degrees of freedom irrespective of frequency. A second method is to rewrite the total partition function as a product of partition functions corresponding to a given vibrational frequency and include degeneracy at a given frequency resulting in the
corresponding partition function being raised to the power of the degeneracy:

$$(q_V)_{Total} = \sum_{i=1}^{n'} (q_V)_i^{g_i} \tag{14.51}$$

where n' is the total number of unique vibrational frequencies indexed by i. It is important to note that n' is *not* the number of vibrational degrees of freedom! Carbon dioxide serves as a classic example of vibrational degeneracy, as Example Problem 14.11
illustrates.

EXAMPLE PROBLEM 14.11

CO_2 has the following vibrational degrees of freedom: 1388, 667.4 (doubly
degenerate), and 2349 cm^{-1}. What is the total vibrational partition function for
this molecule at 1000 K?

Solution
Evaluation of the partition function can be performed by calculating the
individual vibrational partition functions for each unique frequency, then taking

the product of these partition functions raised to the power of the degeneracy at a given frequency:

$$(q_V)_{1388} = \cfrac{1}{1 - \exp\left[-\cfrac{(6.626 \times 10^{-34}\,\text{J s})(3.00 \times 10^{10}\,\text{cm s}^{-1})(1388\,\text{cm}^{-1})}{(1.38 \times 10^{-23}\,\text{J K}^{-1})(1000\,\text{K})} \right]} = 1.16$$

$$(q_V)_{667.4} = \cfrac{1}{1 - \exp\left[-\cfrac{(6.626 \times 10^{-34}\,\text{J s})(3.00 \times 10^{10}\,\text{cm s}^{-1})(667.4\,\text{cm}^{-1})}{(1.38 \times 10^{-23}\,\text{J K}^{-1})(1000\,\text{K})} \right]} = 1.62$$

$$(q_V)_{2349} = \cfrac{1}{1 - \exp\left[-\cfrac{(6.626 \times 10^{-34}\,\text{J s})(3.00 \times 10^{10}\,\text{cm s}^{-1})(2349\,\text{cm}^{-1})}{(1.38 \times 10^{-23}\,\text{J K}^{-1})(1000\,\text{K})} \right]} = 1.04$$

$$(q_V)_{Total} = \prod_{i=1}^{n'} (q_V)_i^{g_i} = (q_V)_{1388} (q_V)_{667.4}^2 (1.04)_{2349}$$

$$= (1.16)(1.62)^2(1.04) = 3.17$$

14.8 The Equipartition Theorem

In the previous sections regarding rotations and vibrations, equivalence of the high-T and exact expressions for q_R and q_V, respectively, was observed when the temperature was sufficiently large that the thermal energy available to the system was significantly greater than the energy-level spacings. At these elevated temperatures, the quantum nature of the energy levels becomes unimportant and a classical description of the energetics is all that is needed.

The definition of partition function involves summation over quantized energy levels, and one might assume that there is a corresponding classical expression for the partition function in which a classical description of the system energetics is employed. Indeed there is such an expression; however, its derivation is beyond the scope of this text, so we simply state the result here. The expression for the three-dimensional partition function for a molecule consisting of N atoms is

$$q_{classical} = \frac{1}{h^{3N}} \int \dots \int e^{-\beta H} \, dp^{3N} \, dx^{3N} \tag{14.52}$$

In the expression for the partition function, the terms p and x represent the momentum and position coordinates for each particle, respectively, with three Cartesian dimensions available for each term. The integral is multiplied by h^{-3N}, which has units of (momentum $\times$ distance)$^{-3N}$ such that the partition function is unitless.

What does the term $e^{-\beta H}$ represent in $q_{classical}$? The H represents the classical Hamiltonian and, like the quantum Hamiltonian, is the sum of a system's kinetic and potential energy. Therefore, $e^{-\beta H}$ is equivalent to $e^{-\beta \varepsilon}$ in our quantum expression for the molecular partition function. Consider the Hamiltonian for a classical one-dimensional harmonic oscillator with reduced mass μ and force constant k:

$$H = \frac{p^2}{2\mu} + \frac{1}{2}kx^2 \tag{14.53}$$

Using this Hamiltonian, the corresponding classical partition function for the one-dimensional harmonic oscillator is

$$q_{classical} = \frac{1}{h} \int dp \int dx \, e^{-\beta\left(\frac{p^2}{2\mu} + \frac{1}{2}kx^2\right)} = \frac{T}{\Theta_V} \tag{14.54}$$

This result is in agreement with the high-T approximation to q_V derived using the quantum partition function of Equation (14.50). This example illustrates the applicability of classical statistical mechanics to molecular systems when the temperature is sufficiently high such that summation over the quantum states can be replaced by integration. Under these temperature conditions, knowledge of the quantum details of the system is not necessary because, when evaluating Equation (14.54), nothing was implied regarding the quantization of the harmonic oscillator energy levels.

The applicability of classical statistical mechanics to molecular systems at high temperature finds application in an interesting theorem known as the **equipartition theorem.** This theorem states that any term in the classical Hamiltonian that is quadratic with respect to momentum or position (i.e., p^2 or x^2) will contribute $kT/2$ to the average energy. For example, the Hamiltonian for the one-dimensional harmonic oscillator [Equation (14.53)] has both a p^2 and x^2 term such that the average energy for the oscillator by equipartition should be kT (or NkT for a collection of N harmonic oscillators). In the next chapter, the equipartition result will be directly compared to the average energy determined using quantum statistical mechanics. At present, it is important to recognize that the concept of equipartition is a consequence of classical mechanics because, for a given energetic degree of freedom, the change in energy associated with passing from one energy level to the other must be significantly less than kT. As discussed earlier, this is true for translational and rotational degrees of freedom, but is not the case for vibrational degrees of freedom except at relatively high temperatures.

14.1 Variation of q with Temperature

14.9 Electronic Partition Function

Electronic energy levels correspond to the various arrangements of electrons in an atom or molecule. The hydrogen atom provides an excellent example of an atomic system where the orbital energies are given by [see Equation (9.11) in *Quantum Chemistry and Spectroscopy*]:

$$E_n = \frac{-m_e e^4}{8\varepsilon_o^2 h^2 n^2} = 109,737\,\text{cm}^{-1}\frac{1}{n^2} \quad (n = 1,2,3,...) \tag{14.55}$$

This expression demonstrates that the energy of a given orbital in the hydrogen atom is dependent on the quantum number n. In addition, each orbital has a degeneracy of $2n^2$. Using Equation (14.55), the energy levels for the electron in the hydrogen atom can be determined as illustrated in Figure 14.14.

From the perspective of statistical mechanics, the energy levels of the hydrogen atom represent the energy levels for the electronic energetic degree of freedom, with the corresponding partition function derived by summing over the energy levels. However, rather than use the absolute energies as determined in the quantum mechanical solution to the hydrogen atom problem, we will adjust the energy levels such that the energy associated with the $n = 1$ orbital is zero, similar to the adjustment of the ground-state energy of the harmonic oscillator to zero by elimination of the zero point energy. With this redefinition of the orbital energies, the electronic partition for the hydrogen atom becomes

$$q_E = \sum_{n=1}^{\infty} g_n e^{-\beta hc E_n} = 2e^{-\beta hc E_1} + 8e^{-\beta hc E_2} + 18e^{-\beta hc E_3} + ... \tag{14.56}$$

$$= 2e^{-\beta hc(0\ \text{cm}^{-1})} + 8e^{-\beta hc(82,303\ \text{cm}^{-1})} + 18e^{-\beta hc(97,544\ \text{cm}^{-1})} + ...$$

$$= 2 + 8e^{-\beta hc(82,303\ \text{cm}^{-1})} + 18e^{-\beta hc(97,544\ \text{cm}^{-1})} + ...$$

The magnitude of the terms in the partition function corresponding to $n \geq 2$ will depend on the temperature at which the partition function is being evaluated. However, note that

FIGURE 14.14
Orbital energies for the hydrogen atom.
(a) The orbital energies as dictated by
solving the Schrödinger equation for the
hydrogen atom. **(b)** The energy levels
shifted by the addition of 109,737 cm^{-1} of
energy such that the lowest orbital energy
is 0.

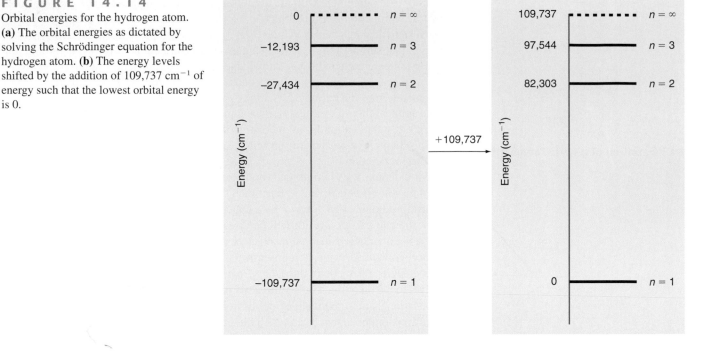

these energies are quite large. Consider defining the "electronic temperature" or Θ_E, in exactly the same way as the rotational and vibrational temperatures were defined:

$$\Theta_E = \frac{hcE_n}{k} = \frac{E_n}{0.695 \text{ cm}^{-1}\text{ K}^{-1}} \tag{14.57}$$

With the definition of $E_1 = 0$, the energy of the $n = 2$ orbital, E_2, is 82,303 cm^{-1} corresponding to $\Theta_E = 118{,}421$ K! This is an extremely high temperature, and this simple calculation illustrates the primary difference between electronic energy levels and the other energetic degrees of freedom discussed thus far. Electronic degrees of freedom are generally characterized by level spacings that are quite large relative to kT. Therefore, only the ground electronic state is populated to a significant extent (although exceptions are known, as presented in the problems at the end of this chapter). Applying this conceptual picture to the hydrogen atom, the terms in the partition function corresponding to $n \geq 2$ should be quite small at 298 K. For example, the term for the $n = 2$ state is as follows:

$$e^{-\beta hcE_2} = \exp\left[\frac{-(6.626\times10^{-34}\text{ J s})(3.00\times10^{10}\text{ cm s}^{-1})(82{,}303\text{ cm}^{-1})}{(1.38\times10^{-23}\text{ J K}^{-1})(298\text{ K})}\right] = e^{-397.3} \approx 0$$

Terms corresponding to higher energy orbitals will also be extremely small such that the electronic partition function for the hydrogen atom is ~2 at 298 K. In general, the contribution of each state to the partition function must be considered, resulting in the following expression for the **electronic partition function:**

$$q_E = \sum_n g_n e^{-\beta hcE_n} \tag{14.58}$$

In the expression for the electronic partition function, the exponential term for each energy level is multiplied by the degeneracy of the level, g_n. If the energy level spacings are very large compared to kT, then $q_E \approx g_O$, or the degeneracy of the ground state. However, certain atoms and molecules may have excited electronic states that are energetically accessible relative to kT, and the contribution of these states must be included in evaluating the partition function. The following problem provides an example of such a system.

EXAMPLE PROBLEM 14.12

The lowest nine energy levels for gaseous vanadium (V) have the following energies and degeneracies:

Level(n)	Energy(cm^{-1})	Degeneracy
0	0	4
1	137.38	6
2	323.46	8
3	552.96	10
4	2112.28	2
5	2153.21	4
6	2220.11	6
7	2311.36	8
8	2424.78	10

What is the value of the electronic partition function for V at 298 K?

Solution

Due to the presence of unpaired electrons in V, the electronic excited states are accessible relative to kT. Therefore, the partition function is not simply equal to the ground-state degeneracy and must instead be determined by writing out the summation explicitly, paying careful attention to the energy and degeneracy of each level:

$$q_E = \sum_n g_n e^{-\beta hcE_n} = g_o e^{-\beta hcE_0} + g_1 e^{-\beta hcE_1} + g_2 e^{-\beta hcE_2} + g_3 e^{-\beta hcE_0} + g_4 e^{-\beta hcE_0} + \dots$$

$$= 4\exp\left[\frac{-0 \text{ cm}^{-1}}{(0.695 \text{ cm}^{-1}\text{ K}^{-1})(298 \text{ K})}\right] + 6\exp\left[\frac{-137.38 \text{ cm}^{-1}}{(0.695 \text{ cm}^{-1}\text{ K}^{-1})(298 \text{ K})}\right]$$

$$+ 8\exp\left[\frac{-323.46 \text{ cm}^{-1}}{(0.695 \text{ cm}^{-1}\text{ K}^{-1})(298 \text{ K})}\right] + 10\exp\left[\frac{-552.96 \text{ cm}^{-1}}{(0.695 \text{ cm}^{-1}\text{ K}^{-1})(298 \text{ K})}\right]$$

$$+ 2\exp\left[\frac{-2112.28 \text{ cm}^{-1}}{(0.695 \text{ cm}^{-1}\text{ K}^{-1})(298 \text{ K})}\right] + \dots$$

Notice in the preceding expression that the energy of state $n = 4$ (2112.28 cm^{-1}) is large with respect to kT (208 cm^{-1}). The exponential term for this state is approximately e^{-10}, or 4.5×10^{-5}. Therefore, the contribution of state $n = 4$ and higher energy states to the partition function will be extremely modest, and these terms can be disregarded when evaluating the partition function. Focusing on the lower energy states that make the dominant contribution to the partition function results in the following:

$$q_E \approx 4\exp\left[\frac{-0 \text{ cm}^{-1}}{(0.695 \text{ cm}^{-1}\text{ K}^{-1})(298 \text{ K})}\right] + 6\exp\left[\frac{-137.38 \text{ cm}^{-1}}{(0.695 \text{ cm}^{-1}\text{ K}^{-1})(298 \text{ K})}\right]$$

$$+ 8\exp\left[\frac{-323.46 \text{ cm}^{-1}}{(0.695 \text{ cm}^{-1}\text{ K}^{-1})(298 \text{ K})}\right] + 10\exp\left[\frac{-552.96 \text{ cm}^{-1}}{(0.695 \text{ cm}^{-1}\text{ K}^{-1})(298 \text{ K})}\right]$$

$$\approx 4 + 6(0.515) + 8(0.211) + 10(0.070)$$

$$\approx 9.49$$

In summary, if the energy of an electronic excited state is sufficiently greater than kT, the contribution of the state to the electronic partition function will be minimal and the state can be disregarded in numerical evaluation of the partition function.

Although the previous discussion focused on atomic systems, similar logic applies to molecules. Molecular electronic energy levels were first introduced in the discussion of molecular orbital (MO) theory in Chapter 14 of *Quantum Chemistry and Spectroscopy*. In MO theory, linear combinations of atomic orbitals are used to construct a new set of electronic orbitals known as molecular orbitals. The molecular orbitals differ in energy, and the electronic configuration of the molecule is determined by placing spin-paired electrons into the orbitals starting with the lowest energy orbital. The highest energy occupied molecular orbital was designated as the HOMO or highest occupied molecular orbital. The molecular orbital energy-level diagram for butadiene is presented in Figure 14.15.

Figure 14.15 presents both the lowest and next highest energy electronic energy states for butadiene, with the difference in energies corresponding to the promotion of an electron from the HOMO to the lowest unoccupied molecular orbital (LUMO). The separation in energy between these two states corresponds to the amount of energy it takes to excite the electron. The wavelength of the lowest energy electronic transition of butadiene is ~220 nm, demonstrating that the separation between the HOMO and LUMO is ~45,000 cm^{-1}, significantly greater than kT at 298 K. As such, only the lowest energy electronic state contributes to the electronic partition function, and $q_E = 1$ for butadiene since the degeneracy of the lowest energy level is one. Typically, the first electronic excited level of molecules will reside 5000 to 50,000 cm^{-1} higher in energy than the lowest level such that at room temperature only this lowest level is considered in evaluating the partition function:

$$q_E = \sum_{n=0} g_n e^{-\beta E_n} \approx g_0 \qquad (14.59)$$

FIGURE 14.15
Depiction of the molecular orbitals for butadiene. The highest occupied molecular orbital (HOMO) is indicated by the yellow rectangle, and the lowest unoccupied molecular orbital (LUMO) is indicated by the pink rectangle. The lowest energy electron configuration is shown on the left, and the next highest energy configuration is shown on the right, corresponding to the promotion of an electron from the HOMO to the LUMO.

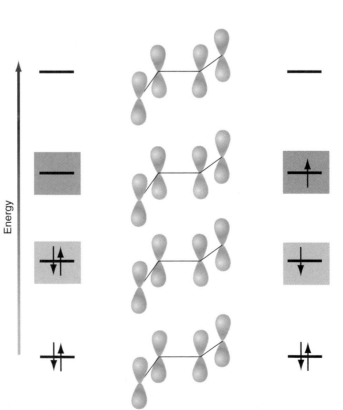

In the absence of degeneracy of the ground electronic state, the electronic partition function will equal one.

14.10 Review

Given the numerous derivations, equations, and examples provided in this chapter, it is important for the reader to focus on the primary concepts developed throughout. There are two primary goals of this chapter: to relate the canonical partition, Q, to the molecular partition function, q, and to express the molecular partition function in terms of the individual energetic degrees of freedom. The relationship between the molecular and canonical partition function depends on whether the individual units comprising the ensemble are distinguishable or indistinguishable such that

$$Q_{Total} = q_{Total}^N \quad \text{(distinguishable)}$$

$$Q_{Total} = \frac{1}{N!} q_{Total}^N \quad \text{(indistinguishable)}$$

Evaluation of the canonical partition function requires knowledge of the total molecular partition function, which is equal to the product of the partition functions for each energetic degree of freedom:

$$q_{Total} = q_T q_R q_V q_E$$

An example will help to illustrate the approach that was developed in this chapter. Imagine that we are interested in the canonical partition function for a mole of a diatomic molecule at an arbitrary temperature. First, gaseous molecules are indistinguishable such that

$$Q = \frac{1}{N_A!} q_{Total}^{N_A}$$

Next, the expression for the total molecular partition function is simply given by

$$q_{Total} = q_T q_R q_V q_E$$

$$= \left(\frac{V}{\Lambda^3}\right)\left(\frac{1}{\sigma \beta hcB}\right)\left(\frac{1}{1 - e^{-\beta hc\tilde{v}}}\right)(g_0)$$

Notice that we have used the appropriate expressions for the rotational and vibrational partition functions for a diatomic molecule. Numerical evaluation of the preceding equation requires knowledge of the specific molecular parameters, the volume of the gas, and the temperature of the ensemble. However, the conceptual approach is universal, and illustrates the connection between the ensemble Q and the microscopic properties of the individual units of the ensemble q. Using this approach, we can connect the microscopic description of molecules to the macroscopic behavior of molecular ensembles, an undertaking we explore in the next chapter.

For Further Reading

Chandler, D., *Introduction to Modern Statistical Mechanics.* Oxford, New York, 1987.

Hill, T., *Statistical Mechanics. Principles and Selected Applications.* Dover, New York, 1956.

McQuarrie, D., *Statistical Mechanics.* Harper & Row, New York, 1973.

Nash, L. K., "On the Boltzmann Distribution Law." *J. Chemical Education* 59 (1982), 824.

Nash, L. K., *Elements of Statistical Thermodynamics.* Addison-Wesley, San Francisco, 1972.

Noggle, J. H., *Physical Chemistry.* HarperCollins, New York, 1996.

Townes, C. H., and A. L. Schallow, *Microwave Spectroscopy.* Dover, New York, 1975. (This book contains an excellent appendix of spectroscopic constants.)

Widom, B., *Statistical Mechanics.* Cambridge University Press, Cambridge, 2002.

Vocabulary

canonical ensemble	high-temperature (high-T) limit	rotational temperature
canonical partition function	molecular partition function	symmetry number
diatomic	*ortho*-hydrogen	thermal de Broglie wavelength
electronic partition function	*para*-hydrogen	thermal wavelength
energetic degrees of freedom	polyatomic	translational partition function
ensemble	rotational constant	vibrational partition function
equipartition theorem	rotational partition function	vibrational temperature

Questions on Concepts

Q14.1 What is the canonical ensemble? What properties are held constant in this ensemble?

Q14.2 What is the relationship between Q and q? How does this relationship differ if the particles of interest are distinguishable versus indistinguishable?

Q14.3 List the atomic and/or molecular energetic degrees of freedom discussed in this chapter. For each energetic degree of freedom, briefly summarize the corresponding quantum mechanical model.

Q14.4 For which energetic degrees of freedom are the spacings between energy levels small relative to kT at room temperature?

Q14.5 For the translational and rotational degrees of freedom, evaluation of the partition function involved replacement of the summation by integration. Why could

integration be performed? How does this relate back to the discussion of probability distributions of discrete variables treated as continuous?

Q14.6 What is the high-T approximation for rotations and vibrations? For which degree of freedom do you expect this approximation to be generally valid at room temperature?

Q14.7 State the equipartition theorem. Why is this theorem inherently classical?

Q14.8 Why is the electronic partition function generally equal to the degeneracy of the ground electronic state?

Q14.9 What is q_{Total}, and how is it constructed using the partition functions for each energetic degree of freedom discussed in this chapter?

Q14.10 Why is it possible to set the energy of the ground vibrational and electronic energy level to zero?

Problems

Problem numbers in **red** indicate that the solution to the problem is given in the *Student's Solutions Manual*.

P14.1 Evaluate the translational partition function for H_2 confined to a volume of 100 cm^3 at 298 K. Perform the same calculation for N_2 under identical conditions (*Hint:* Do you need to reevaluate the full expression for q_T?).

P14.2 Evaluate the translational partition function for Ar confined to a volume of 1000 cm^3 at 298 K. At what temperature will the translational partition function of Ne be identical to that of Ar at 298 K confined to the same volume?

P14.3 At what temperature are there Avogadro's number of translational states available for O_2 confined to a volume of 1000 cm^3?

P14.4 Imagine gaseous Ar at 298 K confined to move in a two-dimensional plane of area 1.00 cm^2. What is the value of the translational partition function?

P14.5 For N_2 at 77.3 K, 1 atm, in a 1-cm^3 container, calculate the translational partition function and ratio of this

partition function to the number of N_2 molecules present under these conditions.

P14.6 What is the symmetry number for the following molecules?

a. $^{35}Cl^{37}Cl$

b. $^{35}Cl^{35}Cl$

c. H_2O

d. C_6H_6

e. CH_2Cl_2

P14.7 Which species will have the largest rotational partition function: H_2, HD, or D_2? Which of these species will have the largest translational partition function assuming that volume and temperature are identical? When evaluating the rotational partition functions, you can assume that the high-temperature limit is valid.

P14.8 Consider *para*-H_2 ($B = 60.853$ cm^{-1}) for which only even-J levels are available. Evaluate the rotational partition function for this species at 50 K. Perform this same calculation for HD ($B = 45.655$ cm^{-1}).

P14.9 For which of the following diatomic molecules is the high-temperature expression for the rotational partition function valid if $T = 40$ K?

a. DBr ($B = 4.24$ cm^{-1})

b. DI ($B = 3.25$ cm^{-1})

c. CsI ($B = 0.0236$ cm^{-1})

d. F^{35}Cl ($B = 0.516$ cm^{-1})

P14.10 Calculate the rotational partition function for SO_2 at 298 K where $B_A = 2.03$ cm^{-1}, $B_B = 0.344$ cm^{-1}, and $B_C = 0.293$ cm^{-1}.

P14.11 Calculate the rotational partition function for ClNO at 500 K where $B_A = 2.84$ cm^{-1}, $B_B = 0.187$ cm^{-1}, and $B_C = 0.175$ cm^{-1}.

P14.12

a. In the rotational spectrum of H^{35}Cl ($I = 2.65 \times 10^{-47}$ kg m^2), the transition corresponding to the $J = 4$ to $J = 5$ transition is the most intense. At what temperature was the spectrum obtained?

b. At 1000 K, which rotational transition of H^{35}Cl would you expect to demonstrate the greatest intensity?

c. Would you expect the answers for parts (a) and (b) to change if the spectrum were of H^{37}Cl?

P14.13 Calculate the rotational partition function for oxygen ($B = 1.44$ cm^{-1}) at its boiling point, 90.2 K, using the high-temperature approximation and by discrete summation. Why should only odd values of J should be included in this summation?

P14.14

a. Calculate the percent population of the first 10 rotational energy levels for HBr ($B = 8.46$ cm^{-1}) at 298 K.

b. Repeat this calculation for HF assuming that the bond length of this molecule is identical to that of HBr.

P14.15 In general, the high-temperature limit for the rotational partition function is appropriate for almost all molecules at temperatures above the boiling point. Hydrogen is an exception to this generality because the moment of inertia is small due to the small mass of H. Given this, other molecules with H may also represent exceptions to this general rule. For example, methane (CH_4) has relatively modest moments of inertia ($I_A = I_B = I_C = 5.31 \times 10^{-40}$ g cm^2) and has a relatively low boiling point of $T = 112$ K.

a. Determine B_A, B_B, and B_C for this molecule.

b. Use the answer from part (a) to determine the rotational partition function. Is the high-temperature limit valid?

P14.16 Calculate the vibrational partition function for H^{35}Cl at 300 and 3000 K. What fraction of molecules will be in the ground vibrational state at these temperatures?

P14.17 For IF ($\tilde{\nu} = 610$ cm^{-1}) calculate the vibrational partition function and populations in the first three vibrational energy levels for $T = 300$ and 3000 K. Repeat this calculation for IBr ($\tilde{\nu} = 269$ cm^{-1}). Compare the probabilities for IF and IBr. Can you explain the differences between the probabilities of these molecules?

P14.18 Evaluate the vibrational partition function for H_2O at 2000 K where the vibrational frequencies are 1615, 3694, and 3802 cm^{-1}.

P14.19 Evaluate the vibrational partition function for SO_2 at 298 K where the vibrational frequencies are 519, 1151, and 1361 cm^{-1}.

P14.20 Evaluate the vibrational partition function for NH_3 at 1000 K for which the vibrational frequencies are 950, 1627.5 (doubly degenerate), 3335, and 3414 cm^{-1} (doubly degenerate). Are there any modes that you can disregard in this calculation? Why or why not?

P14.21 In deriving the vibrational partition function, a mathematical expression for the series expression for the partition function was employed. However, what if one performed integration instead of summation to evaluate the partition function? Evaluate the following expression for the vibrational partition function:

$$q_v = \sum_{n=0}^{\infty} e^{-\beta hcn\tilde{\nu}} \approx \int_0^{\infty} e^{-\beta hcn\tilde{\nu}} dn$$

Under what conditions would you expect the resulting expression for q_V to be applicable?

P14.22 You have in your possession the first vibrational spectrum of a new diatomic molecule, X_2, obtained at 1000 K. From the spectrum you determine that the fraction of molecules occupying a given vibrational energy state n is as follows:

n	0	1	2	3	>3
Fraction	0.352	0.184	0.0963	0.050	0.318

What are the vibrational energy spacings for X_2?

P14.23

a. In this chapter, the assumption was made that the harmonic oscillator model is valid such that anharmonicity can be neglected. However, anharmonicity can be included in the expression for vibrational energies. The energy levels for an anharmonic oscillator are given by

$$E_n = hc\tilde{\nu}\left(n + \frac{1}{2}\right) - hc\tilde{\chi}\tilde{\nu}\left(n + \frac{1}{2}\right)^2 + ...$$

Neglecting zero point energy, the energy levels become $E_n = hc\tilde{\nu}n - hc\tilde{\chi}\tilde{\nu}n^2 + ...$ Using the preceding expression, demonstrate that the vibrational partition function for the anharmonic oscillator is

$$q_{V,anharmonic} = q_{V,harm}(1 + \beta hc\tilde{\chi}\tilde{\nu}q_{V,harm}^2(e^{-2\beta\tilde{\nu}n} + e^{-\beta\tilde{\nu}n}))$$

In deriving the preceding result, the following series relationship will prove useful:

$$\sum_{n=0}^{\infty} n^2 x^n = \frac{x^2 + x}{(1-x)^3}$$

b. For H_2, $\tilde{\nu} = 4401.2$ cm^{-1} and $\tilde{\chi}\tilde{\nu} = 121.3$ cm^{-1}. Use the result from part (a) to determine the percent error in q_V if anharmonicity is ignored.

P14.24 Consider a particle free to translate in one dimension. The classical Hamiltonian is $H = p^2/2m$.

a. Determine $q_{classical}$ for this system. To what quantum system should you compare it in order to determine the equivalence of the classical and quantum statistical mechanical treatments?

b. Derive $q_{classical}$ for a system with translational motion in three dimensions for which $H = (p_x^2 + p_y^2 + p_z^2)/2m$.

P14.25 Evaluate the electronic partition function for atomic Fe at 298 K given the following energy levels. Terms in parenthesis are divided by $2m$.

Level(n)	Energy(cm^{-1})	Degeneracy
0	0	9
1	415.9	7
2	704.0	5
3	888.1	3
4	978.1	1

P14.26

a. Evaluate the electronic partition function for atomic Si at 298 K given the following energy levels:

Level(n)	Energy (cm^{-1})	Degeneracy
0	0	1
1	77.1	3
2	223.2	5
3	6298	5

b. At what temperature will the $n = 3$ energy level contribute 0.1 to the electronic partition function?

P14.27 NO is a well-known example of a molecular system in which excited electronic energy levels are readily accessible at room temperature. Both the ground and excited electronic states are doubly degenerate and are separated by 121.1 cm^{-1}.

a. Evaluate the electronic partition function for this molecule at 298 K.

b. Determine the temperature at which $q_E = 3$.

P14.28 Determine the total molecular partition function for I_2, confined to a volume of 1000 cm^3 at 298 K. Other information you will find useful: $B = 0.0374$ cm^{-1}, $\tilde{\nu} = 208$ cm^{-1}, and the ground electronic state is nondegenerate.

P14.29 Determine the total molecular partition function for gaseous H_2O at 1000 K confined to a volume of 1 cm^3. The rotational constants for water are $B_A = 27.8$ cm^{-1}, $B_B = 14.5$ cm^{-1}, and $B_C = 9.95$ cm^{-1}. The vibrational frequencies are 1615, 3694, and 3802 cm^{-1}. The ground electronic state is nondegenerate.

P14.30 The effect of symmetry on the rotational partition function for H_2 was evaluated by recognizing that each hydrogen is a spin 1/2 particle and is, therefore, a fermion. However, this development is not limited to fermions, but is also applicable to bosons. Consider CO_2 in which rotation by 180° results in the interchange of two spin 0 particles.

a. Because the overall wave function describing the interchange of two bosons must be symmetric with respect to exchange, to what J levels is the summation limited to in evaluating q_R for CO_2?

b. The rotational constant for CO_2 is 0.390 cm^{-1}. Calculate q_R at 298 K. Do you have to evaluate q_R by summation of the allowed rotational energy levels? Why or why not?

Web-Based Simulations, Animations, and Problems

W14.1 In this web-based simulation, the variation in q_T, q_R, and q_V with temperature is investigated for three diatomic molecules: HF, H^{35}Cl, and ^{35}ClF. Comparisons of q_T and q_R are performed to illustrate the mass and temperature dependence of these partition functions. Also, the expected dependence of q_T, q_R, and q_V on temperature in the high-temperature limit is investigated.

CHAPTER 15

Statistical Thermodynamics

With the central concepts of statistical mechanics in hand, the relationship between statistical mechanics and classical thermodynamics can be explored. In this chapter, the microscopic viewpoint of matter is connected to fundamental thermodynamic quantities such as internal energy, entropy, and Gibbs free energy using statistical mechanics. As will be shown, the statistical perspective is not only capable of reproducing thermodynamic properties of matter, but it also provides critical insight into the microscopic details behind these properties. We will see that the insight into the behavior of chemical systems gained from the statistical perspective is remarkable. ■

15.1 Energy

We begin our discussion of statistical thermodynamics by returning to the canonical ensemble (Section 14.1) and considering the **average energy** content of an ensemble unit, $\langle \varepsilon \rangle$, which is simply the **total energy** of the ensemble, E, divided by the number of units in the ensemble, N:

$$\langle \varepsilon \rangle = \frac{E}{N} = \frac{\sum_n \varepsilon_n a_n}{N} = \sum_n \varepsilon_n \frac{a_n}{N} \tag{15.1}$$

In this equation, ε_n is the level energy and a_n is the occupation number for the level. Consistent with our development in Chapter 14, the ensemble is partitioned such that there is one atom or molecule in each unit. The Boltzmann distribution for a series of nondegenerate energy levels is

$$\frac{a_n}{N} = \frac{e^{-\beta \varepsilon_n}}{q} \tag{15.2}$$

In this expression, q is the molecular partition function. Substituting Equation (15.2) into Equation (15.1) yields

$$\langle \varepsilon \rangle = \sum_n \varepsilon_n \frac{a_n}{N} = \frac{1}{q} \sum_n \varepsilon_n e^{-\beta \varepsilon_n} \tag{15.3}$$

As a final step in deriving $\langle \varepsilon \rangle$, consider the derivative of the molecular partition function with respect to β, which is given by

$$\frac{-dq}{d\beta} = \sum_n \varepsilon_n e^{-\beta \varepsilon_n} \tag{15.4}$$

Using Equation (15.4), Equation (15.3) can be rewritten to obtain the following expressions for the average unit energy and total ensemble energy:

$$\langle \varepsilon \rangle = \frac{-1}{q}\left(\frac{dq}{d\beta}\right) = -\left(\frac{d\ln q}{d\beta}\right) \tag{15.5}$$

$$E = N\langle \varepsilon \rangle = \frac{-N}{q}\left(\frac{dq}{d\beta}\right) = -N\left(\frac{d\ln q}{d\beta}\right) \tag{15.6}$$

At times, Equations 15.6 and 15.6 are easier to evaluate by taking the derivative with respect to T rather than β. Using the definition $\beta = (kT)^{-1}$,

$$\frac{d\beta}{dT} = \frac{d}{dT}(kT)^{-1} = -k(kT)^{-2} = -\frac{1}{kT^2} \tag{15.7}$$

Using Equation (15.7), the expressions for average and total energy can be written as follows:

$$\langle \varepsilon \rangle = kT^2\left(\frac{d\ln q}{dT}\right) \tag{15.8}$$

$$E = NkT^2\left(\frac{d\ln q}{dT}\right) \tag{15.9}$$

Equation (15.9) demonstrates that E will change with temperature, a result we are familiar with from our study of thermodynamics. Example Problem 15.1 involves an ensemble comprised of particles with two energy levels, commonly referred to as a **two-level system** (see Figure 15.1), to illustrate the variation of E with T. To derive the functional form of E for this ensemble, q must be constructed and used to derive an expression for E, as shown next.

FIGURE 15.1
Depiction of the two-level system.

EXAMPLE PROBLEM 15.1

Determine the total energy of an ensemble consisting of N particles that have only two energy levels separated by energy $h\nu$.

Solution
The energy levels for the particles are illustrated in Figure 15.1. As mentioned, systems with only two energy levels are commonly referred to as two-level systems. To determine the average energy, the partition function describing this system must be evaluated. The partition function consists of a sum of two terms as follows:

$$q = 1 + e^{-\beta h\nu}$$

The derivative of the partition function with respect to β is

$$\frac{dq}{d\beta} = \frac{d}{d\beta}(1 + e^{-\beta h\nu})$$
$$= -h\nu e^{-\beta h\nu}$$

Using this result, the total energy is

$$E = \frac{-N}{q}\left(\frac{dq}{d\beta}\right) = \frac{-N}{(1 + e^{-\beta h\nu})}(-h\nu^{-\beta h\nu})$$
$$= \frac{Nh\nu e^{-\beta h\nu}}{1 + e^{-\beta h\nu}}$$
$$= \frac{Nh\nu}{e^{\beta h\nu} + 1}$$

In the final step of this example, the expression for E was multiplied by unity in the form of $\exp(\beta h\nu)/\exp(\beta h\nu)$ to facilitate numerical evaluation.

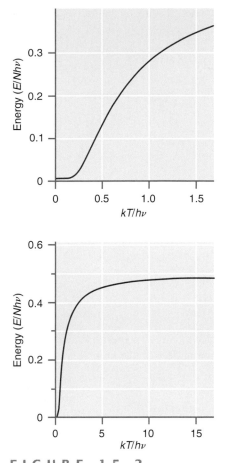

FIGURE   15.2

Total energy as a function of temperature is presented for an ensemble consisting of units that have two energy levels separated by an amount $h\nu$. Temperature is described relative to the energy gap through $kT/h\nu$. The plot on the top provides an expanded view of the low $kT/h\nu$ region plotted on the bottom.

Figure 15.2 presents the evolution in E with T for the two-level system in Figure 15.1. The total energy as presented in the figure is divided by the number of particles in the ensemble, N, and the energy-level spacing, $h\nu$. In addition, temperature is expressed as kT divided by the energy spacing between levels.

Two interesting trends are evident in Figure 15.2. First, at the lowest temperatures the total energy does not change appreciably as the temperature is increased until $kT/h\nu \approx 0.2$, at which point a significant increase in E is observed. Second, the total energy reaches a limiting value at high temperature, and further increases in T do not affect the total energy. Why does this occur? Recall that the change in energy of a system correlates with a change in occupation number, with higher energy levels being more readily populated as the temperature is increased. For the two-level system just presented, the probability of occupying the excited energy level is

$$p_1 = \frac{a_1}{N} = \frac{e^{-\beta h\nu}}{q} = \frac{e^{-\beta h\nu}}{1 + e^{-\beta h\nu}} = \frac{1}{e^{\beta h\nu} + 1} \qquad (15.10)$$

At $T = 0$, the exponential term in the denominator is infinite such that $p_1 = 0$. The probability of occupying the excited state becomes finite when the amount of thermal energy available, kT, is comparable to the separation energy of the states, $h\nu$. At low temperatures $kT \ll h\nu$, and the probability of occupying the excited state is quite small such that the occupation number for the excited state, a_1, is essentially zero. Inspection of Equation (15.10) demonstrates that p_1 increases until $h\nu \ll kT$. At these elevated temperatures, the exponential term in the denominator approaches unity, and p_1 approaches its limiting value of 0.5. Because this is a two-level system, the probability of occupying the lower energy state must also be 0.5. When the probability of occupying all energy states is equal, the occupation numbers do not change. The exact result for the evolution of p_1 as a function of temperature is presented in Figure 15.3. Again, the evolution in temperature is described as $kT/h\nu$ such that kT is measured with respect to the energy gap between the two states. Notice that the change in p_1 exactly mimics the behavior observed for the total energy depicted in Figure 15.2, demonstrating that changes in energy correspond to changes in occupation number.

15.1.1 Energy and the Canonical Partition Function

Equation (15.9) allows one to calculate the ensemble energy, but to what thermodynamic quantity is this energy related? Recall that we are interested in the canonical ensemble in which N, V, and T are held constant. Because V is constant, there can be no P-V-type work, and by the first law of thermodynamics any change in internal energy must occur by heat flow, q_V. Using the first law, the change in heat is related to the change in system **internal energy** at constant volume by

$$U - U_o = q_V \qquad (15.11)$$

In Equation (15.11), q_V is heat, not the molecular partition function. The energy as expressed by Equation (15.11) is the difference in internal energy at some finite temperature to that at 0 K. If there is residual, internal energy present at 0 K, it must be included to determine the overall energy of the system. However, by convention U_o is generally set to zero. For example, in the two-level example described earlier, the internal energy will be zero at 0 K since the energy of the ground state is defined as zero.

The second important relationship to establish is that between the internal energy and the canonical partition function. Fortunately, we have already encountered this relationship. For an ensemble of indistinguishable noninteracting particles, the canonical partition function is given by

$$Q = \frac{q^N}{N!} \qquad (15.12)$$

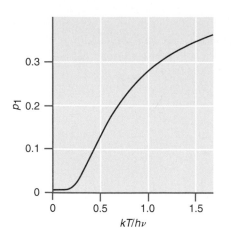

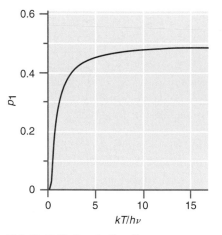

FIGURE 15.3
The probability of occupying the excited state in a two-level system is shown as a function of temperature. The evolution in temperature is plotted as $kT/h\nu$.

In Equation (15.12), q is the molecular partition function. Taking the natural log of Equation (15.12),

$$\ln Q = \ln\left(\frac{q^N}{N!}\right) = N \ln q - \ln N! \tag{15.13}$$

Finally, taking the derivative of Equation (15.13) with respect to β and recognizing that $\ln(N!)$ is constant in the canonical ensemble,

$$\frac{d \ln Q}{d\beta} = \frac{d}{d\beta}(N \ln q) - \frac{d}{d\beta}(\ln N!) \tag{15.14}$$

$$= N \frac{d \ln q}{d\beta}$$

The last relationship is simply the total energy; therefore, the relationship between the canonical partition function and total internal energy, U, is simply

$$U = -\left(\frac{d \ln Q}{d\beta}\right)_V \tag{15.15}$$

EXAMPLE PROBLEM 15.2

For an ensemble consisting of a mole of particles having two energy levels separated by $h\nu = 1.00 \times 10^{-20}$ J, at what temperature will the internal energy of this system equal 1.00 kJ?

Solution

Using the expression for total energy and recognizing that $N = nN_a$,

$$U = -\left(\frac{d \ln Q}{d\beta}\right)_V = -nN_A\left(\frac{d \ln q}{d\beta}\right)_V$$

Evaluating the preceding expression and paying particular attention to units, we get

$$U = -nN_A\left(\frac{d}{d\beta}\ln q\right)_V = -\frac{nN_A}{q}\left(\frac{dq}{d\beta}\right)_V$$

$$\frac{U}{nN_A} = \frac{-1}{(1+e^{-\beta h\nu})}\left(\frac{d}{d\beta}(1+e^{-\beta h\nu})\right)_V$$

$$= \frac{h\nu e^{-\beta h\nu}}{1+e^{-\beta h\nu}} = \frac{h\nu}{e^{\beta h\nu}+1}$$

$$\frac{nN_A h\nu}{U} - 1 = e^{\beta h\nu}$$

$$\ln\left(\frac{nN_A h\nu}{U} - 1\right) = \beta h\nu = \frac{h\nu}{kT}$$

$$T = \frac{h\nu}{k \ln\left(\frac{nN_A h\nu}{U} - 1\right)}$$

$$= \frac{1.00 \times 10^{-20}\,\text{J}}{(1.38 \times 10^{-23}\,\text{J}\,\text{K}^{-1})\ln\left(\dfrac{(1\,\text{mol})(6.022 \times 10^{23}\,\text{mol}^{-1})(1.00 \times 10^{-20}\,\text{J})}{(1.00 \times 10^3\,\text{J})} - 1\right)}$$

$$= 449\,\text{K}$$

15.2 Energy and Molecular Energetic Degrees of Freedom

In the previous chapter, the relationship between the total molecular partition function (q_{Total}) and the partition functions corresponding to an individual energetic degree of freedom was derived under the assumption that the energetic degrees of freedom are not coupled:

$$q_{Total} = q_T q_R q_V q_E \tag{15.16}$$

In this expression, the subscripts T, R, V, and E refer to translational, rotational, vibrational, and electronic energetic degrees of freedom, respectively. In a similar fashion, the internal energy can be decomposed into the contributions from each energetic degree of freedom:

$$U = -\left(\frac{d \ln Q}{d\beta}\right)_V = -N\left(\frac{d \ln q}{d\beta}\right)_V \tag{15.17}$$

$$= -N\left(\frac{d \ln(q_T q_R q_V q_E)}{d\beta}\right)_V$$

$$= -N\left(\frac{d}{d\beta}\left(\ln q_T + \ln q_R + \ln q_V + \ln q_E\right)\right)_V$$

$$= -N\left[\left(\frac{d \ln q_T}{d\beta}\right)_V + \left(\frac{d \ln q_R}{d\beta}\right)_V + \left(\frac{d \ln q_V}{d\beta}\right)_V + \left(\frac{d \ln q_E}{d\beta}\right)_V\right]$$

$$= U_T + U_R + U_V + U_E$$

The last line in this expression demonstrates a very intuitive result—that the total internal energy is simply the sum of contributions from each molecular energetic degree of freedom. This result also illustrates the connection between the macroscopic property of the ensemble (internal energy) and the microscopic details of the units themselves (molecular energy levels). To relate the total internal energy to the energetic degrees of freedom, expressions for the energy contribution from each energetic degree of freedom (U_T, U_R, and so on) are needed. The remainder of this section is dedicated to deriving these relationships.

15.2.1 Translations

The contribution to the system internal energy from translational motion is

$$U_T = \frac{-N}{q_T}\left(\frac{dq_T}{d\beta}\right)_V \tag{15.18}$$

In Equation (15.18), q_T is the translational partition function, which in three dimensions is given by

$$q_T = \frac{V}{\Lambda^3} \quad \text{with} \ \Lambda^3 = \left(\frac{h^2 \beta}{2\pi m}\right)^{3/2} \tag{15.19}$$

With this partition function, the translational contribution to the internal energy becomes

$$U_T = \frac{-N}{q_T}\left(\frac{dq_T}{d\beta}\right)_V = \frac{-N\Lambda^3}{V}\left(\frac{d}{d\beta}\frac{V}{\Lambda^3}\right)_V$$

$$= -N\Lambda^3\left(\frac{d}{d\beta}\frac{1}{\Lambda^3}\right)_V$$

$$= -N\Lambda^3\left(\frac{d}{d\beta}\left(\frac{2\pi m}{h^2\beta}\right)^{3/2}\right)_V$$

$$= -N\Lambda^3\left(\frac{2\pi m}{h^2}\right)^{3/2}\left(\frac{d}{d\beta}\beta^{-3/2}\right)_V$$

$$= -N\Lambda^3\left(\frac{2\pi m}{h^2}\right)^{3/2}\frac{-3}{2}\beta^{-5/2}$$

$$= \frac{3}{2}N\Lambda^3\left(\frac{2\pi m}{h^2\beta}\right)^{3/2}\beta^{-1}$$

$$= \frac{3}{2}N\beta^{-1}$$

$$U_T = \frac{3}{2}NkT = \frac{3}{2}nRT \tag{15.20}$$

Equation (15.20) should look familiar. Recall from Chapter 2 that the internal energy of an ideal monatomic gas is $nC_V\Delta T = 3/2(nRT)$ with $T_{initial} = 0$ K, identical to the result just obtained. The convergence between the thermodynamic and statistical mechanical descriptions of monatomic-gas systems is remarkable. It is also important to note that the starting point in deriving the preceding relationship was a quantum mechanical description of translational motion and the partition function.

It is also interesting to note that the contribution of translational motion to the internal energy is equal to that predicted by the equipartition theorem (Section 14.8). The equipartition theorem states that any term in the classical Hamiltonian that is quadratic with respect to momentum or position will contribute $kT/2$ to the energy. The Hamiltonian corresponding to three-dimensional translational motion of an ideal monatomic gas is

$$H_{trans} = \frac{1}{2m}(p_x^2 + p_y^2 + p_z^2) \tag{15.21}$$

Each p^2 term in Equation (15.21) will contribute $1/2\ kT$ to the energy by equipartition such that the total contribution will be $3/2\ kT$, or $3/2\ RT$ for a mole of particles that is identical to the result derived using the quantum mechanical description of translational motion. This agreement is not surprising given the small energy gap between translational energy levels such that classical behavior is expected.

15.2.2 Rotations

Within the rigid rotor approximation, the rotational partition function for a diatomic molecule in the high-temperature limit is given by

$$q_R = \frac{1}{\sigma\beta hcB} \tag{15.22}$$

With this partition function, the contribution to the internal energy from rotational motion is

$$U_R = \frac{-N}{q_R}\left(\frac{dq_R}{d\beta}\right)_V = -N\sigma\beta hcB\left(\frac{d}{d\beta}\frac{1}{\sigma\beta hcB}\right)_V$$

$$= -N\beta\left(\frac{d}{d\beta}\beta^{-1}\right)_V$$

$$= -N\beta(-\beta^{-2})$$

$$= N\beta^{-1}$$

$$\boxed{U_R = NkT = nRT} \tag{15.23}$$

Recall from Chapter 14 that the rotational partition function employed in this derivation is for a diatomic in which the rotational temperature, Θ_R, is much less than kT. In this limit, a significant number of rotational states are accessible. If Θ_R is not small relative to kT, then full evaluation of the sum form of the rotational partition function is required.

In the high-temperature limit, the rotational energy can be thought of as containing contributions of $1/2\ kT$ from each nonvanishing moment of inertia. This partitioning of energy is analogous to the case of translational energy discussed earlier within the context of the equipartition theorem. The concept of equipartition can be used to extend the result for U_R obtained for a diatomic molecule to linear and nonlinear polyatomic molecules. Because each nonvanishing moment of inertia will provide $1/2\ kT$ to the rotational energy equipartition, we can state that

$$U_R = nRT \quad \text{(linear polyatomic)} \tag{15.24}$$

$$U_R = \frac{3}{2}nRT \quad \text{(nonlinear polyatomic)} \tag{15.25}$$

The result for nonlinear polyatomic molecules can be confirmed by evaluating the expression for the average rotational energy employing the partition function for a nonlinear polyatomic that was presented in the previous chapter.

15.2.3 Vibrations

Unlike translational and rotational degrees of freedom, vibrational energy-level spacings are typically greater than kT such that the equipartition theorem is usually not applicable to this energetic degree of freedom. Fortunately, within the harmonic oscillator model, the regular energy-level spacings provide for a relatively simple expression for the vibrational partition function:

$$q_V = (1 - e^{-\beta hc\tilde{\nu}})^{-1} \tag{15.26}$$

The term $\tilde{\nu}$ represents vibrational frequency in units of cm^{-1}. Using this partition function, the vibrational contribution to the average energy is

$$U_V = \frac{-N}{q_V}\left(\frac{dq_V}{d\beta}\right)_V = -N(1 - e^{-\beta hc\tilde{\nu}})\left(\frac{d}{d\beta}(1 - e^{-\beta hc\tilde{\nu}})^{-1}\right)_V$$

$$= -N(1 - e^{-\beta hc\tilde{\nu}})(-hc\tilde{\nu}e^{-\beta hc\tilde{\nu}})(1 - e^{-\beta hc\tilde{\nu}})^{-2}$$

$$= \frac{Nhc\tilde{\nu}e^{-\beta hc\tilde{\nu}}}{(1 - e^{-\beta hc\tilde{\nu}})}$$

$$\boxed{U_V = \frac{Nhc\tilde{\nu}}{e^{\beta hc\tilde{\nu}} - 1}} \tag{15.27}$$

The temperature dependence of U_V/Nhc for a vibrational degree of freedom with $\tilde{\nu} = 1000$ cm^{-1} is presented in Figure 15.4. First, note that at lowest temperatures U_V is zero, which is reminiscent of the two-level example presented earlier in this chapter. At

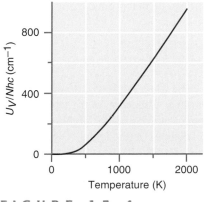

FIGURE 15.4

The variation in average vibrational energy as a function of temperature where $\tilde{\nu} = 1000$ cm^{-1}.

TABLE 15.1

Molecular Vibrational Temperatures

Molecule	Θ_V (K)	Molecule	Θ_V (K)
I_2	309	N_2	3392
Br_2	468	CO	3121
Cl_2	807	O_2	2274
F_2	1329	H_2	6338

low temperatures, $kT << hc\tilde{v}$ such that there is insufficient thermal energy to populate the first excited vibrational state to an appreciable extent. However, for temperatures ≥ 1000 K the average energy increases linearly with temperature, identical to the behavior observed for translational and rotational energy. This observation suggests that a high-temperature expression for U_V also exists.

To derive the high-temperature limit expression for the vibrational energy, the exponential term in q_V is written using the following series expression:

$$e^x = 1 + x + \frac{x^2}{2!} + \cdots$$

Because $x = \beta hc\tilde{v} = hc\tilde{v} / kT$, when $kT >> hc\tilde{v}$, the series is approximately equal to $1 + x$ yielding

$$U_V = \frac{Nhc\tilde{v}}{e^{\beta hc\tilde{v}} - 1} = \frac{Nhc\tilde{v}}{(1 + \beta hc\tilde{v}) - 1}$$

$$= \frac{N}{\beta}$$

$$U_V = NkT = nRT \tag{15.28}$$

When the temperature is sufficiently high so that the high-temperature limit is applicable, the vibrational contribution to the internal energy is nRT, identical to the prediction of the equipartition theorem. Note that this is the contribution for a single vibrational degree of freedom, and that the overall contribution will be the sum over all vibrational degrees of freedom. Furthermore, the applicability of the high-temperature approximation is dependent on the details of the vibrational energy spectrum such that while the high-temperature approximation may be applicable for some low-energy vibrations, it may not be applicable to higher energy vibrations.

The applicability of the high-temperature limit can be determined using the vibrational temperature defined previously:

$$\Theta_V = \frac{hc\tilde{v}}{k} \tag{15.29}$$

It was stated that the high-temperature limit is applicable when $T \geq 10\,\Theta_V$. Some examples of Θ_V for diatomic molecules are provided in Table 15.1. Inspection of the table demonstrates that relatively elevated temperatures must be reached before the high-temperature limit is applicable to the vast majority of vibrational degrees of freedom.

15.2.4 Electronic

Because electronic energy-level spacings are generally quite large compared to kT, the partition function is simply equal to the ground-state degeneracy. Since the degeneracy is a constant, the derivative of this quantity with respect to β must be zero and

$$U_E = 0 \tag{15.30}$$

Exceptions to this result do exist. In particular, for systems in which electronic energy levels are comparable to kT, it is necessary to evaluate the full partition function.

EXAMPLE PROBLEM 15.3

The ground state of O_2 is $^3\Sigma_g^-$. When O_2 is electronically excited, emission from the excited state ($^1\Delta_g$) to the ground state is observed at 1263 nm. Calculate q_E and determine the electronic contribution to U for a mole of O_2 at 500 K.

Solution

The first step to solving this problem is construction of the electronic partition function. The ground state is threefold degenerate, and the excited state is nondegenerate. The energy of the excited state relative to the ground state is determined using the emission wavelength:

$$\varepsilon = \frac{hc}{\lambda} = \frac{(6.626\times10^{-34}\,\text{J s})(3.00\times10^8\,\text{m s}^{-1})}{1.263\times10^{-6}\,\text{m}} = 1.57\times10^{-19}\,\text{J}$$

Therefore, the electronic partition function is

$$q_E = g_0 + g_1 e^{-\beta\varepsilon} = 3 + e^{-\beta(1.57\times10^{-19}\,\text{J})}$$

With the electronic partition function, U_E is readily determined:

$$U_E = \frac{-nN_A}{q_E}\left(\frac{\partial q_E}{\partial\beta}\right)_V = \frac{(1\,\text{mol})N_A(1.57\times10^{-19}\,\text{J})e^{-\beta(1.57\times10^{-19}\,\text{J})}}{3 + e^{-\beta(1.57\times10^{-19}\,\text{J})}}$$

$$= \frac{94.5kJ\exp\left[-\dfrac{1.57\times10^{-19}\,\text{J}}{(1.38\times10^{-23}\,\text{J K}^{-1})(500\,\text{K})}\right]}{3 + \exp\left[-\dfrac{1.57\times10^{-19}\,\text{J}}{(1.38\times10^{-23}\,\text{J K}^{-1})(500\,\text{K})}\right]}$$

$$= 4.1\times10^{-6}\,\text{kJ}$$

Notice that U_E is quite small, reflecting the fact that even at 500 K the $^1\Delta_g$ excited state of O_2 is not readily populated. Therefore, the contribution to the internal energy from electronic degrees of freedom for O_2 at 500 K is negligible.

15.2.5 Review

At the beginning of this section, we established that the total average energy was simply the sum of average energies for each energetic degree of freedom. Applying this logic to a diatomic system, the total energy is given by

$$U_{Total} = U_T + U_R + U_V + U_E$$
$$= \frac{3}{2}NkT + NkT + \frac{Nhc\tilde{v}}{e^{\beta hc\tilde{v}} - 1} + 0$$
$$= \frac{5}{2}NkT + \frac{Nhc\tilde{v}}{e^{\beta hc\tilde{v}} - 1} \tag{15.31}$$

In arriving at Equation (15.31), the assumption has been made that the rotational degrees of freedom are in the high-temperature limit and that degeneracy of the ground electronic level makes the lone contribution to the electronic partition function. Although the internal energy is dependent on molecular details (B, $\tilde{v}$, and so on), it is important to realize that the total energy can be decomposed into contributions from each molecular degree of freedom.

15.3 Heat Capacity

As discussed in Chapter 2, the thermodynamic definition of the **heat capacity** at constant volume (C_V) is

$$C_V = \left(\frac{\partial U}{\partial T}\right)_V = -k\beta^2 \left(\frac{\partial U}{\partial \beta}\right)_V \tag{15.32}$$

Because the internal energy can be decomposed into contributions from each energetic degree of freedom, the heat capacity can also be decomposed in a similar fashion. In the previous section, the average internal energy contribution from each energetic degree of freedom was determined. Correspondingly, the heat capacity is given by the derivative of the internal energy with respect to temperature for a given energetic degree of freedom. In the remainder of this section, we use this approach to determine the translational, vibrational, rotational, and electronic contributions to the constant volume heat capacity.

EXAMPLE PROBLEM 15.4

Determine the heat capacity for an ensemble consisting of units that have only two energy levels separated by an arbitrary amount of energy $h\nu$.

Solution

This is the same system discussed in Example Problem 15.1 where the partition function was determined to be $q = 1 + e^{-\beta h\nu}$. The corresponding average energy calculated using this partition function is $U = Nh\nu/e^{\beta h\nu} + 1$. Given the functional form of the average energy, the heat capacity is most easily determined by taking the derivative with respect to β as follows:

$$C_v = -k\beta^2 \left(\frac{\partial U}{\partial \beta}\right)_V = -Nk\beta^2 \left(\frac{\partial}{\partial \beta} h\nu (e^{\beta h\nu} + 1)^{-1}\right)_V$$

$$= \frac{Nk\beta^2 (h\nu)^2 e^{\beta h\nu}}{(e^{\beta h\nu} + 1)^2}$$

The functional form of the heat capacity is rather complex and is graphically represented in Figure 15.5. As observed previously, limiting behavior is observed at both low and high temperatures. At the lowest temperature, C_V is zero, then increases to a maximum value after which further increases in temperature result in a decrease in the heat capacity. This behavior is reminiscent of the evolution in energy as a function of temperature (Figure 15.2) and can be traced back to the same origin, the evolution in state populations as a function of temperature. Heat capacity is a measure of the ability a system to absorb energy from the surroundings. At the lowest temperatures, there is insufficient energy to access the excited state such that the two-level system is incapable of absorbing energy. As the available thermal energy is increased, the excited state becomes accessible and the heat capacity increases. Finally, at the highest temperatures the populations in the ground and excited states reach their limiting values of 0.5. Here, the system is again incapable of absorbing energy, and the heat capacity approaches zero.

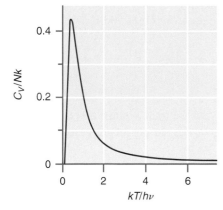

FIGURE 15.5
Constant volume heat capacity for a two-level system as a function of temperature. Note that the heat capacity has been divided by the product of Boltzmann's constant and the number of particles.

15.3.1 Translational Heat Capacity

Translational energy-level spacings are extremely small such that as long as the molecules remain in the gas phase, the high-temperature approximation is valid. The contribution of translational motion to the average molecular energy for an ideal gas is

$$U_T = \frac{3}{2}NkT \tag{15.33}$$

Therefore, the translational contribution to C_V is readily determined by taking the derivative of the preceding expression with respect to temperature:

$$(C_V)_T = \left(\frac{dU_T}{dT}\right)_V = \frac{3}{2}Nk \qquad (15.34)$$

This result demonstrates that the contribution of translations to the overall constant volume heat capacity is simply a constant, with no temperature dependence. The constant value of the translational contribution to C_V down to very low temperatures is a consequence of the dense manifold of closely spaced translational states.

15.3.2 Rotational Heat Capacity

Assuming that the rotational energy spacings are small relative to kT, the high-temperature expression for the rotational contribution to the average energy can be used to determine the rotational contribution to C_V. The internal energy is dependent on molecular geometry:

$$U_R = NkT \quad \text{(linear)} \qquad (15.35)$$

$$U_R = \frac{3}{2}NkT \quad \text{(nonlinear)} \qquad (15.36)$$

Using these two equations, the rotational contribution to C_V becomes

$$(C_V)_R = Nk \quad \text{(linear)} \qquad (15.37)$$

$$(C_V)_R = \frac{3}{2}Nk \quad \text{(nonlinear)} \qquad (15.38)$$

Note that these expressions are correct in the high-temperature limit. What will occur at low temperature? At the lowest temperatures, there is insufficient thermal energy to provide for population of excited rotational energy levels. Therefore, the heat capacity approaches zero. As temperature is increased, the heat capacity increases until the high-temperature limit is reached. At these intermediate temperatures, the heat capacity must be determined by evaluating the summation form of q_R.

15.3.3 Vibrational Heat Capacity

In contrast to translations and rotations, the high-temperature limit is generally not applicable to the vibrational degrees of freedom. Therefore, the exact functional form of the energy must be evaluated to determine the vibrational contribution to C_V. The derivation is somewhat more involved, but still straightforward. First, recall that the vibrational contribution to U is [Equation (15.27)]

$$U_V = \frac{Nhc\tilde{\nu}}{e^{\beta hc\tilde{\nu}} - 1} \qquad (15.39)$$

Given this equation, the vibrational contribution to the constant volume heat capacity is

$$(C_V)_{Vib} = \left(\frac{dU_V}{dT}\right)_V = -k\beta^2\left(\frac{dU_V}{d\beta}\right)_V$$

$$= -Nk\beta^2 hc\tilde{\nu}\left(\frac{d}{d\beta}(e^{\beta hc\tilde{\nu}} - 1)^{-1}\right)_V$$

$$= -Nk\beta^2 hc\tilde{\nu}(-hc\tilde{\nu}e^{\beta hc\tilde{\nu}}(e^{\beta hc\tilde{\nu}} - 1)^{-2})_V$$

$$(C_V)_{Vib} = Nk\beta^2(hc\tilde{\nu})^2 \frac{e^{\beta hc\tilde{\nu}}}{(e^{\beta hc\tilde{\nu}} - 1)^2} \qquad (15.40)$$

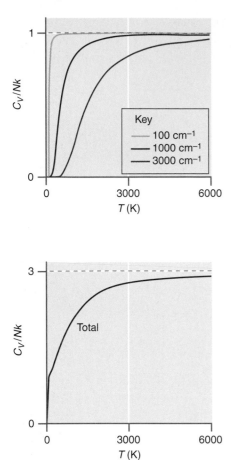

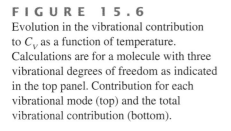

FIGURE 15.6

Evolution in the vibrational contribution to C_V as a function of temperature. Calculations are for a molecule with three vibrational degrees of freedom as indicated in the top panel. Contribution for each vibrational mode (top) and the total vibrational contribution (bottom).

15.1 Vibrational Heat Capacity Simulation

The heat capacity can also be cast in terms of the vibrational temperature Θ_V as follows

$$(C_V)_{Vib} = Nk \left(\frac{\Theta_V}{T} \right)^2 \frac{e^{\Theta_V/T}}{(e^{\Theta_V/T} - 1)^2} \qquad (15.41)$$

As mentioned in Chapter 14, a polyatomic molecule has $3N - 6$ or $3N - 5$ vibrational degrees of freedom, respectively. Each vibrational degree of freedom contributes to the overall vibrational constant volume heat capacity such that

$$(C_V)_{Vib,Total} = \sum_{m=1}^{3N-6 \ or \ 3N-5} (C_V)_{Vib,m} \qquad (15.42)$$

How does the vibrational heat capacity change as a function of T? The contribution of a given vibrational degree of freedom depends on the spacing between vibrational levels relative to kT. Therefore, at lowest temperatures we expect the vibrational contribution to C_V to be zero. As the temperature increases, the lowest energy vibrational modes have spacings comparable to kT such that these modes will contribute to the heat capacity. Finally, the highest energy vibrational modes contribute only at high temperature. In summary, we expect the vibrational contribution to C_V to demonstrate a significant temperature dependence.

Figure 15.6 presents the vibrational contribution to C_V as a function of temperature for a nonlinear triatomic molecule having three nondegenerate vibrational modes with $\tilde{\nu}$ equal to 100, 1000, and 3000 cm^{-1}. As expected, the lowest frequency mode contributes to the heat capacity at lowest temperatures, reaching a constant value of Nk for temperatures greatly in excess of Θ_V for a given vibration. As the temperature increases, each successively higher energy vibrational mode begins to contribute to the heat capacity and reaches the same limiting value of k at highest temperatures. Finally, the total vibrational heat capacity, which is simply the sum of the contributions from the individual modes, approaches a limiting value of $3 Nk$.

The behavior illustrated in Figure 15.6 demonstrates that for temperatures where $kT >> hc\tilde{\nu}$, the heat capacity approaches a constant value, suggesting that, like translations and rotations, there is a high-temperature limit for the vibrational contribution to C_V. Recall that the high-temperature approximation for the average energy for an individual vibrational degree of freedom is

$$U_V = \frac{N}{\beta} = NkT \qquad (15.43)$$

Differentiation of this expression with respect to temperature reveals that the vibrational contribution to the C_V is indeed equal to Nk per vibrational mode in the high-temperature limit as illustrated in Figure 15.6 and as expected from the equipartition theorem.

15.3.4 Electronic Heat Capacity

Because the partition function for the energetic degree of freedom is generally equal to the ground-state degeneracy, the resulting average energy is zero. Therefore, there is no contribution to the constant volume heat capacity from these degrees of freedom. However, for systems with electronic excited states that are comparable to kT, the contribution to C_V from electronic degrees of freedom can be finite and must be determined using the summation form of the partition function.

15.3.5 Review

Similar to energy, the overall constant volume heat capacity is simply the sum of the contributions to the heat capacity from each individual degree of freedom. For example, for a diatomic molecule with translational and rotational degrees of freedom well described by the high-temperature approximation, we have

FIGURE 15.7
The constant volume heat capacity for gaseous HCl as a function of temperature. The contributions of translational (yellow), rotational (orange), and vibrational (light blue) degrees of freedom to the heat capacity are shown.

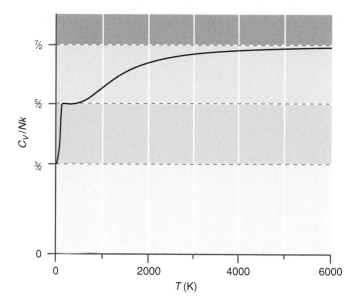

$$(C_V)_{Total} = (C_V)_T + (C_V)_R + (C_V)_{Vib}$$

$$= \frac{3}{2}Nk + Nk + Nk\beta^2(hc\tilde{\nu})^2 \frac{e^{\beta hc\tilde{\nu}}}{(e^{\beta hc\tilde{\nu}} - 1)^2}$$

$$= \frac{5}{2}Nk + Nk\beta^2(hc\tilde{\nu})^2 \frac{e^{\beta hc\tilde{\nu}}}{(e^{\beta hc\tilde{\nu}} - 1)^2} \tag{15.44}$$

The theoretical prediction for the heat capacity for gaseous HCl as a function of temperature is presented in Figure 15.7. The figure illustrates that at the lowest temperatures the heat capacity is $3/2Nk$ due to the contribution of translational motion. As the temperature is increased, the rotational contribution to the heat capacity increases, reaching the high-temperature limit at ~150 K, which is approximately 10-fold greater than the rotational temperature of 15.2 K. The temperature dependence of the rotational contribution was determined by numerical evaluation, a tedious procedure that must be performed when the high-temperature limit is not applicable. Finally, the vibrational contribution to the heat capacity becomes significant at highest temperatures consistent with the high vibrational temperature of this molecule (~4000 K). The curve depicted in the figure is theoretical and does not take into account molecular dissociation or phase transitions.

15.3.6 The Einstein Solid

The **Einstein solid** model was developed to describe the thermodynamic properties of atomic crystalline systems. In this model, each atom is envisioned to occupy a lattice site where the restoring potential is described as a three-dimensional harmonic oscillator. All of the harmonic oscillators are assumed to be separable such that motion of the atom in one dimension does not affect the vibrational motion (i.e., energy) in orthogonal dimensions. Finally, the harmonic oscillators are assumed to be isoenergetic such that they are characterized by the same frequency. For a crystal containing N atoms, there are $3N$ vibrational degrees of freedom, and the total heat capacity will simply be the sum of contributions from each vibrational degree of freedom. Therefore,

$$(C_V)_{Total} = 3Nk\left(\frac{\Theta_V}{T}\right)^2 \frac{e^{\Theta_V/T}}{(e^{\Theta_V/T} - 1)^2} \tag{15.45}$$

This total heat capacity is identical to that of a collection of $3N$ harmonic oscillators. The important aspect to note about Equation (15.45) is that C_V of the crystal at a given temperature is predicted to depend only on the vibrational frequency. This suggests

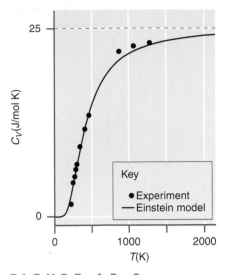

FIGURE 15.8
Comparison of C_V for diamond to the theoretical prediction of the Einstein solid model. The classical limit of 24.91 J/mol K is shown as the dashed line.

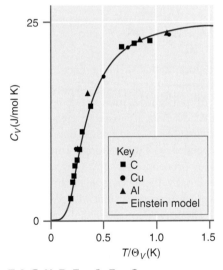

FIGURE 15.9
Comparison of C_V versus T/Θ_V for C, Cu, and Al.

that measurement of the constant volume heat capacity of an atomic crystal as a function of temperature allows for a determination of the characteristic vibrational temperature and corresponding vibrational frequency. This was one of the earliest quantum mechanical models used to describe a thermodynamic observable, and it has proved quite useful. For example, a comparison of the Einstein model and experiment for diamond ($\Theta_V = 1320$ K) is presented in Figure 15.8, and the agreement is remarkably good. Note that the model predicts that the heat capacity should reach a limiting value of 24.91 J/mol K, or $3R$, at high temperature. This limiting value is known as the **Dulong and Petit law,** and represents the high-temperature or classical prediction for the heat capacity of such systems. The elegance of the Einstein model is that it provides a reason for the failure of classical models at low temperatures where significant deviations from the Dulong and Petit law are observed. As has been seen, the quantum nature of vibrational degrees of freedom becomes exceedingly important when $\Theta_V > T$, and the Einstein model provided a critical view into the microscopic nature of matter manifesting itself in macroscopic behavior that classical mechanics cannot explain.

A second prediction of the Einstein model is that a plot of C_V versus T/Θ_V can describe the variation in heat capacity for all solids. Figure 15.9 presents a comparison of this prediction to the experimental values for three atomic crystals. The agreement is nothing short of astonishing and suggests that one only need measure C_V at a single temperature to determine the characteristic Θ_V for a given solid! Furthermore, with Θ_V in hand, the heat capacity at all other temperatures is defined! In short, starting with a quantum model for vibrations and the mathematics of probability theory, a unifying theory of heat capacity for atomic solids has been developed. This is quite an accomplishment.

The agreement between the Einstein model and experiment at high temperatures is excellent; however, discrepancies between this theory and experiment have been noted. First, the model is applicable only to atomic crystals, and it does not accurately reproduce the temperature dependence of C_V for molecular solids. Second, the Einstein model predicts that the heat capacity should demonstrate exponential dependence at low temperature; however, experimentally the heat capacity demonstrates T^3 dependence. The discrepancy between the Einstein model and experiment lies in the fact that the normal frequencies of the crystal are not due to vibrations of single atoms, but due to concerted harmonic motion of all the atoms. These lattice vibrations are not all characterized by the same frequency. More advanced treatments of atomic crystals, such as the Debye model, are capable of quantitatively reproducing crystalline heat capacities.

15.4 Entropy

Entropy is perhaps the most misunderstood thermodynamic property of matter. Introductory chemistry courses generally describe the entropic driving force for a reaction as an inherent desire for the system to increase its "randomness" or "disorder." With statistical mechanics, we will see that the tendency of an isolated system to evolve toward the state of maximum entropy is a direct consequence of statistics. To illustrate this point, recall from Chapter 13 that a system approaches equilibrium by achieving the configuration of energy with the maximum weight. In other words, the configuration of energy that will be observed at equilibrium corresponds to W_{max}. The tendency of a system to maximize W and entropy, S, suggests that a relationship exists between these quantities. Boltzmann expressed this relationship in following equation, known as **Boltzmann's formula:**

$$S = k \ln W \qquad (15.46)$$

This relationship states that entropy is directly proportional to $\ln(W)$, with the Boltzmann constant serving as the proportionality constant. Equation (15.46) makes it clear that a maximum in W will correspond to a maximum in S. Although at first view the Boltzmann formula appears to be a rather ad hoc statement, this formula is in fact equivalent to the thermodynamic definition of entropy discussed in Chapter 5. To illustrate this equivalence, consider the energy of an ensemble of particles. This energy is equal to the sum of the product of level energies and the occupation numbers for these levels such that

$$E = \sum_n \varepsilon_n a_n \tag{15.47}$$

The total differential of E is given by

$$dE = \sum_n \varepsilon_n da_n + \sum_n a_n d\varepsilon_n \tag{15.48}$$

Because volume is constant in the canonical ensemble, the level energies are also constant. Therefore, the second term in the preceding expression is zero, and the change in energy is related exclusively to changes in level occupation numbers:

$$dE = \sum_n \varepsilon_n da_n \tag{15.49}$$

The constraint of constant volume also dictates that there is no P-V-type work; therefore, by the first law of thermodynamics the change in energy must be due to heat flow:

$$dE = dq_{rev} = \sum_n \varepsilon_n da_n \tag{15.50}$$

where dq_{rev} is the reversible heat exchanged between system and surroundings. The thermodynamic definition of entropy is

$$dS = \frac{dq_{rev}}{T} \tag{15.51}$$

Comparison of the last two equations provides for the following definition of entropy:

$$dS = \frac{1}{T} \sum_n \varepsilon_n da_n = k\beta \sum_n \varepsilon_n da_n \tag{15.52}$$

We have rewritten T^{-1} as $k\beta$ given our need for β in the upcoming steps of the derivation. In the derivation of the Boltzmann distribution presented in Chapter 13, the following relationship was obtained:

$$\left(\frac{d \ln W}{da_n}\right) + \alpha - \beta\varepsilon_n = 0 \tag{15.53}$$

$$\beta\varepsilon_n = \left(\frac{d \ln W}{da_n}\right) + \alpha$$

where α and β are constants (i.e., the Lagrange multipliers in the derivation of the Boltzmann distribution). Employing Equation (15.53), our expression for entropy becomes

$$dS = k\beta \sum_n \varepsilon_n da_n = k \sum_n \left(\frac{d \ln W}{da_n}\right) da_n + k \sum_n \alpha da_n$$

$$= k \sum_n \left(\frac{d \ln W}{da_n}\right) da_n + k\alpha \sum_n da_n$$

$$= k \sum_n \left(\frac{d \ln W}{da_n}\right) da_n$$

$$= k(d \ln W) \tag{15.54}$$

Equation (15.54) is simply an alternative form of the Boltzmann formula of Equation (15.46). The definition of entropy provided by the Boltzmann formula provides insight into the underpinnings of entropy, but how does one calculate the entropy for a molecular system? The answer to this question requires a bit more work. Because the partition function provides a measure of energy state accessibility, it can be assumed that a relationship between the partition function and entropy exists. This relationship can be derived by applying the Boltzmann formula as follows:

$$S = k \ln W = k \ln \left(\frac{N!}{\prod_n a_n!} \right) \tag{15.55}$$

which can be reduced as follows:

$$
\begin{aligned}
S &= k \ln \left(\frac{N!}{\prod_n a_n!} \right) \\
&= k \ln N! - k \ln \prod_n a_n! \\
&= k \ln N! - k \sum_n \ln a_n \\
&= k(N \ln N - N) - k \sum_n (a_n \ln a_n - a_n) \\
&= k \left(N \ln N - \sum_n a_n \ln a_n \right) \tag{15.56}
\end{aligned}
$$

In the last step the following definition for N was used:

$$N = \sum_n a_n \tag{15.57}$$

Using this definition again,

$$
\begin{aligned}
S &= k \left(N \ln N - \sum_n a_n \ln a_n \right) \\
&= k \left(\sum_n a_n \ln N - \sum_n a_n \ln a_n \right) \\
&= -k \sum_n a_n \ln \frac{a_n}{N} \\
&= -k \sum_n a_n \ln p_n \tag{15.58}
\end{aligned}
$$

In the last line of this equation, p_n is the probability of occupying energy level n as defined previously. The Boltzmann distribution law can be used to rewrite the probability term in Equation (15.58) as follows:

$$\ln p_n = \ln \left(\frac{e^{-\beta \varepsilon_n}}{q} \right) = -\beta \varepsilon_n - \ln q \tag{15.59}$$

The expression for entropy then becomes

$$
\begin{aligned}
S &= -k \sum_n a_n \ln p_n \\
&= -k \left(\sum_n a_n (-\beta \varepsilon_n - \ln q) \right) \\
&= k\beta \sum_n a_n \varepsilon_n + k \sum_n a_n \ln q \\
&= k\beta E + kN \ln q \\
&= \frac{E}{T} + k \ln q^N \\
S &= \frac{E}{T} + k \ln Q = \frac{U}{T} + k \ln Q \tag{15.60}
\end{aligned}
$$

In this expression, U is the internal energy of the system and Q is the canonical partition function. In Equation (15.60), E is replaced by U consistent with the previous discussion regarding internal energy. The internal energy is related to the canonical partition function by

$$U = -\left(\frac{d\ln Q}{d\beta}\right)_V = kT^2\left(\frac{d\ln Q}{dT}\right)_V \tag{15.61}$$

Using this relationship, we arrive at a very compact expression for entropy:

$$S = \frac{U}{T} + k\ln Q = kT\left(\frac{d\ln Q}{dT}\right)_V + k\ln Q$$

$$S = \left(\frac{d}{dT}(kT\ln Q)\right)_V \tag{15.62}$$

15.4.1 Entropy of an Ideal Monatomic Gas

What is the general expression for the molar entropy of an ideal monatomic gas? A monatomic gas is a collection of indistinguishable particles. Assuming that the electronic partition function is unity (i.e., the ground electronic energy level is nondegenerate), only translational degrees of freedom remain to be evaluated resulting in

$$S = \left(\frac{d}{dT}(kT\ln Q)\right)_V = \frac{U}{T} + k\ln Q$$

$$= \frac{1}{T}\left(\frac{3}{2}NkT\right) + k\ln\frac{q^N_{trans}}{N!}$$

$$= \frac{3}{2}Nk + Nk\ln q_{trans} - k(N\ln N - N)$$

$$= \frac{5}{2}Nk + Nk\ln q_{trans} - Nk\ln N$$

$$= \frac{5}{2}Nk + Nk\ln\frac{V}{\Lambda^3} - Nk\ln N$$

$$= \frac{5}{2}Nk + Nk\ln V - Nk\ln\Lambda^3 - Nk\ln N$$

$$= \frac{5}{2}Nk + Nk\ln V - Nk\ln\left(\frac{h^2}{2\pi mkT}\right)^{3/2} - Nk\ln N$$

$$= \frac{5}{2}Nk + Nk\ln V + \frac{3}{2}Nk\ln T - Nk\ln\left(\frac{N^{2/3}h^2}{2\pi mk}\right)^{3/2}$$

$$= \frac{5}{2}nR + nR\ln V + \frac{3}{2}nR\ln T - nR\ln\left(\frac{n^{2/3}N_A^{2/3}h^2}{2\pi mk}\right) \tag{15.63}$$

The final line of Equation (15.63) is a version of the **Sackur–Tetrode equation,** which can be written in the more compact form:

$$S = nR\ln\left[\frac{RTe^{5/2}}{\Lambda^3 N_A P}\right] \quad \text{where } \Lambda^3 = \left(\frac{h^2}{2\pi mkT}\right)^{3/2} \tag{15.64}$$

The Sackur–Tetrode equation reproduces many of the classical thermodynamics properties of ideal monatomic gases encountered previously. For example, consider the isothermal expansion of an ideal monatomic gas from an initial volume V_1 to a final volume V_2. Inspection of the expanded form of the Sackur–Tetrode equation [Equation (15.63)] demonstrates that all of the terms in this expression are unchanged except for the second term involving volume such that

$$\Delta S = S_{final} - S_{initial} = nR \ln \frac{V_2}{V_1} \tag{15.65}$$

This is the same result obtained from classical thermodynamics. What if the entropy change were initiated by isochoric ($\Delta V = 0$) heating? Using the difference in temperature between initial (T_1) and final (T_2) states, Equation (15.63) yields

$$\Delta S = S_{final} - S_{initial} = \frac{3}{2} nR \ln \frac{T_2}{T_1} = nC_V \ln \frac{T_2}{T_1} \tag{15.66}$$

Recognizing that $C_V = 3/2R$ for an ideal monatomic gas, we again arrive at a result first encountered in thermodynamics.

Does the Sackur–Tetrode equation provide any information not available from thermodynamics? Indeed, note the first and fourth terms in Equation (15.63). These terms are simply constants, with the latter varying with the atomic mass. Classical thermodynamics is entirely incapable of explaining the origin of these terms, and only through empirical studies could the presence of these terms be determined. However, their contribution to entropy appears naturally (and elegantly) when using the statistical perspective.

EXAMPLE PROBLEM 15.5

Determine the standard molar entropy of Ne and Kr under standard thermodynamics conditions.

Solution

Beginning with the expression for entropy derived in the text:

$$S = \frac{5}{2}R + R\ln\left(\frac{V}{\Lambda^3}\right) - R\ln N_A$$

$$= \frac{5}{2}R + R\ln\left(\frac{V}{\Lambda^3}\right) - 54.75R$$

$$= R\ln\left(\frac{V}{\Lambda^3}\right) - 52.25R$$

The conventional standard state is defined by $T = 298$ K and $V_m = 24.4$ l (0.0244 m³). The thermal wavelength for Ne is

$$\Lambda = \left(\frac{h^2}{2\pi mkT}\right)^{1/2} = \left(\frac{(6.626\times 10^{-34}\,\text{J s})^2}{2\pi\left(\dfrac{0.02018\text{ kg mol}^{-1}}{N_A}\right)(1.38\times 10^{-23}\,\text{J K}^{-1})(298\text{ K})}\right)^{1/2}$$

$$= 2.25\times 10^{-11}\text{ m}$$

Using this value for the thermal wavelength, the entropy becomes

$$S = R\ln\left(\frac{0.0244\text{ m}^3}{(2.25\times 10^{-11}\text{ m})^3}\right) - 52.25R$$

$$= 69.83R - 52.25R = 17.59R = 146\text{ J mol}^{-1}\text{ K}^{-1}$$

The experimental value is 146.48 J/mol K. Rather than determining the entropy of Kr directly, it is easier to determine the difference in entropy relative to Ne:

$$\Delta S = S_{Kr} - S_{Ne} = S = R \ln\left(\frac{V}{\Lambda_{Kr}^3}\right) - R \ln\left(\frac{V}{\Lambda_{Ne}^3}\right)$$

$$= R \ln\left(\frac{\Lambda_{Ne}}{\Lambda_{Kr}}\right)^3$$

$$= 3R \ln\left(\frac{\Lambda_{Ne}}{\Lambda_{Kr}}\right)$$

$$= 3R \ln\left(\frac{m_{Kr}}{m_{Ne}}\right)^{1/2}$$

$$= \frac{3}{2} R \ln\left(\frac{m_{Kr}}{m_{Ne}}\right) = \frac{3}{2} R \ln(4.15)$$

$$= 17.7 \ \text{J mol}^{-1} \text{K}^{-1}$$

Using this difference, the standard molar entropy of Kr becomes

$$S_{Kr} = \Delta S + S_{Ne} = 164 \ \text{J mol}^{-1} \text{K}^{-1}$$

The experimental value is 163.89 J/mol K in excellent agreement with the calculated value.

When calculating the entropy for an ideal gas consisting of diatomic or polyatomic molecules, it is best to start with the general expression for entropy [Equation (15.62)] and express the canonical partition function in terms of the product of molecular partition functions for each energetic degree of freedom. In addition to the translational entropy term derived earlier for an ideal monatomic gas, contributions from rotational, vibrational, and electronic degrees of freedom will also be included when calculating the entropy.

15.5 Residual Entropy

As illustrated in Example Problem 15.5, when the entropy calculated using statistical mechanics is compared to experiment, good agreement is observed for a variety of atomic and molecular systems. However, for many molecular systems, this agreement is less than ideal. A famous example of such a system is carbon monoxide where the calculated entropy at thermodynamic standard temperature and pressure is 197.9 J mol^{-1} K^{-1} and the experimental value is only 193.3 J mol^{-1} K^{-1}. In this and other systems, the calculated entropy is always greater than that observed experimentally.

The reason for the systematic discrepancy between calculated and experimental entropies for such systems is **residual entropy,** or entropy associated with molecular orientation in the molecular crystal at low temperature. Using CO as an example, the weak electric dipole moment of the molecule dictates that dipole–dipole interactions do not play a dominant role in determining the orientation of one CO molecule relative to neighboring molecules in a crystal. Therefore, each CO can assume one of two orientations as illustrated in Figure 15.10. The solid corresponding to the possible orientations of CO will have an inherent randomness to it. Because each CO molecule can assume one of two possible orientations, the entropy associated with this orientational disorder is

$$S = k \ln W = k \ln 2^N = Nk \ln 2 = nR \ln 2 \qquad (15.67)$$

FIGURE 15.10
The origin of residual entropy for CO. Each CO molecule in the solid can have one of two possible orientations as illustrated by the central CO. Each CO will have two possible directions such that the total number of arrangements possible is 2^N where N is the number of CO molecules.

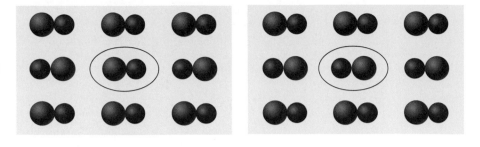

In Equation (15.67), W is the total number of CO arrangements possible, and it is equal to 2^N where N is the number of CO molecules. For a system consisting of 1 mol of CO, the residual entropy is predicted to be $R \ln 2$ or 5.76 J mol^{-1} K^{-1}, roughly equal to the difference between the experimental and calculated entropy values.

Finally, note that the concept of residual entropy sheds light on the origin of the third law of thermodynamics. As discussed in Chapter 5, the third law states that the entropy of a pure and crystalline substance is zero at 0 K. By "pure and crystalline," the third law means that the system must be pure with respect to both composition (i.e., a single component) and orientation in the solid at 0 K. For such a pure system, $W = 1$ and correspondingly $S = 0$ by Equation (15.67). Therefore, the definition of zero entropy provided by the third law is a natural consequence of the statistical nature of matter.

EXAMPLE PROBLEM 15.6

The Van der Waals radii of H and F are similar such that steric effects on molecular ordering in the crystal are minimal. Do you expect the residual molar entropies for crystalline 1,2-difluorobenzene and 1,4-difluorobenzene to be the same?

Solution

The structures of 1,2-difluorobenzene and 1,4-difluorobenzene are shown here:

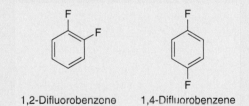

1,2-Difluorobenzene   1,4-Difluorobenzene

In crystalline 1,2-difluorobenzene, there will be six possible arrangements that can be visualized by rotation of the molecule about the C_6 symmetry axis of the molecule. Therefore, $W = 6^N$ and

$$S = k \ln W = k \ln 6^{N_A} = N_A k \ln 6 = R \ln 6$$

Similarly, for 1,4-difluorobenzene, there are three possible arrangements such that $W = 3^N$ and $S = R \ln 3$. The residual molar entropies are expected to differ for molecular crystals involving these species.

15.6 Other Thermodynamic Functions

The discussion thus far illustrates the convergence of the statistical and thermodynamics perspectives, and the utility of the statistical perspective in illustrating the underlying contributions to macroscopic properties of matter is impressive. The relationship between the canonical partition function and other thermodynamic quantities can be derived using the following familiar thermodynamics relationships for enthalpy, H, Helmholtz energy, A, and Gibbs energy, G:

$$H = U + PV \tag{15.68}$$

$$A = U - TS \tag{15.69}$$

$$G = H - TS \tag{15.70}$$

Using these expressions, relationships between the canonical partition function and the previous thermodynamic quantities can be derived.

15.6.1 Helmholtz Energy

Beginning with the thermodynamic definition of A from Equation (15.69), the relationship of this quantity to the canonical partition function becomes

$$A = U - TS$$

$$= U - T\left(\frac{U}{T} + k\ln Q\right)$$

$$A = -kT\ln Q \tag{15.71}$$

The **Helmholtz energy** provides an interesting pathway to obtain a well-known relationship: the ideal gas law. From thermodynamics, pressure is related to the Helmholtz energy by

$$P = \left(\frac{-\partial A}{\partial V}\right)_T \tag{15.72}$$

Substituting the definition for A provided in Equation (15.71) into the preceding equation,

$$P = \left(\frac{-\partial}{\partial V}(-kT\ln Q)\right)_T$$

$$= kT\left(\frac{\partial}{\partial V}\ln Q\right)_T \tag{15.73}$$

The canonical partition function for an ideal gas is

$$Q = \frac{q^N}{N!}$$

Substituting this expression for the canonical partition function into the preceding equation,

$$P = kT\left(\frac{\partial}{\partial V}\ln\frac{q^N}{N!}\right)_T$$

$$= kT\left(\frac{\partial}{\partial V}\ln q^N - \frac{\partial}{\partial V}\ln N!\right)_T$$

$$= kT\left(\frac{\partial}{\partial V}\ln q^N\right)_T$$

$$= NkT\left(\frac{\partial}{\partial V}\ln q\right)_T \tag{15.74}$$

For a monatomic gas,

$$q = \frac{V}{\Lambda^3} \tag{15.75}$$

such that

$$P = NkT \left(\frac{\partial}{\partial V} \ln \frac{V}{\Lambda^3} \right)_T$$

$$= NkT \left[\frac{\partial}{\partial V} \ln V - \frac{\partial}{\partial V} \ln \Lambda^3 \right]_T$$

$$= NkT \left(\frac{\partial}{\partial V} \ln V \right)_T$$

$$= \frac{NkT}{V} = \frac{nRT}{V} \tag{15.76}$$

This result is nothing short of remarkable. This relationship was first obtained by empirical measurements, expressed as the laws of Boyle, Charles, Avogadro, and Gay-Lussac. However, in deriving this relationship we have not employed these empirical laws. Instead, starting with a quantum mechanical model for translational motion, we have derived the ideal gas law from a purely theoretical perspective. Once again, statistical mechanics has provided microscopic insight into a macroscopic property. This example illustrates one of the major contributions of statistical mechanics to physical chemistry: the ability to predict relationships that are largely derived from empirical observations and subsequently stated as thermodynamic laws. As a final point, an objection that may be raised is that the preceding derivation was performed for a monatomic gas. However, the skeptical reader is encouraged to demonstrate that this result also holds for polyatomic systems, a task that is provided as a problem at the end of this chapter.

15.6.2 Enthalpy

Using the thermodynamic definition of entropy from Equation (15.68), this quantity can be expressed in terms of the canonical partition function as follows:

$$H = U + PV$$

$$= \left(\frac{-\partial}{\partial \beta} \ln Q \right)_V + V \left(\frac{-\partial A}{\partial V} \right)_T$$

$$= \left(\frac{-\partial}{\partial \beta} \ln Q \right)_V + V \left(\frac{-\partial}{dV} (-kT \ln Q) \right)_T$$

$$= kT^2 \left(\frac{\partial}{\partial T} \ln Q \right)_V + VkT \left(\frac{\partial}{\partial V} \ln Q \right)_T$$

$$H = T \left[kT \left(\frac{\partial}{\partial T} \ln Q \right)_V + Vk \left(\frac{\partial}{\partial V} \ln Q \right)_T \right] \tag{15.77}$$

Although Equation (15.77) is correct, it clearly requires a bit of work to implement. Yet, the statistical perspective has shown that one can relate enthalpy to microscopic molecular details through the partition function. When calculating enthalpy, it is sometimes easier to use a combination of thermodynamic and statistical perspectives, as the following example demonstrates.

EXAMPLE PROBLEM 15.7

What is the enthalpy of 1 mol of an ideal monatomic gas?

Solution
One approach to this problem is to start with the expression for the canonical partition function in terms of the molecular partition function for an ideal

monatomic gas and evaluate the result. However, a more efficient approach is to begin with the thermodynamic definition of enthalpy:

$$H = U + PV$$

Recall that the translational contribution to U is $3/2RT$, and this is the only degree of freedom operative for the monatomic gas. In addition, we can apply the ideal gas law (because we have now demonstrated its validity from a statistical perspective) such that the enthalpy is simply

$$H = U + PV$$

$$= \frac{3}{2}RT + RT$$

$$= \frac{5}{2}RT$$

The interested reader is encouraged to obtain this result through a full evaluation of the statistical expression for enthalpy.

15.6.3 Gibbs Energy

Perhaps the most important state function to emerge from thermodynamics is the **Gibbs energy.** Using this quantity, one can determine if a chemical reaction will occur spontaneously. The statistical expression for the Gibbs energy is also derived starting with the thermodynamic definition of this quantity [Equation (15.70)]:

$$G = A + PV$$

$$= -kT \ln Q + VkT \left(\frac{\partial}{\partial V} \ln Q \right)_T$$

$$G = -kT \left[\ln Q - V \left(\frac{\partial \ln Q}{\partial V} \right)_T \right] \qquad (15.78)$$

Previously derived expressions for the Helmholtz energy and pressure were employed in arriving at the result of Equation (15.78). A more intuitive result can be derived by applying the preceding relationship to an ideal gas such that $PV = nRT = NkT$. With this relationship:

$$G = A + PV$$

$$= -kT \ln Q + NkT$$

$$= -kT \ln \left(\frac{q^N}{N!} \right) + NkT$$

$$= -kT \ln q^N + kT \ln N! + NkT$$

$$= -NkT \ln q + kT(N \ln N - N) + NkT$$

$$= -NkT \ln q + NkT \ln N$$

$$G = -NkT \ln \left(\frac{q}{N} \right) = -nRT \ln \left(\frac{q}{N} \right) \qquad (15.79)$$

This relationship is extremely important because it provides insight into the origin of the Gibbs energy. At constant temperature the nRT prefactor in the expression for G is equivalent for all species; therefore, differences in the Gibbs energy must be due to the partition function. Because the Gibbs energy is proportional to $-\ln(q)$, an increase in the value for the partition function will result in a lower Gibbs energy. The partition function quantifies the number of states that are accessible at a given temperature; therefore, the statistical perspective dictates that species with a comparatively greater number of accessible energy states will have a lower Gibbs energy. This relationship will have profound consequences when discussing chemical equilibria in the next section.

EXAMPLE PROBLEM 15.8

Calculate the Gibbs energy for 1 mol of Ar at 298.15 K and standard pressure (10^5 Pa) assuming that the gas demonstrates ideal behavior.

Solution

Argon is a monatomic gas; therefore, $q = q_{trans}$. Using Equation (15.83),

$$G° = -nRT \ln\left(\frac{q}{N}\right) = -nRT \ln\left(\frac{V}{N\Lambda^3}\right)$$

$$= -nRT \ln\left(\frac{kT}{P\Lambda^3}\right)$$

The superscript on G indicates standard thermodynamic conditions of 298.15 K and 1 bar. In the last step, the ideal gas law was used to express V in terms of P, and the relationships $N = nN_A$ and $R = N_A k$ were employed. The units of pressure must be Pa = J m^{-3}. Solving for the thermal wavelength term, Λ^3, we get

$$\Lambda^3 = \left(\frac{h^2}{2\pi mkT}\right)^{3/2} = \left(\frac{(6.626\times10^{-34}\,\mathrm{J\,s})^2}{2\pi\left(\dfrac{0.040\ \mathrm{kg\ mol^{-1}}}{6.022\times10^{23}\,\mathrm{mol^{-1}}}\right)(1.38\times10^{-23}\,\mathrm{J\,K^{-1}})(298\ \mathrm{K})}\right)^{3/2}$$

$$= 4.09\times10^{-33}\,\mathrm{m^3}$$

With this result, $G°$ becomes

$$G° = -nRT \ln\left(\frac{kT}{P\Lambda^3}\right) = -(1\ \mathrm{mol})(8.314\ \mathrm{J\ mol^{-1}\ K^{-1}})$$

$$\times (298\ \mathrm{K}) \ln\left(\frac{(1.38\times10^{-23}\,\mathrm{J\,K^{-1}})(298\ \mathrm{K})}{(10^5\,\mathrm{Pa})(4.09\times10^{-33}\,\mathrm{m^3})}\right)$$

$$= -3.99\times10^4\,\mathrm{J} = -39.9\ \mathrm{kJ}$$

15.7 Chemical Equilibrium

Consider the following generic reaction:

$$a\mathrm{A} + b\mathrm{B} \Leftrightarrow c\mathrm{C} + d\mathrm{D} \tag{15.80}$$

The change in Gibbs energy for this reaction is related to the Gibbs energy for the associated species as follows:

$$\Delta G° = cG_C° + dG_D° - aG_A° - bG_B° \tag{15.81}$$

In this expression, the superscript indicates standard thermodynamic state. In addition, the equilibrium constant K is given by

$$\Delta G° = -RT \ln K \tag{15.82}$$

In the previous section of this chapter, the Gibbs energy was related to the molecular partition function. Therefore, it should be possible to define $\Delta G°$ and K in terms of partition functions for the various species involved. This development can be initiated by substituting into our expression for $\Delta G°$ from Equation (15.81) the definition for G given in Equation (15.79):

$$\Delta G^{\circ} = c\left(-RT \ln\left(\frac{q^{\circ}_C}{N}\right)\right) + d\left(-RT \ln\left(\frac{q^{\circ}_D}{N}\right)\right) - a\left(-RT \ln\left(\frac{q^{\circ}_A}{N}\right)\right) \quad (15.83)$$

$$- b\left(-RT \ln\left(\frac{q^{\circ}_B}{N}\right)\right)$$

$$= -RT\left(\ln\left(\frac{q^{\circ}_C}{N}\right)^c + \ln\left(\frac{q^{\circ}_D}{N}\right)^d - \ln\left(\frac{q^{\circ}_A}{N}\right)^a - \ln\left(\frac{q^{\circ}_B}{N}\right)^b\right)$$

$$= -RT \ln\left(\frac{\left(\frac{q^{\circ}_C}{N}\right)^c \left(\frac{q^{\circ}_D}{N}\right)^d}{\left(\frac{q^{\circ}_A}{N}\right)^a \left(\frac{q^{\circ}_B}{N}\right)^b}\right)$$

Comparison of the preceding relationship to the thermodynamics definition of ΔG° demonstrates that the equilibrium constant can be defined as follows:

$$K_P = \frac{\left(\frac{q^{\circ}_C}{N}\right)^c \left(\frac{q^{\circ}_D}{N}\right)^d}{\left(\frac{q^{\circ}_A}{N}\right)^a \left(\frac{q^{\circ}_B}{N}\right)^b} \quad (15.84)$$

Although Equation (15.84) is correct as written, there is one final detail to consider. Specifically, imagine taking the preceding relationship to $T = 0$ K such that only the lowest energy states along all energetic degrees of freedom are populated. The translational and rotational ground states for all species are equivalent; however, the vibrational and electronic ground states are not. Figure 15.11 illustrates the origin of this discrepancy. The figure illustrates the ground *vibronic* (vibrational and electronic) potential for a diatomic molecule. The presence of a bond between the two atoms in the molecule lowers the energy of the molecule relative to the separated atomic fragments. Because the energy of the atomic fragments is defined as zero, the ground vibrational state is lower than zero by an amount equal to the **dissociation energy** of the molecule, ε_D. Furthermore, different molecules will have different values for the dissociation energy such that this offset from zero will be molecule specific.

Establishing a common reference state for vibrational and electronic degrees of freedom is accomplished as follows. First, differences in ε_D can be incorporated into the vibrational part of the problem such that the electronic partition function remains identical

FIGURE 15.11

The ground-state potential energy curve for a diatomic molecule. The lowest three vibrational levels are indicated.

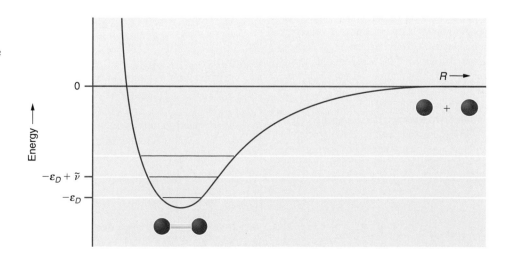
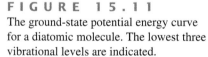

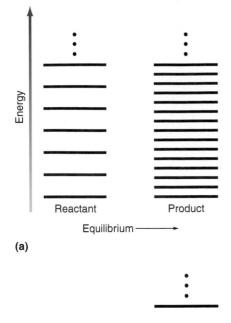

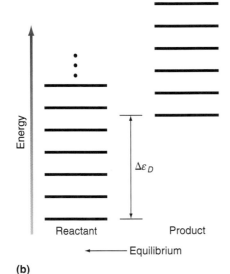

(a)

(b)

FIGURE 15.12
The statistical interpretation of equilibrium. **(a)** Reactant and product species having equal ground-state energies are depicted. However, the energy spacings of the product are less than the reactant such that more product states are available at a given temperature. Therefore, equilibrium will lie with the product. **(b)** Reactant and product species having equal state spacings are depicted. In this case, the product states are higher in energy than those of the reactant such that equilibrium lies with the reactant.

to our earlier definition. Turning to the vibrational problem, the general expression for the vibrational partition function can be written incorporating ε_D as follows:

$$q'_{vib} = \sum_n e^{-\beta \varepsilon_n} = e^{-\beta(-\varepsilon_D)} + e^{-\beta(-\varepsilon_D + \tilde{\nu})} + e^{-\beta(-\varepsilon_D + 2\tilde{\nu})} + \ldots$$

$$= e^{\beta \varepsilon_D}(1 + e^{-\beta \tilde{\nu}} + e^{-2\beta \tilde{\nu}} + \ldots)$$

$$= e^{\beta \varepsilon_D} q_{vib} \tag{15.85}$$

In other words, the corrected form for the vibrational partition function is simply the product of our original q_{vib} (without zero point energy) times a factor that corrects for the offset in energy due to dissociation. With this correction factor in place, the ground vibrational states are all set to zero. Therefore, the final expression for the equilibrium constant is

$$K_P = \frac{\left(\dfrac{q^\circ_C}{N}\right)^c \left(\dfrac{q^\circ_D}{N}\right)^d}{\left(\dfrac{q^\circ_A}{N}\right)^a \left(\dfrac{q^\circ_B}{N}\right)^b} e^{\beta(c\varepsilon_C + d\varepsilon_D - a\varepsilon_A - b\varepsilon_B)} = \frac{\left(\dfrac{q^\circ_C}{N}\right)^c \left(\dfrac{q^\circ_D}{N}\right)^d}{\left(\dfrac{q^\circ_A}{N}\right)^a \left(\dfrac{q^\circ_B}{N}\right)^b} e^{-\beta \Delta \varepsilon} \tag{15.86}$$

The dissociation energies needed to evaluate Equation (15.86) are readily obtained using spectroscopic techniques.

What insight does this expression for the equilibrium constant provide? Equation (15.86) can be viewed as consisting of two parts. The first part is the ratio of partition functions for products and reactants. Because the partition functions quantify the number of energy states available, this ratio dictates that equilibrium will favor those species with the greatest number of available energy states at a given temperature. The second half is the dissociation energy. This term dictates that equilibrium will favor those species with the lowest energy states. This behavior is illustrated in Figure 15.12. The upper part of the figure shows a reactant and product species having equal ground-state energies. The only difference between the two is that the product species has more accessible energy states at a given temperature than the reactant. Another way to envision this relationship is that the partition function describing the product will be greater than that for the reactant at the same temperature such that, at equilibrium, products will be favored ($K > 1$). On the lower part of the figure, the reactant and product energy spacings are equivalent; however, the product states lie higher in energy than those of the reactant. In this case, equilibrium will lie with the reactant ($K < 1$).

EXAMPLE PROBLEM 15.9

What is the general form of the equilibrium constant for the dissociation of a diatomic molecule?

Solution

The dissociation reaction is

$$X_2(g) \rightleftarrows 2X(g)$$

We first need to derive the partition functions that describe the reactants and products. The products are monatomic species such that only translations and electronic degrees of freedom are relevant. The partition function is then

$$q^\circ_X = q^\circ_T q_E = \left(\frac{V^\circ}{\Lambda^3_X}\right) g_o$$

The superscripts indicate standard thermodynamic conditions. With these conditions, $V° = RT/P°$ (for 1 mol) such that

$$q_X° = q_T° q_E = \left(\frac{RT}{\Lambda_X^3 P°} \right) g_o$$

For molar quantities, $N = N_A$ such that

$$\frac{q_X°}{N_A} = \frac{g_o RT}{N_A \Lambda_X^3 P°}$$

The partition function for X_2 will be equivalent to that for X, with the addition of rotational and vibrational degrees of freedom:

$$\frac{q_{X_2}°}{N_A} = \frac{g_o RT}{N_A \Lambda_{X_2}^3 P°} q_R q_V$$

Using the preceding expressions, the equilibrium constant for the dissociation of a diatomic becomes

$$K_P = \frac{\left(\dfrac{q_X°}{N_A} \right)^2}{\dfrac{q_{X_2}°}{N_A}} e^{-\beta \varepsilon_D} = \frac{\left(\dfrac{g_{o,X} RT}{N_A \Lambda_X^3 P°} \right)^2}{\left(\dfrac{g_{o,X_2} RT}{N_A \Lambda_{X_2}^3 P°} \right) q_R q_V} e^{-\beta \varepsilon_D}$$

$$= \left(\frac{g_{o,X}^2}{g_{o,X_2}} \right) \left(\frac{RT}{N_A P°} \right) \left(\frac{\Lambda_{X_2}^3}{\Lambda_X^6} \right) \frac{1}{q_R q_V} e^{-\beta \varepsilon_D}$$

For a specific example, we can use the preceding expression to predict K_P for the dissociation of I_2 at 298 K given the following parameters:

$$g_{o,I} = 4 \quad \text{and} \quad g_{o,I_2} = 1$$
$$\Lambda_I = 3.20 \times 10^{-12} \text{ m} \quad \text{and} \quad \Lambda_{I_2} = 2.26 \times 10^{-12} \text{ m}$$
$$q_R = 2773$$
$$q_V = 1.58$$
$$\varepsilon_D = 12{,}461 \text{ cm}^{-1}$$

Evaluation of the thermal wavelengths, q_R, and q_V was performed as described in Chapter 14. Given these values, K_P becomes

$$K_P = \left(\frac{g_{o,I}^2}{g_{o,I_2}} \right) \left(\frac{RT}{N_A P°} \right) \left(\frac{\Lambda_{I_2}^3}{\Lambda_I^6} \right) \frac{1}{q_R q_V} e^{-\beta \varepsilon_D}$$

$$= (16)(4.06 \times 10^{-26} \text{ m}^3) \left(\frac{(2.26 \times 10^{-12} \text{ m})^3}{(3.20 \times 10^{-12} \text{ m})^6} \right) \frac{1}{(2773)(1.58)} \exp \left[\frac{-hc(12461 \text{ cm}^{-1})}{kT} \right]$$

$$= (1.59 \times 10^6) \exp \left[\frac{-(6.626 \times 10^{-34} \text{ J s})(3.00 \times 10^{10} \text{ cm s}^{-1})(12461 \text{ cm}^{-1})}{(1.38 \times 10^{-23} \text{ J K}^{-1})(298 \text{ K})} \right]$$

$$= 1.10 \times 10^{-20}$$

Using tabulated values for ΔG provided in the back of the text and the relationship $\Delta G = -RT \ln K$, a value of $K = 6 \times 10^{-22}$ is obtained for this reaction.

For Further Reading

Chandler, D., *Introduction to Modern Statistical Mechanics*. Oxford, New York, 1987.

Hill, T., *Statistical Mechanics. Principles and Selected Applications*. Dover, New York, 1956.

Linstrom, P. J., and Mallard, W. G., Eds., *NIST Chemistry Webbook: NIST Standard Reference Database Number 69*. National Institute of Standards and Technology, Gaithersburg, MD, retrieved from http://webbook.nist.gov/chemistry. (This site contains a searchable database of thermodynamic and spectroscopic properties of numerous atomic and molecular species.)

McQuarrie, D., *Statistical Mechanics*. Harper & Row, New York, 1973.

Nash, L. K., "On the Boltzmann Distribution Law," *J. Chemical Education* 59 (1982), 824.

Nash, L. K., *Elements of Statistical Thermodynamics*. Addison-Wesley, San Francisco, 1972.

Noggle, J. H., *Physical Chemistry*. HarperCollins, New York, 1996.

Townes, C. H., and A. L. Schallow, *Microwave Spectroscopy*. Dover, New York, 1975. (This book contains an excellent appendix of spectroscopic constants.)

Widom, B., *Statistical Mechanics*. Cambridge University Press, Cambridge, 2002.

Vocabulary

average energy	Gibbs energy	Sackur–Tetrode equation
Boltzmann's formula	heat capacity	total energy
dissociation energy	Helmholtz energy	translational heat capacity
Dulong and Petit law	internal energy	two-level systems
Einstein solid	residual entropy	vibrational heat capacity
electronic heat capacity	rotational heat capacity	

Questions on Concepts

Q15.1   What is the relationship between ensemble energy and the thermodynamic concept of internal energy?

Q15.2   List the energetic degrees of freedom expected to contribute to the internal energy at 298 K for a diatomic molecule. Given this list, what spectroscopic information do you need to numerically determine the internal energy?

Q15.3   List the energetic degrees of freedom for which the contribution to the internal energy determined by statistical mechanics is equal to the prediction of the equipartition theorem at 289 K.

Q15.4   Write down the contribution to the constant volume heat capacity from translations and rotations for an ideal monatomic, diatomic, and nonlinear polyatomic gas, assuming that the high-temperature limit is appropriate for the rotational degrees of freedom.

Q15.5   When are rotational degrees of freedom expected to contribute R or $3/2R$ (linear and nonlinear, respectively) to the molar constant volume heat capacity? When will a vibrational degree of freedom contribute R to the molar heat capacity?

Q15.6   Why do electronic degrees of freedom generally not contribute to the constant volume heat capacity?

Q15.7   What is the Boltzmann formula, and how can it be used to predict residual entropy?

Q15.8   How does the Boltzmann formula provide an understanding of the third law of thermodynamics?

Q15.9   Which thermodynamic quantity is used to derive the ideal gas law for a monatomic gas? What molecular partition function is employed in this derivation? Why?

Q15.10   What is the definition of "zero" energy employed in constructing the statistical mechanical expression for the equilibrium constant? Why was this definition necessary?

Q15.11   Assume you have an equilibrium expression that involves monatomic species only. What difference in energy between reactants and products would you use in the expression for K_P?

Problems

Problem numbers in **red** indicate that the solution to the problem is given in the *Student's Solutions Manual.*

P15.1 Consider two separate molar ensembles of particles characterized by the following energy-level diagrams:

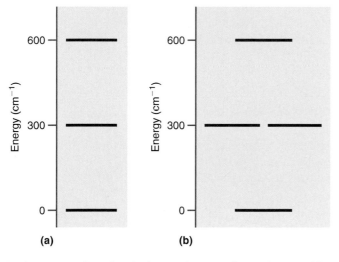

(a)   (b)

Derive expressions for the internal energy for each ensemble. At 298 K, which ensemble is expected to have the greatest internal energy?

P15.2 What is the contribution to the internal energy from translations for an ideal monatomic gas confined to move on a surface? What is the expected contribution from the equipartition theorem?

P15.3 For a system of energy levels, $\varepsilon_m = m^2\alpha$, where α is a constant with units of energy and $m = 0, 1, 2, \ldots, \infty$. What is the internal energy and heat capacity of this system in the high-temperature limit?

P15.4 Consider the following table of diatomic molecules and associated rotational constants:

Molecule	B (cm^{-1})	$\tilde{\nu}$ (cm^{-1})
H^{35}Cl	10.59	2886
^{12}C^{16}O	1.93	2170
^{39}KI	0.061	200
CsI	0.024	120

a. Calculate the rotational temperature for each molecule.

b. Assuming that these species remain gaseous 100 K, for which species is the equipartition theorem prediction for the rotational contribution to the internal energy appropriate?

c. Calculate the vibrational temperature for each molecule.

d. If these species were to remain gaseous at 1000 K, for which species is the equipartition theorem prediction for

the vibrational contribution to the internal energy appropriate?

P15.5 The lowest four energy levels for atomic vanadium (V) have the following energies and degeneracies:

Level (n)	Energy (cm^{-1})	Degeneracy
0	0	4
1	137.38	6
2	323.46	8
3	552.96	10

What is the contribution to the average energy from electronic degrees of freedom for V when $T = 298$ K?

P15.6 Consider an ensemble of units in which the first excited electronic state at energy ε_1 is m_1-fold degenerate, and the energy of the ground state is m_o-fold degenerate with energy ε_0.

a. Demonstrate that if $\varepsilon_0 = 0$, the expression for the electronic partition function is

$$q_E = m_o\left(1 + \frac{m_1}{m_o}e^{-\varepsilon_1/kT}\right)$$

b. Determine the expression for the internal energy U of a ensemble of N such units. What is the limiting value of U as the temperature approaches zero and infinity?

P15.7 Calculate the internal energy of He, Ne, and Ar under standard thermodynamic conditions. Do you need to redo the entire calculation for each species?

P15.8 Determine the internal energy of HCl ($B = 10.59$ cm^{-1} and $\tilde{\nu} = 2886$ cm^{-1}) under standard thermodynamic conditions.

P15.9 Determine the vibrational contribution to C_V for HCl ($\tilde{\nu} = 2886$ cm^{-1}) over a temperature range from 500 to 5000 K in 500-K intervals and plot your result. At what temperature do you expect to reach the high-temperature limit for the vibrational contribution to C_V?

P15.10 Determine the vibrational contribution to C_V for HCN where $\tilde{\nu}_1 = 2041$ cm^{-1}, $\tilde{\nu}_2 = 712$ cm^{-1} (doubly degenerate), and $\tilde{\nu}_3 = 3369$ cm^{-1} at $T = 298$, 500, and 1000 K.

P15.11 Consider the following energy levels and associated degeneracies for atomic Fe:

Level (n)	Energy (cm^{-1})	Degeneracy
0	0	9
1	415.9	7
2	704.0	5
3	888.1	3
4	978.1	1

a. Determine the electronic contribution to C_V for atomic Fe at 150 K assuming that only the first two levels contribute to C_V.

b. How does your answer to part (a) change if the $n = 2$ level is included in the calculation of C_V? Do you need to include the other levels?

P15.12 The speed of sound is given by the relationship

$$c_{sound} = \left(\frac{\frac{C_p}{C_V} RT}{M}\right)^{1/2}$$

where C_p is the constant pressure heat capacity (equal to $C_V + R$), R is the ideal gas constant, T is temperature, and M is molar mass.

a. What is the expression for the speed of sound for an ideal monatomic gas?

b. What is the expression for the speed of sound of an ideal diatomic gas?

c. What is the speed of sound in air at 298 K, assuming that air is mostly made up of nitrogen ($B = 2.00$ cm^{-1} and $\tilde{\nu} = 2359$ cm^{-1})?

P15.13 The measured molar heat capacities for crystalline KCl are as follows at the indicated temperatures:

T (K) /mole	C_V (J/mol K)
50	21.1
100	39.0
175	46.1
250	48.6

a. Explain why the high-temperature limit for C_V is apparently twofold greater than that predicted by the Dulong–Petit law.

b. Determine if the Einstein model is applicable to ionic solids. To do this use the value for C_V at 100 K to determine Θ_V, then use this temperature to determine C_V at 175 K.

P15.14 Inspection of the thermodynamic tables in the back of the text reveals that many molecules have quite similar constant volume heat capacities.

a. The value of $C_{V,M}$ for Ar(g) at standard temperature and pressure is 12.48 J mol^{-1} K^{-1}, identical to gaseous He(g). Using statistical mechanics, demonstrate why this equivalence is expected.

b. The value of $C_{V,M}$ for $N_2(g)$ is 20.81 J mol^{-1} K^{-1}. Is this value expected given your answer to part (a)? For N_2, $\tilde{\nu} = 2359$ cm^{-1} and $B = 2.00$ cm^{-1}.

P15.15 Determine the molar entropy for 1 mol of gaseous Ar at 200, 300, and 500 K and $V = 1000$ cm^3 assuming that Ar can be treated as an ideal gas. How does the result of this calculation change if the gas is Kr instead of Ar?

P15.16 The standard molar entropy of O_2 is 205.14 J mol^{-1} K^{-1}. Using this information, determine the bond length of O_2. For this molecule, $\tilde{\nu} = 1580$ cm^{-1}, and the ground electronic state degeneracy is three.

P15.17 Determine the standard molar entropy of N_2O, a linear triatomic molecule at 298 K. For this molecule, $B = 0.419$ cm^{-1} and $\tilde{\nu}_1 = 1285$ cm^{-1}, $\tilde{\nu}_2 = 589$ cm^{-1} (doubly degenerate), and $\tilde{\nu}_3 = 2224$ cm^{-1}.

P15.18 Determine the standard molar entropy of OClO, a nonlinear triatomic molecule where $B_A = 1.06$ cm^{-1}, $B_B = 0.31$ cm^{-1}, $B_C = 0.29$ cm^{-1} and $\tilde{\nu}_1 = 938$ cm^{-1}, $\tilde{\nu}_2 = 450$ cm^{-1}, and $\tilde{\nu}_3 = 1100$ cm^{-1}.

P15.19 Determine the standard molar entropy of N_2 ($\tilde{\nu} = 2359$ cm^{-1} and $B = 2.00$ cm^{-1}, $g_0 = 1$) and the entropy when $P = 1$ atm but $T = 2500$ K.

P15.20 Determine the standard molar entropy of HCl^{35} at 298 K where $B = 10.58$ cm^{-1}, $\tilde{\nu} = 2886$ cm^{-1}, and the ground-state electronic level degeneracy is one.

P15.21 Derive the expression for the standard molar entropy of a monatomic gas restricted to two-dimensional translational motion (*Hint:* You are deriving the two-dimensional version of the Sackur–Tetrode equation.)

P15.22 The molecule NO has a ground electronic level that is doubly degenerate, and a first excited level at 121.1 cm^{-1} that is also twofold degenerate. Determine the contribution of electronic degrees of freedom to the standard molar entropy of NO. Compare your result to $R\ln(4)$. What is the significance of this comparison?

P15.23 Determine the residual molar entropies for molecular crystals of the following:

a. $^{35}Cl^{37}Cl$ c. CF_2Cl_2

b. $CFCl_3$ d. CO_2

P15.24 Using the Helmholtz energy, demonstrate that the pressure for an ideal polyatomic gas is identical to that derived for an ideal monatomic gas in the text.

P15.25 Derive an expression for the standard molar enthalpy of an ideal monatomic gas by evaluation of the statistical mechanical expression for enthalpy as opposed to the thermodynamic argument provided in Example Problem 15.7.

P15.26 Demonstrate that the molar enthalpy is equal to the molar energy for a collection of one-dimensional harmonic oscillators.

P15.27 Calculate the standard Helmholtz energy for molar ensembles of Ne and Kr at 298 K.

P15.28 What is the vibrational contribution to the Helmholtz and Gibbs energies from a molar ensemble of one-dimensional harmonic oscillators?

P15.29 Determine the standard Gibbs energy for $^{35}Cl^{35}Cl$ where $\tilde{\nu} = 560$ cm^{-1}, $B = 0.244$ cm^{-1}, and the ground electronic state is nondegenerate.

P15.30 Determine the rotational and vibrational contributions to the standard Gibbs energy for N_2O (NNO), a linear triatomic molecule where $B = 0.419 \text{ cm}^{-1}$ and $\tilde{\nu}_1 = 1285 \text{ cm}^{-1}$, $\tilde{\nu}_2 = 589 \text{ cm}^{-1}$ (doubly degenerate), and $\tilde{\nu}_3 = 2224 \text{ cm}^{-1}$.

P15.31 Determine the equilibrium constant for the dissociation of sodium at 298 K: $Na_2(g) \rightleftharpoons 2Na(g)$. For Na_2, $B = 0.155 \text{ cm}^{-1}$, $\tilde{\nu} = 159 \text{ cm}^{-1}$, the dissociation energy is 70.4 kJ/mol, and the ground-state electronic degeneracy for Na is 2.

P15.32 The isotope exchange reaction for Cl_2 is as follows: $^{35}Cl^{35}Cl + {}^{37}Cl^{37}Cl \rightleftharpoons 2\,{}^{37}Cl^{35}Cl$. The equilibrium constant for this reaction is ~4. Furthermore, the equilibrium constant for similar isotope-exchange reactions is also close to this value. Demonstrate why this would be so.

P15.33 In "Direct Measurement of the Size of the Helium Dimer" by F. Luo, C. F. Geise, and W. R. Gentry [*J. Chemical Physics* 104 (1996), 1151], evidence for the helium dimer is presented. As one can imagine, the chemical bond in the dimer is extremely weak, with an estimated value of only 8.3 mJ/mol.

a. An estimate for the bond length of 65 Å is presented in the paper. Using this information, determine the rotational constant for He_2. Using this value for the rotational constant, determine the location of the first rotational state. If correct, you will determine that the first excited rotational level is well beyond the dissociation energy of He_2.

b. Consider the following equilibrium between He_2 and its atomic constituents: $He_2(g) \rightleftharpoons 2He(g)$. If there are no rotational or vibrational states to consider, the equilibrium is determined exclusively by the translational degrees of freedom and the dissociation energy of He_2. Using the dissociation energy provided earlier and $V = 1000 \text{ cm}^3$, determine K_p assuming that $T = 10$ K. The experiments were actually performed at 1 mK; why was such a low temperature employed?

Web-Based Simulations, Animations, and Problems

W15.1 In this simulation, the temperature dependence of C_V for vibrational degrees of freedom is investigated for diatomic and polyatomic molecules. Diatomic molecules are studied, as are polyatomics with and without mode degeneracy. Comparisons of exact values to those expected for the high-temperature limit are performed.

CHAPTER 16

Kinetic Theory of Gases

Gas particle motion is of importance in many aspects of physical chemistry, from transport phenomena to chemical kinetics. In this chapter, the translational motion of gas particles is described. Gas particle motion is characterized by a distribution of velocities and speeds. These distributions, including the Maxwell speed distribution, are derived. Benchmark values for these distributions are presented that provide insight into how the distributions change with temperature and particle mass. Finally, molecular collisions are discussed, including the frequency of collisional events and the distance particles travel between collisions. The concepts presented in this chapter find wide application in the remainder of this text since they provide the first step in understanding gas-phase molecular dynamics. ■

16.1 Kinetic Theory of Gas Motion and Pressure

In this chapter, we expand on the microscopic viewpoint of matter by considering the translational motion of gas particles. **Gas kinetic theory** provides the starting point for this development and represents a central concept of physical chemistry. In this theory, gases are envisioned as a collection of atoms or molecules that we will refer to as *particles.* Gas kinetic theory is applicable when the particle density of the gas is such that the distance between particles is very large in comparison to their size. To illustrate this point, consider a mole of Ar at a temperature of 298 K and 1 atm pressure. Using the ideal gas law, the gas occupies a volume of 24.4 L or 0.0244 m^3. Dividing this volume by Avogadro's number provides an average volume per Ar atom of 4.05×10^{-26} m^3, or 40.5 nm^3. The diameter of Ar is ~0.29 nm corresponding to a particle volume of 0.013 nm^3. Comparison of the particle volume to the average volume of an individual Ar atom demonstrates that on average only 0.03% of the available volume is occupied by the particle. Even for particles with diameters substantially greater than Ar, the difference between the average volume per particle and the volume of the particle itself will be such that the distance between particles in the gas phase is substantial.

Given the large distance between particles in a gas, each particle is envisioned as traveling through space as a separate, unperturbed entity until a collision occurs with another particle or with the walls of the container confining the gas. In gas kinetic theory, particle motion is described using Newton's laws of motion. Although in previous chapters we have seen that classical descriptions fail to capture many of the microscopic

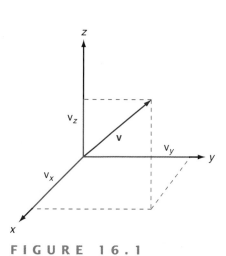

FIGURE 16.1
Cartesian components of velocity. The particle velocity **v** can be decomposed into velocity components along three Cartesian dimensions: v_x, v_y, and v_z.

properties of atoms and molecules, recall from Chapter 14 that the energy spacings between translational energy states is very small relative to kT such that a classical description of translational motion is appropriate.

One hallmark of kinetic theory is its ability to describe the pressure of an ideal gas. As opposed to the thermodynamic derivation of the ideal gas law, which relied on empirical relations between gas variables, the kinetic theory description of pressure relies on classical mechanics and a microscopic description of the system. The pressure exerted by a gas on the container confining the gas arises from collisions of gas particles with the container walls. Because the number of molecules comprising the gas is on the order of Avogadro's number, one might imagine that the number of collisions is also large. To describe pressure using kinetic theory, a molecule is envisioned as traveling through space with a velocity, **v**, that can be decomposed into three Cartesian components: v_x, v_y, and v_z as illustrated in Figure 16.1.

The square of the velocity magnitude (v^2) in terms of the three velocity components is

$$v^2 = \mathbf{v} \bullet \mathbf{v} = v_x^2 + v_y^2 + v_z^2 \tag{16.1}$$

The particle kinetic energy is $1/2\ mv^2$ such that

$$\varepsilon_{Tr} = \frac{1}{2}mv^2 = \frac{1}{2}mv_x^2 + \frac{1}{2}mv_y^2 + \frac{1}{2}mv_z^2 = \varepsilon_{Tr,x} + \varepsilon_{Tr,y} + \varepsilon_{Tr,z} \tag{16.2}$$

where the subscript Tr indicates that the energy corresponds to translational motion of the particle. Furthermore, this equation dictates that the total translational energy is simply the sum of translational energy along each Cartesian dimension.

Pressure arises from the collisions of gas particles with the walls of the container; therefore, to describe pressure using gas kinetic theory we must consider what occurs during a collision between a gas particle and the wall. First, we assume that the collisions with the wall are **elastic** such that translational energy of the particle is conserved. Although the collision is elastic, this does not mean that nothing happens. As a result of the collision, linear momentum is imparted to the wall, which results in pressure. The definition of pressure is force per unit area and, by Newton's second law, force is equal to the product of mass and acceleration. Using these two definitions, pressure is expressed as

$$P = \frac{F}{A} = \frac{ma_i}{A} = \frac{m}{A}\left(\frac{dv_i}{dt}\right) = \frac{1}{A}\left(\frac{dmv_i}{dt}\right) = \frac{1}{A}\left(\frac{dp_i}{dt}\right) \tag{16.3}$$

In Equation (16.3), F is the force of the collision, A is the area of the wall with which the particle has collided, m is the mass of the particle, v_i is the velocity in the i direction ($i = x$, y, or z), and p_i is the particle linear momentum in the i direction. Equation (16.3) illustrates that pressure is related to the change in linear momentum with respect to time, and suggests that momentum is exchanged between the particle and wall during a collision. Due to conservation of momentum, any change in particle linear momentum must result in an equal and opposite change in momentum of the container wall. A single collision is depicted in Figure 16.2. This figure illustrates that the particle linear momentum change in the x direction is $-2mv_x$ (note there is no change in momentum in the y or z direction). Given this, a corresponding momentum change of $2mv_x$ must occur for the wall. (Keep in mind that the wall is very massive; therefore, it does not move as a result of the collision.)

The pressure measured at the container wall corresponds to the sum of collisions involving a large number of particles that occur per unit time. Therefore, the total momentum change that gives rise to the pressure is equal to the product of the momentum change from a single particle collision and the total number of particles that collide with the wall:

$$\Delta p_{Total} = \frac{\Delta p}{\text{molecule}}(\text{number of molecules}) \tag{16.4}$$

How many molecules strike the side of the container in a given period of time? To answer this question, the time over which collisions are counted must be considered. Con-

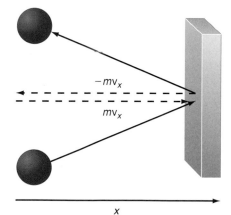

FIGURE 16.2
Collision between a gas particle and a wall. Before the collision, the particle has a momentum of mv_x in the x direction, and after the collision the momentum is $-mv_x$. Therefore, the change in particle momentum resulting from the collision is $-2mv_x$. By conservation of momentum, the change in momentum of the wall must be $2mv_x$.

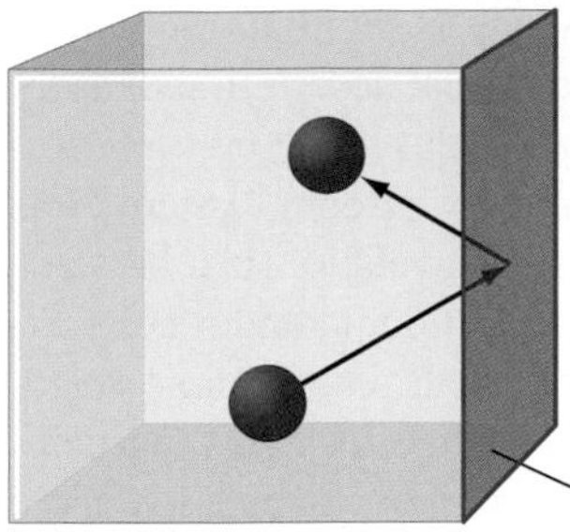

FIGURE 16.3
Volume element to determine the number of collisions with the wall per unit time.

sider a volume element defined by the area of the wall, A, times length, Δx, as illustrated in Figure 16.3. The collisional volume element depicted in Figure 16.3 is given by

$$V = A(\Delta x) \tag{16.5}$$

The length of the box, Δx, is related to the time period over which collisions will be counted, Δt, and the component of particle velocity parallel to the side of the box (taken to be the x direction):

$$\Delta x = \mathrm{v}_x \Delta t$$

In this expression, v_x is for a single particle; however, an average of this quantity will be used when describing the collisions from a collection of particles. Finally, the number of particles that will collide with the container wall in the time interval Δt is equal to the number density, $\tilde{N}$, which is equal to the number of particles in the container, N, divided by the container volume, V, times the collisional volume element depicted in Figure 16.3:

$$N = \tilde{N} \times (A\mathrm{v}_x \Delta t)\left(\frac{1}{2}\right) = \frac{nN_A}{V}(A\mathrm{v}_x \Delta t)\left(\frac{1}{2}\right) \tag{16.6}$$

The factor of 1/2 in Equation (16.6) accounts for the directionality of particle motion. The collisions of interest occur with the container wall defined by area A depicted in Figure 16.3. Because particles travel in either the $+x$ or $-x$ direction with equal probability, only those molecules traveling in the $+x$ direction will strike the area of interest. Therefore, the total number of particles, N, is divided by two to take the direction of particle motion into account. Employing Equation (16.6), the total change in linear momentum of the container wall imparted by particle collisions is given by

$$\Delta p_{Total} = \frac{\Delta p}{\text{molecule}}(\text{number of molecules}) \tag{16.7}$$

$$= (2m\mathrm{v}_x)(N)$$

$$= (2m\mathrm{v}_x)\left(\frac{nN_A}{V}\frac{A\mathrm{v}_x \Delta t}{2}\right)$$

$$= \frac{nN_A}{V}A\Delta t\, m\langle \mathrm{v}_x^2 \rangle$$

In Equation (16.7), angle brackets appear around v_x^2 to indicate that this quantity represents an average value since the particles will demonstrate a distribution of velocities. This distribution is considered in detail later in this chapter. With the total change in linear momentum provided in Equation (16.7), the force and corresponding pressure exerted by the gas on the container wall [Equation (16.3)] are as follows:

$$F = \frac{\Delta p_{Total}}{\Delta t} = \frac{nN_A}{V}Am\langle \mathrm{v}_x^2 \rangle$$

$$P = \frac{F}{A} = \frac{nN_A}{V}m\langle \mathrm{v}_x^2 \rangle \tag{16.8}$$

Equation (16.8) can be converted into a more familiar expression once $1/2\, m\langle \mathrm{v}_x^2 \rangle$ is recognized as the translational energy in the x direction. In Chapter 15, it was shown that the average translational energy for an individual particle in one dimension is

$$\frac{m\langle \mathrm{v}_x^2 \rangle}{2} = \frac{kT}{2}$$

Substituting this result into Equation (16.8) results in the following expression for pressure:

$$P = \frac{nN_A}{V}m\langle \mathrm{v}_x^2 \rangle = \frac{nN_A}{V}kT = \frac{nRT}{V} \tag{16.9}$$

Equation (16.9) is the **ideal gas law.** Although this relationship is familiar, the way in which it was derived is extremely interesting. Employing a classical description of a single molecular collision with the container wall, and then scaling this result up to macroscopic proportions, one of the central results in the chemistry of gaseous systems was derived. A second result derived from gas kinetic theory is the relationship between **root-mean-squared velocity** and temperature. If the particle motion is random, the average velocities along all three Cartesian dimensions are equivalent. The average velocity along any dimension will be zero because there will be just as many particles traveling in both the positive and negative directions. In contrast, the root-mean-squared velocity is given by the following:

$$\langle v^2 \rangle^{1/2} = \langle v_x^2 + v_y^2 + v_z^2 \rangle^{1/2}$$
$$= \langle 3v_x^2 \rangle^{1/2}$$
$$= \left(\frac{3kT}{m} \right)^{1/2} \tag{16.10}$$

Kinetic theory thus predicts that the root-mean-squared speed of the gas particles should increase as the square root of temperature and decrease as the square root of the particle mass.

The success of this approach is tempered by uncertainty regarding the assumptions made during the course of the derivation. For example, it has been assumed that the individual particle velocities can be characterized by some average value. Can we determine what this average value is given the distribution of molecular velocities that exist? Just what does the distribution of particle velocities or speed look like? In addition, the molecules collide with each other as well as the walls of the container. How frequent are such collisions? The frequency of molecular collisions will be important in subsequent chapters describing transport phenomena and the rates of chemical reactions. Therefore, a more critical look at atomic and molecular speed distributions and collisional dynamics is warranted.

16.2 Velocity Distribution in One Dimension

From the previous discussion of statistical thermodynamics, it is clear that a distribution of translational energies and, therefore, velocities will exist for a collection of gaseous particles. What does this distribution of velocities look like?

The variation in particle velocities is described by the **velocity distribution function.** In Chapter 12 the concept of a distribution function was presented. The velocity distribution function describes the probability of a gas particle having a velocity within a given range. In section 14.4 we found that the translational energy-level spacings are sufficiently small that velocity can be treated as a continuous variable. Therefore, the velocity distribution function describes the probability of a particle having a velocity in the range $v_x + dv_x$, $v_y + dv_y$, and $v_z + dv_z$.

To begin the derivation of the velocity distribution function, let $\Omega(v_x, v_y, v_z)$ represent the function that describes the distribution of velocity for an ensemble of gaseous particles. We assume that the distribution function can be decomposed into a product of distribution functions for each Cartesian dimension, and the distribution of velocities in one dimension is independent of the distribution in the other two dimensions. With this assumption, $\Omega(v_x, v_y, v_z)$ is expressed as follows:

$$\Omega(v_x, v_y, v_z) = f(v_x)f(v_y)f(v_z) \tag{16.11}$$

In Equation (16.11), $f(v_x)$ is the velocity distribution for velocity in the x direction, and so forth. We assume that the gas is confined to an isotropic space such that the direction in which the particle moves does not affect the properties of the gas. In this case, the distribution function $\Omega\,(v_x, v_y, v_z)$ only depends on the magnitude of the velocity, or speed (v). The natural log of Equation (16.12) yields

$$\ln \Omega(v) = \ln f(v_x) + \ln f(v_y) + \ln f(v_z)$$

To determine the velocity distribution along a single direction, the partial derivative of $\ln \Omega(v)$ is taken with respect to v_x while keeping the velocity along the other two directions constant:

$$\left(\frac{\partial \ln \Omega(v)}{\partial v_x} \right)_{v_y, v_z} = \frac{d \ln f(v_x)}{dv_x} \tag{16.12}$$

Equation (16.12) can be rewritten using the chain rule for differentiation (see Appendix B, Math Supplement) allowing for the derivative of $\ln \Omega(v)$ with respect to v_x to be written as

$$\left(\frac{d \ln \Omega(v)}{dv} \right)\left(\frac{\partial v}{\partial v_x} \right)_{v_y, v_z} = \frac{d \ln f(v_x)}{dv_x} \tag{16.13}$$

Using Equation (16.1), the second factor on the left-hand side of Equation (16.13) can be readily evaluated:

$$\left(\frac{\partial v}{\partial v_x} \right)_{v_y, v_z} = \left(\frac{\partial}{\partial v_x} (v_x^2 + v_y^2 + v_z^2)^{1/2} \right)_{v_y, v_z} \tag{16.14}$$

$$= \frac{1}{2}(2v_x)(v_x^2 + v_y^2 + v_z^2)^{-1/2}$$

$$= \frac{v_x}{v}$$

Substituting this result into Equation (16.13) and rearranging yields

$$\left(\frac{d \ln \Omega(v)}{dv} \right)\left(\frac{\partial v}{\partial v_x} \right)_{v_y, v_z} = \frac{d \ln f(v_x)}{dv_x}$$

$$\left(\frac{d \ln \Omega(v)}{dv} \right)\left(\frac{v_x}{v} \right) = \frac{d \ln f(v_x)}{dv_x}$$

$$\frac{d \ln \Omega(v)}{v dv} = \frac{d \ln f(v_x)}{v_x dv_x} \tag{16.15}$$

It is important to recall at this point that the velocity distributions along each direction are equivalent. Therefore, the preceding derivation could just as easily have been performed considering v_y or v_z, resulting in the following expressions analogous to Equation (16.15):

$$\frac{d \ln \Omega(v)}{v dv} = \frac{d \ln f(v_y)}{v_y dv_y} \tag{16.16}$$

$$\frac{d \ln \Omega(v)}{v dv} = \frac{d \ln f(v_z)}{v_z dv_z} \tag{16.17}$$

Comparison of Equations (16.15), (16.16), and (16.17) suggests that the following equality exists:

$$\frac{d \ln f(v_x)}{v_x dv_x} = \frac{d \ln f(v_y)}{v_y dv_y} = \frac{d \ln f(v_z)}{v_z dv_z} \tag{16.18}$$

In order for Equation (16.18) to be correct, each of the terms in Equation (16.17) must be equal to a constant, γ, such that

$$\frac{d \ln f(v_j)}{v_j dv} = -\gamma \quad \text{for } j = x, y, z \tag{16.19}$$

In this equation, the negative of γ has been employed recognizing that γ must be a positive quantity to ensure that $f(v_j)$ does not diverge as v_j approaches infinity. Integration of Equation (16.19) results in the following expression for the velocity distribution along one direction:

$$f(v_j) = Ae^{-\gamma v_j^2/2} \tag{16.20}$$

The last step remaining in the derivation is to determine A and γ. To determine A, we refer back to Chapter 12 and the discussion of normalized distribution functions and require that the velocity distribution be normalized. Because a particle can be traveling in either the $+j$ or $-j$ direction, the range of the velocity distribution is $-\infty \le v_j \le \infty$. Applying the normalization condition and integrating over this range,

$$\int_{-\infty}^{\infty} f(v_j)dv_j = 1 = \int_{-\infty}^{\infty} Ae^{-\gamma v_j^2/2}dv_j \tag{16.21}$$

$$1 = A\int_{-\infty}^{\infty} e^{-\gamma v_j^2/2}dv_j$$

$$1 = A\sqrt{\frac{2\pi}{\gamma}}$$

$$\sqrt{\frac{\gamma}{2\pi}} = A$$

In evaluating this integral, the property of even integrands has been used, which says that the integral from $-\infty$ to ∞ is equal to twice the integral from 0 to ∞. With the normalization factor, the velocity distribution in one dimension becomes

$$f(v_j) = \left(\frac{\gamma}{2\pi}\right)^{1/2} e^{-\gamma v_j^2/2} \tag{16.22}$$

All that remains is to evaluate γ. Earlier we encountered the following definition for $\langle v_x^2 \rangle$:

$$\langle v_x^2 \rangle = \frac{kT}{m}$$

Recall that angle brackets around v_x^2 indicate that this quantity is an average over the ensemble of particles. Furthermore, this quantity is equal to the second moment of the velocity distribution; therefore, γ can be determined as follows:

$$\langle v_x^2 \rangle = \frac{kT}{m} = \int_{-\infty}^{\infty} v_x^2 f(v_x)dv_x \tag{16.23}$$

$$= \int_{-\infty}^{\infty} v_x^2 \sqrt{\frac{\gamma}{2\pi}} e^{-\gamma v_x^2/2}dv_x$$

$$= \sqrt{\frac{\gamma}{2\pi}} \int_{-\infty}^{\infty} v_x^2 e^{-\gamma v_x^2/2}dv_x$$

$$= \sqrt{\frac{\gamma}{2\pi}} \left(\frac{1}{\gamma}\sqrt{\frac{2\pi}{\gamma}}\right)$$

$$= \frac{1}{\gamma}$$

$$\frac{m}{kT} = \gamma$$

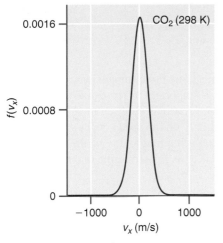

FIGURE 16.4
One-dimensional velocity distribution for CO_2 at 298 K.

The integration was performed using the integral tables provided in the Math Supplement, Appendix B. With the definition of γ, the **Maxwell–Boltzmann velocity distribution** in one dimension becomes

$$f(v_j) = \left(\frac{m}{2\pi kT}\right)^{1/2} e^{(-mv_j^2/2kT)} = \left(\frac{M}{2\pi RT}\right)^{1/2} e^{(-Mv_j^2/2RT)} \qquad (16.24)$$

In Equation (16.24), m is the particle mass in units of kilograms, and M is the molar mass in units of kg mol^{-1}, obtained from the expression involving m using the relationship $R = N_A k$. The velocity distribution is equal to the product of a preexponential factor that is independent of velocity, and an exponential factor that is velocity dependent, and this latter term is very reminiscent of the Boltzmann distribution. The one-dimensional velocity distribution in the x direction for CO_2 at 298 K is presented in Figure 16.4. Notice that the distribution maximum is at 0 ms^{-1}.

EXAMPLE PROBLEM 16.1

Compare $\langle v_x \rangle$ and $\langle v_x^2 \rangle$ for an ensemble of gaseous particles.

Solution
The average velocity is simply the first moment of the velocity distribution function:

$$\langle v_x \rangle = \int_{-\infty}^{\infty} v_x f(v_x) dv_x$$

$$= \int_{-\infty}^{\infty} v_x \left(\frac{m}{2\pi kT}\right)^{1/2} e^{-mv_x^2/2kT} dv_x$$

$$= \left(\frac{m}{2\pi kT}\right)^{1/2} \int_{-\infty}^{\infty} v_x e^{-mv_x^2/2kT} dv_x$$

$$= 0$$

The integral involves the product of odd (v_x) and even (exponential) factors so that the integral over the domain of v_x equals zero (a further description of even and odd functions is provided in the Math Supplement). The fact that the average value of $v_x = 0$ reflects the vectorial character of velocity with particles equally likely to be moving in the $+x$ or $-x$ direction.
The quantity $\langle v_x^2 \rangle$ was determined earlier:

$$\langle v_x^2 \rangle = \int_{-\infty}^{\infty} v_x^2 f(v_x) dv_x = \frac{kT}{m} \qquad (16.25)$$

Notice that the average value for the second moment is greater than zero, reflecting the fact that the square of the velocity must be a positive quantity.

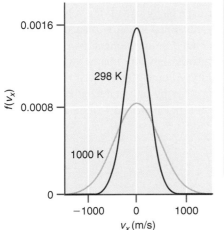

FIGURE 16.5
One-dimensional velocity distributions for Ar at 298 K (red line) and 1000 K (yellow line).

Figure 16.5 presents the one-dimensional velocity distribution functions for Ar at two different temperatures. Notice how the width of the distribution increases with temperature consistent with the increased probability of populating higher energy translational states with correspondingly greater velocities.
Figure 16.6 presents the distribution for Kr, Ar, and Ne at 298 K. The velocity distribution is narrowest for Kr and broadest for Ne reflecting the mass dependence of the distribution.

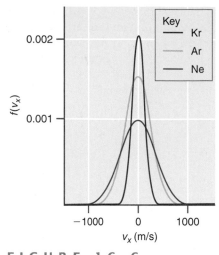

FIGURE 16.6
Velocity distributions for Kr (red line, molar mass = 83.8 g mol⁻¹), Ar (yellow line, molar mass = 39.9 g mol⁻¹), and Ne (blue line, molar mass = 20.2 g mol⁻¹) at 298 K.

16.3 The Maxwell Distribution of Molecular Speeds

With one-dimensional velocity distributions in hand, the three-dimensional distribution of molecular speeds can be determined. First, it is important to recognize that speed is not a vector. The reason we concern ourselves with speed as opposed to velocity is that many physical properties of gases are dependent only on the speed of the gas particles, and not on the direction of motion. Therefore, only the magnitude of the velocity, or speed, is generally of interest. The **particle speed,** v, is related to the one-dimensional velocity Cartesian components by the following:

$$v = (v_x^2 + v_y^2 + v_z^2)^{1/2} \tag{16.26}$$

We are interested in determining the particle speed distribution, $F(v)$, but how can this distribution be derived using the velocity distributions derived in the previous section? We can connect these concepts using the geometric interpretation of velocity presented in Figure 16.7. The figure depicts **velocity space,** which can be understood in analogy to Cartesian space with linear distance (x, y, z) replaced by the Cartesian components of velocity (v_x, v_y, v_z). The figure demonstrates that the molecular velocity **v**, is described by a vector with coordinates v_x, v_y, and v_z in velocity space with length equal to the speed [Equation (16.26)].

Particle speed distribution $F(v)$ is defined in terms of one-dimensional velocity distributions along each direction [Equation (16.24)] as follows:

$$
\begin{aligned}
F(v)dv &= f(v_x)f(v_y)f(v_z)dv_x\,dv_y\,dv_z \\
&= \left[\left(\frac{m}{2\pi kT}\right)^{1/2} e^{-mv_x^2/2kT}\right]\left[\left(\frac{m}{2\pi kT}\right)^{1/2} e^{-mv_y^2/2kT}\right] \\
&\quad \times \left[\left(\frac{m}{2\pi kT}\right)^{1/2} e^{-mv_z^2/2kT}\right]dv_x\,dv_y\,dv_z \\
&= \left(\frac{m}{2\pi kT}\right)^{3/2} e^{\left[-m(v_x^2+v_y^2+v_z^2)\right]/2kT}dv_x\,dv_y\,dv_z
\end{aligned}
\tag{16.27}
$$

Notice that in Equation (16.27), the speed distribution of interest is defined with respect to the Cartesian components of velocity; therefore, the factors involving velocity need to be expressed in terms of speed to obtain $F(v)dv$. This transformation is accomplished as follows. First, the factor $dv_x\,dv_y\,dv_z$ is an infinitesimal volume element in velocity space (Figure 16.7). Similar to the transformation from Cartesian coordinates to spherical coordinates (see the Math Supplement), the velocity volume element can be written as $4\pi v^2 dv$ after integration over angular dimensions leaving only v dependence. In addition, $v_x^2 + v_y^2 + v_z^2$ in the exponent of Equation (16.27) can be written as v^2 [Equation (16.26)] such that

$$F(v)dv = 4\pi\left(\frac{m}{2\pi kT}\right)^{3/2} v^2 e^{-mv^2/2kT}dv \tag{16.28}$$

Equation (16.23) is written in terms of the mass of an individual gas particle. Because the molar mass M is equal to $N_A \times m$, where N_A is Avogadro's number and m is the particle mass, and $R = N_A \times k$, Equation (16.28) can be written in terms of M as follows:

$$F(v)dv = 4\pi\left(\frac{M}{2\pi RT}\right)^{3/2} v^2 e^{-Mv^2/2RT}dv \tag{16.29}$$

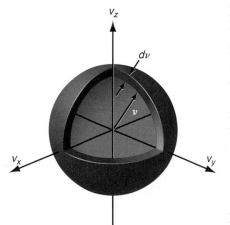

FIGURE 16.7
Illustration of velocity space. The Cartesian components of the particle velocity **v** are given by v_x, v_y, and v_z. The spherical shell represents a differential volume element of velocity space having volume $4\pi v^2 dv$.

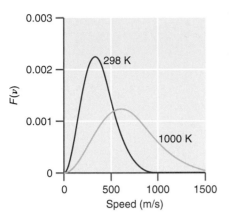

FIGURE 16.8
Speed distributions for Ar at 298 and 1000 K.

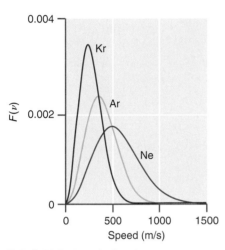

FIGURE 16.9
Speed distributions for Ne (blue line, molar mass = 20.2 g mol^{-1}), Ar (yellow line, molar mass = 39.9 g mol^{-1}), and Kr (red line, molar mass = 83.8 g mol^{-1}) at 298 K.

The **Maxwell speed distribution** is given by Equation (16.28) or (16.29) and represents the probability distribution of a molecule having a speed between v and $v + dv$. Comparison of this distribution to the one-dimensional velocity distribution of Equation (16.24) reveals many similarities, with the main difference being the v^2 dependence that now appears in the preexponential factor in Equation (16.29). A second difference is the range of the distribution. Unlike velocity where negative values are possible since the particle is always free to move in the negative direction with respect to a given coordinate, particle speeds must be greater than or equal to zero so that the range of the distribution is from zero to infinity.

Figure 16.8 presents the Maxwell speed distribution for Ar at 298 and 1000 K and illustrates the dependence of the speed distribution on temperature. Notice that unlike the velocity distribution of Equation (16.20), the speed distribution is not symmetric. This is because the initial increase in probability is due to the v^2 factor in Equations (16.28) or (16.29), but at higher speeds the probability decays exponentially. Second, as temperature increases two trends become evident. The maximum of the distribution shifts to higher speed as temperature increases. This is expected because an increase in temperature corresponds to an increase in kinetic energy and subsequently an increase in particle speed. However, the entire distribution does not simply shift to higher velocity. Instead, the curvature of the distribution changes, with this behavior quite pronounced on the high-speed side of the distribution since an increase in kT will increase the probability of occupying higher energy translational states.

Figure 16.9 presents a comparison of the particle speed distributions for Ne, Ar, and Kr, at 298 K. The speed distribution peaks at lower speeds for heavier particles. This behavior can be understood since the average kinetic energy is 3/2 kT, a quantity that is only dependent on temperature. Because kinetic energy is also equal to 1/2 mv^2, an increase in mass must be offset by a reduction in the root-mean-squared speed. This expectation is reflected by the distributions presented in Figure 16.9.

One of the first detailed experimental verifications of the Maxwell distribution law was provided in 1955 by Miller and Kusch [*Phys. Rev.* 99 (1955), 1314]. A schematic drawing of the apparatus employed in this study is presented in Figure 16.10. An oven was used to create a gas of known temperature, and a hole was placed in the side of the oven through which the gas could emerge. The stream of gas molecules escaping the oven was then directed through spatial apertures to create a beam of particles, which was then directed toward a velocity selector. By changing the rotational speed of the velocity selector, the required gas speed necessary to pass through the cylinder is varied. The number of gas particles passing through the cylinder as a function of rotational speed is then measured to determine the distribution of gas speeds.

FIGURE 16.10
Schematic of an experimental apparatus used to verify the Maxwell speed distribution. Molecules are emitted from the oven and pass through the slits to produce a beam of molecules. This beam reaches a velocity selector consisting of two rotating disks with slots in each disk. After passing through the first disk, the molecules will reach the second disk that will have rotated by angle θ. Only molecules for which the velocity equals $\omega x/\theta$, where ω is the angular rotational velocity of the disks and x is the separation between disks, will pass through the second disk and reach the detector.

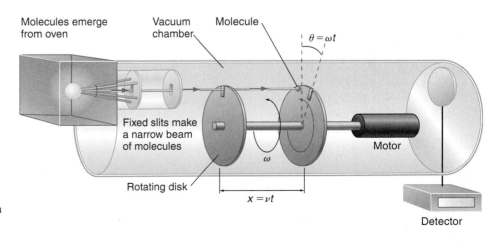

FIGURE 16.11
Experimentally determined distribution of particle speeds for gaseous potassium at 466 ± 2 K (circles). The expected Maxwellian distribution is presented as the red line, and demonstrates excellent agreement with experiment.

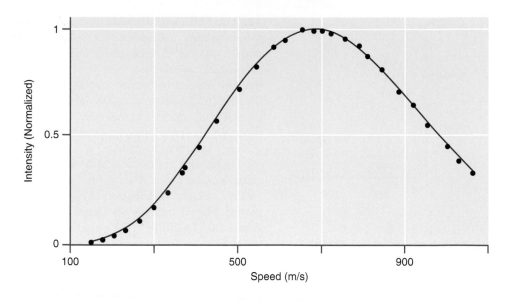

The results of this experiment for potassium at an oven temperature of 466 ± 2 K are presented in Figure 16.11. The comparison between the experimentally measured number of particles with a certain speed versus the theoretical prediction for a Maxwellian distribution is excellent. The interested reader is encouraged to read the Miller and Kusch manuscript for a fuller presentation of this elegant experiment.

16.4 Comparative Values for Speed Distribution: v_{ave}, v_{mp}, and v_{rms}

The Maxwell speed distribution describes the probability of observing a particle within a given range of speeds; however, knowledge of the entire distribution is seldom required when comparing the properties of gases. Instead, representative quantities of this distribution that provide a metric as to how the distribution changes as a function of mass or temperature are sometimes more useful. For example, in Figure 16.9 it is clear that the speed distributions for Ne, Ar, and Kr are different, but can these differences be described without depicting the entire distribution? It would be much more convenient to compare only certain aspects of the distribution, such as the speed at which the distribution is maximized or the average speed.

The first comparative value we consider is the **most probable speed,** or v_{mp}, which is equal to the speed at which $F(v)$ is at a maximum. This quantity is determined by calculating the derivative of $F(v)$ with respect to speed:

$$\frac{dF(v)}{dv} = \frac{d}{dv}\left(4\pi\left(\frac{m}{2\pi kT}\right)^{3/2} v^2 e^{-mv^2/2kT}\right)$$

$$= 4\pi\left(\frac{m}{2\pi kT}\right)^{3/2} \frac{d}{dv}(v^2 e^{-mv^2/2kT})$$

$$= 4\pi\left(\frac{m}{2\pi kT}\right)^{3/2} e^{-mv^2/2kT}\left[2v - \frac{mv^3}{kT}\right]$$

The most probable speed is the speed at which $dF(v)/dv$ is equal to zero, which will be the case when the factor contained in the square brackets in the preceding equation is equal to zero. Recognizing this, v_{mp} is given by

$$2v_{mp} - \frac{mv_{mp}^3}{kT} = 0$$

$$v_{mp} = \sqrt{\frac{2kT}{m}} = \sqrt{\frac{2RT}{M}} \tag{16.30}$$

EXAMPLE PROBLEM 16.2

What is the most probable speed for Ne and Kr at 298 K?

Solution

First, v_{mp} for Ne is readily determined using Equation (16.30):

$$v_{mp} = \sqrt{\frac{2RT}{M}} = \sqrt{\frac{2(8.314 \text{ J mol}^{-1} \text{ K}^{-1})298 \text{ K}}{0.020 \text{ kg mol}^{-1}}} = 498 \text{ m s}^{-1}$$

The corresponding v_{mp} for Kr can be determined in the same manner, or with reference to Ne as follows:

$$\frac{(v_{mp})_{Kr}}{(v_{mp})_{Ne}} = \sqrt{\frac{M_{Ne}}{M_{Kr}}} = \sqrt{\frac{0.020 \text{ kg mol}^{-1}}{0.083 \text{ kg mol}^{-1}}} = 0.491$$

With this result, v_{mp} for Kr is readily determined:

$$(v_{mp})_{Kr} = 0.491(v_{mp})_{Ne} = 244 \text{ m s}^{-1}$$

For the sake of comparison, the speed of sound in dry air at 298 K is 346 m/s, and a typical commercial airliner travels at about 500 miles/hour or 224 m/s.

Average speed can be determined using the Maxwell speed distribution and the definition of average value provided in Chapter 12:

$$v_{ave} = \langle v \rangle = \int_0^\infty v F(v) dv$$

$$= \int_0^\infty v \left(4\pi \left(\frac{m}{2\pi kT} \right)^{3/2} v^2 e^{-mv^2/2kT} \right) dv$$

$$= 4\pi \left(\frac{m}{2\pi kT} \right)^{3/2} \int_0^\infty v^3 e^{-mv^2/2kT} dv$$

$$= 4\pi \left(\frac{m}{2\pi kT} \right)^{3/2} \frac{1}{2} \left(\frac{2kT}{m} \right)^2$$

$$v_{ave} = \left(\frac{8kT}{\pi m} \right)^{1/2} = \left(\frac{8RT}{\pi M} \right)^{1/2} \tag{16.31}$$

A solution to the integral in Equation (16.31) is presented in the Math Supplement.

The final comparative quantity is the **root-mean-squared speed,** or v_{rms}. This quantity is equal to $[\langle v^2 \rangle]^{1/2}$, or simply the square root of the second moment of the distribution:

$$v_{rms} = \left[\langle v^2 \rangle \right]^{1/2} = \left(\frac{3kT}{m} \right)^{1/2} = \left(\frac{3RT}{M} \right)^{1/2} \tag{16.32}$$

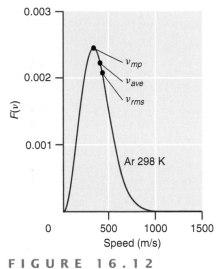

FIGURE 16.12
Comparison of v_{mp}, v_{ave}, and v_{rms} for Ar at 298 K.

Notice that Equation (16.32) is equal to the prediction of kinetic theory of Equation (16.10). The locations of v_{mp}, v_{ave}, and v_{rms} relative to the speed distribution for Ar at 298 K are presented in Figure 16.12. Comparison of Equations (16.30), (16.31), and (16.32) reveals that the only differences between the values are constants, which becomes evident when the ratios of these quantities are considered. Specifically, the ratio of $v_{rms}/v_{mp} = (3/2)^{1/2}$ and $v_{ave}/v_{mp} = (4/\pi)^{1/2}$ so that $v_{rms} > v_{ave} > v_{mp}$. Note also that all three **benchmark values** demonstrate the same dependence on T and particle mass: they increase as the square root of T and decrease as the square root of M.

EXAMPLE PROBLEM 16.3

Determine v_{mp}, v_{ave}, and v_{rms} for Ar at 298 K.

Solution
Using Equations (16.30), (16.31), and (16.32), the benchmark speed values are as follows:

$$v_{mp} = \sqrt{\frac{2RT}{M}} = \sqrt{\frac{2(8.314 \text{ J mol}^{-1} \text{ K}^{-1})(298 \text{ K})}{0.040 \text{ kg mol}^{-1}}} = 352 \text{ m s}^{-1}$$

$$v_{ave} = \sqrt{\frac{8RT}{\pi M}} = \sqrt{\frac{8(8.314 \text{ J mol}^{-1} \text{ K}^{-1})(298 \text{ K})}{\pi(0.040 \text{ kg mol}^{-1})}} = 397 \text{ m s}^{-1}$$

$$v_{rms} = \sqrt{\frac{3RT}{M}} = \sqrt{\frac{3(8.314 \text{ J mol}^{-1} \text{ K}^{-1})(298 \text{ K})}{0.040 \text{ kg mol}^{-1}}} = 432 \text{ m s}^{-1}$$

16.5 Gas Effusion

As described earlier, the experiments that verified the accuracy of the Maxwell speed distribution were performed using a gas escaping through an aperture in the wall of the oven containing the gas (see Figure 16.10). In this technique, the gas confined to the box is at some finite pressure and is separated from a vacuum by a thin wall of the oven containing the aperture. The pressure of the gas and size of the aperture is such that molecules do not undergo collisions near or when passing through the aperture. The process by which a gas passes through an opening under these conditions is called **effusion** and is employed to produce a stream or "beam" of gas particles. For example, this technique is used to create atomic or molecular beams that can collide with beams of other molecules to study chemical reaction dynamics.

The rate at which gas particles will escape through an aperture of a given area can be related to the rate at which particles strike an area on the side of the box or oven. To derive the rate of gas effusion, we proceed in a fashion analogous to that used in Section 16.1 to derive pressure. Let dN_c be the number of particles that hit the wall of the container. The collisional rate, dN_c/dt, is the number of collisions with the wall per unit time. This quantity will be proportional to the area being struck, A. In addition, the collisional rate will depend on particle velocity, with increased velocity resulting in an increased collisional rate. Finally, the collisional rate should be directly proportional to the particle density, $\tilde{N}$, defined as the number of particles per unit volume. Taking these three ideas into account, we can write

$$\frac{dN_c}{dt} = \tilde{N}A \int_0^\infty v_x f(v_x) dv_x \tag{16.33}$$

The integral in Equation (16.33) is simply the average particle velocity in the direction that will result in collision with the area of interest (taken as the positive x direction with corresponding limits of integration from zero to positive infinity). Evaluating this integral yields the following expression for the collision rate:

$$\frac{dN_c}{dt} = \tilde{N}A \int_0^\infty v_x \left(\frac{m}{2\pi kT}\right)^{1/2} e^{-mv_x^2/2kT} dv_x$$

$$= \tilde{N}A \left(\frac{m}{2\pi kT}\right)^{1/2} \int_0^\infty v_x e^{-mv_x^2/2kT} dv_x$$

$$= \tilde{N}A \left(\frac{m}{2\pi kT}\right)^{1/2} \left(\frac{kT}{m}\right)$$

$$= \tilde{N}A \left(\frac{kT}{2\pi m}\right)^{1/2}$$

$$\frac{dN_c}{dt} = \tilde{N}A \frac{1}{4} v_{ave} \tag{16.34}$$

In the final step, the definition of average speed, v_{ave}, provided in Equation (16.31) has been used. The **collisional flux**, Z_c, is defined as the number of collisions per unit time and per unit area. This quantity is equal to collisional rate divided by the area of interest, A:

$$Z_c = \frac{dN_c/dt}{A} = \frac{1}{4} \tilde{N} v_{ave} \tag{16.35}$$

It is sometimes more convenient to express the collisional flux in terms of gas pressure. This is accomplished by rewriting $\tilde{N}$ as follows:

$$\tilde{N} = \frac{N}{V} = \frac{nN_a}{V} = \frac{P}{kT} \tag{16.36}$$

With this definition of $\tilde{N}$, Z_c becomes

$$Z_c = \frac{P}{(2\pi mkT)^{1/2}} = \frac{PN_A}{(2\pi MRT)^{1/2}} \tag{16.37}$$

where m is the particle mass (in kilograms) and M is molar mass (in kg mol^{-1}). Evaluating the preceding expression requires careful attention to units, as Example Problem 16.4 illustrates.

EXAMPLE PROBLEM 16.4

How many collisions per second occur on a container wall with an area of 1 cm^2 for a collection of Ar particles at 1 atm and 298 K?

Solution
Using Equation (16.37):

$$Z_c = \frac{PN_A}{(2\pi MRT)^{1/2}} = \frac{(1.01325 \times 10^5\,\text{Pa})(6.022 \times 10^{23}\,\text{mol}^{-1})}{(2\pi(0.040\,\text{kg mol}^{-1})(8.314\,\text{J mol}^{-1}\,\text{K}^{-1})(298\,\text{K})^{1/2})}$$

$$= 2.45 \times 10^{27}\,\text{m}^{-2}\,\text{s}^{-1}$$

Notice that pressure is in units of Pa (kg m^{-1} s^{-2}) resulting in the appropriate units for Z_c. Then, multiplying the collisional flux by the area of interest yields the collisional rate:

$$\frac{dN_c}{dt} = Z_c A = (2.45 \times 10^{27}\,\text{m}^{-2}\,\text{s}^{-1})(10^{-4}\,\text{m}^2) = 2.45 \times 10^{23}\,\text{s}^{-1}$$

This quantity represents the number of collisions per second with a section of the container wall having an area of 1 cm². This is a rather large quantity, and it demonstrates the substantial number of collisions that occur in a container for a gas under standard temperature and pressure conditions.

Effusion will result in a decrease in gas pressure as a function of time. The change in pressure is related to the change in the number of particles in the container, N, as follows:

$$\frac{dP}{dt} = \frac{d}{dt}\left(\frac{NkT}{V}\right) = \frac{kT}{V}\frac{dN}{dt} \tag{16.38}$$

The quantity dN/dt can be related to the collisional rate [Equation (16.37)] by recognizing the following. First, if the space outside of the container is at a significantly lower pressure than the container and a particle escapes the container, it will not return. Second, each collision corresponds to a particle striking the aperture area so that the number of collisions with the aperture area is equal to the number of molecules lost, resulting in $N_c = N$ in Equation (16.35) and

$$\frac{dN}{dt} = -Z_c A = \frac{-PA}{(2\pi mkT)^{1/2}}$$

where the negative sign is consistent with the expectation that the number of particles in the container will decrease as effusion proceeds. Substituting the preceding result into Equation (16.38), the change in pressure as a function of time becomes

$$\frac{dP}{dt} = \frac{kT}{V}\left(\frac{-PA}{(2\pi mkT)^{1/2}}\right) \tag{16.39}$$

Integration of Equation (16.39) yields the following expression for container pressure as a function of time:

$$P = P_0 \exp\left[-\frac{At}{V}\left(\frac{kT}{2\pi m}\right)^{1/2}\right] \tag{16.40}$$

In Equation (16.40), P_0 is initial container pressure. This result demonstrates that effusion will result in an exponential decrease in container pressure as a function of time.

EXAMPLE PROBLEM 16.5

A 1-L container filled with Ar at 298 K and at an initial pressure of 1.00×10^{-2} atm is allowed to effuse through an aperture having an area of 0.01 μm^2. Will the pressure inside the container be significantly reduced after 1 hour of effusion?

Solution

In evaluating Equation (16.40), it is easiest to first determine the exponential factor and then determine the pressure:

$$\frac{At}{V}\left(\frac{kT}{2\pi m}\right)^{1/2} = \frac{10^{-14}\ m^2 (3600\ s)}{10^{-3}\ m^3}\left(\frac{1.38 \times 10^{-23}\ J\ K^{-1}(298\ K)}{2\pi\left(\dfrac{0.040\ kg\ mol^{-1}}{N_A}\right)}\right)^{1/2}$$

$$= 3.60 \times 10^{-8}\ s\ m^{-1}(99.3\ m\ s^{-1}) = 3.57 \times 10^{-6}$$

The pressure after 1 hour of effusion, therefore, is

$$P = P_0 e^{-3.57 \times 10^{-6}} = (0.01\ atm)e^{-3.57 \times 10^{-6}} \approx 0.01\ atm$$

Given the large volume of the container and the relatively small aperture through which effusion occurs, the pressure inside the container is essentially unchanged.

16.6 Molecular Collisions

Kinetic theory can also be used to determine the collisional frequency between gaseous particles. Recall that one of the primary ideas behind kinetic theory is that the distance between gaseous particles is on average much greater than actual particle volume. However, the particles translate through space, and collisions between particles will occur. How does one think about these collisions with respect to the intermolecular interactions that occur between particles? As discussed in Chapter 7, at high gas pressures, intermolecular forces are substantial and particle interactions become important. Even during collisions at low pressures, the intermolecular forces must be relevant during the collisions. Modeling collisions including the subtleties of intermolecular interactions is beyond the scope of this text. Instead, we adopt a limiting viewpoint of collisions in which we treat the particles as hard spheres. Billiard balls are an excellent example of a hard-sphere particle. Collisions occur when two billiard balls attempt to occupy the same region of space, and this is the only time the particles interact. We will see in upcoming chapters that an understanding of the frequency of molecular collisions is important in describing a variety of chemical phenomena, including the rates of chemical reactions.

How frequent are molecular collisions? To answer this question, we assume that the particle of interest is moving and that all other molecules are stationary (we will relax this assumption shortly). In this picture, the particle of interest sweeps out a collisional cylinder, which determines the number of collisions the particle undergoes per unit time. A depiction of this cylinder is provided in Figure 16.13. Collisions occur between the particle of interest and other particles that are positioned within the cylinder. The area of the collisional cylinder base or **collision diameter,** σ, is dependent on the radii of the gas particles, r, as follows:

$$\sigma = \pi(r_1 + r_2)^2 \tag{16.41}$$

The subscripts 1 and 2 in Equation (16.41) denote the two collisional partners, which may or may not be the same species. The length of the cylinder is given by the product of a given time interval (dt) and v_{ave}, the average speed of the molecule. Therefore, the total cylinder volume is equal to $\sigma v_{ave}(dt)$ as depicted in Figure 16.13.

This derivation of the cylinder volume is not exactly correct because the other molecules are not stationary. To incorporate the motion of other molecules in the derivation, we now introduce the concept of **effective speed.** To illustrate this concept, imagine two particles moving with some average speed as depicted in Figure 16.14. The effective speed at which two particles approach each other is dependent on the relative direction of particle motion. The first case depicted in the figure is of two particles traveling in the same direction where $\langle v_{12} \rangle = \langle v_1 \rangle - \langle v_2 \rangle$. If the particles were identical, the effective speed would be zero since the molecules would travel with

FIGURE 16.13
Schematic of the hard-sphere collisional process.

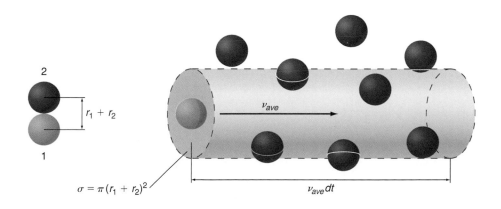

2

$r_1 + r_2$

1

$\sigma = \pi(r_1 + r_2)^2$

v_{ave}

$v_{ave}\,dt$

FIGURE 16.14
Depiction of effective speed in a two-particle collision.

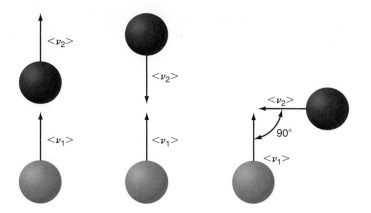

constant separation and never collide. The opposite case is depicted in the middle of the figure, where the effective speed is $\langle v_1 \rangle + \langle v_2 \rangle$ when the particles are traveling directly at each other. Again, if the particles were identical then $\langle v_{12} \rangle = 2\langle v_1 \rangle$.

A full derivation of the effective speed is quite involved, and results in the intuitive result depicted as the third case in Figure 16.14. In this third case, the average approach angle is 90° and, using the Pythagorean theorem, the effective speed is equal to

$$\langle v_{12} \rangle = (\langle v_1 \rangle^2 + \langle v_2 \rangle^2)^{1/2} = \left[\left(\frac{8kT}{\pi m_1} \right) + \left(\frac{8kT}{\pi m_2} \right) \right]^{1/2} \tag{16.42}$$

$$= \left[\frac{8kT}{\pi} \left(\frac{1}{m_1} + \frac{1}{m_2} \right) \right]^{1/2}$$

$$= \left(\frac{8kT}{\pi \mu} \right)^{1/2}$$

where

$$\mu = \frac{m_1 m_2}{m_1 + m_2}$$

With the effective speed as defined by Equation (16.42), the collisional-cylinder volume is now defined. The number of collisional partners in the collisional cylinder is equal to the product of the collisional-partner number density (N_2/V) and the volume of the cylinder, $V_{cyl} = \sigma v_{ave} dt$ (Figure 16.13). We define the individual **particle collisional frequency**, z_{12}, as the number of collisions an individual molecule (denoted by the subscript 1) undergoes with other collisional partners (denoted by the subscript 2) per unit time (dt). This quantity is equal to the number of collisional partners divided by dt:

$$z_{12} = \frac{N_2}{V} \left(\frac{V_{cyl}}{dt} \right) = \frac{N_2}{V} \left(\frac{\sigma v_{ave} dt}{dt} \right) = \frac{N_2}{V} \sigma \left(\frac{8kT}{\pi \mu} \right)^{1/2} \tag{16.43}$$

If the gas is comprised of one type of particle, $\mu = m_1/2$ and Equation (16.43) becomes

$$z_{11} = \frac{N_1}{V} \sigma \sqrt{2} \left(\frac{8kT}{\pi m_1} \right)^{1/2} = \frac{P_1 N_A}{RT} \sigma \sqrt{2} \left(\frac{8RT}{\pi M_1} \right)^{1/2} \tag{16.44}$$

The **total collisional frequency** is defined as the total number of collisions that occurs for all gas particles. The total collisional frequency for a collection of two types of gas molecules, Z_{12}, is given by z_{12} times the number density of species 1:

$$Z_{12} = \frac{N_1}{V} z_{12} = \frac{N_1}{V} \frac{N_2}{V} \sigma \left(\frac{8kT}{\pi\mu} \right)^{1/2} = \left(\frac{P_1 N_A}{RT} \right)\left(\frac{P_2 N_A}{RT} \right) \sigma \left(\frac{8kT}{\pi\mu} \right)^{1/2} \quad (16.45)$$

The units of Z_{12} are collisions per cubic meter, or the total number of collisions per unit volume. The corresponding collisional frequency for a gas consisting of only one type of particle is

$$Z_{11} = \frac{1}{2}\frac{N_1}{V} z_{11} = \frac{1}{\sqrt{2}}\left(\frac{N_1}{V} \right)^2 \sigma \left(\frac{8kT}{\pi m_1} \right)^{1/2} = \frac{1}{\sqrt{2}}\left(\frac{P_1 N_A}{RT} \right)^2 \sigma \left(\frac{8RT}{\pi M_1} \right)^{1/2} \quad (16.46)$$

The factor of 1/2 appears in Equation (16.46) to ensure that each collision is only counted once. Evaluation of Equations (16.43) through (16.46) requires knowledge of the collisional cross sections that are in turn dependent on the effective hard-sphere radii. As we will see in the next chapter, these values can be determined by the measure of various gas properties. Table 16.1 provides the hard-sphere radii for a variety of common gases determined from such measurements. Generally, for monatomic gases and small molecules the radius is on the order of 0.2 nm.

TABLE 16.1

Collisional Parameters for Various Gases

Species	r (nm)	σ (nm²)
He	0.13	0.21
Ne	0.14	0.24
Ar	0.17	0.36
Kr	0.20	0.52
N_2	0.19	0.43
O_2	0.18	0.40
CO_2	0.20	0.52

EXAMPLE PROBLEM 16.6

What is z_{11} for CO_2 at 298 K and 1 atm?

Solution

The question asks for the single-particle collisional frequency of CO_2. Using Equation (16.44) and the collisional cross section provided in Table 16.1, we obtain

$$z_{CO_2} = \frac{P_{CO_2} N_A}{RT} \sigma\sqrt{2}\left(\frac{8RT}{\pi M_{CO_2}} \right)^{1/2}$$

$$= \frac{101,325\ \text{Pa}(6.022\times10^{23}\ \text{mol}^{-1})}{8.314\ \text{J mol}^{-1}\ \text{K}^{-1}(298\ \text{K})}(5.2\times10^{-19}\ \text{m}^2)\sqrt{2}$$

$$\times \left(\frac{8(8.314\ \text{J mol}^{-1}\ \text{K}^{-1})(298\ \text{K})}{\pi(0.044\ \text{kg mol}^{-1})} \right)^{1/2}$$

$$= 6.86\times10^9\ \text{s}^{-1}$$

This calculation demonstrates that a single CO_2 molecule undergoes roughly 7 billion collisions per second under standard temperature and pressure conditions! The inverse of the collisional frequency corresponds to the time between molecular collision or roughly 150 picoseconds (1 ps = 10^{-12} s) between collisions.

EXAMPLE PROBLEM 16.7

What is the total collisional frequency (Z_{ArKr}) at 298 K for a collection of Ar and Kr confined to a 1-cm³ container with partial pressures of 360 Torr for Ar and 400 Torr for Kr?

Solution

Evaluation of Equation (16.45) is best performed by evaluating each factor in the equation separately, then combining factors to calculate the total collisional frequency as follows:

$$\left(\frac{P_{Ar}N_A}{RT}\right) = \left(\frac{47{,}996\ \text{Pa}(6.022\times10^{23}\ \text{mol}^{-1})}{8.314\ \text{J mol}^{-1}\ \text{K}^{-1}(298\ \text{K})}\right) = 1.17\times10^{25}\ \text{m}^{-3}$$

$$\left(\frac{P_{Kr}N_A}{RT}\right) = \left(\frac{53{,}328\ \text{Pa}(6.022\times10^{23}\ \text{mol}^{-1})}{8.314\ \text{J mol}^{-1}\ \text{K}^{-1}(298\ \text{K})}\right) = 1.29\times10^{25}\ \text{m}^{-3}$$

$$\sigma = \pi(r_{Ar}+r_{Kr})^2 = \pi(0.17\ \text{nm}+0.20\ \text{nm})^2 = 4.30\ \text{nm}^2 = 4.30\times10^{-18}\ \text{m}^2$$

$$\mu = \frac{m_{Ar}m_{Kr}}{m_{Ar}+m_{Kr}} = \frac{(0.040\ \text{kg mol}^{-1})(0.084\ \text{kg mol}^{-1})}{(0.040\ \text{kg mol}^{-1})+(0.084\ \text{kg mol}^{-1})}\times\frac{1}{N_A} = 4.48\times10^{-26}\ \text{kg}$$

$$\left(\frac{8kT}{\pi\mu}\right)^{1/2} = \left(\frac{8(1.38\times10^{-23}\ \text{J K}^{-1})(298\ \text{K})}{\pi(4.48\times10^{-26}\ \text{kg})}\right)^{1/2} = 484\ \text{m s}^{-1}$$

$$Z_{ArKr} = \left(\frac{P_{Ar}N_A}{RT}\right)\left(\frac{P_{Kr}N_A}{RT}\right)\sigma\left(\frac{8kT}{\pi\mu}\right)^{1/2}$$

$$= (1.17\times10^{25}\ \text{m}^{-3})(1.29\times10^{25}\ \text{m}^{-3})(4.30\times10^{-18}\ \text{m}^2)(484\ \text{m s}^{-1})$$

$$= 3.14\times10^{35}\ \text{m}^{-3}\ \text{s}^{-1} = 3.14\times10^{32}\ \text{L}^{-1}\ \text{s}^{-1}$$

Comparison of the last two example problems reveals that the total collisional frequency (Z_{12}) is generally much larger than the collisional frequency for an individual molecule (z_{12}). The total collisional frequency incorporates all collisions that occur for a collection of gas particles (consistent with the units of inverse volume); therefore, magnitude of this value relative to the collisional frequency for a single particle is not unexpected.

16.7 The Mean Free Path

The **mean free path** is defined as the average distance a gas particle travels between successive collisions. In a given time interval, dt, the distance a particle will travel is equal to $v_{ave}dt$ where v_{ave} is the average speed of the particle. In addition, the number of collisions the particle undergoes is given by $(z_{11} + z_{12})dt$, where the frequency of collisions with either type of collisional partner in the binary mixture is included. Given these quantities, the mean free path, λ, is given by the average distance traveled divided by the number of collisions:

$$\lambda = \frac{v_{ave}dt}{(z_{11}+z_{12})dt} = \frac{v_{ave}}{(z_{11}+z_{12})} \tag{16.47}$$

If our discussion is limited to a gas with one type of particle, $N_2 = 0$ resulting in $z_{12} = 0$ and the mean free path becomes

$$\lambda = \frac{v_{ave}}{z_{11}} = \frac{v_{ave}}{\left(\dfrac{N_1}{V}\right)\sqrt{2}\sigma v_{ave}} = \left(\frac{RT}{P_1 N_A}\right)\frac{1}{\sqrt{2}\sigma} \tag{16.48}$$

Equation (16.41) demonstrates that the mean free path decreases if the pressure increases or if the collisional cross section of the particle increases. This behavior makes intuitive sense. As particle density increases (i.e., as pressure increases), we would expect the particle to travel a shorter distance between collisions. Also, as the particle size increases, we would expect the probability of collision to also increase, thereby reducing the mean free path.

What does the mean free path tell us about the length scale of collisional events relative to molecular size? Recall that one of the assumptions of kinetic theory is that the distance between particles is large compared to their size. Is the mean free path consistent with this assumption? To answer this question, we return to the first example provided in this chapter, Ar at a pressure of 1 atm and temperature of 298 K, for which the mean free path is

$$\lambda_{Ar} = \left(\frac{RT}{P_{Ar}N_A} \right) \frac{1}{\sqrt{2}\sigma}$$

$$= \left(\frac{(8.314 \text{ J mol}^{-1} \text{ K}^{-1})(298 \text{ K})}{(101,325 \text{ Pa})(6.022 \times 10^{23} \text{ mol}^{-1})} \right) \frac{1}{\sqrt{2}(3.6 \times 10^{-19} \text{ m}^2)}$$

$$= 7.98 \times 10^{-8} \text{ m} \approx 80 \text{ nm}$$

Compared to the 0.29-nm diameter of Ar, the mean free path demonstrates that an Ar atom travels an average distance equal to ~275 times its diameter between collisions. This difference in length scales is consistent with the assumptions of kinetic theory. To provide further insight into the behavior of a collection of gaseous particles, as indicated in Example Problem 16.6, the collisional frequency can be used to determine the timescale between collisions as follows:

$$\frac{1}{z_{11}} = \frac{\lambda}{v_{ave}} = \frac{7.98 \times 10^{-8} \text{ m}}{397 \text{ m s}^{-1}} = 2.01 \times 10^{-10} \text{ s}$$

A picosecond is equal to 10^{-12} s; therefore, an individual Ar atom undergoes on average a collision every 200 ps. As will be discussed in the following chapters, properties such as collisional frequency and mean free path are important in describing transport properties of gases and chemical reaction dynamics involving collisional processes. The physical picture of gas particle motion outlined here will prove critical in understanding these important aspects of physical chemistry.

For Further Reading

Castellan, G. W., *Physical Chemistry,* 3rd ed. Addison Wesley, Reading, MA, 1983.

Hirschfelder, J. O., C. F. Curtiss, and R. B. Bird, *The Molecular Theory of Gases and Liquids.* Wiley, New York, 1954.

Liboff, R. L., *Kinetic Theory: Classical, Quantum, and Relativistic Descriptions.* Springer, New York, 2003.

McQuarrie, D., *Statistical Mechanics.* Harper & Row, New York, 1973.

Vocabulary

average speed

benchmark value

collision diameter

collisional flux

effective speed

effusion

elastic collision

gas kinetic theory

ideal gas law

Maxwell speed distribution

Maxwell–Boltzmann velocity distribution

mean free path

most probable speed

particle collisional frequency

particle speed

root-mean-squared speed

root-mean-squared velocity

total collisional frequency

velocity distribution function

velocity space

Questions on Concepts

Q16.1 Why is probability used to describe the velocity and speed of gas molecules?

Q16.2 Describe pressure using gas kinetic theory. Why would one expect pressure to depend on the inverse of volume in this theory?

Q16.3 What are the inherent assumptions about gas particle interactions in gas kinetic theory?

Q16.4 Provide a physical explanation as to why the Maxwell speed distribution approaches zero at high speeds. Why is $f(v) = 0$ at $v = 0$?

Q16.5 How would the Maxwell speed distributions for He versus Kr compare if the gases were at the same temperature?

Q16.6 How does the average speed of a collection of gas particles vary with particle mass and temperature?

Q16.7 Does the average kinetic energy of a particle depend on particle mass?

Q16.8 Why does the mean free path depend on σ^2? Would an increase in $\tilde{N}$ increase or decrease the mean free path?

Q16.9 What is the typical length scale for a molecular diameter?

Q16.10 What is the difference between z_{11} and z_{12}?

Q16.11 Define the mean free path. How does this quantity vary with number density, particle diameter, and average particle speed?

Problems

Problem numbers in **red** indicate that the solution to the problem is given in the *Student's Solutions Manual*.

P16.1 Consider a collection of gas particles confined to translate in two dimensions (for example, a gas molecule on a surface). Derive the Maxwell speed distribution for such a gas.

P16.2 Determine v_{mp}, v_{ave}, and v_{rms} for the following species at 298 K:

a. Ne

b. Kr

c. CH_4

d. C_6H_6

e. C_{60}

P16.3 Compute v_{mp}, v_{ave}, and v_{rms} for O_2 at 300 and 500 K. How would your answers change for H_2?

P16.4 Compare the average speed and average kinetic energy of O_2 with that of CCl_4 at 298 K.

P16.5

a. What is the average time required for H_2 to travel 1 m at 298 K and 1 atm?

b. How much longer does it take N_2 to travel 1 m on average relative to H_2 under these same conditions?

c. (Challenging) What fraction of N_2 particles will require more than this average time to travel 1 m? Answering this question will require evaluating a definite integral of the speed distribution, which requires using numerical methods such as Simpson's rule.

P16.6 As mentioned in Section 16.3, the only differences between the quantities v_{mp}, v_{ave}, and v_{rms} involve constants.

a. Derive the expressions for v_{ave} and v_{rms} relative to v_{mp} provided in the text.

b. Your result from part (a) will involve quantities that are independent of gas-specific quantities such as mass or temperature. Given this, it is possible to construct a "generic" speed distribution curve for speed in reduced units

of v/v_{mp}. Transform the Maxwell distribution into a corresponding expression involving reduced speed.

P16.7 At what temperature is the v_{rms} of Ar equal to that of SF_6 at 298 K? Perform the same calculation for v_{mp}.

P16.8 The probability that a particle will have a velocity in the x direction in the range of $-v_{x_0}$ and v_{x_0} is given by

$$f(-v_{x_0} \leq v_x \leq v_{x_0}) = \left(\frac{m}{2\pi kT}\right)^{1/2} \int_{-v_{x_0}}^{v_{x_0}} e^{-mv_x^2/2kT} dv_x$$

$$= \left(\frac{2m}{\pi kT}\right)^{1/2} \int_{0}^{v_{x_0}} e^{-mv_x^2/2kT} dv_x$$

The preceding integral can be rewritten using the following substitution: $\xi^2 = mv_x^2/2kT$, resulting in $f(-v_{x_0} \leq v_x \leq v_{x_0}) = 2/\sqrt{\pi} \left(\int_{0}^{\xi_0} e^{-\xi^2} d\xi\right)$, which can be evaluated using the error function defined as $\text{erf}(z) = 2/\sqrt{\pi} \left(\int_{0}^{z} e^{-x^2} dx\right)$. The complementary error function is defined as $\text{erfc}(z) = 1 - \text{erf}(z)$. Finally, a plot of both $\text{erf}(z)$ and $\text{erfc}(z)$ as a function of z is shown here (tabulated values are available in the Math Supplement, Appendix B):

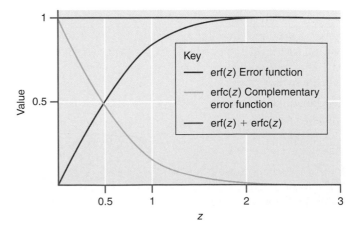

Using this graph of erf(z), determine the probability that $|v_x| \le$ $(2kT/m)^{1/2}$. What is the probability that $|v_x| > (2kT/m)^{1/2}$?

P16.9 The speed of sound is given by $v_{sound} = \sqrt{\gamma kT/m}$ $= \sqrt{\gamma RT/M}$, where $\gamma = C_P/C_v$.

a. What is the speed of sound in Ne, Kr, and Ar at 1000 K?

b. At what temperature will the speed of sound in Kr equal the speed of sound in Ar at 1000 K?

P16.10 For O_2 at 1 atm and 298 K, what fraction of molecules has a speed that is greater than v_{rms}?

P16.11 The escape velocity from the Earth's surface is given by $v_E = (2gR)^{1/2}$ where g is gravitational acceleration (9.80 m s^{-2}) and R is the radius of the Earth (6.37×10^6 m).

a. At what temperature will v_{mp} for N_2 be equal to the escape velocity?

b. How does the answer for part (a) change if the gas of interest is He?

P16.12 For N_2 at 298 K, what fraction of molecules has a speed between 200 and 300 m/s? What is this fraction if the gas temperature is 500 K?

P16.13 Demonstrate that the Maxwell–Boltzmann speed distribution is normalized.

P16.14 (Challenging) Derive the Maxwell–Boltzmann distribution using the Boltzmann distribution introduced in statistical mechanics. Begin by developing the expression for the distribution in translational energy in one dimension and then extend it to three dimensions.

P16.15 Starting with the Maxwell speed distribution, demonstrate that the probability distribution for translational energy for $\varepsilon_{Tr} \gg kT$ is given by

$$f(\varepsilon_{Tr})d\varepsilon_{Tr} = 2\pi \left(\frac{1}{\pi kT} \right)^{3/2} e^{-\varepsilon_{Tr}/kT} \varepsilon_{Tr}^{1/2} d\varepsilon_{Tr}$$

P16.16 Using the distribution of particle translational energy provided in Problem P16.15, derive expressions for the average and most probable translational energies for a collection of gaseous particles.

P16.17 (Challenging) Using the distribution of particle translational energy provided in Problem P16.15, derive an expression for the fraction of molecules that have energy greater than some energy ε^*. The rate of many chemical reactions is dependent on the thermal energy available, kT, versus some threshold energy. Your answer to this question will provide insight into why one might expect the rate of such chemical reactions to vary with temperature.

P16.18 As discussed in Chapter 12, the nth moment of a distribution can be determined as follows: $\langle x^n \rangle = \int x^n f(x)dx$, where integration is over the entire domain of the distribution. Derive expressions for the nth moment of the gas speed distribution.

P16.19 Imagine a cubic container with sides 1 cm in length that contains 1 atm of Ar at 298 K. How many gas–wall collisions are there per second?

P16.20 The vapor pressure of various substances can be determined using effusion. In this process, the material of interest is placed in an oven (referred to as a Knudsen cell) and the mass of material lost through effusion is determined. The mass loss (Δm) is given by $\Delta m = Z_c Am\Delta t$, where Z_c is the collisional flux, A is the area of the aperture through which effusion occurs, m is the mass of one atom, and Δt is the time interval over which the mass loss occurs. This technique is quite useful for determining the vapor pressure of nonvolatile materials. A 1.00-g sample of UF_6 is placed in a Knudsen cell equipped with a 100-μm-radius hole and heated to 18.2°C where the vapor pressure is 100 Torr.

a. The best scale in your lab has an accuracy of ± 0.01g. What is the minimum amount of time you must wait until the mass change of the cell can be determined by your balance?

b. How much UF_6 will remain in the Knudsen cell after 5 min of effusion?

P16.21

a. How many molecules strike a 1-cm^2 surface during 1 min if the surface is exposed to O_2 at 1 atm and 298 K?

b. Ultrahigh vacuum studies typically employ pressures on the order of 10^{-10} Torr. How many collisions will occur at this pressure at 298 K?

P16.22 You are a NASA engineer faced with the task of ensuring that the material on the hull of a spacecraft can withstand puncturing by space debris. The initial cabin air pressure in the craft of 1 atm can drop to 0.7 atm before the safety of the crew is jeopardized. The volume of the cabin is 100 m^3, and the temperature in the cabin is 285 K. Assuming it takes the space shuttle about 8 hours from entry into orbit until landing, what is the largest circular aperture created by a hull puncture that can be safely tolerated assuming that the flow of gas out of the spaceship is effusive? Can the escaping gas from the spaceship be considered as an effusive process? (You can assume that the air is adequately represented by N_2.)

P16.23 Many of the concepts developed in this chapter can be applied to understanding the atmosphere. Because atmospheric air is comprised primarily of N_2 (roughly 78% by volume), approximate the atmosphere as consisting only of N_2 in answering the following questions:

a. What is the single-particle collisional frequency at sea level, with $T = 298$ K and $P = 1$ atm? The corresponding single-particle collisional frequency is reported as 10^{10} s^{-1} in the *CRC Handbook of Chemistry and Physics* (62nd ed., p. F-171).

b. At the tropopause (11 km in altitude), the collisional frequency decreases to 3.16×10^9 s^{-1}, primarily due to a reduction in temperature and barometric pressure (i.e., fewer particles). The temperature at the tropopause is ~220 K. What is the pressure of N_2 at this altitude?

c. At the tropopause, what is the mean free path for N_2?

P16.24

a. Determine the total collisional frequency for CO_2 at 1 atm and 298 K.

b. At what temperature would the collisional frequency be 10% of the value determined in part (a)?

P16.25

a. A standard rotary pump is capable of producing a vacuum on the order of 10^{-3} Torr. What is the single-particle collisional frequency and mean free path for N_2 at this pressure and 298 K?

b. A cryogenic pump can produce a vacuum on the order of 10^{-10} Torr. What is the collisional frequency and mean free path for N_2 at this pressure and 298 K?

P16.26 Determine the mean free path for Ar at 298 K at the following pressures:

a. 0.5 atm   b. 0.005 atm   c. 5×10^{-6} atm

P16.27 Determine the mean free path at 500 K and 1 atm for the following:

a. Ne   b. Kr   c. CH_4

Rather than simply calculating the mean free path for each species separately, instead develop an expression for the ratio of mean free paths for two species and use the calculated value for one species to determine the other two.

P16.28 Consider the following diagram of a molecular beam apparatus:

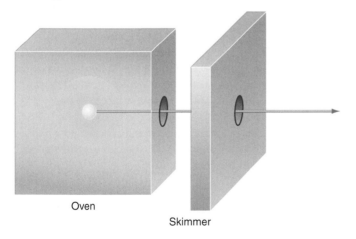

Oven

Skimmer

In the design of the apparatus, it is important to ensure that the molecular beam effusing from the oven does not collide with other particles until the beam is well past the skimmer, a device that selects molecules that are traveling in the appropriate direction, resulting in the creation of a molecular beam. The skimmer is located 10 cm in front of the oven so that a mean free path of 20 cm will ensure that the molecules are well past the skimmer before a collision can occur. If the molecular beam consists of O_2 at a temperature of 500 K, what must the pressure outside the oven be to ensure this mean free path?

P16.29 A comparison of v_{ave}, v_{mp}, and v_{rms} for the Maxwell speed distribution reveals that these three quantities are not equal. Is the same true for the one-dimensional velocity distributions?

P16.30 At 30 km above the Earth's surface (roughly in the middle of the stratosphere), the pressure is roughly 0.013 atm and the gas density is 3.74×10^{23} molecules/m^3. Assuming N_2 is representative of the stratosphere, using the collisional diameter information provided in Table 16.1 determine:

a. The number of collisions a single gas particle undergoes in this region of the stratosphere in 1 s

b. The total number of particles collisions that occur in 1 s

c. The mean free path of a gas particle in this region of the stratosphere.

Web-Based Simulations, Animations, and Problems

W16.1 In this simulation, the variations of gas particle velocity and speed distributions with particle mass and temperature are explored. Specifically, the variations in these distributions with mass are studied and compared to calculations performed by the student. Similar calculations are performed with respect to the distribution variation with temperature.

CHAPTER 17

Transport Phenomena

How will a system respond when it is not at equilibrium? The first steps toward answering this question are provided in this chapter. The study of system relaxation toward equilibrium is known as dynamics. In this chapter, transport phenomena involving the evolution of a system's physical properties such as mass or energy are described. All transport phenomena are connected by one central idea: the rate of change for a system physical property is dependent on the spatial gradient of the property. In this chapter, this underlying idea is first described as a general concept, then applied to mass (diffusion), energy (thermal conduction), linear momentum (viscosity), and charge (ionic conductivity) transport. The timescale for mass transport is discussed and approached from both the macroscopic and microscopic perspective. It is important to note that although the various transport phenomena outlined here look different, the underlying concepts describing these phenomena have a common origin. ∎

17.1 What Is Transport?

To this point we have been concerned with describing system properties at equilibrium. However, consider the application of an external perturbation to a system such that a property of the system is shifted away from equilibrium. Examples of such system properties are mass and energy. Once the external perturbation is removed, the system will evolve to reestablish the equilibrium distribution of the property. **Transport phenomena** involve the evolution a system property in response to a non-equilibrium distribution of the property. The system properties of interest in this chapter are given in Table 17.1, and each property is listed with the corresponding transport process.

In order for a system property to be transported, a spatial distribution of the property must exist that is different from that at equilibrium. For example, consider a collection of gas particles in which the equilibrium spatial distribution of the particles corresponds to a homogeneous particle number density throughout the container. What would happen if the particle number density were greater on one side of the container than the other? The expectation is that gas particles will translate to reestablish a homogeneous number density throughout the container. That is, the system evolves to reestablish a distribution of the system property that is consistent with equilibrium.

TABLE 17.1

Transported Properties and the Corresponding Transport Process

Property Transported	Transport Process
Matter	Diffusion
Energy	Thermal conductivity
Linear momentum	Viscosity
Charge	Ionic conductivity

A central concept in transport phenomena is **flux,** which is defined as a quantity transferred through a given area in a given amount of time. Flux will occur when a spatial imbalance or gradient exists for a system property, and the flux will act in opposition to this gradient. In the example just discussed, imagine dividing the container in two parts with a partition and counting the number of particles that move from one side of the container to the other side, as illustrated in Figure 17.1. The flux in this case is equal to the number of particles that move through the partition per unit time. The theoretical underpinning of *all* of the transport processes listed in Table 17.1 involves flux and the fact that a spatial gradient in a system property will give rise to a corresponding flux. The most basic relationship between flux and the spatial gradient in transported property is as follows:

$$J_x = -\alpha \frac{d(\text{property})}{dx} \tag{17.1}$$

In Equation (17.1), J_x is the flux expressed in units of property area^{-1} time^{-1}. The derivative in Equation (17.1) represents the spatial gradient of the quantity of interest (mass, energy, etc.). The linear relationship between flux and the property spatial gradient is reasonable when the displacement away from equilibrium is modest. The limit of modest displacement of the system property away from equilibrium is assumed in the remainder of this chapter.

The negative sign in Equation (17.1) indicates that the flux occurs in the opposite direction of the gradient; therefore, flux will result in a reduction of the gradient if external action is not taken to maintain the gradient. If the gradient is externally maintained at a constant value, the flux will also remain constant. Again, consider Figure 17.1, which presents a graphical example of the relationship between gradient and flux. The gas density is greatest on the left-hand side of the container such that the particle density increases as one goes from the right side of the container to the left. According to Equation (17.1), particle flux occurs in opposition to the number density gradient in an attempt to make the particle density spatially homogeneous. The final quantity of interest in Equation (17.1) is the factor α. Mathematically, this quantity serves as the proportionality constant between the gradient and flux and is referred to as the **transport coefficient.** In the following sections, we will determine the transport coefficients for the processes listed in Table 17.1 and derive the expressions for flux involving these various transport phenomena. Although the derivations for each transport property will look different, it is important to note that all originate from Equation (17.1). That is, the underlying principle behind all transport phenomena is the relationship between flux and gradient.

FIGURE 17.1
Illustration of flux. The flux J_x of gas particles is in opposition to the gradient in particle number density $\tilde{N}$.

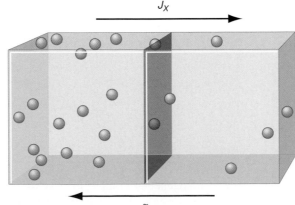

17.2 Mass Transport: Diffusion

Diffusion is the process by which particle density evolves in response to a spatial gradient in concentration. With respect to thermodynamics, this spatial gradient represents a gradient in chemical potential, and the system will relax toward equilibrium by eliminating this gradient. The first case to be considered is diffusion in an ideal gas. Diffusion in liquids will be treated later in this chapter.

Consider a gradient in gas particle number density, $\tilde{N}$, as depicted in Figure 17.2. According to Equation (17.1), there will be a flux of gas particles in opposition to the gradient. The flux is determined by quantifying the flow of particles per unit time through an imaginary plane located at $x = 0$ with area A. We will refer to this plane as the *flux plane*. Two other planes are located one mean free path $\pm\lambda$ away on either side of the flux plane, and the net flux arises from particles traveling from either of these planes to the flux plane. The mean free path is the distance a particle travels on average between collisions as defined in Equation (16.48).

Figure 17.2 demonstrates that a gradient in particle number density exists in the x direction such that

$$\frac{d\tilde{N}}{dx} \neq 0$$

If the gradient in $\tilde{N}$ were equal to zero, flux J_x would also equal zero by Equation (17.1). However, this does not mean that particles are now stationary. Instead, $J_x = 0$ indicates that the flow of particles through the flux plane from left to right is exactly balanced by the flow of particles from right to left. Therefore, the flux expressed in Equation (17.1) represents the net flux, or sum of flux in each direction through the flux plane.

Equation (17.1) provides the relationship between the flux and the spatial gradient in $\tilde{N}$. Solution of this **mass transport** problem involves determining the proportionality constant α. This quantity is referred to as the diffusion coefficient for mass transport. To determine this constant, consider the particle number density at $\pm\lambda$:

$$\tilde{N}(-\lambda) = \tilde{N}(0) - \lambda\left(\frac{d\tilde{N}}{dx}\right)_{x=0} \tag{17.2}$$

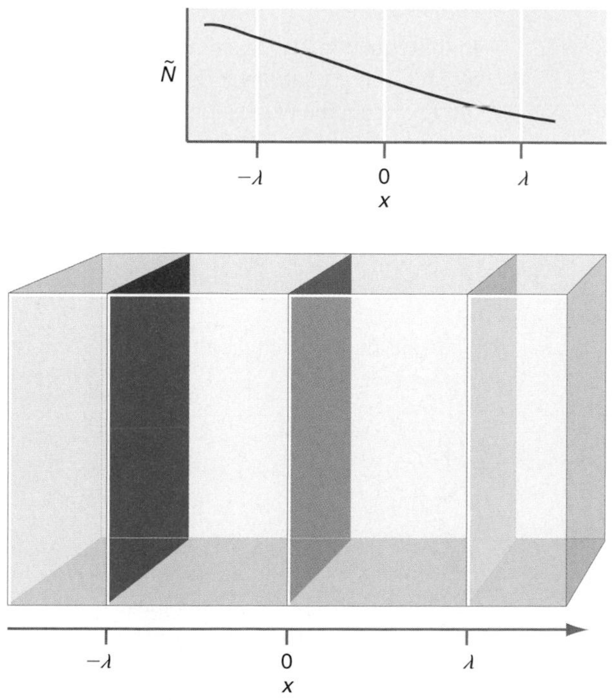

FIGURE 17.2
Model used to describe gas diffusion. The gradient in number density $\tilde{N}$ results in particles diffusing from $-x$ to $+x$. The plane located at $x = 0$ is where the flux of particles in response to the gradient is calculated (the *flux plane*). Two planes, located one mean free path distance away ($\pm\lambda$) are considered with particles traveling from either of these planes to the flux plane. The total flux through the flux plane is equal to the difference in flux from the planes located at $\pm\lambda$.

$$\tilde{N}(\lambda) = \tilde{N}(0) + \lambda \left(\frac{d\tilde{N}}{dx} \right)_{x=0} \qquad (17.3)$$

These expressions state that the value of $\tilde{N}$ away from $x = 0$ is equal to the value of $\tilde{N}$ at $x = 0$ plus a second term representing the change in concentration as one moves toward the planes at $\pm\lambda$. Formally, Equations (17.2) and (17.3) are derived from a Taylor series expansion of the number density with respect to distance, and only the first two terms in the expansion are kept consistent with λ being sufficiently small such that higher order terms of the expansion can be neglected. The diffusion process can be viewed as an effusion of particles through a plane or aperture of area A. Effusion was described in detail in Section 16.5. Proceeding in a similar fashion to the derivation of effusion presented earlier, the number of particles, N, striking a given area per unit time is equal to

$$\frac{dN}{dt} = J_x \times A \qquad (17.4)$$

Consider the flux of particles traveling from the plane at $-\lambda$ to the flux plane (Figure 17.2). We are interested in the number of particles striking the flux plane per unit time; therefore, we need to count only those particles traveling toward the flux plane. Again, this same problem was encountered in the section on gas effusion where the number of particles traveling toward the wall was equal to the product of number density and the average velocity in the $+x$ direction. Taking the identical approach, the flux in the $+x$ direction is given by

$$J_x = \tilde{N} \int_0^\infty v_x f(v_x) dv_x \qquad (17.5)$$

$$= \tilde{N} \int_0^\infty v_x \left(\frac{m}{2\pi kT} \right)^{1/2} e^{-mv_x^2/2kT} dv_x$$

$$= \tilde{N} \left(\frac{kT}{2\pi m} \right)^{1/2}$$

$$= \frac{\tilde{N}}{4} v_{ave}$$

In this equation, we have employed the definition of v_{ave} as defined in Equation (16.31). Substituting Equations (17.2) and (17.3) into the expression of Equation (17.5) for J_x, the flux from the planes located at $-\lambda$ and λ is given by

$$J_{-\lambda,0} = \frac{1}{4} v_{ave} \tilde{N}(-\lambda) = \frac{1}{4} v_{ave} \left[\tilde{N}(0) - \lambda \left(\frac{d\tilde{N}}{dx} \right)_{x=0} \right] \qquad (17.6)$$

$$J_{\lambda,0} = \frac{1}{4} v_{ave} \tilde{N}(\lambda) = \frac{1}{4} v_{ave} \left[\tilde{N}(0) + \lambda \left(\frac{d\tilde{N}}{dx} \right)_{x=0} \right] \qquad (17.7)$$

The total flux through the flux plane is simply the difference in flux from the planes at $\pm\lambda$:

$$J_{Total} = J_{-\lambda,0} - J_{\lambda,0} = \frac{1}{4} v_{ave} \left(-2\lambda \left(\frac{d\tilde{N}}{dx} \right)_{x=0} \right) \qquad (17.8)$$

$$= -\frac{1}{2} v_{ave} \lambda \left(\frac{d\tilde{N}}{dx} \right)_{x=0}$$

One correction remains before the derivation is complete. We have assumed that the particles move from the planes located at $\pm\lambda$ to the flux plane directly along the x axis. However, Figure 17.3 illustrates that if the particle trajectory is not aligned with the x axis, the particle will not reach the flux plane after traveling one mean free path. At this point, collisions with other particles can occur, resulting in postcollision particle trajectories away from the flux plane; therefore, these particles will not contribute to the

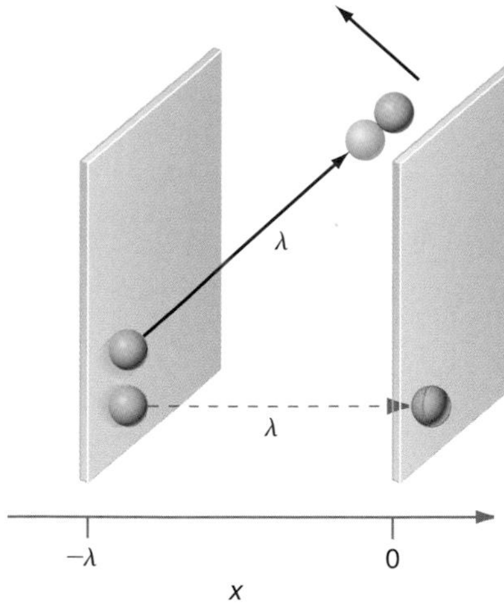

FIGURE 17.3
Particle trajectories aligned with the x axis (dashed line) result in the particle traveling between planes without collision. However, trajectories not aligned with the x axis (solid line) result in the particle not reaching the flux plane before a collision with another particle occurs. This collision may result in the particle being directed away from the flux plane.

flux. Inclusion of these trajectories requires one to take the orientational average of the mean free path, and this averaging results in a reduction in total flux as expressed by Equation (17.8) by a factor of 2/3. With this averaging, the total flux becomes

$$J_{Total} = -\frac{1}{3} v_{ave} \lambda \left(\frac{d\tilde{N}}{dx} \right)_{x=0} \tag{17.9}$$

Equation (17.9) is identical to Equation (17.1), with the diffusion proportionality constant, or simply the **diffusion coefficient,** defined as follows:

$$D = \frac{1}{3} v_{ave} \lambda \tag{17.10}$$

The diffusion coefficient has units of $m^2 \ s^{-1}$ in SI units. With this definition of the diffusion coefficient, Equation (17.9) becomes:

$$J_{Total} = -D \left(\frac{d\tilde{N}}{dx} \right)_{x=0} \tag{17.11}$$

Equation (17.11) is referred to as **Fick's first law** of diffusion. It is important to note that the diffusion coefficient is defined using parameters derived from gas kinetic theory first encountered in Chapter 16, namely, the average speed of the gas and the mean free path. Example Problem 17.1 illustrates the dependence of the diffusion coefficient on these parameters.

EXAMPLE PROBLEM 17.1

Determine the diffusion coefficient for Ar at 298 K and a pressure of 1.00 atm.

Solution
Using Equation (17.10) and the collisional cross section for Ar provided in Table 16.1:

$$D_{Ar} = \frac{1}{3} v_{ave,Ar} \lambda_{Ar}$$

$$= \frac{1}{3} \left(\frac{8RT}{\pi M_{Ar}} \right)^{1/2} \left(\frac{RT}{PN_A \sqrt{2} \sigma_{Ar}} \right)$$

$$= \frac{1}{3} \left(\frac{8(8.314 \ \text{J mol}^{-1} \ \text{K}^{-1})298 \ \text{K}}{\pi(0.040 \ \text{kg mol}^{-1})} \right)^{1/2} \left(\frac{(8.314 \ \text{J mol}^{-1} \ \text{K}^{-1})298 \ \text{K}}{(101,325 \ \text{Pa})(6.022 \times 10^{23} \text{mol}^{-1})} \right.$$
$$\left. \times \frac{1}{\sqrt{2}(3.6 \times 10^{-19} \text{m}^2)} \right)$$

$$= \frac{1}{3} (397 \ \text{m s}^{-1})(7.98 \times 10^{-8} \text{m})$$

$$= 1.06 \times 10^{-5} \text{m}^2 \ \text{s}^{-1}$$

Transport properties of gases can be described using concepts derived from gas kinetic theory. For example, the diffusion coefficient is dependent on the mean free path, which is in turn dependent on the collisional cross section. One criticism of this approach is that parameters such as average velocity are derived using an equilibrium distribution, yet these concepts are now applied in a nonequilibrium context when discussing transport phenomena. The development presented here is performed under the assumption that the displacement of the system away from equilibrium is modest; therefore, equilibrium-based quantities remain relevant. That said, transport phenomena can be described using nonequilibrium distributions; however, the mathematical complexity of this approach is beyond the scope of this text.

With the expression for the diffusion coefficient in hand [Equation (17.10)], the relationship between this quantity and the details of the gas particles are clear. This relationship suggests that transport properties such as diffusion can be used to determine particle parameters such as effective size as described by the collisional cross section. Example Problem 17.2 illustrates the connection between the diffusion coefficient and particle size.

EXAMPLE PROBLEM 17.2

Under identical temperature and pressure conditions, the diffusion coefficient of He is roughly four times larger than that of Ar. Determine the ratio of the collisional cross sections.

Solution

Using Equation (17.10), the ratio of diffusion coefficients (after canceling the 1/3 constant term) can be written in terms of the average speed and mean free path as follows:

$$\frac{D_{He}}{D_{Ar}} = 4 = \frac{v_{ave,He}\lambda_{He}}{v_{ave,Ar}\lambda_{Ar}}$$

$$= \frac{\left(\dfrac{8RT}{\pi M_{He}}\right)^{1/2}\left(\dfrac{RT}{P_{He}N_A\sqrt{2}\sigma_{He}}\right)}{\left(\dfrac{8RT}{\pi M_{Ar}}\right)^{1/2}\left(\dfrac{RT}{P_{Ar}N_A\sqrt{2}\sigma_{Ar}}\right)}$$

$$= \left(\frac{M_{Ar}}{M_{He}}\right)^{1/2}\left(\frac{\sigma_{Ar}}{\sigma_{He}}\right)$$

$$\left(\frac{\sigma_{He}}{\sigma_{Ar}}\right) = \frac{1}{4}\left(\frac{M_{Ar}}{M_{He}}\right)^{1/2} = \frac{1}{4}\left(\frac{39.9\ \text{g mol}^{-1}}{4.00\ \text{g mol}^{-1}}\right)^{1/2} = 0.79$$

Recall from Section 16.6 that the collisional cross section for a pure gas is equal to πd^2 where d is the diameter of a gas particle. The ratio of collisional cross sections determined using the diffusion coefficients is consistent with a He diameter that is 0.89 smaller than that of Ar. However, the diameter of Ar as provided in tables of atomic radii is roughly 2.5 times greater than that of He. The origin of this discrepancy can be traced to the hard-sphere approximation for interparticle interactions.

17.3 The Time Evolution of a Concentration Gradient

As illustrated in the previous section, the existence of a concentration gradient results in particle diffusion. What is the timescale for diffusion, and how far can a particle diffuse in a given amount of time? These questions are addressed by the diffusion equation, which can be derived as follows. Beginning with Fick's first law, the particle flux is given by

$$J_x = -D\left(\frac{d\tilde{N}(x)}{dx}\right) \tag{17.12}$$

The quantity J_x in Equation (17.12) is the flux through a plane located at x as illustrated in Figure 17.4. The flux at location $x + dx$ can also be written using Fick's first law:

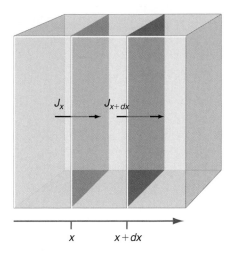

FIGURE 17.4
Depiction of flux through two separate planes. If $J_x = J_{x+dx}$, then the concentration between the planes will not change. However, if the fluxes are unequal, then the concentration will change with time.

17.1 Fick's Second Law

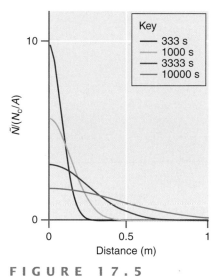

FIGURE 17.5
The spatial variation in particle number density, $\tilde{N}(x,t)$, as a function of time. The number density is defined with respect to N_0/A, the number of particles confined to a plane located at $x = 0$ of area A. In this example, $D = 10^{-5}$ m^2 s^{-1}, a typical value for a gas at 1 atm and 298 K (see Example Problem 17.1). The corresponding diffusion time for a given concentration profile is indicated.

$$J_{x+dx} = -D\left(\frac{d\tilde{N}(x+dx)}{dx}\right) \quad (17.13)$$

The particle density at $(x + dx)$ is related to the corresponding value at x as follows:

$$\tilde{N}(x+dx) = \tilde{N}(x) + dx\left(\frac{d\tilde{N}(x)}{dx}\right) \quad (17.14)$$

This equation is derived by keeping the first two terms in the Taylor series expansion of number density with distance equivalent to the procedure described earlier for obtaining Equations (17.2) and (17.3). Substituting Equation (17.14) into Equation (17.13), the flux through the plane at $(x + dx)$ becomes

$$J_{x+dx} = -D\left(\frac{d\tilde{N}(x)}{dx} + \left(\frac{d^2\tilde{N}(x)}{dx^2}\right)dx\right) \quad (17.15)$$

Consider the space between the two flux planes illustrated in Figure 17.4. The change in particle density in this region of space depends on the difference in flux through the two planes. If the fluxes are identical, the particle density will remain constant in time. However, a difference in flux will result in an evolution in particle number density as a function of time. The flux is equal to the number of particles that pass through a given area per unit time. If the area of the two flux planes is equivalent, then the difference in flux is directly proportional to the difference in number density. This relationship between the time dependence of the number density and the difference in flux is expressed as follows:

$$\frac{d\tilde{N}(x,t)}{dt} = \frac{d(J_x - J_{x+dx})}{dx}$$

$$= \frac{d}{dx}\left[-D\left(\frac{d\tilde{N}(x,t)}{dx}\right) - \left(-D\left(\left(\frac{d\tilde{N}(x,t)}{dx}\right) + \left(\frac{d^2\tilde{N}(x,t)}{dx^2}\right)dx\right)\right)\right]$$

$$\frac{d\tilde{N}(x,t)}{dt} = D\frac{d^2\tilde{N}(x,t)}{dx^2} \quad (17.16)$$

Equation (17.16) is called the **diffusion equation,** and is also known as **Fick's second law of diffusion.** Notice that $\tilde{N}$ depends on both position x and time t. Equation (17.16) demonstrates that the time evolution of the concentration gradient is proportional to the second derivative of the spatial gradient in concentration. That is, the greater the "curvature" of the concentration gradient, the faster the relaxation will proceed. Equation (17.16) is a differential equation that can be solved using standard techniques and a set of initial conditions (see Crank entry in the Further Reading section at the end of the chapter) resulting in the following expression for $\tilde{N}(x,t)$:

$$\tilde{N}(x,t) = \frac{N_0}{2A(\pi Dt)^{1/2}}e^{-x^2/4Dt} \quad (17.17)$$

In this expression, N_0 represents the initial number of molecules confined to a plane at $t = 0$, A is the area of this plane, x is distance away from the plane, and D is the diffusion coefficient. Equation (17.17) can be viewed as a distribution function that describes the probability of finding a particle at time $= t$ at a plane located a distance x away from the initial plane at $t = 0$. An example of the spatial variation in $\tilde{N}$ versus time is provided in Figure 17.5 for a species with $D = 10^{-5}$ m^2 s^{-1}, roughly equivalent to the diffusion coefficient of Ar at 298 K and 1 atm. The figure demonstrates that with an increase in time, $\tilde{N}$ increases at distances farther away from the initial plane (located at 0 m in Figure 17.5).

Similar to other distribution functions encountered thus far, it is more convenient to use a metric or benchmark value that provides a measure of $\tilde{N}(x,t)$ as opposed to describing the entire distribution. The primary metric employed to describe $\tilde{N}(x,t)$ is the root-mean-square (rms) displacement, determined using what should by now be a familiar approach:

$$x_{rms} = \langle x^2 \rangle^{1/2} = \left[\frac{A}{N_0} \int_{-\infty}^{\infty} x^2 \tilde{N}(x,t)dx \right]^{1/2}$$

$$= \left[\frac{A}{N_0} \int_{-\infty}^{\infty} x^2 \frac{N_0}{A2(\pi Dt)^{1/2}} e^{-x^2/4Dt} dx \right]^{1/2}$$

$$= \left[\frac{1}{2(\pi Dt)^{1/2}} \int_{-\infty}^{\infty} x^2 e^{-x^2/4Dt} dx \right]^{1/2}$$

$$x_{rms} = \sqrt{2Dt} \qquad (17.18)$$

Notice that the rms displacement increases as the square root of both the diffusion coefficient and time. Equation (17.18) represents the rms displacement in a single dimension. For diffusion in three dimensions, the corresponding term, r_{rms}, can be determined using the Pythagorean theorem under the assumption that diffusion is equivalent in all three dimensions:

$$r_{rms} = \sqrt{6Dt} \qquad (17.19)$$

The diffusion relationships derived in this section and the previous section involved gases and employed concepts from gas kinetic theory. However, these relationships are also applicable to diffusion in solution, as will be demonstrated later in this chapter.

EXAMPLE PROBLEM 17.3

Determine x_{rms} for a particle where $D = 10^{-5}$ m^2 s^{-1} for diffusion times of 1000 and 10,000 s.

Solution
Employing Equation (17.18):

$$x_{rms,1000s} = \sqrt{2Dt} = \sqrt{2(10^{-5}\text{m}^2\text{ s}^{-1})(1000\text{ s})} = 0.141\text{ m}$$

$$x_{rms,10,000s} = \sqrt{2Dt} = \sqrt{2(10^{-5}\text{m}^2\text{ s}^{-1})(10,000\text{ s})} = 0.447\text{ m}$$

The diffusion coefficient employed in this example is equivalent to that used in Figure 17.5, and the rms displacements determined here can be compared to the spatial variation in $\tilde{N}(x,t)$ depicted in the figure to provide a feeling for the x_{rms} distance versus the overall distribution of particle diffusion distances versus time.

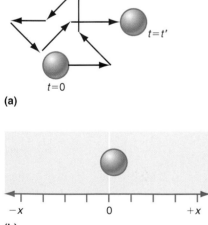

(a)

(b)

FIGURE 17.6
(a) Illustration of a random walk. Diffusion of the particle is modeled as a series of discrete steps (each arrow); the length of each step is the same, but the direction of the step is random.
(b) Illustration of the one-dimensional random walk model.

SUPPLEMENTAL

17.4 Statistical View of Diffusion

In deriving Fick's first law of diffusion, a gas particle was envisioned to move a distance equal to the mean free path before colliding with another particle. After this collision, memory of the initial direction of motion is lost and the particle is free to move in the same or a new direction until the next collision occurs. This conceptual picture of particle motion is mathematically described by the statistical approach to diffusion. In the statistical approach, illustrated in Figure 17.6, particle diffusion is also modeled as a series of discrete displacements or steps, with the direction of one step being uncorrelated with that of the previous step. That is, once the particle has taken a step, the direction of the next step is random. A series of such steps is referred to as a **random walk.**

In the previous section we determined the probability of finding a particle at a distance x away from the origin after a certain amount of time. The statistical model of diffusion can be connected directly to this idea using the random walk model. Consider a particle undergoing a random walk along a single dimension x such that the particle moves one step in either the $+x$ or $-x$ direction (Figure 17.6). After a certain number of steps, the particle will have taken Δ total steps with Δ_- steps in the $-x$ direction and Δ_+ steps in the $+x$ direction. The probability that the particle will have traveled a distance X from the origin is related to the weight associated with that distance, as given by the following expression:

$$W = \frac{\Delta!}{\Delta_+!\Delta_-!} = \frac{\Delta!}{\Delta_+!(\Delta-\Delta_+)!} \tag{17.20}$$

Equation (17.20) is identical to the weight associated with observing a certain number of heads after tossing a coin Δ times, as discussed in Chapter 13. This similarity is not a coincidence—the one-dimensional random walk model is very much like tossing a coin. Each outcome of a coin toss is independent of the previous outcome, and only one of two outcomes is possible per toss. Evaluation of Equation (17.20) requires an expression for Δ_+ in terms of Δ. This relationship can be derived by recognizing that X is equal to the difference in the number of steps in the $+x$ and $-x$ direction:

$$X = \Delta_+ - \Delta_- = \Delta_+ - (\Delta - \Delta_+) = 2\Delta_+ - \Delta$$

$$\frac{X+\Delta}{2} = \Delta_+$$

With this definition for Δ_+, the expression for W becomes

$$W = \frac{\Delta!}{\left(\dfrac{\Delta+X}{2}\right)!\left(\Delta - \dfrac{\Delta+X}{2}\right)!} = \frac{\Delta!}{\left(\dfrac{\Delta+X}{2}\right)!\left(\dfrac{\Delta-X}{2}\right)!} \tag{17.21}$$

The probability of the particle being a distance X away from the origin is given by the weight associated with this distance divided by the total weight, 2^Δ, such that:

$$P = \frac{W}{W_{Total}} = \frac{W}{2^\Delta} \propto e^{-X^2/2\Delta} \tag{17.22}$$

The final proportionality can be derived by evaluation of Equation (17.21) using Stirling's approximation. Recall from the solution to the diffusion equation in the previous section that the distance a particle diffuses away from the origin was also proportional to an exponential term:

$$\tilde{N}(x,t) \propto e^{-x^2/4Dt} \tag{17.23}$$

In Equation (17.23), x is the actual diffusion distance, D is the diffusion coefficient, and t is time. For these two pictures of diffusion to converge on the same physical result, the exponents must be equivalent such that

$$\frac{x^2}{4Dt} = \frac{X^2}{2\Delta} \tag{17.24}$$

At this point, the random walk parameters X and Δ must be expressed in terms of the actual quantities of diffusion distance x and total diffusion time t. The total number of random walk steps is expressed as the total diffusion, t, time divided by the time per random walk step, τ:

$$\Delta = \frac{t}{\tau} \tag{17.25}$$

In addition, x can be related to the random walk displacement X by using a proportionality constant, x_o, that represents the average distance in physical space a particle traverses between collisions such that

$$x = Xx_o \qquad (17.26)$$

With these definitions for Δ and X, substitution into Equation (17.24) results in the following definition of D:

$$D = \frac{x_o^2}{2\tau} \qquad (17.27)$$

Equation (17.27) is the **Einstein–Smoluchowski equation.** The importance of this equation is that it relates a macroscopic quantity, D, to microscopic aspects of the diffusion as described by the random walk model. For reactions in solution, x_o is generally taken to be the particle diameter. Using this definition and the experimental value for D, the timescale associated with each random walk event can be determined.

EXAMPLE PROBLEM 17.4

The diffusion coefficient of liquid benzene is 2.2×10^{-5} cm^2 s^{-1}. (Liquids typically demonstrate diffusion coefficients on the order of 1 cm^2 s^{-1} under standard conditions.) Given an estimated molecular diameter of 0.3 nm, what is the timescale for a random walk?

Solution

Rearranging Equation (17.27), the time per random walk step is

$$\tau = \frac{x_o^2}{2D} = \frac{(0.3 \times 10^{-9} \text{ m})^2}{2(2.2 \times 10^{-9} \text{ m}^2 \text{ s}^{-1})} = 2.0 \times 10^{-11} \text{ s}$$

This is an extremely short time, only 20 ps! This example illustrates that, on average, the diffusional motion of a benzene molecule in the liquid phase is characterized by short-range translational motion between frequent collisions with neighboring molecules.

The Einstein–Smoluchowski equation can also be related to gas diffusion described previously. If we equate x_o with the mean free path λ and define the time per step as the average time it takes a gas particle to translate one mean free path λ / v_{ave}, then the diffusion coefficient is given by

$$D = \frac{\lambda^2}{2\left(\dfrac{\lambda}{v_{ave}}\right)} = \frac{1}{2}\lambda v_{ave}$$

This is exactly the same expression for D derived from gas kinetic theory in the absence of the 2/3 correction for particle trajectories as discussed earlier. That is, the statistical and kinetic theory viewpoints of diffusion provide equivalent descriptions of gas diffusion.

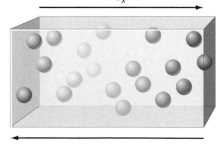

J_x

Temperature

FIGURE 17.7
Temperature gradient in a collection of gas particles. Regions containing high kinetic energy particles are red and low kinetic energy regions are blue. The gradient in kinetic energy, and therefore temperature, is indicated, and the flux in energy in response to this gradient, J_x, is also shown.

17.5 Thermal Conduction

Thermal conduction is the transport process in which energy migrates in response to a gradient in temperature. Figure 17.7 depicts a collection of identical gas particles for which a gradient in temperature exists. Note that the gradient is with respect to temperature only, and that the particle number density is the same throughout the box. Equilibrium is reached when the system has an identical temperature in all regions of the box. Because temperature and kinetic energy are related, relaxation toward equilibrium will involve the transport of kinetic energy from the high-temperature side of the box to the lower temperature side.

Thermal conduction occurs through a variety of mechanisms, depending on the phase of matter. In this derivation we assume that energy transfer occurs during particle collisions and that equilibrium with the energy gradient is established after each collision, thereby ensuring that the particles are at equilibrium with the gradient after each collision. This collisional picture of energy transfer is easy to envision for a gas; however, it can also be applied to liquids and solids. In these phases, molecules do not translate freely, yet the molecular energy can be transferred through collisions with nearby molecules, resulting in energy transfer. In addition to collisional energy transfer, energy can also be transferred through convection or radiative transfer. In convection, differences in density resulting from the temperature gradient can produce convection currents. Although energy is still transferred through collisional events when the particles in the currents collide with other particles, particle migration is not random such that transfer through convection is physically distinct from collisional transport as we have defined it. In radiative transfer, matter is treated as a blackbody that is capable of emitting and absorbing electromagnetic radiation. Radiation from higher temperature matter is absorbed by lower temperature matter, resulting in energy transfer. Both convection and radiative transfer are assumed to be negligible here.

The problem of thermal conduction is approached in exactly the same manner as was done for diffusion. The process begins by considering energy transfer for an ideal monatomic gas. Recall from statistical mechanics and thermodynamics that the average translational energy of a monatomic gas particle is $3/2 \, kT$. In addition to translational energy, diatomic and polyatomic molecules can carry more energy in rotational and vibrational energetic degrees of freedom, and our treatment will be extended to these more complex molecules shortly. Kinetic energy is transferred when the flux plane is struck by particles from one side of the plane. In this collision, energy is transmitted from one side of the flux plane to the other. It is important to note that the particle number density is equivalent throughout the box such that mass transfer does not occur. Energy transport and the relaxation toward thermal equilibrium are accomplished through molecular collisions, not diffusion.

Using techniques identical to those employed in Section 17.2 to describe diffusion, the energy at planes located $\pm\lambda$ away from the flux plane (Figure 17.8) is

$$\varepsilon(-\lambda) = \varepsilon(0) - \lambda\left(\frac{d\varepsilon}{dx}\right)_{x=0} \tag{17.28}$$

FIGURE 17.8
Model used to determine the thermal conductivity of a gas. The gradient in temperature will result in the transfer of kinetic energy from regions of high temperature (indicated by the red plane) to regions of lower temperature (indicated by the blue plane). The plane located at $x = 0$ is the location at which the flux of kinetic energy in response to the gradient is determined. Two planes, located one mean free path away ($\pm\lambda$) from the flux plane, are established, with particles traveling from one of these planes to the flux plane. Energy is transferred via collisional events at the flux plane.

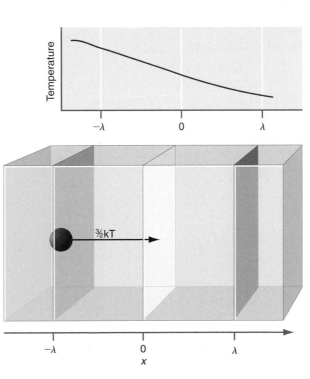

$$\varepsilon(\lambda) = \varepsilon(0) + \lambda \left(\frac{d\varepsilon}{dx} \right)_{x=0} \tag{17.29}$$

Proceeding in a manner identical to the derivation in Section 17.2 for diffusion, the flux is defined as

$$J_x = \tilde{N}\varepsilon(x) \int_0^\infty v_x f(v_x)\, dv_x = \frac{1}{4} \tilde{N}\varepsilon(x) v_{ave} \tag{17.30}$$

In Equation (17.30), $\tilde{N}$ is the number density of gas particles, which is constant (recall that there is no spatial gradient in number density, only energy). Substituting Equations (17.28) and (17.29) into Equation (17.30) yields the following expression for the flux in energy from planes located $\pm\lambda$ from the flux plane:

$$J_{-\lambda,0} = \frac{1}{4} v_{ave} \tilde{N}\varepsilon(-\lambda) \tag{17.31}$$

$$J_{\lambda,0} = \frac{1}{4} v_{ave} \tilde{N}\varepsilon(\lambda) \tag{17.32}$$

The total energy flux is given by the difference between Equations (17.31) and (17.32):

$$J_{Total} = J_{-\lambda,0} - J_{\lambda,0} \tag{17.33}$$

$$= \frac{1}{4} v_{ave} \tilde{N}(\varepsilon(-\lambda) - \varepsilon(\lambda))$$

$$= \frac{1}{4} v_{ave} \tilde{N} \left(-2\lambda \left(\frac{d\varepsilon}{dx} \right)_{x=0} \right)$$

$$= -\frac{1}{2} v_{ave} \tilde{N}\lambda \left(\left(\frac{d\left(\frac{3}{2}kT \right)}{dx} \right)_{x=0} \right)$$

$$= -\frac{3}{4} k v_{ave} \tilde{N}\lambda \left(\frac{dT}{dx} \right)_{x=0}$$

After multiplying the total flux by 2/3 for orientational averaging of the particle trajectories, the expression for total flux through the flux plane becomes

$$J_{Total} = -\frac{1}{2} k v_{ave} \tilde{N}\lambda \left(\frac{dT}{dx} \right)_{x=0} \tag{17.34}$$

Equation (17.34) describes the total energy flux for an ideal monatomic gas, where the energy transferred per collision is $3/2\, kT$. The expression for total flux can be defined in terms of molar heat capacity, $C_{V,m}$, by recognizing that the constant volume molar heat capacity for a monatomic gas is $3/2\, R$, therefore:

$$\frac{C_{V,m}}{N_A} = \frac{3}{2}k \quad \text{(for an ideal monatomic gas)}$$

With this substitution, the flux can be expressed in terms of the $C_{V,m}$ for the species of interest and is not restricted to monatomic species. Employing this identity, the final expression for total energy flux is

$$J_{Total} = -\frac{1}{3} \frac{C_{V,m}}{N_A} v_{ave} \tilde{N}\lambda \left(\frac{dT}{dx} \right)_{x=0} \tag{17.35}$$

Comparison of Equation (17.35) to the general relationship between flux and gradient of Equation (17.1) dictates that the proportionality constant for energy transfer, referred to as the **thermal conductivity,** κ, be given by

$$\kappa = \frac{1}{3}\frac{C_{V,m}}{N_A}v_{ave}\tilde{N}\lambda \qquad (17.36)$$

Unit analysis of Equation (17.36) reveals that κ has units of $J\ K^{-1}\ m^{-1}\ s^{-1}$. Every factor in the expression for thermal conductivity is known from thermodynamics or gas kinetic theory such that the evaluation of κ is relatively straightforward for gases. For fluids, thermal conductivity still serves as the proportionality constant between the net flux in energy and the gradient in temperature; however, a simple expression similar to Equation (17.36) does not exist. The appearance of the mean free path in Equation (17.36) demonstrates that similar to the diffusion coefficient, the thermal conductivity can be used to estimate the hard-sphere radius of gases, as Example Problem 17.5 illustrates.

EXAMPLE PROBLEM 17.5

The thermal conductivity of Ar at 300 K and 1 atm pressure is $0.0177\ J\ K^{-1}\ m^{-1}\ s^{-1}$. What is the collisional cross section of Ar assuming ideal gas behavior?

Solution

The collisional cross section is contained in the mean free path. Rearranging Equation (17.25) to isolate the mean free path, we obtain

$$\lambda = \frac{3\kappa}{\left(\dfrac{C_{V,m}}{N_A}\right)v_{ave}\tilde{N}}$$

The thermal conductivity is provided, and the other terms needed to calculate λ are as follows:

$$\frac{C_{V,m}}{N_A} = \frac{\frac{3}{2}R}{N_A} = \frac{3}{2}k = 2.07\times10^{-23}\ J\ K^{-1}$$

$$v_{ave} = \left(\frac{8RT}{\pi M}\right)^{1/2} = 399\ m\ s^{-1}$$

$$\tilde{N} = \frac{N}{V} = \frac{N_A P}{RT} = 2.45\times10^{25}\ m^{-3}$$

With these quantities, the mean free path is

$$\lambda = \frac{3(0.0177\ J\ K^{-1}\ m^{-1}\ s^{-1})}{(2.07\times10^{-23}\ J\ K^{-1})(399\ m\ s^{-1})(2.44\times10^{25}\ m^{-3})} = 2.61\times10^{-7}\ m$$

Using the definition of the mean free path (Section 16.7), the collisional cross section is

$$\sigma = \frac{1}{\sqrt{2}\tilde{N}\lambda} = \frac{1}{\sqrt{2}(2.45\times10^{25}\ m^{-3})(2.61\times10^{-7}\ m)}$$

$$= 1.11\times10^{-19}\ m^2$$

This value for the collisional cross section corresponds to a particle diameter of 188 pm ($pm = 10^{-12}$ m), which is remarkably close to the 194-pm value provided by the tabulated atomic radii of Ar.

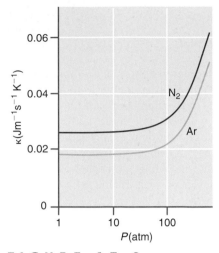

FIGURE 17.9
Thermal conductivity as a function of pressure for N_2 and Ar at 300 K. Note that the pressure scale is logarithmic. The figure demonstrates that κ is roughly independent of pressure up to 50 atm.

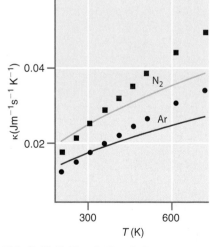

FIGURE 17.10
Variation of κ with temperature for N_2 and Ar at 1 atm. Experimental data are indicated by the squares, and the predicted $T^{1/2}$ dependence as the solid lines. The calculated κ was set equal to the experimental value at 300 K, and then $T^{1/2}$ dependence was applied to generate the predicted variation in κ with temperature.

The thermal conductivity of a material is dependent on v_{ave} $\tilde{N}$, and λ, quantities that are dependent on temperature and pressure. Therefore, κ will also demonstrate temperature and pressure dependence. Predictions regarding the dependence of κ on T and P can be made within the gas kinetic theory approach used to describe the gas particles. With respect to pressure, λ is inversely proportional to $\tilde{N}$ such that the mean free path will decrease with pressure:

$$\lambda = \left(\frac{RT}{PN_A}\right)\frac{1}{\sqrt{2}\sigma} = \left(\frac{V}{N}\right)\frac{1}{\sqrt{2}\sigma} = \frac{1}{\tilde{N}\sqrt{2}\sigma}$$

The product of the mean free path and $\tilde{N}$ results in the absence of $\tilde{N}$ dependence in Equation (17.36), leading to the surprising result that κ is predicted to be independent of pressure. Figure 17.9 illustrates the pressure dependence of κ for N_2 and Ar at 300 K. The figure demonstrates that κ is indeed essentially independent of pressure up to 50 atm! At elevated pressures, the intermolecular forces between gas molecules become appreciable, resulting in failure of the hard-sphere model, and the expressions derived using gas kinetic theory are not applicable. This region is indicated by the dramatic rise in κ at high pressures. The pressure dependence of κ is also observed at very low pressure, when the mean free path is greater than the dimensions of the container. In this case, the energy is transported from one side of the container to the other by wall–particle collisions such that κ increases with pressure in this low-pressure regime.

With respect to the temperature dependence of κ, the cancellation of $\tilde{N}$ with the mean free path results in only v_{ave} and $C_{V,m}$ carrying temperature dependence. For a monatomic gas, $C_{V,m}$ is expected to demonstrate minimal temperature dependence. Therefore, the v_{ave} term in Equation (17.36) predicts that κ is proportional to $T^{1/2}$. Identical behavior is predicted for diatomic and polyatomic molecules in which the temperature dependence of $C_{V,m}$ is minimal. Figure 17.10 presents the variation in κ with temperature for N_2 and Ar at 1 atm. Also presented is the predicted $T^{1/2}$ dependence of κ. The comparison between the experimental and predicted temperature dependence demonstrates that κ increases more rapidly with temperature than the predicted $T^{1/2}$ dependence due to the presence of intermolecular interactions that are neglected in the hard-sphere approximation employed in gas kinetic theory. Notice that differences between the predicted and experimental κ are evident at low temperature as well, as opposed to the good agreement between the predicted and experimental behavior of κ with pressure evident in Figure 17.9.

17.6 Viscosity of Gases

The third transport phenomenon considered is that of linear momentum. Practical experience provides an intuitive guide with respect to this area of transport. Consider the flow of a gas through a pipe under pressure. Some gases will flow more easily than others, and the property that characterizes resistance of flow is **viscosity,** represented by the symbol η (lowercase Greek eta). What does viscosity have to do with linear momentum? Figure 17.11 provides a cutaway view of a gas flowing between two plates. It can be shown experimentally that the velocity of the gas, v_x, is greatest midway between the plates and decreases as the gas approaches either plate with $v_x = 0$ at the fluid–plate boundary. Therefore, a gradient in v_x exists along the coordinate orthogonal to the direction of flow (z in Figure 17.11). Because the linear momentum in the x direction is mv_x, a gradient in linear momentum must also exist.

We assume that the gas flow is **laminar flow,** meaning that the gas can be decomposed into layers of constant speed as illustrated in Figure 17.11. This regime will exist for most gases and some liquids provided the flow rate is not too high (see Problem P17.15 at the end of the chapter). At high flow rates, the **turbulent flow** regime is reached where the layers are intermixed such that a clear dissection of the gas in terms

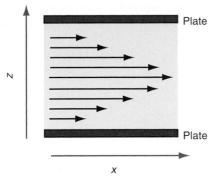

FIGURE 17.11
Cross section of a fluid flowing between two plates. The fluid is indicated by the blue area between the plates, with arrow lengths representing the speed of the fluid.

of layers of the same speed cannot be performed. The discussion presented here is limited to conditions of laminar flow.

The analysis of linear momentum transport proceeds in direct analogy to diffusion and thermal conductivity. As illustrated in Figure 17.11, a gradient in linear momentum exists in the z direction; therefore, planes of similar linear momentum are defined parallel to the direction of fluid flow as illustrated in Figure 17.12. The transfer of linear momentum occurs by a particle from one momentum layer colliding with the flux plane and thereby transferring its momentum to the adjacent layer.

To derive the relationship between flux and the gradient in linear momentum, we proceed in a fashion analogous to that used to derive diffusion (Section 17.2) and thermal conductivity (Section 17.5). First, the linear momentum, p, at $\pm\lambda$ is given by

$$p(-\lambda) = p(0) - \lambda\left(\frac{dp}{dz}\right)_{z=0} \tag{17.37}$$

$$p(\lambda) = p(0) + \lambda\left(\frac{dp}{dz}\right)_{z=0} \tag{17.38}$$

Proceeding just as before for diffusion and thermal conduction, the flux in linear momentum from each plane located at $\pm\lambda$ to the flux plane is

$$J_{-\lambda,0} = \frac{1}{4}v_{ave}\tilde{N}p(-\lambda) \tag{17.39}$$

$$J_{\lambda,0} = \frac{1}{4}v_{ave}\tilde{N}p(\lambda) \tag{17.40}$$

The total flux is the difference between Equations (17.39) and (17.40):

$$J_{Total} = \frac{1}{4}v_{ave}\tilde{N}\left(-2\lambda\left(\frac{dp}{dz}\right)_{z=0}\right) \tag{17.41}$$

$$= -\frac{1}{2}v_{ave}\tilde{N}\lambda\left(\frac{d(mv_x)}{dz}\right)_{z=0}$$

$$= -\frac{1}{2}v_{ave}\tilde{N}\lambda m\left(\frac{dv_x}{dz}\right)_{z=0}$$

FIGURE 17.12
Parameterization of the box model used to derive viscosity. Planes of identical particle velocity (v_x) are given by the blue planes, with the magnitude of velocity given by the arrows. The gradient in linear momentum will result in momentum transfer from regions of high momentum (darker blue plane) to regions of lower momentum (indicated by the light blue plane). The plane located at $z = 0$ is the location at which the flux of linear momentum in response to the gradient is determined.

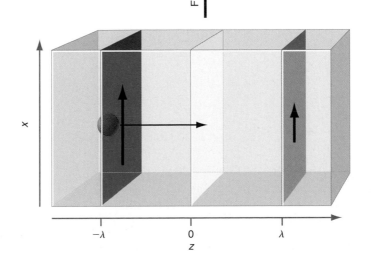

Finally, Equation (17.41) is multiplied by 2/3 as a result of orientational averaging of the particle trajectories, resulting in the final expression for the total flux in linear momentum:

$$J_{Total} = -\frac{1}{3} v_{ave} \tilde{N} \lambda m \left(\frac{dv_x}{dz}\right)_0 \qquad (17.42)$$

Comparison to Equation (17.1) and Equation (17.28) indicates that the proportionality constant between flux and the gradient in velocity is defined as

$$\eta = \frac{1}{3} v_{ave} \tilde{N} \lambda m \qquad (17.43)$$

Equation (17.43) represents the viscosity of the gas, η, given in terms of parameters derived from gas kinetic theory. The units of viscosity are the poise (P), or $0.1 \text{ kg m}^{-1} \text{ s}^{-1}$. Notice the quantity 0.1 in the conversion to SI units. Viscosities are generally reported in μP (10^{-6} P) for gases and cP (10^{-2} P) for liquids.

EXAMPLE PROBLEM 17.6

The viscosity of Ar is 227 μP at 300 K and 1 atm. What is the collisional cross section of Ar assuming ideal gas behavior?

Solution

Because the collisional cross section is related to the mean free path, Equation (17.43) is first rearranged as follows:

$$\lambda = \frac{3\eta}{v_{ave} \tilde{N} m}$$

Next, we evaluate each term separately, then use these terms to calculate the mean free path:

$$v_{ave} = \left(\frac{8RT}{\pi M}\right)^{1/2} = 399 \text{ m s}^{-1}$$

$$\tilde{N} = \frac{PN_A}{RT} = 2.45 \times 10^{25} \text{ m}^{-3}$$

$$m = \frac{M}{N_A} = \frac{0.040 \text{ kg mol}^{-1}}{6.022 \times 10^{23} \text{ mol}^{-1}} = 6.64 \times 10^{-26} \text{ kg}$$

$$\lambda = \frac{3\eta}{v_{ave} \tilde{N} m} = \frac{3(227 \times 10^{-6} \text{P})}{(399 \text{ m s}^{-1})(2.45 \times 10^{25} \text{ m}^{-3})(6.64 \times 10^{-26} \text{ kg})}$$

$$= \frac{3(227 \times 10^{-7} \text{ kg m}^{-1} \text{ s}^{-1})}{(399 \text{ m s}^{-1})(2.45 \times 10^{25} \text{ m}^{-3})(6.64 \times 10^{-26} \text{ kg})} = 1.05 \times 10^{-7} \text{ m}$$

Note the conversion of poise to SI units in the last step. Using the definition of the mean free path, the collisional cross section can be determined:

$$\sigma = \frac{1}{\sqrt{2} \tilde{N} \lambda} = \frac{1}{\sqrt{2}(2.45 \times 10^{25} \text{ m}^{-3})(1.05 \times 10^{-7} \text{ m})} = 2.75 \times 10^{-19} \text{ m}^2$$

In Example Problem 17.5, the thermal conductivity of Ar provided an estimate for the collisional cross section that was roughly 1/2 this value. The discrepancy between collisional cross-section values determined using two different measured transport properties illustrates the approximate nature of the treatment presented here. Specifically, we have assumed that the intermolecular interactions are well modeled by the hard-sphere approximation. The difference in σ suggests that attractive forces are also important in describing the interaction of particles during collisions.

Similar to thermal conductivity, viscosity is dependent on v_{ave}, $\tilde{N}$, and λ, quantities that are both temperature and pressure dependent. Both $\tilde{N}$ and λ demonstrate pressure dependence, but the product of $\tilde{N}$ and λ contains no net $\tilde{N}$ dependence; therefore, η is predicted to be independent of pressure. Figure 17.13 presents the pressure dependence of η for N_2 and Ar at 300 K. The behavior is almost identical that observed for thermal conductivity (Figure 17.9), with η demonstrating little pressure dependence until ~$P = 50$ atm. At elevated pressures, intermolecular interactions become important, and the increased interaction between particles gives rise to a substantial increase in η for $P > 50$ atm, similar to the case for thermal conductivity.

With respect to the temperature dependence of η, the v_{ave} term in Equation (17.43) dictates that η should increase as $T^{1/2}$, identical to the temperature dependence of thermal conductivity (Figure 17.10). This result is perhaps a bit surprising since liquids demonstrate a decrease in η with an increase in temperature. However, the predicted increase in η with temperature for a gas is born out by experiment as illustrated in Figure 17.14 where the variation in η with temperature for N_2 and Ar at 1 atm is presented. Also shown is the predicted $T^{1/2}$ dependence. The prediction of increased viscosity with temperature is a remarkable confirmation of gas kinetic theory. However, comparison of the experimental and predicted temperature dependence demonstrates that η increases more rapidly than predicted due to the presence of intermolecular interactions that are neglected in the hard-sphere model, similar to the discussion of the temperature dependence of κ provided earlier. The increase in η with temperature for a gas is consistent with the increase in velocity accompanying a rise in temperature, and corresponding increase in momentum flux.

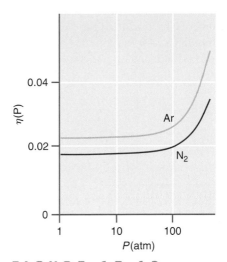

FIGURE 17.13
Pressure dependence of η for gaseous N_2 and Ar at 300 K.

17.7 Measuring Viscosity

Viscosity is a measure of a fluid's resistance to flow; therefore, it is not surprising that the viscosity of gases and liquids is measured using flow. Viscosity is typically measured by monitoring the flow of a fluid through a tube with the underlying idea that the greater the viscosity, the smaller the flow through the tube. The following equation was derived by Poiseuille to describe flow of a liquid through a round tube under conditions of laminar flow:

$$\frac{\Delta V}{\Delta t} = \frac{\pi r^4}{8\eta}\left(\frac{P_2 - P_1}{x_2 - x_1}\right) \tag{17.44}$$

This equation is referred to as **Poiseuille's law.** In Equation (17.44), $\Delta V/\Delta t$ represents the volume of fluid, ΔV, that passes through the tube in a specific amount of time, Δt; r is the radius of the tube through which the fluid flows; η is the fluid viscosity; and the factor in parentheses represents the macroscopic pressure gradient over the tube length. Notice that the fluid flow rate is dependent on the radius of the tube and is also inversely proportional to fluid viscosity. As anticipated, the more viscous the fluid, the smaller the flow rate. The flow of an ideal gas through a tube is given by

$$\frac{\Delta V}{\Delta t} = \frac{\pi r^4}{16\eta L P_0}(P_2^2 - P_1^2) \tag{17.45}$$

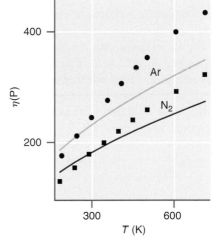

FIGURE 17.14
Temperature dependence of η for N_2 and Ar at 1 atm. Experimental values are given by the squares and circles, and the predicted $T^{1/2}$ dependence is given as the solid line.

where L is the length of the tube, P_2 and P_1 are the pressures at the entrance and exit of the tube, respectively, and P_0 is the pressure at which the volume is measured (and is equal to P_1 if the volume is measured at the end of the tube).

EXAMPLE PROBLEM 17.7

Gas cylinders of CO_2 are sold in terms of weight of CO_2. A cylinder contains 50 lb (22.7 kg) of CO_2. How long can this cylinder be used in an experiment that requires flowing CO_2 at 293 K ($\eta = 146\ \mu P$) through a 1.00-m-long tube (diameter = 0.75 mm) with an input pressure of 1.05 atm and output pressure of 1.00 atm? The flow is measured at the tube output.

Solution

Using Equation (17.45), the gas flow rate $\Delta V / \Delta t$ is

$$\frac{\Delta V}{\Delta t} = \frac{\pi r^4}{16 \eta L P_0}(P_2^2 - P_1^2)$$

$$= \frac{\pi (0.375 \times 10^{-3}\,\mathrm{m})^4}{16(1.46 \times 10^{-5}\ \mathrm{kg\ m^{-1}\ s^{-1}})(1.00\ \mathrm{m})(101{,}325\ \mathrm{Pa})}$$

$$\times ((106{,}391\ \mathrm{Pa})^2 - (101{,}325\ \mathrm{Pa})^2)$$

$$= 2.76 \times 10^{-6}\,\mathrm{m^3\ s^{-1}}$$

Converting the CO_2 contained in the cylinder to the volume occupied at 298 K and 1 atm pressure, we get

$$n_{CO_2} = 22.7\ \mathrm{kg}\left(\frac{1}{0.044\ \mathrm{kg\ mol^{-1}}}\right) = 516\ \mathrm{mol}$$

$$V = \frac{nRT}{P} = 1.24 \times 10^4\,\mathrm{L}\left(\frac{10^{-3}\,\mathrm{m^3}}{\mathrm{L}}\right) = 12.4\ \mathrm{m^3}$$

Given the effective volume of CO_2 contained in the cylinder, the duration over which the cylinder can be used is

$$\frac{12.4\ \mathrm{m^3}}{2.76 \times 10^{-6}\,\mathrm{m^3\ s^{-1}}} = 4.49 \times 10^6\,\mathrm{s}$$

This time corresponds to roughly 52 days.

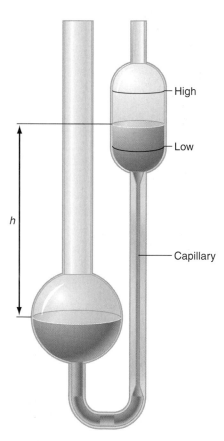

h

High

Low

Capillary

FIGURE 17.15
An Ostwald viscometer. The time for a volume of fluid (ΔV) to flow from the "High" level mark to the "Low" level mark is measured and then used to determine the viscosity of the fluid by means of Equation (17.47).

A convenient tool for measuring the viscosity of liquids is an **Ostwald viscometer** (Figure 17.15). To determine the viscosity, one measures the time it takes for the liquid level to fall from the "high" level mark to the "low" level mark, and the fluid flows through a thin capillary that ensures laminar flow. The pressure driving the liquid through the capillary is $\rho g h$, where ρ is the fluid density, g is the acceleration due to gravity, and h is the difference in liquid levels in the two sections of the viscometer, as illustrated in the figure. Because the height difference will evolve as the fluid flows, h represents an average height. Because $\rho g h$ is the pressure generating flow, Equation (17.44) can be rewritten as follows:

$$\frac{\Delta V}{\Delta t} = \frac{\pi r^4}{8\eta}\frac{\rho g h}{(x_2 - x_1)} = \frac{\pi r^4}{8\eta}\frac{\rho g h}{l} \tag{17.46}$$

where l is the length of the capillary. This equation can be rearranged:

$$\eta = \left(\frac{\pi r^4}{8}\frac{gh}{\Delta V l}\right)\rho \Delta t = A \rho \Delta t \tag{17.47}$$

In Equation (17.47), A is known as the viscometer constant and is dependent on the geometry of the viscometer. All of the viscometer parameters can be determined through careful measurement of the viscometer dimensions; however, the viscometer constant is generally determined by calibration using a fluid of known density and viscosity.

17.8 Diffusion in Liquids and Viscosity of Liquids

The concept of random motion resulting in particle diffusion was evidenced in the famous microscopy experiments of Robert Brown performed in 1827. [An excellent account of this work can be found in *The Microscope* 40 (1992), 235–241.] In these experiments, Brown took ~5-μm-diameter pollen grains suspended in water, and using a microscope he was able to see that the particles were "very evidently in motion." After performing experiments to show that the motion was not from convection or evaporation, Brown concluded that the motion was associated with the particle itself. This apparently random motion of the pollen grain is referred to as **Brownian motion,** and this motion is actually the diffusion of a large particle in solution.

Consider Figure 17.16 where a particle of mass m is embedded in a liquid having viscosity η. The motion of the particle is driven by collisions with liquid particles, which will provide a time-varying force, $F(t)$. We decompose $F(t)$ into its directional components and focus on the component in the x direction, $F_x(t)$. Motion of the particle in the x direction will result in a frictional force due to the liquid's viscosity:

$$F_{fr,x} = -fv_x = -f\left(\frac{dx}{dt}\right) \tag{17.48}$$

The negative sign in Equation (17.48) indicates that the frictional force is in opposition to the direction of motion. In Equation (17.48), f is referred to as the friction coefficient and is dependent on both the geometry of the particle and the viscosity of the fluid. The total force on the particle is simply the sum of the collisional and frictional forces:

$$F_{total,x} = F_x(t) + F_{fr,x} \tag{17.49}$$

$$m\left(\frac{d^2x}{dt^2}\right) = F_x(t) - f\left(\frac{dx}{dt}\right)$$

This differential equation was studied by Einstein in 1905, who demonstrated that when averaged over numerous fluid–particle collisions the average square displacement of the particle in a specific amount of time t is given by

$$\langle x^2 \rangle = \frac{2kTt}{f} \tag{17.50}$$

where k is Boltzmann's constant and T is temperature. If the particle undergoing diffusion is spherical, the frictional coefficient is given by

$$f = 6\pi\eta r \tag{17.51}$$

FIGURE 17.16

Illustration of Brownian motion. A spherical particle with radius r is embedded in a liquid of viscosity η. The particle undergoes collisions with solvent molecules (the red dot in the right-hand figure) resulting in a time-varying force, $F(t)$, that initiates particle motion. Particle motion is opposed by a frictional force that is dependent on the solvent viscosity.

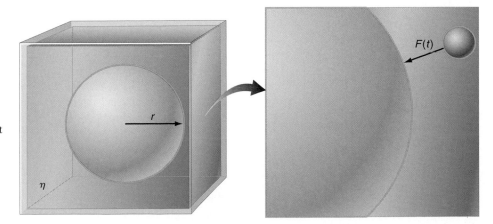

such that

$$\langle x^2 \rangle = 2\left(\frac{kT}{6\pi\eta r}\right)t \tag{17.52}$$

Comparison of this result to the value of $\langle x^2 \rangle$ determined by the diffusion equation [Equation (17.18)] dictates that the term in parentheses is the diffusion coefficient, D, given by

$$D = \frac{kT}{6\pi\eta r} \tag{17.53}$$

Equation (17.53) is the **Stokes–Einstein equation** for diffusion of a spherical particle. This equation states that the diffusion coefficient is dependent on the viscosity of the medium, the size of the particle, and the temperature. This form of the Stokes–Einstein equation is applicable to diffusion in solution when the radius of the diffusing particle is significantly greater than the radius of a liquid particle. For diffusion of a particle in a fluid of similar size, experimental data demonstrate that Equation (17.53) should be modified as follows:

$$D = \frac{kT}{4\pi\eta r} \tag{17.54}$$

Equation (17.54) can also be used for **self-diffusion,** in which the diffusion of a single solvent molecule in the solvent itself occurs.

EXAMPLE PROBLEM 17.8

Hemoglobin is a protein responsible for oxygen transport. The diffusion coefficient of human hemoglobin in water at 298 K ($\eta = 0.891$ cP) is 6.9×10^{-11} m^2 s^{-1}. Assuming this protein can be approximated as spherical, what is the radius of hemoglobin?

Solution

Rearranging Equation (17.53) and paying close attention to units, the radius is

$$r = \frac{kT}{6\pi\eta D} = \frac{(1.38 \times 10^{-23}\,\text{J K}^{-1})(298\,\text{K})}{6\pi(0.891 \times 10^{-3}\,\text{kg m}^{-1}\,\text{s}^{-1})(6.9 \times 10^{-11}\,\text{m}^2\,\text{s}^{-1})}$$

$$= 3.55\,\text{nm}$$

X-ray crystallographic studies of hemoglobin have shown that this globular protein can be approximated as a sphere having a radius of 2.75 nm. One reason for this discrepancy is that in aqueous solution, hemoglobin will have associated water, which is expected to increase the effective size of the particle relative to the crystallographic measurement.

Unlike gases, the dependence of fluid viscosity on temperature is difficult to explain theoretically. The primary reason for this is that, on average, the distance between particles in a liquid is small, and intermolecular interactions become important when describing particle interactions. As a general rule, the stronger the intermolecular interactions in a liquid, the greater the viscosity of the liquid. With respect to temperature, liquid viscosities increase as the temperature decreases. The reason for this behavior is that as the temperature is reduced, the kinetic energy of the particles is also reduced, and the particles have less kinetic energy to overcome the potential energy arising from intermolecular interactions, resulting in the fluid being more resistant to flow.

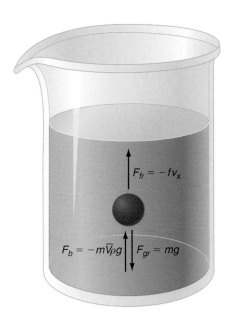

FIGURE 17.17
Illustration of the forces involved in sedimentation of a particle; F_{fr} is the frictional force, F_{gr} is the gravitational force, and F_b is the buoyant force.

SUPPLEMENTAL

17.9 Sedimentation and Centrifugation

An important application of transport phenomena in liquids is **sedimentation.** Sedimentation can be used with diffusion to determine the molecular weights of macromolecules. Figure 17.17 depicts a molecule with mass m undergoing sedimentation in a liquid of density ρ under the influence of the Earth's gravitational field. Three distinct forces are acting on the particle:

1. The frictional force: $F_{fr} = -fv_x$
2. The gravitational force: $F_{gr} = mg$
3. The buoyant force: $F_b = -m\bar{V}\rho g$

In the expression for the buoyant force, $\bar{V}$ is the **specific volume** of the solute, equal to the change in solution volume per mass of solute with units of $cm^3 \ g^{-1}$.

Imagine placing the particle at the top of the solution and then letting it fall. Initially, the downward velocity (v_x) is zero, but the particle will accelerate and the velocity will increase. Eventually, a particle velocity will be reached where the frictional and buoyant forces are balanced by the gravitational force. This velocity is known as the terminal velocity, and when this velocity is reached the particle acceleration is zero. Using Newton's second law,

$$F_{total} = ma = F_{fr} + F_{gr} + F_b \tag{17.55}$$

$$0 = -fv_{x,ter} + mg - m\bar{V}\rho g$$

$$v_{x,ter} = \frac{mg(1-\bar{V}\rho)}{f}$$

$$\bar{s} = \frac{v_{x,ter}}{g} = \frac{m(1-\bar{V}\rho)}{f}$$

The **sedimentation coefficient,** $\bar{s}$, is defined as the terminal velocity divided by the acceleration due to gravity. Sedimentation coefficients are generally reported in the units of Svedbergs (S) with $1 \ S = 10^{-13}$ s; however, we will use units of seconds to avoid confusion with other units in upcoming sections of this chapter.

Sedimentation is generally not performed using acceleration due to the Earth's gravity. Instead, acceleration of the particle is accomplished using a **centrifuge,** with ultracentrifuges capable of producing accelerations on the order of 10^5 times the acceleration due to gravity. In centrifugal sedimentation, the acceleration is equal to $\omega^2 x$ where ω is the angular velocity (radians s^{-1}) and x is the distance of the particle from the center of rotation. During centrifugal sedimentation, the particles will also reach a terminal velocity that depends on the acceleration due to centrifugation, and the sedimentation coefficient is expressed as

$$\bar{s} = \frac{v_{x,ter}}{\omega^2 x} = \frac{m(1-\bar{V}\rho)}{f} \tag{17.56}$$

EXAMPLE PROBLEM 17.9

The sedimentation coefficient of lysozyme (M = 14,100 g mol^{-1}) in water at 20°C is 1.91×10^{-13} s, and the specific volume is 0.703 cm^3 g^{-1}. The density of water at this temperature is 0.998 g cm^{-3} and $\eta = 1.002$ cP. Assuming lysozyme is spherical, what is the radius of this protein?

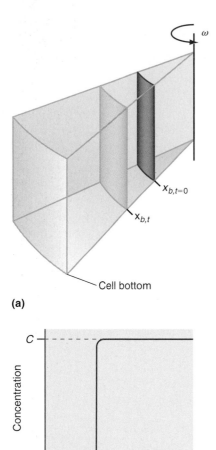

(a)

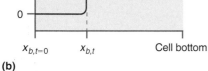

(b)

FIGURE 17.18

Determination of sedimentation coefficient by centrifugation. **(a)** Schematic drawing of the centrifuge cell, which is rotating with angular velocity ω. The blue plane at $x_{b,t=0}$ is the location of the solution meniscus before centrifugation. As the sample is centrifuged, a boundary between the solution with increased molecular concentration versus the solvent is produced. This boundary represented by the yellow plane at $x_{b,t}$. **(b)** As centrifugation proceeds, the boundary layer will move toward the cell bottom. A plot of $\ln(x_{b,t}/x_{b,t=0})$ versus time will yield a straight line with slope equal to ω^2 times the sedimentation coefficient.

Solution

The frictional coefficient is dependent on molecular radius. Solving Equation (17.56) for f, we obtain

$$f = \frac{m(1-\bar{V}\rho)}{\bar{s}} = \frac{\dfrac{(14,100 \text{ g mol}^{-1})}{(6.022\times10^{23} \text{ mol}^{-1})}(1-(0.703 \text{ mL g}^{-1})(0.998 \text{ g mL}^{-1}))}{1.91\times10^{-13} \text{ s}}$$

$$= 3.66\times10^{-8} \text{ g s}^{-1}$$

The frictional coefficient is related to the radius for a spherical particle by Equation (17.51) such that

$$r = \frac{f}{6\pi\eta} = \frac{3.66\times10^{-8}\text{g s}^{-1}}{6\pi(1.002 \text{ g m}^{-1}\text{ s}^{-1})} = 1.94\times10^{-9} \text{ m} = 1.93 \text{ nm}$$

One method for measurement of macromolecular sedimentation coefficients is by centrifugation, as illustrated in Figure 17.18. In this process, an initially homogeneous solution of macromolecules is placed in a centrifuge and spun. Sedimentation occurs, resulting in regions of the sample farther away from the axis of rotation experiencing an increase in macromolecule concentrations, and a corresponding reduction in concentration for the sample regions closest to the axis of rotation. A boundary between these two concentration layers will be established, and this boundary will move away from the axis of rotation with time in the centrifuge. If we define the x_b as the midpoint of the **boundary layer** (Figure 17.18), the following relationship exists between the location of the boundary layer and centrifugation time:

$$\bar{s} = \frac{V_{x,ter}}{\omega^2 x} = \frac{\dfrac{dx_b}{dt}}{\omega^2 x_b}$$

$$\omega^2 \bar{s}\int_0^t dt = \int_{x_{b,t=0}}^{x_{b,t}} \frac{dx_b}{x_b}$$

$$\omega^2 \bar{s}t = \ln\left(\frac{x_{b,t}}{x_{b,t=0}}\right) \tag{17.57}$$

Equation (17.57) suggests that a plot of $\ln(x_b/x_{b,t=0})$ versus time will yield a straight line with slope equal to ω^2 times the sedimentation coefficient. The determination of sedimentation coefficients by boundary centrifugation is illustrated in Example Problem 17.10.

EXAMPLE PROBLEM 17.10

The sedimentation coefficient of lysozyme is determined by centrifugation at 55,000 rpm in water at 20°C. The following data were obtained regarding the location of the boundary layer as a function of time:

Time (min)	x_b (cm)
0	6.00
30	6.07
60	6.14
90	6.21
120	6.28
150	6.35

Using these data, determine the sedimentation coefficient of lysozyme in water at 20°C.

Solution

First, we transform the data to determine $\ln(x_b/x_{b,t=0})$ as a function of time:

Time (min)	x_b (cm)	$(x_b/x_{b,t=0})$	$\ln(x_b/x_{b,t=0})$
0	6.00	1	0
30	6.07	1.01	0.00995
60	6.14	1.02	0.01980
90	6.21	1.03	0.02956
120	6.28	1.04	0.03922
150	6.35	1.05	0.04879

The plot of $\ln(x_b/x_{b,t=0})$ versus time is shown here:

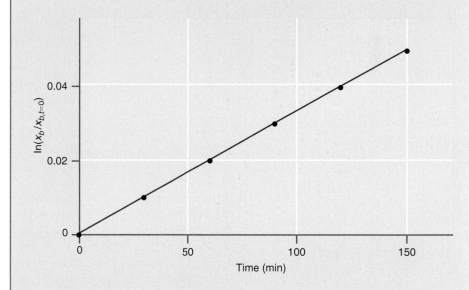

The slope of the line in the preceding plot is 3.75×10^{-4} min^{-1}, which is equal to ω^2 times the sedimentation coefficient:

$$3.75\times10^{-4}\ \mathrm{min}^{-1} = 6.25\times10^{-6}\ \mathrm{s}^{-1} = \omega^2\overline{s}$$

$$\overline{s} = \frac{6.25\times10^{-6}\ \mathrm{s}^{-1}}{\omega^2} = \frac{6.25\times10^{-6}\ \mathrm{s}^{-1}}{((55{,}000\ \mathrm{rev\ min}^{-1})(2\pi\ \mathrm{rad\ rev}^{-1})(0.0167\ \mathrm{min\ s}^{-1}))^2}$$

$$\overline{s} = 1.88\times10^{-13}\mathrm{s}$$

Finally with knowledge of the sedimentation coefficient and diffusion coefficient, the molecular weight of a macromolecule can be determined. Equation (17.56) can be rearranged to isolate the frictional coefficient:

$$f = \frac{m(1-\overline{V}\rho)}{\overline{s}} \tag{17.58}$$

The frictional coefficient can also be expressed in terms of the diffusion coefficient [Equations (17.51) and (17.53)]:

$$f = \frac{kT}{D} \tag{17.59}$$

Setting Equations (17.58) and (17.59) equal, the weight of the molecule is given by

$$m = \frac{kT\,\bar{s}}{D(1-\bar{V}\rho)} \quad \text{or} \quad M = \frac{RT\,\bar{s}}{D(1-\bar{V}\rho)} \tag{17.60}$$

Equation (17.60) dictates that with the sedimentation and diffusion coefficients, as well as the specific volume of the molecule, the molecular weight of a macromolecule can be determined.

17.10 Ionic Conduction

Ionic conduction is a transport phenomenon in which electrical charge in the form of electrons or ions migrates under the influence of an electrical potential. The amount of charge that migrates is equal to the electrical current, I, which is defined as the amount of charge, Q, migrating in a given time interval:

$$I = \frac{dQ}{dt} \tag{17.61}$$

The unit used for current is the ampere (A), which is equal to one coulomb of charge per second:

$$1\,A = 1\,C\,s^{-1} \tag{17.62}$$

Recall that the charge on an electron is 1.60×10^{-19} C. Therefore, 1.00 A of current corresponds to the flow of 6.25×10^{18} electrons in one second. Current can also be quantified by the current density, j, defined as the amount of current that flows through a conductor of cross-sectional area A (Figure 17.19):

$$j = \frac{I}{Area} \tag{17.63}$$

In the remainder of this section, the cross-sectional area will be written as *Area* rather than A to avoid confusion with amperes.

The migration of charge is initiated by the presence of an electrical force, created by an electrical field. In Figure 17.19, the electric field across the conductor is generated using a battery. The current density is proportional to the magnitude of the electric field, E, and the proportionality constant is known as the **electrical conductivity,** κ:

$$j = \kappa E \tag{17.64}$$

FIGURE 17.19
Cross section of a current-carrying conductor. The conductor cross-sectional area is indicated by the shaded area. The direction of the applied electric field, E_x, created by the battery is also shown.

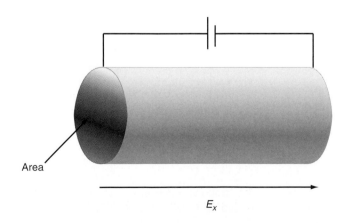

Area

E_x

Unfortunately, by convention thermal conductivity and electrical conductivity are both denoted by the symbol κ. Therefore, we will use the terms *thermal* or *electrical* when discussing conductivity to avoid confusion. Electrical conductivity has units of siemens per meter, or S m^{-1}, where 1 S equals 1 Ω^{-1} (ohm^{-1}). Finally, the **resistivity, ρ,** of a material is the inverse of the electrical conductivity:

$$\rho = \frac{1}{\kappa} = \frac{E}{j} \tag{17.65}$$

Resistivity is expressed in units of Ω m. Consider Figure 17.19 in which an electric field E is applied to a cylindrical conductor with a cross-sectional area and length l. If the cylinder is homogeneous in composition, the electric field and current density will be equivalent throughout the conductor allowing us to write:

$$E = \frac{V}{l} \tag{17.66}$$

$$j = \frac{I}{Area} \tag{17.67}$$

In Equation (17.66), V is the magnitude of the electric potential difference between the ends of the conductor, commonly referred to as the voltage. Substituting Equations (17.66) and (17.67) into the expression for resistivity, Equation (17.65), yields the following:

$$\rho = \frac{1}{\kappa} = \left(\frac{V}{I}\right)\frac{Area}{l} = R\frac{Area}{l} \tag{17.68}$$

In Equation (17.68), Ohm's law was used to rewrite V/I as R, the resistance. This equation demonstrates that the resistivity is proportional to the resistance of the material, R, in units of Ω with a proportionality constant that is equal to the area A of the conductor cross section divided by the conductor length l.

How is electrical conductivity related to transport? Again, consider the application of a battery to a conductor as shown in Figure 17.19. The battery will produce an electric field inside the conductor, and the direction of the electric field is taken to be the x direction. The electric field is related to the gradient in the electrical potential as follows:

$$E_x = -\frac{d\phi}{dx} \tag{17.69}$$

Using this definition, Equation (17.64) becomes

$$\frac{1}{Area}\frac{dQ}{dt} = -\kappa\frac{d\phi}{dx} \tag{17.70}$$

The left-hand side of Equation (17.70) is simply the flux of charge through the conductor in response to the gradient in electrical potential. Comparison of Equation (17.70) to Equation (17.1) demonstrates that ionic conductivity is described exactly as other transport processes. The proportionality constant between the potential gradient and flux in charge is the electrical conductivity.

Insight into the nature of charge transport in solution can be obtained by measuring the conductivity of ion-containing solutions. Measurements are usually performed using a conductivity cell, which is simply a container in which two electrodes of well-defined area and spacing are placed, and into which a solution of electrolyte is added (Figure 17.20). The resistance of the conductivity cell is generally measured using alternating currents to limit the buildup of electrolysis products at the electrodes. The resistance of the cell is measured using a Wheatstone bridge as illustrated in Figure 17.20. The operating principle behind this circuit is that when the two "arms" of the bridge have equal resistance, no current will flow between points A and B. In the

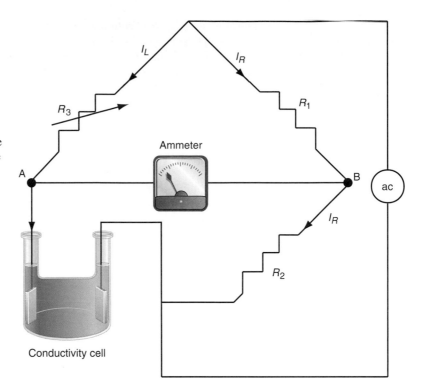

experiment, the resistance of R_3 is varied until this occurs. Under these conditions, the resistance of the cell is given by

$$R_{cell} = \frac{R_2 R_3}{R_1} \tag{17.71}$$

If the resistance of the cell is measured, the electrical conductivity of the solution can be determined using Equation (17.68).

The electrical conductivity of a solution depends on the number of ions present. The conductivity of ionic solutions is expressed as the **molar conductivity,** Λ_m, which is defined as follows:

$$\Lambda_m = \frac{\kappa}{c} \tag{17.72}$$

where c is the molar concentration of electrolyte. The molar conductivity has units of $S \, m^2 \, mol^{-1}$. If κ is linearly proportional to electrolyte concentration, then the molar conductivity will be independent of concentration. Figure 17.21 presents the measured molar conductivity for a variety of electrolytes as a function of concentration in water at 25°C. Two trends are immediately evident. First, none of the electrolytes demonstrates a concentration-independent molar conductivity. Second, the concentration dependence of the various species indicates that electrolytes can be divided into two categories. In the first category are **strong electrolytes,** which demonstrate a modest decrease in molar conductivity with an increase in concentration. In Figure 17.21, HCl, NaCl, KCl, and $CuSO_4$ are all strong electrolytes. In the second category are **weak electrolytes** represented by CH_3COOH in Figure 17.21. The molar conductivity of a weak electrolyte is comparable to a strong electrolyte at low concentration, but decreases rapidly with increased concentration. Note that a given species can behave as either a weak or strong electrolyte depending on the solvent employed. In the remainder of this section, we discuss the underlying physics behind the conductivity of strong and weak electrolytes.

FIGURE 17.21

Molar conductivity, Λ_m, of various compounds as a function of concentration, c. Measurements were performed in water at 25°C. Notice the mild reduction in Λ_m as a function of concentration for HCl, KCl, NaCl, and $CuSO_4$. This behavior is characteristic of a strong electrolyte. In contrast, CH_3COOH demonstrates a substantial reduction in Λ_m with increased concentration, characteristic of a weak electrolyte.

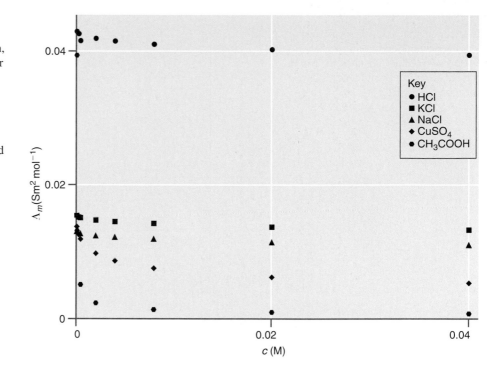

17.10.1 Strong Electrolytes

The central difference between weak and strong electrolytes is the extent of ion formation in solution. Strong electrolytes, including ionic solids (e.g., KCl) and strong acids/bases (e.g., HCl), exist as ionic species in solution and the electrolyte concentration is directly related to the concentration of ionic species in solution.

In 1900, Kohlrausch demonstrated that the concentration dependence of the molar conductivity for strong electrolytes is well described by the following equation:

$$\Lambda_m = \Lambda_m^0 - K\sqrt{\frac{c}{c_0}} \qquad (17.73)$$

This equation, known as **Kohlrausch's law,** states that the molar conductivity for strong electrolytes decreases as the square root of concentration, with slope given by the quantity K. The slope is found to depend more on the stoichiometry of the electrolyte (AB, AB_2, etc.) than composition. In Equation (17.73), Λ_m^0 is the molar conductivity at infinite dilution and represents the expected molar conductivity if interionic interactions were entirely absent. Clearly, any measurement at "infinite" dilution is impossible to perform; however, Kohlrausch's law demonstrates that Λ_m^0 can be readily determined by plotting Λ_m as a function of the square root of concentration (referenced to a standard concentration, c_0, generally 1 M), and then extrapolating to $c = 0$. Figure 17.22 presents the molar conductivity of HCl, KCl, and NaCl originally presented in Figure 17.21, but now shown with Λ_m plotted as a function of $(c/c_0)^{1/2}$. The solid line is a fit to the data using Kohlrausch's law, and it demonstrates the ability of this equation to reproduce the concentration dependence of Λ_m for strong electrolytes.

Kohlrausch was also able to demonstrate that for strong electrolytes, the molar conductivity at infinite dilution can be described in terms of contributions from the individual ionic constituents as follows:

$$\Lambda_m^0 = v_+\lambda_+ + v_-\lambda_- \qquad (17.74)$$

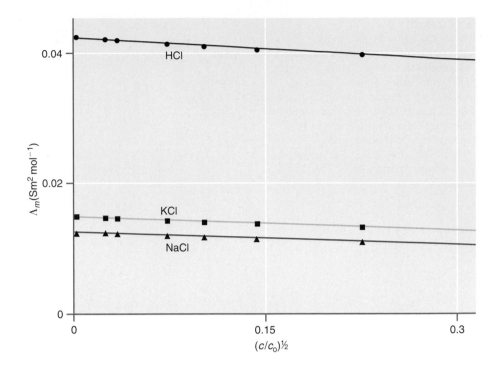

In Equation (17.74), v_+ and v_- are the number of positively and negatively charged species, respectively, in the formula weight of the electrolyte. For example, $v_+ = v_- = 1$ for NaCl, corresponding to Na^+ and Cl^-. The **ionic equivalent conductance** of the cations and anions is denoted by λ_+ and λ_-, respectively, and represents the conductivity of an individual ionic constituent of the electrolyte. Equation (17.74) is known as the **law of independent migration of ions** and states that under dilute conditions the molecular conductivity of an electrolyte is equal to the sum of the conductivities of the ionic constituents. Table 17.2 provides the molar conductivities of various ions in aqueous solution.

EXAMPLE PROBLEM 17.11

What is the expected molar conductivity at infinite dilution for $MgCl_2$?

Solution
Using Equation (17.44) and the data provided in Table 17.2:

$$\Lambda_m^0(MgCl_2) = 1\lambda(Mg^{2+}) + 2\lambda(Cl^-)$$
$$= (0.0106 \text{ S m}^2 \text{ mol}^{-1}) + 2(0.0076 \text{ S m}^2 \text{ mol}^{-1})$$
$$= 0.0258 \text{ S m}^2 \text{ mol}^{-1}$$

As determined by a plot of conductivity versus $(c/c_0)^{1/2}$, Λ_m^0 for $MgCl_2$ is equal to 0.212 S m^2 mol^{-1}. Comparison of this value to the expected molar conductivity calculated earlier demonstrates that $MgCl_2$ behaves as a strong electrolyte in aqueous solution.

What is the origin of the $c^{1/2}$ dependence stated in Kohlrausch's law? In Sections 10.3 and 10.4, it was shown that the activity of electrolytes demonstrates a similar concentration dependence, and this dependence was related to dielectric screening as expressed by the Debye–Hückel limiting law. This law states that the natural log of the ion activity is proportional to the square root of the solution ionic strength, or ion concentration. An ion in solution is characterized not only by the ion itself, but also by the ionic atmosphere surrounding the ion as discussed in Chapter 10,

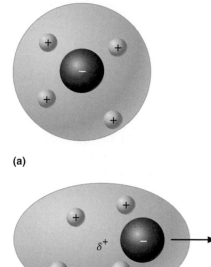

(a)

(b)

FIGURE 17.23
Negative ion in solution and associated ionic atmosphere. **(a)** In the absence of an electric field, the center of the positive and negative charge is identical. **(b)** However, when an electric field is present, the motion of the negative ion results in displacement of the center of the positive and negative charge. This local electric field is in opposition to the applied electric field such that the rate of ion migration is reduced.

TABLE 17.2

Ionic Equivalent Conductance Values for Representative Ions

Ion	λ (S m^2 mol^{-1})	Ion	λ (S m^2 mol^{-1})
H$^+$	0.0350	OH$^-$	0.0199
Na$^+$	0.0050	Cl$^-$	0.0076
K$^+$	0.0074	Br$^-$	0.0078
Mg^{+2}	0.0106	F$^-$	0.0054
Cu^{+2}	0.0107	NO$_3^-$	0.0071
Ca^{+2}	0.0119	CO$_3^{2-}$	0.0139

and depicted in Figure 17.23. In the figure, a negative ion is surrounded by a few close-lying positive ions comprising the ionic atmosphere. In the absence of an electric field, the ionic atmosphere will be spherically symmetric, and the centers of positive and negative charge will be identical. However, application of a field will result in ion motion, and this motion will create two effects. First, consider the situation where the negative ion in Figure 17.23 migrates and the atmosphere follows this migrating ion. The ionic atmosphere is not capable of responding to the motion of the negative ion instantaneously, and the lag in response will result in displacement of the centers of positive and negative charge. This displacement will give rise to a local electric field in opposition to the applied field. Correspondingly, the migration rate of the ion, and subsequently the conductivity, will be reduced. This effect is called the **relaxation effect,** and this term implies that a time dependence is associated with this phenomenon. In this case, it is the time it takes the ionic atmosphere to respond to the motion of the negative ion.

Information regarding the timescale for ionic atmosphere relaxation can be gained by varying the frequency of the ac electric field to study ionic conductivity. A second effect that reduces the migration rate of the ion is the **electrophoretic effect.** The negatively and positively charged ions will migrate in opposite directions such that the viscous drag the ion experiences will increase. The reduction in ion mobility accompanying the increase in viscous drag will decrease the conductivity.

17.10.2 Weak Electrolytes

Weak electrolytes undergo only fractional conversion into their ionic constituents in solution. Acetic acid, a weak acid, is a classic example of a weak electrolyte. Weak acids are characterized by incomplete conversion to H_3O^+ and the conjugate base. The dependence of Λ_m on concentration for a weak electrolyte can be understood by considering the extent of ionization. For the case of a weak acid (HA) with dissociation in water to form H_3O^+ and the conjugate base (A^-), incomplete ionization can be factored into the ionic and electrolyte concentrations as follows:

$$[H_3O^+] = [A^-] = \alpha c \tag{17.75}$$

$$[HA] = (1-\alpha)c \tag{17.76}$$

In Equations (17.75) and (17.76), α is the degree of ionization (ranging in value from 0 to 1), and c is the initial concentration of electrolyte. Substituting Equations (17.75) and (17.76) into the expression for the equilibrium constant of a weak acid results in Equation (17.77):

$$K_a = \frac{\alpha^2 c}{(1-\alpha)} \tag{17.77}$$

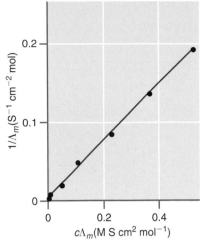

FIGURE 17.24

Comparison of experimental and predicted conductivity of acetic acid (CH_3COOH) as a function of concentration using the Ostwald dilution law of Equation (17.80). The y intercept on the graph is equal to $1/\Lambda_m^0$.

Solving Equation (17.77) for α yields

$$\alpha = \frac{K_a}{2c}\left(\left(1+\frac{4c}{K_a}\right)^{1/2}-1\right) \qquad (17.78)$$

The molar conductivity of a weak electrolyte at a given electrolyte concentration, Λ_m, will be related to the molar conductivity at infinite dilution, Λ_m^0, by

$$\Lambda_m = \alpha\Lambda_m^0 \qquad (17.79)$$

Using Equations (17.78) and (17.79), the concentration dependence of the molar conductivity can be determined and compared to experiment. The relationship between Λ_m and c for a weak electrolyte is described by the **Ostwald dilution law:**

$$\frac{1}{\Lambda_m} = \frac{1}{\Lambda_m^0} + \frac{c\Lambda_m}{K_a(\Lambda_m^0)^2} \qquad (17.80)$$

The comparison of the predicted behavior given by Equation (17.80) to experiment for acetic acid ($K_a = 1.8\times10^{-5}$, $\Lambda_m^0 = 003957$ S m^{-2} mol^{-1}) is presented in Figure 17.24, and excellent agreement is observed. The Ostwald dilution law can be used to determine the molar conductivity at infinite dilution for a weak acid.

For Further Reading

Bird, R. B., W. E. Stewart, and E. N. Lightfoot, *Transport Phenomena.* Wiley, New York, 1960.

Cantor, C. R., and P. R. Schimmel, *Biophysical Chemistry. Part II: Techniques for the Study of Biological Structure and Function.* W. H. Freeman, San Francisco, 1980.

Castellan, G. W., *Physical Chemistry.* Addison-Wesley, Reading, MA, 1983.

Crank, J., *The Mathematics of Diffusion.* Clarendon Press, Oxford, 1975.

Hirschfelder, J. O., C. F. Curtiss, and R. B. Bird, *The Molecular Theory of Gases and Liquids.* Wiley, New York, 1954.

Reid, R. C., J. M. Prausnitz, and T. K. Sherwood, *The Properties of Gases and Liquids.* McGraw-Hill, New York, 1977.

Vargaftik, N. B., *Tables on the Thermophysical Properties of Liquids and Gases.* Wiley, New York, 1975.

Welty, J. R., C. E. Wicks, and R. E. Wilson, *Fundamentals of Momentum, Heat, and Mass Transfer.* Wiley, New York, 1969.

Vocabulary

boundary layer

Brownian motion

centrifuge

diffusion

diffusion coefficient

diffusion equation

Einstein–Smoluchowski equation

electrical conductivity

electrophoretic effect

Fick's first law

Fick's second law of diffusion

flux

ionic conduction

ionic equivalent conductance

Kohlrausch's law

laminar flow

law of independent migration of ions

mass transport

molar conductivity

Ostwald dilution law

Ostwald viscometer

Poiseuille's law

random walk

relaxation effect

resistivity

sedimentation

sedimentation coefficient

self-diffusion

specific volume

Stokes–Einstein equation

strong electrolytes

thermal conduction

thermal conductivity

transport coefficient

transport phenomena

turbulent flow

viscosity

weak electrolytes

Questions on Concepts

Q17.1 What is the general relationship between the spatial gradient in a system property and the flux of that property?

Q17.2 What is the expression for the diffusion coefficient, D, in terms of gas kinetic theory parameters? How is D expected to vary with an increase in molecular mass or collisional cross section?

Q17.3 Particles are confined to a plane and then allowed to diffuse. How does the number density vary with distance away from the initial plane?

Q17.4 How does the root-mean-square diffusion distance vary with the diffusion coefficient? How does this quantity vary with time?

Q17.5 What is the expression for thermal conductivity in terms of particle parameters derived from gas kinetic theory?

Q17.6 Why is the thermal conductivity for an ideal gas expected to be independent of pressure? Why does the thermal conductivity for an ideal gas increase as $T^{1/2}$?

Q17.7 In describing viscosity, what system quantity was transported? What is the expression for viscosity in terms of particle parameters derived from gas kinetic theory?

Q17.8 What observable is used to measure the viscosity of a gas or liquid?

Q17.9 What is Brownian motion?

Q17.10 In the Stokes–Einstein equation that describes particle diffusion for a spherical particle, how does the diffusion coefficient depend on fluid viscosity and particle size?

Q17.11 What is the difference between a strong and weak electrolyte?

Q17.12 According to Kohlrausch's law, how will the molar conductivity for a strong electrolyte change with concentration?

Problems

Problem numbers in **red** indicate that the solution to the problem is given in the *Student's Solutions Manual*.

P17.1 The diffusion coefficient for CO_2 at 273 K and 1 atm is 1.00×10^{-5} m^2 s^{-1}. Estimate the collisional cross section of CO_2 given this diffusion coefficient.

P17.2

a. The diffusion coefficient for Xe at 273 K and 1 atm is 0.5×10^{-5} m^2 s^{-1}. What is the collisional cross section of Xe?

b. The diffusion coefficient of N_2 is threefold greater than that of Xe under the same pressure and temperature conditions. What is the collisional cross section of N_2?

P17.3

a. The diffusion coefficient of sucrose in water at 298 K is $0.522 \ 10^{-9}$ m^2 s^{-1}. Determine the time it will take a sucrose molecule on average to diffuse an rms distance of 1 mm.

b. If the molecular diameter of sucrose is taken to be 0.8 nm, what is the time per random walk step?

P17.4

a. The diffusion coefficient of the protein lysozyme (MW = 14.1 kg/mol) is 0.104×10^{-5} cm^2 s^{-1}. How long will it take this protein to diffuse an rms distance of 1 μm? Model the diffusion as a three-dimensional process.

b. You are about to perform a microscopy experiment in which you will monitor the fluorescence from a single lysozyme molecule. The spatial resolution of the microscope is 1 μm. You intend to monitor the diffusion using a camera that is capable of one image every 60 s. Is the imaging rate of the

camera sufficient to detect the diffusion of a single lysozyme protein over a length of 1 μm?

c. Assume that in the microscopy experiment of part (b) you use a thin layer of water such that diffusion is constrained to two dimensions. How long will it take a protein to diffuse an rms distance of 1 μm under these conditions?

P17.5 A solution consisting of 1 g of sucrose in 10 mL of water is poured into a 1-L graduated cylinder with a radius of 2.5 cm. Then the cylinder is filled with pure water.

a. The diffusion of sucrose can be considered diffusion in one dimension. Derive an expression for the average distance of diffusion, x_{ave}.

b. Determine x_{ave} and x_{rms} for sucrose for time periods of 1 s, 1 min, and 1 h.

P17.6 A thermopane window consists of two sheets of glass separated by a volume filled with air (which we will model as N_2 where $\kappa = 0.0240$ J K^{-1} m^{-1} s^{-1}). For a thermopane window that is 1 m^2 in area with a separation between glass sheets of 3 cm, what is the loss of energy when:

a. the exterior of the window is at a temperature of 10°C and the interior of the window is at a temperature of 22°C?

b. the exterior of the window is at a temperature of -20°C and the interior of the window is at a temperature of 22°C?

c. the same temperature differential as in part (b) is used, but the window is filled with Ar ($\kappa = 0.0163$ J K^{-1} m^{-1} s^{-1}) rather than N_2?

P17.7 Two parallel metal plates separated by 1 cm are held at 300 and 298 K, respectively. The space between the plates is filled with N_2 ($\sigma = 0.430$ nm² and $C_{V,m} = 5/2\ R$). Determine the heat flow between the two plates in units of W cm⁻².

P17.8 Determine the thermal conductivity of the following species at 273 K and 1 atm:

a. Ar ($\sigma = 0.36$ nm²)

b. Cl_2 ($\sigma = 0.93$ nm²)

c. SO_2 ($\sigma = 0.58$ nm², geometry: bent)

You will need to determine $C_{V,m}$ for the species listed. You can assume that the translational and rotational degrees of freedom are in the high-temperature limit, and that the vibrational contribution to $C_{V,m}$ can be ignored at this temperature.

P17.9 The thermal conductivity of Kr is 0.0087 J K⁻¹ m⁻¹ s⁻¹ at 273 K and 1 atm. Estimate the collisional cross section of Kr.

P17.10 The thermal conductivity of Kr is roughly half that of Ar under identical pressure and temperature conditions. Both gases are monatomic such that $C_{V,m} = 3/2\ R$.

a. Why would one expect the thermal conductivity of Kr to be less than that of Ar?

b. Determine the ratio of collisional cross sections for Ar relative to Kr assuming identical pressure and temperature conditions.

c. For Kr at 273 K at 1 atm, $\kappa = 0.0087$ J K⁻¹ m⁻¹ s⁻¹. Determine the collisional cross section of Kr.

P17.11

a. Determine the ratio of thermal conductivity for N_2 ($\sigma = 0.43$ nm²) at sea level ($T = 300$ K, $P = 1$ atm) versus the lower stratosphere ($T = 230$ K, $P = 0.25$ atm).

b. Determine the ratio of thermal conductivity for N_2 at sea level if $P = 1$ atm, but the temperature is 100 K. Which energetic degrees of freedom will be operative at the lower temperature, and how will this affect $C_{V,m}$?

P17.12 The thermal conductivities of acetylene (C_2H_2) and N_2 at 273 K and 1 atm are 0.01866 and 0.0240 J m⁻¹ s⁻¹ K⁻¹, respectively. Based on these data, what is the ratio of the collisional cross section of acetylene relative to N_2?

P17.13

a. The viscosity of Cl_2 at 293 K and 1 atm is 132 μP. Determine the collisional cross section of this molecule based on the viscosity.

b. Given your answer in part (a), estimate the thermal conductivity of Cl_2 under the same pressure and temperature conditions.

P17.14

a. The viscosity of O_2 at 293 K and 1 atm is 204 μP. What is the expected flow rate through a tube having a radius of 2 mm, length of 10 cm, input pressure of 765 Torr, output

pressure of 760 Torr, with the flow measured at the output end of the tube?

b. If Ar were used in the apparatus ($\eta = 223\ \mu P$) of part (a), what would be the expected flow rate? Can you determine the flow rate without evaluating Poiseuille's equation?

P17.15 The Reynolds' number (Re) is defined as $Re = \rho \langle v_x \rangle d / \eta$, where ρ and η are the fluid density and viscosity, respectively; d is the diameter of the tube through which the fluid is flowing; and $\langle v_x \rangle$ is the average velocity. Laminar flow occurs when Re < 2000, the limit in which the equations for gas viscosity were derived in this chapter. Turbulent flow occurs when Re > 2000. For the following species, determine the maximum value of $\langle v_x \rangle$ for which laminar flow will occur:

a. Ne at 293 K ($\eta = 313\ \mu P$, $\rho =$ that of an ideal gas) through a 2-mm-diameter pipe.

b. Liquid water at 293 K ($\eta = 0.891$ cP, $\rho = 0.998$ g mL⁻¹) through a 2-mm-diameter pipe.

P17.16 The viscosity of H_2 at 273 K at 1 atm is 84 μP. Determine the viscosities of D_2 and HD.

P17.17 An Ostwald viscometer is calibrated using water at 20°C ($\eta = 1.0015$ cP, $\rho = 0.998$ g mL⁻¹). It takes 15 s for the fluid to fall from the upper to the lower level of the viscometer. A second liquid is then placed in the viscometer and it takes 37 s for the fluid to fall between the levels. Finally, 100 mL of the second liquid weighs 76.5 g. What is the viscosity of the liquid?

P17.18 How long will it take to pass 200 mL of H_2 at 273 K through a 10-cm-long capillary tube of 0.25 mm if the gas input and output pressures are 1.05 and 1.00 atm, respectively?

P17.19

a. Derive the general relationship between the diffusion coefficient and viscosity for a gas.

b. Given that the viscosity of Ar is 223 μP at 293 K and 1 atm, what is the diffusion coefficient?

P17.20

a. Derive the general relationship between the thermal conductivity and viscosity.

b. Given that the viscosity of Ar is 223 μP at 293 K and 1 atm, what is the thermal conductivity?

c. Given that the viscosity of Ar is 223 μP at 293 K and 1 atm, what is the thermal conductivity of Ne under these same conditions?

P17.21 As mentioned in the text, the viscosity of liquids decreases with increasing temperature. The empirical equation $\eta(T) = Ae^{E/RT}$ provides the relationship between viscosity and temperature for a liquid. In this equation, A and E are constants, with E being referred to as the activation energy for flow.

a. How can one use the equation provided to determine A and E given a series of viscosity versus temperature measurements?

b. Use your answer in part (a) to determine A and E for liquid benzene given the following data:

T (°C)	η (cP)
5	0.826
40	0.492
80	0.318
120	0.219
160	0.156

P17.22 Myoglobin is a protein that participates in oxygen transport. For myoglobin in water at 20°C, $\bar{s} = 2.04 \times 10^{-13}$ s, $D = 1.13 \times 10^{-11}$ m² s⁻¹, and $\bar{V} = 0.740$ cm³ g⁻¹. The density of water is 0.998 g cm³ and the viscosity is 1.002 cP at this temperature.

a. Using the information provided, estimate the size of myoglobin.

b. What is the molecular weight of myoglobin?

P17.23 You are interested in purifying a sample containing the protein alcohol dehydrogenase obtained from horse liver; however, the sample also contains a second protein, catalase. These two proteins have the following transport properties:

	Catalase	Alcohol Dehydrogenase
$\bar{s}$ (s)	11.3×10^{-13}	4.88×10^{-13}
D (m² s⁻¹)	4.1×10^{-11}	6.5×10^{-11}
$\bar{V}$ (cm³ g⁻¹)	0.715	0.751

a. Determine the molecular weight of catalase and alcohol dehydrogenase.

b. You have access to a centrifuge that can provide angular velocities up to 35,000 rpm. For the species you expect to travel the greatest distance in the centrifuge tube, determine the time it will take to centrifuge until a 3-cm displacement of the boundary layer occurs relative to the initial 5-cm location of the boundary layer relative to the centrifuge axis.

c. To separate the proteins, you need a separation of at least 1.5 cm between the boundary layers associated with each protein. Using your answer to part (b), will it be possible to separate the proteins by centrifugation?

P17.24 Boundary centrifugation is performed at an angular velocity of 40,000 rpm to determine the sedimentation coefficient of cytochrome c (M = 13,400 g mol⁻¹) in water at 20°C ($\rho = 0.998$ g cm³, $\eta = 1.002$ cP). The following data are obtained on the position of the boundary layer as a function of time:

Time (h)	x_b (cm)
0	4.00
2.5	4.11
5.2	4.23
12.3	4.57
19.1	4.91

a. What is the sedimentation coefficient for cytochrome c under these conditions?

b. The specific volume of cytochrome c is 0.728 cm³ g⁻¹. Estimate the size of cytochrome c.

P17.25 A current of 2.00 A is applied to a metal wire for 30 s. How many electrons pass through a given point in the wire during this time?

P17.26 Use the following data to determine the conductance at infinite dilution for $NaNO_3$:

$$\Lambda_m^0 (KCl) = 0.0149 \text{ S m}^2 \text{ mol}^{-1}$$
$$\Lambda_m^0 (NaCl) = 0.0127 \text{ S m}^2 \text{ mol}^{-1}$$
$$\Lambda_m^0 (KNO_3) = 0.0145 \text{ S m}^2 \text{ mol}^{-1}$$

P17.27 The following molar conductivity data are obtained for an electrolyte:

Concentration (M)	Λ_m (S m² mol⁻¹)
0.0005	0.01245
0.001	0.01237
0.005	0.01207
0.01	0.01185
0.02	0.01158
0.05	0.01111
0.1	0.01067

Determine if the electrolyte is strong or weak, and the conductivity of the electrolyte at infinite dilution.

P17.28 The molar conductivity of sodium acetate, CH_3COONa, is measured as a function of concentration in water at 298 K, and the following data are obtained:

Concentration (M)	Λ_m (S m² mol⁻¹)
0.0005	0.00892
0.001	0.00885
0.005	0.00857
0.01	0.00838
0.02	0.00812
0.05	0.00769
0.1	0.00728

Is sodium acetate a weak or strong electrolyte? Determine Λ_m^0 using appropriate methodology depending on your answer.

P17.29 Starting with Equations (17.47) and (17.48), derive the Ostwald dilution law.

P17.30 For a one-dimensional random walk, determine the probability that the particle will have moved six steps in either the $+x$ or $-x$ direction after 10, 20, and 100 steps.

P17.31 In the early 1990s, fusion involving hydrogen dissolved in palladium at room temperature, or *cold fusion*, was proposed as a new source of energy. This process relies on the diffusion of H_2 into palladium. The diffusion of hydrogen gas through a 0.005-cm-thick piece of palladium foil with a cross section of 0.750 cm² is measured. On one side of the foil, a volume of gas maintained at 298 K and 1 atm is applied, while a vacuum is applied to the other side of the foil. After 24 h, the volume of hydrogen has decreased by 15.2 cm³. What is the diffusion coefficient of hydrogen gas in palladium?

P17.32 In the determination of molar conductivities, it is convenient to define the cell constant, K, as $K = l/A$, where l is the separation between the electrodes in the conductivity cell, and A is the area of the electrodes.

a. A standard solution of KCl (conductivity or $\kappa = 1.06296 \times 10^{-6}$ S m⁻¹ at 298 K) is employed to standardize the cell, and a resistance of 4.2156 Ω is measured. What is the cell constant?

b. The same cell is filled with a solution of HCl and a resistance of 1.0326 Ω is measured. What is the conductivity of the HCl solution?

P17.33 Conductivity measurements were one of the first methods used to determine the autoionization constant of water. The autoionization constant of water is given by the following equation:

$$K_w = a_{H^+}a_{OH^-} = \left(\frac{[H^+]}{1M}\right)\left(\frac{[OH^-]}{1M}\right)$$

where a is the activity of the species, which is equal to the actual concentration of the species divided by the standard state concentration at infinite dilution. This substitution of concentrations for activities is a reasonable approximation given the small concentrations of H^+ and OH^- that result from autoionization.

a. Using the expression provided, show that the conductivity of pure water can be written as.

$$\Lambda_m(H_2O) = (1M)K_w^{1/2}(\lambda(H^+) + \lambda(OH^-))$$

b. Kohlrausch and Heydweiller measured the conductivity of water in 1894 and determined that $\Lambda_m(H_2O) = 5.5 \times 10^{-6}$ S m⁻¹ at 298 K. Using the information in Table 17.2, determine K_w.

Web-Based Simulations, Animations, and Problems

W17.1 In this problem, concentration time dependence in one dimension is depicted as predicted using Fick's second law of diffusion. Specifically, variation of particle number density as a function of distance with time and diffusion constant D is investigated. Comparisons of the full distribution to x_{rms} are performed to illustrate the behavior of x_{rms} with time and D.

CHAPTER 18

Elementary Chemical Kinetics

In Chapters 18 and 19, the evolution of a system toward equilibrium with respect to chemical composition is discussed. The central question of interest in chemical kinetics is perhaps the first question one asks about any chemical reaction: just how do the reactants become products? In chemical kinetics, this question is answered by determining the timescale and mechanism of a chemical reaction. The basic tools of this field are presented in this chapter. First, reaction rates and methods for their determination are discussed. Next, the concept of reaction mechanisms in terms of elementary reaction steps is presented. Approaches to describing elementary reaction steps, including integrated rate law expressions and numerical methods, are outlined. Basic reaction types, such as unidirectional, sequential, and parallel reactions, are described. The possibility of back or reverse reactions is introduced, and these reactions are used to establish the kinetic definition of equilibrium. Finally, reaction dynamics are described by means of potential energy surfaces, and the concepts of activated complex theory are introduced. The ideas presented here provide a kinetic "toolkit" that will be employed to describe complex reactions in Chapter 19. ■

18.1 Introduction to Kinetics

In the previous chapter, the evolution of a system's physical properties toward equilibrium was discussed. In these transport phenomena, the system undergoes relaxation without a change in chemical composition. Transport phenomena are sometimes referred to as physical kinetics to indicate that the physical properties of the system are evolving and not the system composition. In this chapter and the next, we focus on chemical kinetics where the composition of the system evolves with time.

Chemical kinetics involves the study of the rates and mechanisms of chemical reactions. This area bridges an important gap in our discussion of chemical reactions to this point. Thermodynamic descriptions of chemical reactions involved the Gibbs or Helmholtz energy for a reaction and the corresponding equilibrium constant. These quantities are sufficient to predict the reactant and product concentrations at equilibrium, but are of little use in determining the timescale over which the reaction occurs. That is, thermodynamics may dictate that a reaction is spontaneous, but does it occur in 10^{-15} s (a femtosecond, the timescale for the fastest chemical processes known to date) or 10^{15} s (the age of the universe)? The answer to this question lies in the domain of chemical kinetics.

FIGURE 18.1

Concentration as a function of time for the conversion of reactant A into product B. The concentration of A at $t = 0$ is $[A]_0$, and the concentration of B is zero. As the reaction proceeds, the loss of A results in the production of B.

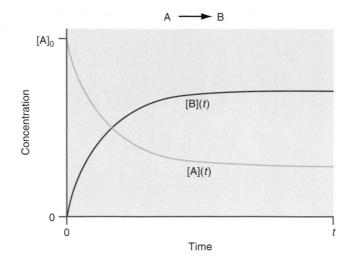

In the course of a chemical reaction, concentrations will change with time as "reactants" become "products." Figure 18.1 presents possibly the first chemical reaction you were introduced to in introductory chemistry: the conversion of reactant A into product B. The figure illustrates that, as the reaction proceeds, a decrease in reactant concentration and a corresponding increase in product concentration is observed. One way to describe this process is to define the rate of concentration change with time, a quantity that is referred to as the reaction rate.

The central ideal behind chemical kinetics is this: by monitoring the rate at which chemical reactions occur and determining the dependence of this rate on system parameters such as temperature, concentration, and pressure, we can gain insight into the mechanism of the reaction. Experimental chemical kinetics includes the development of techniques that allow for the study of chemical reactions including the measurement and analysis of chemical reaction dynamics. In addition to experiments, a substantial amount of theoretical work has been performed to understand reaction mechanisms and the underlying physics that govern the rates of chemical transformations. The synergy between experiment and theoretical chemical kinetics has provided for dramatic advances in this field. Finally, the importance of chemical kinetics is evidenced by its application in nearly every area of chemistry. Found in areas such as enzyme catalysis, materials processing, and atmospheric chemistry, chemical kinetics is clearly an important aspect of physical chemistry.

18.2 Reaction Rates

Consider the following "generic" chemical reaction:

$$a\text{A} + b\text{B} + ... \rightarrow c\text{C} + d\text{D} + ... \tag{18.1}$$

In Equation (18.1), uppercase letters indicate a chemical species and lowercase letters represent the stoichiometric coefficient for the species in the balanced reaction. The species on the left-hand and right-hand sides of the arrow are referred to as *reactants* and *products,* respectively. The number of moles of a species during any point of the reaction is given by

$$n_i = n_i^o + \nu_i \xi \tag{18.2}$$

where n_i is the number of moles of species i at any given time during the reaction, n_i^o is the number of moles of species i present before initiation of the reaction, ν_i is related to the stoichiometric coefficient of species i, and ξ represents the advancement of the reaction and is equal to zero at the start of the reaction. The advancement variable allows us to quantify the rate of the reaction with respect to all species, irrespective of stoichiometry (see later discussion). For the reaction depicted in Equation (18.1), reactants will be consumed and products formed during the reaction. To ensure that this behavior is reflected in Equation (18.2), v_i is set equal to -1 times the stoichiometric coefficient for reactants, and is set equal to the stoichiometric coefficient for products.

The time evolution of the reactant and product concentrations is quantified by differentiating both sides of Equation (18.2) with respect to time:

$$\frac{dn_i}{dt} = v_i \frac{d\xi}{dt} \tag{18.3}$$

The **reaction rate** is defined as the change in the advancement of the reaction with time:

$$Rate = \frac{d\xi}{dt} \tag{18.4}$$

With this definition, the rate of reaction with respect to the change in the number of moles of a given species with time is

$$Rate = \frac{1}{v_i} \frac{dn_i}{dt} \tag{18.5}$$

As an example of how the rate of reaction is defined relative to the change in moles of reactant or product with time, consider the following reaction:

$$4NO_2(g) + O_2(g) \rightarrow 2N_2O_5(g) \tag{18.6}$$

The rate of reaction can be expressed with respect to any species in Equation (18.6):

$$Rate = -\frac{1}{4} \frac{dn_{NO_2}}{dt} = -\frac{dn_{O_2}}{dt} = \frac{1}{2} \frac{dn_{N_2O_5}}{dt} \tag{18.7}$$

Notice the sign convention of the coefficient with respect to reactants and products: negative for reactants and positive for products. Also, notice that the rate of reaction can be defined with respect to both reactants and products. In our example, 4 mol of NO_2 react with 1 mol of O_2 to produce 2 mol of N_2O_5 product. Therefore, the **rate of conversion** of NO_2 will be four times greater than the rate of O_2 conversion. Although the conversion rates are different, the reaction rate defined with respect to either species will be the same. Furthermore, because both NO_2 and O_2 are reactants, the change in the moles of these species with respect to time is negative. However, by using a negative stoichiometric coefficient, the reaction rate defined with respect to the reactants is still a positive quantity.

In applying Equation (18.5) to define a rate of reaction, a set of stoichiometric coefficients must be employed; however, these coefficients are not unique. For example, if we multiply both sides of Equation (18.6) by a factor of 2, the expression for the rate of conversion must also change. Generally, one decides on a given set of coefficients for a balanced reaction and uses these coefficients consistently throughout a given kinetics problem.

In our present definition, the rate of reaction as written is an extensive property; therefore, it will depend on the system size. The rate can be made intensive by dividing Equation (18.5) by the volume of the system:

$$R - \frac{Rate}{V} = \frac{1}{V} \left(\frac{1}{v_i} \frac{dn_i}{dt} \right) = \frac{1}{v_i} \frac{d[i]}{dt} \tag{18.8}$$

In Equation (18.8), R is the intensive reaction rate. The last equality in Equation (18.8) is performed recognizing that moles of species i per unit volume is simply the molarity of species i, or $[i]$. Equation (18.8) is the definition for the rate of reaction at constant volume. For species in solution, the application of Equation (18.8) in defining the rate of reaction is clear, but it can also be used for gases, as Example Problem 18.1 illustrates.

EXAMPLE PROBLEM 18.1

The decomposition of acetaldehyde is given by the following balanced reaction:

$$CH_3COH(g) \rightarrow CH_4(g) + CO(g)$$

Define the rate of reaction with respect to the pressure of the reactant.

Solution

Beginning with Equation (18.2) and focusing on the acetaldehyde reactant, we obtain

$$n_{CH_3COH} = n^o_{CH_3COH} - \xi$$

Using the ideal gas law, the pressure of acetaldehyde is expressed as

$$P_{CH_3COH} = \frac{n_{CH_3COH}}{V} RT = [CH_3COH]RT$$

Therefore, the pressure is related to the concentration by the quantity RT. Substituting this result into Equation (18.8) with $v_i = 1$ yields

$$R = \frac{Rate}{V} = -\frac{1}{v_{CH_3COH}} \frac{d[CH_3COH]}{dt}$$

$$= -\frac{1}{RT} \frac{dP_{CH_3COH}}{dt}$$

18.3 Rate Laws

We begin our discussion of rate laws with a few important definitions. The rate of a reaction will generally depend on the temperature, pressure, and concentrations of species involved in the reaction. In addition, the rate may depend on the phase or phases in which the reaction occurs. Homogeneous reactions occur in a single phase, whereas heterogeneous reactions involve more than one phase. Reactions that involve a surface are classic examples of heterogeneous reactions. We will limit our initial discussion to homogeneous reactions, with heterogeneous reactivity discussed in Chapter 19. For the majority of homogeneous reactions, an empirical relationship between reactant concentrations and the rate of a chemical reaction can be written. This relationship is known as a **rate law,** and for the reaction shown in Equation (18.1) it is written as

$$R = k[A]^\alpha[B]^\beta... \tag{18.9}$$

where [A] is the concentration of reactant A, [B] is the concentration of reactant B, and so forth. The constant α is known as the **reaction order** with respect to species A, β the reaction order with respect to species B, and so forth. The overall reaction order is equal to the sum of the individual reaction orders ($\alpha + \beta + ...$). Finally, the constant k is referred to as the **rate constant** for the reaction. The rate constant is independent of concentration, but dependent on pressure and temperature, as discussed later in Section 18.9.

The reaction order dictates the concentration dependence of the reaction rate. The reaction order may be integer, zero, or fractional. *It cannot be overemphasized that reaction orders have no relation to stoichiometric coefficients, and they are determined by experiment.* For example, reconsider the reaction of nitrogen dioxide with molecular oxygen [Equation (18.6)]:

$$4NO_2(g) + O_2(g) \rightarrow 2N_2O_5(g)$$

The experimentally determined rate law expression for this reaction is

$$R = k[NO_2]^2[O_2]$$

TABLE 18.1

Relationship between Rate Law, Order, and the Rate Constant, k^*

Rate Law	Order	Units of k
Rate $= k$	Zero	$M\ s^{-1}$
Rate $= k[A]$	First order with respect to A First order overall	s^{-1}
Rate $= k[A]^2$	Second order with respect to A Second order overall	$M^{-1}\ s^{-1}$
Rate $= k[A][B]$	First order with respect to A First order with respect to B Second order overall	$M^{-1}\ s^{-1}$
Rate $= k[A][B][C]$	First order with respect to A First order with respect to B First order with respect to C Third order overall	$M^{-2}\ s^{-1}$

*In the units of k, M represents mol L^{-1} or moles per liter.

That is, the reaction is second order with respect to NO_2, first order with respect to O_2, and third order overall. Notice that the reaction orders are not equal to the stoichiometric coefficients. *All rate laws must be determined experimentally with respect to each reactant,* and there is no insight to be gained by considering the stoichiometry of the reaction.

In the rate law expression of Equation (18.9), the rate constant serves as the proportionality constant between the concentration of the various species and the reaction rate. Inspection of Equation (18.8) demonstrates that the reaction rate will *always* have units of concentration time^{-1}. Therefore, the units of k must change with respect to the overall order of the reaction to ensure that the reaction rate has the correct units. The relationship between the rate law expression, order, and the units of k is presented in Table 18.1.

18.3.1 Measuring Reaction Rates

With the definitions for the reaction rate and rate law provided by Equations (18.8) and (18.9), the question of how one measures the rate of reaction becomes important. To illustrate this point, consider the following reaction:

$$A \xrightarrow{\ k\ } B \tag{18.10}$$

The rate of this reaction in terms of [A] is given by

$$R = -\frac{d[A]}{dt} \tag{18.11}$$

Furthermore, suppose experiments demonstrate that the reaction is first order in A, first order overall, and $k = 40\ s^{-1}$ so that

$$R = k[A] = (40\ s^{-1})[A] \tag{18.12}$$

Equation (18.11) states that the rate of the reaction is equal to the negative of the time derivative of [A]. Imagine that we perform an experiment in which [A] is measured as a function of time as shown in Figure 18.2. The derivative in Equation (18.11) is simply the slope of the tangent for the concentration curve at a specific time. Therefore, the reaction rate will depend on the time at which the rate is determined. Figure 18.2 presents a measurement of the rate at two time points, $t = 0$ ms (1 ms $= 10^{-3}$ s) and $t = 30$ ms.

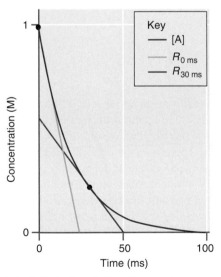

FIGURE 18.2

Measurement of the reaction rate. The concentration of reactant A as a function of time is presented. The rate R is equal to the slope of the tangent of this curve. This slope depends on the time at which the tangent is determined. The tangent determined 30 ms into the reaction is presented as the blue line, and the tangent at $t = 0$ is presented as the yellow line.

At $t = 0$ ms, the reaction rate is given by the negative of the slope of the line corresponding to the change in [A] with time, per Equation (18.11):

$$R_{t=0} = -\frac{d[A]}{dt} = 40 \text{ M s}^{-1}$$

However, when measured at 30 ms the rate is

$$R_{t=30\,\text{ms}} = -\frac{d[A]}{dt} = 12 \text{ M s}^{-1}$$

Notice that the reaction rate is decreasing with time. This behavior is a direct consequence of the change of [A] as a function of time, as expected from the rate law of Equation (18.8). Specifically, at $t = 0$,

$$R_{t=0} = 40 \text{ s}^{-1}[A]_{t=0} = 40 \text{ s}^{-1} \ (1 \text{ M}) = 40 \text{ M s}^{-1}$$

However, by $t = 30$ ms the concentration of A has decreased to 0.3 M so that the rate is

$$R_{t=30\,\text{ms}} = 40 \text{ s}^{-1}[A]_{t=30\,\text{ms}} = 40 \text{ s}^{-1}(0.3 \text{ M}) = 12 \text{ M s}^{-1}$$

This difference in rates brings to the forefront an important issue in kinetics: how does one define a reaction rate if the rate changes with time? One convention is to define the rate before the reactant concentrations have undergone any substantial change from their initial values. The reaction rate obtained under such conditions is known as the **initial rate.** The initial rate in the previous example is that determined at $t = 0$. In the remainder of our discussion of kinetics, the rate of reaction is taken to be synonymous with initial rate. However, the rate constant is independent of concentration; therefore, if the rate constant, concentrations, and order dependence of the reaction rate are known, the reaction rate can be determined at any time.

18.3.2 Determining Reaction Orders

Consider the following reaction:

$$A + B \xrightarrow{\ k\ } C \qquad (18.13)$$

The rate law expression for this reaction is

$$R = k[A]^\alpha[B]^\beta \qquad (18.14)$$

How can one determine the order of the reaction with respect to A and B? First, note that the measurement of the rate under a single set of concentrations for A and B will not by itself provide a measure of α and β because one would have only one equation [Equation (18.14)], but two unknown quantities. Therefore, the determination of reaction order will involve the measurement of the reaction rate under various concentration conditions. This approach assumes that one can vary the reactant concentrations, but for some reactions this is not possible. In this case, another approach must be taken to determine the order dependence of the reaction, and this approach will be discussed shortly. For now, assume that the reactant concentrations can be varied. The question then becomes "What set of concentrations should be used to determine the reaction rate?" One answer to this question is known as the **isolation method.** In this approach, the reaction is performed with all species but one in excess. Under these conditions, only the concentration of one species will vary to a significant extent during the reaction. For example, consider the example A + B reaction shown in Equation (18.13). Imagine performing the experiment where the initial concentration of A is 1.00 M and the concentration of B is 0.01 M. The rate of the reaction will be zero when all of reactant B has been used; however, the concentration of A will have been reduced to 0.99 M, only a slight reduction from the initial concentration. This simple example demonstrates that the concentration of species present in excess will be essentially constant with time. This time independence simplifies the reaction rate expression because the reaction rate will depend only on the concentration of the nonexcess species. In our example reaction where A is in excess, Equation (18.14) simplifies to

$$R = k'[B]^\beta \tag{18.15}$$

In Equation (18.15), k' is the product of the original rate constant and $[A]^\alpha$, both of which are time independent. Isolation results in the dependence of the reaction rate on [B] exclusively, and the reaction order with respect to B is determined by measuring the reaction rate as [B] is varied. Of course, the isolation method could just as easily be applied to determine α by performing measurements with B in excess.

A second strategy employed to determine reaction rates is referred to as the **method of initial rates.** In this approach, the concentration of a single reactant is changed while holding all other concentrations constant, and the initial rate of the reaction determined. The variation in the initial rate as a function of concentration is then analyzed to determine the order of the reaction with respect to the reactant that is varied. Consider the reaction depicted by Equation (18.13). To determine the order of the reaction for each reactant, the reaction rate is measured as [A] is varied and the concentration of B is held constant. The reaction rate at two different values of [A] is then analyzed to determine the order of the reaction with respect to [A] as follows:

$$\frac{R_1}{R_2} = \frac{k[A]_1^\alpha [B]_0^\beta}{k[A]_2^\alpha [B]_0^\beta} = \left(\frac{[A]_1}{[A]_2} \right)^\alpha$$

$$\ln\left(\frac{R_1}{R_2} \right) = \alpha \ln\left(\frac{[A]_1}{[A]_2} \right) \tag{18.16}$$

Notice that [B] and k are constant in each measurement; therefore, they cancel when one evaluates the ratio of the measured reaction rates. Using Equation (18.16), the order of the reaction with respect to A is readily determined. A similar experiment to determine β can be performed where [A] is held constant and the dependence of the reaction rate on [B] is measured.

EXAMPLE PROBLEM 18.2

Using the following data for the reaction illustrated in Equation (18.13), determine the order of the reaction with respect to A and B, and the rate constant for the reaction:

[A] (M)	[B] (M)	Initial Rate (M s^{-1})
2.30×10^{-4}	3.10×10^{-5}	5.25×10^{-4}
4.60×10^{-4}	6.20×10^{-5}	4.20×10^{-3}
9.20×10^{-4}	6.20×10^{-5}	1.70×10^{-2}

Solution
Using the last two entries in the table, the order of the reaction with respect to A is

$$\ln\left(\frac{R_1}{R_2} \right) = \alpha \ln\left(\frac{[A]_1}{[A]_2} \right)$$

$$\ln\left(\frac{4.20 \times 10^{-3}}{1.70 \times 10^{-2}} \right) = \alpha \ln\left(\frac{4.60 \times 10^{-4}}{9.20 \times 10^{-4}} \right)$$

$$-1.398 = \alpha(-0.693)$$

$$2 = \alpha$$

Using this result and the first two entries in the table, the order of the reaction with respect to B is given by

$$\frac{R_1}{R_2} = \frac{k[A]_1^2[B]_1^\beta}{k[A]_2^2[B]_2^\beta} = \frac{[A]_1^2[B]_1^\beta}{[A]_2^2[B]_2^\beta}$$

$$\left(\frac{5.25 \times 10^{-4}}{4.20 \times 10^{-3}}\right) = \left(\frac{2.30 \times 10^{-4}}{4.60 \times 10^{-4}}\right)^2 \left(\frac{3.10 \times 10^{-5}}{6.20 \times 10^{-5}}\right)^\beta$$

$$0.500 = (0.500)^\beta$$

$$1 = \beta$$

Therefore, the reaction is second order in A, first order in B, and third order overall. Using any row from the table, the rate constant is readily determined:

$$R = k[A]^2[B]$$

$$5.2 \times 10^{-4}\,\text{Ms}^{-1} = k(2.3 \times 10^{-4}\,\text{M})^2(3.1 \times 10^{-5}\,\text{M})$$

$$3.17 \times 10^8\,\text{M}^{-2}\text{s}^{-1} = k$$

Having determined k, the overall rate law is

$$R = (3.17 \times 10^8\,\text{M}^{-2}\text{s}^{-1})[A]^2[B]$$

The remaining question to address is how one experimentally determines the rate of a chemical reaction. Measurement techniques are usually separated into one of two categories: chemical and physical. As the name implies, **chemical methods** in kinetics studies rely on chemical processing to determine the progress of a reaction with respect to time. In this method, a chemical reaction is initiated, and samples are removed from the reaction and manipulated such that the reaction in the sample is terminated. Termination of the reaction is accomplished by rapidly cooling the sample or by adding a chemical species that depletes one of the reactants. After stopping the reaction, the sample contents are analyzed. By performing this analysis on a series of samples removed from the original reaction container as a function of time after initiation of the reaction, the kinetics of the reaction can be determined. Chemical methods are generally cumbersome to use and are limited to reactions that occur on slow timescales.

The majority of modern kinetics experiments involve **physical methods.** In these methods, a physical property of the system is monitored as the reaction proceeds. For some reactions, the system pressure or volume provides a convenient physical property for monitoring the progress of a reaction. For example, consider the thermal decomposition of PCl_5:

$$PCl_5(g) \longrightarrow PCl_3(g) + Cl_2(g)$$

As the reaction proceeds, for every gaseous PCl_5 molecule that decays, two gaseous product molecules are formed. Therefore, the total system pressure will increase as the reaction proceeds. Measurement of this pressure increase as a function of time provides information on the reaction kinetics.

More complex physical methods involve techniques that are capable of monitoring the concentration of an individual species as a function of time. Many of the spectroscopic techniques described in Engel's *Quantum Chemistry and Spectroscopy* are extremely useful for such measurements. For example, electronic absorption measurements can be performed in which the concentration of a species is monitored using the electronic absorption of a molecule and the Beer–Lambert law. Vibrational spectroscopic measurements using infrared absorption and Raman scattering (Chapter 8 of *Quantum Chemistry and Spectroscopy*, Sections 8.7 and 8.8) can be employed to monitor vibrational transitions of reactants or products providing information on their consumption or production. Finally, NMR spectroscopy (Chapter 18 of *Quantum Chemistry and Spectroscopy*) is a useful technique for following the reaction kinetics of complex systems.

The challenge in chemical kinetics is to perform measurements with sufficient time resolution to monitor the chemistry of interest. If the reaction is slow (seconds or longer), then the chemical methods just described can be used to monitor the kinetics. However, many chemical reactions occur on timescales as short as picoseconds (10^{-12} s) and femtoseconds (10^{-15} s). Reactions occurring on these short timescales are most easily studied using physical methods.

For reactions that occur on timescales as short at 1 ms (10^{-3} s), **stopped-flow techniques** provide a convenient method by which to measure solution phase reactions. These techniques are exceptionally popular for biochemical studies. A stopped-flow experiment is illustrated in Figure 18.3. Two reactants (A and B) are held in reservoirs connected to a syringe pump. The reaction is initiated by depressing the reactant syringes, and the reactants are mixed at the junction indicated in the figure. The reaction is monitored by observing the change in absorbance of the reaction mixture as a function of time. The temporal resolutions of stopped-flow techniques are generally limited by the time it takes for the reactants to mix.

Reactions that can be triggered by light are studied using **flash photolysis techniques.** In flash photolysis, the sample is exposed to a temporal pulse of light that initiates the reaction. Light pulses as short as 10 femtoseconds ($10\ \text{fs} = 10^{-14}\ \text{s}$) in the visible region of the electromagnetic spectrum are available such that reaction dynamics on this extremely short or ultrafast timescale can be studied. For reference, a 3000-cm^{-1} vibrational mode has a period of roughly 10 fs. Therefore, reactions can be initiated on the same timescale as vibrational molecular motion, and this capability has opened up many exciting fields in chemical kinetics. This capability has been used to determine the ultrafast reaction kinetics associated with vision, photosynthesis, atmospheric processes, and charge-carrier dynamics in semiconductors. Recent references to some of this work are included in the "For Further Reading" section at the end of this chapter. Short optical pulses can be used to perform vibrational spectroscopic measurements (infrared absorption or Raman) on the 100-fs timescale. Finally, NMR techniques, as well as optical absorption and vibrational spectroscopy, can be used to study reactions that occur on the microsecond (10^{-6} s) and longer timescale.

Another approach to studying chemical kinetics is that of **perturbation-relaxation methods.** In this approach, a chemical system initially at equilibrium is perturbed such that the system is no longer at equilibrium. By following the relaxation of the system back toward equilibrium, the rate constants for the reaction can be determined. Any system variable that affects the position of the equilibrium such as pressure, pH, or temperature can be used in a perturbation-relaxation experiment. Temperature perturbation or T-jump experiments are the most common type of perturbation experiment and are described in detail later in this chapter.

In summary, the measurement technique chosen for reaction rate determination will depend on both the specifics of the reaction as well as the timescale over which the reaction occurs. In any event, the determination of reaction rates is an experimental exercise and must be accomplished through careful measurements involving well-designed experiments.

FIGURE 18.3

Schematic of a stopped-flow experiment. Two reactants are rapidly introduced into the mixing chamber by syringes. After mixing, the reaction kinetics are monitored by observing the change in sample concentration versus time, in this example by measuring the absorption of light as a function of time after mixing.

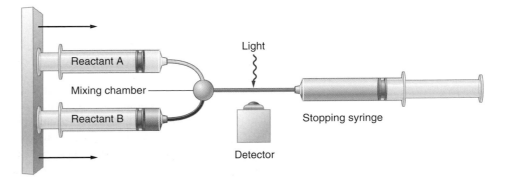

18.4 Reaction Mechanisms

As discussed in the previous section, the order of a reaction with respect to a given reactant is not determined by the stoichiometry of the reaction. The reason for the inequivalence of the stoichiometric coefficient and reaction order is that the balanced chemical reaction provides no information with respect to the mechanism of the chemical reaction. A **reaction mechanism** is defined as the collection of individual kinetic processes or elementary steps involved in the transformation of reactants into products. The rate law expression for a chemical reaction, including the order of the reaction, is entirely dependent on the reaction mechanism. In contrast, the Gibbs energy for a reaction is dependent on the equilibrium concentration of reactants and products. Just as the study of concentrations as a function of reaction conditions provides information on the thermodynamics of the reaction, the study of reaction rates as a function of reaction conditions provides information on the reaction mechanism.

All reaction mechanisms consist of a series of **elementary reaction steps,** or a chemical process that occurs in a single step. The **molecularity** of a reaction step is the stoichiometric quantity of reactants involved in the reaction step. For example, unimolecular reactions involve a single reactant species. An example of a unimolecular reaction step is the decomposition of a diatomic molecule into its atomic fragments:

$$I_2 \xrightarrow{k_d} 2I \tag{18.17}$$

Although Equation (18.17) is referred to as a unimolecular reaction, enthalpy changes accompanying the reaction generally involve the transfer of this heat through collisions with other, neighboring molecules. The role of collisional energy exchange with surrounding molecules will figure prominently in the discussion of unimolecular dissociation reactions in the following chapter (Section 19.3), but these energy-exchange processes are suppressed in this discussion. Bimolecular reaction steps involve the interaction of two reactants. For example, the reaction of nitric oxide with ozone is a biomolecular reaction:

$$NO + O_3 \xrightarrow{k_r} NO_2 + O_2 \tag{18.18}$$

The importance of elementary reaction steps is that the corresponding rate law expression for the reaction can be written based on the molecularity of the reaction. For the unimolecular reaction, the rate law expression is that of a first-order reaction. For the unimolecular decomposition of I_2 presented in Equation (18.18), the rate law expression for this elementary step is

$$R = -\frac{d[I_2]}{dt} = k_d[I_2] \tag{18.19}$$

Likewise, the rate law expression for the bimolecular reaction of NO and O_3 is

$$R = -\frac{d[NO]}{dt} = k_r[NO][O_3] \tag{18.20}$$

Comparison of the rate law expressions with their corresponding reactions demonstrates that the order of the reaction is equal to the stoichiometric coefficient. For elementary reactions, the order of the reaction can be inferred from the molecularity of the reaction. Keep in mind that *the equivalence of order and molecularity is only true for elementary reaction steps.*

A common problem in kinetics is identifying which of a variety of proposed reaction mechanisms is the "correct" mechanism. The design of kinetic experiments to differentiate between proposed mechanisms is quite challenging. Due to the complexity of many reactions, it is often difficult to experimentally differentiate between several potential mechanisms. A general rule of kinetics is that although it may be possible to rule out a proposed mechanism, it is never possible to prove unequivocally that a given mechanism is correct. The following example illustrates the origins of this rule. Consider the following reaction:

$$A \longrightarrow P \tag{18.21}$$

As written, the reaction is a simple first-order transformation of reactant A into product P, and it may occur through a single elementary step. However, what if the reaction were to occur through two elementary steps as follows:

$$A \xrightarrow{k_1} I \tag{18.22}$$
$$I \xrightarrow{k_2} P$$

In this mechanism, the decay of reactant A results in the formation of an intermediate species, I, that undergoes subsequent decay to produce the reaction product, P. One way to validate this mechanism is to observe the formation of the intermediate species. However, if the rate of the second reaction step is fast compared to the rate of the first step, the concentration of [I] will be quite small such that detection of the intermediate may be difficult. As will be seen later, in this limit the product formation kinetics will be consistent with the single elementary step mechanism, and verification of the two-step mechanism is not possible. It is usually assumed that the simplest mechanism consistent with the experimentally determined order dependence is correct until proven otherwise. In this example, a simple single-step mechanism would be considered "correct" until a clever chemist discovered a set of reaction conditions that demonstrates the reaction must occur by a sequential mechanism.

In order for a reaction mechanism to be valid, the order of the reaction predicted by the mechanism must be in agreement with the experimentally determined rate law. In evaluating a reaction mechanism, one must express the mechanism in terms of elementary reaction steps. The remainder of this chapter involves an investigation of various elementary reaction processes and derivations of the rate law expressions for these elementary reactions. The techniques developed in this chapter can be readily employed in the evaluation of complex kinetic problems, as illustrated in Chapter 19.

18.5 Integrated Rate Law Expressions

The rate law determination methods described in the previous section assume that one has a substantial amount of control over the reaction. Specifically, application of the initial rates method requires that the reactant concentrations be controlled and mixed in any proportion desired. In addition, this method requires that the rate of reaction be measured immediately after initiation of the reaction. Unfortunately, many reactions cannot be studied by this technique due to the instability of the reactants involved, or the timescale of the reaction of interest. In this case, other approaches must be employed.

One approach is to assume that the reaction occurs with a given order dependence and then determine how the concentrations of reactants and products will vary as a function of time. The predictions of the model are compared to experiment to determine if the model provides an appropriate description of the reaction kinetics. **Integrated rate law expressions** provide the predicted temporal evolution in reactant and product concentrations for reactions having an assumed order dependence. In this section these expressions are derived. For many elementary reactions, integrated rate law expressions can be derived, and some of those cases are considered in this section. However, more complex reactions may be difficult to approach using this technique, and one must resort to numerical methods to evaluate the kinetic behavior associated with a given reaction mechanism. Numerical techniques are discussed in Section 18.6.

18.5.1 First-Order Reactions

Consider the following elementary reaction step where reactant A decays, resulting in the formation of product P:

$$A \xrightarrow{k} P \tag{18.23}$$

If the reaction is first order with respect to [A], the corresponding rate law expression is

$$R = k[A] \tag{18.24}$$

where k is the rate constant for the reaction. The reaction rate can also be written in terms of the time derivative of [A]:

$$R = -\frac{d[A]}{dt} \tag{18.25}$$

Because the reaction rates given by Equations (18.24) and (18.25) are the same, we can write

$$\frac{d[A]}{dt} = -k[A] \tag{18.26}$$

Equation (18.26) is known as a differential rate expression. It relates the time derivative of A to the rate constant and concentration dependence of the reaction. It is also a standard differential equation that can be integrated as follows:

$$\int_{[A]_0}^{[A]} \frac{d[A]}{[A]} = \int_0^t -k\,dt$$

$$\ln\left(\frac{[A]}{[A]_0}\right) = -kt$$

$$[A] = [A]_0 e^{-kt} \tag{18.27a}$$

The limits of integration employed in obtaining Equation (18.27a) correspond to the initial concentration of reactant when the reaction is initiated ($[A] = [A]_0$ at $t = 0$) and the concentration of reactant at a given time after the reaction has started. If only the reactant is present at $t = 0$, the sum of reactant and product concentrations at any time must be equal to $[A]_0$. Using this idea, the concentration of product with time for this **first-order reaction** is

$$[P] + [A] = [A]_0$$

$$[P] = [A]_0 - [A]$$

$$[P] = [A]_0(1 - e^{-kt}) \tag{18.27b}$$

Equation (18.27a) demonstrates that for a first-order reaction, the concentration of A will undergo exponential decay with time. A graphically convenient version of Equation (18.27a) for comparison to experiment is obtained by taking the natural log of the equation:

$$\ln[A] = \ln[A]_0 - kt \tag{18.28}$$

Equation (18.28) predicts that for a first-order reaction, a plot of the natural log of the reactant concentration versus time will be a straight line of slope $-k$ and y intercept equal to the natural log of the initial concentration. Figure 18.4 provides a comparison of the concentration dependences predicted by Equations (18.27) and Equation (18.28) for first-order reactions. It is important to note that the comparison of experimental data to an integrated rate law expression requires that the variation in concentration with time be accurately known over a wide range of reaction times to determine if the reaction indeed follows a certain order dependence.

18.5.2 Half-Life and First-Order Reactions

The time it takes for the reactant concentration to decrease to one-half of its initial value is called the **half-life** of the reaction and is denoted as $t_{1/2}$. For a first-order reaction, substitution of the definition for $t_{1/2}$ into Equation (18.28) results in the following:

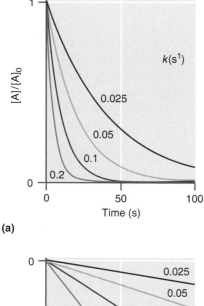

(a)

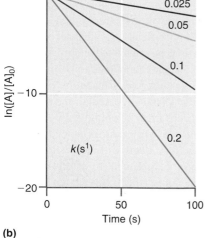

(b)

FIGURE 18.4

Reactant concentration as a function of time for a first-order chemical reaction as given by Equation (18.27). **(a)** Plots of [A] as a function of time for various rate constants k. The rate constant of a given curve is provided in the figure. **(b)** The natural log of reactant concentration as a function of time for a first-order chemical reaction as given by Equation (18.28).

$$-kt_{1/2} = \ln\left(\frac{[A]_0/2}{[A]_0}\right) = -\ln 2$$

$$t_{1/2} = \frac{\ln 2}{k} \qquad (18.29)$$

Notice that the half-life for a first-order reaction is independent of the initial concentration, and only the rate constant of the reaction influences $t_{1/2}$.

EXAMPLE PROBLEM 18.3

The decomposition of N_2O_5 is an important process in tropospheric chemistry. The half-life for the first-order decomposition of this compound is 2.05×10^4 s. How long will it take for a sample of N_2O_5 to decay to 60% of its initial value?

Solution
Using Equation (18.29), the rate constant for the decay reaction is determined using the half-life as follows:

$$k = \frac{\ln 2}{t_{1/2}} = \frac{\ln 2}{2.05 \times 10^4 \text{ s}} = 3.38 \times 10^{-5} \text{ s}^{-1}$$

The time at which the sample has decayed to 60% of its initial value is then determined using Equation (18.27a):

$$[N_2O_5] = 0.6[N_2O_5]_0 = [N_2O_5]_0 e^{-(3.38 \times 10^{-5} \text{ s}^{-1})t}$$

$$0.6 = e^{-(3.38 \times 10^{-5} \text{ s}^{-1})t}$$

$$\frac{-\ln(0.6)}{3.38 \times 10^{-5} \text{ s}^{-1}} = t = 1.51 \times 10^4 \text{ s}$$

Radioactive decay of unstable nuclear isotopes is an important example of a first-order process. The decay rate is usually stated as the half-life. Example Problem 18.4 demonstrates the use of radioactive decay in determining the age of a carbon-containing material.

EXAMPLE PROBLEM 18.4

Carbon-14 is a radioactive nucleus with a half-life of 5760 years. Living matter exchanges carbon with its surroundings (for example, through CO_2) so that a constant level of ^{14}C is maintained, corresponding to 15.3 decay events per minute. Once living matter has died, carbon contained in the matter is not exchanged with the surroundings, and the amount of ^{14}C that remains in the dead material decreases with time due to radioactive decay. Consider a piece of fossilized wood that demonstrates 2.4 ^{14}C decay events per minute. How old is the wood?

Solution
The ratio of decay events yields the amount of ^{14}C present currently versus the amount that was present when the tree died:

$$\frac{[^{14}C]}{[^{14}C]_0} = \frac{2.40 \text{ min}^{-1}}{15.3 \text{ min}^{-1}} = 0.157$$

The rate constant for isotope decay is related to the half-life as follows:

$$k = \frac{\ln 2}{t_{1/2}} = \frac{\ln 2}{5760 \text{ years}} = \frac{\ln 2}{1.82 \times 10^{11} \text{ s}} = 3.81 \times 10^{-12} \text{ s}^{-1}$$

With the rate constant and ratio of isotope concentrations, the age of the fossilized wood is readily determined:

$$\frac{[^{14}C]}{[^{14}C]_0} = e^{-kt}$$

$$\ln\left(\frac{[^{14}C]}{[^{14}C]_0}\right) = -kt$$

$$-\frac{1}{k}\ln\left(\frac{[^{14}C]}{[^{14}C]_0}\right) = -\frac{1}{3.81\times10^{-12}\,\text{s}}\ln(0.157) = t$$

$$4.86\times10^{11}\,\text{s} = t$$

This time corresponds to an age of roughly 15,400 years.

18.5.3 Second-Order Reaction (Type I)

Consider the following reaction, which is second order with respect to the reactant A:

$$2A \xrightarrow{\ k\ } P \tag{18.30}$$

Second-order reactions involving a single reactant species are referred to as **type I.** Another reaction that is second order overall involves two reactants, A and B, with a rate law that is first order with respect to each reactant. Such reactions are referred to as **second-order reactions of type II.** We focus first on the type I case. For this reaction, the corresponding rate law expression is

$$R = k[A]^2 \tag{18.31}$$

The rate as expressed as the derivative of reactant concentration is

$$R = -\frac{1}{2}\frac{d[A]}{dt} \tag{18.32}$$

The rates in the preceding two expressions are equivalent such that

$$-\frac{d[A]}{dt} = 2k[A]^2 \tag{18.33}$$

Generally, the quantity $2k$ is written as an effective rate constant, denoted as k_{eff}. With this substitution, integration of Equation (18.33) yields

$$-\int_{[A]_0}^{[A]}\frac{d[A]}{[A]^2} = \int_0^t k_{eff}\,dt$$

$$\frac{1}{[A]} - \frac{1}{[A]_0} = k_{eff}t$$

$$\boxed{\frac{1}{[A]} = \frac{1}{[A]_0} + k_{eff}t} \tag{18.34}$$

Equation (18.34) demonstrates that for a second-order reaction, a plot of the inverse of reactant concentration versus time will result in a straight line having a slope of k_{eff} and y intercept of $1/[A]_0$. Figure 18.5 presents a comparison between [A] versus time for a second-order reaction and $1/[A]$ versus time. The linear behavior predicted by Equation (18.34) is evident.

18.5.4 Half-Life and Reactions of Second Order (Type I)

Recall that the definition of half-life is when the concentration of a reactant is half of its initial value. With this definition, the half-life for a type I second-order reaction is

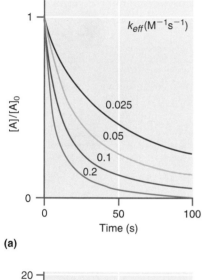

$k_{eff}(\text{M}^{-1}\text{s}^{-1})$

0.025
0.05
0.1
0.2

(a)

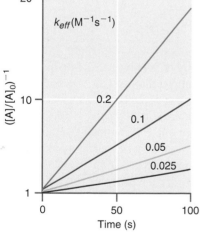

$k_{eff}(\text{M}^{-1}\text{s}^{-1})$

0.2
0.1
0.05
0.025

(b)

FIGURE 18.5
Reactant concentration as a function of time for a type I second-order chemical reaction. **(a)** Plots of [A] as a function of time for various rate constants. The rate constant of a given curve is provided in the figure. **(b)** The inverse of reactant concentration as a function of time as given by Equation (18.34).

$$t_{1/2} = \frac{1}{k_{eff}[A]_0} \tag{18.35}$$

In contrast to first-order reactions, the half-life for a second-order reaction is dependent on the initial concentration of reactant, with an increase in initial concentration resulting in a decrease in $t_{1/2}$. This behavior is consistent with a first-order reaction occurring through a unimolecular process, whereas the second-order reaction involves a bimolecular process in which the concentration dependence of the reaction rate is anticipated.

18.5.5 Second-Order Reaction (Type II)

Second-order reactions of type II involves two different reactants, A and B, as follows:

$$A + B \xrightarrow{k} P \tag{18.36}$$

Assuming that the reaction is first order in both A and B, the reaction rate is

$$R = k[A][B] \tag{18.37}$$

In addition, the rate with respect to the time derivative of the reactant concentrations is

$$R = -\frac{d[A]}{dt} = -\frac{d[B]}{dt} \tag{18.38}$$

Notice that the loss rate for the reactants is equal such that

$$[A]_0 - [A] = [B]_0 - [B]$$
$$[B]_0 - [A]_0 + [A] = [B]$$
$$\Delta + [A] = [B] \tag{18.39}$$

Equation (18.39) provides a definition for [B] in terms of [A] and the difference in initial concentration, $[B]_0 - [A]_0$, denoted as Δ. With this definition, the integrated rate law expression can be solved as follows. First, setting Equations (18.37) and (18.38) equal, the following expression is obtained:

$$\frac{d[A]}{dt} = -k[A][B] = -k[A](\Delta + [A])$$

$$\int_{[A]_0}^{[A]} \frac{d[A]}{[A](\Delta + [A])} = -\int_0^t k\,dt \tag{18.40}$$

Next, solution to the integral involving [A] is given by

$$\int \frac{dx}{x(c+x)} = -\frac{1}{c}\ln\left(\frac{c+x}{x}\right)$$

Using this solution to the integral, the integrated rate law expression becomes

$$-\frac{1}{\Delta}\ln\left(\frac{\Delta + [A]}{[A]}\right)\Bigg|_{[A]_0}^{[A]} = -kt$$

$$\frac{1}{\Delta}\left[\ln\left(\frac{\Delta + [A]}{[A]}\right) - \ln\left(\frac{\Delta + [A]_0}{[A]_0}\right)\right] = kt$$

$$\frac{1}{\Delta}\left[\ln\left(\frac{[B]}{[A]}\right) - \ln\left(\frac{[B]_0}{[A]_0}\right)\right] = kt$$

$$\frac{1}{[B]_0 - [A]_0}\ln\left(\frac{[B]/[B]_0}{[A]/[A]_0}\right) = kt \tag{18.41}$$

Equation (18.41) is not applicable in the case for which the initial concentrations are equivalent, that is, when $[B]_0 = [A]_0$. For this specific case, the concentrations of [A] and [B] reduce to the expression for a second-order reaction of type I with $k_{eff} = k$. The time evolution in reactant concentrations depends on the amount of each reactant present. Finally, the concept of half-life does not apply to second-order reactions of type II. Unless the reactants are mixed in stoichiometric proportions (1:1 for the case discussed in this section), the concentrations of both species will not be 1/2 their initial concentrations at the identical time.

SUPPLEMENTAL

18.6 Numerical Approaches

For the simple reactions outlined in the preceding section, an integrated rate law expression can be readily determined. However, there is a wide variety of kinetic problems for which an integrated rate law expression cannot be obtained. How can one compare a kinetic model with experiment in the absence of an integrated rate law? In such cases, numerical methods provide another approach by which to determine the time evolution in concentrations predicted by a kinetic model. To illustrate this approach, consider the following first-order reaction:

$$A \xrightarrow{k} P \tag{18.42}$$

The differential rate expression for this reaction is

$$\frac{d[A]}{dt} = -k[A] \tag{18.43}$$

The time derivative corresponds to the change in [A] for a time duration that is infinitesimally small. Using this idea, we can state that for a finite time duration, Δt, the change in [A] is given by

$$\frac{\Delta[A]}{\Delta t} = -k[A]$$

$$\Delta[A] = -\Delta t(k[A]) \tag{18.44}$$

In Equation (18.44), [A] is the concentration of [A] at a specific time. Therefore, we can use this equation to determine the change in the concentration of A, or $\Delta[A]$, over a time period Δt and then use this concentration change to determine the concentration at the end of the time period. This new concentration can be used to determine the subsequent change in [A] over the next time period, and this process is continued until the reaction is complete. Mathematically,

$$[A]_{t+\Delta t} = [A]_t + \Delta[A]$$
$$= [A]_t + \Delta t(-k[A]_t)$$
$$= [A]_t - k\Delta t[A]_t \tag{18.45}$$

In Equation (18.45), $[A]_t$ is the concentration at the beginning of the time interval, and $[A]_{t+\Delta t}$ is the concentration at the end of the time interval. This process is illustrated in Figure 18.6. In the figure, the initial concentration is used to determine $\Delta[A]$ over the time interval Δt. The concentration at this next time point, $[A]_1$, is used to determine $\Delta[A]$ over the next time interval, resulting in concentration $[A]_2$. This process is continued until the entire concentration profile is evaluated.

The specific example discussed here is representative of the general approach to numerically integrating differential equations, known as **Euler's method.** Application of Euler's method requires some knowledge of the timescale of interest, and then selection of a time interval, Δt, that is sufficiently small to capture the evolution in concentration.

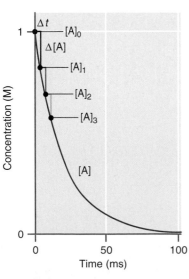

FIGURE 1 8 . 6
Schematic representation of the numerical evaluation of a rate law.

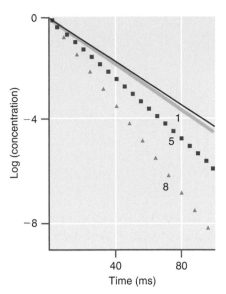

FIGURE 18.7

Comparison of the numerical approximation method to the integrated rate law expression for a first-order reaction. The rate constant for the reaction is 0.1 m s^{-1}. The time evolution in reactant concentration determined by the integrated rate law expression of Equation (18.27) is shown as the solid red line. Comparison to three numerical approximations is given, and the size of the time step (in ms) employed for each approximation is indicated. Notice the improvement in the numerical approximation as the time step is decreased.

Figure 18.7 presents a comparison of the reactant concentration determined using the integrated rate law expression for a first-order reaction to that determined numerically for three different choices for Δt. The figure illustrates that the accuracy of this method is highly dependent on an appropriate choice for Δt. In practice, convergence of the numerical model is demonstrated by reducing Δt and observing that the predicted evolution in concentrations does not change.

The numerical method can be applied to any kinetic process for which differential rate expressions can be prescribed. Euler's method provides the most straightforward way by which to predict how reactant and product concentrations will vary for a specific kinetic scheme. However, this method is "brute force" in that a sufficiently small time step must be chosen to accurately capture the slope of the concentration, and the time steps may be quite small, requiring a large number of iterations in order to reproduce the full time course of the reaction. As such, Euler's method can be computationally demanding. More elegant approaches, such as the Runge–Kutta method, exist that allow for larger time steps to be performed in numerical evaluations, and the interested reader is encouraged to investigate these approaches.

18.7 Sequential First-Order Reactions

Many chemical reactions occur in a series of steps in which reactants are transformed into products through multiple sequential elementary reaction steps. For example, consider the following **sequential reaction** scheme:

$$A \xrightarrow{k_A} I \xrightarrow{k_I} P \tag{18.46}$$

In this scheme, the reactant A decays to form **intermediate** I, and this intermediate undergoes subsequent decay resulting in the formation of product P. Species I is known as an intermediate. The sequential reaction scheme illustrated in Equation (18.46) involves a series of elementary first-order reactions. Recognizing this, the differential rate expressions for each species can be written as follows:

$$\frac{d[A]}{dt} = -k_A[A] \tag{18.47}$$

$$\frac{d[I]}{dt} = k_A[A] - k_I[I] \tag{18.48}$$

$$\frac{d[P]}{dt} = k_I[I] \tag{18.49}$$

These expressions follow naturally from the elementary reaction steps in which a given species participates. For example, the decay of A occurs in the first step of the reaction. The decay is a standard first-order process, consistent with the differential rate expression in Equation (18.47). The formation of product P is also a first-order process per Equation (18.49). The expression of Equation (18.48) for intermediate I reflects the fact that I is involved in both elementary reaction steps, the decay of A ($k_A[A]$), and the formation of P ($-k_I[I]$). Correspondingly, the differential rate expression for [I] is the sum of the rates associated with these two reaction steps. To determine the concentrations of each species as a function of time, we begin with Equation (18.47), which can be readily integrated given a set of initial concentrations. Let only the reactant A be present at $t = 0$ such that

$$[A]_0 \neq 0 \quad [I]_0 = 0 \quad [P]_0 = 0 \tag{18.50}$$

With these initial conditions, the expression for [A] is exactly that derived previously:

$$[A] = [A]_0 e^{-k_A t} \tag{18.51}$$

The expression for [A] given by Equation (18.51) can be substituted into the differential rate expression for I resulting in

$$\frac{d[\text{I}]}{dt} = k_A[\text{A}] - k_I[\text{I}] \tag{18.52}$$

$$= k_A[\text{A}]_0 e^{-k_A t} - k_I[\text{I}]$$

Equation (18.52) is a differential equation that when solved yields the following expression for [I]:

$$[\text{I}] = \frac{k_A}{k_I - k_A}(e^{-k_A t} - e^{-k_I t})[\text{A}]_0 \tag{18.53}$$

Finally, the expression for [P] is readily determined using the initial conditions of the reaction, with the initial concentration of A, $[\text{A}]_0$, equal to the sum of all concentrations for $t > 0$:

$$[\text{A}]_0 = [\text{A}] + [\text{I}] + [\text{P}]$$

$$[\text{P}] = [\text{A}]_0 - [\text{A}] - [\text{I}] \tag{18.54}$$

Substituting Equations (18.51) and (18.53) into Equation (18.54) results in the following expression for [P]:

$$[\text{P}] = \left(\frac{k_A e^{-k_I t} - k_I e^{-k_A t}}{k_I - k_A} + 1\right)[\text{A}]_0 \tag{18.55}$$

FIGURE 18.8

Concentration profiles for a sequential reaction in which the reactant (A, blue line) forms an intermediate (I, yellow line) that undergoes subsequent decay to form the product (P, red line) where **(a)** $k_A = 2k_I = 0.1$ s^{-1} and **(b)** $k_A = 8k_I = 0.4$ s^{-1}. Notice that both the maximal amount of I in addition to the time for the maximum is changed relative to the first panel. **(c)** $k_A = 0.025k_I = 0.0125$ s^{-1}. In this case, very little intermediate is formed, and the maximum in [I] is delayed relative to the first two examples.

Although the expressions for [I] and [P] look complicated, the temporal evolution in concentration predicted by these equations is intuitive as shown in Figure 18.8. Figure 18.8a presents the evolution in concentration when $k_A = 2k_I$. Notice that A undergoes exponential decay resulting in the production of I. The intermediate in turn undergoes subsequent decay to form the product. The temporal evolution of [I] is extremely dependent on the relative rate constants for the production, k_A, and decay, k_I. Figure 18.8b presents the case where $k_A \gg k_I$. Here, the maximum intermediate concentration is greater than in the first case. The opposite limit is illustrated in Figure 18.8c, where $k_A < k_I$ and the maximum in intermediate concentration is significantly reduced. This behavior is consistent with intuition: if the intermediate undergoes decay at a faster rate than the rate at which it is being formed, then the intermediate concentration will be small. Of course, the opposite logic holds as evidenced by the $k_A \gg k_I$ example presented in the Figure 18.8b.

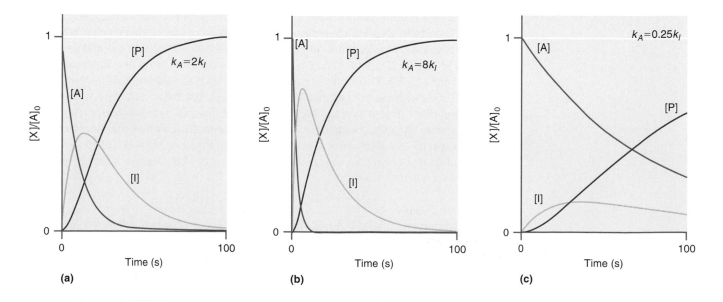

(a) Time (s)

(b) Time (s)

(c) Time (s)

18.7.1 Maximum Intermediate Concentration

Inspection of Figure 18.8 demonstrates that the time at which the concentration of the intermediate species will be at a maximum depends on the rate constants for its production and decay. Can we predict when [I] will be at a maximum? The maximum intermediate concentration has been reached when the derivative of [I] with respect to time is equal to zero:

$$\left(\frac{d[I]}{dt}\right)_{t=t_{max}} = 0 \tag{18.56}$$

Using the expression for [I] given in Equation (18.53) in the preceding equation, the time at which [I] is at a maximum, t_{max}, is

$$t_{max} = \frac{1}{k_A - k_I} \ln\left(\frac{k_A}{k_I}\right) \tag{18.57}$$

EXAMPLE PROBLEM 18.5

Determine the time at which [I] is at a maximum for $k_A = 2k_I = 0.1 \text{ s}^{-1}$.

Solution

This is the first example illustrated in Figure 18.8 where $k_A = 0.1 \text{ s}^{-1}$ and $k_I = 0.05 \text{ s}^{-1}$. Using these rate constants and Equation (18.57), t_{max} is determined as follows:

$$t_{max} = \frac{1}{k_A - k_I} \ln\left(\frac{k_A}{k_I}\right) = \frac{1}{0.1 \text{ s}^{-1} - 0.05 \text{ s}^{-1}} \ln\left(\frac{0.1 \text{ s}^{-1}}{0.05 \text{ s}^{-1}}\right) = 13.9 \text{ s}$$

18.7.2 Rate-Determining Steps

In the preceding subsection, the rate of product formation in a sequential reaction was found to depend on the timescale for production and decay of the intermediate species. Two limiting situations can be envisioned at this point. The first limit is where the rate constant for intermediate decay is much greater than the rate constant for production, that is, where $k_I \gg k_A$ in Equation (18.46). In this limit, any intermediate formed will rapidly go on to product, and the rate of product formation depends on the rate of reactant decay. The opposite limit occurs when the rate constant for intermediate production is significantly greater than the intermediate decay rate constant, that is, where $k_A \gg k_I$ in Equation (18.46). In this limit, reactants quickly produce intermediate, but the rate of product formation depends on the rate of intermediate decay. These two limits give rise to one of the most important approximations made in the analysis of kinetic problems, that of the **rate-determining step** or the rate-limiting step. The central idea behind this approximation is as follows: if one step in the sequential reaction is much slower than any other step, this slow step will control the rate of product formation and is therefore the rate-determining step.

   Consider the sequential reaction illustrated in Equation (18.46) when $k_A \gg k_I$. In this limit, the kinetic step corresponding to the decay of intermediate I is the rate-limiting step. Because $k_A \gg k_I$, $e^{-k_A t} \ll e^{-k_I t}$ and the expression for [P] of Equation (18.55) becomes

$$\lim_{k_A \to \infty} [P] = \lim_{k_A \to \infty} \left(\left(\frac{k_A e^{-k_I t} - k_I e^{-k_A t}}{k_I - k_A} + 1\right)[A]_0\right) = (1 - e^{-k_I t})[A]_0 \tag{18.58}$$

The time dependence of [P] when k_I is the rate-limiting step is identical to that predicted for first-order decay of I resulting in product formation. The other limit occurs when $k_I \gg k_A$, where $e^{-k_I t} \ll e^{-k_A t}$, and the expression for [P] becomes

$$\lim_{k_I \to \infty} [P] = \lim_{k_I \to \infty} \left(\left(\frac{k_A e^{-k_I t} - k_I e^{-k_A t}}{k_I - k_A} + 1\right)[A]_0\right) = (1 - e^{-k_A t})[A]_0 \tag{18.59}$$

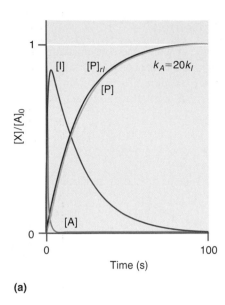

(a)

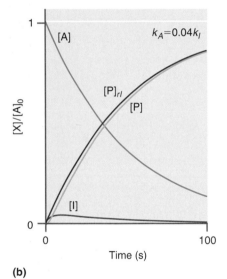

(b)

FIGURE 18.9
Rate-limiting step behavior in sequential reactions. **(a)** $k_A = 20k_I = 1\ s^{-1}$ such that the rate-limiting step is the decay of intermediate I. In this case, the reduction in [I] is reflected by the appearance of [P]. The time evolution of [P] predicted by the sequential mechanism is given by the yellow line, and the corresponding evolution assuming rate-limiting step behavior, $[P]_{rl}$, is given by the red curve. **(b)** The opposite case from part (a) in which $k_A = 0.04k_I = 0.02\ s^{-1}$ such that the rate-limiting step is the decay of reactant A.

In this limit, the time dependence of [P] is identical to that predicted for the first-order decay of the reactant A, resulting in product formation.

When is the rate-determining step approximation appropriate? For the two-step reaction under consideration, 20-fold differences between rate constants are sufficient to ensure that the smaller rate constant will be rate determining. Figure 18.9 presents a comparison for [P] determined using the exact result from Equation (18.55) and the rate-limited prediction of Equations (18.58) and (18.59), for the case where $k_A = 20k_I = 1\ s^{-1}$ and where $k_A = 0.04k_I = 0.02\ s^{-1}$. In Figure 18.9a, decay of the intermediate is the rate-limiting step in product formation. Notice the rapid reactant decay, resulting in an appreciable intermediate concentration, with the subsequent decay of the intermediate reflected by a corresponding increase in [P]. The similarity of the exact and rate-limiting curves for [P] demonstrates the validity of the rate-limiting approximation for this ratio of rate constants. The opposite limit is presented in Figure 18.9b. In this case, decay of the reactant is the rate-limiting step in product formation. When reactant decay is the rate-limiting step, very little intermediate is produced. In this case, the loss of [A] is mirrored by an increase in [P]. Again, the agreement between the exact and rate-limiting descriptions of [P] demonstrates the validity of the rate-limiting approximation when a substantial difference in rate constants for intermediate production and decay exists.

18.7.3 The Steady-State Approximation

Consider the following sequential reaction scheme:

$$A \xrightarrow{k_A} I_1 \xrightarrow{k_1} I_2 \xrightarrow{k_2} P \tag{18.60}$$

In this reaction, product formation results from the formation and decay of two intermediate species, I_1 and I_2. The differential rate expressions for this scheme are as follows:

$$\frac{d[A]}{dt} = -k_A[A] \tag{18.61}$$

$$\frac{d[I_1]}{dt} = k_A[A] - k_1[I_1] \tag{18.62}$$

$$\frac{d[I_2]}{dt} = k_1[I_1] - k_2[I_2] \tag{18.63}$$

$$\frac{d[P]}{dt} = k_2[I_2] \tag{18.64}$$

A determination of the time-dependent concentrations for the species involved in this reaction by integration of the differential rate expressions is not trivial; therefore, how can the concentrations be determined? One approach is to use Euler's method to determine numerically the concentrations as a function of time. The result of this approach for $k_A = 0.02\ s^{-1}$ and $k_1 = k_2 = 0.2\ s^{-1}$ is presented in Figure 18.10. Notice that the relative magnitude of the rate constants results in only modest intermediate concentrations.

Inspection of Figure 18.10 illustrates that in addition to the modest intermediate concentrations, $[I_1]$ and $[I_2]$ change very little with time such that the time derivative of these concentrations can be set approximately equal to zero:

$$\frac{d[I]}{dt} = 0 \tag{18.65}$$

Equation (18.65) is known as the **steady-state approximation.** This approximation is used to evaluate the differential rate expressions by simply setting the time derivative of

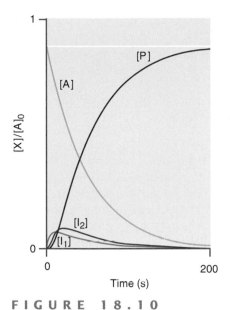

FIGURE 18.10
Concentrations determined by numerical evaluation of the sequential reaction scheme presented in Equation (18.44) where $k_A = 0.02$ s^{-1} and $k_1 = k_2 = 0.2$ s^{-1}.

all intermediates to zero. This approximation is particularly good when the decay rate of the intermediate is greater than the rate of production so that the intermediates are present at very small concentrations during the reaction (as in the case illustrated in Figure 18.10). Applying the steady-state approximation to I_1 in our example reaction results in the following expression for $[I_1]$:

$$\frac{d[I_1]_{ss}}{dt} = 0 = k_A[A] - k_1[I_1]_{ss}$$

$$[I_1]_{ss} = \frac{k_A}{k_1}[A] = \frac{k_A}{k_1}[A]_0 e^{-k_A t} \tag{18.66}$$

where the subscript ss indicates that the concentration is that predicted using the steady-state approximation. The final equality in Equation (18.66) results from integration of the differential rate expression for [A] with the initial conditions that $[A]_0 \neq 0$ and all other initial concentrations are zero. The corresponding expression for $[I_2]$ under the steady-state approximation is

$$\frac{d[I_2]_{ss}}{dt} = 0 = k_1[I_1]_{ss} - k_2[I_2]_{ss}$$

$$[I_2]_{ss} = \frac{k_1}{k_2}[I_1]_{ss} = \frac{k_A}{k_2}[A]_0 e^{-k_A t} \tag{18.67}$$

Finally, the differential expression for P is

$$\frac{d[P]_{ss}}{dt} = k_2[I_2] = k_A[A]_0 e^{-k_A t} \tag{18.68}$$

Integration of Equation (18.68) results in the now familiar expression for [P]:

$$[P]_{ss} = [A]_0 (1 - e^{-k_A t}) \tag{18.69}$$

Equation (18.69) demonstrates that within the steady-state approximation, [P] is predicted to demonstrate appearance kinetics consistent with the first-order decay of A.

When is the steady-state approximation valid? The approximation requires that the concentration of intermediate be constant as a function of time. Consider the concentration of the first intermediate under the steady-state approximation. The time derivative of $[I_1]_{ss}$ is

$$\frac{d[I_1]_{ss}}{dt} = \frac{d}{dt}\left(\frac{k_A}{k_1}[A]_0 e^{-k_A t}\right) = -\frac{k_A^2}{k_1}[A]_0 e^{-k_A t} \tag{18.70}$$

The steady-state approximation is valid when Equation (18.70) is equal to zero, which is true when $k_1 \gg k_A^2[A]_0$. In other words, k_1 must be sufficiently large such that $[I_1]$ is small at all times. Similar logic applies to I_2 for which the steady-state approximation is valid when $k_2 \gg k_A^2[A]_0$.

Figure 18.11 presents a comparison between the numerically determined concentrations and those predicted using the steady-state approximation for the two-intermediate sequential reaction where $k_A = 0.02$ s^{-1} and $k_1 = k_2 = 0.2$ s^{-1}. Notice that even for these conditions where the steady-state approximation is expected to be valid, the discrepancy between [P] determined by numerical evaluation versus the steady-state approximation value, $[P]_{ss}$, is evident. For the examples presented here, the steady-state approximation is relatively easy to implement; however, for many reactions the approximation of constant intermediate concentration with time is not appropriate. In addition, the steady-state approximation is difficult to implement if the intermediate concentrations are not isolated to one or two of the differential rate expressions derived from the mechanism of interest.

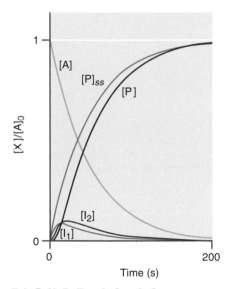

FIGURE 18.11
Comparison of the numerical and steady-state concentration profiles for the sequential reaction scheme presented in Equation (18.44) where $k_A = 0.02$ s^{-1} and $k_1 = k_2 = 0.2$ s^{-1}. Curves corresponding to the steady-state approximation are indicated by the subscript ss.

EXAMPLE PROBLEM 18.6

Consider the following sequential reaction scheme:

$$A \xrightarrow{k_A} I \xrightarrow{k_I} P$$

Assuming that only reactant A is present at $t = 0$, what is the expected time dependence of [P] using the steady-state approximation?

Solution

The differential rate expressions for this reaction were provided in Equations (18.47), (18.48), and (18.49):

$$\frac{d[A]}{dt} = -k_A[A]$$

$$\frac{d[I]}{dt} = k_A[A] - k_I[I]$$

$$\frac{d[P]}{dt} = k_I[I]$$

Applying the steady-state approximation to the differential rate expression for I and substituting in the integrated expression for [A] of Equation (18.51) yield

$$\frac{d[I]}{dt} = 0 = k_A[A] - k_I[I]$$

$$\frac{k_A}{k_I}[A] = \frac{k_A}{k_I}[A]_0 e^{-k_A t} = [I]$$

Substituting the preceding expression for [I] into the differential rate expression for the product and integrating yield

$$\frac{d[P]}{dt} = k_I[I] = \frac{k_A}{k_I}(k_I[A]_0 e^{-k_A t})$$

$$\int_0^{[P]} d[P] = k_A[A]_0 \int_0^t e^{-k_A t}$$

$$[P] = k_A[A]_0 \left[\frac{1}{k_A}(1 - e^{-k_A t}) \right]$$

$$[P] = [A]_0 (1 - e^{-k_A t})$$

This expression for [P] is identical to that derived in the limit that the decay of A is the rate-limiting step in the sequential reaction [Equation (18.59)].

18.1 Sequential Reaction Kinetics

18.8 Parallel Reactions

In the reactions discussed thus far, reactant decay results in the production of only a single species. However, in many instances a single reactant can become a variety of products. Such reactions are referred to as **parallel reactions.** Consider the following reaction in which the reactant A can form one of two products, B or C:

(18.71)

$$\begin{array}{ccc} & & B \\ & \nearrow^{k_b} & \\ A & & \\ & \searrow_{k_c} & \\ & & C \end{array}$$

The differential rate expressions for the reactant and products are

$$\frac{d[A]}{dt} = -k_B[A] - k_C[A] = -(k_B + k_C)[A] \tag{18.72}$$

$$\frac{d[B]}{dt} = k_B[A] \tag{18.73}$$

$$\frac{d[C]}{dt} = k_C[A] \tag{18.74}$$

Integration of the preceding expression involving [A] with the initial conditions $[A]_0 \neq 0$ and $[B] = [C] = 0$ yields

$$[A] = [A]_0 e^{-(k_B + k_C)t} \tag{18.75}$$

The product concentrations can be determined by substituting the expression for [A] into the differential rate expressions and integrating, which results in

$$[B] = \frac{k_B}{k_B + k_c}[A]_0(1 - e^{-(k_B + k_c)t}) \tag{18.76}$$

$$[C] = \frac{k_C}{k_B + k_c}[A]_0(1 - e^{-(k_B + k_c)t}) \tag{18.77}$$

Figure 18.12 provides an illustration of the reactant and product concentrations for this branching reaction where $k_B = 2k_C = 0.1 \text{ s}^{-1}$. A few general trends demonstrated by branching reactions are evident in the figure. First, notice that the decay of A occurs with an apparent rate constant equal to $k_B + k_C$, the sum of rate constants for each reaction branch. Second, the ratio of product concentrations is independent of time. That is, at any time point the ratio [B]/[C] is identical. This behavior is consistent with Equations (18.76) and (18.77) where this ratio of product concentrations is predicted to be

$$\frac{[B]}{[C]} = \frac{k_B}{k_C} \tag{18.78}$$

Equation (18.78) is a very interesting result. The equation states that as the rate constant for one of the reaction branches increases relative to the other, the greater the final concentration of the corresponding product will be. Furthermore, there is no time dependence in Equation (18.78); therefore, the product ratio remains constant with time.

Equation (18.78) demonstrates that the extent of product formation in a parallel reaction is dependent on the rate constants. Another way to view this behavior is with respect to probability; the larger the rate constant for a given process, the more likely that product will be formed. The **yield**, Φ, is defined as the probability that a given product will be formed by decay of the reactant:

$$\Phi_i = \frac{k_i}{\sum_n k_n} \tag{18.79}$$

In Equation (18.79), k_i is the rate constant for the reaction leading to formation of the product of interest indicated by the subscript i. The denominator is the sum over all rate constants for the reaction branches. The total yield is the sum of the yields for forming each product, and it is normalized such that

$$\sum_i \Phi_i = 1 \tag{18.80}$$

In the example reaction depicted in Figure 18.12 where $k_B = 2k_C$, the yield for the formation of product C is

$$\Phi_C = \frac{k_C}{k_B + k_C} = \frac{k_C}{(2k_C) + k_C} = \frac{1}{3} \tag{18.81}$$

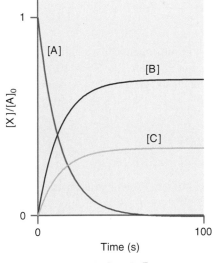

1

[A]

[B]

[C]

$[X]/[A]_0$

0

0   Time (s)   100

FIGURE 18.12
Concentrations for a parallel reaction where $k_B = 2k_C = 0.1 \text{ s}^{-1}$.

WWW

18.2 Parallel Reaction Kinetics

Because there are only two branches in this reaction, $\Phi_B = 2/3$. Inspection of Figure 18.12 reveals that $[B] = 2[C]$, which is consistent with the calculated yields.

EXAMPLE PROBLEM 18.7

In acidic conditions, benzyl penicillin (BP) undergoes the following parallel reaction:

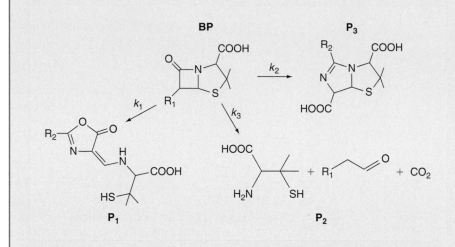

In the molecular structures, R_1 and R_2 indicate alkyl substituents. In a solution where pH = 3, the rate constants for the processes at 22°C are $k_1 = 7.0 \times 10^{-4}\,s^{-1}$, $k_2 = 4.1 \times 10^{-3}\,s^{-1}$, and $k_3 = 5.7 \times 10^{-3}\,s^{-1}$. What is the yield for P_1 formation?

Solution

Using Equation (18.79),

$$\Phi_{P_1} = \frac{k_1}{k_1 + k_2 + k_3} = \frac{7.0 \times 10^{-4}\,s^{-1}}{7.0 \times 10^{-4}\,s^{-1} + 4.1 \times 10^{-3}\,s^{-1} + 5.7 \times 10^{-3}\,s^{-1}} = 0.067$$

Of the BP that undergoes acid-catalyzed dissociation, 6.7% will result in the formation of P_1.

18.9 Temperature Dependence of Rate Constants

As mentioned at the beginning of this chapter, rate constants k are generally temperature-dependent quantities. Experimentally, it is observed that for many reactions a plot of $\ln(k)$ versus T^{-1} demonstrates linear or close to linear behavior. The following empirical relationship between temperature and k, first proposed by Arrhenius in the late 1800s, is known as the **Arrhenius expression:**

$$k = Ae^{-E_a/RT} \tag{18.82}$$

In Equation (18.82), the constant A is referred to as the **frequency factor** or **Arrhenius preexponential factor,** and E_a is the **activation energy** for the reaction. The units of the preexponential factor are identical to those of the rate constant and will vary depending on the order of the reaction. The activation energy is in units of energy mol^{-1} (for example kJ mol^{-1}). The natural log of Equation (18.82) results in the following expression:

$$\ln(k) = \ln(A) - \frac{E_a}{R}\frac{1}{T} \tag{18.83}$$

Equation (18.83) predicts that a plot of $\ln(k)$ versus T^{-1} will yield a straight line with slope equal to $-E_a/R$ and y intercept equal to $\ln(A)$. Example Problem 18.8 provides an example of the application of Equation (18.83) to determine the Arrhenius parameters for a reaction.

EXAMPLE PROBLEM 18.8

The temperature dependence of the acid-catalyzed hydrolysis of penicillin (illustrated in Example Problem 18.7) is investigated, and the dependence of k_1 on temperature is given in the following table. What is the activation energy and Arrhenius preexponential factor for this branch of the hydrolysis reaction?

Temperature (°C)	k_1 (s⁻¹)
22.2	7.0×10^{-4}
27.2	9.8×10^{-4}
33.7	1.6×10^{-3}
38.0	2.0×10^{-3}

Solution

A plot of $\ln(k_1)$ versus T^{-1} is shown here:

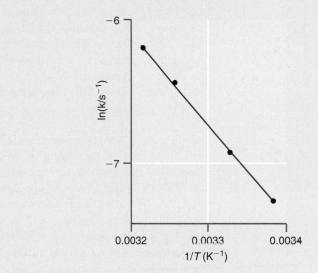

The data are indicated by the points, and the solid line corresponds to the linear least-squares fit to the data. The equation for the line is

$$\ln(k) = (-6306.3 \text{ K})\frac{1}{T} + 14.1$$

As shown in Equation (18.83), the slope of the line is equal to $-E_a/R$ such that

$$6306.3 \text{ K} = \frac{E_a}{R} \Rightarrow E_a = 52,400 \text{ J mol}^{-1} = 52.4 \text{ kJ mol}^{-1}$$

The y intercept is equal to $\ln(A)$ such that

$$A = e^{14.1} = 1.33 \times 10^6 \text{ s}^{-1}$$

The origin of the energy term in the Arrhenius expression can be understood as follows. The activation energy corresponds to the energy needed for the chemical reaction to occur. Conceptually, we envision a chemical reaction as occurring along an energy profile as illustrated in Figure 18.13. If the reactants have an energy greater than the activation energy, the reaction can proceed. The exponential dependence on the activation energy is consistent with Boltzmann statistics, with $\exp(-E_a/RT)$ representing the

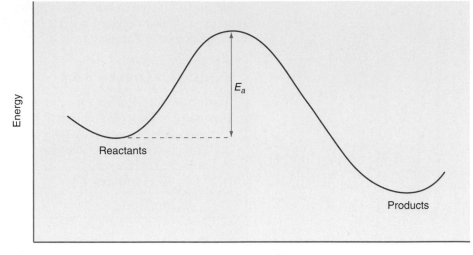

fraction of molecules with sufficient kinetic energy to undergo reaction (Chapter 13). As the activation energy increases, the fraction of molecules with sufficient energy to react will decrease as will the reaction rate.

Not all chemical reactions demonstrate Arrhenius behavior. Specifically, the inherent assumption in Equation (18.83) is that both E_a and A are temperature-independent quantities. However, there are many reactions for which a plot of $\ln(k)$ versus T^{-1} does not yield a straight line, consistent with the temperature dependence of one or both of the Arrhenius parameters. Modern theories of reaction rates predict that the rate constant will demonstrate the following behavior:

$$k = aT^m e^{-E'/RT}$$

where a and E' are temperature-independent quantities, and m can assume values such as 1, 1/2, and $-1/2$ depending on the details of the theory used to predict the rate constant. For example, in the upcoming section on activated complex theory (Section 18.14), a value of $m = 1$ is predicted. With this value for m, a plot of $\ln(k/T)$ versus T^{-1} should yield a straight line with slope equal to $-E'/R$ and the y intercept equal to $\ln(a)$. Although the limitations of the Arrhenius expression are well known, this relationship still provides an adequate description of the temperature dependence of reaction rate constants for a wide variety of reactions.

18.10 Reversible Reactions and Equilibrium

In the kinetic models discussed in earlier sections, it was assumed that once reactants form products, the opposite or "back" reaction does not occur. However, the **reaction coordinate** presented in Figure 18.14 suggests that, depending on the energetics of the reaction, such reactions can indeed occur. Specifically, the figure illustrates that reactants form products if they have sufficient energy to overcome the activation energy for the reaction. But what if the reaction coordinate is viewed from the product's perspective? Can the coordinate be followed in reverse, with products returning to reactants by overcoming the activation energy barrier from the product side, E'_a, of the coordinate? Such **reversible reactions** are discussed in this section.

Consider the following reaction in which the forward reaction is first order in A, and the back reaction is first order in B:

$$A \underset{k_B}{\overset{k_A}{\rightleftharpoons}} B \tag{18.84}$$

The forward and back rate constants are k_A and k_B, respectively. Integrated rate law expressions can be obtained for this reaction starting with the differential rate expressions for the reactant and product:

FIGURE 18.14

Reaction coordinate demonstrating the activation energy for reactants to form products, E_a, and the back reaction in which products form reactants, E_a'.

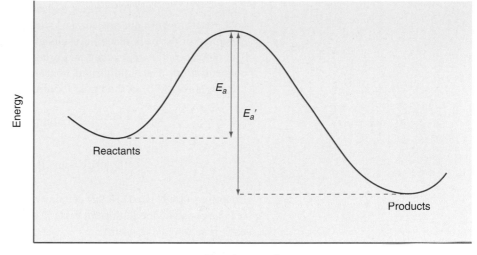

$$\frac{d[A]}{dt} = -k_A[A] + k_B[B] \tag{18.85}$$

$$\frac{d[B]}{dt} = k_A[A] - k_B[B] \tag{18.86}$$

Equation (18.85) should be contrasted with the differential rate expression for first-order reactant decay given in Equation (18.26). Reactant decay is included through the $-k_A[A]$ term similar to first-order decay discussed earlier; however, a second term involving the formation of reactant by product decay, $k_B[B]$, is now included. The initial conditions are identical to those employed in previous sections. Only reactant is present at $t = 0$, and the concentration of reactant and product for $t > 0$ must be equal to the initial concentration of reactant:

$$[A]_0 = [A] + [B] \tag{18.87}$$

With these initial conditions, Equation (18.59) can be integrated as follows:

$$\frac{d[A]}{dt} = -k_A[A] + k_B[B] \tag{18.88}$$

$$= -k_A[A] + k_B([A]_0 - [A])$$

$$= -[A](k_A + k_B) + k_B[A]_0$$

$$\int_{[A]_0}^{[A]} \frac{d[A]}{[A](k_A + k_B) - k_B[A]_0} = -\int_0^t dt$$

Equation (18.88) can be evaluated using the following standard integral:

$$\int \frac{dx}{(a + bx)} = \frac{1}{b} \ln(a + bx)$$

Using this relationship with the initial conditions specified earlier, the concentrations of reactant and products are

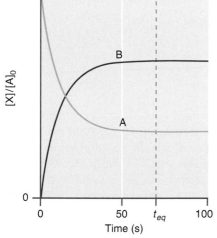

FIGURE 18.15

Time-dependent concentrations in which both forward and back reactions exist between reactant A and product B. In this example, $k_A = 2k_B = 0.06 \text{ s}^{-1}$. Note that the concentrations reach a constant value at longer times ($t \geq t_{eq}$) at which point the reaction reaches equilibrium.

$$[A] = [A]_0 \frac{k_B + k_A e^{-(k_A + k_B)t}}{k_A + k_B} \tag{18.89}$$

$$[B] = [A]_0 \left(1 - \frac{k_B + k_A e^{-(k_A + k_B)t}}{k_A + k_B} \right) \tag{18.90}$$

Figure 18.15 presents the time dependence of [A] and [B] for the case where $k_A = 2k_B = 0.06 \text{ s}^{-1}$. Note that [A] undergoes exponential decay with an apparent rate constant equal to $k_A + k_B$, and [B] appears exponentially with an equivalent rate constant. If the back reaction were not present, [A] would be expected to decay to zero; however, the existence of the back reaction results in both [A] and [B] being nonzero at long times. The concentration of reactant and product at long times is defined as the equilibrium concentration. The equilibrium concentrations are equal to the limit of Equations (18.89) and (18.90) as time goes to infinity:

$$[A]_{eq} = \lim_{t \to \infty}[A] = [A]_0 \frac{k_B}{k_A + k_B} \tag{18.91}$$

$$[B]_{eq} = \lim_{t \to \infty}[B] = [A]_0 \left(1 - \frac{k_B}{k_A + k_B}\right) \tag{18.92}$$

Equations (18.91) and (18.92) demonstrate that the reactant and product concentrations reach a constant or equilibrium value that depends on the relative size of the forward and back reaction rate constants.

Theoretically, one must wait an infinite amount of time before equilibrium is reached. In practice, there will be a time after which the reactant and product concentrations are sufficiently close to equilibrium, and the change in these concentrations with time is so modest, that approximating the system as having reached equilibrium is reasonable. This time is indicated by t_{eq} in Figure 18.15, where inspection of the figure demonstrates that the concentrations are near their equilibrium values for times after t_{eq}. After equilibrium has been established, the time independence of the reactant and product concentrations can be expressed as

$$\frac{d[A]_{eq}}{dt} = \frac{d[B]_{eq}}{dt} = 0 \tag{18.93}$$

The subscripts in Equation (18.93) indicate that equality applies only after equilibrium has been established. A common misconception is that Equation (18.93) states that at equilibrium the forward and back reaction rates are zero. Instead, at equilibrium the forward and back reaction rates are equal, but not zero, such that the macroscopic concentration of reactant or product does not evolve with time. That is, the forward and back reactions still occur, but they occur with equal rates at equilibrium. Using Equation (18.93) in combination with the differential rate expressions for the reactant [Equation (18.85)], we arrive at what is hopefully a familiar relationship:

$$\frac{d[A]_{eq}}{dt} = \frac{d[B]_{eq}}{dt} = 0 = -k_A[A]_{eq} + k_B[B]_{eq}$$

$$\frac{k_A}{k_B} = \frac{[B]_{eq}}{[A]_{eq}} = K_c \tag{18.94}$$

In this equation, K_c is the equilibrium constant defined in terms of concentration. This quantity is identical to that first encountered in thermodynamics (Chapter 6) and statistical mechanics (Chapter 15). We now have a definition of equilibrium from the kinetic perspective; therefore, Equation (18.94) is a remarkable result in which the concept of equilibrium as described by these three different perspectives is connected into one deceptively simple equation. From the kinetic standpoint, K_c is related to the ratio of forward and back rate constants for the reaction. The greater the forward rate constant relative to that for the back reaction, the more equilibrium will favor products over reactants.

Figure 18.16 illustrates the methodology by which forward and back rate constants can be determined. Specifically, measurement of the reactant decay kinetics (or equivalently the product formation kinetics) provides a measure of the apparent rate constant, $k_A + k_B$. The measurement of K_c, or the reactant and product concentrations at equilibrium, provides a measure of the ratio of the forward and backward rate constants. Together, these measurements represent a system of two equations and two unknowns that can be readily solved to determine k_A and k_B.

FIGURE 18.16

Methodology for determining forward and back rate constants. The apparent rate constant for reactant decay is equal to the sum of forward, k_A, and back, k_B, rate constants. The equilibrium constant is equal to k_A/k_B. These two measurements provide a system of two equations and two unknowns that can be readily evaluated to produce k_A and k_B.

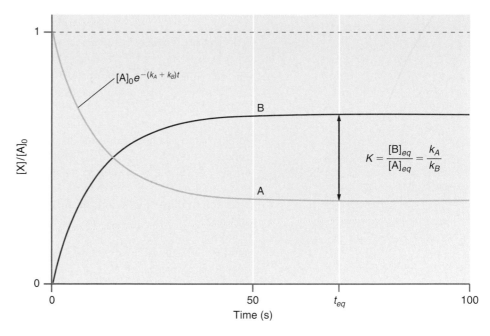

EXAMPLE PROBLEM 18.9

Consider the interconversion of the "boat" and "chair" conformations of cyclohexane:

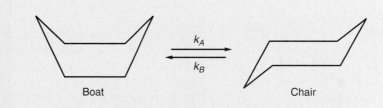

The reaction is first order in each direction, with an equilibrium constant of 10^4. The activation energy for the conversion of the chair conformer to the boat conformer is 42 kJ/mol. Assuming an Arrhenius preexponential factor of 10^{12} s^{-1}, what is the expected observed reaction rate constant at 298 K if one were to initiate this reaction starting with only the boat conformer?

Solution

Using the Arrhenius expression of Equation (18.56), k_B is given by

$$k_B = Ae^{-E_a/RT} = 10^{12}\,\text{s}^{-1}\exp\left[\frac{-42{,}000\ \text{J mol}^{-1}}{(8.314\ \text{J mol}^{-1}\ \text{K}^{-1})(298\ \text{K})}\right]$$

$$= 4.34 \times 10^4\,\text{s}^{-1}$$

Using the equilibrium constant, k_A can be determined as follows:

$$K_c = 10^4 = \frac{k_A}{k_B}$$

$$k_A = 10^4 k_B = 10^4(4.34 \times 10^4\,\text{s}^{-1}) = 4.34 \times 10^8\,\text{s}^{-1}$$

Finally, the apparent rate constant is simply the sum of k_A and k_B:

$$k_{app} = k_A + k_B \approx 4.34 \times 10^8\,\text{s}^{-1}$$

18.11 Perturbation-Relaxation Methods

The previous section demonstrated that for reactions with appreciable forward and backward rate constants, concentrations approaching those at equilibrium will be established at some later time after initiation of the reaction. The forward and backward rate constants for such reactions can be determined by monitoring the evolution in reactant or product concentrations as equilibrium is approached, and by measuring the concentrations at equilibrium. But what if the initial conditions for the reaction cannot be controlled? For example, what if it is impossible to sequester the reactants such that initiation of the reaction at a specified time is impossible? In such situations, application of the methodology described in the preceding section to determine forward and back rate constants is not possible. However, if one can perturb the system by changing temperature, pressure, or concentration, the system will no longer be at equilibrium and will evolve until a new equilibrium is established. If the perturbation occurs on a timescale that is rapid compared to the system relaxation, the kinetics of the relaxation can be monitored and related to the forward and back rate constants. This is the conceptual idea behind perturbation methods and their application to chemical kinetics.

There are many perturbation techniques; however, the focus here is on **temperature jump** (or T-jump) methods to illustrate the type of information available using perturbation techniques. Consider again the following reaction in which both the forward and back reactions are first order:

$$A \underset{k_B}{\overset{k_A}{\rightleftharpoons}} B \qquad (18.95)$$

Next, a rapid change in temperature occurs such that the forward and back rate constants are altered in accord with the Arrhenius expression of Equation (18.82), and a new equilibrium is established:

$$A \underset{k_B^+}{\overset{k_A^+}{\rightleftharpoons}} B \qquad (18.96)$$

The superscript $+$ in this expression indicates that the rate constants correspond to the conditions after the temperature jump. Following the temperature jump, the concentrations of reactants and products will evolve until the new equilibrium concentrations are reached. At the new equilibrium, the differential rate expression for the reactant is equal to zero so that

$$\frac{d[A]_{eq}}{dt} = 0 = -k_A^+[A]_{eq} + k_B^+[B]_{eq} \qquad (18.97)$$

$$k_A^+[A]_{eq} = k_B^+[B]_{eq}$$

The subscripts on the reactant and product concentrations represent the new equilibrium concentrations after the temperature jump. The evolution of reactant and product concentrations from the pre-temperature to post-temperature jump values can be expressed using a coefficient of reaction advancement (Section 18.2). Specifically, let the variable ξ represent the extent to which the pre-temperature jump concentration is shifted away from the concentration for the post-temperature jump equilibrium:

$$[A] - \xi = [A]_{eq} \qquad (18.98)$$

$$[B] + \xi = [B]_{eq} \qquad (18.99)$$

Immediately after the temperature jump, the concentrations will evolve until equilibrium is reached. Using this idea, the differential rate expression describing the extent of reaction advancement is as follows:

$$\frac{d\xi}{dt} = -k_A^+[A] + k_B^+[B]$$

Notice in this equation that the forward and back rate constants are the post-temperature jump values. Substitution of Equations (18.98) and (18.99) into the differential rate expression yields the following:

$$\frac{d\xi}{dt} = -k_A^+(\xi + [A]_{eq}) + k_B^+(-\xi + [B]_{eq})$$
$$= -k_A^+[A]_{eq} + k_B^+[B]_{eq} - \xi(k_A^+ + k_B^+)$$
$$= -\xi(k_A^+ + k_B^+) \tag{18.100}$$

In the final step of the preceding equation, derivation was performed recognizing that at equilibrium the first two terms cancel in accord with Equation (18.97). The relaxation time, τ, is defined as follows:

$$\tau = (k_A^+ + k_B^+)^{-1} \tag{18.101}$$

Employing the relaxation time, Equation (18.100) is readily evaluated:

$$\frac{d\xi}{dt} = -\frac{\xi}{\tau}$$
$$\int_{\xi_0}^{\xi} \frac{d\xi}{\xi} = -\frac{1}{\tau}\int_0^t dt$$
$$\xi = \xi_0 e^{-t/\tau} \tag{18.102}$$

Equation (18.102) demonstrates that for this reaction, the concentrations will change exponentially, and that the relaxation time is the time it takes for the coefficient of reaction advancement to decay to e^{-1} of its initial value. The timescale for relaxation after the temperature jump is related to the sum of the forward and back rate constants. This information in combination with the equilibrium constant (given by measurement of $[A]_{eq}$ and $[B]_{eq}$) can be used to determine the individual values for the rate constants. Figure 18.17 presents a schematic of this process.

FIGURE 18.17
Example of a temperature-jump experiment for a reaction in which the forward and back rate processes are first order. The yellow and blue portions of the graph indicate times before and after the temperature jump, respectively. After the temperature jump, [A] decreases with a time constant related to the sum of the forward and back rate constants. The change between the pre-jump and post-jump equilibrium concentrations is given by ξ_0.

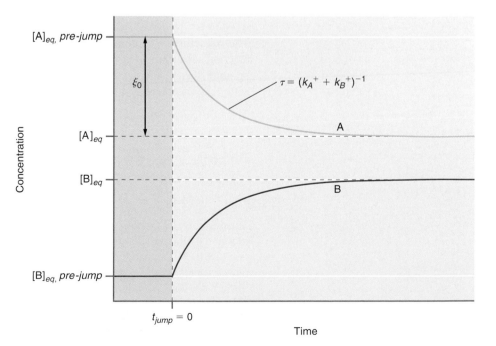

18.12 The Autoionization of Water: A T-Jump Example

In the autoionization of water, the equilibrium of interest is the following:

$$H_2O(aq) \underset{k_r}{\overset{k_f}{\rightleftharpoons}} H^+(aq) + OH^-(aq) \qquad (18.103)$$

The reaction is first order in the forward direction, and second order in the reverse direction. The differential rate expressions that describe the temporal evolution of H_2O and H^+ concentrations are as follows:

$$\frac{d[H_2O]}{dt} = -k_f[H_2O] + k_r[H^+][OH^-] \qquad (18.104)$$

$$\frac{d[H^+]}{dt} = k_f[H_2O] - k_r[H^+][OH^-] \qquad (18.105)$$

Following a temperature jump to 298 K, the measured relaxation time constant was 37 μs. In addition, the pH of the solution is 7. Given this information, the forward and back rate constants can be determined as follows. First, the equilibrium constant after the temperature jump is

$$\frac{k_f^+}{k_r^+} = \frac{[H^+]_{eq}[OH^-]_{eq}}{[H_2O]_{eq}} = K_c \qquad (18.106)$$

The differential rate expression for the extent of reaction advancement after the perturbation is given by

$$\frac{d\xi}{dt} = -k_f^+[H_2O] + k_r^+[H^+][OH^-] \qquad (18.107)$$

$$= -k_f^+(\xi + [H_2O]_{eq}) + k_r^+(\xi - [H^+]_{eq})(\xi - [OH^-]_{eq})$$

$$= -k_f^+\left(\xi + \frac{k_r^+}{k_f^+}[H^+]_{eq}[OH^-]_{eq}\right) + k_r^+(\xi - [H^+]_{eq})(\xi - [OH^-]_{eq})$$

$$= -k_f^+\xi - k_r^+\xi([H^+]_{eq} + [OH^-]_{eq}) + O(\xi^2)$$

The last term in Equation (18.107) represents terms on the order of ξ^2. If the extent of reaction advancement is small, corresponding to a small perturbation of the system temperature, then this term can be neglected resulting in the following expression for the reaction advancement as a function of time:

$$\frac{d\xi}{dt} = -\xi(k_f^+ + k_r^+([H^+]_{eq} + [OH^-]_{eq})) \qquad (18.108)$$

Proceeding as before, the relaxation time is defined as

$$\frac{1}{\tau} = (k_f^+ + k_r^+([H^+]_{eq} + [OH^-]_{eq})) \qquad (18.109)$$

Substitution of Equation (18.109) into Equation (18.108) and integration yields an expression for the post-temperature jump evolution identical to that derived earlier in Equation (18.102). The parameters needed to determine the autoionization forward and back rate constants are the expression for the relaxation time [Equation (18.109)] and the equilibrium constant. Recall that the experimental relaxation time was 37 μs (1 μs = 10^{-6} s) such that

$$\frac{1}{3.7 \times 10^{-5}\,s} = (k_f^+ + k_r^+([H^+]_{eq} + [OH^-]_{eq})) \qquad (18.110)$$

In addition, the pH at equilibrium is 7 such that $[H^+] = [OH^-] = 1 \times 10^{-7}$ M. Finally, the concentration of water at 298 K is 55.4 M. Using this information, the ratio of the forward to back rate constants becomes

$$\frac{k_f^+}{k_r^+} = \frac{[H^+]_{eq}[OH^-]_{eq}}{[H_2O]_{eq}} = \frac{(1 \times 10^{-7}\,M)(1 \times 10^{-7}\,M)}{55.4\,M} = 1.81 \times 10^{-16}\,M \quad (18.111)$$

Substitution of Equation (18.111) into Equation (18.110) yields the following value for the reverse rate constant:

$$\frac{1}{3.7 \times 10^{-5}\,s} = (k_f^+ + k_r^+([H^+]_{eq} + [OH^-]_{eq}))$$

$$= (1.81 \times 10^{-16}\,M(k_r^+) + k_r^+(2 \times 10^{-7}\,M))$$

$$\frac{1}{(3.7 \times 10^{-5}\,s)(2 \times 10^{-7}\,M)} \cong k_r^+$$

$$1.35 \times 10^{11}\,M^{-1}s^{-1} \cong k_r^+$$

Finally, the forward rate constant is

$$k_f^+ = (k_r^+)1.83 \times 10^{-16}\,M = (1.35 \times 10^{11}\,M^{-1}s^{-1})(1.83 \times 10^{-16}\,M) = 2.47 \times 10^{-5}\,s^{-1}$$

Notice the substantial difference between the forward and back rate constants, consistent with the modest amount of autoionized species in water. In addition, the forward and reverse rate constants are temperature dependent, and the autoionization constant also demonstrates temperature dependence.

18.13 Potential Energy Surfaces

In the discussion of the Arrhenius equation, the energetics of the reaction were identified as an important factor determining the rate of a reaction. This connection between reaction kinetics and energetics is central to the concept of the potential energy surface. To illustrate this concept, consider the following bimolecular reaction:

$$AB + C \longrightarrow A + BC \quad (18.112)$$

The diatomic species AB and BC are stable, but we will assume that the triatomic species ABC and the diatomic species AC are not formed during the course of the reaction. This reaction can be viewed as the interaction of three atoms, and the potential energy of this collection of atoms can be defined with respect to the relative positions in space. The geometric relationship between these species is generally defined with respect to the distance between two of the three atoms (R_{AB} and R_{BC}), and the angle formed between these two distances, as illustrated in Figure 18.18.

The potential energy of the system can be expressed as a function of these coordinates. The variation of the potential energy with a change along these coordinates can then be presented as a graph or surface referred to as a **potential energy surface.** Formally, for our example reaction this surface would be four dimensional (the three geometric coordinates and energy). The dimensionality of the problem can be reduced by considering the energetics of the reaction at a fixed value for one of the geometric coordinates. In the example reaction, the centers of A, B, and C must be aligned during the reaction such that $\theta = 180°$. With this requirement, the potential energy is reduced to a three-dimensional problem, and plots of the potential energy surface for the case where $\theta = 180°$ are presented in Figure 18.19. The graphs represent the variation in energy with displacement along R_{AB} and R_{BC} with the arrows indicating the direction of increased separation.

Figures 18.19a and b illustrate the three-dimensional potential energy surface and the two minima in this surface corresponding to the stable diatomic molecules AB and BC. A more convenient way to view the potential energy surface is to use a two-dimensional

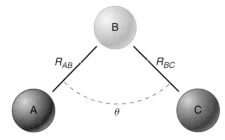

FIGURE 18.18
Definition of geometric coordinates for the AB + C → A + BC reaction.

FIGURE 18.19

Illustration of a potential energy surface for the AB + C reaction at a colinear geometry ($\theta = 180°$ in Figure 18.18). **(a, b)** Three-dimensional views of the surface. **(c)** Contour plot of the surface with contours of equipotential energy. The curved dashed line represents the path of a reactive event, corresponding to the reaction coordinate. The transition state for this coordinate is indicated by the symbol ‡. **(d, e)** Cross sections of the potential energy surface along the lines $a'–a$ and $b'–b$, respectively. These two graphs correspond to the potential for two-body interactions of B with C, and A with B. [Adapted from J. H. Noggle, *Physical Chemistry*, Harper Collins, New York, 1996.]

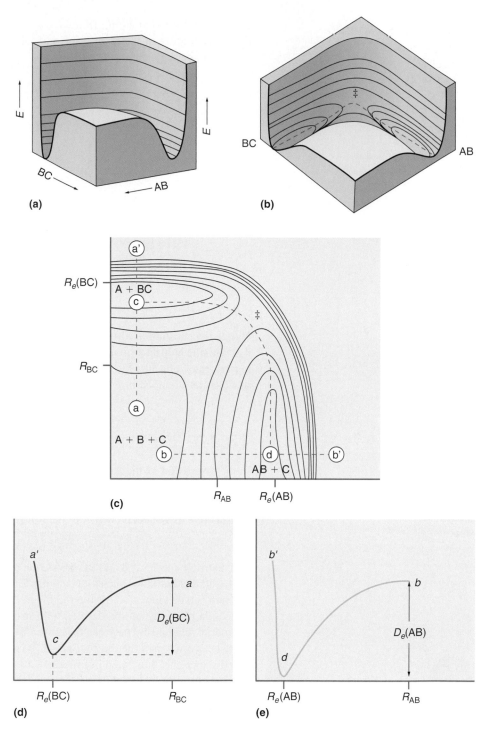

contour plot, as illustrated in Figure 18.19c. One can think of this plot as a view straight down onto the three-dimensional surface presented in Figure 18.19b. The lines on the contour plot connect regions of equal energy. On the lower left-hand region of the surface is a broad energetic plateau that corresponds to the energy when the three atoms are separated, or the dissociated state A + B + C. The pathway corresponding to the reaction of B + C to form BC is indicated by the dashed line between points a and a'. The cross section of the potential energy surface along this line is presented in Figure 18.19d, and this contour is simply the potential energy diagram for the diatomic molecule BC. The depth of the potential is equal to the dissociation energy of the diatomic, $D_e(\text{BC})$, and the minimum along R_{BC} corresponds to the equilibrium bond length of the diatomic. Figure 18.19e presents the corresponding diagram for the diatomic AB, as indicated by the dashed line between points b and b' in Figure 18.19c.

FIGURE 18.20
Reaction coordinates involving an activated complex and a reactive intermediate. The graph corresponds to the reaction coordinate derived from the dashed line between points c and d on the contour plot of Figure 18.19c. The maximum in energy along this coordinate corresponds to the transition state, and the species at this maximum is referred to as an activated complex.

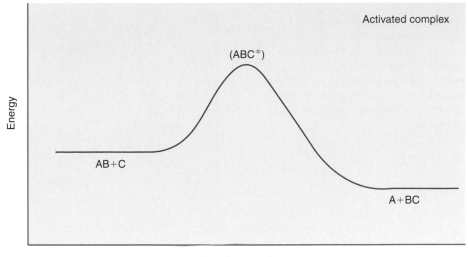

The dashed line between points c and d in Figure 18.19c represents the system energy as C approaches AB and reacts to form BC and A under the constraint that $\theta = 180°$. This pathway represents the AB + C → A + BC reaction. The maximum in energy along this pathway is referred to as the **transition state** and is indicated by the double dagger symbol, ‡. The variation in energy as one proceeds from reacting to products along this reactive pathway can be plotted to construct a reaction coordinate as presented in Figure 18.20. Note that the transition state corresponds to a maximum along this coordinate; therefore, the activated complex is not a stable species (i.e., an intermediate) along the reaction coordinate.

The discussion of potential energy surfaces just presented suggests that the kinetics and product yields will depend on the energy content of the reactants and the relative orientation of reactants. This sensitivity can be explored using techniques of crossed-molecular beams. In this approach, reactants with well-defined energies are seeded into a molecular beam that intersects another beam of reactants at well-defined beam geometries. The products formed in the reaction can be analyzed in terms of their energetics, spatial distribution of the products, and beam geometry. This experimental information is then used to construct a potential energy surface (following a substantial amount of analysis). Crossed-molecular beam techniques have provided much insight into the nature of reactive pathways, and detailed, introductory references to this important area of research are presented at the end of this chapter.

18.14 Activated Complex Theory

The concept of equilibrium is central to a theoretical description of reaction rates developed principally by Henry Eyring in the 1930s. This theory, known as **activated complex theory** or **transition state theory,** provides a theoretical description of reaction rates. To illustrate the conceptual ideas behind activated complex theory, consider the following bimolecular reaction:

$$A + B \xrightarrow{k} P \tag{18.113}$$

Figure 18.21 illustrates the reaction coordinate for this process, where A and B react to form an activated complex that undergoes decay, resulting in product formation. The **activated complex** represents the system at the transition state. This complex is not stable and has a lifetime on the order of one or a few vibrational periods ($\sim 10^{-14}$ s). When this theory was first proposed, experiments were incapable of following reaction dynamics on such short timescales such that evidence for an activated complex corresponding to the transition state was not available. However, recent developments in

FIGURE 18.21
Illustration of transition state theory.
Similar to reaction coordinates depicted
previously, the reactants (A and B) and
products (P) are separated by an energy
barrier. The transition state is an activated
reactant complex envisioned to exist at the
free-energy maximum along the reaction
coordinate.

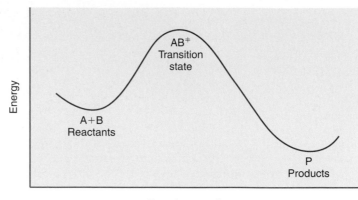

experimental kinetics have allowed for the investigation of these transient species, and
a few references to this work are provided at the end of this chapter.

Activated complex theory involves a few major assumptions that are important to
acknowledge before proceeding. The primary assumption is that an equilibrium exists
between the reactants and the activated complex. It is also assumed that the reaction coordinate describing decomposition of the activated complex can be mapped onto a single energetic degree of freedom of the activated complex. For example, if product
formation involves the breaking of a bond, then the vibrational degree of freedom corresponding to bond stretching is taken to be the reactive coordinate.

With these approximations in mind, we can take the kinetic methods derived earlier
in this chapter and develop an expression for the rate of product formation. For the example bimolecular reaction from Equation (18.113), the kinetic mechanism corresponding to the activated complex model described earlier is

$$A + B \underset{k_{-1}}{\overset{k_1}{\rightleftharpoons}} AB^{\ddagger} \tag{18.114}$$

$$AB^{\ddagger} \xrightarrow{k_2} P \tag{18.115}$$

Equation (18.114) represents the equilibrium between reactants and the activated complex,
and Equation (18.115) represents the decay of the activated complex to form product. With
the assumption of an equilibrium between the reactants and the activated complex, the differential rate expression for one of the reactants (A in this case) is set equal to zero consistent with equilibrium, and an expression for $[AB^{\ddagger}]$ is obtained as follows:

$$\frac{d[A]}{dt} = 0 = -k_1[A][B] + k_{-1}[AB^{\ddagger}]$$

$$[AB^{\ddagger}] = \frac{k_1}{k_{-1}}[A][B] = \frac{K_c^{\ddagger}}{c^{\circ}}[A][B] \tag{18.116}$$

In Equation (18.116), $K_c^{\ddagger}$ is the equilibrium constant involving the reactants and the activated complex, and it can be expressed in terms of the molecular partition functions of
these species as described in Chapter 15. In addition, c° is the standard state concentration (typically 1 M), which appears in the following definition for $K_c^{\ddagger}$:

$$K_c^{\ddagger} = \frac{[AB^{\ddagger}]/c^{\circ}}{([A]/c^{\circ})([B]/c^{\circ})} = \frac{[AB^{\ddagger}]c^{\circ}}{[A][B]}$$

The rate of the reaction is equal to the rate of product formation, which by Equation (18.115) is equal to

$$R = \frac{d[P]}{dt} = k_2[AB^{\ddagger}] \tag{18.117}$$

Substitution into Equation (18.117) the expression for $[AB^\ddagger]$ provided in Equation (18.116) yields the following expression for the reaction rate:

$$R = \frac{d[P]}{dt} = \frac{k_2 K_c^\ddagger}{c^\circ}[A][B] \qquad (18.118)$$

Further evaluation of the reaction rate expression requires that k_2 be defined. This rate constant is associated with the rate of activated complex decay. Imagine that product formation requires the dissociation of a bond in the activated complex. The activated complex is not stable; therefore, the dissociating bond must be relatively weak and the complex can dissociate with the initial motion along the corresponding bond-stretching coordinate. Therefore, k_2 is related to the vibrational frequency associated with bond stretching, ν. The rate constant is equal to ν only if every time an activated complex is formed, it dissociates resulting in product formation. However, it is possible that the activated complex will instead revert back to reactants. If the reverse reaction can occur, then only a fraction of the activated complexes that are formed will continue along the reaction coordinate and result in product formation. To account for this possibility, a term referred to as the transmission coefficient, κ, is included in the definition of k_2:

$$k_2 = \kappa\nu \qquad (18.119)$$

With this definition of k_2, the reaction rate becomes

$$R = \frac{\kappa\nu K_c^\ddagger}{c^\circ}[A][B] \qquad (18.120)$$

As stated earlier, one can express $K_c^\ddagger$ in terms of the partition function of reactants and the activated complex using the techniques outlined in Chapter 15. In addition, the partition function for the activated complex can decompose into a product of partition functions corresponding to the reactive coordinate and the remaining energetic degrees of freedom. Removing the partition function for the reactive coordinate from the expression of $K_c^\ddagger$ yields

$$K_c^\ddagger = q_{rc}\overline{K}_c^\ddagger = \frac{k_B T}{h\nu}\overline{K}_c^\ddagger \qquad (18.121)$$

where q_{rc} is the partition function associated with the reactive coordinate, $\overline{K}_c^\ddagger$ is the remainder of the original equilibrium constant in the absence of q_{rc}, and k_B is Boltzmann's constant. The final equality in Equation (18.121) is made by recognizing that the reactive vibrational coordinate corresponds to a weak bond for which $h\nu \ll kT$, and the high-temperature approximation for q_{rc} is valid. Substituting Equation (18.121) into Equation (18.120) yields the following expression for k_2:

$$k_2 = \kappa\frac{k_B T}{hc^\circ}\overline{K}_c^\ddagger \qquad (18.122)$$

Equation (18.122) is the central result of activated complex theory, and it provides a connection between the rate constant for product formation and the molecular parameters for species involved in the reaction. Evaluation of this rate expression requires that one determine $\overline{K}_c^\ddagger$, which is related to the partition functions of the activated complex and reactants (Chapter 15). The partition functions for the reactants can be readily determined using the techniques discussed in Chapters 14 and 15; however, the partition function for the activated complex requires some thought.

The translational partition function for the complex can also be determined using the techniques described earlier, but determination of the rotational and vibrational partition functions requires some knowledge of the structure of the activated complex. The determination of the vibrational partition function is further complicated by the requirement that one of the vibrational degrees of freedom be designated as the reactive coordinate; however, identification of this coordinate may be far from trivial for an activated complex with more than one weak bond. At times, computational techniques, as discussed in

Chapter 16 of Engel's *Quantum Chemistry and Spectroscopy,* can be used to provide insight into the structure of the activated complex and assist in determination of the partition function for this species. With these complications acknowledged, Equation (18.122) represents an important theoretical accomplishment in chemical reaction kinetics.

Note that the presentation of activated complex theory provided here is a very rudimentary description of this field. Work continues to the present day to advance and refine this theory, and references are provided at the end of this chapter to review articles describing the significant advances in this field.

We end this discussion by connecting the results of activated complex theory to earlier thermodynamics descriptions of chemical reactions. Recall from thermodynamics that equilibrium constant $K_c^{\ddagger}$ is related to the corresponding change in Gibbs energy using the following thermodynamic definition:

$$\Delta G^{\ddagger} = -RT \ln K_c^{\ddagger} \tag{18.123}$$

In this definition, $\Delta G^{\ddagger}$ is the difference in Gibbs energy between the transition state and the reactants. With this definition for $K_c^{\ddagger}$, k_2 becomes (with $\kappa = 1$ for convenience)

$$k_2 = \frac{k_B T}{hc^{\circ}} e^{-\Delta G^{\ddagger}/RT} \tag{18.124}$$

In addition, $\Delta G^{\ddagger}$ can be related to the corresponding changes in enthalpy and entropy using

$$\Delta G^{\ddagger} = \Delta H^{\ddagger} - T\Delta S^{\ddagger} \tag{18.125}$$

Substituting Equation (18.125) into Equation (18.124) yields

$$k_2 = \frac{k_B T}{hc^{\circ}} e^{\Delta S^{\ddagger}/R} e^{-\Delta H^{\ddagger}/RT} \tag{18.126}$$

Equation (18.126) is known as the **Eyring equation.** Notice that the temperature dependence of the reaction rate constant predicted by transition state theory is different than that assumed by the Arrhenius expression of Equation (18.82). In particular, the preexponential term in the Eyring equation demonstrates temperature dependence as opposed to the assumed temperature independence of the corresponding term in the Arrhenius expression. However, both the Eyring equation and the Arrhenius expression provide an expression for the temperature dependence of rate constants; therefore, one might expect that the parameters in the Eyring equation ($\Delta H^{\ddagger}$ and $\Delta S^{\ddagger}$) can be related to corresponding parameters in the Arrhenius expression (E_a and A). To derive this relationship, we begin with a modification of Equation (18.82) where the Arrhenius activation energy is written as

$$E_a = RT^2 \left(\frac{d \ln k}{dT} \right) \tag{18.127}$$

Substituting for k the expression for k_2 given in Equation (18.122) yields

$$E_a = RT^2 \left(\frac{d}{dT} \ln \left(\frac{kT}{hc^{\circ}} \bar{K}_c^{\ddagger} \right) \right) = RT + RT^2 \left(\frac{d \ln \bar{K}_c^{\ddagger}}{dT} \right)$$

From thermodynamics (Chapter 6), the temperature derivative of $\ln(K_c)$ is equal to $\Delta U/RT^2$. Employing this definition to the previous equation results in the following:

$$E_a = RT + \Delta U^{\ddagger}$$

We also make use of the thermodynamic definition of enthalpy, $H = U + PV$, to write

$$\Delta U^{\ddagger} = \Delta H^{\ddagger} - \Delta(PV)^{\ddagger} \tag{18.128}$$

In Equation (18.128), the $\Delta(PV)^{\ddagger}$ term is related to the difference in the product PV with respect to the activated complex and reactants. For a solution-phase reaction, P is constant and the change in V is negligible such that $\Delta U^{\ddagger} \cong \Delta H^{\ddagger}$ and the activation energy in terms of $\Delta H^{\ddagger}$ becomes

$$E_a = \Delta H^{\ddagger} + RT \quad \text{(solutions)} \tag{18.129}$$

Comparison of this result with Equation (18.126) demonstrates that the Arrhenius pre-exponential factor in this case is

$$A = \frac{ek_B T}{hc^{\circ}} e^{\Delta S^{\ddagger}/R} \quad \text{(solutions, bimolecular)} \tag{18.130}$$

For solution-phase unimolecular reactions, $\Delta U^{\ddagger} \cong \Delta H^{\ddagger}$, and the activation energy for a unimolecular solution phase reaction is identical to Equation (18.129). All that changes relative to the bimolecular case is the Arrhenius preexponential factor, resulting in

$$A = \frac{ek_B T}{h} e^{\Delta S^{\ddagger}/R} \quad \text{(solutions, unimolecular)} \tag{18.131}$$

For a gas-phase reaction, $\Delta(PV)^{\ddagger}$ in Equation (18.128) is proportional to the difference in the number of moles between the transition state and reactants. For a unimolecular ($\Delta n^{\ddagger} = 0$), bimolecular ($\Delta n^{\ddagger} = -1$), and trimolecular ($\Delta n^{\ddagger} = -2$) reaction, E_a and A are given by

$$gas, uni \qquad E_a = \Delta H^{\ddagger} + RT \qquad A = \frac{ek_B T}{h} e^{\Delta S^{\ddagger}/R} \tag{18.132}$$

$$gas, bi \qquad E_a = \Delta H^{\ddagger} + 2RT \qquad A = \frac{e^2 k_B T}{hc^{\circ}} e^{\Delta S^{\ddagger}/R} \tag{18.133}$$

$$gas, tri \qquad E_a = \Delta H^{\ddagger} + 3RT \qquad A = \frac{e^3 k_B T}{h(c^{\circ})^2} e^{\Delta S^{\ddagger}/R} \tag{18.134}$$

Notice now that both the Arrhenius activation energy and preexponential terms are expected to demonstrate temperature dependence. If $\Delta H^{\ddagger} \gg RT$, then the temperature dependence of E_a will be modest. Also notice that if the enthalpy of the transition state is lower than that of the reactants, then the reaction rate may become faster as temperature is decreased! However, the entropy difference between the transition state and reactants is also important in determining the rate. If this entropy difference is positive and the activation energy is near zero, then the reaction rate is determined by entropic rather than enthalpic factors.

EXAMPLE PROBLEM 18.10

The thermal decomposition reaction of nitrosyl halides is important in tropospheric chemistry. For example, consider the decomposition of NOCl:

$$2NOCl(g) \longrightarrow 2NO(g) + Cl_2(g)$$

The Arrhenius parameters for this reaction are $A = 1.00 \times 10^{13} \text{ M}^{-1} \text{ s}^{-1}$ and $E_a = 104 \text{ kJ mol}^{-1}$. Calculate $\Delta H^{\ddagger}$ and $\Delta S^{\ddagger}$ for this reaction with $T = 300$ K.

Solution

This is a bimolecular reaction such that

$$\Delta H^{\ddagger} = E_a - 2RT = 104 \text{ kJ mol}^{-1} - 2(8.314 \text{ J mol}^{-1} \text{ K}^{-1})(300 \text{ K})$$

$$= 104 \text{ kJ mol}^{-1} - (4.99 \times 10^3 \text{ J mol}^{-1})\left(\frac{1 \text{ kJ}}{1000 \text{ J}}\right) = 99.0 \text{ kJ mol}^{-1}$$

$$\Delta S^{\ddagger} = R \ln\left(\frac{Ahc^{\circ}}{e^2 kT}\right)$$

$$= (8.314 \text{ J mol}^{-1}\text{K}^{-1}) \ln\left(\frac{(1.00 \times 10^{13} \text{ M}^{-1} \text{ s}^{-1})(6.626 \times 10^{-34} \text{ J s})(1 \text{ M})}{e^2 (1.38 \times 10^{-23} \text{ J K}^{-1})(300 \text{ K})}\right)$$

$$= -12.7 \text{ J mol}^{-1}\text{K}^{-1}$$

One of the utilities of this calculation is that the sign and magnitude of $\Delta S^{\ddagger}$ provide information on the structure of the activated complex at the transition state relative to the reactants. The negative value in this example illustrates that the activated complex has a lower entropy (or is more ordered) than the reactants. This observation is consistent with a mechanism in which the two NOCl reactants form a complex that eventually decays to produce NO and Cl.

For Further Reading

Brooks, P. R., "Spectroscopy of Transition Region Species," *Chemical Reviews* 87 (1987), 167.

Callender, R. H., R. B. Dyer, R. Blimanshin, and W. H. Woodruff, "Fast Events in Protein Folding: The Time Evolution of a Primary Process," *Annual Review of Physical Chemistry* 49 (1998), 173.

Castellan, G. W., *Physical Chemistry*. Addison-Wesley, Reading, MA, 1983.

Eyring, H., S. H. Lin, and S. M. Lin, *Basic Chemical Kinetics*. Wiley, New York, 1980.

Frost, A. A., and R. G. Pearson, *Kinetics and Mechanism*. Wiley, New York, 1961.

Hammes, G. G., *Thermodynamics and Kinetics for the Biological Sciences*. Wiley, New York, 2000.

Laidler, K. J., *Chemical Kinetics*. Harper & Row, New York, 1987.

Martin, J.-L., and M. H. Vos, "Femtosecond Biology," *Annual Review of Biophysical and Biomolecular Structure* 21 (1992), 1999.

Pannetier, G., and P. Souchay, *Chemical Kinetics*. Elsevier, Amsterdam, 1967.

Schoenlein, R. W., L. A. Peteanu, R. A. Mathies, and C. V. Shank, "The First Step in Vision: Femtosecond Isomerization of Rhodopsin," *Science* 254 (1991), 412.

Steinfeld, J. I., J. S. Francisco, and W. L. Hase. *Chemical Kinetics and Dynamics*. Prentice-Hall, Upper Saddle River, NJ, 1999.

Truhlar, D. G., W. L. Hase, and J. T. Hynes, "Current Status in Transition State Theory," *Journal of Physical Chemistry* 87 (1983), 2642.

Vos, M. H., F. Rappaport, J.-C. Lambry, J. Breton, and J.-L. Martin, "Visualization of Coherent Nuclear Motion in a Membrane Protein by Femtosecond Spectroscopy," *Nature* 363 (1993), 320.

Zewail, H., "Laser Femtochemistry," *Science* 242 (1988), 1645.

Vocabulary

activated complex

activated complex theory

activation energy

Arrhenius expression

Arrhenius preexponential factor

chemical kinetics

chemical methods

contour plot

elementary reaction step

Euler's method

Eyring equation

first-order reaction

flash photolysis techniques

frequency factor

half-life

initial rate

integrated rate law expression

intermediate

isolation method

method of initial rates

molecularity

parallel reaction

perturbation-relaxation methods

physical methods

potential energy surface

rate constant

rate-determining step

rate law

rate of conversion

reaction coordinate

reaction mechanism

reaction order

reaction rate

reversible reaction

second-order reaction (type I)

second-order reaction (type II)

sequential reaction

steady-state approximation

stopped-flow techniques

temperature jump

transition state

transition state theory

yield

Questions on Concepts

Q18.1 Why is the stoichiometry of a reaction generally not sufficient to determine reaction order?

Q18.2 What is an elementary chemical step, and how is one used in kinetics?

Q18.3 What is the difference between chemical and physical methods for studying chemical kinetics?

Q18.4 What is the method of initial rates, and why is it used in chemical kinetics studies?

Q18.5 What is a rate law expression, and how is it determined?

Q18.6 What is the difference between a first-order reaction and a second-order reaction?

Q18.7 What is a half-life? Is the half-life for a first-order reaction dependent on concentration?

Q18.8 In a sequential reaction, what is an intermediate?

Q18.9 What is meant by the rate-determining step in a sequential reaction?

Q18.10 What is the steady-state approximation, and when is this approximation employed?

Q18.11 In a parallel reaction in which two products can be formed from the same reactant, what determines the extent to which one product will be formed over another?

Q18.12 What is the kinetic definition of equilibrium?

Q18.13 In a temperature-jump experiment, why does a change in temperature result in a corresponding change in equilibrium?

Q18.14 What is a transition state? How is the concept of a transition state used in activated complex theory?

Q18.15 What is the relationship between the parameters in the Arrhenius equation and in the Eyring equation?

Problems

Problem numbers in **red** indicate that the solution to the problem is given in the *Student's Solutions Manual*.

P18.1 Express the rate of reaction with respect to each species in the following reactions:

a. $2NO(g) + O_2(g) \longrightarrow N_2O_4(g)$

b. $H_2(g) + I_2(g) \longrightarrow 2HI(g)$

c. $ClO(g) + BrO(g) \longrightarrow ClO_2(g) + Br(g)$

P18.2 Consider the first-order decomposition of cyclobutane at 438°C at constant volume:

$C_4H_8(g) \longrightarrow 2C_2H_4(g)$

a. Express the rate of the reaction in terms of the change in total pressure as a function of time.

b. The rate constant for the reaction is 2.48×10^{-4} s^{-1}. What is the half-life?

c. After initiation of the reaction, how long will it take for the initial pressure of C_4H_8 to drop to 90% of its initial value?

P18.3 As discussed in the text, the total system pressure can be used to monitor the progress of a chemical reaction. Consider the following reaction: $SO_2Cl_2(g) \longrightarrow SO_2(g) + Cl_2(g)$. The reaction is initiated, and the following data are obtained:

Time (h)	0	3	6	9	12	15
P_{Total} (kPa)	11.07	14.79	17.26	18.90	19.99	20.71

a. Is the reaction first or second order with respect to SO_2Cl_2?

b. What is the rate constant for this reaction?

P18.4 Consider the following reaction involving bromophenol blue (BPB) and OH$^-$: BPB(aq) + OH$^-(aq) \longrightarrow$ BPBOH$^-(aq)$. The concentration of BPB can be monitored by following the absorption of this species and using the Beer–Lambert law. In this law, absorption, A, and concentration are linearly related.

a. Express the reaction rate in terms of the change in absorbance as a function of time.

b. Let A_o be the absorbance due to BPB at the beginning of the reaction. Assuming that the reaction is first order with respect to both reactants, how is the absorbance of BPB expected to change with time?

c. Given your answer to part (b), what plot would you construct to determine the rate constant for the reaction?

P18.5 For the following rate expressions, state the order of the reaction with respect to each species, the total order of the reaction, and the units of the rate constant, k:

a. $Rate = k[ClO][BrO]$

b. $Rate = k[NO]^2[O_2]$

c. $Rate = k\dfrac{[HI]^2[O_2]}{[H^+]^{1/2}}$

P18.6 What is the overall order of the reaction corresponding to the following rate constants?

a. $k = 1.63 \times 10^{-4}$ M^{-1} s^{-1}

b. $k = 1.63 \times 10^{-4}$ M^{-2} s^{-1}

c. $k = 1.63 \times 10^{-4}$ M$^{-1/2}$ s^{-1}

P18.7 The reaction rate as a function of initial reactant pressures was investigated for the reaction $2NO(g) + 2H_2(g) \longrightarrow N_2(g) + 2H_2O(g)$, and the following data were obtained:

Run	P_o H$_2$ (kPa)	P_o N$_2$ (kPa)	Rate (kPa s^{-1})
1	53.3	40.0	0.137
2	53.3	20.3	0.033
3	38.5	53.3	0.213
4	19.6	53.3	0.105

What is the rate law expression for this reaction?

P18.8 (Challenging) The first-order thermal decomposition of chlorocyclohexane is as follows: $C_6H_{11}Cl(g) \longrightarrow C_6H_{10}(g) + HCl(g)$. For a constant volume system the following total pressures were measured as a function of time:

Time (s)	P (torr)	Time (s)	P (torr)
3	237.2	24	332.1
6	255.3	27	341.1
9	271.3	30	349.3
12	285.8	33	356.9
15	299.0	36	363.7
18	311.2	39	369.9
21	322.2	42	375.5

a. Derive the following relationship for a first-order reaction: $P(t_2) - P(t_1) = (P(t_\infty) - P(t_0))e^{-kt_1}(1 - e^{-k(t_2 - t_1)})$. In this relation, $P(t_1)$ and $P(t_2)$ are the pressures at two specific times; $P(t_0)$ is the initial pressure when the reaction is initiated, $P(t_\infty)$ is the pressure at the completion of the reaction, and k is the rate constant for the reaction. To derive this relationship:

   i. Given the first-order dependence of the reaction, write the expression for the pressure of chlorocyclohexane at a specific time, t_1.

   ii. Write the expression for the pressure at another time, t_2, which is equal to $t_1 + \Delta$ where delta is a fixed quantity of time.

   iii. Write expressions for $P(t_\infty) - P(t_1)$ and $P(t_\infty) - P(t_2)$.

   iv. Subtract the two expressions from part (iii).

b. Using the natural log of the relationship from part (a) and the data provided in the table given earlier in this problem, determine the rate constant for the decomposition of chlorocyclohexane. (*Hint:* Transform the data in the table by defining $t_2 - t_1$ to be a constant value, for example, 9 s.)

P18.9   You are given the following data for the decomposition of acetaldehyde:

Initial Concentration (M)	9.72×10^{-3}	4.56×10^{-3}
Half-Life (s)	328	572

Determine the order of the reaction and the rate constant for the reaction.

P18.10   Consider the schematic reaction $A \xrightarrow{k} P$.

a. If the reaction is one-half order with respect to [A], what is the integrated rate law expression for this reaction?

b. What plot would you construct to determine the rate constant k for the reaction?

c. What would be the half-life for this reaction? Will it depend on initial concentration of the reactant?

P18.11   A certain reaction is first order, and 540 s after initiation of the reaction, 32.5% of the reactant remains.

a. What is the rate constant for this reaction?

b. At what time after initiation of the reaction will 10% of the reactant remain?

P18.12   The half-life of ^{238}U is 4.5×10^9 years. How many disintegrations occur in 1 min for a 10-mg sample of this element?

P18.13   You are performing an experiment using ^3H (half-life $= 4.5 \times 10^3$ days) labeled phenylalanine in which the five aromatic hydrogens are labeled. To perform the experiment, the initial activity cannot be lower than 10% of the initial activity when the sample was received. How long after receiving the sample can you wait before performing the experiment?

P18.14   A convenient source of gamma rays for radiation chemistry research is ^{60}Co, which undergoes the following decay process: $^{60}_{27}Co \xrightarrow{k} {}^{60}_{28}Ni + \beta^- + \gamma$. The half-life of ^{60}Co is 1.9×10^3 days.

a. What is the rate constant for the decay process?

b. How long will it take for a sample of ^{60}Co to decay to half of its original concentration?

P18.15   The growth of a bacterial colony can be modeled as a first-order process in which the probability of cell division is linear with respect to time such that $dN/N = \zeta dt$, where dN is the number of cells that divide in the time interval dt, and ζ is a constant.

a. Use the preceding expression to show that the number of cells in the colony is given by $N = N_0 e^{\zeta t}$, where N is the number of cell colonies and N_0 is the number of colonies present at $t = 0$.

b. The generation time is the amount of time it takes for the number of cells to double. Using the answer to part (a), derive an expression for the generation time.

c. In milk at 37°C, the bacteria lactobacillus acidophilus has a generation time of about 75 min. Construct a plot of the acidophilus concentration as a function of time for time intervals of 15, 30, 45, 60, 90, 120, and 150 min after a colony of size N_0 is introduced to a container of milk.

P18.16   Show that the ratio of the half-life to the three-quarter life, $t_{1/2}/t_{3/4}$, for a reaction that is nth order ($n > 1$) in reactant A can be written as a function of n alone (that is, there is no concentration dependence in the ratio).

P18.17   Given the following kinetic scheme and associated rate constants, determine the concentration profiles of all species using Euler's method. Assume that the reaction is initiated with only the reactant A present at an initial concentration of 1 M. To perform this calculation, you may want to use a spreadsheet program such as Excel.

$$A \xrightarrow{k=1.5\times10^{-3}\text{s}^{-1}} B$$
$$A \xrightarrow{k=2.5\times10^{-3}\text{s}^{-1}} C \xrightarrow{k=1.8\times10^{-3}\text{s}^{-1}} D$$

P18.18   For the sequential reaction $A \xrightarrow{k_A} B \xrightarrow{k_B} C$, the rate constants are $k_A = 5 \times 10^6$ s^{-1} and $k_B = 3 \times 10^6$ s^{-1}. Determine the time at which [B] is at a maximum.

P18.19   For the sequential reaction $A \xrightarrow{k_A} B \xrightarrow{k_B} C$, $k_A = 1.00 \times 10^{-3}$ s^{-1}. Using a computer spreadsheet program such as Excel, plot the concentration of each species for cases where $k_B = 10k_A$, $k_B = 1.5k_A$, and $k_B = 0.1k_A$. Assume that only the reactant is present when the reaction is initiated.

P18.20 (Challenging) For the sequential reaction in Problem P18.19, plot the concentration of each species for the case where $k_B = k_A$. Can you use the analytical expression for [B] in this case?

P18.21 For a type II second-order reaction, the reaction is 60% complete in 60 seconds when $[A]_0 = 0.1$ M and $[B]_0 = 0.5$ M.

a. What is the rate constant for this reaction?

b. Will the time for the reaction to reach 60% completion change if the initial reactant concentrations are decreased by a factor of 2?

P18.22 Bacteriorhodopsin is a protein found in *Halobacterium halobium* that converts light energy into a transmembrane proton gradient that is used for ATP synthesis. After light is absorbed by the protein, the following initial reaction sequence occurs:

$$Br \xrightarrow{k_1 = 2.0 \times 10^{12} s^{-1}} J \xrightarrow{k_2 = 3.3 \times 10^{11} s^{-1}} K$$

a. At what time will the maximum concentration of the intermediate J occur?

b. Construct plots of the concentration of each species versus time.

P18.23 Bananas are somewhat radioactive due to the presence of substantial amounts of potassium. Potassium-40 decays by two different paths:

$$^{40}_{19}K \rightarrow {}^{40}_{20}Ca + \beta \quad (89.3\%)$$
$$^{40}_{19}K \rightarrow {}^{40}_{18}Ar + \beta^+ \quad (10.7\%)$$

The half-life for potassium decay is 1.3×10^9 years. Determine the rate constants for the individual channels.

P18.24 In the stratosphere, the rate constant for the conversion of ozone to molecular oxygen by atomic chlorine is $Cl + O_3 \rightarrow ClO + O_2$ $[(k=1.7 \times 10^{10}$ M^{-1} s$^{-1})e^{-260K/T}]$.

a. What is the rate of this reaction at 20 km where [Cl] = 5 $\times 10^{-17}$ M, $[O_3] = 8 \times 10^{-9}$ M, and $T = 220$ K?

b. The actual concentrations at 45 km are [Cl] = 3×10^{-15} M and $[O_3] = 8 \times 10^{-11}$ M. What is the rate of the reaction at this altitude where $T = 270$ K?

c. (Optional) Given the concentrations in part (a), what would you expect the concentrations at 45 km to be assuming that the gravity represents the operative force defining the potential energy?

P18.25 An experiment is performed on the following parallel reaction:

Two things are determined: (1) The yield for B at a given temperature is found to be 0.3 and (2) the rate constants are described well by an Arrhenius expression with the activation to B and C formation being 27 and 34 kJ mol^{-1}, respectively, and with identical preexponential factors. Demonstrate that these two statements are inconsistent with each other.

P18.26 A standard "rule of thumb" for thermally activated reactions is that the reaction rate doubles for every 10 K increase in temperature. Is this statement true independent of the activation energy (assuming that the activation energy is positive and independent of temperature)?

P18.27 Calculate the ratio of rate constants for two thermal reactions that have the same Arrhenius preexponential term, but with activation energies that differ by 1, 10, and 30 kJ/mol.

P18.28 The rate constant for the reaction of hydrogen with iodine is 2.45×10^{-4} M^{-1} s^{-1} at 302°C and 0.950 M^{-1} s^{-1} at 508°C.

a. Calculate the activation energy and Arrhenius preexponential factor for this reaction.

b. What is the value of the rate constant at 400°C?

P18.29 Consider the thermal decomposition of 1 atm of $(CH_3)_3COOC(CH_3)_3$ to acetone $(CH_3)_2CO$ and ethane (C_2H_6), which occurs with a rate constant of 0.0019 s^{-1}. After initiation of the reaction, at what time would you expect the pressure to be 1.8 atm?

P18.30 At 552.3 K, the rate constant for the thermal decomposition of SO_2Cl_2 is 1.02×10^{-6} s^{-1}. If the activation energy is 210 kJ mol^{-1}, calculate the Arrhenius preexponential factor and determine the rate constant at 600 K.

P18.31 The melting of double-strand DNA into two single strands can be initiated using temperature-jump methods. Derive the expression for the T-jump relaxation time for the following equilibrium involving double-strand (DS) and single-strand (SS) DNA:

$$DS \underset{k_r}{\overset{k_f}{\rightleftharpoons}} 2SS$$

P18.32 Consider the reaction

$$A + B \underset{k'}{\overset{k}{\rightleftharpoons}} P$$

A temperature-jump experiment is performed where the relaxation time constant is measured to be 310 μs, resulting in an equilibrium where $K_{eq} = 0.7$ with $[P]_{eq} = 0.2$ M. What are k and k'? (Watch the units!)

P18.33 The unimolecular decomposition of urea in aqueous solution is measured at two different temperatures and the following data are observed:

Trial Number	Temperature (°C)	k (s^{-1})
1	60.0	1.2×10^{-7}
2	71.5	4.40×10^{-7}

a. Determine the Arrhenius parameters for this reaction.

b. Using these parameters, determine $\Delta H^{\ddagger}$ and $\Delta S^{\ddagger}$ as described by the Eyring equation.

P18.34 The gas-phase decomposition of ethyl bromide is a first-order reaction, occurring with a rate constant that demonstrates the following dependence on temperature:

Trial Number	Temperature (K)	$k\ (s^{-1})$
1	800	0.036
2	900	1.410

a. Determine the Arrhenius parameters for this reaction.

b. Using these parameters, determine $\Delta H^{\ddagger}$ and $\Delta S^{\ddagger}$ as described by the Eyring equation.

P18.35 Hydrogen abstraction from hydrocarbons by atomic chlorine is a mechanism for Cl loss in the atmosphere. Consider the reaction of Cl with ethane:

$$C_2H_6(g) + Cl(g) \longrightarrow C_2H_5(g) + HCl$$

This reaction was studied in the laboratory, and the following data was obtained:

T (K)	$10^{-10}\ k\ (M^{-2}s^{-1})$
270	3.43
370	3.77
470	3.99
570	4.13
670	4.23

a. Determine the Arrhenius parameters for this reaction.

b. At the tropopause (the boundary between the troposphere and stratosphere located approximately 11 km above the surface of the Earth), the temperature is roughly 220 K. What do you expect the rate constant to be at this temperature?

c. Using the Arrhenius parameters obtained in part (a), determine the Eyring parameters $\Delta H^{\ddagger}$ and $\Delta S^{\ddagger}$ for this reaction.

P18.36 Consider the "unimolecular" isomerization of methylcyanide, a reaction that will be discussed in detail in Chapter 19: $CH_3NC(g) \longrightarrow CH_3CN(g)$. The Arrhenius parameters for this reaction are $A = 2.5 \times 10^{16}\ s^{-1}$ and $E_a = 272\ kJ\ mol^{-1}$. Determine the Eyring parameters $\Delta H^{\ddagger}$ and $\Delta S^{\ddagger}$ for this reaction with $T = 300$ K.

P18.37 Reactions involving hydroxyl radical (OH) are extremely important in atmospheric chemistry. The reaction of hydroxyl radical with molecular hydrogen is as follows:

$$OH{\bullet}(g) + H_2(g) \longrightarrow H_2O(g) + H{\bullet}(g)$$

Determine the Eyring parameters $\Delta H^{\ddagger}$ and $\Delta S^{\ddagger}$ for this reaction where $A = 8 \times 10^{13}\ M^{-1}\ s^{-1}$ and $E_a = 42\ kJ\ mol^{-1}$.

P18.38 Chlorine monoxide (ClO) demonstrates three bimolecular self-reactions:

$$Rxn_1:\quad ClO{\bullet}(g) + ClO{\bullet}(g) \xrightarrow{k_1} Cl_2(g) + O_2(g)$$

$$Rxn_2:\quad ClO{\bullet}(g) + ClO{\bullet}(g) \xrightarrow{k_2} Cl{\bullet}(g) + ClOO{\bullet}(g)$$

$$Rxn_3:\quad ClO{\bullet}(g) + ClO{\bullet}(g) \xrightarrow{k_3} Cl{\bullet}(g) + OClO{\bullet}(g)$$

The following table provides the Arrhenius parameters for this reaction:

	$A\ (M^{-1}\ s^{-1})$	E_a (kJ/mol)
Rxn_1	6.08×10^8	13.2
Rxn_2	1.79×10^{10}	20.4
Rxn_3	2.11×10^8	11.4

a. For which reaction is $\Delta H^{\ddagger}$ greatest and by how much relative to the next closest reaction?

b. For which reaction is $\Delta S^{\ddagger}$ the smallest and by how much relative to the next closest reaction?

Web-Based Simulations, Animations, and Problems

W18.1 In this problem, concentration profiles as a function of rate constant are explored for the following sequential reaction scheme:

$$A \xrightarrow{k_a} B \xrightarrow{k_b} C$$

Students vary the rate constants k_a and k_b and explore the following behavior:

a. The variation in concentrations as the rate constants are varied

b. Comparison of the maximum intermediate concentration time determined by simulation and through computation

c. Visualization of the conditions under which the rate-limiting step approximation is valid.

W18.2 In this simulation, the kinetic behavior of the following parallel reaction is studied:

$$A \xrightarrow{k_b} B$$
$$A \xrightarrow{k_c} C$$

The variation in concentrations as a function of k_b and k_c is studied. In addition, the product yields for the reaction determined based on the relative values of k_b and k_c are compared to the simulation result.

CHAPTER 19

Complex Reaction Mechanisms

In this chapter, the kinetic techniques developed in the previous chapter are applied to complex reactions. Reaction mechanisms and their use in predicting reaction rate law expressions are explored. The preequilibrium approximation is presented and used in the evaluation of catalytic reactions including enzyme catalysis. In addition, homogeneous and heterogeneous catalytic processes are described. Reactions involving radicals including polymerization and radical-initiated explosions are discussed. The chapter concludes with an introduction to photochemistry. The unifying theme behind these apparently different topics is that all of the reaction mechanisms for these phenomena can be developed using the techniques of elementary chemical kinetics. Seemingly complex reactions can be decomposed into a series of well-defined kinetic steps, thereby providing substantial insight into the underlying chemical reaction dynamics. ■

19.1 Reaction Mechanisms and Rate Laws

Reaction mechanisms are defined as the collection of individual kinetic processes or steps involved in the transformation of reactants into products. The rate law expression for a chemical reaction, including the order of the reaction, is entirely dependent on the reaction mechanism. For a reaction mechanism to be valid, the order of the reaction predicted by the mechanism must agree with the experimentally determined order. Consider the following reaction:

$$2N_2O_5 \longrightarrow 4NO_2 + O_2 \tag{19.1}$$

One possible mechanism for this reaction is a single step consisting of a bimolecular collision between two N_2O_5 molecules. A reaction mechanism that consists of a single elementary step is known as a **simple reaction**. The rate law predicted by this mechanism is second order with respect to N_2O_5. However, the experimentally determined rate law for this reaction is first order in N_2O_5, not second order. Therefore, the single-step mechanism cannot be correct. To explain the observed order dependence of the reaction rate, the following mechanism was proposed:

$$2\left\{ N_2O_5 \underset{k_{-1}}{\overset{k_1}{\rightleftharpoons}} NO_2 + NO_3 \right\} \tag{19.2}$$

$$NO_2 + NO_3 \overset{k_2}{\longrightarrow} NO_2 + O_2 + NO \tag{19.3}$$

$$NO + NO_3 \overset{k_3}{\longrightarrow} 2NO_2 \tag{19.4}$$

This mechanism is an example of a **complex reaction**, defined as a reaction that occurs in two or more elementary steps. In this mechanism, the first step, Equation (19.2), represents an equilibrium between N_2O_5 to NO_2 and NO_3. In the second step, Equation (19.3), the bimolecular reaction of NO_2 and NO_3 results in the dissociation of NO_3 to product NO and O_2. In the final step of the reaction, Equation (19.4), NO and NO_3 undergo a bimolecular reaction to produce $2NO_2$.

In addition to the reactant (N_2O_5) and overall reaction products (NO_2 and O_2), two other species appear in the mechanism (NO and NO_3) that are not in the overall reaction of Equation (19.1). These species are referred to as **reaction intermediates**. Reaction intermediates that are formed in one step of the mechanism must be consumed in a subsequent step. Given this requirement, step 1 of the reaction must occur twice in order to balance the NO_3 that appears in steps 2 and 3. Therefore, we have multiplied this reaction by two in Equation (19.2) to emphasize that step a must occur twice for every occurrence of steps 2 and 3. The number of times a given step occurs in a reaction mechanism is referred to as the **stoichiometric number**. In the mechanism under discussion, step 1 has a stoichiometric number of two, whereas the other two steps have stoichiometric numbers of one. With correct stoichiometric numbers, the sum of the elementary reaction steps will produce an overall reaction that is stoichiometrically equivalent to the reaction of interest.

For a reaction mechanism to be considered valid, the mechanism must be consistent with the experimentally determined rate law. Using the mechanism depicted by Equations (19.2) through (19.4), the rate of the reaction is

$$R = -\frac{1}{2}\frac{d[N_2O_5]}{dt} = \frac{1}{2}(k_1[N_2O_5] - k_{-1}[NO_2][NO_3]) \tag{19.5}$$

Notice that the stoichiometric number of the reaction is not included in the differential rate expression. Equation (19.5) corresponds to the loss of N_2O_5 by unimolecular decay and production by the bimolecular reaction of NO_2 with NO_3. As discussed earlier, NO and NO_3 are reaction intermediates. Writing the differential rate expression for these species and applying the steady-state approximation to the concentrations of both intermediates (Section 18.7.3) yields

$$\frac{d[NO]}{dt} = 0 = k_2[NO_2][NO_3] - k_3[NO][NO_3] \tag{19.6}$$

$$\frac{d[NO_3]}{dt} = 0 = k_1[N_2O_5] - k_{-1}[NO_2][NO_3] - k_2[NO_2][NO_3] - k_3[NO][NO_3] \tag{19.7}$$

Equation (19.6) can be rewritten to produce the following expression for [NO]:

$$[NO] = \frac{k_2[NO_2]}{k_3} \tag{19.8}$$

Substituting this result into Equation (19.7) yields

$$0 = k_1[N_2O_5] - k_{-1}[NO_2][NO_3] - k_2[NO_2][NO_3] - k_3\left(\frac{k_2[NO_2]}{k_3}\right)[NO_3]$$

$$0 = k_1[N_2O_5] - k_{-1}[NO_2][NO_3] - 2k_2[NO_2][NO_3]$$

$$\frac{k_1[N_2O_5]}{k_{-1} + 2k_2} = [NO_2][NO_3] \tag{19.9}$$

Substituting Equation (19.9) into Equation (19.5) results in the following predicted rate law expression for this mechanism:

$$R = \frac{1}{2}(k_1[N_2O_5] - k_{-1}[NO_2][NO_3]) \tag{19.10}$$

$$= \frac{1}{2}\left(k_1[N_2O_5] - k_{-1}\left(\frac{k_1[N_2O_5]}{k_{-1} + 2k_2}\right)\right)$$

$$= \frac{k_1k_2}{k_{-1} + 2k_2}[N_2O_5] = k_{eff}[N_2O_5]$$

In Equation (19.10), the collection of rate constants multiplying $[N_2O_5]$ has been renamed k_{eff}. Equation (19.10) demonstrates that the mechanism is consistent with the experimentally observed first-order dependence on $[N_2O_5]$. However, as discussed in Chapter 18, the consistency of a reaction mechanism with the experimental order dependence of the reaction is not proof that the mechanism is absolutely correct, but instead demonstrates that the mechanism is consistent with the experimentally determined order dependence. It is quite possible that an alternative mechanism could be constructed that is also consistent with the experimental order dependence.

The example just presented illustrates how reaction mechanisms are used to explain the order dependence of the reaction rate. A theme that reoccurs throughout this chapter is the relation between reaction mechanisms and elementary reaction steps. As we will see, the mechanisms for many complex reactions can be decomposed into a series of elementary steps, and the techniques developed in the previous chapter can be readily employed in the evaluation of these complex kinetic problems.

19.2 The Preequilibrium Approximation

The preequilibrium approximation is a central concept employed in the evaluation of reaction mechanisms. This approximation is used when equilibrium among a subset of species is established before product formation occurs. In this section, the preequilibrium approximation is defined. This approximation will prove to be extremely useful in subsequent sections.

19.2.1 General Solution

Consider the following reaction:

$$A + B \underset{k_r}{\overset{k_f}{\rightleftharpoons}} I \overset{k_p}{\longrightarrow} P \tag{19.11}$$

In Equation (19.11), forward and back rate constants link the reactants A and B with an intermediate species, I. Decay of I results in the formation of product, P. If the forward and back reactions involving the reactants and intermediate are more rapid than the decay of the intermediate to form products, then the reaction of Equation (19.11) can be envisioned as occurring in two distinct steps:

1. First, equilibrium between the reactants and the intermediate is maintained during the course of the reaction.
2. The intermediate undergoes decay to form product.

This description of events is referred to as the **preequilibrium approximation.** This approximation is appropriate when the rate of the backward reaction of I to form reactants is much greater than the rate of product formation, a condition that occurs when $k_r \gg k_p$. The application of the preequilibrium approximation in evaluating reaction mechanisms containing equilibrium steps is performed as follows. The differential rate expression for the product is

$$\frac{d[P]}{dt} = k_p[I] \tag{19.12}$$

In Equation (19.12), [I] can be rewritten by recognizing that this species is in equilibrium with the reactants; therefore,

$$\frac{[I]}{[A][B]} = \frac{k_f}{k_r} = K_c \tag{19.13}$$

$$[I] = K_c[A][B] \tag{19.14}$$

In the preceding equations, K_c is the equilibrium constant expressed in terms of reactant and product concentrations. Substituting the definition of [I] provided by Equation (19.14) into Equation (19.12), the differential rate expression for the product becomes

$$\frac{d[P]}{dt} = k_p[I] = k_p K_c[A][B] = k_{eff}[A][B] \tag{19.15}$$

Equation (19.15) demonstrates that with the preequilibrium approximation, the predicted rate law is second order overall and first order with respect to both reactants (A and B). Finally, the rate constant for product formation is not simply k_p, but is instead the product of this rate constant with the equilibrium constant, which is in turn equal to the ratio of the forward and back rate constants.

19.2.2 A Preequilibrium Example

The reaction of NO and O_2 to form product NO_2 provides an example in which the preequilibrium approximation provides insight into the mechanism of NO_2 formation. The specific reaction of interest is

$$2NO(g) + O_2(g) \longrightarrow 2NO_2(g) \tag{19.16}$$

One possible mechanism for this reaction is that of a single elementary step corresponding to a trimolecular reaction of two NO molecules and one O_2 molecule. The experimental rate law for this reaction is second order in NO and first order in O_2, consistent with this mechanism. However, this mechanism was further evaluated by measuring the temperature dependence of the reaction rate. If correct, raising the temperature will increase the number of collisions, and the reaction rate should increase. However, as the temperature is increased, a *reduction* in the reaction rate is observed, proving that the trimolecular-collisional mechanism is incorrect. In contrast, these experimental observations were explained using the following mechanism:

$$2NO \underset{k_r}{\overset{k_f}{\rightleftharpoons}} N_2O_2 \tag{19.17}$$

$$N_2O_2 + O_2 \xrightarrow{k_p} 2NO_2 \tag{19.18}$$

In the first step of this mechanism, Equation (19.17), an equilibrium between NO and the dimer, N_2O_2 is established rapidly compared to the rate of product formation. In the second step, Equation (19.18), a bimolecular reaction involving the dimer and O_2 results in the production of the NO_2 product. The stoichiometric number for each step is one. To evaluate this mechanism, the preequilibrium approximation is applied to step 1 of the mechanism, and the concentration of N_2O_2 is expressed as

$$[N_2O_2] = \frac{k_f}{k_r}[NO]^2 = K_c[NO]^2 \tag{19.19}$$

Using the second step of the mechanism, the reaction rate is written as

$$R = \frac{1}{2}\frac{d[NO_2]}{dt} = k_p[N_2O_2][O_2] \tag{19.20}$$

Substitution of Equation (19.19) into the differential rate expression for $[NO_2]$ yields

$$R = k_p[N_2O_2][O_2] = k_p K_c[NO]^2[O_2] \tag{19.21}$$
$$= k_{eff}[NO]^2[O_2]$$

The rate law predicted by this mechanism is second order in NO and first order in O_2, consistent with experiment. Furthermore, the preequilibrium approximation provides an explanation for the temperature dependence of product formation. Specifically, the formation of N_2O_2 is an exothermic process such that an increase in temperature shifts the equilibrium between NO and N_2O_2 toward NO. As such, there is less N_2O_2 to react with O_2, and this is reflected by a reduction in the rate of NO_2 formation with increased temperature.

19.3 The Lindemann Mechanism

The **Lindemann mechanism** for **unimolecular reactions** provides an elegant example of the relationship between kinetics and reaction mechanisms. This mechanism was developed to describe the observed concentration dependence in unimolecular dissociation reactions of the form:

$$A \rightarrow \textit{fragments} \tag{19.22}$$

In this reaction, a reactant molecule undergoes decomposition when the energy content of a vibrational mode or modes is sufficient for decomposition to occur. The question is how does the reactant acquire sufficient energy to undergo decomposition? One possibility is that the reactant acquires sufficient energy to react through a bimolecular collision. Experimentally, however, the rate of decomposition demonstrates only first-order behavior at high reactant concentrations, and not second order as expected for a single-step bimolecular mechanism. Lindemann proposed another mechanism to explain the order dependence of the reaction with respect to reactant concentration.

The Lindemann mechanism involves two steps. First, reactants acquire sufficient energy to undergo reaction through a bimolecular collision:

$$A + A \xrightarrow{k_1} A^* + A \tag{19.23}$$

In this reaction, A^* is the "activated" reactant that has received sufficient energy to undergo decomposition. The collisional partner of the activated reactant molecule leaves the collision with insufficient energy to decompose. In the second step of the Lindemann mechanism, the **activated reactant** undergoes one of two reactions: collision resulting in deactivation or decomposition resulting in product formation:

$$A^* + A \xrightarrow{k_{-1}} A + A \tag{19.24}$$

$$A^* \xrightarrow{k_2} P \tag{19.25}$$

The separation of the reaction into two steps is the key conceptual contribution of the Lindemann mechanism. Specifically, the mechanism implies that a separation in timescale exists between activation and deactivation/product formation. Inspection of the mechanism described by Equations (19.24) and (19.25) demonstrates that the only process resulting in product formation is the final decomposition step of Equation (19.25); therefore, the rate of product production is written as

$$\frac{d[P]}{dt} = k_2[A^*] \tag{19.26}$$

Evaluation of Equation (19.26) requires an expression for $[A^*]$ in terms of reactant concentration, $[A]$. Because A^* is an intermediate species, the relationship between $[A^*]$ and $[A]$ is obtained by writing the differential rate expression for $[A^*]$ and applying the steady-state approximation:

$$\frac{d[A^*]}{dt} = k_1[A]^2 - k_{-1}[A][A^*] - k_2[A^*] = 0$$

$$[A^*] = \frac{k_1[A]^2}{(k_{-1}[A] + k_2)} \tag{19.27}$$

Substituting Equation (19.27) into Equation (19.26) results in the final differential rate expression for [P]:

$$\frac{d[P]}{dt} = \frac{k_1 k_2 [A]^2}{k_{-1}[A] + k_2} \tag{19.28}$$

Equation (19.28) is the central result of the Lindemann mechanism. It states that the observed order dependence on [A] depends on the relative magnitude of $k_{-1}[A]$ versus k_2. At high reactant concentrations, $k_{-1}[A] > k_2$ and Equation (19.28) reduces to

$$\frac{d[P]}{dt} = \frac{k_1 k_2}{k_{-1}}[A] \tag{19.29}$$

Equation (19.29) demonstrates that at high reactant concentrations or pressures (recall that $P_A/RT = n_A/V = [A]$) the rate of product formation will be first order in [A], consistent with experiment. Mechanistically, at high pressures activated molecules will be produced faster than decomposition occurs such that the rate of decomposition is the rate-limiting step in product formation. At low reactant concentrations $k_2 > k_{-1}[A]$ and Equation (19.28) becomes

$$\frac{d[P]}{dt} = k_1 [A]^2 \tag{19.30}$$

Equation (19.30) demonstrates that at low pressures the formation of activated complex becomes the rate-limiting step in the reaction and the rate of product formation is second order in [A].

The Lindemann mechanism can be generalized to describe a variety of unimolecular reactions through the following generic scheme:

$$A + M \underset{k_{-1}}{\overset{k_1}{\rightleftharpoons}} A* + M \tag{19.31}$$

$$A* \overset{k_2}{\longrightarrow} P \tag{19.32}$$

In this mechanism, M is a collisional partner that can be the reactant itself (A) or some other species such as a nonreactive buffer gas added to the reaction. If the concentration of all species remains approximately constant during the reaction (for example, the products of the decomposition reaction can serve as species M in the creation of the activated reactant), then the rate of product formation can be written as follows:

$$R = \frac{d[P]}{dt} = \frac{k_1 k_2 [A][M]}{k_{-1}[M] + k_2} = k_{uni}[A] \tag{19.33}$$

In Equation (19.33), k_{uni} is the apparent rate constant for the reaction defined as

$$k_{uni} = \frac{k_1 k_2 [M]}{k_{-1}[M] + k_2} = \frac{k_1 k_2}{k_{-1} + \dfrac{k_2}{[M]}} \tag{19.34}$$

In the limit of high M concentrations, $k_{-1}[M] \gg k_2$ and $k_{uni} = k_1 k_2 / k_{-1}$ resulting in an apparent rate constant that is independent of [M]. As [M] decreases, k_{uni} will decrease until $k_2 > k_{-1}[M]$, at which point $k_{uni} = k_1[M]$ and the apparent rate demonstrates first-order dependence on M. Figure 19.1 presents a plot of the observed rate constant for the isomerization of methyl isocyanide versus pressure measured at 230.4°C by Schneider and Rabinovitch. The figure demonstrates that the predicted linear relationship between k_{uni} and pressure at low pressure, which is consistent with the corresponding limiting behavior of Equation (19.34), is observed for this reaction. In addition, at high pressure k_{uni} reaches a constant value, which is also consistent with the limiting behavior of Equation (19.34).

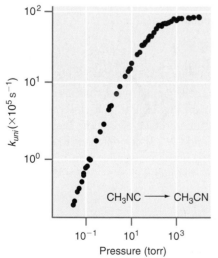

FIGURE 19.1

Pressure dependence of the observed rate constant for the unimolecular isomerization of methyl isocyanide. [Data from Schneider and Rabinovitch, *J. American Chemical Society* 84 (1962), 4225.]

www

19.1 The Lindemann Mechanism

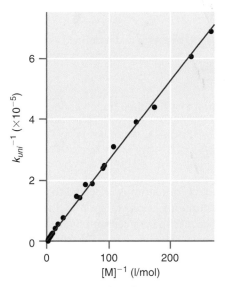

FIGURE 19.2

Plot of k_{uni}^{-1} versus $[M]^{-1}$ for the unimolecular isomerization of methyl isocyanide at 230°C. The solid line is the best fit to the data.

The Lindemann mechanism provides a detailed prediction of how the rate constant for a unimolecular reaction will vary with pressure or concentration. Inverting Equation (19.34), the relationship between k_{uni} and reactant concentration becomes

$$\frac{1}{k_{uni}} = \frac{k_{-1}}{k_1 k_2} + \left(\frac{1}{k_1}\right)\frac{1}{[M]}$$

(19.35)

Equation (19.35) predicts that a plot of k_{uni}^{-1} versus $[M]^{-1}$ should yield a straight line with slope $1/k_1$ and y intercept of $k_{-1}/k_1 k_2$. A plot of the data presented in Figure 19.1 employing Equation (19.35) is presented in Figure 19.2. The figure demonstrates that the expected linear relationship between k_{uni}^{-1} and $[M]^{-1}$ that is observed for this reaction is consistent with the Lindemann mechanism. The solid line in the figure is the best fit to the data by a straight line. The slope of this line provides a value for k_1 of 4.16×10^6 M^{-1} s^{-1}, and the y intercept in combination with the value for k_1 dictates that $k_{-1}/k_2 = 1.76 \times 10^5$ M^{-1}.

19.4 Catalysis

A **catalyst** is a substance that participates in chemical reactions by increasing the rate of reaction, yet the catalyst itself remains intact after the reaction is complete. The general function of a catalyst is to provide an additional mechanism by which reactants are converted to products. The presence of a new reaction mechanism involving the catalyst results in a second reaction coordinate that connects reactants and products. The activation energy along this second reaction coordinate will be lower in comparison to the reaction coordinate for the uncatalyzed reaction; therefore, the overall reaction rate will increase. For example, consider Figure 19.3 in which a reaction involving the conversion of reactant A to product B with and without a catalyst is depicted. In the absence of a catalyst, the rate of product formation is given by rate = r_0. In the presence of the catalyst, a second pathway is created, and the reaction rate is now the sum of the original rate plus the rate for the catalyzed reaction, or $r_0 + r_c$.

An analogy for a catalyzed reaction is found in the electrical circuits depicted in Figure 19.3. In the "catalyzed" electrical circuit, a second, parallel pathway for current flow has been added, allowing for increased total current when compared to the "uncatalyzed" circuit. By analogy, the addition of the second, parallel pathway is equivalent to the alternative reaction mechanism involving the catalyst.

To be effective, a catalyst must combine with one or more of the reactants or with an intermediate species involved in the reaction. After the reaction has taken place, the catalyst is freed, and can combine with another reactant or intermediate in a subsequent reaction. The catalyst is not consumed during the reaction so that a small amount of catalyst can participate in numerous reactions. The simplest mechanism describing a catalytic process is as follows:

$$S + C \underset{k_{-1}}{\overset{k_1}{\rightleftarrows}} SC$$

(19.36)

$$SC \overset{k_2}{\longrightarrow} P + C$$

(19.37)

where S represents the reactant or substrate, C is the catalyst, and P is the product. The **substrate–catalyst complex** is represented by SC and is an intermediate species in this mechanism. The differential rate expression for product formation is

$$\frac{d[P]}{dt} = k_2[SC]$$

(19.38)

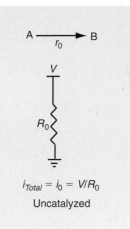

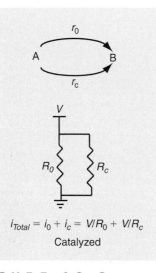

FIGURE 19.3

Illustration of catalysis. In the uncatalyzed reaction, the rate of reaction is given by r_0. In the catalyzed case, a new pathway is created by the presence of the catalyst with corresponding rate r_c. The total rate of reaction for the catalyzed case is $r_0 + r_c$. The analogous electrical circuits are also presented for comparison.

Because SC is an intermediate, we write the differential rate expression for this species and apply the steady-state approximation:

$$\frac{d[SC]}{dt} = k_1[S][C] - k_{-1}[SC] - k_2[SC] = 0 \tag{19.39}$$

$$[SC] = \frac{k_1[S][C]}{k_{-1} + k_2} = \frac{[S][C]}{K_m}$$

In Equation (19.39), K_m is referred to as the **composite constant** and is defined as follows:

$$K_m = \frac{k_{-1} + k_2}{k_1} \tag{19.40}$$

Substituting the expression for [SC] into Equation (19.20), the rate of product formation becomes

$$\frac{d[P]}{dt} = \frac{k_2[S][C]}{K_m} \tag{19.41}$$

Equation (19.41) illustrates that the rate of product formation is expected to increase linearly with both substrate and catalyst concentrations. This equation is difficult to evaluate over the entire course of the reaction because the concentrations of substrate and catalyst given in Equation (19.41) correspond to species not in the SC complex, and these concentrations can be quite difficult to measure. A more convenient measurement is to determine how much substrate and catalyst are present at the beginning of the reaction. Conservation of mass dictates the following relationship between these initial concentrations and the concentrations of all species present after the reaction is initiated:

$$[S]_0 = [S] + [SC] + [P] \tag{19.42}$$

$$[C]_0 = [C] + [SC] \tag{19.43}$$

Rearrangement of Equations (19.42) and (19.43) yields the following definitions for [S] and [C]:

$$[S] = [S]_0 - [SC] - [P] \tag{19.44}$$

$$[C] = [C]_0 - [SC] \tag{19.45}$$

Substituting these expressions into Equation (19.39) yields

$$K_m[SC] = [S][C] = ([S]_0 - [SC] - [P])([C]_0 - [SC]) \tag{19.46}$$

$$0 = \left[[C]_0([S]_0 - [P])\right] - [SC]([S]_0 + [C]_0 - [P] + K_m) + [SC]^2$$

Equation (19.46) can be evaluated as a quadratic equation to determine [SC]. However, two assumptions are generally employed at this point to simplify matters. First, through control of the initial substrate and catalyst concentrations, conditions can be employed such that [SC] is small. Therefore, the $[SC]^2$ term in Equation (19.46) can be neglected. Second, we confine ourselves to early stages of the reaction when little product has been formed; therefore, terms involving [P] can also be neglected. With these two approximations, Equation (19.46) is readily evaluated, providing the following expression for [SC]:

$$[SC] = \frac{[S]_0[C]_0}{[S]_0 + [C]_0 + K_m} \tag{19.47}$$

Substituting Equation (19.47) into Equation (19.38), the rate of the reaction becomes

$$R_0 = \frac{d[P]}{dt} = \frac{k_2[S]_0[C]_0}{[S]_0 + [C]_0 + K_m} \tag{19.48}$$

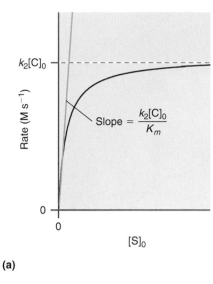

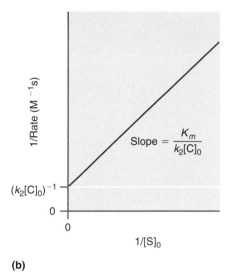

(a)

(b)

FIGURE 19.4
Illustration of the variation in the reaction rate with substrate concentration under Case 1 conditions as described in the text. **(a)** Plot of the initial reaction rate with respect to substrate concentration [Equation (19.49)]. At low substrate concentrations, the reaction rate increases linearly with substrate concentration. At high substrate concentrations, a maximum reaction rate of $k_2[C]_0$ is reached. **(b)** Reciprocal plot where the inverse of the reaction rate is plotted with respect to the inverse of substrate concentration [Equation (19.50)]. The y intercept of this line is equal to the inverse of the maximum reaction rate, or $(k_2[C]_0)^{-1}$. The slope of the line is equal to $K_m(k_2[C]_0)^{-1}$; therefore, with the slope and y intercept, K_m can be determined.

In Equation (19.48), the subscript on the rate indicates that this expression applies to the early-time or initial reaction rate. We next consider two limiting cases of Equation (19.48).

19.4.1 Case 1: $[C]_0 \ll [S]_0$

In the $[C]_0 \ll [S]_0$ case, the most common case in catalysis, much more substrate is present in comparison to catalyst. In this limit $[C]_0$ can be neglected in the denominator of Equation (19.48) and the rate becomes

$$R_0 = \frac{k_2[S]_0[C]_0}{[S]_0 + K_m} \tag{19.49}$$

For substrate concentrations where $[S]_0 < K_m$, the reaction rate should increase linearly with substrate concentration, with a slope equal to $k_2[C]_0/K_m$. Although parameters such as k_2 and K_m can be obtained by comparing experimental reaction rates to Equation (19.49), another approach is generally taken. Specifically, inverting Equation (19.49) provides the following relationship between the reaction rate and initial substrate concentration:

$$\frac{1}{R_0} = \left(\frac{K_m}{k_2[C]_0}\right)\frac{1}{[S]_0} + \frac{1}{k_2[C]_0} \tag{19.50}$$

Equation (19.50) demonstrates that a plot of the inverse of the initial reaction rate versus $[S]_0^{-1}$, referred to as a **reciprocal plot**, should yield a straight line. The y intercept and slope of this line provide a measure of K_m and k_2, assuming $[C]_0$ is known.

At elevated concentrations of substrate where $[S]_0 \gg K_m$, the denominator in Equation (19.49) can be approximated as $[S]_0$, resulting in the following expression for the reaction rate:

$$R_0 = k_2[C]_0 = R_{max} \tag{19.51}$$

In other words, the rate of reaction will reach a limiting value where the rate becomes zero order in substrate concentration. In this limit, the reaction rate can only be enhanced by increasing the amount of catalyst. An illustration of the variation in the reaction rate with initial substrate concentration predicted by Equations (19.49) and (19.50) is provided in Figure 19.4.

19.4.2 Case 2: $[C]_0 \gg [S]_0$

In the $[C]_0 \gg [S]_0$ limit, Equation (19.48) becomes

$$R_0 = \frac{k_2[S]_0[C]_0}{[C]_0 + K_m} \tag{19.52}$$

In this concentration limit, the reaction rate is first order in $[S]_0$, but can be first or zero order in $[C]_0$ depending on the size of $[C]_0$ relative to K_m. In catalysis studies, this limit is generally avoided because the insight to be gained regarding the rate constants for the various reaction steps are more easily evaluated for the previously discussed Case 1. In addition, good catalysts can be expensive; therefore, employing excess catalyst in a reaction is not cost effective.

19.4.3 Michaelis–Menten Enzyme Kinetics

Enzymes are protein molecules that serve as catalysts in a wide variety of chemical reactions. Enzymes are noted for their reaction specificity, with nature having developed specific catalysts to facilitate the vast majority of biological reactions required for organism survival. An illustration enzyme with associated substrate is presented in

FIGURE 19.5
Space-filling model of the enzyme phospholipase A_2 (white) containing a bound substrate analogue (red). The substrate analogue contains a stable phosphonate group in place of the enzyme-susceptible ester; therefore, the substrate analogue is resistant to enzymatic hydrolysis and the enzyme–substrate complex remains stable in the complex during the X-ray diffraction structure determination process. [Structural data from Scott, White, Browning , Rosa, Gelb, and Sigler, *Science* 5034 (1991), 1007.]

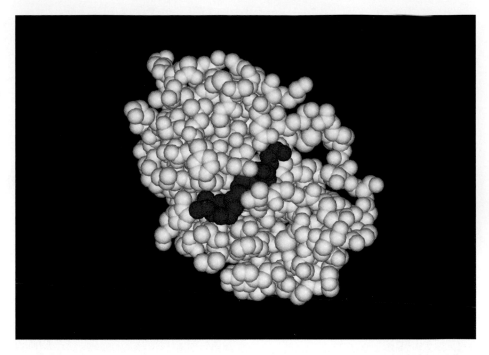

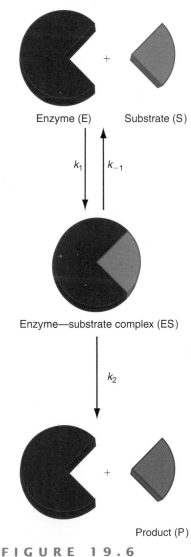

Enzyme (E)   Substrate (S)

k_1 k_{-1}

Enzyme—substrate complex (ES)

k_2

Product (P)

FIGURE 19.6
Schematic of enzyme catalysis.

Figure 19.5. The figure presents a space-filling model derived from a crystal structure of phospholipase A_2 (white) containing a bound substrate analogue (red). This enzyme catalyzes the hydrolysis of esters in phospholipids. The substrate analogue contains a stable phosphonate group in place of the enzyme-susceptible ester. The substrate analogue is resistant to enzymatic hydrolysis so that it does not suffer chemical breakdown during the structure determination process. With reactive substrate, ester hydrolysis occurs and the products of the reaction are released from the enzyme, resulting in regeneration of the free enzyme.

The kinetic mechanism of phospholipase A_2 catalysis can be described using the **Michaelis–Menten mechanism** of enzyme activity illustrated in Figure 19.6. The figure depicts the "lock-and-key" model for enzyme reactivity in which the substrate is bound to the active site of the enzyme where the reaction is catalyzed. The enzyme and substrate form the enzyme–substrate complex, which dissociates into product and uncomplexed enzyme. The interactions involved in creation of the enzyme–substrate complex are enzyme specific. For example, the active site may bind the substrate in more than one location, thereby creating geometric strain that promotes product formation. The enzyme may orient the substrate so that the reaction geometry is optimized. In summary, the details of enzyme-mediated chemistry are highly dependent on the reaction of interest. Rather than an exhaustive presentation of enzyme kinetics, our motivation here is to describe enzyme kinetics within the general framework of catalyzed reactions.

A schematic description of the mechanism illustrated in Figure 19.6 is as follows:

$$E + S \underset{k_{-1}}{\overset{k_1}{\rightleftharpoons}} ES \overset{k_2}{\longrightarrow} E + P \tag{19.53}$$

In this mechanism, E is enzyme, S is substrate, ES is the complex, and P is product. Comparison of the mechanism of Equation (19.53) to the general catalytic mechanism described earlier in Equations (19.36) and (19.37) demonstrates that this mechanism is identical to the general catalysis mechanism except that the catalyst C is now the enzyme E. In the limit where the initial substrate concentration is substantially greater than that of the enzyme ($[S]_0 \gg [E]_0$ or Case 1 conditions as described previously), the rate of product formation is given by

$$R_0 = \frac{k_2 [S]_0 [E]_0}{[S]_0 + K_m} \tag{19.54}$$

The composite constant, K_m, in Equation (19.54) is referred to as the **Michaelis constant** in enzyme kinetics, and Equation (19.54) is referred to as the **Michaelis–Menten rate law**. When $[S]_0 \gg K_m$, the Michaelis constant can be neglected, resulting in the following expression for the rate:

$$R_0 = k_2 [E]_0 = R_{max} \tag{19.55}$$

Equation (19.55) demonstrates that the rate of product formation will plateau at some maximum value equal to the product of initial enzyme concentration and k_2, the rate constant for product formation, consistent with the behavior depicted in Figure 19.4. A reciprocal plot of the reaction rate can also be constructed by inverting Equation (19.54), which results in the **Lineweaver–Burk equation**:

$$\frac{1}{R_0} = \frac{1}{R_{max}} + \frac{K_m}{R_{max}} \frac{1}{[S]_0} \tag{19.56}$$

For the Michaelis–Menten mechanism to be consistent with experiment, a plot of the inverse of the initial rate with respect to $[S]_0^{-1}$ should yield a straight line from which the y intercept and slope can be used to determine the maximum reaction rate and the Michaelis constant. This reciprocal plot is referred to as **Lineweaver–Burk plot**. In addition, because $[E]_0$ is readily determined experimentally, the maximum rate can be used to determine k_2, referred to as the **turnover number** of the enzyme [Equation (19.55)]. The turnover number is the maximum number of substrate molecules per unit time that can be converted into product, with most enzymes demonstrating turnover numbers between 1 and $10^5 \ s^{-1}$ under physiological conditions.

EXAMPLE PROBLEM 19.1

DeVoe and Kistiakowsky [*J. American Chemical Society* 83 (1961), 274] studied the kinetics of CO_2 hydration catalyzed by the enzyme carbonic anhydrase:

$$CO_2 + H_2O \rightleftharpoons HCO_3^- + H^+$$

In this reaction, CO_2 is converted to bicarbonate ion. Bicarbonate is transported in the bloodstream and converted back to CO_2 in the lungs, a reaction that is also catalyzed by carbonic anhydrase. The following initial reaction rates for the hydration reaction were obtained for an initial enzyme concentration of 2.3 nM and temperature of 0.5°C:

Rate (M s^{-1})	[CO$_2$] (mM)
2.78×10^{-5}	1.25
5.00×10^{-5}	2.5
8.33×10^{-5}	5.0
1.67×10^{-4}	20.0

Determine K_m and k_2 for the enzyme at this temperature.

Solution

The Lineweaver–Burk plot of the rate^{-1} versus $[CO_2]^{-1}$ is shown here:

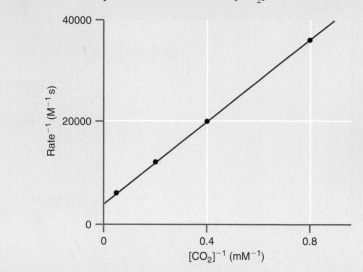

The y intercept for the best fit line to the data is 4000 M^{-1} s corresponding to $rate_{max} = 2.5 \times 10^{-4}$ M s^{-1}. Using this value and $[E]_0 = 2.3$ nM, k_2 is

$$k_2 = \frac{R_{max}}{[E]_0} = \frac{2.5 \times 10^{-4}\,M\,s^{-1}}{2.3 \times 10^{-9}\,M} = 1.1 \times 10^5\,s^{-1}$$

Notice that the units of k_2, the turnover number, are consistent with a first-order process, in agreement with the Michaelis–Menten mechanism. The slope of the best fit line is 40 s such that, per Equation (19.56), K_m is given by

$$K_m = slope \times R_{max} = (40\,s)(2.5 \times 10^{-4}\,M\,s^{-1})$$
$$= 10\ mM$$

In addition to the Lineweaver–Burk plot, K_m can be estimated if the maximum rate is known. Specifically, if the initial rate is equal to one-half the maximum rate, Equation (19.54) reduces to

$$R_0 = \frac{k_2[S]_0[E]_0}{[S]_0 + K_m} = \frac{R_{max}[S]_0}{[S]_0 + K_m}$$
$$\frac{R_{max}}{2} = \frac{R_{max}[S]_0}{[S]_0 + K_m}$$
$$[S]_0 + K_m = 2[S]_0$$
$$K_m = [S]_0 \tag{19.57}$$

● ● ●
WWW
● ● ●

**19.2 Michaelis-Menten
Enzyme Kinetics**

Equation (19.57) demonstrates that when the initial rate is half the maximum rate, K_m is equal to the initial substrate concentration. Therefore, K_m can be determined by viewing a substrate saturation curve, as illustrated in Figure 19.7 for the carbonic-anhydrase catalyzed hydration of CO_2 discussed in Example Problem 19.1. The figure demonstrates that the initial rate is equal to half the maximum rate when $[S]_0 = 10$ mM. Therefore, the value of K_m determined in this relatively simple approach is in excellent agreement with that determined from the Lineweaver–Burk plot. Notice in Figure 19.7 that the maximum rate depicted was that employed using the Lineweaver–Burk analysis as shown in the example problem. When employing this method to determine K_m, the high-substrate-concentration limit must be carefully explored to ensure that the reaction rate is indeed at a maximum.

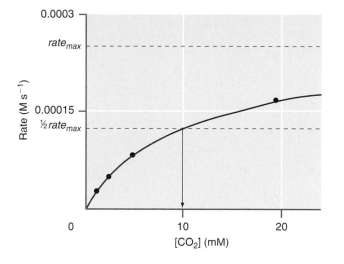

FIGURE 19.7
Determination of K_m for the carbonic-anhydrase catalyzed hydration of CO_2. The substrate concentration at which the rate of reaction is equal to half that of the maximum rate is equal to K_m.

19.4.4 Competitive Inhibition in Enzyme Catalysis

The activity of an enzyme can be affected by the introduction of species that structurally resemble the substrate and that can occupy the enzyme active site; however, once bound to the active site the molecules are nonreactive. Such molecules are referred to as **competitive inhibitors.** The phosphonated substrate bound to phospholipase A_2 in Figure 19.5 is an example of a competitive inhibitor. Competitive inhibition can be described using the following mechanism:

$$E + S \underset{k_{-1}}{\overset{k_1}{\rightleftharpoons}} ES \qquad (19.58)$$

$$ES \overset{k_2}{\longrightarrow} E + P \qquad (19.59)$$

$$E + I \underset{k_{-3}}{\overset{k_3}{\rightleftharpoons}} EI \qquad (19.60)$$

In this mechanism, I is the inhibitor, EI is the enzyme–inhibitor complex, and the other species are identical to those employed in the standard enzyme kinetic scheme of Equation (19.53). How does the rate of reaction differ from the noninhibited case discussed earlier? To answer this question, we first define the initial enzyme concentration:

$$[E]_0 = [E] + [EI] + [ES] \qquad (19.61)$$

Next, assuming that k_1, k_{-1}, k_3, and $k_{-3} \gg k_2$ the preequilibrium approximation is applied using Equations (19.58) and (19.60) yielding

$$K_s = \frac{[E][S]}{[ES]} \approx K_m \qquad (19.62)$$

$$K_i = \frac{[E][I]}{[EI]} \qquad (19.63)$$

In Equation (19.62), the equilibrium constant describing the enzyme and substrate is equivalent to K_m when $k_{-1} \gg k_2$. With these relationships, Equation (19.61) can be written as

$$
\begin{aligned}
[E]_0 &= \frac{K_m[ES]}{[S]} + \frac{[E][I]}{K_i} + [ES] \\
&= \frac{K_m[ES]}{[S]} + \left(\frac{K_m[ES]}{[S]} \right) \frac{[I]}{K_i} + [ES] \\
&= [ES]\left(\frac{K_m}{[S]} + \frac{K_m[I]}{[S]K_i} + 1 \right)
\end{aligned}
\qquad (19.64)
$$

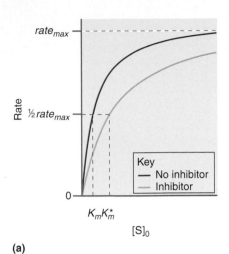

(a)

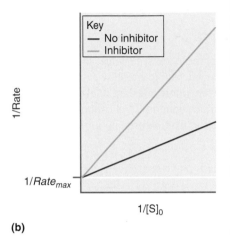

(b)

FIGURE 19.8
Comparison of enzymatic reaction rates in the presence and absence of a competitive inhibitor. **(a)** Plot of rate versus initial substrate concentration. The location of K_m and K_m^* is indicated. **(b)** Reciprocal plots ($1/Rate$ versus $1/[S]_0$). Notice that $1/Rate_{max}$ is identical in the presence and absence of a competitive inhibitor.

Solving Equation (19.64) for [ES] yields

$$[ES] = \frac{[E]_0}{1 + \dfrac{K_m}{[S]} + \dfrac{K_m[I]}{[S]K_i}} \qquad (19.65)$$

Finally, the rate of product formation is given by

$$R_0 = \frac{d[P]}{dt} = k_2[ES] = \frac{k_2[E]_0}{1 + \dfrac{K_m}{[S]} + \dfrac{K_m[I]}{[S]K_i}} = \frac{k_2[S][E]_0}{[S] + K_m\left(1 + \dfrac{[I]}{K_i}\right)}$$

$$R_0 \cong \frac{k_2[S]_0[E]_0}{[S]_0 + K_m\left(1 + \dfrac{[I]}{K_i}\right)} \qquad (19.66)$$

In Equation (19.66), the assumption that [ES] and [P] << [S] has been employed so that $[S] \cong [S]_0$, consistent with the previous treatment of uninhibited catalysis. Comparison of Equation (19.66) to the corresponding expression for the uninhibited case of Equation (19.54) illustrates that with competitive inhibition, a new apparent Michaelis constant can be defined:

$$K_m^* = K_m\left(1 + \frac{[I]}{K_i}\right) \qquad (19.67)$$

Notice that K_m^* reduces to K_m in the absence of inhibitor ([I] = 0). Next, using the definition of maximum reaction rate defined earlier in Equation (19.55), the reaction rate in the case of competitive inhibition can be written as

$$R_0 \cong \frac{R_{max}[S]_0}{[S]_0 + K_m^*} \qquad (19.68)$$

In the presence of inhibitor, $K_m^* > K_m$, and more substrate is required to reach half the maximum rate in comparison to the uninhibited case. The effect of inhibition can also be observed in a Lineweaver–Burk plot of the following form:

$$\frac{1}{R_0} = \frac{1}{R_{max}} + \frac{K_m^*}{R_{max}}\frac{1}{[S]_0} \qquad (19.69)$$

Because $K_m^* > K_m$, the slope of the Lineweaver–Burk plot will be greater with inhibitor compared to the slope without inhibitor. Figure 19.8 presents an illustration of this effect.

Competitive inhibition has been used in drug design for antiviral, antibacterial, and antitumor applications. These drugs are molecules that serve as competitive inhibitors for enzymes required for viral, bacterial, or cellular replication. For example sulfanilamide (Figure 19.9) is a powerful antibacterial drug. This compound is similar to p-aminobenzoic acid, the substrate for the enzyme dihydropteroate synthetase that participates in the production of folate. When present, the enzyme in bacteria cannot produce folate, and the bacteria die. However, humans can obtain folate from other sources; therefore, sulfanilamide is not toxic.

19.4.5 Homogeneous and Heterogeneous Catalysis

A **homogeneous catalyst** is a catalyst that exists in the same phase as the species involved in the reaction, and heterogeneous catalysts exist in a different phase. Enzymes serve as an example of a homogeneous catalyst; they exist in solution and catalyze reactions that occur in solution. A famous example of gas-phase catalysis is the catalytic depletion of stratospheric ozone by atomic chlorine. In the mid-1970s, F. Sherwood Rowland and Mario Molina proposed that Cl atoms catalyze the decomposition of stratospheric ozone by the following mechanism:

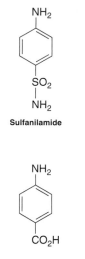

Sulfanilamide

p-Aminobenzoic acid

FIGURE 19.9
Structural comparison of the antibacterial drug sulfanilamide, a competitive inhibitor of the enzyme dihydropteroate synthetase, and the active substrate, p-aminobenzoic acid. The change in functional group from $-CO_2H$ to $-SO_2NH_2$ is such that sulfanilamide cannot be used by bacteria to synthesize folate, and the bacterium starves.

$$Cl + O_3 \xrightarrow{k_1} ClO + O_2 \tag{19.70}$$

$$ClO + O \xrightarrow{k_2} Cl + O_2 \tag{19.71}$$

$$O_3 + O \longrightarrow 2O_2 \tag{19.72}$$

In this mechanism, Cl reacts with ozone to produce chlorine monoxide (ClO) and molecular oxygen. The ClO undergoes a second reaction with atomic oxygen, resulting in the reformation of Cl and the product of O_2. The sum of these reactions leads to the net conversion of O_3 and O to $2O_2$. Notice that the Cl is not consumed in the net reaction.

The catalytic efficiency of Cl can be determined using standard techniques in kinetics. The experimentally determined rate law expression for the uncatalyzed reaction of Equation (19.72) is

$$R_{nc} = k_{nc}[O][O_3] \tag{19.73}$$

The stratospheric temperature where this reaction occurs is roughly 220 K, at which temperature k_{nc} has a value of 3.30×10^5 M^{-1} s^{-1}. For the Cl catalyzed decomposition of ozone, the rate constants at this temperature are $k_1 = 1.56 \times 10^{10}$ M^{-1} s^{-1} and $k_2 = 2.44 \times 10^{10}$ M^{-1} s^{-1}. To employ these rates in determining the overall rate of reaction, the rate law expression for the catalytic mechanism must be determined. Notice that both Cl and ClO are intermediates in this mechanism. Applying the steady-state approximation, the concentration of intermediates is taken to be a constant such that

$$[X] = [Cl] + [ClO] \tag{19.74}$$

where [X] is defined as the sum of reaction intermediate concentrations, a definition that will prove useful in deriving the rate law. In addition, the steady-state approximation is applied in evaluating the differential rate expression for [Cl] as follows:

$$\frac{d[Cl]}{dt} = 0 = -k_1[Cl][O_3] + k_2[ClO][O]$$

$$k_1[Cl][O_3] = k_2[ClO][O]$$

$$\frac{k_1[Cl][O_3]}{k_2[O]} = [ClO] \tag{19.75}$$

Substituting Equation (19.75) into Equation (19.74) yields the following expression for [Cl]:

$$[Cl] = \frac{k_2[X][O]}{k_1[O_3] + k_2[O]} \tag{19.76}$$

Using Equation (19.76), the rate law expression for the catalytic mechanism is determined as follows:

$$R_{cat} = -\frac{d[O_3]}{dt} = k_1[Cl][O_3] = \frac{k_1k_2[X][O][O_3]}{k_1[O_3] + k_2[O]} \tag{19.77}$$

The composition of the stratosphere is such that $[O_3] \gg [O]$. Taken in combination with the numerical values for k_1 and k_2 presented earlier, the $k_2[O]$ term in the denominator of Equation (19.77) can be neglected, and the rate law expression for the catalyzed reaction becomes

$$R_{cat} = k_2[X][O] \tag{19.78}$$

The ratio of catalyzed to uncatalyzed reaction rates is

$$\frac{R_{cat}}{R_{uc}} = \frac{k_2[X]}{k_{nc}[O_3]} \tag{19.79}$$

In the stratosphere $[O_3]$ is roughly 10^3 greater than $[X]$, and Equation (19.79) becomes

$$\frac{R_{cat}}{R_{uc}} = \frac{k_2}{k_{nc}} \times 10^{-3} = \frac{2.44 \times 10^{10}\,M^{-1}\,s^{-1}}{3.30 \times 10^5\,M^{-1}\,s^{-1}} \times 10^{-3} \approx 74$$

Therefore, through Cl-mediated catalysis, the rate of O_3 loss is roughly two orders of magnitude greater than the loss through the bimolecular reaction of O_3 and O directly.

Where does stratospheric Cl come from? Rowland and Molina proposed that a major source of Cl was from the photolysis of chlorofluorocarbons such as $CFCl_3$ and CF_2Cl_2, anthropogenic compounds that were common refrigerants at the time. These molecules are extremely robust, and when released into the atmosphere, they readily survive transport through the troposphere and into the stratosphere. Once in the stratosphere, these molecules can absorb a photon of light with sufficient energy to dissociate the C–Cl bond, and Cl is produced. This proposal served as the impetus to understand the details of stratospheric ozone depletion, and led to the Montreal Protocol in which the vast majority of nations agreed to phase out the industrial use of chlorofluorocarbons.

Heterogeneous catalysts are extremely important in industrial chemistry. The majority of industrial catalysts are solids. For example, the synthesis of NH_3 from N_2 and H_2 is catalyzed using Fe. This is an example of heterogeneous catalysis because the reactant products are in the gas phase, but the catalyst is a solid. An important step in reactions involving solid catalysis is the adsorption of one or more of the reactants to the solid surface. First, we assume that the particles adsorb to the surface without changing their internal bonding, a process referred to as **physisorption.** A dynamic equilibrium exists between the free and surface-adsorbed species or adsorbate, and information regarding the kinetics of surface adsorption and desorption can be obtained by studying this equilibrium as a function of reactant pressure over the surface of the catalyst. A critical parameter in evaluating surface adsorption is the **fractional coverage,** θ, defined as

$$\theta = \frac{\text{Number of adsorption sites occupied}}{\text{Total number of adsorption sites}}$$

Figure 19.10 provides an illustration of a surface with a series of adsorption sites. Reactant molecules (given by the blue spheres) can exist in either the gas phase or be adsorbed to one of these sites. The fractional coverage is simply the fraction of adsorption sites occupied.

The fractional coverage can also be defined as $\theta = V_{adsorbed}/V_m$ where $V_{adsorbed}$ is the volume of absorbate at a specific pressure, and V_m is the volume of absorbate in the high-pressure limit corresponding to monolayer coverage.

Studies of adsorption involve measuring the extent of adsorption or θ as a function of reactant-gas pressure at a specific temperature. The variation in θ with pressure at fixed temperature is called an **adsorption isotherm.** The simplest kinetic model describing the adsorption process is known as the **Langmuir model,** where adsorption is described by the following mechanism:

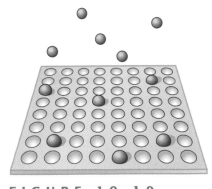

FIGURE 19.10
Illustration of fractional coverage, θ. The surface (orange parallelogram) contains a series of adsorption sites (white circles). The reactant (blue spheres) exists in an equilibrium between free reactants and adsorbates. The fractional coverage is the number of occupied adsorption sites divided by the total number of sites on the surface.

$$R(g) + M(surface) \underset{k_d}{\overset{k_a}{\rightleftharpoons}} RM(surface) \qquad (19.80)$$

In Equation (19.80), R is reagent, M is an unoccupied absorption site on the surface of the catalyst, RM is an occupied adsorption site, k_a is the rate constant for adsorption, and k_d is the rate constant for desorption. Three approximations are employed in the Langmuir model:

1. Adsorption is complete once monolayer coverage has been reached.

2. All adsorption sites are equivalent and the surface is uniform.

3. Adsorption and desorption are uncooperative processes. The occupancy state of the adsorption site will not affect the probability of adsorption or desorption for adjacent sites.

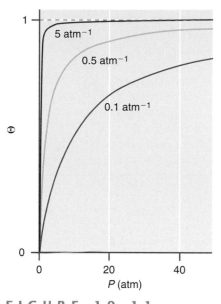

FIGURE 19.11
Langmuir isotherms for a range of k_a/k_d.

With these approximations, the rate of change in θ will depend on the rate constant for absorption, k_a, reagent pressure, P, and the number of vacant sites, which is equal to the total number of adsorption sites, N, times the fraction of sites that are open $(1 - \theta)$:

$$\frac{d\theta}{dt} = k_a PN(1-\theta) \tag{19.81}$$

The corresponding change in θ due to desorption is related to the rate constant for desorption, k_d, and the number of occupied adsorption sites, $N\theta$, as follows:

$$\frac{d\theta}{dt} = -k_d N\theta \tag{19.82}$$

At equilibrium, the change in fractional coverage with time is equal to zero so we can write

$$\frac{d\theta}{dt} = 0 = k_a PN(1-\theta) - k_d N\theta$$

$$(k_a PN + k_d N)\theta = k_a PN$$

$$\theta = \frac{k_a P}{k_a P + k_d}$$

$$\theta = \frac{KP}{KP + 1} \tag{19.83}$$

where K is the equilibrium constant defined as k_a/k_d. Equation (19.83) is the equation for the **Langmuir isotherm.** Figure 19.11 presents Langmuir isotherms for various values of k_a/k_d. Notice that as the rate constant for desorption increases relative to adsorption, higher pressures must be employed to reach $\theta = 1$, and this behavior can be understood based on the competition between the kinetics for adsorption and desorption. Correspondingly, if the rate constant for desorption is small, the coverage becomes independent of pressure for lower values of pressure.

In many instances adsorption is accompanied by dissociation of the adsorbate, a process that is referred to as **chemisorption** and described by the following mechanism:

$$R_2(g) + 2M(\textit{surface}) \underset{k_d}{\overset{k_a}{\rightleftharpoons}} 2RM(\textit{surface})$$

Kinetic analysis of this mechanism (see the end-of-chapter problems) yields the following expression for θ:

$$\theta = \frac{(KP)^{1/2}}{1 + (KP)^{1/2}} \tag{19.84}$$

Inspection of Equation (19.84) reveals that the extent of surface coverage should demonstrate weaker pressure dependence compared to physisorption. Figure 19.12 provides a comparison of the isotherms predicted using nondissociative and dissociative mechanisms corresponding to the same ratio of k_a/k_d. Finally, different Langmuir isotherms can be collected and evaluated as a range of temperatures to determine K as a function of T. With this information, a van't Hoff plot of $\ln K$ versus $1/T$ should provide a straight line of slope $-\Delta H_{ads}/R$. Through this analysis, the enthalpy of adsorption, ΔH_{ads}, can be determined.

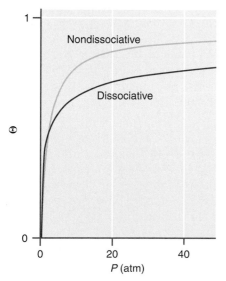

FIGURE 19.12
Comparison of Langmuir isotherms for nondissociative and dissociative adsorption with $k_a/k_d = 0.5$ atm^{-1}.

The assumptions employed in the Langmuir model may not be rigorously obeyed in real heterogeneous systems. First, surfaces are generally not uniform, resulting in the presence of more than one type of adsorption site. Second, the rate of adsorption and desorption may depend on the occupation state of nearby adsorption sites. Finally, it has been established that adsorbed molecules can diffuse on the surface and then adsorb corresponding to a kinetic process of adsorption that is more complicated than the Langmuir mechanism envisions.

EXAMPLE PROBLEM 19.2

The following data were obtained for the adsorption of Kr on charcoal at 193.5 K. Using the Langmuir model, construct the absorption isotherm, determine V_m and the equilibrium constant for adsorption/desorption.

V_{ads} (cm³ g⁻¹)	P (Torr)
5.98	2.45
7.76	3.5
10.1	5.2
12.35	7.2
16.45	11.2
18.05	12.8
19.72	14.6
21.1	16.1

Solution

The fractional coverage is related to the experimentally measured V_{ads}. The adsorption isotherm is given by a plot of V_{ads} versus P, which can be compared to the behavior predicted by Equation (19.83) as illustrated here:

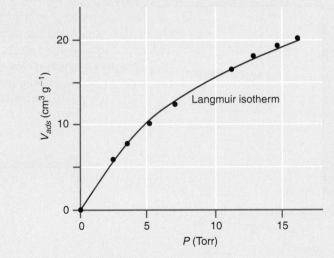

Although the above comparison of the adsorption isotherm to Equation (19.83) illustrates that the Langmuir model is consistent with the adsorption of Kr on charcoal, determination of the Langmuir parameters is difficult because parameters such as V_m are unknown. This information is more readily determined by using the reciprocal of Equation (19.83):

$$\frac{1}{V_{ads}} = \left(\frac{1}{KV_m}\right)\frac{1}{P} + \frac{1}{V_m}$$

This equation demonstrates that a plot of $(V_{ads})^{-1}$ versus $(P)^{-1}$ should yield a straight line with slope equal to $(KV_m)^{-1}$ and y intercept equal to $(V_m)^{-1}$. A plot of the data in reciprocal form with the best fit line is shown next:

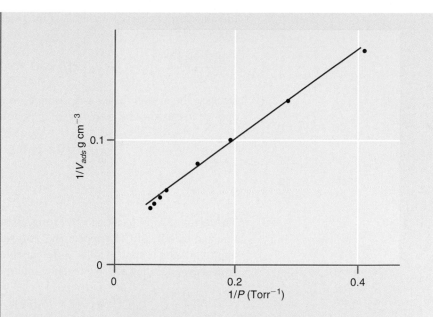

The y intercept obtained from the best fit line is 0.0293 g cm^{-3} such that $V_m =$ 34.1 cm^3 g^{-1}. The slope of the best fit line is 0.3449 Torr g cm^{-3}. Using V_m determined from the y intercept, K is found to be 8.38×10^{-2} Torr^{-1}.

19.3 The Langmuir Isotherm

19.5 Radical-Chain Reactions

Radicals are chemical species that contain an unpaired electron. Due to the presence of the unpaired electron, radicals tend to be extremely reactive. In 1934, Rice and Herzfeld were able to demonstrate that the kinetic behavior of many organic reactions was consistent with the presence of radicals in the reaction mechanism. An example of a radical-mediated reaction is the thermal decomposition of ethane:

$$C_2H_6(g) \longrightarrow C_2H_4(g) + H_2(g) \tag{19.85}$$

Small amounts of methane (CH$_4$) are also produced in the decomposition. The decomposition mechanism proposed by Rice and Herzfeld is as follows:

$$\text{Initiation} \quad C_2H_6 \xrightarrow{k_1} 2CH_3\bullet \tag{19.86}$$

$$\text{Propagation} \quad CH_3\bullet + C_2H_6 \xrightarrow{k_2} CH_4 + C_2H_5\bullet \tag{19.87}$$

$$C_2H_5\bullet \xrightarrow{k_3} C_2H_4 + H\bullet \tag{19.88}$$

$$H\bullet + C_2H_6 \xrightarrow{k_4} C_2H_5\bullet + H_2 \tag{19.89}$$

$$\text{Termination} \quad H\bullet + C_2H_5\bullet \xrightarrow{k_5} C_2H_6 \tag{19.90}$$

In this section, we include a dot ($\bullet$) in the formula of a compound to indicate that the species is a radical. The first elementary step in the mechanism involves the creation of two methyl radicals, referred to as an **initiation step** [Equation (19.86)]. In an initiation step, radicals are produced from a precursor species. The next three steps in the mechanism [Equations (19.87) through (19.89)] are referred to as **propagation steps** in which a radical reacts with another species to produce a radical and nonradical products, and the radical products go on to participate in subsequent reactions. The final step in the mechanism is a **termination step** in which two radicals recombine to produce a nonradical product.

Although the mechanism of Equations (19.86) through (19.90) is clearly complex, the rate law expression predicted by this mechanism is remarkably simple. To derive this rate law, we begin with the differential rate expression for the disappearance of ethane:

$$-\frac{d[C_2H_6]}{dt} = k_1[C_2H_6] + k_2[C_2H_6][CH_3\bullet] + k_4[C_2H_6][H\bullet] - k_5[C_2H_5\bullet][H\bullet] \quad (19.91)$$

Because the methyl radical is a reactive intermediate, the steady-state approximation is applied to this species, resulting in the following expression for $[CH_3]$:

$$\frac{d[CH_3\bullet]}{dt} = 0 = 2k_1[C_2H_6] - k_2[C_2H_6][CH_3\bullet] \quad (19.92)$$

$$[CH_3\bullet] = \frac{2k_1}{k_2} \quad (19.93)$$

The factor of 2 in Equation (19.92) originates from the relationship between the reaction rate and the rate of $CH_3\bullet$ appearance as discussed in Chapter 18. Next, the steady-state approximation is applied to the differential rate expressions for ethyl radical and atomic hydrogen since they are also reaction intermediates:

$$\frac{d[C_2H_5\bullet]}{dt} = 0 = k_2[CH_3\bullet][C_2H_6] - k_3[C_2H_5\bullet] + k_4[C_2H_6][H\bullet] - k_5[C_2H_5\bullet][H\bullet] \quad (19.94)$$

$$\frac{d[H\bullet]}{dt} = 0 = k_3[C_2H_5\bullet] - k_4[C_2H_6][H\bullet] - k_5[C_2H_5\bullet][H\bullet] \quad (19.95)$$

Adding Equations (19.94), (19.95), and (19.92) yields the following expression for $[H\bullet]$:

$$0 = 2k_1[C_2H_6] - 2k_5[C_2H_5\bullet][H\bullet]$$

$$[H\bullet] = \frac{k_1[C_2H_6]}{k_5[C_2H_5\bullet]} \quad (19.96)$$

Substituting Equation (19.96) into Equation (19.95) yields

$$0 = k_3[C_2H_5\bullet] - k_4[C_2H_6][H\bullet] - k_5[C_2H_5\bullet][H\bullet]$$

$$0 = k_3[C_2H_5\bullet] - k_4[C_2H_6]\left(\frac{k_1[C_2H_6]}{k_5[C_2H_5\bullet]}\right) - k_5[C_2H_5\bullet]\left(\frac{k_1[C_2H_6]}{k_5[C_2H_5\bullet]}\right)$$

$$0 = k_3[C_2H_5\bullet] - \frac{k_4k_1[C_2H_6]^2}{k_5[C_2H_5\bullet]} - k_1[C_2H_6]$$

$$0 = [C_2H_5\bullet]^2 - \frac{k_4k_1[C_2H_6]^2}{k_5k_3} - \frac{k_1}{k_3}[C_2H_6][C_2H_5\bullet] \quad (19.97)$$

The last expression is a quadratic equation in $[C_2H_5\bullet]$ for which the solution yields the following expression for $[C_2H_5\bullet]$:

$$[C_2H_5\bullet] = [C_2H_6]\left[\frac{k_1}{2k_3} + \left(\left(\frac{k_1}{2k_3}\right)^2 + \left(\frac{k_1k_4}{k_3k_5}\right)\right)^{1/2}\right] \quad (19.98)$$

Because the rate constant for initiation, k_1, is quite small, only the lowest power term in this quantity is kept when evaluating Equation (19.98) such that

$$[C_2H_5\bullet] = \left(\frac{k_1k_4}{k_3k_5}\right)^{1/2}[C_2H_6] \quad (19.99)$$

With Equation (19.99), $[H\bullet]$ from Equation (19.96) becomes

$$[H\bullet] = \frac{k_1[C_2H_6]}{k_5[C_2H_5\bullet]} = \frac{k_1}{k_5}\left(\frac{k_3k_5}{k_1k_4}\right)^{1/2} = \left(\frac{k_1k_3}{k_4k_5}\right)^{1/2} \quad (19.100)$$

With the preceding definitions for $[H\bullet]$ and $[C_2H_5\bullet]$ in hand, the differential rate expression for the disappearance of ethane [Equation (19.91)] becomes

$$-\frac{d[C_2H_6]}{dt} = \left(k_2[CH_3\bullet] + \left(\frac{k_1k_3k_4}{k_5} \right)^{1/2} \right)[C_2H_6] \qquad (19.101)$$

Finally, using the definition of $[CH_3\bullet]$ in Equation (19.93) and ignoring higher powers of k_1, the final differential rate expression for $[C_2H_6]$ is

$$-\frac{d[C_2H_6]}{dt} = \left(\frac{k_1k_3k_4}{k_5} \right)^{1/2}[C_2H_6] \qquad (19.102)$$

Equation (19.102) predicts that the decay of ethane should be first order with respect to ethane, as is observed experimentally. The remarkable aspect of this result is that from a very complex mechanism a relatively simple rate expression is derived. In general, even the most complex Rice–Herzfeld radical mechanisms will yield orders of 1/2, 1, 3/2, and 2.

EXAMPLE PROBLEM 19.3

Consider the following reaction of methane with molecular chlorine:

$$CH_4(g) + Cl_2(g) \longrightarrow CH_3Cl(g) + HCl(g)$$

Experimental studies have shown that the rate law for this reaction is one-half order with respect to Cl_2. Is the following mechanism consistent with this behavior?

$$Cl_2 \xrightarrow{k_1} 2Cl\bullet$$

$$Cl\bullet + CH_4 \xrightarrow{k_2} HCl + CH_3\bullet$$

$$CH_3\bullet + Cl_2 \xrightarrow{k_3} CH_3Cl + Cl\bullet$$

$$Cl\bullet + Cl\bullet \xrightarrow{k_4} Cl_2$$

Solution

The rate of reaction in terms of product HCl is given by

$$R = \frac{d[HCl]}{dt} = k_2[Cl\bullet][CH_4]$$

Because $Cl\bullet$ is a reaction intermediate, it cannot appear in the final rate law expression, and $[Cl\bullet]$ must be expressed in terms of $[CH_4]$ and $[Cl_2]$. The differential rate expressions for $[Cl\bullet]$ and $[CH_3\bullet]$ are

$$\frac{d[Cl\bullet]}{dt} = 2k_1[Cl_2] - k_2[Cl\bullet][CH_4] + k_3[CH_3\bullet][Cl_2] - 2k_4[Cl\bullet]^2$$

$$\frac{d[CH_3\bullet]}{dt} = k_2[Cl\bullet][CH_4] - k_3[CH_3\bullet][Cl_2]$$

Applying the steady-state approximation to the expression for $[CH_3\bullet]$ yields

$$[CH_3\bullet] = \frac{k_2[Cl\bullet][CH_4]}{k_3[Cl_2]}$$

Next, we substitute this definition of $[CH_3\bullet]$ into the differential rate expression for $[Cl\bullet]$ and apply the steady-state approximation:

$$0 = 2k_1[Cl_2] - k_2[Cl\bullet][CH_4] + k_3[CH_3\bullet][Cl_2] - 2k_4[Cl\bullet]^2$$

$$0 = 2k_1[Cl_2] - k_2[Cl\bullet][CH_4] + k_3\left(\frac{k_2[Cl\bullet][CH_4]}{k_3[Cl_2]} \right)[Cl_2] - 2k_4[Cl\bullet]^2$$

$$0 = 2k_1[Cl_2] - 2k_4[Cl\bullet]^2$$

$$[Cl\bullet] = \left(\frac{k_1}{k_4}[Cl_2] \right)^{1/2}$$

With this result, the predicted rate law expression becomes

$$R = k_2[Cl\bullet][CH_4] = k_2\left(\frac{k_1}{k_4}\right)^{1/2}[CH_4][Cl_2]^{1/2}$$

The mechanism is consistent with the experimentally observed one-half order dependence on $[Cl_2]$.

19.6 Radical-Chain Polymerization

A very important class of radical reactions are **radical polymerization** reactions. In these processes, a monomer is activated through the reaction with a radical initiator, creating a monomer radical. Next, the monomer radical reacts with another monomer to create a radical dimer. The radical dimer then reacts with another monomer, and the process continues, resulting in formation of a **polymer** chain. The mechanism for chain polymerization is as follows. First, the activated monomer must be created in an initiation step:

$$I \xrightarrow{k_i} 2R\bullet \tag{19.103}$$

$$R\bullet + M \xrightarrow{k_1} M_1\bullet \tag{19.104}$$

In this step, the initiator, I, is transformed into radicals, $R\bullet$, that react with a monomer to form an activated monomer, $M_1\bullet$. The next mechanistic step is propagation where the activated monomer reacts with another monomer to form activated dimer, and the dimer undergoes subsequent reaction as follows:

$$M_1\bullet + M \xrightarrow{k_p} M_2\bullet \tag{19.105}$$

$$M_2\bullet + M \xrightarrow{k_p} M_3\bullet \tag{19.106}$$

$$M_{n-1}\bullet + M \xrightarrow{k_p} M_n\bullet \tag{19.107}$$

In the preceding equations, the subscript indicates the number of monomers contained in the polymer chain, and the rate constant for propagation, k_p, is assumed to be independent of polymer size. The final step in the mechanism is termination in which two radical chains undergo reaction:

$$M_m\bullet + M_n\bullet \xrightarrow{k_t} M_{m+n} \tag{19.108}$$

The rate of activated monomer production is related to the rate of radical ($R\bullet$) formation as follows:

$$\left(\frac{d[M\bullet]}{dt}\right)_{production} = 2\phi k_i[I] \tag{19.109}$$

where ϕ represents the probability that the initiator-generated radical will create a radical chain. The rate of activated monomer loss is equal to the rate of termination:

$$\left(\frac{d[M\bullet]}{dt}\right)_{decay} = -2k_t[M\bullet]^2 \tag{19.110}$$

The total differential rate expression for $[M\bullet]$ is given by the sum of Equations (19.109) and (19.110):

$$\frac{d[M\bullet]}{dt} = 2\phi k_i[I] - 2k_t[M\bullet]^2 \tag{19.111}$$

Because $M\bullet$ is an intermediate species, the steady-state approximation is applied to Equation (19.111), yielding the following:

$$[\text{M}\bullet] = \left(\frac{\phi k_i}{k_t} \right)^{1/2} [\text{I}]^{1/2} \tag{19.112}$$

Finally, the monomer consumption is dominated by the propagation compared to radical activation such that the differential rate expression for [M] becomes

$$\frac{d[\text{M}]}{dt} = -k_p[\text{M}\bullet][\text{M}] \tag{19.113}$$

$$= -k_p \left(\frac{\phi k_i}{k_t} \right)^{1/2} [\text{I}]^{1/2}[\text{M}]$$

Equation (19.113) demonstrates that the rate of monomer is predicted to overall 3/2 order, 1/2 order in initiator concentration, and first order in monomer concentration.

One measure of the efficiency of polymerization is the kinetic chain length, ν. This quantity is equal to the rate of monomer unit consumption divided by the rate of active center production:

$$\nu = \frac{k_p[\text{M}\bullet][\text{M}]}{2\phi k_i[\text{I}]} = \frac{k_p[\text{M}\bullet][\text{M}]}{2k_t[\text{M}\bullet]^2} = \frac{k_p[\text{M}]}{2k_t[\text{M}\bullet]} \tag{19.114}$$

Substitution of Equation (19.112) into Equation (19.114) provides the final expression for the kinetic chain length:

$$\nu = \frac{k_p[\text{M}]}{2k_t[\text{M}\bullet]} = \frac{k_p[\text{M}]}{2k_t \left(\frac{\phi k_i}{k_t} \right)^{1/2} [\text{I}]^{1/2}} = \frac{k_p[\text{M}]}{2(\phi k_i k_t)^{1/2}[\text{I}]^{1/2}} \tag{19.115}$$

Equation (19.115) predicts that the kinetic chain length will increase as the rate constants for chain initiation or termination or the concentration of initiator are reduced. Therefore, polymerization is usually carried out at minimal initiator concentrations such that the number of activated monomers is kept small.

19.7 Explosions

Consider a highly exothermic reaction in which a significant amount of heat is liberated during the reaction. The reaction will proceed with a certain initial rate, but if the heat liberated during the reaction is not dissipated, the system temperature will rise, as will the rate of reaction. The final result of this process is a thermal explosion. A second type of explosion involves chain-branching reactions. In the previous section, the concentration of radical intermediate species was determined by applying the steady-state approximation. However, what if the concentrations of radical intermediate species were not constant with time? Two limits can be envisioned: either a significant reduction or increase in radical intermediate concentration as time proceeds. If the concentration of radical intermediates decreases with time, then the reaction will terminate. What if the concentration of reactive intermediates increases rapidly with time? From the mechanisms discussed thus far, an increase in radical intermediate concentration would lead to the creation of more radical species. In this case, the number of radical chains increases exponentially with time leading to explosion.

A standard introductory chemistry demonstration is the ignition of a balloon containing hydrogen and oxygen. The balloon is ignited, and if the two gases are present in the correct stoichiometric ratio, an explosion occurs as evidenced by a loud bang and a number of startled students. The specific reaction is

$$2\text{H}_2(g) + \text{O}_2(g) \longrightarrow 2\text{H}_2\text{O}(g) \tag{19.116}$$

The reaction is deceptively simple in that the mechanism of the reaction is not fully understood. Important mechanistic components of this reaction are

$$H_2 + O_2 \longrightarrow 2OH\bullet \tag{19.117}$$

$$H_2 + OH\bullet \longrightarrow H\bullet + H_2O \tag{19.118}$$

$$H\bullet + O_2 \longrightarrow OH\bullet + \bullet O\bullet \tag{19.119}$$

$$\bullet O\bullet + H_2 \longrightarrow OH\bullet + H\bullet \tag{19.120}$$

$$H\bullet + O_2 + M \longrightarrow HO_2\bullet + M^* \tag{19.121}$$

The first step, Equation (19.117), is an initiation step in which two hydroxyl radicals (OH•) are created. The radical is propagated in the second step, Equation (19.118). Steps three and four, Equations (19.119) and (19.120), are referred to as **branching reactions**, in which a single radical species reacts to produce two radical species. As such the number of reactive radicals increases twofold in these branching steps. These branching steps lead to a chain-branching explosion because the concentration of the reactive species grows rapidly in time.

The occurrence of explosions for this reaction is dependent on temperature and pressure, as shown in Figure 19.13. First, explosion will occur only if the temperature is greater than 460°C. At lower temperatures, the rates for the various radical-producing reactions are insufficient to support appreciable chain branching. In addition to temperature, the pressure must also be sufficiently high to support chain branching. If the pressure is low, radicals that are produced can diffuse to the walls of the vessel where they are destroyed. Under these conditions, the rates of radical production and decay are balanced such that branching is not prevalent and explosion does not occur. As the pressure is increased, the first explosion regime is reached in which the radicals can participate in branching reactions before reaching the container walls. A further increase in pressure results in the reaction returning to a controlled regime where the pressure is so great that radical–radical reactions reduce the number of reactive species present. The final reaction in the mechanism under discussion is an example of such a reaction. In this step, H• and O_2 react to produce $HO_2\bullet$, which does not contribute to the reaction. The formation of this species requires a three-body collision, with the third species, M, taking away excess energy such that the $HO_2\bullet$ radical is stable. Such reactions will occur only at elevated pressures. Finally, at highest pressures another explosive regime is encountered. This is a thermal explosive regime where the dissipation of heat is insufficient to keep the system temperature from increasing rapidly, providing for explosive behavior.

The likelihood of undergoing an explosion is highly dependent on radical concentration. A generic scheme for chain-branching reactions is as follows:

$$A + B \xrightarrow{k_i} R\bullet \tag{19.122}$$

$$R\bullet \xrightarrow{k_b} \phi R\bullet + P_1 \tag{19.123}$$

$$R\bullet \xrightarrow{k_t} P_2 \tag{19.124}$$

In this scheme, the first step involves the reaction of reactants A and B, resulting in the formation of radical R•. The second step is chain branching in which the radical undergoes reaction to produce other radicals with a branching efficiency given by ϕ. For example, $\phi = 2$ is equivalent to the H• + O_2 and produces the OH• + •O• reaction discussed in the previous example. The final step of the mechanism is termination. Finally, the species P_1 and P_2 represent nonreactive products. The differential rate expression for [R•] consistent with this mechanism is as follows:

$$\frac{d[R\bullet]}{dt} = k_i[A][B] - k_b[R\bullet] + \phi k_b[R\bullet] - k_t[R\bullet]$$

$$= \Gamma + k_{eff}[R\bullet] \tag{19.125}$$

In Equation (19.125), the following definitions have been employed:

$$\Gamma = k_i[A][B]$$
$$k_{eff} = k_b(\phi - 1) - k_t$$

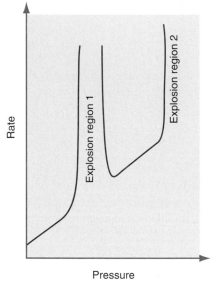

FIGURE 19.13

Schematic of an explosive reaction. As the pressure is increased, two explosive regimes are encountered. The region at lower pressures is due to chain-branching reactions, and the higher pressure region corresponds to a thermal explosion.

Rate (vertical axis) *Pressure* (horizontal axis)

Explosion region 1

Explosion region 2

Equation (19.125) can be solved for [R•] to yield

$$[R\bullet] = \frac{\Gamma}{k_{eff}}(e^{k_{eff}t} - 1) \tag{19.126}$$

Equation (19.126) demonstrates that [R•] is dependent on k_{eff}. Two cases can be envisioned depending on the magnitude of k_t in comparison to $k_b(\phi - 1)$. In the limit where $k_t \gg k_b(\phi - 1)$, termination dominates and Equation (19.126) becomes

$$\lim_{k_t \gg k_b(\phi-1)} [R\bullet] = \frac{\Gamma}{k_t}(1 - e^{-k_t t}) \tag{19.127}$$

Equation (19.127) demonstrates that in this limit the radical concentration will reach a limiting value of Γ/k_t at $t = \infty$. The interpretation of this limiting behavior is that the [R•] will never become large enough to support branching, and explosion will not occur. The second limit occurs when $k_b(\phi - 1) \gg k_t$ and branching dominates. In this limit, Equation (19.126) becomes

$$\lim_{k_b(\phi-1) \gg k_t} [R\bullet] = \frac{\Gamma}{k_b(\phi - 1)}(e^{k_b(\phi-1)t} - 1) \tag{19.128}$$

Equation (19.128) demonstrates that [R•] is predicted to increase exponentially corresponding to explosion. This simple mechanism illustrates the importance of efficient propagation/branching in promoting explosions in chain-branching reactions.

19.8 Photochemistry

Photochemical processes involve the initiation of a chemical reaction through the absorption of a photon by an atom or molecule. In these reactions, photons can be thought of as reactants, and initiation of the reaction occurs when the photon is absorbed. Photochemical reactions are important in a wide variety of areas. The primary event in vision involves the absorption of a photon by the visual pigment rhodopsin. Photosynthesis involves the conversion of light energy into chemical energy by plants and bacteria. Finally, numerous photochemical reactions occur in the atmosphere (e.g., ozone production and decomposition) that are critical to life on Earth. As illustrated by these examples, photochemical reactions are an extremely important area of chemistry, and are explored in this section.

19.8.1 Photophysical Processes

When a molecule absorbs a photon of light, the energy contained in the photon is transferred to the molecule. The amount of energy contained by a photon is given by the Planck equation:

$$E_{photon} = h\nu = \frac{hc}{\lambda} \tag{19.129}$$

In Equation (19.129), h is Planck's constant (6.626×10^{-34} J s), c is the speed of light in a vacuum (3.00×10^8 m s^{-1}), ν is the frequency of light, and λ is the corresponding wavelength of light. A mole of photons is referred to as an Einstein, and the energy contained by an Einstein of photons is Avogadro's number times E_{photon}. The intensity of light is generally stated as energy per unit area per unit time. Because one joule per second is a watt, a typical intensity unit is W cm^{-2}.

The simplest photochemical process is the absorption of a photon by a reactant resulting in product formation:

$$A \xrightarrow{h\nu} P \tag{19.130}$$

The rate of reactant photoexcitation is given by

$$Rate = -\frac{d[A]}{dt} = \frac{I_{abs}1000}{l} \tag{19.131}$$

In Equation (19.131), I_{abs} is the intensity of absorbed light in units of Einstein $cm^{-2} s^{-1}$, l is the path length of the sample in centimeters, and 1000 represents the conversion from cubic centimeters to liters such that the rate has appropriate units of $M s^{-1}$. In Equation (19.131), it is assumed that reactant excitation occurs through the absorption of a single photon. According to the Beer–Lambert law introduced in Chapter 15 of Engel's *Quantum Chemistry and Spectroscopy*, the intensity of light transmitted through a sample (I_{trans}) is given by

$$I_{trans} = I_0 10^{-\varepsilon l[A]} \tag{19.132}$$

where I_0 is the intensity of incident radiation, ε is the **molar absorptivity** of species A, and [A] is the concentration of reactant. Recall that the molar absorptivity will vary with excitation wavelength. Because $I_{abs} = I_0 - I_{trans}$,

$$I_{abs} = I_0(1 - 10^{-\varepsilon l[A]}) \tag{19.133}$$

The series expansion of the exponential term in Equation (19.133) is

$$10^{-\varepsilon l[A]} = 1 - 2.303\varepsilon l[A] + \frac{(2.303\varepsilon l[A])^2}{2!} - ... \tag{19.134}$$

If the concentration of reactant is kept small, only the first two terms in Equation (19.134) are appreciable, and substitution into Equation (19.133) yields

$$I_{abs} = I_0(2.303)\varepsilon l[A] \tag{19.135}$$

Substitution of Equation (19.135) into the rate expression for reactant photoexcitation of Equation (19.131) and integration yield the following expression for [A]:

$$[A] = [A]_0 e^{-I_0(2303)\varepsilon t} = [A]_0 e^{-kt} \tag{19.136}$$

Equation (19.136) demonstrates that the absorption of light will result in the decay of reactant concentration consistent with first-order kinetic behavior. Most photochemical reactions are first order in reactant concentration such that Equation (19.136) describes the evolution in reactant concentration for the majority of photochemical processes. At times it is more useful to discuss photochemical processes with respect to the number of molecules as opposed to concentration. This is precisely the limit one encounters when considering the **photochemistry** of individual molecules as presented later. In this case, Equation (19.131) becomes

$$-\frac{dA}{dt} = I_0 \frac{2303\varepsilon}{N_A} A \tag{19.137}$$

where A represents the number of molecules of reactant and N_A is Avogadro's number. Integrating Equation (19.137), we obtain

$$A = A_0 e^{-I_0(2303\varepsilon/N_A)t} = A_0 e^{-I_0\sigma_A t} \tag{19.138}$$

where σ_A is known as the **absorption cross section**, and the rate constant for excitation, k_a, is equal to $I_0\sigma_A$ with I_0 in units of photons $cm^{-2} s^{-1}$.

As discussed in Chapter 15 of *Quantum Chemistry and Spectroscopy*, the absorption of light may occur when the photon energy is equal to the energy difference between two energy states of the molecule. A schematic of the processes that occur following photon absorption resulting in an electronic energy-level transition (or "electronic transition") is given in Figure 19.14. Such diagrams are referred to as **Jablonski diagrams** after Aleksander Jablonski, a Polish physicist who developed these diagrams for describing kinetic processes initiated by electronic transitions. In a Jablonski diagram, the vertical axis represents increasing energy. The electronic states depicted are the ground-state singlet, S_0, first excited singlet, S_1, and triplet, T_1. In the singlet states, the electrons are spin

FIGURE 19.14
A Jablonski diagram depicting various photophysical processes, where S_0 is the ground electronic singlet state, S_1 is the first excited singlet state, and T_1 is the first excited triplet state. Radiative processes are indicated by the straight lines. The nonradiative processes of intersystem crossing (ISC), internal conversion (IC), and vibrational relaxation (VR) are indicated by the wavy lines.

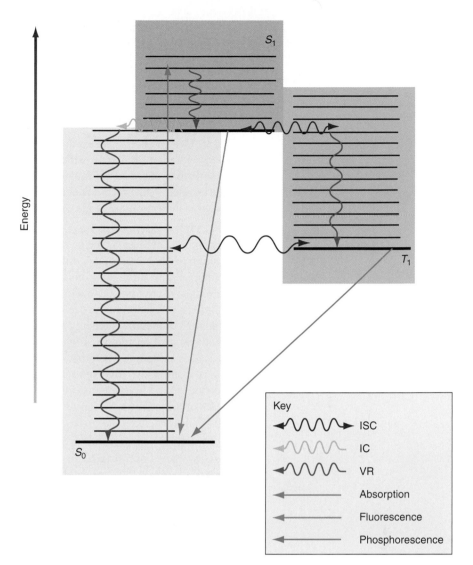

paired such that the spin multiplicity is one (i.e., a "singlet"), and in the triplet state two electrons are unpaired such that the spin multiplicity is three (a "triplet"). The subscripts indicate the energy ordering of the states. Because triplets are generally formed by electronic excitation, the lowest energy triplet state is labeled T_1 as opposed to T_0 (the lowest energy spin configuration of molecular oxygen is a triplet, a famous exception to this generality). Finally, the lowest vibrational level for each electronic state is indicated by dark horizontal lines, with higher vibrational levels indicated by the lighter horizontal lines. In addition, a manifold of rotational states will exist for each vibrational level; however, the rotational energy levels have been suppressed for clarity in Figure 19.14.

The solid and wavy lines in Figure 19.14 represent a variety of processes that couple the electronic states. These processes, including the absorption of light and subsequent energetic relaxation pathways, are referred to as **photophysical processes** because the structure of the molecule remains unchanged. In fact, many of the processes of interest in "photochemistry" do not involve photochemical transformation of the reactant at all, but are instead photophysical in nature. The absorption of light decreases the population in the lowest energy singlet state, S_0, referred to as a depletion. Correspondingly, the population in the first excited singlet, S_1, is increased. The absorption transition depicted in Figure 19.14 is to a higher vibrational level in S_1, with the probability of transition to a specific vibrational level determined by the Franck–Condon factor (Chapter 15 of *Quantum Chemistry and Spectroscopy*) between the lowest energy vibrational level in S_0 and the vibrational states in S_1.

After populating S_1, thermal equilibrium of the vibrational energy will occur, a process referred to as vibrational relaxation. Vibrational relaxation is extremely rapid (~100 fs), and when complete the vibrational state population in S_1 will be governed by the Boltzmann distribution. The vibrational energy-level spacings are assumed to be sufficiently large such that only the lowest vibrational level of S_1 is populated to a significant extent after equilibration. Decay of S_1 resulting in repopulation of S_0 can occur through one of three paths:

1. *Path 1*: Loss of excess electronic energy through the emission of a photon. Such processes are referred to as **radiative transitions.** The process by which photons are emitted in the radiative transitions between S_1 and S_0 is referred to as **fluorescence.** This process is equivalent to the spontaneous emission process discussed in Section 8.2 of *Quantum Chemistry and Spectroscopy.*

2. *Path 2*: **Intersystem crossing** (ISC in Figure 19.14) resulting in population of T_1. This process involves a change in spin state, a process that is forbidden by quantum mechanics. As such, intersystem crossing is significantly slower than vibrational relaxation, but is competitive with fluorescence in systems where the triplet state is populated to a significant extent. Following intersystem crossing, vibrational relaxation in the triplet vibrational manifold occurs, resulting in population of the lowest energy vibrational level. From this level, a second radiative transition can occur where S_0 is populated and the excess energy is released as a photon. This process is referred to as **phosphorescence.** Because the T_1–S_0 transition also involves a change in spin, it is also forbidden by spin selection rules. Therefore, the rate for this process is slow and phosphorescence occurs over longer timescales (~10^{-6} s to seconds) as compared to fluorescence (~10^{-9} s).

3. *Path 3*: Rather than undergoing a radiative transition, decay from S_1 to a high vibrational level of S_0 can occur followed by rapid vibrational relaxation. This process is referred to as **internal conversion** or nonradiative decay. Nonradiative decay can also occur through the triplet state by crossing to S_0 followed by vibrational relaxation.

From the viewpoint of kinetics, the absorption of light and subsequent relaxation processes can be viewed as a collection of reactions with corresponding rates. Figure 19.15 presents a modified version of the Jablonski diagram that focuses on these processes and

FIGURE 19.15

Kinetic description of photophysical processes. Rate constants are indicated for absorption (k_a), fluorescence (k_f), internal conversion (k_{ic}), intersystem crossing from S_1 to T_1 (k^S_{isc}), intersystem crossing from T_1 to S_0 (k^T_{isc}), and phosphorescence (k_p).

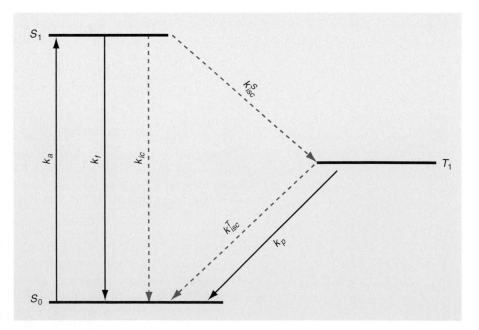

TABLE 19.1

Photophysical Reactions and Corresponding Rate Expressions

Process	Reaction	Rate
Absorption/excitation	$S_0 + h\nu \longrightarrow S_1$	$k_a[S_0]$ $(k_a = I_0\sigma_A)$
Fluorescence	$S_1 \longrightarrow S_0 + h\nu$	$k_f[S_1]$
Internal conversion	$S_1 \longrightarrow S_0$	$k_{ic}[S_1]$
Intersystem crossing	$S_1 \longrightarrow T_1$	$k_{isc}^S[S_1]$
Phosphorescence	$T_1 \longrightarrow S_0 + h\nu$	$k_p[T_1]$
Intersystem crossing	$T_1 \longrightarrow S_0$	$k_{isc}^T[T_1]$

corresponding rate constants. The individual processes, reactions, and notation for the reaction rates are provided in Table 19.1.

19.8.2 Fluorescence and Fluorescence Quenching

The photophysical processes outlined in Table 19.1 are present for any molecular system. To study excited state lifetimes, another photophysical process is introduced: **collisional quenching**. In this process, a collision occurs between a species, Q, and a molecule populating an excited electronic state. The result of the collision is the removal of energy from the molecule with the accompanying conversion of the molecule from S_1 to S_0:

$$S_1 + Q \xrightarrow{k_q} S_0 + Q \tag{19.139}$$

The rate expression for this process is

$$R_q = k_q[S_1][Q] \tag{19.140}$$

By studying the rate of collisional quenching as a function of [Q], it is possible to determine the k_f. To demonstrate this procedure, we begin by recognizing that in the kinetic scheme illustrated in Figure 19.15, S_1 can be considered an intermediate species. Under constant illumination, the concentration of this intermediate will not change. Therefore, we can write the differential rate expression for S_1 and apply the steady-state approximation:

$$\frac{d[S_1]}{dt} = 0 = k_a[S_0] - k_f[S_1] - k_{ic}[S_1] - k_{isc}^S[S_1] - k_q[S_1][Q] \tag{19.141}$$

The **fluorescence lifetime**, τ_f, is defined as

$$\frac{1}{\tau_f} = k_f + k_{ic} + k_{isc}^S + k_q[Q] \tag{19.142}$$

Using this definition of τ_f, Equation (19.141) becomes

$$\frac{d[S_1]}{dt} = 0 = k_a[S_0] - \frac{[S_1]}{\tau_f} \tag{19.143}$$

Equation (19.143) is readily solved for $[S_1]$:

$$[S_1] = k_a[S_0]\tau_f \tag{19.144}$$

The fluorescence intensity, I_f, depends on the rate of fluorescence given by

$$I_f = k_f[S_1] \tag{19.145}$$

Substituting Equation (19.144) into Equation (19.145) results in

$$I_f = k_a[S_0]k_f\tau_f \tag{19.146}$$

Inspection of the last two factors in Equation (19.146) illustrates the following relationship:

$$k_f \tau_f = \frac{k_f}{k_f + k_{ic} + k_{ics}^S + k_q[Q]} = \Phi_f \qquad (19.147)$$

The product of the fluorescence rate constant and fluorescence lifetime is equivalent to the radiative rate constant divided by the sum of rate constants for all processes leading to the decay of S_1. In effect, S_1 decay can be viewed as a branching reaction, and the ratio of rate constants contained in Equation (19.147) can be rewritten as the quantum yield for fluorescence, Φ_f, similar to the definition of reaction yield provided in Section 18.8. The fluorescence quantum yield is also defined as the number of photons emitted as fluorescence divided by the number of photons absorbed. Comparison of this definition to Equation (19.147) demonstrates that the fluorescence quantum yield will be large for molecules in which k_f is significantly greater than other rate constants corresponding to S_1 decay. Inverting Equation (19.146) and using the definition of τ_f, the following expression is obtained:

$$\frac{1}{I_f} = \frac{1}{k_a[S_0]}\left(1 + \frac{k_{ic} + k_{ics}^S}{k_f}\right) + \frac{k_q[Q]}{k_a[S_0]k_f} \qquad (19.148)$$

In fluorescence quenching experiments, fluorescence intensity is measured as a function of [Q]. Measurements are generally performed by referencing to the fluorescence intensity observed in the absence of quencher, I_f^0, such that

$$\frac{I_f^0}{I_f} = 1 + \frac{k_q}{k_f}[Q] \qquad (19.149)$$

Equation (19.149) reveals that a plot of the fluorescence intensity ratio as a function of [Q] will yield a straight line, with slope equal to k_q/k_f. Such plots are referred to as **Stern–Volmer plots**, an example of which is shown in Figure 19.16.

19.8.3 Measurement of τ_f

In the development presented in the preceding subsection, it was assumed that the system of interest was subjected to continuous irradiation so that the steady-state approximation could be applied to $[S_1]$. However, it is often more convenient to photoexcite the system with a temporally short burst of photons, or pulse of light. If the temporal duration of the pulse is short compared to the rate of S_1 decay, the decay of this state can be measured directly by monitoring the fluorescence intensity as a function of time. Optical pulses as short as 4 fs (4×10^{-15} s) can be produced that provide excitation on a timescale that is significantly shorter than decay time of S_1.

After excitation by a temporally short optical pulse, the concentration of molecules in $[S_1]$ will be finite. In addition, the rate constant for excitation is zero because $I_0 = 0$; therefore, the differential rate expression for S_1 becomes

$$\frac{d[S_1]}{dt} = -k_f[S_1] - k_{ic}[S_1] - k_{isc}^S[S_1] - k_q[Q][S_1]$$

$$\frac{d[S_1]}{dt} = -\frac{[S_1]}{\tau_f} \qquad (19.150)$$

Equation (19.150) can be solved for $[S_1]$ resulting in

$$[S_1] = [S_1]_0 e^{-t/\tau_f} \qquad (19.151)$$

Because the fluorescence intensity is linearly proportional to $[S_1]$ per Equation (19.145), Equation (19.151) predicts that the fluorescence intensity will undergo exponential decay with time constant τ_f. In the limit where $k_f \gg k_{ic}$ and $k_f \gg k_{isc}^S$, τ_f can be approximated as follows:

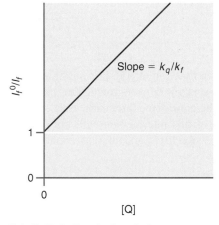

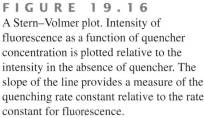

FIGURE 19.16
A Stern–Volmer plot. Intensity of fluorescence as a function of quencher concentration is plotted relative to the intensity in the absence of quencher. The slope of the line provides a measure of the quenching rate constant relative to the rate constant for fluorescence.

$$\lim_{k_f \gg k_{ic}, k_{isc}^S} \tau_f = \frac{1}{k_f + k_q[Q]} \qquad (19.152)$$

In this limit, measurement of the fluorescence lifetime at a known quencher concentration combined with the slope from a Stern–Volmer plot is sufficient to uniquely determine k_f and k_q. Taking the reciprocal of Equation (19.152), we obtain

$$\frac{1}{\tau_f} = k_f + k_q[Q] \qquad (19.153)$$

Equation (19.153) demonstrates that a plot of $(\tau_f)^{-1}$ versus [Q] will yield a straight line with y intercept equal to k_f and slope equal to k_q.

EXAMPLE PROBLEM 19.4

Thomaz and Stevens (in *Molecular Luminescence*, Lim, 1969) studied the fluorescence quenching of pyrene in solution. Using the following information, determine k_f and k_q for pyrene in the presence of the quencher Br_6C_6.

$[Br_6C_6]$ (M)	τ_f (s)
0.0005	2.66×10^{-7}
0.001	1.87×10^{-7}
0.002	1.17×10^{-7}
0.003	8.50×10^{-8}
0.005	5.51×10^{-8}

Solution
Using Equation (19.153), a plot of $(\tau_f)^{-1}$ versus [Q] for this system is as follows:

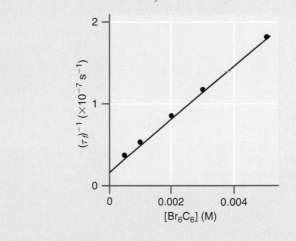

The best fit to the data by a straight line corresponds to a slope of 3.00×10^9 s^{-1}, which is equal to k_q by Equation (19.153), and a y intercept of 1.98×10^6 s^{-1}, which is equal to k_f.

19.8.4 Single-Molecule Fluorescence

Equation (19.151) describes how the population of molecules in S_1 will evolve as a function of time, and the fluorescence intensity is predicted to demonstrate exponential decay. This predicted behavior is for a collection, or ensemble, of molecules; however,

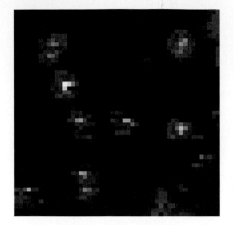

FIGURE 19.17
Microscope image of single Rhodamine B dye molecules on glass. Image was obtained using a confocal scanning microscope with the bright spots in the image corresponding to molecular fluorescence. The image dimension is 5 μm by 5 μm.

recent spectroscopic techniques and advances in light detection have allowed for the detection of fluorescence from a single molecule. Figure 19.17 presents an image of single molecules obtained using a confocal scanning microscope. In a confocal microscope, the excitation source and image occur at identical focal distances such that fluorescence from sample areas not directly in focus can be rejected. Using this technique in combination with laser excitation and efficient detectors, it is possible to observe the fluorescence from a single molecule. In Figure 19.17, the bright features represent fluorescence from single molecules. The spatial dimension of these features is determined by the diameter of the light beam at the sample (~300 nm).

What does the fluorescence from a single molecule look like as a function of time? Instead of a population of molecules in S_1 being responsible for the emission, the fluorescence is derived from a single molecule. Figure 19.18 presents the observed fluorescence intensity from a single molecule with continuous photoexcitation. Fluorescence is observed after the light field is turned on, and the molecule cycles between S_0 and S_1 due to photoexcitation and subsequent relaxation via fluorescence. This regime of constant fluorescence intensity continues until the fluorescence abruptly stops. At this point, depopulation of the S_1 state results in the production of T_1 or some other state of the molecule that does not fluoresce. Eventually, fluorescence is again observed at later times corresponding to the eventual recovery of S_0 by relaxation from these other states, with excitation resulting in the repopulation of S_1 followed by fluorescence. This pattern continues until a catastrophic event occurs in which the structural integrity of the molecules is lost. This catastrophic event is referred to as photodestruction, and it results in irreversible photochemical conversion of the molecule to another, nonemissive species.

Clearly, the fluorescence behavior observed in Figure 19.18 is dramatically different than the behavior predicted for an ensemble of molecules. Current interest in this field involves the application of single-molecule techniques to elucidate behavior that is not reflected by the ensemble. Such studies are extremely useful for isolating molecular dynamics from an ensemble of molecules having inherently inhomogeneous behavior. In addition, molecules can be studied in isolation of the bulk, thereby providing a window into the connection between molecular and ensemble behavior.

19.8.5 Photochemical Processes

As discussed earlier, photochemical processes are distinct from photophysical processes in that the absorption of a photon results in chemical transformation of the reactant. For a photochemical process that occurs through the first excited singlet state,

FIGURE 19.18
Fluorescence from a single Rhodamine B dye molecule. Steady illumination of the single molecule occurs at t_{on} resulting in fluorescence, I_f. The fluorescence continues until decay of the S_1 state leads to population of a nonfluorescent state. At the end of the time axis, a brief period of fluorescence is observed corresponding to decay of the nonfluorescent state to populate S_0 followed by photoexcitation resulting in the population of S_1 and fluorescence. However, this second period of fluorescence ends abruptly due to photodestruction of the molecule as evidenced by the absence of fluorescence after the decay event (t_{pd}).

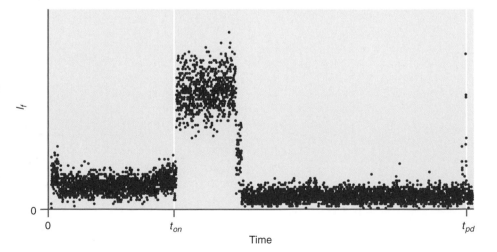

S_1, a photochemical reaction can be viewed kinetically as another reaction branch resulting in decay of S_1. The corresponding expression for the rate corresponding to this photochemical reaction branch is

$$R_{photochem.} = k_{photo}^S [S_1] \qquad (19.154)$$

where k_{photo} is the rate constant for the photochemical reaction. For photochemical processes occurring through T_1, a rate expression similar to Equation (19.154) can be constructed as follows:

$$R_{photochem.} = k_{photo}^T [T_1] \qquad (19.155)$$

The absorption of a photon can also provide sufficient energy to initiate a chemical reaction. However, given the range of photophysical processes that occurs, absorption of a photon is not sufficient to guarantee that the photochemical reaction will occur. The extent of photochemistry is quantified by the overall **quantum yield, ϕ**, which is defined as the number of reactant molecules consumed in photochemical processes per photon absorbed. The overall quantum yield can be greater than one, as demonstrated by the photoinitiated decomposition of HI that proceeds by the following mechanism:

$$HI + h\nu \longrightarrow H\bullet + I\bullet \qquad (19.156)$$
$$H\bullet + HI \longrightarrow H_2 + I\bullet \qquad (19.157)$$
$$I\bullet + I\bullet \longrightarrow I_2 \qquad (19.158)$$

In this mechanism, absorption of a photon results in the loss of two HI molecules such that $\phi = 2$. In general, the overall quantum yield can be determined experimentally by comparing the molecules of reactant lost to the number of photons absorbed as illustrated in Example Problem 19.5.

EXAMPLE PROBLEM 19.5

The reactant 1,3 cyclohexadiene can be photochemically converted to *cis*-hexatriene. In an experiment, 2.5 mmol of cyclohexadiene are converted to *cis*-hexatriene when irradiated with 100 W of 280-nm light for 27 s. All of the light is absorbed by the sample. What is the overall quantum yield for this photochemical process?

Solution

First, the total photon energy absorbed by the sample, E_{abs}, is

$$E_{abs} = (power)\Delta t = (100 \text{ J s}^{-1})(27 \text{ s}) = 2.7 \times 10^3 \text{ J}$$

Next, the photon energy at 280 nm is

$$E_{ph} = \frac{hc}{\lambda} = \frac{(6.626 \times 10^{-34} \text{ J s})(3 \times 10^8 \text{ m s}^{-1})}{2.80 \times 10^{-7} \text{ m}} = 7.10 \times 10^{-19} \text{ J}$$

The total number of photons absorbed by the sample is therefore

$$\frac{E_{abs}}{E_{ph}} = \frac{2.7 \times 10^3 \text{ J}}{7.10 \times 10^{-19} \text{ J photon}^{-1}} = 3.80 \times 10^{21} \text{ photons}$$

Dividing this result by Avogadro's number results in 6.31×10^{-3} Einsteins or moles of photons. Therefore, the overall quantum yield is

$$\phi = \frac{moles_{react}}{moles_{photon}} = \frac{2.50 \times 10^{-3} \text{ mol}}{6.31 \times 10^{-3} \text{ mol}} = 0.396 \approx 0.4$$

For Further Reading

CRC Handbook of Photochemistry and Photophysics. CRC Press, Boca Raton, FL, 1990.

Eyring, H., S. H. Lin, and S. M. Lin, *Basic Chemical Kinetics.* Wiley, New York, 1980.

Pannetier, G., and P. Souchay, *Chemical Kinetics.* Elsevier, Amsterdam, 1967.

Hammes, G. G., *Thermodynamics and Kinetics for the Biological Sciences.* Wiley, New York, 2000.

Fersht, A., *Enzyme Structure and Mechanism.* W. H. Freeman, New York, 1985.

Fersht, A., *Structure and Mechanism in Protein Science.* W. H. Freeman, New York, 1999.

Laidler, K. J., *Chemical Kinetics.* Harper & Row, New York, 1987.

Robinson, P. J., and K. A. Holbrook. *Unimolecular Reactions.* Wiley, New York, 1972.

Turro, N. J., *Modern Molecular Photochemistry.* Benjamin Cummings, Menlo Park, CA, 1978.

Simons, J. P., *Photochemistry and Spectroscopy.* Wiley, New York, 1971.

Lim, E. G., *Molecular Luminescence.* Benjamin Cummings, Menlo Park, CA, 1969.

Noggle, J. H., *Physical Chemistry.* HarperCollins, New York, 1996.

Vocabulary

absorption cross section

activated reactant

adsorption isotherm

branching reaction

catalyst

chemisorption

collisional quenching

competitive inhibitor

complex reaction

composite constant

enzymes

fluorescence

fluorescence lifetime

fractional coverage

heterogeneous catalysts

homogeneous catalyst

initiation step

internal conversion

intersystem crossing

Jablonski diagram

Langmuir isotherm

Langmuir model

Lindemann mechanism

Lineweaver–Burk equation

Lineweaver–Burk plot

Michaelis constant

Michaelis–Menten mechanism

Michaelis–Menten rate law

molar absorptivity

phosphorescence

photochemistry

photophysical processes

physisorption

polymer

preequilibrium approximation

propagation step

quantum yield

radiative transitions

radical

radical polymerization

reaction intermediate

reaction mechanism

reciprocal plot

simple reaction

Stern–Volmer plots

stoichiometric number

substrate–catalyst complex

termination step

turnover number

unimolecular reactions

Questions on Concepts

Q19.1   How is a simple reaction different from a complex reaction?

Q19.2   For a reaction mechanism to be considered correct, what property must it demonstrate?

Q19.3   What is a reaction intermediate? Can an intermediate be present in the rate law expression for the overall reaction?

Q19.4   What is the preequilibrium approximation, and under what conditions is it considered valid?

Q19.5   What is the one main assumption in the Lindemann mechanism for unimolecular reactions?

Q19.6   How is a catalyst defined, and how does such a species increase the reaction rate?

Q19.7   What is an enzyme? What is the general mechanism describing enzyme catalysis?

Q19.8   What is the Michaelis–Menten rate law? What is the maximum reaction rate predicated by this rate law?

Q19.9 How is the standard enzyme kinetic scheme modified to incorporate competitive inhibition? What plot is used to establish competitive inhibition and to determine the kinetic parameters associated with inhibition?

Q19.10 What is the difference between a homogeneous and a heterogeneous catalyst?

Q19.11 What are the inherent assumptions in the Langmuir model of surface adsorption?

Q19.12 What is a radical? What elementary steps are involved in a reaction mechanism involving radicals?

Q19.13 In what ways are radical polymerization reactions similar to radical reactions in general?

Q19.14 What is photochemistry? How does one calculate the energy of a photon?

Q19.15 What depopulation pathways occur from the first excited singlet state? For the first excited triplet state?

Q19.16 What is the expected variation in excited state lifetime with quencher concentration in a Stern–Volmer plot?

Problems

Problem numbers in **red** indicate that the solution to the problem is given in the *Student's Solutions Manual.*

P19.1 A proposed mechanism for the formation of N_2O_5 from NO_2 and O_3 is

$$NO_2 + O_3 \xrightarrow{k_1} NO_3 + O_2$$

$$NO_2 + NO_3 + M \xrightarrow{k_2} N_2O_5 + M$$

Determine the rate law expression for the production of N_2O_5 given this mechanism.

P19.2 The Rice-Herzfeld mechanism for the thermal decomposition of acetaldehyde (CH_3CO) is

$$CH_3CHO \xrightarrow{k_1} CH_3\bullet + CHO\bullet$$

$$CH_3\bullet + CH_3CHO \xrightarrow{k_2} CH_4 + CH_2CHO\bullet$$

$$CH_2CHO\bullet \xrightarrow{k_3} CO + CH_3\bullet$$

$$CH_3\bullet + CH_3\bullet \xrightarrow{k_4} C_2H_6$$

Using the steady-state approximation, determine the rate of methane (CH_4) formation.

P19.3 Consider the following mechanism for ozone thermal decomposition:

$$O_3 \underset{k_{-1}}{\overset{k_1}{\rightleftharpoons}} O_2 + O$$

$$O_3 + O \xrightarrow{k_2} 2O_2$$

a. Derive the rate law expression for the loss of O_3.

b. Under what conditions will the rate law expression for O_3 decomposition be first order with respect to O_3?

P19.4 The hydrogen-bromine reaction corresponds to the production of HBr from H_2 and Br_2 as follows: $H_2 + Br_2 \rightleftharpoons 2HBr$. This reaction is famous for its complex rate law, determined by Bodenstein and Lind in 1906:

$$\frac{d[HBr]}{dt} = \frac{k[H_2][Br_2]^{1/2}}{1 + \frac{m[HBr]}{[Br_2]}}$$

where k and m are constants. It took 13 years for the correct mechanism of this reaction to be proposed, and this feat was accomplished simultaneously by Christiansen, Herzfeld, and Polyani. The mechanism is as follows:

$$Br_2 \underset{k_{-1}}{\overset{k_1}{\rightleftharpoons}} 2Br\bullet$$

$$Br\bullet + H_2 \xrightarrow{k_2} HBr + H\bullet$$

$$H\bullet + Br_2 \xrightarrow{k_3} HBr + Br\bullet$$

$$HBr + H\bullet \xrightarrow{k_4} H_2 + Br\bullet$$

Construct the rate law expression for the hydrogen-bromine reaction by performing the following steps:

a. Write down the differential rate expression for [HBr].

b. Write down the differential rate expressions for [Br] and [H].

c. Because Br and H are reaction intermediates, apply the steady-state approximation to the result of part (b).

d. Add the two equations from part (c) to determine [Br] in terms of $[Br_2]$.

e. Substitute the expression for [Br] back into the equation for [H] derived in part (c) and solve for [H].

f. Substitute the expressions for [Br] and [H] determined in part (e) into the differential rate expression for [HBr] to derive the rate law expression for the reaction.

P19.5

a. For the hydrogen-bromine reaction presented in Problem P19.4 imagine initiating the reaction with only Br_2 and H_2 present. Demonstrate that the rate law expression at $t = 0$ reduces to

$$\left(\frac{d[HBr]}{dt} \right)_{t=0} = 2k_2 \left(\frac{k_1}{k_5} \right)^{1/2} [H_2]_0 [Br_2]_0^{1/2}$$

b. The activation energies for the rate constants are as follows:

Rate Constant	ΔE_a (kJ/mol)
k_1	192
k_2	0
k_5	74

What is the overall activation energy for this reaction?

c. How much will the rate of the reaction change if the temperature is increased to 400 K from 298 K?

P19.6 For the reaction $I^-(aq) + OCl^-(aq) \rightleftharpoons OI^-(aq) + Cl^-(aq)$ occurring in aqueous solution, the following mechanism has been proposed:

$$OCl^- + H_2O \underset{k_{-1}}{\overset{k_1}{\rightleftharpoons}} HOCl + OH^-$$

$$I^- + HOCl \overset{k_2}{\longrightarrow} HOI + Cl^-$$

$$HOI + OH^- \overset{k_3}{\longrightarrow} H_2O + OI^-$$

a. Derive the rate law expression for this reaction based on this mechanism. (*Hint:* $[OH^-]$ should appear in the rate law.)

b. The initial rate of reaction was studied as a function of concentration by Chia and Connick [*J. Physical Chemistry* 63 (1959), 1518], and the following data were obtained:

$[I^-]_0$ (M)	$[OCl^-]_0$ (M)	$[OH^-]_0$ (M)	Initial Rate (M s^{-1})
2.0×10^{-3}	1.5×10^{-3}	1.00	1.8×10^{-4}
4.0×10^{-3}	1.5×10^{-3}	1.00	3.6×10^{-4}
2.0×10^{-3}	3.0×10^{-3}	2.00	1.8×10^{-4}
4.0×10^{-3}	3.0×10^{-3}	1.00	7.2×10^{-4}

Is the predicted rate law expression derived from the mechanism consistent with these data?

P19.7 Using the preequilibrium approximation, derive the predicted rate law expression for the following mechanism:

$$A_2 \underset{k_{-1}}{\overset{k_1}{\rightleftharpoons}} 2A$$

$$A + B \overset{k_2}{\longrightarrow} P$$

P19.8 Consider the following mechanism, which results in the formation of product P:

$$A \underset{k_{-1}}{\overset{k_1}{\rightleftharpoons}} B \underset{k_{-2}}{\overset{k_2}{\rightleftharpoons}} C$$

$$B \overset{k_3}{\longrightarrow} P$$

If only the species A is present at $t = 0$, what is the expression for the concentration of P as a function of time? You can apply the preequilibrium approximation in deriving your answer.

P19.9 Consider the gas-phase isomerization of cyclopropane:

$$\begin{array}{c} CH_2 \\ CH_2 - CH_2 \end{array} \longrightarrow CH_3CH=CH_2$$

Are the following data of the observed rate constant as a function of cyclopropane pressure consistent with the Lindemann mechanism?

P (Torr)	k (10^4 s^{-1})	P (Torr)	k (10^4 s^{-1})
84.1	2.98	1.37	1.30
34.0	2.82	0.569	0.857
11.0	2.23	0.170	0.486
6.07	2.00	0.120	0.392
2.89	1.54	0.067	0.303

P19.10 In the discussion of the Lindemann mechanism, it was assumed that the rate of activation by collision with another reactant molecule, A, was the same as collision with a nonreactant molecule, M, such as a buffer gas. What if the rates of activation for these two processes are different? In this case, the mechanism becomes

$$A + M \underset{k_{-1}}{\overset{k_1}{\rightleftharpoons}} A* + M$$

$$A + A \underset{k_{-2}}{\overset{k_2}{\rightleftharpoons}} A* + A$$

$$A* \overset{k_3}{\longrightarrow} P$$

a. Demonstrate that the rate law expression for this mechanism is

$$R = \frac{k_3(k_1[A][M] + k_2[A]^2)}{k_{-1}[M] + k_{-2}[A] + k_{-3}}$$

b. Does this rate law reduce to the expected form when $[M] = 0$?

P19.11 In the unimolecular isomerization of cyclobutane to butylene, the following values for k_{uni} as a function of cyclobutane pressure were measured:

P_0 (Torr)	110	210	390	760
k_{uni} (s^{-1})	9.58	10.3	10.8	11.1

Assuming that the Lindemann mechanism accurately describes this reaction, determine k_1 and the ratio k_{-1}/k_2.

P19.12 The enzyme fumarase catalyzes the hydrolysis of fumarate: Fumarate + $H_2O \longrightarrow$ L-malate. The turnover number for this enzyme is 2.5×10^3 s^{-1}, and the Michaelis constant is 4.2×10^{-6} M. What is the rate of fumarate conversion if the initial enzyme concentration is 1×10^{-6} M and the fumarate concentration is 2×10^{-4} M?

P19.13 The enzyme catalase catalyzes the decomposition of hydrogen peroxide. The following data are obtained regarding the rate of reaction as a function of substrate concentration:

$[H_2O_2]_0$ (M)	0.001	0.002	0.005
Initial Rate (M s^{-1})	1.38×10^{-3}	2.67×10^{-3}	6.00×10^{-3}

The concentration of catalase is 3.5×10^{-9} M. Use these data to determine $rate_{max}$, K_m, and the turnover number for this enzyme.

P19.14 Protein tyrosine phosphatases (PTPases) are a general class of enzymes that are involved in a variety of disease processes including diabetes and obesity. In a study by Z.-Y. Zhang and coworkers [*J. Medicinal Chemistry* 43 (2000), 146], computational techniques were used to identify potential competitive inhibitors of a specific PTPase known as PTP1B. The structure of one of the identified potential competitive inhibitors is shown here:

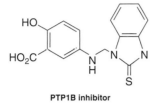

PTP1B inhibitor

The reaction rate was determined in the presence and absence of inhibitor, I, and revealed the following initial reaction rates as a function of substrate concentration:

[S] (μM)	$Rate_0$ (μM s^{-1}), [I] = 0	$Rate_0$ (μM s^{-1}) [I] = 200 μM
0.299	0.071	0.018
0.500	0.100	0.030
0.820	0.143	0.042
1.22	0.250	0.070
1.75	0.286	0.105
2.85	0.333	0.159
5.00	0.400	0.200
5.88	0.500	0.250

a. Determine K_m and $rate_{max}$ for PTP1B.

b. Demonstrate that the inhibition is competitive, and determine K_i.

P19.15 The rate of reaction can be determined by measuring the change in optical rotation of the sample as a function of time if a reactant or product is chiral. This technique is especially useful for kinetic studies of enzyme catalysis involving sugars. For example, the enzyme invertase catalyzes the hydrolysis of sucrose, an optically active sugar. The initial reaction rates as a function of sucrose concentration are as follows:

[Sucrose]$_0$ (M)	Rate (M s^{-1})
0.029	0.182
0.059	0.266
0.088	0.310
0.117	0.330
0.175	0.372
0.234	0.371

Use these data to determine the Michaelis constant and the turnover number for invertase.

P19.16 The enzyme glycogen synthase kinase 3β (GSK-3β) plays a central role in Alzheimer's disease. The onset of Alzheimer's disease is accompanied by the production of highly phosphorylated forms of a protein referred to as "τ." GSK-3β contributes to the hyperphosphorylation of τ such that inhibiting the activity of this enzyme represents a pathway for the development of an Alzheimer's drug. A compound known as Ro 31-8220 is a competitive inhibitor of GSK-3β. The following data were obtained for the rate of GSK-3β activity in the presence and absence of Ro 31-8220 [A. Martinez *et al.*, *J. Medicinal Chemistry* 45 (2002), 1292]:

[S] (μM)	$Rate_0$ (μM s^{-1}), [I] = 0	$Rate_0$ (μM s^{-1}) [I] = 200 μM
66.7	4.17×10^{-8}	3.33×10^{-8}
40.0	3.97×10^{-8}	2.98×10^{-8}
20.0	3.62×10^{-8}	2.38×10^{-8}
13.3	3.27×10^{-8}	1.81×10^{-8}
10.0	2.98×10^{-8}	1.39×10^{-8}
6.67	2.31×10^{-8}	1.04×10^{-8}

Determine K_m and $rate_{max}$ for GSK-3β and, using the data with the inhibitor, determine $K_m{}^*$ and K_i.

P19.17 In the Michaelis–Menten mechanism, it is assumed that the formation of product from the enzyme–substrate complex is irreversible. However, consider the following modified version in which the product formation step is reversible:

$$E + S \underset{k_{-1}}{\overset{k_1}{\rightleftharpoons}} ES \underset{k_{-2}}{\overset{k_2}{\rightleftharpoons}} E + P$$

Derive the expression for the Michaelis constant for this mechanism in the limit where $[S]_0 \gg [E]_0$.

P19.18 Determine the predicted rate law expression for the following radical-chain reaction:

$$A_2 \overset{k_1}{\longrightarrow} 2A\bullet$$
$$A\bullet \overset{k_2}{\longrightarrow} B\bullet + C$$
$$A\bullet + B\bullet \overset{k_3}{\longrightarrow} P$$
$$A\bullet + P \overset{k_4}{\longrightarrow} B\bullet$$

P19.19 The overall reaction for the halogenation of a hydrocarbon (RH) using Br as the halogen is $RH + Br_2 \longrightarrow RBr + HBr$. The following mechanism has been proposed for this process:

$$Br_2 \overset{k_1}{\longrightarrow} 2Br\bullet$$
$$Br\bullet + RH \overset{k_2}{\longrightarrow} R\bullet + HBr$$
$$R\bullet + Br_2 \overset{k_3}{\longrightarrow} RBr + Br\bullet$$
$$Br\bullet + R\bullet \overset{k_4}{\longrightarrow} RBr$$

Determine the rate law predicted by this mechanism.

P19.20 The chlorination of vinyl chloride, $C_2H_3Cl + Cl_2 \longrightarrow C_2H_3Cl_3$, is believed to proceed by the following mechanism:

$$Cl_2 \xrightarrow{k_1} 2Cl\bullet$$

$$Cl\bullet + C_2H_3Cl \xrightarrow{k_2} C_2H_3Cl_2\bullet$$

$$C_2H_3Cl_2\bullet + Cl_2 \xrightarrow{k_3} C_2H_3Cl_3 + Cl\bullet$$

$$C_2H_3Cl_2\bullet + C_2H_3Cl_2\bullet \xrightarrow{k_4} \text{stable species}$$

Derive the rate law expression for the chlorination of vinyl chloride based on this mechanism.

P19.21 Determine the expression for fractional coverage as a function of pressure for the dissociative adsorption mechanism described in the text in which adsorption is accompanied by dissociation:

$$R_2(g) + 2M(surface) \underset{k_d}{\overset{k_a}{\rightleftharpoons}} 2RM(surface)$$

P19.22 The adsorption of ethyl chloride on a sample of charcoal at 0°C measured at several different pressures is as follows:

$P_{C_2H_5Cl}$ (Torr)	V_{ads} (mL)
20	3.0
50	3.8
100	4.3
200	4.7
300	4.8

Using the Langmuir isotherm, determine the fractional coverage at each pressure, and V_m.

P19.23 Given the limitations of the Langmuir model, many other empirical adsorption isotherms have been proposed to better reproduce observed adsorption behavior. One of these empirical isotherms is the Temkin isotherm: $V_{adsorbed} = r \ln sP$, where V is the volume of gas adsorbed, P is pressure, and r and s are empirical constants.

a. Given the Temkin isotherm provided, what type of plot is expected to give a straight line?

b. Use your answer from part (a) to determine r and s for the data presented in Problem P19.22.

P19.24 Use the following data to determine the Langmuir adsorption parameters for nitrogen on mica:

V_{ads} (cm³ g⁻¹)	P (Torr)
0.494	2.1×10^{-3}
0.782	4.60×10^{-3}
1.16	1.30×10^{-2}

P19.25 Many surface reactions require the adsorption of two or more different gases. For the case of two gases,

assuming that the adsorption of a gas simply limits the number of surface sites available for adsorption, derive expressions for the fractional coverage of each gas.

P19.26 Sunburn is caused primarily by sunlight in what is known as the UVB band, or the wavelength range from 290 to 320 nm. The minimum dose of radiation needed to create a sunburn (erythema) is known as a MED (minimum erythema dose). The MED for a person of average resistance to burning is 50 mJ cm⁻².

a. Determine the number of 290-nm photons corresponding to the MED assuming each photon is absorbed. Repeat this calculation for 320-nm photons.

b. At 20° latitude, the solar flux in the UVB band at the surface of the earth is 1.45 mW cm⁻². Assuming that each photon is absorbed, how long would a person with unprotected skin be able to stand in the sun before acquiring one MED?

P19.27 A likely mechanism for the photolysis of acetaldehyde is

$$CH_3CHO + h\nu \longrightarrow CH_3\bullet + CHO\bullet$$

$$CH_3\bullet + CH_3CHO \xrightarrow{k_1} CH_4 + CH_3CO\bullet$$

$$CH_3CO\bullet \xrightarrow{k_2} CO + CH_3\bullet$$

$$CH_3\bullet + CH_3\bullet \xrightarrow{k_3} C_2H_6$$

Derive the rate law expression for the formation of CO based on this mechanism.

P19.28 If $\tau_f = 1 \times 10^{-10}$ s and $k_{ic} = 5 \times 10^8$ s⁻¹, what is ϕ_f? Assume that the rate constants for intersystem crossing and quenching are sufficiently small that these processes can be neglected.

P19.29 The quantum yield for CO production in the photolysis of gaseous acetone is unity for wavelengths between 250 and 320 nm. After 20 min of irradiation at 313 nm, 18.4 cm³ of CO (measured at 1008 Pa and 22°C) is produced. Calculate the number of photons absorbed, and the absorbed intensity in J s⁻¹.

P19.30 If 10% of the energy of a 100-W incandescent bulb is in the form of visible light having an average wavelength of 600 nm, how many quanta of light are emitted per second from the light bulb?

P19.31 For phenanthrene, the measured lifetime of the triplet state, τ_p, is 3.3 s, the fluorescence quantum yield is 0.12, and the phosphorescence quantum yield is 0.13 in an alcohol-ether glass at 77 K. Assume that no quenching and no internal conversion from the singlet state occurs. Determine k_p, k_{isc}^T, and k_{isc}^S / k_r.

P19.32 In this problem you will investigate the parameters involved in a single-molecule fluorescence experiment. Specifically, the incident photon power needed to see a single molecule with a reasonable signal-to-noise ratio will be determined.

a. Rhodamine dye molecules are typically employed in such experiments because their fluorescence quantum yields are large. What is the fluorescence quantum yield for Rhodamine B (a specific rhodamine dye) where $k_r = 1 \times 10^9 \text{ s}^{-1}$ and $k_{ic} = 1 \times 10^8 \text{ s}^{-1}$? You can ignore intersystem crossing and quenching in deriving this answer.

b. If care is taken in selecting the collection optics and detector for the experiment, a detection efficiency of 10% can be readily achieved. Furthermore, detector dark noise usually limits these experiments, and dark noise on the order of 10 counts s^{-1} is typical. If we require a signal-to-noise ratio of 10:1, then we will need to detect 100 counts s^{-1}. Given the detection efficiency, a total emission rate of 1000 fluorescence photons s^{-1} is required. Using the fluorescence quantum yield and a molar extinction coefficient for Rhodamine B of ~40,000 $\text{M}^{-1} \text{ cm}^{-1}$, what is the intensity of light needed in this experiment in terms of photons $\text{cm}^{-2} \text{ s}^{-1}$?

c. The smallest diameter focused spot one can obtain in a microscope using conventional refractive optics is one-half the wavelength of incident light. Studies of Rhodamine B generally employ 532-nm light such that the focused-spot diameter is ~270 nm. Using this diameter, what incident power in watts is required for this experiment? Don't be surprised if this value is relatively modest.

P19.33 A central issue in the design of aircraft is improving the lift of aircraft wings. To assist in the design of more efficient wings, wind-tunnel tests are performed in which the pressures at various parts of the wing are measured generally using only a few localized pressure sensors. Recently, pressure-sensitive paints have been developed to provide a more detailed view of wing pressure. In these paints, a luminescent molecule is dispersed into an oxygen-permeable paint and the aircraft wing is painted. The wing is placed into an airfoil, and luminescence from the paint is measured. The variation in O_2 pressure is measured by monitoring the luminescence intensity, with lower intensity demonstrating areas of higher O_2 pressure due to quenching.

a. The use of platinum octaethylprophyrin (PtOEP) as an oxygen sensor in pressure-sensitive paints was described by Gouterman and coworkers [*Review of Scientific Instruments* 61 (1990), 3340]. In this work, the following relationship between luminescence intensity and pressure was derived: $I_0/I = A + B(P/P_0)$, where I_0 is the fluorescence intensity at ambient pressure P_0, and I is the fluorescence intensity at an arbitrary pressure P. Determine coefficients A and B in the preceding

expression using the Stern–Volmer equation: $k_{total} = 1/\tau_l = k_r + k_q[Q]$. In this equation τ_l is the luminescence lifetime, k_r is the luminescent rate constant, and k_q is the quenching rate constant. In addition, the luminescent intensity ratio is equal to the ratio of luminescence quantum yields at ambient pressure, Φ_0, and an arbitrary pressure, Φ: $\Phi_0/\Phi = I_0/I$.

b. Using the following calibration data of the intensity ratio versus pressure observed for PtOEP, determine A and B:

I_0/I	P/P_0	I_0/I	P/P_0
1.0	1.0	0.65	0.46
0.9	0.86	0.61	0.40
0.87	0.80	0.55	0.34
0.83	0.75	0.50	0.28
0.77	0.65	0.46	0.20
0.70	0.53	0.35	0.10

c. At an ambient pressure of 1 atm, $I_0 = 50,000$ (arbitrary units) and 40,000 at the front and back of the wing. The wind tunnel is turned on to a speed of Mach 0.36 and the measured luminescence intensity is 65,000 and 45,000 at the respective locations. What is the pressure differential between the front and back of the wing?

P19.34 Oxygen sensing is important in biological studies of many systems. The variation in oxygen content of sapwood trees was measured by del Hierro and coworkers [*J. Experimental Biology* 53 (2002), 559] by monitoring the luminescence intensity of $[\text{Ru}(\text{dpp})_3]^{2+}$ immobilized in a sol-gel that coats the end of an optical fiber implanted into the tree. As the oxygen content of the tree increases, the luminescence from the ruthenium complex is quenched. The quenching of $[\text{Ru}(\text{dpp})_3]^{2+}$ by O_2 was measured by Bright and coworkers [*Applied Spectroscopy* 52 (1998), 750] and the following data were obtained:

I_0/I	% O_2
3.6	12
4.8	20
7.8	47
12.2	100

a. Construct a Stern–Volmer plot using the data supplied in the table. For $[\text{Ru}(\text{dpp})_3]^{2+}$ $k_r = 1.77 \times 10^5 \text{ s}^{-1}$, what is k_q?

b. Comparison of the Stern–Volmer prediction to the quenching data led the authors to suggest that some of the $[\text{Ru}(\text{dpp})_3]^{2+}$ molecules are located in sol-gel environments that are not equally accessible to O_2. What led the authors to this suggestion?

Web-Based Simulations, Animations, and Problems

W19.1   In this problem, the Lindemann mechanism for unimolecular rearrangements is investigated. Students will investigate the dependence of the reaction rate on reactant concentration and on the relative rate constants for the reaction. The turnover of the rate law from second order to first order as the reactant concentration is increased is explored. Also, dependence of this turnover on the relative magnitudes of $k_{-1}[A]$ and k_2 is investigated.

W19.2   In this problem, Michaelis–Menten enzyme kinetics are investigated, specifically, the variation in reaction rate with substrate concentration for three enzymes having significantly different Michaelis–Menten kinetic parameters.

Students will investigate how the maximum reaction rate and overall kinetics depend on K_m and turnover number. Finally, hand calculations of enzyme kinetic parameters performed by the students are compared to the results obtained by simulation.

W19.3   In this problem, the Langmuir isotherms for nondissociative and dissociative adsorption to a surface are investigated. Students study the dependence of the isotherms for these two adsorption processes on the adsorption/desorption rate constants and on adsorbate pressure.

Data Tables

Data tables referenced throughout the text are listed here and appear either on the page number in parentheses or, if no page number is listed, in this appendix.

Sources of Data

The most extensive databases for thermodynamic data (and the abbreviations listed with the tables) are as follows:

HCP Lide, D. R., Ed., *Handbook of Chemistry and Physics,* 83rd ed. CRC Press, Boca Raton, FL, 2002.

NIST Chemistry Webbook Linstrom, P. J., and W. G. Mallard, Eds., *NIST Chemistry Webbook: NIST Standard Reference Database Number 69.* National Institute of Standards and Technology, Gaithersburg, MD, retrieved from http://webbook.nist.gov.

Additional data sources used in the tables include the following:

Bard Bard, A. J., R. Parsons, and J. Jordan, *Standard Potentials in Aqueous Solution.* Marcel Dekker, New York, 1985.

DAL Blachnik, R., Ed., *D'Ans Lax Taschenbuch für Chemiker und Physiker,* 4th ed. Springer, Berlin, 1998.

HP Benenson, W., J. W. Harris, H. Stocker, and H. Lutz, *Handbook of Physics.* Springer, New York, 2002.

HTTD Lide, D. R., Ed., *CRC Handbook of Thermophysical and Thermochemical Data.* CRC Press, Boca Raton, FL, 1994.

TDOC Pedley, J. B., R. D. Naylor, and S. P. Kirby, *Thermochemical Data of Organic Compounds.* Chapman and Hall, London, 1977.

TDPS Barin, I., *Thermochemical Data of Pure Substances.* VCH Press, Weinheim, 1989.

AS Alberty, R. A., and R. S. Silbey, *Physical Chemistry.* John Wiley & Sons, New York, 1992.

TABLE 2.2
Physical Properties of Selected Elements

Densities are shown for nongaseous elements under standard conditions.

Substance	Atomic Weight	Melting Point (K)	Boiling Point (K)	$\rho°$ (kg m^{-3})	$C°_{P,m}$ (J K^{-1} mol^{-1})	Oxidation States
Aluminum	26.982	933.47	2792.15	2698.9	25.4	3
Argon	39.948	83.79 tp (69 kPa)	87.30	—	20.79	
Barium	137.33	1000.15	2170.15	3620	28.07	2
Boron	10.811	2348.15	4273.15	2340	11.1	3
Bromine	79.904	265.95	331.95	3103	36.05	1, 3, 4, 5, 6
Calcium	40.078	1115.15	1757.15	1540	25.9	2
Carbon	12.011	4713.15 (12.4 GPa) 4762.15 tp (10.3 MPa)	4098.15 (graphite)	3513 (diamond) 2250 (graphite)	6.113 (diamond) 8.527 (graphite)	2, 4
Cesium	132.91	301.65	944.15	1930	32.20	1
Chlorine	35.453	171.65	239.11	—	33.95	1, 3, 4, 5, 6, 7
Copper	63.546	1357.77	2835.15	8960	24.4	1, 2
Fluorine	18.998	53.48 tp	85.03	—	31.30	1
Gold	196.97	1337.33	3129.15	19320	25.42	1, 3
Helium	4.0026	0.95	4.22	—	20.79	
Hydrogen	1.0079	13.81	20.28	—	28.84	1
Iodine	126.90	386.85	457.55	4933	54.44	1, 3, 5, 7
Iron	55.845	1811.15	3134.15	7874	25.10	2, 3
Krypton	83.80	115.77 tp (73.2 kPa)	119.93	—	20.79	
Lead	207.2	600.61	2022.15	11343	26.44	2, 4
Lithium	6.941	453.65	1615.15	534	24.77	1
Magnesium	24.305	923.15	1363.15	1740	24.89	2
Manganese	54.938	1519.15	2334.15	7300	26.3	2, 3, 4, 6, 7
Mercury	200.59	234.31	629.88	13534	27.98	1, 2
Molybdenum	95.94	2896.15	4912.15	10222	23.90	2, 3, 4, 5, 6
Neon	20.180	24.54 tp 43 kPa)	27.07	—	20.79	
Nickel	58.693	1728.15	3186	8902	26.07	2, 3
Nitrogen	14.007	63.15	77.36	—	29.12	1, 2, 3, 4, 5
Oxygen	15.999	54.36	90.20	—	29.38	2
Palladium	106.42	1828.05	3236.15	11995	25.98	2, 4
Phosphorus (white)	30.974	317.3	553.65	1823	23.84	3, 5
Platinum	195.08	2041.55	4098.15	21500	25.85	2, 4, 6
Potassium	39.098	336.65	1032.15	890	29.58	1
Rhenium	186.21	3459.15	5869.15	20800	25.31	2, 4, 5, 6, 7
Rhodium	102.91	2237.15	3968.15	12410	24.98	2, 3, 4
Ruthenium	101.07	2607.15	4423.15	12100	24.04	3, 4, 5, 6, 8
Silicon	28.086	1687.15	3538.15	2330	20.00	4
Silver	107.87	1234.93	2435.15	10500	25.35	1
Sodium	22.990	370.95	1156.15	971	28.24	1
Sulfur	32.066	388.36	717.75	1819	22.76	2, 4, 6
Tin	118.71	505.08	2879	7310	26.99	2, 4
Titanium	47.867	1941.15	3560.15	4506	25.05	2, 3, 4
Vanadium	50.942	2183.15	3680.15	6000	24.89	2, 3, 4, 5
Xenon	131.29	161.36 tp (81.6 kPa)	165.03	—	20.79	2, 4, 6, 8
Zinc	65.39	692.68	1180.15	7135	25.40	2

Sources: HCP and DAL.

TABLE 2.3

Physical Properties of Selected Compounds

Densities are shown for nongaseous compounds under standard conditions.

Formula	Name	Molecular Weight	Melting Point (K)	Boiling Point (K)	Density $\rho°$ (kg m^{-3})	Heat Capacity $C°_{P,m}$ (J K^{-1} mol^{-1})
CO (g)	Carbon monoxide	28.01	68.13	81.65	—	29.1
COCl$_2$ (g)	Phosgene	98.92	145.4	281	—	
CO$_2$ (g)	Carbon dioxide	44.01	216.6 tp	194.75	—	37.1
D$_2$O (l)	Deuterium oxide	20.03	277	374.6	1108	
HCl (g)	Hydrogen chloride	36.46	158.98	188.15	—	29.1
HF (g)	Hydrogen fluoride	20.01	189.8	293.15	—	
H$_2$O	Water	18.02	273.15	373.15	998(l) 20°C 917(s) 0°C	75.3(l) 36.2(s)
H$_2$O$_2$ (l)	Hydrogen peroxide	34.01	272.72	423.35	1440	43.1
H$_2$SO$_4$ (l)	Sulfuric acid	98.08	283.46	610.15	1800	
KBr (s)	Potassium bromide	119.00	1007.15	1708.15	2740	52.3
KCl (s)	Potassium chloride	74.55	1044.15		1988	51.3
KI (s)	Potassium iodide	166.0	954.15	1596.15	3120	52.9
NaCl (s)	Sodium chloride	58.44	1073.85	1738.15	2170	50.5
NH$_3$ (g)	Ammonia	17.03	195.42	239.82	—	35.1
SO$_2$ (g)	Sulfur dioxide	64.06	197.65	263.10	—	39.9
CCl$_4$ (l)	Carbon tetrachloride	153.82	250.3	349.8	1594	131.3
CH$_4$ (g)	Methane	16.04	90.75	111.65	—	35.7
HCOOH (l)	Formic acid	46.03	281.45	374.15	1220	99.04
CH$_3$OH (l)	Methanol	32.04	175.55	337.75	791.4	81.1
CH$_3$CHO (l)	Acetaldehyde	44.05	150.15	293.25	783.4	89.0
CH$_3$COOH (l)	Acetic acid	60.05	289.6	391.2	1044.6	123.1
CH$_3$COCH$_3$ (l)	Acetone	58.08	178.5	329.2	789.9	125.45
C$_2$H$_5$OH (l)	Ethanol	46.07	158.8	351.5	789.3	112.3
C$_3$H$_7$OH (l)	1-Propanol	60.10	147.05	370.35	799.8	156.5
C$_4$H$_{11}$OH (l)	1-Butanol	74.12	183.35	390.85	809.8	176.9
C$_5$H$_5$NH$_2$ (l)	Pyridine	79.10	231.55	388.35	981.9	193
C$_5$H$_{12}$ (l)	Pentane	72.15	143.45	309.15	626.2	167.2
C$_5$H$_{11}$OH (l)	1-Pentanol	88.15	194.25	411.05	814.4	207.5
C$_6$H$_{12}$ (l)	Cyclohexane	84.16	279.6	353.9	773.9	156.0
C$_6$H$_5$CHO (l)	Benzaldehyde	106.12	247.15	452.15	1041.5	172.0
C$_6$H$_5$COOH (s)	Benzoic acid	122.12	395.55	522.35	1265.9	147.8
C$_6$H$_5$CH$_3$ (l)	Toluene	92.14	178.2	383.8	866.9	157.1
C$_6$H$_5$NH$_2$ (l)	Aniline	93.13	267	457	1021.7	194.1
C$_6$H$_5$OH (s)	Phenol	94.11	314.05	454.95	1057.6	127.2
C$_6$H$_6$ (l)	Benzene	78.11	278.6	353.3	876.5	135.7
1,2-(CH$_3$)$_2$C$_6$H$_5$ (l)	o-Xylene	106.17	248	417.6	880.2	187.7
C$_8$H$_{18}$ (l)	Octane	114.23	216.35	398.75	698.6	254.7

Sources: HCP and TDOC.

TABLE 2.4

Molar Heat Capacity, $C_{P,m}$, of Gases in the Range 298–800 K

Given by

$$C_{P,m}(\text{J K}^{-1}\text{ mol}^{-1}) = A(1) + A(2)\frac{T}{\text{K}} + A(3)\frac{T^2}{\text{K}^2} + A(4)\frac{T^3}{\text{K}^3}$$

Note that $C_{P,m}$ for solids and liquids at 298.15 K is listed in Tables 2.1 and 2.2.

Name	Formula	C_P° (298.15 K) in J K^{-1}mol^{-1}	A(1)	A(2)	A(3)	A(4)
All monatomic gases	He, Ne, Ar, Xe, O, H, among others	20.79	20.79			
Bromine	Br$_2$	36.05	30.11	0.03353	-5.5009×10^{-5}	3.1711×10^{-8}
Chlorine	Cl$_2$	33.95	22.85	0.06543	-1.2517×10^{-4}	1.1484×10^{-7}
Carbon monoxide	CO	29.14	31.08	-0.01452	3.1415×10^{-5}	-1.4973×10^{-8}
Carbon dioxide	CO$_2$	37.14	18.86	0.07937	-6.7834×10^{-5}	2.4426×10^{-8}
Fluorine	F$_2$	31.30	23.06	0.03742	-3.6836×10^{-5}	1.351×10^{-8}
Hydrogen	H$_2$	28.84	22.66	0.04381	-1.0835×10^{-4}	1.1710×10^{-7}
Water	H$_2$O	33.59	33.80	-0.00795	2.8228×10^{-5}	-1.3115×10^{-8}
Hydrogen bromide	HBr	29.13	29.72	-0.00416	7.3177×10^{-6}	
Hydrogen chloride	HCl	29.14	29.81	-0.00412	6.2231×10^{-6}	
Hydrogen fluoride	HF	29.14	28.94	0.00152	-4.0674×10^{-6}	$\times 3.8970 \times 10^{-9}$
Ammonia	NH$_3$	35.62	29.29	0.01103	4.2446×10^{-5}	-2.7706×10^{-8}
Nitrogen	N$_2$	29.13	30.81	-0.01187	2.3968×10^{-5}	-1.0176×10^{-8}
	NO	29.86	33.58	-0.02593	5.3326×10^{-5}	-2.7744×10^{-8}
	NO$_2$	37.18	32.06	-0.00984	1.3807×10^{-4}	-1.8157×10^{-7}
Oxygen	O$_2$	29.38	32.83	-0.03633	1.1532×10^{-4}	-1.2194×10^{-7}
Sulfur dioxide	SO$_2$	39.83	26.07	0.05417	2.6774×10^{-5}	
Methane	CH$_4$	35.67	30.65	-0.01739	1.3903×10^{-4}	8.1395×10^{-8}
Methanol	CH$_3$OH	44.07	26.53	0.03703	9.451×10^{-5}	-7.2006×10^{-8}
Ethyne	C$_2$H$_2$	44.05	10.82	0.15889	-1.8447×10^{-4}	8.5291×10^{-8}
Ethene	C$_2$H$_4$	42.86	8.39	0.12453	-2.5224×10^{-5}	-1.5679×10^{-8}
Ethane	C$_2$H$_6$	52.38	6.82	0.16840	-5.2347×10^{-5}	
Propane	C$_3$H$_8$	73.52	0.56	0.27559	-1.0355×10^{-4}	
Butane	C$_4$H$_{10}$	101.01	172.02	-1.08574	4.4887×10^{-3}	-6.5539×10^{-6}
Pentane	C$_5$H$_{12}$	120.11	2.02	0.44729	-1.7174×10^{-4}	
Benzene	C$_6$H$_6$	82.39	-46.48	0.53735	-3.8303×10^{-4}	1.0184×10^{-7}
Hexane	C$_6$H$_{12}$	142.13	-13.27	0.61995	-3.5408×10^{-4}	7.6704×10^{-8}

Sources: HCP and HTTD.

TABLE 2.5

Molar Heat Capacity, $C_{P,m}$, of Solids

$$C_{P,m}(\text{J K}^{-1}\text{ mol}^{-1}) = A(1) + A(2)\frac{T}{\text{K}} + A(3)\frac{T^2}{\text{K}^2} + A(4)\frac{T^3}{\text{K}^3} + A(5)\frac{T^4}{\text{K}^4}$$

Formula	Name	$C_{P,m}$ (298.15 K) (J K^{-1} mol^{-1})	A(1)	A(2)	A(3)	A(4)	A(5)	Range (K)
Ag	Silver	25.35	26.12	−0.0110	3.826×10^{-5}	3.750×10^{-8}	1.396×10^{-11}	290–800
Al	Aluminum	24.4	6.56	0.1153	-2.460×10^{-4}	1.941×10^{-7}		200–450
Au	Gold	25.4	34.97	−0.0768	2.117×10^{-4}	-2.350×10^{-7}	9.500×10^{-11}	290–800
CsCl	Cesium chloride	52.5	43.38	0.0467	-8.973×10^{-5}	1.421×10^{-7}	-8.237×10^{-11}	200–600
CuSO$_4$	Copper sulfate	98.9	−13.81	0.7036	-1.636×10^{-3}	2.176×10^{-6}	-1.182×10^{-9}	200–600
Fe	Iron	25.1	−10.99	0.3353	-1.238×10^{-3}	2.163×10^{-6}	-1.407×10^{-9}	200–450
NaCl	Sodium chloride	50.5	25.19	0.1973	-6.011×10^{-4}	8.815×10^{-7}	-4.765×10^{-10}	200–600
Si	Silicon	20.0	−6.25	0.1681	-3.437×10^{-4}	2.494×10^{-7}	6.667×10^{-12}	200–450
C (graphite)	C	8.5	−12.19	0.1126	-1.947×10^{-4}	1.919×10^{-7}	-7.800×10^{-11}	290–600
C$_6$H$_5$OH	Phenol	127.4	−5.97	1.0380	-6.467×10^{-3}	2.304×10^{-5}	-2.658×10^{-8}	100–314
C$_{10}$H$_8$	Naphthalene	165.7	−6.16	1.0383	-5.355×10^{-3}	1.891×10^{-5}	-2.053×10^{-8}	100–353
C$_{14}$H$_{10}$	Anthracene	210.5	11.10	0.5816	2.790×10^{-4}			100–488

Sources: HCP and HTTD.

TABLE 4.1
Thermodynamic Data for Inorganic Compounds

Substance	ΔH_f° (kJ mol^{-1})	ΔG_f° (kJ mol^{-1})	S° (J mol^{-1} K^{-1})	C_P° (J mol^{-1} K^{-1})	Atomic or Molecular Weight (amu)
Aluminum					
Al(s)	0	0	28.3	24.4	26.98
Al$_2$O$_3$(s)	−1675.7	−1582.3	50.9	79.0	101.96
Al^{3+}(aq)	−538.4	−485.0	−325		26.98
Antimony					
Sb(s)	0	0	45.7	25.2	121.75
Argon					
Ar(g)	0	0	154.8	20.8	39.95
Barium					
Ba(s)	0	0	62.5	28.1	137.34
BaO(s)	−548.0	−520.3	72.1	47.3	153.34
BaCO$_3$(s)	−1216.3	−1137.6	112.1	85.4	197.35
BaCl$_2$(s)	−855.0	−806.7	123.7	75.1	208.25
BaSO$_4$(s)	−1473.2	−1362.3	132.2	101.8	233.40
Ba^{2+}(aq)	−537.6	−560.8	9.6		137.34
Bromine					
Br$_2$(l)	0	0	152.2	75.7	159.82
Br$_2$(g)	30.9	3.1	245.5	36.0	159.82
Br(g)	111.9	82.4	175.0	20.8	79.91
HBr(g)	−36.3	−53.4	198.7	29.1	90.92
Br$^-$(aq)	−121.6	−104.0	82.4		79.91
Calcium					
Ca(s)	0	0	41.6	25.9	40.08
CaCO$_3$(s) calcite	−1220.0	−1081.4	110.0	81.9	100.09
CaCl$_2$(s)	−795.4	−748.8	108.4	72.9	110.99
CaO(s)	−634.9	−603.3	38.1	42.0	56.08
CaSO$_4$(s)	−1434.5	−1322.0	106.5	99.7	136.15
Ca^{2+}(aq)	−542.8	−553.6	−53.1		40.08
Carbon					
Graphite(s)	0	0	5.74	8.52	12.011
Diamond(s)	1.89	2.90	2.38	6.12	12.011
C(g)	716.7	671.2	158.1	20.8	12.011
CO(g)	−110.5	−137.2	197.7	29.1	28.011
CO$_2$(g)	−393.5	−394.4	213.8	37.1	44.010
HCN(g)	135.5	124.7	201.8	35.9	27.03
CN$^-$(aq)	150.6	172.4	94.1		26.02
HCO$_3^-$(aq)	−692.0	−586.8	91.2		61.02
CO$_3^{2-}$(aq)	−675.2	−527.8	−50.0		60.01
Chlorine					
Cl$_2$(g)	0	0	223.1	33.9	70.91
Cl(g)	121.3	105.7	165.2	21.8	35.45
HCl(g)	−92.3	−95.3	186.9	29.1	36.46
ClO$_2$(g)	102.5	120.5	256.8	42.0	67.45

TABLE 4.1

(Continued)

Substance	ΔH_f° (kJ mol^{-1})	ΔG_f° (kJ mol^{-1})	S° (J mol^{-1} K^{-1})	C_P° (J mol^{-1} K^{-1})	Atomic or Molecular Weight (amu)
ClO$_4^-$(aq)	−128.1	−8.52	184.0		99.45
Cl$^-$(aq)	−167.2	−131.2	56.5		35.45
Copper					
Cu(s)	0	0	33.2	24.4	63.54
CuCl$_2$(s)	−220.1	−175.7	108.1	71.9	134.55
CuO(s)	−157.3	−129.7	42.6	42.3	79.54
Cu$_2$O(s)	−168.6	−146.0	93.1	63.6	143.08
CuSO$_4$(s)	−771.4	−662.2	109.2	98.5	159.62
Cu$^+$(aq)	71.7	50.0	40.6		63.54
Cu^{2+}(aq)	64.8	65.5	−99.6		63.54
Deuterium					
D$_2$(g)	0	0	145.0	29.2	4.028
HD(g)	0.32	−1.46	143.8	29.2	3.022
D$_2$O(g)	−249.2	−234.5	198.3	34.3	20.028
D$_2$O(l)	−294.6	−243.4	75.94	84.4	20.028
HDO(g)	−246.3	−234.5	199.4	33.8	19.022
HDO(l)	−289.9	−241.9	79.3		19.022
Fluorine					
F$_2$(g)	0	0	202.8	31.3	38.00
F(g)	79.4	62.3	158.8	22.7	19.00
HF(g)	−273.3	−275.4	173.8	29.1	20.01
F$^-$(aq)	−332.6	−278.8	−13.8		19.00
Gold					
Au(s)	0	0	47.4	25.4	196.97
Au(g)	366.1	326.3	180.5	20.8	197.97
Hydrogen					
H$_2$(g)	0	0	130.7	28.8	2.016
H(g)	218.0	203.3	114.7	20.8	1.008
OH(g)	39.0	34.2	183.7	29.9	17.01
H$_2$O(g)	−241.8	−228.6	188.8	33.6	18.015
H$_2$O(l)	−285.8	−237.1	70.0	75.3	18.015
H$_2$O(s)			48.0	36.2 (273 K)	18.015
H$_2$O$_2$(g)	−136.3	−105.6	232.7	43.1	34.015
H$^+$(aq)	0	0	0		1.008
OH$^-$(aq)	−230.0	−157.24	−10.9		17.01
Iodine					
I$_2$(s)	0	0	116.1	54.4	253.80
I$_2$(g)	62.4	19.3	260.7	36.9	253.80
I(g)	106.8	70.2	180.8	20.8	126.90
I$^-$(aq)	−55.2	−51.6	111.3		126.90
Iron					
Fe(s)	0	0	27.3	25.1	55.85
Fe(g)	416.3	370.7	180.5	25.7	55.85

(continued)

TABLE 4.1

(Continued)

Substance	ΔH_f° (kJ mol^{-1})	ΔG_f° (kJ mol^{-1})	S° (J mol^{-1} K^{-1})	C_P° (J mol^{-1} K^{-1})	Atomic or Molecular Weight (amu)
Fe$_2$O$_3$(s)	−824.2	−742.2	87.4	103.9	159.69
Fe$_3$O$_4$(s)	−1118.4	−1015.4	146.4	143.4	231.54
FeSO$_4$(s)	−928.4	−820.8	107.5	100.6	151.92
Fe^{2+}(aq)	−89.1	−78.9	−137.7		55.85
Fe^{3+}(aq)	−48.5	−4.7	−315.9		55.85
Lead					
Pb(s)	0	0	64.8	26.4	207.19
Pb(g)	195.2	162.2	175.4	20.8	207.19
PbO$_2$(s)	−277.4	−217.3	68.6	64.6	239.19
PbSO$_4$(s)	−920.0	−813.20	148.5	103.2	303.25
Pb^{2+}(aq)	0.92	−24.4	18.5		207.19
Lithium					
Li(s)	0	0	29.1	24.8	6.94
Li(g)	159.3	126.6	138.8	20.8	6.94
LiH(s)	−90.5	−68.3	20.0	27.9	7.94
LiH(g)	140.6	117.8	170.9	29.7	7.94
Li$^+$(aq)	−278.5	−293.3	13.4		6.94
Magnesium					
Mg(s)	0	0	32.7	24.9	24.31
Mg(g)	147.1	112.5	148.6	20.8	24.31
MgO(s)	−601.6	−569.3	27.0	37.2	40.31
MgSO$_4$(s)	−1284.9	−1170.6	91.6	96.5	120.38
MgCl$_2$(s)	−641.3	−591.8	89.6	71.4	95.22
MgCO$_3$(s)	−1095.8	−1012.2	65.7	75.5	84.32
Mg^{2+}(aq)	−466.9	−454.8	−138.1		24.31
Manganese					
Mn(s)	0	0	32.0	26.3	54.94
Mn(g)	280.7	238.5	173.7	20.8	54.94
MnO$_2$(s)	−520.0	−465.1	53.1	54.1	86.94
Mn^{2+}(aq)	−220.8	−228.1	−73.6		54.94
MnO$_4^-$(aq)	−541.4	−447.2	191.2		118.94
Mercury					
Hg(l)	0	0	75.9	28.0	200.59
Hg(g)	61.4	31.8	175.0	20.8	200.59
Hg$_2$Cl$_2$(s)	−265.4	−210.7	191.6	101.9	472.09
Hg^{2+}(aq)	170.2	164.4	−36.2		401.18
Hg$_2^{2+}$(aq)	166.9	153.5	65.7		401.18
Nickel					
Ni(s)	0	0	29.9	26.1	58.71
Ni(g)	429.7	384.5	182.2	23.4	58.71
NiCl$_2$(s)	−305.3	−259.0	97.7	71.7	129.62
NiO(s)	−239.7	−211.5	38.0	44.3	74.71
NiSO$_4$(s)	−872.9	−759.7	92.0	138.0	154.77
Ni^{2+}(aq)	−54.0	−45.6	−128.9		58.71

(continued)

TABLE 4.1

(Continued)

Substance	ΔH_f° (kJ mol^{-1})	ΔG_f° (kJ mol^{-1})	S° (J mol^{-1} K^{-1})	C_P° (J mol^{-1} K^{-1})	Atomic or Molecular Weight (amu)
Nitrogen					
N$_2$(g)	0	0	191.6	29.1	28.013
N(g)	472.7	455.5	153.3	20.8	14.007
NH$_3$(g)	−45.9	−16.5	192.8	35.1	17.03
NO(g)	91.3	87.6	210.8	29.9	30.01
N$_2$O(g)	81.6	103.7	220.0	38.6	44.01
NO$_2$(g)	33.2	51.3	240.1	37.2	46.01
NOCl(g)	51.7	66.1	261.7	44.7	65.46
N$_2$O$_4$(g)	11.1	99.8	304.4	79.2	92.01
N$_2$O$_4$(l)	−19.5	97.5	209.2	142.7	92.01
HNO$_3$(l)	−174.1	−80.7	155.6	109.9	63.01
HNO$_3$(g)	−133.9	−73.5	266.9	54.1	63.01
NO$_3^-$(aq)	−207.4	−111.3	146.4		62.01
NH$_4^+$(aq)	−132.5	−79.3	113.4		18.04
Oxygen					
O$_2$(g)	0	0	205.2	29.4	31.999
O(g)	249.2	231.7	161.1	21.9	15.999
O$_3$(g)	142.7	163.2	238.9	39.2	47.998
OH(g)	39.0	34.22	183.7	29.9	17.01
OH$^-$(aq)	−230.0	−157.2	−10.9		17.01
Phosphorus					
P(s) white	0	0	41.1	23.8	30.97
P(s) red	−17.6	−12.1	22.8	21.2	30.97
P$_4$(g)	58.9	24.4	280.0	67.2	123.90
PCl$_5$(g)	−374.9	−305.0	364.6	112.8	208.24
PH$_3$(g)	5.4	13.5	210.2	37.1	34.00
H$_3$PO$_4$(l)	−1284.4	−1124.3	110.5	106.1	94.97
PO$_4^{3-}$(aq)	−1277.4	−1018.7	−220.5		91.97
HPO$_4^{2-}$(aq)	−1299.0	−1089.2	−33.5		92.97
H$_2$PO$_4^-$(aq)	−1302.6	−1130.2	92.5		93.97
Potassium					
K(s)	0	0	64.7	29.6	39.10
K(g)	89.0	60.5	160.3	20.8	39.10
KCl(s)	−436.5	−408.5	82.6	51.3	74.56
K$_2$O(s)	−361.5	−322.8	102.0	77.4	94.20
K$_2$SO$_4$(s)	−1437.8	−1321.4	175.6	131.5	174.27
K$^+$(aq)	−252.4	−283.3	102.5		39.10
Silicon					
Si(s)	0	0	18.8	20.0	28.09
Si(g)	450.0	405.5	168.0	22.3	28.09
SiCl$_4$(l)	−687.0	−619.8	239.7	145.3	169.70
SiO$_2$(quartz)	−910.7	−856.3	41.5	44.4	60.09

(continued)

TABLE 4.1
(Continued)

Substance	ΔH_f° (kJ mol^{-1})	ΔG_f° (kJ mol^{-1})	S° (J mol^{-1} K^{-1})	C_P° (J mol^{-1} K^{-1})	Atomic or Molecular Weight (amu)
Silver					
Ag(s)	0	0	42.6	25.4	107.87
Ag(g)	284.9	246.0	173.0	20.8	107.87
AgCl(s)	−127.0	−109.8	96.3	50.8	143.32
AgNO$_2$(s)	−124.4	−33.3	140.6	93.0	153.88
AgNO$_3$(s)	−124.4	−33.4	140.9	93.1	169.87
Ag$_2$SO$_4$(s)	−715.9	−618.4	200.4	131.4	311.80
Ag$^+$(aq)	105.6	77.1	72.7		107.87
Sodium					
Na(s)	0	0	51.3	28.2	22.99
Na(g)	107.5	77.0	153.7	20.8	22.99
NaCl(s)	−411.2	−384.1	72.1	50.5	58.44
NaOH(s)	−425.8	−379.7	64.4	59.5	40.00
Na$_2$SO$_4$(s)	−1387.1	−1270.2	149.6	128.2	142.04
Na$^+$(aq)	−240.1	−261.9	59.0		22.99
Sulfur					
S($rhombic$)	0	0	32.1	22.6	32.06
SF$_6$(g)	−1220.5	−1116.5	291.5	97.3	146.07
H$_2$S(g)	−20.6	−33.4	205.8	34.2	34.09
SO$_2$(g)	−296.8	−300.1	248.2	39.9	64.06
SO$_3$(s)	−454.5	−374.2	70.7		80.06
SO$_3^{2-}$(aq)	−635.5	−486.6	−29.3		80.06
SO$_4^{2-}$(aq)	−909.3	−744.5	20.1		96.06
Tin					
Sn($white$)	0	0	51.2	27.0	118.69
Sn(g)	301.2	266.2	168.5	21.3	118.69
SnO$_2$(s)	−577.6	−515.8	49.0	52.6	150.69
Sn^{2+}(aq)	−8.9	−27.2	−16.7		118.69
Titanium					
Ti(s)	0	0	30.7	25.0	47.87
Ti(g)	473.0	428.4	180.3	24.4	47.87
TiCl$_4$(l)	−804.2	−737.2	252.4	145.2	189.69
TiO$_2$(s)	−944.0	−888.8	50.6	55.0	79.88
Xenon					
Xe(g)	0	0	169.7	20.8	131.30
XeF$_4$(s)	−261.5	−123	146	118	207.29
Zinc					
Zn(s)	0	0	41.6	25.4	65.37
ZnCl$_2$(s)	−415.1	−369.4	111.5	71.3	136.28
ZnO(s)	−350.5	−320.5	43.7	40.3	81.37
ZnSO$_4$(s)	−982.8	−871.5	110.5	99.2	161.43
Zn^{2+}(aq)	−153.9	−147.1	−112.1		65.37

Sources: HCP, HTTD, and TDPS.

TABLE 4.2

Thermodynamic Data for Selected Organic Compounds

Substance	Formula	Molecular Weight	ΔH_f° (kJ mol^{-1})	$\Delta H_{combustion}^\circ$ (kJ mol^{-1})	ΔG_f° (kJ mol^{-1})	S° (J mol^{-1} K^{-1})	C_P° (J mol^{-1} K^{-1})
Carbon (graphite)	C	12.011	0	−393.5	0	5.74	8.52
Carbon (diamond)	C	12.011	1.89	−395.4	2.90	2.38	6.12
Carbon monoxide	CO	28.01	−110.5	−283.0	−137.2	197.7	29.1
Carbon dioxide	CO_2	44.01	−393.5		−394.4	213.8	37.1
Acetaldehyde (*l*)	C_2H_4O	44.05	−192.2	−1166.9	−127.6	160.3	89.0
Acetic acid (*l*)	$C_2H_4O_2$	60.05	−484.3	−874.2	−389.9	159.8	124.3
Acetone (*l*)	C_3H_6O	58.08	−248.4	−1790	−155.2	199.8	126.3
Benzene (*l*)	C_6H_6	78.12	49.1	−3268	124.5	173.4	136.0
Benzene (*g*)	C_6H_6	78.12	82.9	−3268	129.7	269.2	82.4
Benzoic acid (*s*)	$C_7H_6O_2$	122.13	−385.2	−3227	−245.5	167.6	146.8
1,3-Butadiene (*g*)	C_4H_6	54.09	110.0	−2541			79.8
n-Butane (*g*)	C_4H_{10}	58.13	−125.7	−2878	−17.0	310.2	97.5
1-Butene (*g*)	C_4H_8	56.11	−0.63	−2718	71.1	305.7	85.7
Carbon disulfide (*g*)	CS_2	76.14	116.9	−1112	66.8	238.0	45.7
Carbon tetrachloride (*l*)	CCl_4	153.82	−128.2	−360	−62.5	214.4	133.9
Carbon tetrachloride (*g*)	CCl_4	153.82	−95.7		−58.2	309.7	83.4
Cyclohexane (*l*)	C_6H_{12}	84.16	−156.4	−3920	26.8	204.5	154.9
Cyclopentane(*l*)	C_5H_{10}	70.13	−105.1	−3291	38.8	204.5	128.8
Cyclopropane (*g*)	C_3H_6	42.08	53.3	−2091	104.5	237.5	55.6
Dimethyl Ether (*g*)	C_2H_6O	131.6	−184.1	−1460	−112.6	266.4	64.4
Ethane (*g*)	C_2H_6	30.07	−84.0	−1561	−32.0	229.2	52.5
Ethanol (*l*)	C_2H_6O	46.07	−277.6	−1367	−174.8	160.7	112.3
Ethanol (*g*)	C_2H_6O	46.07	−234.8	−1367	−167.9	281.6	65.6
Ethene (*g*)	C_2H_4	28.05	52.4	−1411	68.4	219.3	42.9
Ethyne (*g*)	C_2H_2	26.04	227.4	−1310	209.2	200.9	44
Formaldehyde (*g*)	CH_2O	30.03	−108.6	−571	−102.5	218.8	35.4
Formic acid (*l*)	CH_2O_2	46.03	−425.0	−255	−361.4	129.0	99.0
Formic acid (*g*)	CH_2O_2	46.03	378.7	−256	−351.0	248.7	45.2
α-D-Glucose (*s*)	$C_6H_{12}O_6$	180.16	−1273.1	−2805	−910.6	209.2	219.2
n-Hexane (*l*)	C_6H_{14}	86.18	−198.7	−4163	−4.0	296.0	195.6
Hydrogen cyanide (*l*)	HCN	27.03	108.9		125.0	112.8	70.6
Hydrogen cyanide (*g*)	HCN	27.03	135.5		124.7	201.8	35.9
Methane (*g*)	CH_4	16.04	−74.6	−891	−50.5	186.3	35.7
Methanol (*l*)	CH_4O	32.04	−239.2	−726	−166.6	126.8	81.1
Methanol (*g*)	CH_4O	32.04	−201.0	−764	−162.3	239.9	44.1
Oxalic acid (*g*)	$C_2H_2O_4$	90.04	−731.8	−246	−662.7	320.6	86.2
n-Pentane (*g*)	C_5H_{12}	72.15	−146.9	−3509	−8.2	349.1	120.1
Phenol (*s*)	C_6H_6O	94.11	−165.1	−3054	−50.2	144.0	127.4
Propane (*g*)	C_3H_8	44.10	−103.8	−2219	−23.4	270.3	73.6
Propene (*g*)	C_3H_6	42.08	20.0	−2058	62.7	266.9	64.0
Propyne (*g*)	C_3H_4	40.07	184.9	−2058	194.5	248.2	60.7
Pyridine (*l*)	C_5H_5N	79.10	100.2	−2782		177.9	132.7
Sucrose (*s*)	$C_{12}H_{22}O_{11}$	342.3	−2226.1	−5643	−1544.6	360.2	424.3
Thiophene (*l*)	C_4H_4S	84.14	80.2	−2829		181.2	123.8
Toluene (*g*)	C_7H_8	92.14	50.5	−3910	122.3	320.8	104
Urea (*s*)	$C_2H_4N_2O$	60.06	−333.1	−635	−197.4	104.3	92.8

Sources: HCP, HTTD, TDPS, and TDOC.

TABLE 7.1

Second Virial Coefficients for Selected Gases in Units of cm³ mol⁻¹

	200 K	300 K	400 K	500 K	600 K	700 K
Benzene		−1453	−712	−429	−291	−211
Cl_2		−299	−166	−97	−59	−36
CO_2		−126	−61.7	−30.5	−12.6	−1.18
H_2O		−1126	−356	−175	−104	−67
Heptane		−2782	−1233	−702	−452	−304
Kr	−117	−51.0	−23.0	−7.75	1.78	8.33
N_2	−34.4	−3.91	9.17	16.3	20.8	23.8
Octane		−4042	−1704	−936	−583	−375

Source: Calculated from HTTD.

TABLE 7.2

Critical Constants of Selected Substances

Substance	Formula	T_c (K)	P_c (bar)	$10^3 V_c$ (L)	$z_c = \dfrac{P_c V_c}{RT_c}$
Ammonia	NH_3	405.40	113.53	72.47	0.244
Argon	Ar	150.86	48.98	74.57	0.291
Benzene	C_6H_6	562.05	48.95	256.00	0.268
Bromine	Br_2	588.00	103.40	127.00	0.268
Carbon dioxide	CO_2	304.13	73.75	94.07	0.274
Carbon monoxide	CO	132.91	34.99	93.10	0.295
Ethane	C_2H_6	305.32	48.72	145.50	0.279
Ethanol	C_2H_5OH	513.92	61.37	168.00	0.241
Ethene	C_2H_4	282.34	50.41	131.1	0.281
Ethyne	C_2H_2	308.30	61.38	112.20	0.269
Fluorine	F_2	144.30	51.72	66.20	0.285
Hydrogen	H_2	32.98	12.93	64.20	0.303
Methane	CH_4	190.56	45.99	98.60	0.286
Methanol	CH_3OH	512.50	80.84	117.00	0.221
Nitrogen	N_2	126.20	33.98	90.10	0.292
Oxygen	O_2	154.58	50.43	73.37	0.288
Pentane	C_5H_{12}	469.70	33.70	311.00	0.268
Propane	C_3H_8	369.83	42.48	200.00	0.276
Pyridine	C_5H_5N	620.00	56.70	243.00	0.267
Tetrachloromethane	CCl_4	556.60	45.16	276.00	0.269
Water	H_2O	647.14	220.64	55.95	0.229
Xenon	Xe	289.74	58.40	118.00	0.286

Sources: HCP and DAL.

TABLE 7.4

van der Waals and Redlich-Kwong Parameters for Selected Gases

Substance	Formula	van der Waals		Redlich-Kwong	
		a (dm^6 bar mol^{-2})	b (dm^3 mol^{-1})	a (dm^6 bar mol^{-2} K$^{1/2}$)	b (dm^3 mol^{-1})
Ammonia	NH_3	4.225	0.0371	86.12	0.02572
Argon	Ar	1.355	0.0320	16.86	0.02219
Benzene	C_6H_6	18.82	0.1193	452.0	0.08271
Bromine	Br_2	9.75	0.0591	236.5	0.04085
Carbon dioxide	CO_2	3.658	0.0429	64.63	0.02971
Carbon monoxide	CO	1.472	0.0395	17.20	0.02739
Ethane	C_2H_6	5.580	0.0651	98.79	0.04514
Ethanol	C_2H_5OH	12.56	0.0871	287.7	0.06021
Ethene	C_2H_4	4.612	0.0582	78.51	0.04034
Ethyne	C_2H_2	4.533	0.05240	80.65	0.03632
Fluorine	F_2	1.171	0.0290	14.17	0.01993
Hydrogen	H_2	0.2452	0.0265	1.427	0.01837
Methane	CH_4	2.303	0.0431	32.20	0.02985
Methanol	CH_3OH	9.476	0.0659	217.1	0.04561
Nitrogen	N_2	1.370	0.0387	15.55	0.02675
Oxygen	O_2	1.382	0.0319	17.40	0.02208
Pentane	C_5H_{12}	19.09	0.1448	419.2	0.1004
Propane	C_3H_8	9.39	0.0905	182.9	0.06271
Pyridine	C_5H_5N	19.77	0.1137	498.8	0.07877
Tetrachloromethane	CCl_4	20.01	0.1281	473.2	0.08793
Water	H_2O	5.537	0.0305	142.6	0.02113
Xenon	Xe	4.192	0.0516	72.30	0.03574

Source: Calculated from critical constants.

TABLE 8.1

Triple Point Pressures and Temperatures of Selected Substances

Formula	Name	T_{tp} (K)	P_{tp} (Pa)
Ar	Argon	83.806	68950
Br_2	Bromine	280.4	5879
Cl_2	Chlorine	172.17	1392
HCl	Hydrogen chloride	158.8	
H_2	Hydrogen	13.8	7042
H_2O	Water	273.16	611.73
H_2S	Hydrogen sulfide	187.67	23180
NH_3	Ammonia	195.41	6077
Kr	Krypton	115.8	72920
NO	Nitrogen oxide	109.54	21916
O_2	Oxygen	54.36	146.33
SO_3	Sulfur trioxide	289.94	21130
Xe	Xenon	161.4	81590
CH_4	Methane	90.694	11696
CO	Carbon monoxide	68.15	15420
CO_2	Carbon dioxide	216.58	518500
C_3H_6	Propene	87.89	9.50×10^{-4}

Sources: HCP, HTTP, and DAL.

TABLE 8.3
Vapor Pressure and Boiling Temperature of Liquids

$$\ln \frac{P(T)}{\text{Pa}} = A(1) - \frac{A(2)}{\frac{T}{\text{K}} + A(3)} \qquad T_b(P) = \frac{A(2)}{A(1) - \ln \frac{P}{\text{Pa}}} - A(3)$$

Molecular Formula	Name	T_b (K)	A(1)	A(2)	A(3)	$10^{-3}P(298.15 \text{ K})(\text{Pa})$	Range (K)
Ar	Argon	87.28	22.946	1.0325×10^3	3.130	—	73–90
Br$_2$	Bromine	331.9	20.729	2.5782×10^3	−51.77	28.72	268–354
HF	Hydrogen fluoride	292.65	22.893	3.6178×10^3	25.627	122.90	273–303
H$_2$O	Water	373.15	23.195	3.8140×10^3	−46.290		353–393
SO$_2$	Sulfur dioxide	263.12	21.661	2.3024×10^3	−35.960		195–280
CCl$_4$	Tetrachloromethane	349.79	20.738	2.7923×10^3	−46.6667	15.28	287–350
CHCl$_3$	Trichloromethane	334.33	20.907	2.6961×10^3	−46.926	26.24	263–335
HCN	Hydrogen cyanide	298.81	22.226	3.0606×10^3	−12.773	98.84	257–316
CH$_3$OH	Methanol	337.70	23.593	3.6971×10^3	−31.317	16.94	275–338
CS$_2$	Carbon disulfide	319.38	20.801	2.6524×10^3	−33.40	48.17	255–320
C$_2$H$_5$OH	Ethanol	351.45	23.58	3.6745×10^3	−46.702	7.87	293–366
C$_3$H$_6$	Propene	225.46	20.613	1.8152×10^3	−25.705	1156.6	166–226
C$_3$H$_8$	Propane	231.08	20.558	1.8513×10^3	−26.110	948.10	95–370
C$_4$H$_9$Br	1-Bromobutane	374.75	17.076	1.5848×10^3	−11.188	5.26	195–300
C$_4$H$_9$Cl	1-Chlorobutane	351.58	20.612	2.6881×10^3	−55.725	13.68	256–352
C$_5$H$_{11}$OH	1-Pentanol	411.133	20.729	2.5418×10^3	−134.93	0.29	410–514
C$_6$H$_5$Cl	Chlorobenzene	404.837	20.964	3.2969×10^3	−55.515	1.57	335–405
C$_6$H$_5$I	Iodobenzene	461.48	21.088	3.8136×10^3	−62.654	0.13	298–462
C$_6$H$_6$	Benzene	353.24	20.767	2.7738×10^3	−53.08	12.69	294–378
C$_6$H$_{14}$	Hexane	341.886	20.749	2.7081×10^3	−48.251	20.17	286–343
C$_6$H$_5$CHO	Benzaldehyde	451.90	21.213	3.7271×10^3	−67.156	0.17	311–481
C$_6$H$_5$CH$_3$	Toluene	383.78	21.600	3.6266×10^3	−23.778	3.80	360–580
C$_{10}$H$_8$	Naphthalene	491.16	21.100	4.0526×10^3	−67.866	0.01	353–453
C$_{14}$H$_{10}$	Anthracene	614.0	21.965	5.8733×10^3	−51.394		496–615

Sources: HCP and HTTP.

TABLE 8.4
Sublimation Pressure of Solids

$$\ln \frac{P(T)}{\text{Pa}} = A(1) - \frac{A(2)}{\frac{T}{\text{K}} + A(3)}$$

Molecular Formula	Name	A(1)	A(2)	A(3)	Range (K)
CCl$_4$	Tetrachloromethane	17.613	1.6431×10^3	−95.250	232–250
C$_6$H$_{14}$	Hexane	31.224	4.8186×10^3	−23.150	168–178
C$_6$H$_5$COOH	Benzoic acid	14.870	4.7196×10^3		293–314
C$_{10}$H$_8$	Naphthalene	31.143	8.5750×10^3		270–305
C$_{14}$H$_{10}$	Anthracene	31.620	1.1378×10^4		353–400

Sources: HCP and HTTP.

TABLE 10.2
Dielectric Constants, ε_r, of Selected Liquids

Substance	Dielectric Constant	Substance	Dielectric Constant
Acetic acid	6.2	Heptane	1.9
Acetone	21.0	Isopropyl alcohol	20.2
Benzaldehyde	17.8	Methanol	33.0
Benzene	2.3	Nitrobenzene	35.6
Carbon tetrachloride	2.2	o-Xylene	2.6
Cyclohexane	2.0	Phenol	12.4
Ethanol	25.3	Toluene	2.4
Glycerol	42.5	Water (273 K)	88.0
1-Hexanol	13.0	Water (373 K)	55.3

Source: HCP.

TABLE 10.3
Mean Activity Coefficients in Terms of Molalities at 298 K

Substance	0.1m	0.2m	0.3m	0.4m	0.5m	0.6m	0.7m	0.8m	0.9m	1.0m
$AgNO_3$	0.734	0.657	0.606	0.567	0.536	0.509	0.485	0.464	0.446	0.429
$BaCl_2$	0.500	0.444	0.419	0.405	0.397	0.391	0.391	0.391	0.392	0.395
$CaCl_2$	0.518	0.472	0.455	0.448	0.448	0.453	0.460	0.470	0.484	0.500
$CuCl_2$	0.508	0.455	0.429	0.417	0.411	0.409	0.409	0.410	0.413	0.417
$CuSO_4$	0.150	0.104	0.0829	0.0704	0.0620	0.0559	0.0512	0.0475	0.0446	0.0423
HCl	0.796	0.767	0.756	0.755	0.757	0.763	0.772	0.783	0.795	0.809
HNO_3	0.791	0.754	0.735	0.725	0.720	0.717	0.717	0.718	0.721	0.724
H_2SO_4	0.2655	0.2090	0.1826		0.1557		0.1417			0.1316
KCl	0.770	0.718	0.688	0.666	0.649	0.637	0.626	0.618	0.610	0.604
KOH	0.798	0.760	0.742	0.734	0.732	0.733	0.736	0.742	0.749	0.756
$MgCl_2$	0.529	0.489	0.477	0.475	0.481	0.491	0.506	0.522	0.544	0.570
$MgSO_4$	0.150	0.107	0.0874	0.0756	0.0675	0.0616	0.0571	0.0536	0.0508	0.0485
NaCl	0.778	0.735	0.710	0.693	0.681	0.673	0.667	0.662	0.659	0.657
NaOH	0.766	0.727	0.708	0.697	0.690	0.685	0.681	0.679	0.678	0.678
$ZnSO_4$	0.150	0.140	0.0835	0.0714	0.0630	0.0569	0.0523	0.0487	0.0458	0.0435

Source: HCP.

TABLE 11.1
Standard Reduction Potentials in Alphabetical Order

Reaction	$E°$ (V)	Reaction	$E°$ (V)
$Ag^+ + e^- \rightarrow Ag$	0.7996	$Fe^{2+} + 2e^- \rightarrow Fe$	−0.447
$Ag^{2+} + e^- \rightarrow Ag^+$	1.980	$Fe^{3+} + 3e^- \rightarrow Fe$	−0.030
$AgBr + e^- \rightarrow Ag + Br^-$	0.07133	$Fe^{3+} + e^- \rightarrow Fe^{2+}$	0.771
$AgCl + e^- \rightarrow Ag + Cl^-$	0.22233	$[Fe(CN)_6]^{3-} + e^- \rightarrow [Fe(CN)_6]^{4-}$	0.358
$AgCN + e^- \rightarrow Ag + CN^-$	−0.017	$2H^+ + 2e^- \rightarrow H_2$	0
$AgF + e^- \rightarrow Ag + F^-$	0.779	$HBrO + H^+ + e^- \rightarrow 1/2Br_2 + H_2O$	1.574
$Ag_4[Fe(CN)_6] + 4e^- \rightarrow 4Ag + [Fe(CN)_6]^{4-}$	0.1478	$HClO + H^+ + e^- \rightarrow 1/2Cl_2 + H_2O$	1.611
$AgI + e^- \rightarrow Ag + I^-$	−0.15224	$HClO_2 + 3H^+ + 3e^- \rightarrow 1/2Cl_2 + 2H_2O$	1.628
$AgNO_2 + e^- \rightarrow Ag + NO_2^-$	0.564	$HO_2 + H^+ + e^- \rightarrow H_2O_2$	1.495
$Al^{3+} + 3e^- \rightarrow Al$	−1.662	$HO_2^- + H_2O + 2e^- \rightarrow 3OH^-$	0.878
$Au^+ + e^- \rightarrow Au$	1.692	$2H_2O + 2e^- \rightarrow H_2 + 2OH^-$	−0.8277
$Au^{3+} + 2e^- \rightarrow Au^+$	1.401	$H_2O_2 + 2H^+ + 2e^- \rightarrow 2H_2O$	1.776
$Au^{3+} + 3e^- \rightarrow Au$	1.498	$H_3PO_4 + 2H^+ + 2e^- \rightarrow H_3PO_3 + H_2O$	−0.276
$AuBr_2 + e^- \rightarrow Au + 2Br^-$	0.959	$Hg^{2+} + 2e^- \rightarrow Hg$	0.851
$AuCl_4 + 3e^- \rightarrow Au + 4Cl^-$	1.002	$Hg_2^{2+} + 2e^- \rightarrow 2Hg$	0.7973
$Ba^{2+} + 2e^- \rightarrow Ba$	−2.912	$Hg_2Cl_2 + 2e^- \rightarrow 2Hg + 2Cl^-$	0.26808
$Be^{2+} + 2e^- \rightarrow Be$	−1.847	$Hg_2SO_4 + 2e^- \rightarrow 2Hg + SO_4^{2-}$	0.6125
$Bi^{3+} + 3e^- \rightarrow Bi$	0.20	$I_2 + 2e^- \rightarrow 2I^-$	0.5355
$Br_2(aq) + 2e^- \rightarrow 2Br^-$	1.0873	$I_3^- + 2e^- \rightarrow 3I^-$	0.536
$BrO^- + H_2O + 2e^- \rightarrow Br^- + 2OH^-$	0.761	$In^+ + e^- \rightarrow In$	−0.14
$Ca^+ + e^- \rightarrow Ca$	−3.80	$In^{2+} + e^- \rightarrow In^+$	−0.40
$Ca^{2+} + 2e^- \rightarrow Ca$	−2.868	$In^{3+} + 3e^- \rightarrow In$	−0.3382
$Cd^{2+} + 2e^- \rightarrow Cd$	−0.4030	$K^+ + e^- \rightarrow K$	−2.931
$Cd(OH)_2 + 2e^- \rightarrow Cd + 2OH^-$	−0.809	$Li^+ + e^- \rightarrow Li$	−3.0401
$CdSO_4 + 2e^- \rightarrow Cd + SO_4^{2-}$	−0.246	$Mg^{2+} + 2e^- \rightarrow Mg$	−2.372
$Ce^{3+} + 3e^- \rightarrow Ce$	−2.483	$Mg(OH)_2 + 2e^- \rightarrow Mg + 2OH^-$	−2.690
$Ce^{4+} + e^- \rightarrow Ce^{3+}$	1.61	$Mn^{2+} + 2e^- \rightarrow Mn$	−1.185
$Cl_2(g) + 2e^- \rightarrow 2Cl^-$	1.35827	$Mn^{3+} + e^- \rightarrow Mn^{2+}$	1.5415
$ClO_4^- + 2H^+ + 2e^- \rightarrow ClO_3^- + H_2O$	1.189	$MnO_2 + 4H^+ + 2e^- \rightarrow Mn^{2+} + 2H_2O$	1.224
$ClO^- + H_2O + 2e^- \rightarrow Cl^- + 2OH^-$	.81	$MnO_4^- + 4H^+ + 3e^- \rightarrow MnO_2 + 2H_2O$	1.679
$ClO_4^- + H_2O + 2e^- \rightarrow ClO_3^- + 2OH^-$	0.36	$MnO_4^{2-} + 2H_2O + 2e^- \rightarrow MnO_2 + 4OH^-$	0.595
$Co^{2+} + 2e^- \rightarrow Co$	−0.28	$MnO_4^- + 8H^+ + 5e^- \rightarrow Mn^{2+} + 4H_2O$	1.507
$Co^{3+} + e^- \rightarrow Co^{2+}$ (2 mol / l H_2SO_4)	1.83	$MnO_4^- + e^- \rightarrow MnO_4^{2-}$	0.558
$Cr^{2+} + 2e^- \rightarrow Cr$	−0.913	$MnO_4^{2-} + 2H_2O + 2e^- \rightarrow MnO_2 + 4OH^-$	0.60
$Cr^{3+} + e^- \rightarrow Cr^{2+}$	−0.407	$2NO + 2H^+ + 2e^- \rightarrow N_2O + H_2O$	1.591
$Cr^{3+} + 3e^- \rightarrow Cr$	−0.744	$HNO_2 + H^+ + e^- \rightarrow NO + H_2O$	0.983
$Cr_2O_7^{2-} + 14H^+ + 6e^- \rightarrow 2Cr^{3+} + 7H_2O$	1.232	$NO_2 + H_2O + 3e^- \rightarrow NO + 2OH^-$	−0.46
$Cs^+ + e^- \rightarrow Cs$	−2.92	$NO_3^- + 4H^+ + 3e^- \rightarrow NO + 2H_2O$	0.957
$Cu^+ + e^- \rightarrow Cu$	0.521	$NO_3^- + 2H^+ + e^- \rightarrow NO_2^- + H_2O$	0.835
$Cu^{2+} + e^- \rightarrow Cu^+$	0.153	$NO_3^- + H_2O + 2e^- \rightarrow NO_2^- + 2OH^-$	0.10
$Cu(OH)_2 + 2e^- \rightarrow Cu + 2OH^-$	−0.222	$Na^+ + e^- \rightarrow Na$	−2.71
$F_2 + 2H^+ + 2e^- \rightarrow 2HF$	3.053	$Ni^{2+} + 2e^- \rightarrow Ni$	−0.257
$F_2 + 2e^- \rightarrow 2F^-$	2.866	$NiO_2 + 2H_2O + 2e^- \rightarrow Ni(OH)_2 + 2OH^-$	0.49

(continued)

TABLE 11.1
(Continued)

Reaction	$E°$ (V)	Reaction	$E°$ (V)
$Ni(OH)_2 + 2e^- \rightarrow Ni + 2OH^-$	-0.72	$[PtCl_4]^{2-} + 2e^- \rightarrow Pt + 4Cl^-$	0.755
$NiO_2 + 4H^+ + 2e^- \rightarrow Ni^{2+} + 2H_2O$	1.678	$[PtCl_6]^{2-} + 2e^- \rightarrow [PtCl_4]^{2-} + 2Cl^-$	0.68
$NiOOH + H_2O + e^- \rightarrow Ni(OH)_2 + OH^-$	$+0.52$	$Pt(OH)_2 + 2e^- \rightarrow Pt + 2OH^-$	0.14
$O_2 + e^- \rightarrow O_2^-$	-0.56	$Rb^+ + e^- \rightarrow Rb$	-2.98
$O_2 + 2H^+ + 2e^- \rightarrow H_2O_2$	0.695	$Re^{3+} + 3e^- \rightarrow Re$	0.300
$O_2 + 4H^+ + 4e^- \rightarrow 2H_2O$	1.229	$S + 2e^- \rightarrow S^{2-}$	-0.47627
$O_2 + 2H_2O + 2e^- \rightarrow H_2O_2 + 2OH^-$	-0.146	$S + 2H^+ + 2e^- \rightarrow H_2S(aq)$	0.142
$O_2 + 2H_2O + 4e^- \rightarrow 4OH^-$	0.401	$S_2O_6^{2-} + 4H^+ + 2e^- \rightarrow 2H_2SO_3$	0.564
$O_2 + H_2O + 2e^- \rightarrow HO_2^- + OH^-$	-0.076	$S_2O_6^{2-} + 2e^- + 2H^+ \rightarrow 2HSO_3^-$	0.464
$O_3 + 2H^+ + 2e^- \rightarrow O_2 + H_2O$	2.076	$S_2O_8^{2-} + 2e^- \rightarrow 2SO_4^{2-}$	2.010
$O_3 + H_2O + 2e^- \rightarrow O_2 + 2OH^-$	1.24	$2H_2SO_3 + H^+ + 2e^- \rightarrow H_2SO_4^- + 2H_2O$	-0.056
$Pb^{2+} + 2e^- \rightarrow Pb$	-0.1262	$H_2SO_3 + 4H^+ + 4e^- \rightarrow S + 3H_2O$	0.449
$Pb^{4+} + 2e^- \rightarrow Pb^{2+}$	1.67	$Sn^{2+} + 2e^- \rightarrow Sn$	-0.1375
$PbBr_2 + 2e^- \rightarrow Pb + 2Br^-$	-0.284	$Sn^{4+} + 2e^- \rightarrow Sn^{2+}$	0.151
$PbCl_2 + 2e^- \rightarrow Pb + 2Cl^-$	-0.2675	$Ti^{2+} + 2e^- \rightarrow Ti$	-1.630
$PbO + H_2O + 2e^- \rightarrow Pb + 2OH^-$	-0.580	$Ti^{3+} + 2e^- \rightarrow Ti^{2+}$	-0.368
$PbO_2 + 4H^+ + 2e^- \rightarrow Pb^{2+} + 2H_2O$	1.455	$TiO_2 + 4H^+ + 2e^- \rightarrow Ti^{2+} + 2H_2O$	-0.502
$PbO_2 + SO_4^{2-} + 4H^+ + 2e^- \rightarrow PbSO_4 + 2H_2O$	1.6913	$Zn^{2+} + 2e^- \rightarrow Zn$	-0.7618
$PbSO_4 + 2e^- \rightarrow Pb + SO_4^{2-}$	-0.3505	$ZnO_2^{2-} + 2H_2O + 2e^- \rightarrow Zn + 4OH^-$	-1.215
$Pd^{2+} + 2e^- \rightarrow Pd$	0.951	$Zr(OH)_2 + H_2O + 4e^- \rightarrow Zr + 4OH^-$	-2.36
$Pt^{2+} + 2e^- \rightarrow Pt$	1.118		

Source: HCP and Bard.

TABLE 11.2
Standard Reduction Potentials Ordered by Reduction Potential

Reaction	$E°$ (V)	Reaction	$E°$ (V)
$Ca^+ + e^- \rightarrow Ca$	-3.80	$O_2 + 2H_2O + 2e^- \rightarrow H_2O_2 + 2OH^-$	-0.146
$Li^+ + e^- \rightarrow Li$	-3.0401	$In^+ + e^- \rightarrow In$	-0.14
$Rb^+ + e^- \rightarrow Rb$	-2.98	$Sn^{2+} + 2e^- \rightarrow Sn$	-0.1375
$K^+ + e^- \rightarrow K$	-2.931	$Pb^{2+} + 2e^- \rightarrow Pb$	-0.1262
$Cs^+ + e^- \rightarrow Cs$	-2.92	$O_2 + H_2O + 2e^- \rightarrow HO_2^- + OH^-$	-0.076
$Ba^{2+} + 2e^- \rightarrow Ba$	-2.912	$2H_2SO_3 + H^+ + 2e^- \rightarrow H_2SO_4^- + 2H_2O$	-0.056
$Ca^{2+} + 2e^- \rightarrow Ca$	-2.868	$Fe^{3+} + 3e^- \rightarrow Fe$	-0.030
$Na^+ + e^- \rightarrow Na$	-2.71	$AgCN + e^- \rightarrow Ag + CN^-$	-0.017
$Mg(OH)_2 + 2e^- \rightarrow Mg + 2OH^-$	-2.690	$2H^+ + 2e^- \rightarrow H_2$	0
$Ce^{3+} + 3e^- \rightarrow Ce$	-2.483	$AgBr + e^- \rightarrow Ag + Br^-$	0.07133
$Mg^{2+} + 2e^- \rightarrow Mg$	-2.372	$NO_3^- + H_2O + 2e^- \rightarrow NO_2^- + 2OH^-$	0.10
$Zr(OH)_2 + H_2O + 4e^- \rightarrow Zr + 4OH^-$	-2.36	$Pt(OH)_2 + 2e^- \rightarrow Pt + 2OH^-$	0.14
$Be^{2+} + 2e^- \rightarrow Be$	-1.847	$S + 2H^+ + 2e^- \rightarrow H_2S(aq)$	0.142
$Al^{3+} + 3e^- \rightarrow Al$	-1.662	$Ag_4[Fe(CN)_6] + 4e^- \rightarrow 4Ag + [Fe(CN)_6]^{4-}$	0.1478
$Ti^{2+} + 2e^- \rightarrow Ti$	-1.630	$Sn^{4+} + 2e^- \rightarrow Sn^{2+}$	0.151
$ZnO_2^{2-} + 2H_2O + 2e^- \rightarrow Zn + 4OH^-$	-1.215	$Cu^{2+} + e^- \rightarrow Cu^+$	0.153
$Mn^{2+} + 2e^- \rightarrow Mn$	-1.185	$Bi^{3+} + 3e^- \rightarrow Bi$	0.20
$Cr^{2+} + 2e^- \rightarrow Cr$	-0.913	$AgCl + e^- \rightarrow Ag + Cl^-$	0.22233
$2H_2O + 2e^- \rightarrow H_2 + 2OH^-$	-0.8277	$Hg_2Cl_2 + 2e^- \rightarrow 2Hg + 2Cl^-$	0.26808
$Cd(OH)_2 + 2e^- \rightarrow Cd + 2OH^-$	-0.809	$Re^{3+} + 3e^- \rightarrow Re$	0.300
$Zn^{2+} + 2e^- \rightarrow Zn$	-0.7618	$[Fe(CN)_6]^{3-} + e^- \rightarrow [Fe(CN)_6]^{4-}$	0.358
$Cr^{3+} + 3e^- \rightarrow Cr$	-0.744	$ClO_4^- + H_2O + 2e^- \rightarrow ClO_3^- + 2OH^-$	0.36
$Ni(OH)_2 + 2e^- \rightarrow Ni + 2OH^-$	-0.72	$O_2 + 2H_2O + 4e^- \rightarrow 4OH^-$	0.401
$PbO + H_2O + 2e^- \rightarrow Pb + 2OH^-$	-0.580	$H_2SO_3 + 4H^+ + 4e^- \rightarrow S + 3H_2O$	0.449
$O_2 + e^- \rightarrow O_2^-$	-0.56	$S_2O_6^{2-} + 2e^- + 2H^+ \rightarrow 2HSO_3^-$	0.464
$TiO_2 + 4H^+ + 2e^- \rightarrow Ti^{2+} + 2H_2O$	-0.502	$NiO_2 + 2H_2O + 2e^- \rightarrow Ni(OH)_2 + 2OH^-$	0.49
$S + 2e^- \rightarrow S^{2-}$	-0.47627	$NiOOH + H_2O + e^- \rightarrow Ni(OH)_2 + OH^-$	$+0.52$
$NO_2 + H_2O + 3e^- \rightarrow NO + 2OH^-$	-0.46	$Cu^+ + e^- \rightarrow Cu$	0.521
$Fe^{2+} + 2e^- \rightarrow Fe$	-0.447	$I_2 + 2e^- \rightarrow 2I^-$	0.5355
$Cr^{3+} + e^- \rightarrow Cr^{2+}$	-0.407	$I_3^- + 2e^- \rightarrow 3I^-$	0.536
$Cd^{2+} + 2e^- \rightarrow Cd$	-0.4030	$MnO_4^- + e^- \rightarrow MnO_4^{2-}$	0.558
$In^{2+} + e^- \rightarrow In^+$	-0.40	$AgNO_2 + e^- \rightarrow Ag + NO_2^-$	0.564
$Ti^{3+} + 2e^- \rightarrow Ti^{2+}$	-0.368	$S_2O_6^{2-} + 4H^+ + 2e^- \rightarrow 2H_2SO_3$	0.564
$PbSO_4 + 2e^- \rightarrow Pb + SO_4^{2-}$	-0.3505	$MnO_4^{2-} + 2H_2O + 2e^- \rightarrow MnO_2 + 4OH^-$	0.595
$In^{3+} + 3e^- \rightarrow In$	-0.3382	$MnO_4^{2-} + 2H_2O + 2e^- \rightarrow MnO_2 + 4OH^-$	0.60
$PbBr_2 + 2e^- \rightarrow Pb + 2Br^-$	-0.284	$Hg_2SO_4 + 2e^- \rightarrow 2Hg + SO_4^{2-}$	0.6125
$Co^{2+} + 2e^- \rightarrow Co$	-0.28	$[PtCl_6]^{2-} + 2e^- \rightarrow [PtCl_4]^{2-} + 2Cl^-$	0.68
$H_3PO_4 + 2H^+ + 2e^- \rightarrow H_3PO_3 + H_2O$	-0.276	$O_2 + 2H^+ + 2e^- \rightarrow H_2O_2$	0.695
$PbCl_2 + 2e^- \rightarrow Pb + 2Cl^-$	-0.2675	$[PtCl_4]^{2-} + 2e^- \rightarrow Pt + 4Cl^-$	0.755
$Ni^{2+} + 2e^- \rightarrow Ni$	-0.257	$BrO^- + H_2O + 2e^- \rightarrow Br^- + 2OH^-$	0.761
$CdSO_4 + 2e^- \rightarrow Cd + SO_4^{2-}$	-0.246	$Fe^{3+} + e^- \rightarrow Fe^{2+}$	0.771
$Cu(OH)_2 + 2e^- \rightarrow Cu + 2OH^-$	-0.222	$AgF + e^- \rightarrow Ag + F^-$	0.779
$AgI + e^- \rightarrow Ag + I^-$	-0.15224	$Hg_2^{2+} + 2e^- \rightarrow 2Hg$	0.7973

(continued)

TABLE 11.2

(Continued)

Reaction	$E°$ (V)	Reaction	$E°$ (V)
$Ag^+ + e^- \rightarrow Ag$	0.7996	$Au^{3+} + 3e^- \rightarrow Au$	1.498
$ClO^- + H_2O + 2e^- \rightarrow Cl^- + 2OH^-$	0.81	$MnO_4^- + 8H^+ + 5e^- \rightarrow Mn^{2+} + 4H_2O$	1.507
$NO_3^- + 2H^+ + e^- \rightarrow NO_2^- + H_2O$	0.835	$Mn^{3+} + e^- \rightarrow Mn^{2+}$	1.5415
$Hg^{2+} + 2e^- \rightarrow Hg$	0.851	$HBrO + H^+ + e^- \rightarrow 1/2Br_2 + H_2O$	1.574
$HO_2 + H_2O + 2e^- \rightarrow 3OH^-$	0.878	$2NO + 2H^+ + 2e^- \rightarrow N_2O + H_2O$	1.591
$Pd^{2+} + 2e^- \rightarrow Pd$	0.951	$Ce^{4+} + e^- \rightarrow Ce^{3+}$	1.61
$NO_3^- + 4H^+ + 3e^- \rightarrow NO + 2H_2O$	0.957	$HClO + H^+ + e^- \rightarrow 1/2Cl_2 + H_2O$	1.611
$AuBr_2 + e^- \rightarrow Au + 2Br^-$	0.959	$HClO_2 + 3H^+ + 3e^- \rightarrow 1/2Cl_2 + 2H_2O$	1.628
$HNO_2 + H^+ + e^- \rightarrow NO + H_2O$	0.983	$Pb^{4+} + 2e^- \rightarrow Pb^{2+}$	1.67
$AuCl_4 + 3e^- \rightarrow Au + 4Cl^-$	1.002	$NiO_2 + 4H^+ + 2e^- \rightarrow Ni^{2+} + 2H_2O$	1.678
$Br_2(aq) + 2e^- \rightarrow 2Br^-$	1.0873	$MnO_4^- + 4H^+ + 3e^- \rightarrow MnO_2 + 2H_2O$	1.679
$Pt^{2+} + 2e^- \rightarrow Pt$	1.118	$PbO_2 + SO_4^{2-} + 4H^+ + 2e^- \rightarrow PbSO_4 + 2H_2O$	1.6913
$ClO_4^- + 2H^+ + 2e^- \rightarrow ClO_3^- + H_2O$	1.189	$Au^+ + e^- \rightarrow Au$	1.692
$MnO_2 + 4H^+ + 2e^- \rightarrow Mn^{2+} + 2H_2O$	1.224	$H_2O_2 + 2H^+ + 2e^- \rightarrow 2H_2O$	1.776
$O_2 + 4H^+ + 4e^- \rightarrow 2H_2O$	1.229	$Co^{3+} + e^- \rightarrow Co^{2+} \ (2 \ mol \, / \, l \ H_2SO_4)$	1.83
$Cr_2O_7^{2-} + 14H^+ + 6e^- \rightarrow 2Cr^{3+} + 7H_2O$	1.232	$Ag^{2+} + e^- \rightarrow Ag^+$	1.980
$O_3 + H_2O + 2e^- \rightarrow O_2 + 2OH^-$	1.24	$S_2O_8^{2-} + 2e^- \rightarrow 2SO_4^{2-}$	2.010
$Cl_2(g) + 2e^- \rightarrow 2Cl^-$	1.35827	$O_3 + 2H^+ + 2e^- \rightarrow O_2 + H_2O$	2.076
$Au^{3+} + 2e^- \rightarrow Au^+$	1.401	$F_2 + 2e^- \rightarrow 2F^-$	2.866
$PbO_2 + 4H^+ + 2e^- \rightarrow Pb^{2+} + 2H_2O$	1.455	$F_2 + 2H^+ + 2e^- \rightarrow 2HF$	3.053
$HO_2 + H^+ + e^- \rightarrow H_2O_2$	1.495		

Sources: HCP and Bard.

Math Supplement

B.1 Working with Complex Numbers and Complex Functions

Imaginary numbers can be written in the form

$$z = a + ib \tag{B.1}$$

where a and b are real numbers and $i = \sqrt{-1}$. It is useful to represent complex numbers in the complex plane shown in Figure B.1. The vertical and horizontal axes correspond to the imaginary and real parts of z, respectively.

In the representation shown in Figure B.1, a complex number corresponds to a point in the complex plane. Note the similarity to the polar coordinate system. Because of this analogy, a complex number can be represented either as the pair (a,b), or by the radius vector r and the angle θ. From Figure B.1, it can be seen that

$$r = \sqrt{a^2 + b^2} \quad \text{and} \quad \theta = \cos^{-1}\frac{a}{r} = \sin^{-1}\frac{b}{r} = \tan^{-1}\frac{b}{a} \tag{B.2}$$

Using the relations between a, b, and r as well as the Euler relation $e^{i\theta} = \cos\theta + i\sin\theta$, a complex number can be represented in either of two equivalent ways:

$$a + ib = r\cos\theta + r\sin\theta = re^{i\theta} = \sqrt{a^2 + b^2}\,\exp[i\tan^{-1}(b/a)] \tag{B.3}$$

If a complex number is represented in one way, it can easily be converted to the other way. For example, we express the complex number $6 - 7i$ in the form $re^{i\theta}$. The magnitude of the radius vector r is given by $\sqrt{6^2 + 7^2} = \sqrt{85}$. The phase is given by $\tan\theta = (-7/6)$ or $\theta = \tan^{-1}(-7/6)$. Therefore, we can write $6 - 7i$ as $\sqrt{85}\,\exp[i\tan^{-1}(-7/6)]$.

In a second example, we convert the complex number $2e^{i\pi/2}$, which is in the $re^{i\theta}$ notation, to the $a + ib$ notation. Using the relation $e^{i\alpha} = \exp(i\alpha) = \cos\alpha + i\sin\alpha$, we can write $2e^{i\pi/2}$ as

$$2\left(\cos\frac{\pi}{2} + i\sin\frac{\pi}{2}\right) = 2(0 + i) = 2i$$

The complex conjugate of a complex number z is designated by z^* and is obtained by changing the sign of i, wherever it appears in the complex number. For example, if $z = (3 - \sqrt{5}i)e^{i\sqrt{2}\phi}$, then $z^* = (3 + \sqrt{5}i)e^{-i\sqrt{2}\phi}$. The magnitude of a complex number is defined by $\sqrt{zz^*}$ and is always a real number. This is the case for the previous example:

$$zz^* = (3 - \sqrt{5}i)e^{i\sqrt{2}\phi}(3 + \sqrt{5}i)e^{-i\sqrt{2}\phi} = (3 - \sqrt{5}i)(3 + \sqrt{5}i)e^{i\sqrt{2}\phi - i\sqrt{2}\phi} = 14 \tag{B.4}$$

Note also that $zz^* = a^2 + b^2$.

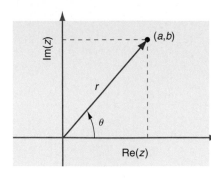

FIGURE B.1

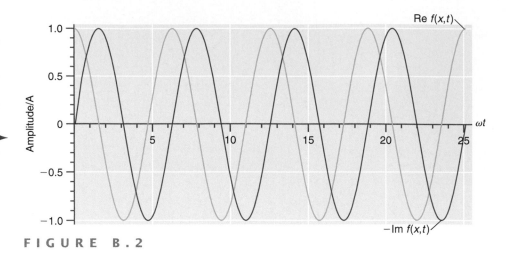

FIGURE B.2

Complex numbers can be added, multiplied, and divided just like real numbers. A few examples follow:

$$(3+\sqrt{2}i)+(1-\sqrt{3}i) = [4+(\sqrt{2}-\sqrt{3})i]$$

$$(3+\sqrt{2}i)(1-\sqrt{3}i) = 3-3\sqrt{3}i+\sqrt{2}i-\sqrt{6}i^2 = (3+\sqrt{6})+(\sqrt{2}-3\sqrt{3})i$$

$$\frac{(3+\sqrt{2}i)}{(1-\sqrt{3}i)} = \frac{(3+\sqrt{2}i)}{(1-\sqrt{3}i)}\frac{(1+\sqrt{3}i)}{(1+\sqrt{3}i)} = \frac{3+3\sqrt{3}i+\sqrt{2}i+\sqrt{6}i^2}{4} = \frac{(3-\sqrt{6})+(3\sqrt{3}+\sqrt{2})i}{4}$$

Functions can depend on a complex variable. It is convenient to represent a plane traveling wave usually written in the form

$$\psi(x,t) = A\sin(kx-\omega t) \tag{B.5}$$

in the complex form

$$Ae^{i(kx-\omega t)} = A\cos(kx-\omega t) - iA\sin(kx-\omega t) \tag{B.6}$$

Note that

$$\psi(x,t) = -\operatorname{Im} Ae^{i(kx-\omega t)} \tag{B.7}$$

The reason for working with the complex form rather than the real form of a function is that calculations such as differentiation and integration can be carried out more easily. Waves in classical physics have real amplitudes, because their amplitudes are linked directly to observables. For example, the amplitude of a sound wave is the local pressure that arises from the expansion or compression of the medium through which the wave passes. However, in quantum mechanics, observables are related to $|\psi(x,t)|^2$ rather than $\psi(x,t)$. Because $|\psi(x,t)|^2$ is always real, $\psi(x,t)$ can be complex, and the observables associated with the wave function are still real.

For the complex function $f(x,t) = Ae^{i(kx-\omega t)}$, $zz^* = \psi(x,t)\psi^*(x,t) = Ae^{i(kx-\omega t)}$ $A^*e^{-i(kx-\omega t)} = AA^*$, so that the magnitude of the function is a constant and does not depend on t or x. As Figure B.2 shows, the real and imaginary parts of $Ae^{i(kx-\omega t)}$ depend differently on the variables x and t; they are phase shifted by $\pi/2$. The figure shows the amplitudes of the real and imaginary parts as a function of ωt for $x = 0$.

B.2 Differential Calculus

B.2.1   The First Derivative of a Function

The derivative of a function has as its physical interpretation the slope of the function evaluated at the position of interest. For example, the slope of the function $y = x^2$ at the point $x = 1.5$ is indicated by the line tangent to the curve shown in Figure B.3.

Mathematically, the first derivative of a function $f(x)$ is denoted $f'(x)$ or $df(x)/dx$. It is defined by

$$\frac{df(x)}{dx} = \lim_{h \to 0} \frac{f(x+h) - f(x)}{h} \tag{B.8}$$

For the function of interest,

$$\frac{df(x)}{dx} = \lim_{h \to 0} \frac{(x+h)^2 - (x)^2}{h} = \lim_{h \to 0} \frac{2hx + h^2}{h} = \lim_{h \to 0} 2x + h = 2x \tag{B.9}$$

In order for $df(x)/dx$ to be defined over an interval in x, $f(x)$ must be continuous over the interval.

Based on this example, $df(x)/dx$ can be calculated if $f(x)$ is known. Several useful rules for differentiating commonly encountered functions are listed next:

$$\frac{d(ax^n)}{dx} = anx^{n-1}, \quad \text{where } a \text{ is a constant and } n > 0 \tag{B.10}$$

For example, $d(\sqrt{3}x^{4/3})/dx = (4/3)\sqrt{3}x^{1/3}$

$$\frac{d(ae^{bx})}{dx} = abe^{bx}, \quad \text{where } a \text{ and } b \text{ are constants} \tag{B.11}$$

For example, $d(5e^{3\sqrt{2}x})/dx = 15\sqrt{2}e^{3\sqrt{2}x}$

$$\frac{d(ae^{bx})}{dx} = abe^{bx}, \quad \text{where } a \text{ and } b \text{ are constants}$$

$$\frac{d(a\sin x)}{dx} = a\cos x, \quad \text{where } a \text{ is a constant} \tag{B.12}$$

$$\frac{d(a\cos x)}{dx} = -a\sin x, \quad \text{where } a \text{ is a constant}$$

Two useful rules in evaluating the derivative of a function that is itself the sum or product of two functions are as follows:

$$\frac{d[f(x) + g(x)]}{dx} = \frac{df(x)}{dx} + \frac{dg(x)}{dx} \tag{B.13}$$

For example,

$$\frac{d(x^3 + \sin x)}{dx} = \frac{dx^3}{dx} + \frac{d\sin x}{dx} = 3x^2 + \cos x$$

$$\frac{d[f(x)g(x)]}{dx} = g(x)\frac{df(x)}{dx} + f(x)\frac{dg(x)}{dx} \tag{B.14}$$

For example,

$$\frac{d[\sin(x)\cos(x)]}{dx} = \cos(x)\frac{d\sin(x)}{dx} + \sin(x)\frac{d\cos(x)}{dx}$$

$$= \cos^2 x - \sin^2 x$$

B.2.2 The Reciprocal Rule and the Quotient Rule

How is the first derivative calculated if the function to be differentiated does not have a simple form such as those listed in the preceding section? In many cases, the derivative can be found by using the product and quotient rules stated here:

$$\frac{d\left(\dfrac{1}{f(x)}\right)}{dx} = -\frac{1}{[f(x)]^2}\frac{df(x)}{dx} \tag{B.15}$$

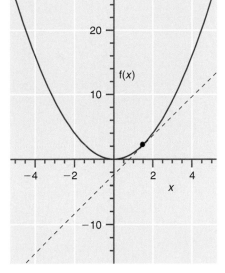

FIGURE B.3

For example,

$$\frac{d\left(\dfrac{1}{\sin x}\right)}{dx} = -\frac{1}{\sin^2 x}\frac{d\sin x}{dx} = \frac{-\cos x}{\sin^2 x}$$

$$\frac{d\left[\dfrac{f(x)}{g(x)}\right]}{dx} = \frac{g(x)\dfrac{df(x)}{dx} - f(x)\dfrac{dg(x)}{dx}}{[g(x)]^2} \tag{B.16}$$

For example,

$$\frac{d\left(\dfrac{x^2}{\sin x}\right)}{dx} = \frac{2x\sin x - x^2\cos x}{\sin^2 x}$$

B.2.3   The Chain Rule

In this section, we deal with the differentiation of more complicated functions. Suppose that $y = f(u)$ and $u = g(x)$. From the previous section, we know how to calculate $df(u)/du$. How do we calculate $df(u)/dx$? The answer to this question is stated as the chain rule:

$$\frac{df(u)}{dx} = \frac{df(u)}{du}\frac{du}{dx} \tag{B.17}$$

Several examples illustrating the chain rule follow:

$$\frac{d\sin(3x)}{dx} = \frac{d\sin(3x)}{d(3x)}\frac{d(3x)}{dx} = 3\cos(3x)$$

$$\frac{d\ln(x^2)}{dx} = \frac{d\ln(x^2)}{d(x^2)}\frac{d(x^2)}{dx} = \frac{2x}{x^2} = \frac{2}{x}$$

$$\frac{d\left(x+\dfrac{1}{x}\right)^{-4}}{dx} = \frac{d\log\left(x+\dfrac{1}{x}\right)^{-4}}{d\left(x+\dfrac{1}{x}\right)}\frac{d\left(x+\dfrac{1}{x}\right)}{dx} = -4\left(x+\frac{1}{x}\right)^{-5}\left(1-\frac{1}{x^2}\right)$$

$$\frac{d\exp(ax^2)}{dx} = \frac{d\exp(ax^2)}{d(ax^2)}\frac{d(ax^2)}{dx} = 2ax\exp(ax^2), \quad \text{where } a \text{ is a constant}$$

B.2.4   Higher Order Derivatives: Maxima, Minima, and Inflection Points

A function $f(x)$ can have higher order derivatives in addition to the first derivative. The second derivative of a function is the slope of a graph of the slope of the function versus the variable. Mathematically,

$$\frac{d^2 f(x)}{dx^2} = \frac{d}{dx}\left(\frac{df(x)}{dx}\right) \tag{B.18}$$

For example,

$$\frac{d^2\exp(ax^2)}{dx^2} = \frac{d}{dx}\left[\frac{d\exp(ax^2)}{dx}\right] = \frac{d[2ax\exp(ax^2)]}{dx}$$

$$= 2a\exp(ax^2) + 4a^2x^2\exp(ax^2), \quad \text{where } a \text{ is a constant}$$

The second derivative is useful in identifying where a function has its minimum or maximum value within a range of the variable, as shown next.

Because the first derivative is zero at a local maximum or minimum, $df(x)/dx = 0$ at the values x_{max} and x_{min}. Consider the function $f(x) = x^3 - 5x$ shown in Figure B.4 over the range $-2.5 \le x \le 2.5$.

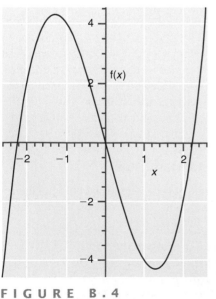

F I G U R E   B . 4

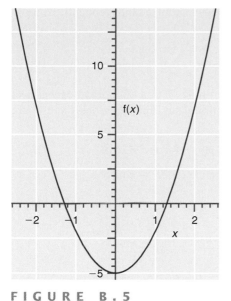

FIGURE B.5

By taking the derivative of this function and setting it equal to zero, we find the minima and maxima of this function in the range

$$\frac{d(x^3 - 5x)}{dx} = 3x^2 - 5 = 0, \quad \text{which has the solutions } x = \pm\sqrt{\frac{5}{3}} = 1.291$$

The maxima and minima can also be determined by graphing the derivative and finding the zero crossings as shown in Figure B.5.

Graphing the function clearly shows that the function has one maximum and one minimum in the range specified. What criterion can be used to distinguish between these extrema if the function is not graphed? The sign of the second derivative, evaluated at the point for which the first derivative is zero, can be used to distinguish between a maximum and a minimum:

$$\frac{d^2 f(x)}{dx^2} = \frac{d}{dx}\left[\frac{df(x)}{dx}\right] < 0 \quad \text{for a maximum} \tag{B.19}$$

$$\frac{d^2 f(x)}{dx^2} = \frac{d}{dx}\left[\frac{df(x)}{dx}\right] > 0 \quad \text{for a minimum}$$

We return to the function graphed earlier and calculate the second derivative:

$$\frac{d^2(x^3 - 5x)}{dx^2} = \frac{d}{dx}\left[\frac{d(x^3 - 5x)}{dx}\right] = \frac{d(3x^2 - 5)}{dx} = 6x$$

By evaluating

$$\frac{d^2 f(x)}{dx^2} \quad \text{at} \quad x = \pm\sqrt{\frac{5}{3}} = \pm 1.291$$

we see that $x = 1.291$ corresponds to the minimum, and $x = -1.291$ corresponds to the maximum.

If a function has an inflection point in the interval of interest, then

$$\frac{df(x)}{dx} = 0 \quad \text{and} \quad \frac{d^2 f(x)}{dx^2} = 0 \tag{B.20}$$

An example for an inflection point is $x = 0$ for $f(x) = x^3$. A graph of this function in the interval $-2 \le x \le 2$ is shown in Figure B.6. As you can verify,

$$\frac{dx^3}{dx} = 3x^2 = 0 \text{ at } x = 0 \quad \text{and} \quad \frac{d^2(x^3)}{dx^2} = 6x = 0 \text{ at } x = 0$$

B.2.5 Maximizing a Function Subject to a Constraint

A frequently encountered problem is that of maximizing a function relative to a constraint. We first outline how to carry out a constrained maximization, and subsequently apply the method to maximizing the volume of a cylinder while minimizing its area. The theoretical framework for solving this problem originated with the French mathematician Lagrange, and the method is known as Lagrange's method of undetermined multipliers. We wish to maximize the function $f(x,y)$ subject to the constraint that $\phi(x,y) - C = 0$, where C is a constant. For example, you may want to maximize the area, A, of a rectangle while minimizing its circumference, C. In this case, $f(x, y) = A(x, y) = xy$ and $\phi(x, y) = C(x, y) = 2(x + y)$, where x and y are the length and width of the rectangle. The total differentials of these functions are given by Equation (B.21):

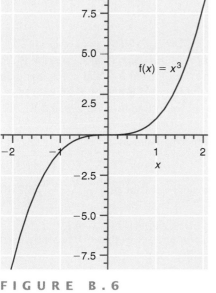

FIGURE B.6

$$df = \left(\frac{\partial f}{\partial x}\right)_y dx + \left(\frac{\partial f}{\partial y}\right)_x dy = 0 \quad \text{and} \quad d\phi = \left(\frac{\partial \phi}{\partial x}\right)_y dx + \left(\frac{\partial \phi}{\partial y}\right)_x dy = 0 \tag{B.21}$$

If x and y were independent variables (there is no constraining relationship), the maximization problem would be identical to those dealt with earlier. However, because $d\phi = 0$ also needs to be satisfied, x and y are not independent variables. In this case, Lagrange found that the appropriate function to minimize is $f - \lambda\phi$, where λ is an undetermined multiplier. He showed that each of the expressions in the square brackets in the differential given by Equation (B.22) can be maximized independently. A separate multiplier is required for each constraint:

$$df = \left[\left(\frac{\partial f}{\partial x} \right)_y - \lambda \left(\frac{\partial \phi}{\partial x} \right)_y \right] dx + \left[\left(\frac{\partial f}{\partial y} \right)_x - \lambda \left(\frac{\partial \phi}{\partial y} \right)_x \right] dy \qquad (B.22)$$

We next use this method to maximize the volume, V, of a cylindrical can subject to the constraint that its exterior area, A, be minimized. The functions f and ϕ are given by

$$V = f(r,h) = \pi r^2 h \quad \text{and} \quad A = \phi(r,h) = 2\pi r^2 + 2\pi rh \qquad (B.23)$$

Calculating the partial derivatives and using Equation (B.22), we have

$$\left(\frac{\partial f(r,h)}{r} \right)_h = 2\pi rh \quad \left(\frac{\partial f(r,h)}{h} \right)_r = \pi r^2 \qquad (B.24)$$

$$\left(\frac{\partial \phi(r,h)}{r} \right)_h = 4\pi r + 2\pi h \quad \left(\frac{\partial \phi(r,h)}{h} \right)_r = 2\pi r$$

$$(2\pi rh - \lambda[4\pi r + 2\pi h])dr = 0 \quad \text{and} \quad (\pi r^2 - \lambda 2\pi r)dh = 0$$

Eliminating λ from these two equations gives

$$\frac{2\pi rh}{4\pi r + 2\pi h} = \frac{\pi r^2}{2\pi r} \qquad (B.25)$$

Solving for h in terms of r gives the result $h = 2r$. Note that there is no need to determine the value of multiplier λ. Perhaps you have noticed that beverage cans do not follow this relationship between r and h. Can you think of factors other than minimizing the amount of metal used in the can that might be important in this case?

B.3 Series Expansions of Functions

B.3.1 Convergent Infinite Series

Physical chemists often express functions of interest in the form of an infinite series. For this application, the series must converge. Consider the series

$$a_0 + a_1 x + a_2 x^2 + a_3 x^3 + \ldots a_n x^n + \ldots \qquad (B.26)$$

How can we determine if such a series converges? A useful convergence criterion is the ratio test. If the absolute ratio of successive terms (designated u_{n-1} and u_n) is less than one as $n \to \infty$, the series converges. We consider the series of Equation B.26 with (a) $a_n = n!$ and (b) $a_n = 1/n!$, and apply the ratio test as shown in Equations B.27a and B.27b.

$$(a) \lim_{n \to \infty} \left| \frac{u_n}{u_{n-1}} \right| = \left| \frac{n! x^n}{(n-1)! x^{n-1}} \right| = \lim_{n \to \infty} |nx| > 1 \text{ unless } x = 0 \qquad (B.27a)$$

$$(b) \lim_{n \to \infty} \left| \frac{u_n}{u_{n-1}} \right| = \left| \frac{x^n/n!}{x^{n-1}/(n-1)!} \right| = \lim_{n \to \infty} \left| \frac{x}{n} \right| < 1 \text{ for all } x \qquad (B.27b)$$

We see that the infinite series converges if $a_n = 1/n!$ but diverges if $a_n = n!$.

The power series is a particularly important form of a series that is frequently used to fit experimental data to a functional form. It has the form

$$a_0 + a_1 x + a_2 x^2 + a_3 x^3 + a_1 x + a_4 x^4 + \ldots = \sum_{n=0}^{\infty} a_n x^n \qquad (B.28)$$

Fitting a data set to a series with a large number of terms is impractical, and to be useful, the series should contain as few terms as possible to satisfy the desired accuracy. For example, the function $\sin x$ can be fit to a power series over the interval $0 \le x \le 1.5$ by the following truncated power series

$$\sin x \approx -1.20835 \times 10^{-3} + 1.02102x - 0.0607398x^2 - 0.11779x^3 \tag{B.29}$$

$$\sin x \approx -8.86688 \times 10^{-5} + 0.996755x + 0.0175769x^2 - 0.200644x^3 - 0.027618x^4$$

The coefficients in Equation B.29 have been determined using a least squares fitting routine. The first series includes terms in x up to x^3, and is accurate to within 2% over the interval. The second series includes terms up to x^4, and is accurate to within 0.1% over the interval. Including more terms will increase the accuracy further.

A special case of a power series is the geometric series, in which successive terms are related by a constant factor. An example of a geometric series and its sum is given in Equation B.30. Using the ratio criterion of Equation B.27, convince yourself that this series converges for $|x| < 1$.

$$a(1 + x + x^2 + x^3 + \ldots) = \frac{a}{1-x}, \quad \text{for } |x| < 1 \tag{B.30}$$

B.3.2 Representing Functions in the Form of Infinite Series

Assume that you have a function in the form $f(x)$ and wish to express it as a power series in x of the form

$$f(x) = a_0 + a_1 x + a_2 x^2 + a_3 x^3 + \ldots \tag{B.31}$$

To do so, we need a way to find the set of coefficients $(a_0, a_1, a_2, a_3, \ldots)$. How can this be done?

If the functional form $f(x)$ is known, the function can be expanded about a point of interest using the Taylor-Mclaurin expansion. In the vicinity of $x = a$, the function can be expanded in the series

$$f(x) = f(a) + \left(\frac{df(x)}{dx}\right)_{x=a} (x-a) + \frac{1}{2!}\left(\frac{d^2 f(x)}{dx^2}\right)_{x=a} (x-a)^2 \tag{B.32}$$

$$+ \frac{1}{3!}\left(\frac{d^3 f(x)}{dx^3}\right)_{x=a} (x-a)^3 + \ldots$$

For example, consider the expansion of $f(x) = e^x$ about $x = 0$. Because $(d^n e^x / dx^n)_{x=0} = 1$ for all values of n, the Taylor-Mclaurin expansion for e^x about $x = 0$ is

$$f(x) = 1 + x + \frac{1}{2!}x^2 + \frac{1}{3!}x^3 + \ldots \tag{B.33}$$

Similarly, the Taylor-Mclaurin expansion for $\ln(1+x)$ is found by evaluating the derivatives in turn:

$$\frac{d \ln(1+x)}{dx} = \frac{1}{1+x}$$

$$\frac{d^2 \ln(1+x)}{dx^2} = \frac{d}{dx}\frac{1}{(1+x)} = -\frac{1}{(1+x)^2}$$

$$\frac{d^3 \ln(1+x)}{dx^3} = -\frac{d}{dx}\frac{1}{(1+x)^2} = \frac{2}{(1+x)^3}$$

$$\frac{d^4 \ln(1+x)}{dx^4} = \frac{d}{dx}\frac{2}{(1+x)^3} = \frac{-6}{(1+x)^4}$$

Each of these derivatives must be evaluated at $x = 0$.

Using these results, the Taylor-Mclaurin expansion for $\ln(1+x)$ about $x = 0$ is

$$f(x) = x - \frac{x^2}{2!} + \frac{2x^3}{3!} - \frac{6x^4}{4!} + \ldots = x - \frac{x^2}{2} + \frac{x^3}{4} - \frac{x^4}{4} + \ldots \tag{B.34}$$

The number of terms that must be included to adequately represent the function depends on the value of x. For $x \ll 1$, the series converges rapidly and, to a very good approximation, we can truncate the Taylor-Mclaurin series after the first one or two terms

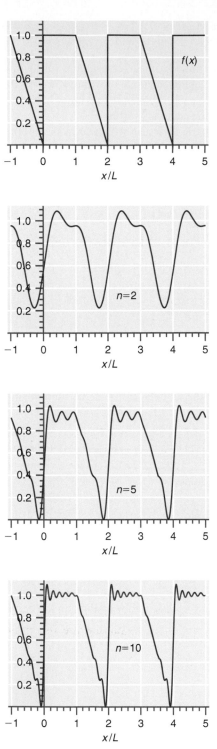

FIGURE B.7

involving the variable. For the two functions just considered, it is reasonable to write $e^x \approx 1 + x$ and $\ln(1 \pm x) \approx \pm x$ if $x \ll 1$.

A second widely used series is the Fourier sine and cosine series. This series can be used to expand functions that are periodic over an interval $-L \leq x \leq L$ by the series

$$f(x) = \frac{1}{2} b_0 + \sum_n b_n \cos \frac{n\pi x}{L} + \sum_n a_n \sin \frac{n\pi x}{L} \tag{B.35}$$

A Fourier series is an infinite series, and the coefficients a_n and b_n can be calculated using the equations

$$a_n = \frac{1}{L} \int_{-L}^{+L} f(x) \sin \frac{n\pi x}{L} dx \quad \text{and} \quad b_n = \frac{1}{L} \int_{-L}^{+L} f(x) \cos \frac{n\pi x}{L} dx \tag{B.36}$$

The usefulness of the Fourier series is that a function can often be approximated by a few terms, depending on the accuracy desired.

For functions that are either even or odd with respect to the variable x, only either the sine or the cosine terms will appear in the series. For even functions, $f(-x) = f(x)$, and for odd functions, $f(-x) = -f(x)$. Because $\sin(-x) = -\sin(x)$ and $\cos(-x) = \cos(x)$, all coefficients a_n are zero for an even function, and all coefficients b_n are zero for an odd function. Note that Equations B.29 are not odd functions of x because the function was only fit over the interval $0 \leq x \leq 1.5$.

Whereas the coefficients for the Taylor-Mclaurin series can be readily calculated, those for the Fourier series require more effort. To avoid mathematical detail here, the Fourier coefficients a_n and b_n are not explicitly calculated for a model function. It is much easier to calculate the coefficients using a program such as *Mathematica* than to calculate them by hand. Our focus here is to show that periodic functions can be approximated to a reasonable degree by using the first few terms in a Fourier series, rather than to carry out the calculations.

To demonstrate the usefulness of expanding a function in a Fourier series, consider the function

$$f(x) = 1 \quad \text{for } 0 \leq x \leq L \tag{B.37}$$
$$f(x) = -x \quad \text{for } -L \leq x \leq 0$$

which is periodic in the interval $-L \leq x \leq L$, in a Fourier series. This function is a demanding function to expand in a Fourier series because the function is discontinuous at $x = 0$ and the slope is discontinuous at $x = 0$ and $x = 1$. The function and the approximate functions obtained by truncating the series at $n = 2$, $n = 5$, and $n = 10$ are shown in Figure B.7. The agreement between the truncated series and the function is reasonably good for $n = 10$. The oscillations seen near $x/L = 0$ are due to the discontinuity in the function. More terms in the series are required to obtain a good fit, because of the discontinuities in the function and its slope.

B.4 Integral Calculus

B.4.1 Definite and Indefinite Integrals

In many areas of physical chemistry, the property of interest is the integral of a function over an interval in the variable of interest. For example, the total probability of finding a particle within an interval $0 \leq x \leq a$ is the integral of the probability density $P(x)$ over the interval

$$P_{total} = \int_0^a P(x) dx \tag{B.38}$$

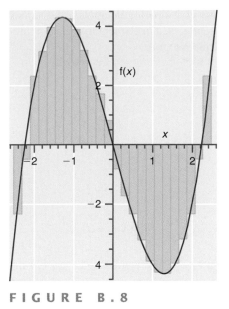

FIGURE B.8

Geometrically, the integral of a function over an integral is the area under the curve describing the function. For example, the integral $\int_{-2.3}^{2.3}(x^3 - 5x)dx$ is the sum of the areas of the individual rectangles in Figure B.8 in the limit within which the width of the rectangles approaches zero. If the rectangles lie below the zero line, the incremental area is negative; if the rectangles lie above the zero line, the incremental area is positive. In this case the total area is zero because the total negative area equals the total positive area. This is the case because $f(x)$ is an odd function of x.

The integral can also be understood as an antiderivative. From this point of view, the integral symbol is defined by the relation

$$f(x) = \int \frac{df(x)}{dx}dx \tag{B.39}$$

and the function that appears under the integral sign is called the integrand. Interpreting the integral in terms of area, we evaluate a definite integral, and the interval over which the integration occurs is specified. The interval is not specified for an indefinite integral.

The geometrical interpretation is often useful in obtaining an integral from experimental data when the functional form of the integrand is not known. For our purposes, the interpretation of the integral as an antiderivative is more useful. The value of the indefinite integral $\int (x^3 - 5x)dx$ is that function which, when differentiated, gives the integrand. Using the rules for differentiation discussed earlier, you can verify that

$$\int (x^3 - 5x)dx = \frac{x^4}{4} - \frac{5x^2}{2} + C \tag{B.40}$$

Note the constant that appears in the evaluation of every indefinite integral. By differentiating the function obtained upon integration, you should convince yourself that any constant will lead to the same integrand. In contrast, a definite integral has no constant of integration. If we evaluate the definite integral

$$\int_{-2.3}^{2.3} (x^3 - 5x)dx = \left(\frac{x^4}{4} - \frac{5x^2}{2} + C \right)_{x=2.3} - \left(\frac{x^4}{4} - \frac{5x^2}{2} + C \right)_{x=-2.3} \tag{B.41}$$

we see that the constant of integration cancels. Because the function obtained upon integration is an even function of x, $\int_{-2.3}^{2.3}(x^3 - 5x)dx = 0$, just as we saw in the geometric interpretation of the integral.

It is useful for the student of physical chemistry to commit the integrals listed next to memory, because they are encountered frequently. These integrals are directly related to the derivatives discussed in Section B.2:

$$\int df(x) = f(x) + C$$

$$\int x^n dx = \frac{x^{n+1}}{n+1} + C$$

$$\int \frac{dx}{x} = \ln x + C$$

$$\int e^{ax} = \frac{e^{ax}}{a} + C, \quad \text{where } a \text{ is a constant}$$

$$\int \sin x dx = -\cos x + C$$

$$\int \cos x dx = \sin x + C$$

However, the primary tool for the physical chemist in evaluating integrals is a good set of integral tables. The integrals that are most frequently used in elementary quantum mechanics are listed here; the first group lists indefinite integrals:

$$\int (\sin \ ax)dx = -\frac{1}{a} \ \cos ax + C$$

$$\int (\cos \ ax)dx = \frac{1}{a} \ \sin ax + C$$

$$\int (\sin^2 \ ax)dx = \frac{1}{2}x - \frac{1}{4a} \ \sin 2ax + C$$

$$\int (\cos^2 \ ax)dx = \frac{1}{2}x + \frac{1}{4a} \ \sin 2ax + C$$

$$\int (x^2 \sin^2 \ ax)dx = \frac{1}{6}x^3 - \left(\frac{1}{4a}x^2 - \frac{1}{8a^3} \right)\sin 2 \ ax - \frac{1}{4a^2}x \cos 2ax + C$$

$$\int (x^2 \cos^2 \ ax) \ dx = \frac{1}{6}x^3 + \left(\frac{1}{4a}x^2 - \frac{1}{8a^3} \right)\sin 2 \ ax + \frac{1}{4a^2}x \cos 2ax + C$$

$$\int x^m e^{ax} dx = \frac{x^m e^{ax}}{a} - \frac{m}{a} \int x^{m-1}e^{ax} dx + C$$

$$\int \frac{e^{ax}}{x^m} dx = -\frac{1}{m-1}\frac{e^{ax}}{x^{m-1}} + \frac{a}{m-1} \int \frac{e^{ax}}{x^{m-1}} \ dx + C$$

The following group lists definite integrals.

$$\int_0^a \sin\left(\frac{n\pi x}{a} \right) \times \sin\left(\frac{m\pi x}{a} \right)dx = \int_0^a \cos\left(\frac{n\pi x}{a} \right) \times \cos\left(\frac{m\pi x}{a} \right)dx = \frac{a}{2}\delta_{mn}$$

$$\int_0^a \left[\sin\left(\frac{n\pi x}{a} \right) \right] \times \left[\cos\left(\frac{n\pi x}{a} \right) \right] dx = 0$$

$$\int_0^\pi \sin^2 mx \ dx = \int_0^\pi \cos^2 mx \ dx = \frac{\pi}{2}$$

$$\int_0^\infty \frac{\sin x}{\sqrt{x}} \ dx = \int_0^\infty \frac{\cos x}{\sqrt{x}} \ dx = \sqrt{\frac{\pi}{2}}$$

$$\int_0^\infty x^n e^{-ax} dx = \frac{n!}{a^{n+1}} \quad (a > 0, \ n \text{ positive integer})$$

$$\int_0^\infty x^{2n} e^{-ax^2} dx = \frac{1 \cdot 3 \cdot 5 \cdots (2n-1)}{2^{n+1}a^n}\sqrt{\frac{\pi}{a}} \quad (a > 0, \ n \text{ positive integer})$$

$$\int_0^\infty x^{2n+1} e^{-ax^2} dx = \frac{n!}{2 \ a^{n+1}} \quad (a > 0, \ n \text{ positive integer})$$

$$\int_0^\infty e^{-ax^2} dx = \left(\frac{\pi}{4a} \right)^{1/2}$$

In the first integral above, $\delta_{mn} = 1$ if $m = n$, and 0 if $m \neq n$.

B.4.2 Multiple Integrals and Spherical Coordinates

In the previous section, integration with respect to a single variable was discussed. Often, however, integration occurs over two or three variables. For example, the wave functions for the particle in a two-dimensional box are given by

$$\psi_{n_x n_y}(x, y) = N \sin \frac{n_x \pi x}{a} \sin \frac{n_y \pi y}{b} \tag{B.42}$$

In normalizing a wave function, the integral of $\left|\psi_{n_x n_y}(x, y)\right|^2$ is required to equal one over the range $0 \le x \le a$ and $0 \le y \le b$. This requires solving the double integral

$$\int_0^b dy \int_0^a \left(N \sin \frac{n_x \pi x}{a} \sin \frac{n_y \pi y}{b} \right)^2 dx = 1 \tag{B.43}$$

to determine the normalization constant N. We sequentially integrate over the variables x and y or vice versa using the list of indefinite integrals from the previous section.

$$\int_0^b dy \int_0^a \left(N \sin \frac{n_x \pi x}{a} \sin \frac{n_y \pi y}{b} \right)^2 dx = \left[\frac{1}{2}x - \frac{a}{4n\pi} \sin \frac{2n_x \pi x}{a} \right]_{x=0}^{x=a} \times N^2 \int_0^b \left(\sin \frac{n_y \pi y}{b} \right)^2 d$$

$$1 = \left[\frac{1}{2}a - \frac{a}{4n\pi}(\sin 2n_x \pi - 0) \right] \times N^2 \int_0^b \left(\sin \frac{n_y \pi y}{b} \right)^2 dy$$

$$1 = N^2 \left[\frac{1}{2}a - \frac{a}{4n\pi}(\sin 2n_x \pi - 0) \right] \times \left[\frac{1}{2}b - \frac{a}{4n\pi}(\sin 2n_y \pi - 0) \right] = \frac{N^2 ab}{4}$$

$$N = \frac{2}{\sqrt{ab}}$$

Convince yourself that the normalization constant for the wave functions of the three-dimensional particle in the box

$$\psi_{n_x n_y n_z}(x, y, z) = N \sin \frac{n_x \pi x}{a} \sin \frac{n_y \pi y}{b} \sin \frac{n_z \pi z}{c} \tag{B.44}$$

has the value $N = 2\sqrt{2}/\sqrt{abc}$.

Up to this point, we have considered functions of a single variable. This restricts us to dealing with a single spatial dimension. The extension to three independent variables becomes important in describing three-dimensional systems. The three-dimensional system of most importance to us is the atom. Closed-shell atoms are spherically symmetric, so we might expect atomic wave functions to be best described by spherical coordinates. Therefore, you should become familiar with integrations in this coordinate system. In transforming from spherical coordinates r, θ, and ϕ to Cartesian coordinates x, y, and z, the following relationships are used:

$$\begin{aligned} x &= r \sin \theta \cos \phi \\ y &= r \sin \theta \sin \phi \\ z &= r \cos \theta \end{aligned} \tag{B.45}$$

These relationships are depicted in Figure B.9. For small increments in the variables r, θ, and ϕ, the volume element depicted in this figure is a rectangular solid of volume

$$dV = (r \sin \theta d\phi)(dr)(rd\theta) = r^2 \sin \theta \, dr \, d\theta \, d\phi \tag{B.46}$$

Note in particular that the volume element in spherical coordinates is not $dr \, d\theta \, d\phi$ in analogy with the volume element $dxdydz$ in Cartesian coordinates.

In transforming from Cartesian coordinates x, y, and z to the spherical coordinates r, θ, and ϕ, these relationships are used:

$$r = \sqrt{x^2 + y^2 + z^2}, \quad \theta = \cos^{-1} \frac{z}{\sqrt{x^2 + y^2 + z^2}}, \quad \text{and} \quad \phi = \tan^{-1} \frac{y}{x} \tag{B.47}$$

What is the appropriate range of variables to integrate over all space in spherical coordinates? If we imagine the radius vector scanning over the range $0 \le \theta \le \pi$; $0 \le \phi \le 2\pi$, the

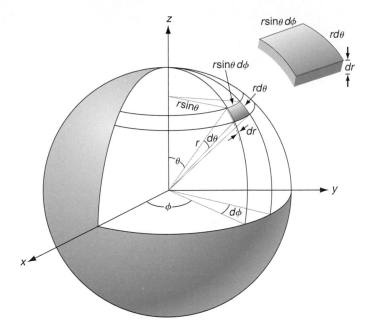

whole angular space is scanned. If we combine this range of θ and ϕ with $0 \le r \le \infty$, all of the three-dimensional space is scanned. Note that $r = \sqrt{x^2 + y^2 + z^2}$ is always positive.

To illustrate the process of integration in spherical coordinates, we normalize the function $e^{-r} \cos \theta$ over the interval $0 \le r \le \infty$; $0 \le \theta \le \pi$; $0 \le \phi \le 2\pi$:

$$N^2 \int_0^{2\pi} d\phi \int_0^{\pi} \sin \theta d\theta \int_0^{\infty} (e^{-r} \cos \theta)^2 r^2 dr = N^2 \int_0^{2\pi} d\phi \int_0^{\pi} \cos^2 \theta \sin \theta d\theta \int_0^{\infty} r^2 e^{-2r} dr = 1$$

It is most convenient to integrate first over ϕ, giving

$$2\pi N^2 \int_0^{\pi} \cos^2 \theta \sin \theta d\theta \int_0^{\infty} r^2 e^{-2r} dr = 1$$

We next integrate over θ, giving

$$2\pi N^2 \left[\frac{-\cos^3 \pi + \cos^3 0}{3} \right] \times \int_0^{\infty} r^2 e^{-2r} dr = \frac{4\pi N^2}{3} \int_0^{\infty} r^2 e^{-2r} dr = 1$$

We finally integrate over r using the standard integral

$$\int_0^{\infty} x^n e^{-ax} dx = \frac{n!}{a^{n+1}} \quad (a > 0, \ n \text{ positive integer})$$

The result is

$$\frac{4\pi N^2}{3} \int_0^{\infty} r^2 e^{-2r} dr = \frac{4\pi N^2}{3} \frac{2!}{8} = 1 \quad \text{or} \quad N = \sqrt{\frac{3}{\pi}}$$

We conclude that the normalized wave function is $\sqrt{3/\pi} \ e^{-r} \cos \theta$.

B.5 Vectors

The use of vectors occurs frequently in physical chemistry. Consider circular motion of a particle at constant speed in two dimensions, as depicted in Figure B.10. The particle is moving in a counterclockwise direction on the ring-like orbit. At any instant in time, its po-

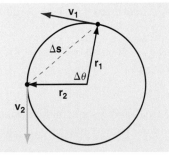

(a)

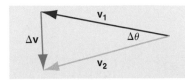

(b)

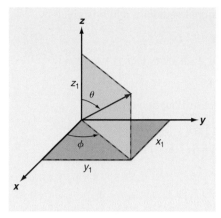

FIGURE B.11

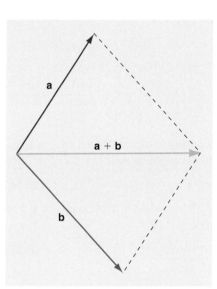

FIGURE B.12

sition, velocity, and acceleration can be measured. The two aspects to these measurements are the magnitude and the direction of each of these observables. Whereas a scalar quantity such as speed has only a magnitude, a vector has both a magnitude and a direction.

For the particular case under consideration, the position vectors $\mathbf{r}_1$ and $\mathbf{r}_2$ extend outward from the origin and terminate at the position of the particle. The velocities $\mathbf{v}_1$ and $\mathbf{v}_2$ are related to the position vector as $\mathbf{v} = \lim_{\Delta t \to 0} [\mathbf{r}(t+\Delta t) - \mathbf{r}(t)]/\Delta t$. Therefore, the velocity vector is perpendicular to the position vector. The acceleration vector is defined by $\mathbf{a} = \lim_{\Delta t \to 0} [\mathbf{v}(t+\Delta t) - \mathbf{v}(t)]/\Delta t$. As we see in part (b) of Figure B.10, $\mathbf{a}$ is perpendicular to $\mathbf{v}$, and is antiparallel to $\mathbf{r}$. As this example of a relatively simple motion shows, vectors are needed to describe the situation properly by keeping track of both the magnitude and direction of each of the observables of interest. For this reason, it is important to be able to work with vectors.

In three–dimensional Cartesian coordinates, any vector can be written in the form

$$\mathbf{r} = x_1\mathbf{i} + y_1\mathbf{j} + z_1\mathbf{k} \tag{B.48}$$

where $\mathbf{i}, \mathbf{j},$ and $\mathbf{k}$ are the mutually perpendicular vectors of unit length along the x, y, and z axes, respectively, and x_1, y_1, and z_1 are numbers. The length of a vector is defined by the equation

$$|\mathbf{r}| = \sqrt{x_1^2 + y_1^2 + z_1^2} \tag{B.49}$$

This vector is depicted in the three-dimensional coordinate system shown in Figure B.11.

By definition, the angle θ is measured from the z axis, and the angle ϕ is measured in the x–y plane from the x axis. The angles θ and ϕ are related to x_1, y_1, and z_1 by

$$\theta = \cos^{-1} \frac{z_1}{\sqrt{x_1^2 + y_1^2 + z_1^2}} \quad \text{and} \quad \phi = \tan^{-1} \frac{y_1}{x_1} \tag{B.50}$$

We next consider the addition and subtraction of two vectors. Two vectors $\mathbf{a} = x\mathbf{i} + y\mathbf{j} + z\mathbf{k}$ and $\mathbf{b} = x'\mathbf{i} + y'\mathbf{j} + z'\mathbf{k}$ can be added or subtracted according to the equations

$$\mathbf{a} \pm \mathbf{b} = (x \pm x')\mathbf{i} + (y \pm y')\mathbf{j} + (z + z')\mathbf{k} \tag{B.51}$$

The addition and subtraction of vectors can also be depicted graphically, as done in Figure B.12.

The multiplication of two vectors can occur in either of two forms. Scalar multiplication of $\mathbf{a}$ and $\mathbf{b}$, also called the dot product of $\mathbf{a}$ and $\mathbf{b}$, is defined by

$$\mathbf{a} \cdot \mathbf{b} = |\mathbf{a}||\mathbf{b}| \cos \alpha \tag{B.52}$$

where α is the angle between the vectors. For $\mathbf{a} = 3\mathbf{i} + 1\mathbf{j} - 2\mathbf{k}$ and $\mathbf{b} = 2\mathbf{i} + -1\mathbf{j} + 4\mathbf{k},$ the vectors in the previous equation can be expanded in terms of their unit vectors:

$$\mathbf{a} \cdot \mathbf{b} = (3\mathbf{i} + 1\mathbf{j} - 2\mathbf{k}) \cdot (2\mathbf{i} + -1\mathbf{j} + 4\mathbf{k})$$
$$= 3\mathbf{i} \cdot 2\mathbf{i} + 3\mathbf{i} \cdot (-1\mathbf{j}) + 3\mathbf{i} \cdot 4\mathbf{k} + 1\mathbf{j} \cdot 2\mathbf{i} + 1\mathbf{j} \cdot (-1\mathbf{j}) + 1\mathbf{j} \cdot 4\mathbf{k} - 2\mathbf{k} \cdot 2\mathbf{i} - 2\mathbf{k} \cdot (-1\mathbf{j}) - 2\mathbf{k} \cdot 4\mathbf{k}$$

However, because $\mathbf{i}$, $\mathbf{j}$, and $\mathbf{k}$ are mutually perpendicular vectors of unit length, $\mathbf{i} \cdot \mathbf{i} = \mathbf{j} \cdot \mathbf{j} = \mathbf{k} \cdot \mathbf{k} = 1$ and $\mathbf{i} \cdot \mathbf{j} = \mathbf{i} \cdot \mathbf{k} = \mathbf{j} \cdot \mathbf{k} = 0$. Therefore, $\mathbf{a} \cdot \mathbf{b} = 3\mathbf{i} \cdot 2\mathbf{i} + 1\mathbf{j} \cdot (-1\mathbf{j}) - 2\mathbf{k} \cdot 4\mathbf{k} = -3$.

The other form in which vectors are multiplied is the vector product, also called the cross product. The vector multiplication of two vectors results in a vector, whereas the scalar multiplication of two vectors results in a scalar. The cross product is defined by the equation

$$\mathbf{a} \times \mathbf{b} = \mathbf{c}|\mathbf{a}||\mathbf{b}| \sin \alpha \tag{B.53}$$

Note that $\mathbf{a} \times \mathbf{b} = -\mathbf{b} \times \mathbf{a}$ as shown in Figure B.13. By contrast, $\mathbf{a} \cdot \mathbf{b} = \mathbf{b} \cdot \mathbf{a}$.

In Equation (B.53), $\mathbf{c}$ is a vector of unit length that is perpendicular to the plane containing $\mathbf{a}$ and $\mathbf{b}$ and has a positive direction found by using the right-hand rule and α is the angle between $\mathbf{a}$ and $\mathbf{b}$.

The cross product between two three-dimensional vectors **a** and **b** is given by

$$\mathbf{a} \times \mathbf{b} = (a_x\mathbf{i} + a_y\mathbf{j} + a_z\mathbf{k}) \times (b_x\mathbf{i} + b_y\mathbf{j} + b_z\mathbf{k}) \tag{B.54}$$

$$= a_x\mathbf{i} \times b_x\mathbf{i} + a_x\mathbf{i} \times b_y\mathbf{j} + a_x\mathbf{i} \times b_z\mathbf{k} + a_y\mathbf{j} \times b_x\mathbf{i} + a_y\mathbf{j} \times b_y\mathbf{j} + a_y\mathbf{j} \times b_z\mathbf{k}$$

$$+\ a_z\mathbf{k} \times b_x\mathbf{i} + a_z\mathbf{k} \times b_y\mathbf{j} + a_z\mathbf{k} \times b_z\mathbf{k}$$

However, using the definition of the cross product in Equation (B.53),

$$\mathbf{i} \times \mathbf{i} = \mathbf{j} \times \mathbf{j} = \mathbf{k} \times \mathbf{k} = 0, \quad \mathbf{i} \times \mathbf{j} = \mathbf{k}, \quad \mathbf{i} \times \mathbf{k} = -\mathbf{j}$$

$$\mathbf{j} \times \mathbf{i} = -\mathbf{k}, \quad \mathbf{j} \times \mathbf{k} = \mathbf{i}, \quad \mathbf{k} \times \mathbf{i} = \mathbf{j}, \quad \mathbf{k} \times \mathbf{j} = -\mathbf{i}$$

Therefore, Equation (B.54) simplifies to

$$\mathbf{a} \times \mathbf{b} = (a_y b_z - a_z b_y)\mathbf{i} + (a_z b_x - a_x b_z)\mathbf{j} + (a_x b_y - a_y b_x)\mathbf{k} \tag{B.55}$$

As we will see in Section B.7, there is a simple way to calculate cross products using determinants.

The angular momentum $\mathbf{l} = \mathbf{r} \times \mathbf{p}$ is of particular interest in quantum chemistry, because s, p, and d electrons are distinguished by their orbital angular momentum. For the example of the particle rotating on a ring depicted at the beginning of this section, the angular momentum vector is pointing upward in a direction perpendicular to the plane of the page. In analogy to Equation (B.55),

$$\mathbf{l} = \mathbf{r} \times \mathbf{p} = (yp_z - zp_y)\mathbf{i} + (zp_x - xp_z)\mathbf{j} + (xp_y - yp_x)\mathbf{k}$$

B.6 Partial Derivatives

In this section, we discuss the differential calculus of functions that depend on several independent variables. Consider the volume of a cylinder of radius r and height h, for which

$$V = f(r,h) = \pi r^2 h \tag{B.56}$$

where V can be written as a function of the two variables r and h. The change in V with a change in r or h is given by the partial derivatives

$$\left(\frac{\partial V}{\partial r}\right)_h = \lim_{\Delta r \to 0} \frac{V(r + \Delta r, h) - V(r,h)}{\Delta r} = 2\pi rh \tag{B.57}$$

$$\left(\frac{\partial V}{\partial h}\right)_r = \lim_{\Delta h \to 0} \frac{V(r, h + \Delta h) - V(r,h)}{\Delta h} = \pi r^2$$

The subscript h in $(\partial V/\partial r)_h$ reminds us that h is being held constant in the differentiation. The partial derivatives in Equation (B.57) allow us to determine how a function changes when one of the variables changes. How does V change if the values of both variables change? In this case, V changes to $V + dV$ where

$$dV = \left(\frac{\partial V}{\partial r}\right)_h dr + \left(\frac{\partial V}{\partial h}\right)_r dh \tag{B.58}$$

These partial derivatives are useful in calculating the error in the function that results from errors in measurements of the individual variables. For example, the relative error in the volume of the cylinder is given by

$$\frac{dV}{V} = \frac{1}{V}\left[\left(\frac{\partial V}{\partial r}\right)_h dr + \left(\frac{\partial V}{\partial h}\right)_r dh\right] = \frac{1}{\pi r^2 h}[2\pi rh\, dr + \pi r^2 dh] = \frac{2dr}{r} + \frac{dh}{h}$$

This equation shows that a given relative error in r generates twice the relative error in V as a relative error in h of the same size.

We can also take second or higher derivatives with respect to either variable. The mixed second partial derivatives are of particular interest. The mixed partial derivatives of V are given by

$$\left(\frac{\partial}{\partial h}\left(\frac{\partial V}{\partial r}\right)_h\right)_r = \left(\partial\left(\frac{\partial[\pi r^2 h]}{\partial r}\right)_h \bigg/ \partial h\right)_r = \left(\frac{\partial[2\pi r h]}{\partial h}\right)_r = 2\pi r \qquad \text{(B.59)}$$

$$\left(\frac{\partial}{\partial r}\left(\frac{\partial V}{\partial h}\right)_r\right)_h = \left(\partial\left(\frac{\partial[\pi r^2 h]}{\partial h}\right)_r \bigg/ \partial r\right)_h = \left(\frac{\partial[\pi r^2]}{\partial r}\right)_h = 2\pi r$$

For the specific case of V, the order in which the function is differentiated does not affect the outcome. Such a function is called a state function. Therefore, for any state function f of the variables x and y,

$$\left(\frac{\partial}{\partial y}\left(\frac{\partial f(x,y)}{\partial x}\right)_y\right)_x = \left(\frac{\partial}{\partial x}\left(\frac{\partial f(x,y)}{\partial y}\right)_x\right)_y \qquad \text{(B.60)}$$

Because Equation (B.60) is satisfied by all state functions, f, it can be used to determine if a function f is a state function.

We demonstrate how to calculate the partial derivatives

$$\left(\frac{\partial f}{\partial x}\right)_y, \left(\frac{\partial f}{\partial y}\right)_x, \left(\frac{\partial^2 f}{\partial x^2}\right)_y, \left(\frac{\partial^2 f}{\partial y^2}\right)_x, \left(\partial\left(\frac{\partial f}{\partial x}\right)_y \bigg/ \partial y\right)_x, \text{ and } \left(\partial\left(\frac{\partial f}{\partial y}\right)_x \bigg/ \partial x\right)_y$$

for the function $f(x,y) = ye^{ax} + xy\cos x + y\ln xy$, where a is a real constant:

$$\left(\frac{\partial f}{\partial x}\right)_y = aye^{ax} + \frac{y}{x} + y\cos x - xy\sin x, \left(\frac{\partial f}{\partial y}\right)_x = 1 + e^{ax} + x\cos x + \ln xy$$

$$\left(\frac{\partial^2 f}{\partial x^2}\right)_y = a^2 ye^{ax} - \frac{y}{x^2} - 2y\sin x - xy\cos x, \left(\frac{\partial^2 f}{\partial y^2}\right)_x = \frac{1}{y}$$

$$\left(\partial\left(\frac{\partial f}{\partial x}\right)_y \bigg/ \partial y\right)_x = ae^{ax} + \frac{1}{x} + \cos x - x\sin x, \left(\partial\left(\frac{\partial f}{\partial y}\right)_x \bigg/ \partial x\right)_y = ae^{ax} + \frac{1}{x} + \cos x - x\sin x$$

Because we have shown that

$$\left(\partial\left(\frac{\partial f}{\partial x}\right)_y \bigg/ \partial y\right)_x = \left(\partial\left(\frac{\partial f}{\partial y}\right)_x \bigg/ \partial x\right)_y$$

$f(x,y)$ is a state function of the variables x and y.

Whereas the partial derivatives tell us how the function changes if the value of one of the variables is changed, the total differential tells us how the function changes when all of the variables are changed simultaneously. The total differential of the function $f(x,y)$ is defined by

$$df = \left(\frac{\partial f}{\partial x}\right)_y dx + \left(\frac{\partial f}{\partial y}\right)_x dy \qquad \text{(B.61)}$$

The total differential of the function used earlier is calculated as follows:

$$df = \left(aye^{ax} + \frac{y}{x} + y\cos x - xy\sin x\right)dx + (1 + e^{ax} + x\cos x + \log xy)dy$$

Two other important results from multivariate differential calculus are used frequently. For a function $z = f(x, y)$, which can be rearranged to $x = g(y, z)$ or $y = h(x, z)$,

$$\left(\frac{\partial x}{\partial y}\right)_z = \frac{1}{\left(\dfrac{\partial y}{\partial x}\right)_z} \tag{B.62}$$

The other important result that is used frequently is the cyclic rule:

$$\left(\frac{\partial x}{\partial y}\right)_z \left(\frac{\partial y}{\partial z}\right)_x \left(\frac{\partial z}{\partial x}\right)_y = -1 \tag{B.63}$$

Consider an additional example of calculating partial derivatives for a function encountered in quantum mechanics. The Schrödinger equation for the hydrogen atom takes the form

$$-\frac{\hbar^2}{2\mu}\left[\frac{1}{r^2}\frac{\partial}{\partial r}\left(r^2 \frac{\partial \psi(r,\theta,\phi)}{\partial r}\right) + \frac{1}{r^2 \sin\theta}\frac{\partial}{\partial\theta}\left(\sin\theta \frac{\partial \psi(r,\theta,\phi)}{\partial\theta}\right) + \frac{1}{r^2 \sin\theta}\frac{\partial^2 \psi(r,\theta,\phi)}{\partial\phi^2}\right]$$

$$-\frac{e^2}{4\pi\varepsilon_0 r}\psi(r,\theta,\phi) = E\psi(r,\theta,\phi)$$

Note that each of the first three terms on the left side of the equation involves partial differentiation with respect to one of the variables r, θ, and ϕ in turn. Two of the solutions to this differential equation are $(r/a_0)e^{-r/2a_0}\sin\theta e^{\pm i\phi}$. Each of these terms is evaluated separately to demonstrate how partial derivatives are taken in quantum mechanics. Although this is a more complex exercise than those presented earlier, it provides good practice in partial differentiation. For the first term, the partial derivative is taken with respect to r:

$$-\frac{\hbar^2}{2\mu}\frac{1}{\sqrt{64\pi}}\left(\frac{1}{a_0}\right)^{3/2}\left[\frac{1}{r^2}\frac{\partial}{\partial r}\left(r^2 \frac{\partial\left(\dfrac{r}{a_0}e^{-r/2a_0}\sin\theta e^{\pm i\phi}\right)}{\partial r}\right)\right]$$

$$= -\frac{\hbar^2}{2\mu}\frac{1}{\sqrt{64\pi}}\left(\frac{1}{a_0}\right)^{3/2}\sin\theta e^{\pm i\phi}\left[\frac{1}{r^2}\frac{\partial}{\partial r}\left(r^2 \frac{\partial\left(\dfrac{r}{a_0}e^{-r/2a_0}\right)}{\partial r}\right)\right]$$

$$= -\frac{\hbar^2}{2\mu}\frac{1}{\sqrt{64\pi}}\left(\frac{1}{a_0}\right)^{3/2}\sin\theta e^{\pm i\phi}\left[\frac{1}{r^2}\frac{\partial}{\partial r}\left(r^2\left(\frac{1}{a_0}e^{-r/2a_0} - \left(r/2a_0^2\right)e^{-r/2a_0}\right)\right)\right]$$

$$= -\frac{\hbar^2}{2\mu}\frac{1}{\sqrt{64\pi}}\left(\frac{1}{a_0}\right)^{3/2}\sin\theta e^{\pm i\phi}\frac{1}{r^2}\left[\begin{array}{c}-r^2\dfrac{e^{-r/2a_0}}{a_0^2} + r^3\dfrac{e^{-r/2a_0}}{4a_0^3} + 2r\dfrac{e^{-r/2a_0}}{a_0}\\[2mm] -2r^2\dfrac{e^{-r/2a_0}}{a_0^2}\end{array}\right]$$

$$= -\frac{\hbar^2}{2\mu}\frac{1}{\sqrt{64\pi}}\left(\frac{1}{a_0}\right)^{3/2}\sin\theta e^{\pm i\phi}e^{-r/2a_0}\frac{(8a_0^2 - 8a_0 r + r^2)}{4a_0^3 r}$$

Partial differentiation with respect to θ is easier, because the terms that depend on r and ϕ are constant:

$$-\frac{\hbar^2}{2\mu}\frac{1}{\sqrt{64\pi}}\left(\frac{1}{a_0}\right)^{3/2}\left[\frac{1}{r^2 \sin\theta}\frac{\partial}{\partial\theta}\left(\sin\theta \frac{\partial\left(\dfrac{r}{a_0}e^{-r/2a_0}\sin\theta e^{\pm i\phi}\right)}{\partial\theta}\right)\right]$$

$$= -\frac{\hbar^2}{2\mu}\frac{1}{\sqrt{64\pi}}\left(\frac{1}{a_0}\right)^{3/2}\frac{r}{a_0}e^{-r/2a_0}e^{\pm i\phi}\left[\frac{1}{r^2 \sin\theta}\frac{\partial}{\partial\theta}\left(\sin\theta \frac{\partial(\sin\theta)}{\partial\theta}\right)\right]$$

$$= -\frac{\hbar^2}{2\mu}\frac{1}{\sqrt{64\pi}}\left(\frac{1}{a_0}\right)^{3/2}\frac{r}{a_0}e^{-r/2a_0}e^{\pm i\phi}\left[\frac{1}{r^2\sin\theta}\frac{\partial}{\partial\theta}(\sin\theta\cos\theta)\right]$$

$$= -\frac{\hbar^2}{2\mu}\frac{1}{\sqrt{64\pi}}\left(\frac{1}{a_0}\right)^{3/2}\frac{r}{a_0}e^{-r/2a_0}e^{\pm i\phi}\left[\frac{1}{r^2\sin\theta}(\cos^2\theta-\sin^2\theta)\right]$$

Partial differentiation with respect to ϕ is also not difficult, because the terms that depend on r and θ are constant:

$$-\frac{\hbar^2}{2\mu}\frac{1}{\sqrt{64\pi}}\left(\frac{1}{a_0}\right)^{3/2}\left[\frac{1}{r^2\sin\theta}\frac{\partial^2\frac{r}{a_0}e^{-r/2a_0}\sin\theta\,e^{\pm i\phi}}{\partial\phi^2}\right]$$

$$= -\frac{\hbar^2}{2\mu}\frac{1}{\sqrt{64\pi}}\left(\frac{1}{a_0}\right)^{3/2}\frac{r}{a_0}e^{-r/2a_0}\frac{1}{r^2}\left[\frac{\partial^2 e^{\pm i\phi}}{\partial\phi^2}\right]$$

$$= \frac{\hbar^2}{2\mu}\frac{1}{\sqrt{64\pi}}\left(\frac{1}{a_0}\right)^{3/2}\frac{r}{a_0}e^{-r/2a_0}\frac{1}{r^2}\left[e^{\pm i\phi}\right]$$

B.7 Working with Determinants

A determinant of nth order is a square $n \times n$ array of numbers symbolically enclosed by vertical lines. A fifth-order determinant is shown here with the conventional indexing of the elements of the array:

$$\begin{vmatrix} a_{11} & a_{12} & a_{13} & a_{14} & a_{15} \\ a_{21} & a_{22} & a_{23} & a_{24} & a_{25} \\ a_{31} & a_{32} & a_{33} & a_{34} & a_{35} \\ a_{41} & a_{42} & a_{43} & a_{44} & a_{45} \\ a_{51} & a_{52} & a_{53} & a_{54} & a_{55} \end{vmatrix} \tag{B.64}$$

A 2×2 determinant has a value that is defined in Equation (B.65). It is obtained by multiplying the elements in the diagonal connected by a line with a negative slope and subtracting from this the product of the elements in the diagonal connected by a line with a positive slope.

$$\begin{vmatrix} a_{11} & a_{12} \\ a_{21} & a_{22} \end{vmatrix} = a_{11}a_{22} - a_{12}a_{21} \tag{B.65}$$

The value of a higher order determinant is obtained by expanding the determinant in terms of determinants of lower order. This is done using the method of cofactors. We illustrate the use of method of cofactors by reducing a 3×3 determinant to a sum of 2×2 determinants. Any row or column can be used in the reduction process. We use the first row of the determinant in the reduction. The recipe is spelled out in this equation:

$$\begin{vmatrix} a_{11} & a_{12} & a_{13} \\ a_{21} & a_{22} & a_{23} \\ a_{31} & a_{32} & a_{33} \end{vmatrix} = (-1)^{1+1}a_{11}\begin{vmatrix} a_{22} & a_{23} \\ a_{32} & a_{33} \end{vmatrix} + (-1)^{1+2}a_{12}\begin{vmatrix} a_{21} & a_{23} \\ a_{31} & a_{33} \end{vmatrix} \tag{B.66}$$

$$+ (-1)^{1+3}a_{13}\begin{vmatrix} a_{21} & a_{22} \\ a_{31} & a_{32} \end{vmatrix}$$

$$= a_{11}\begin{vmatrix} a_{22} & a_{23} \\ a_{32} & a_{33} \end{vmatrix} - a_{12}\begin{vmatrix} a_{21} & a_{23} \\ a_{31} & a_{33} \end{vmatrix} + a_{13}\begin{vmatrix} a_{21} & a_{22} \\ a_{31} & a_{32} \end{vmatrix}$$

Each term in the sum results from the product of one of the three elements of the first row, $(-1)^{m+n}$, where m and n are the indices of the row and column designating the element, respectively, and the 2×2 determinant obtained by omitting the entire row and column to which the element used in the reduction belongs. The product $(-1)^{m+n}$ and the 2×2 determinant are called the cofactor of the element used in the reduction. For example, the value of the following 3×3 determinant is found using the cofactors of the second row:

$$\begin{vmatrix} 1 & 3 & 4 \\ 2 & -1 & 6 \\ -1 & 7 & 5 \end{vmatrix} = (-1)^{2+1}2\begin{vmatrix} 3 & 4 \\ 7 & 5 \end{vmatrix} + (-1)^{2+2}(-1)\begin{vmatrix} 1 & 4 \\ -1 & 5 \end{vmatrix} + (-1)^{2+3}6\begin{vmatrix} 1 & 3 \\ -1 & 7 \end{vmatrix}$$

$$= -1 \times 2 \times (-13) + 1 \times (-1) \times 9 + (-1) \times 6 \times 10 = -43$$

If the initial determinant is of a higher order than 3, multiple sequential reductions as outlined earlier will reduce it in order by one in each step until a sum of 2×2 determinants is obtained.

The main usefulness for determinants is in solving a system of linear equations. Such a system of equations is obtained in evaluating the energies of a set of molecular orbitals obtained by combining a set of atomic orbitals. Before illustrating this method, we list some important properties of determinants that we will need in solving a set of simultaneous equations.

Property I   The value of a determinant is not altered if each row in turn is made into a column or vice versa as long as the original order is kept. By this we mean that the nth row becomes the nth column. This property can be illustrated using 2×2 and 3×3 determinants:

$$\begin{vmatrix} 2 & 1 \\ 3 & -1 \end{vmatrix} = \begin{vmatrix} 2 & 3 \\ 1 & -1 \end{vmatrix} = -5 \text{ and } \begin{vmatrix} 1 & 3 & 4 \\ 2 & -1 & 6 \\ -1 & 7 & 5 \end{vmatrix} = \begin{vmatrix} 1 & 2 & -1 \\ 3 & -1 & 7 \\ 4 & 6 & 5 \end{vmatrix} = -43$$

Property II   If any two rows or columns are interchanged, the sign of the value of the determinant is changed. For example,

$$\begin{vmatrix} 2 & 1 \\ 3 & -1 \end{vmatrix} = -5, \text{ but } \begin{vmatrix} 1 & 2 \\ -1 & 3 \end{vmatrix} = +5 \text{ and } \begin{vmatrix} 1 & 3 & 4 \\ 2 & -1 & 6 \\ -1 & 7 & 5 \end{vmatrix} = -43, \text{ but } \begin{vmatrix} 2 & -1 & 6 \\ 1 & 3 & 4 \\ -1 & 7 & 5 \end{vmatrix} = +43$$

Property III   If two rows or columns of a determinant are identical, the value of the determinant is zero. For example,

$$\begin{vmatrix} 2 & 1 \\ 2 & 1 \end{vmatrix} = 2 - 2 = 0 \text{ and } \begin{vmatrix} 1 & 1 & 4 \\ 2 & 2 & 6 \\ -1 & -1 & 5 \end{vmatrix} = (-1)^{2+1}2\begin{vmatrix} 1 & 4 \\ -1 & 5 \end{vmatrix} + (-1)^{2+2}2\begin{vmatrix} 1 & 4 \\ -1 & 5 \end{vmatrix}$$

$$+ (-1)^{2+3}6\begin{vmatrix} 1 & 1 \\ -1 & -1 \end{vmatrix}$$

$$= -1 \times 2 \times 9 + 1 \times 2 \times 9 + (-1) \times 6 \times 0 = 0$$

Property IV   If each element of a row or column is multiplied by a constant, the value of the determinant is multiplied by that constant. For example,

$$\begin{vmatrix} 2 & 1 \\ 3 & -1 \end{vmatrix} = -5 \text{ and } \begin{vmatrix} 8 & 4 \\ 3 & -1 \end{vmatrix} = -20 \text{ and } \begin{vmatrix} 1 & 2 & -1 \\ 3 & -1 & 7 \\ 4 & 6 & 5 \end{vmatrix} = -43 \text{ and } \begin{vmatrix} 1 & 3\sqrt{2} & 4 \\ 2 & -\sqrt{2} & 6 \\ -1 & 7\sqrt{2} & 5 \end{vmatrix} = -43\sqrt{2}$$

Property V   The value of a determinant is unchanged if a row or column multiplied by an arbitrary number is added to another row or column. For example,

$$\begin{vmatrix} 2 & 1 \\ 3 & -1 \end{vmatrix} = \begin{vmatrix} 2+1 & 1 \\ 3-1 & -1 \end{vmatrix} = \begin{vmatrix} 2 & 1 \\ 3 & -1 \end{vmatrix} = -5 \text{ and}$$

$$\begin{vmatrix} 1 & 3 & 4 \\ 2 & -1 & 6 \\ -1 & 7 & 5 \end{vmatrix} = \begin{vmatrix} 1 & 3 & 4 \\ 2-1 & -1+7 & 6+5 \\ -1 & 7 & 5 \end{vmatrix} = \begin{vmatrix} 1 & 3 & 4 \\ 2 & -1 & 6 \\ -1 & 7 & 5 \end{vmatrix} = -43$$

How are determinants useful? This question can be answered by illustrating how determinants can be used to solve a set of linear equations:

$$\begin{aligned} x + y + z &= 10 \\ 3x + 4y - z &= 12 \\ -x + 2y + 5z &= 26 \end{aligned} \tag{B.67}$$

This set of equations is solved by first constructing the 3×3 determinant that is the array of the coefficients of x, y, and z:

$$\mathbf{D}_{coefficients} = \begin{vmatrix} 1 & 1 & 1 \\ 3 & 4 & -1 \\ -1 & 2 & 5 \end{vmatrix} \tag{B.68}$$

Now imagine that we multiply the first column by x. This changes the value of the determinant as stated in Property IV:

$$\begin{vmatrix} 1x & 1 & 1 \\ 3x & 4 & -1 \\ -1x & 2 & 5 \end{vmatrix} = x\mathbf{D}_{coefficients} \tag{B.69}$$

We next add to the first column of $x\mathbf{D}_{coefficients}$ the second column of $\mathbf{D}_{coefficients}$ multiplied by y, and the third column multiplied by z. According to Properties IV and V, the value of the determinant is unchanged. Therefore,

$$\mathbf{D}_{c1} = \begin{vmatrix} 1 & 1 & 1 \\ 3 & 4 & -1 \\ -1 & 2 & 5 \end{vmatrix} = \begin{vmatrix} x+y+z & 1 & 1 \\ 3x+4y-z & 4 & -1 \\ -x+2y+5z & 2 & 5 \end{vmatrix} = \begin{vmatrix} 10 & 1 & 1 \\ 12 & 4 & -1 \\ 26 & 2 & 5 \end{vmatrix} = x\mathbf{D}_{coefficients} \tag{B.70}$$

To obtain the third determinant in the previous equation, the individual equations in Equation (B.67) are used to substitute the constants for the algebraic expression in the preceding determinants. From the previous equation, we conclude that

$$x = \frac{\mathbf{D}_{c1}}{\mathbf{D}_{coefficients}} = \frac{\begin{vmatrix} 10 & 1 & 1 \\ 12 & 4 & -1 \\ 26 & 2 & 5 \end{vmatrix}}{\begin{vmatrix} 1 & 1 & 1 \\ 3 & 4 & -1 \\ -1 & 2 & 5 \end{vmatrix}} = 3$$

To determine y and z, the exact same procedure can be followed, but we substitute instead in columns 2 and 3, respectively. The first step in each case is to multiply all elements of the second (third) row by $y(z)$. If we do so, we obtain the determinants $\mathbf{D}_{c2}$ and $\mathbf{D}_{c3}$:

$$\mathbf{D}_{c2} = \begin{vmatrix} 1 & 10 & 1 \\ 3 & 12 & -1 \\ -1 & 26 & 5 \end{vmatrix} \text{ and } \mathbf{D}_{c3} = \begin{vmatrix} 1 & 1 & 10 \\ 3 & 4 & 12 \\ -1 & 2 & 26 \end{vmatrix}$$

and we conclude that

$$y = \frac{\mathbf{D}_{c2}}{\mathbf{D}_{coefficients}} = \frac{\begin{vmatrix} 1 & 10 & 1 \\ 3 & 12 & -1 \\ -1 & 26 & 5 \end{vmatrix}}{\begin{vmatrix} 1 & 1 & 1 \\ 3 & 4 & -1 \\ -1 & 2 & 5 \end{vmatrix}} = 2 \quad \text{and} \quad z = \frac{\mathbf{D}_{c3}}{\mathbf{D}_{coefficients}} = \frac{\begin{vmatrix} 1 & 1 & 10 \\ 3 & 4 & 12 \\ -1 & 2 & 26 \end{vmatrix}}{\begin{vmatrix} 1 & 1 & 1 \\ 3 & 4 & -1 \\ -1 & 2 & 5 \end{vmatrix}} = 5$$

This method of solving a set of simultaneous linear equations is known as Cramer's method. If the constants in the set of equations are all zero, as in Equations B.71a and B.71b,

$$x + y + z = 0$$
$$3x + 4y - z = 0 \tag{B.71a}$$
$$-x + 2y + 5z = 0$$

$$3x - y + 2z = 0$$
$$-x + y - z = 0 \tag{B.71b}$$
$$(1+\sqrt{2})x + (1-\sqrt{2})y + \sqrt{2z} = 0$$

the determinants $\mathbf{D}_{c1}$, $\mathbf{D}_{c2}$, and $\mathbf{D}_{c3}$ all have the value zero. An obvious set of solutions is $x = 0$, $y = 0$, and $z = 0$. For most problems in physics and chemistry, this set of solutions is not physically meaningful and is referred to as the set of trivial solutions. A set of nontrivial solutions only exists if the equation $\mathbf{D}_{coefficients} = 0$ is satisfied. There is no nontrivial solution to the set of equation B.71a because $\mathbf{D}_{coefficients} \neq 0$. There is a set of nontrivial colutions to the set of equations B71.b, because $\mathbf{D}_{coefficients} = 0$ in this case.

Determinants offer a convenient way to calculate the cross product of two vectors, as discussed in Section B.5. The following recipe is used:

$$\mathbf{a} \times \mathbf{b} = \begin{vmatrix} \mathbf{i} & \mathbf{j} & \mathbf{k} \\ a_x & a_y & a_z \\ b_x & b_y & b_z \end{vmatrix} = \mathbf{i} \begin{vmatrix} a_y & a_z \\ b_y & b_z \end{vmatrix} - \mathbf{j} \begin{vmatrix} a_x & a_z \\ b_x & b_z \end{vmatrix} + \mathbf{k} \begin{vmatrix} a_x & a_y \\ b_x & b_y \end{vmatrix} \tag{B.72}$$

$$= (a_y b_z - a_z b_y)\mathbf{i} + (a_z b_x - a_x b_z)\mathbf{j} + (a_x b_y - a_y b_x)\mathbf{k}$$

Note that by referring to Property II, you can show that $\mathbf{b} \times \mathbf{a} = -\mathbf{a} \times \mathbf{b}$.

B.8 Working with Matrices

Physical chemists find widespread use for matrices. Matrices can be used to represent symmetry operations in the application of group theory to problems concerning molecular symmetry. They can also be used to obtain the energies of molecular orbitals formed through the linear combination of atomic orbitals. We next illustrate the use of matrices for representing the rotation operation that is frequently encountered in molecular symmetry considerations.

Consider the rotation of a three-dimensional vector about the z axis. Because the z component of the vector is unaffected by this operation, we need only consider the effect of the rotation operation on the two-dimensional vector formed by the projection of the three-dimensional vector on the x–y plane. The transformation can be represented by $(x_1, y_1, z_1) \rightarrow (x_2, y_2, z_1)$. The effect of the operation on the x and y components of the vector is shown in Figure B.14.

Next, relationships are derived among (x_1, y_1, z_1), (x_2, y_2, z_1), the magnitude of the radius vector r, and the angles α and β, based on the preceding figure. The magnitude of the radius vector r is

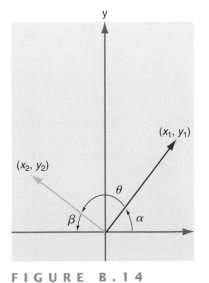

FIGURE B.14

$$r = \sqrt{x_1^2 + y_1^2 + z_1^2} = \sqrt{x_2^2 + y_2^2 + z_1^2} \tag{B.73}$$

Although the values of x and y change in the rotation, r is unaffected by this operation. The relationships between x, y, r, α, and β are given by

$$\theta = 180° - \alpha - \beta \qquad (B.74)$$

$$x_1 = r \cos \alpha, \qquad y_1 = r \sin \alpha$$

$$x_2 = -r \cos \beta, \qquad y_1 = r \sin \beta$$

In the following discussion, these identities are used:

$$\cos(\alpha \pm \beta) = \cos \alpha \cos \beta \mp \sin \alpha \sin \beta \qquad (B.75)$$

$$\sin(\alpha \pm \beta) = \sin \alpha \cos \beta \pm \cos \alpha \sin \beta$$

From Figure B.14, the following relationship between x_2 and x_1 and y_1 can be derived using the identities of Equation (B.75):

$$x_2 = -r \cos \beta = -r \cos(180° - \alpha - \theta) \qquad (B.76)$$

$$= r \sin 180° \sin(-\theta - \alpha) - r \cos 180° \cos(-\theta - \alpha)$$

$$= r \cos(-\theta - \alpha) = r \cos(\theta + \alpha) = r \cos \theta \cos \alpha - r \sin \theta \sin \alpha$$

$$= x_1 \cos \theta - y_1 \sin \theta$$

Using the same procedure, the following relationship between y_2 and x_1 and y_1 can be derived:

$$y_2 = x_1 \sin \theta + y_1 \cos \theta \qquad (B.77)$$

Next, these results are combined to write the following equations relating x_2, y_2, and z_2 to x_1, y_1, and z_1:

$$x_2 = x_1 \cos \theta - y_1 \sin \theta \qquad (B.78)$$

$$y_2 = x_1 \sin \theta + y_1 \cos \theta$$

$$z_2 = 0x_1 + 0y_1 + z_1$$

At this point, the concept of a matrix can be introduced. An $n \times m$ matrix is an array of numbers, functions, or operators that can undergo mathematical operations such as addition and multiplication with one another. The operation of interest to us in considering rotation about the z axis is matrix multiplication. We illustrate how matrices, which are designated in bold script, such as **A,** are multiplied using 2×2 matrices as an example.

$$\mathbf{AB} = \begin{pmatrix} a_{11} & a_{12} \\ a_{21} & a_{22} \end{pmatrix} \begin{pmatrix} b_{11} & b_{12} \\ b_{21} & b_{22} \end{pmatrix} = \begin{pmatrix} a_{11}b_{11} + a_{12}b_{21} & a_{11}b_{12} + a_{12}b_{22} \\ a_{21}b_{11} + a_{22}b_{21} & a_{21}b_{12} + a_{22}b_{22} \end{pmatrix} \qquad (B.79)$$

Using numerical examples,

$$\begin{pmatrix} 2 & 1 \\ -3 & 4 \end{pmatrix} \begin{pmatrix} 1 & 6 \\ 2 & -1 \end{pmatrix} = \begin{pmatrix} 4 & 11 \\ 5 & -22 \end{pmatrix} \quad \text{and} \quad \begin{pmatrix} 1 & 6 \\ 2 & -1 \end{pmatrix} \begin{pmatrix} 1 \\ -1 \end{pmatrix} = \begin{pmatrix} -5 \\ 3 \end{pmatrix}$$

Now consider the initial and final coordinates (x_1, y_1, z_1) and (x_2, y_2, z_1) as 3×1 matrices (x_1, y_1, z_1) and (x_2, y_2, z_2). In that case, the set of simultaneous equations of Equation (B.78) can be written as

$$\begin{pmatrix} x_2 \\ y_2 \\ z_2 \end{pmatrix} = \begin{pmatrix} \cos \theta & -\sin \theta & 0 \\ \sin \theta & \cos \theta & 0 \\ 0 & 0 & 1 \end{pmatrix} \begin{pmatrix} x_1 \\ y_1 \\ z_1 \end{pmatrix} \qquad (B.80)$$

We see that we can represent the operator for rotation about the z axis, R_z, as the following 3×3 matrix:

$$\mathbf{R}_z = \begin{pmatrix} \cos \theta & -\sin \theta & 0 \\ \sin \theta & \cos \theta & 0 \\ 0 & 0 & 1 \end{pmatrix} \qquad (B.81)$$

The rotation operator for 180° and 120° rotation can be obtained by evaluating the sine and cosine functions at the appropriate values of θ. These rotation operators have the form

$$\begin{pmatrix} -1 & 0 & 0 \\ 0 & -1 & 0 \\ 0 & 0 & 1 \end{pmatrix} \quad \text{and} \quad \begin{pmatrix} 1/2 & -\sqrt{3}/2 & 0 \\ \sqrt{3}/2 & 1/2 & 0 \\ 0 & 0 & 1 \end{pmatrix}, \quad \text{respectively} \qquad \text{(B.82)}$$

One special matrix, the identity matrix designated **I**, deserves additional mention. The identity matrix corresponds to an operation in which nothing is changed. The matrix that corresponds to the transformation $(x_1, y_1, z_1) \rightarrow (x_1, y_1, z_1)$ expressed in equation form as

$$\begin{aligned} x_2 &= x_1 + 0y_1 + 0z_1 \\ y_2 &= 0x_1 + y_1 + 0z_1 \\ z_2 &= 0x_1 + 0y_1 + z_1 \end{aligned} \qquad \text{(B.83)}$$

is the identity matrix

$$\mathbf{I} = \begin{pmatrix} 1 & 0 & 0 \\ 0 & 1 & 0 \\ 0 & 0 & 1 \end{pmatrix}$$

The identity matrix is an example of a diagonal matrix. It has this name because only the diagonal elements are nonzero. In the identity matrix of order $n \times n$, all diagonal elements have the value one.

The operation that results from the sequential operation of two individual operations represented by matrices **A** and **B** is the products of the matrices: $\mathbf{C} = \mathbf{AB}$. An interesting case illustrating this relationship is counterclockwise rotation through an angle θ followed by clockwise rotation through the same angle, which corresponds to rotation by $-\theta$. Because $\cos(-\theta) = \cos\theta$ and $\sin\theta = -\sin\theta$, the rotation matrix for $-\theta$ must be

$$\mathbf{R}_{-z} = \begin{pmatrix} \cos\theta & \sin\theta & 0 \\ -\sin\theta & \cos\theta & 0 \\ 0 & 0 & 1 \end{pmatrix} \qquad \text{(B.84)}$$

Because the sequential operations leave the vector unchanged, it must be the case that $\mathbf{R}_z\mathbf{R}_{-z} = \mathbf{R}_{-z}\mathbf{R}_z = \mathbf{I}$. We verify here that the first of these relations is obeyed:

$$\mathbf{R}_{-z} = \begin{pmatrix} \cos\theta & -\sin\theta & 0 \\ \sin\theta & \cos\theta & 0 \\ 0 & 0 & 1 \end{pmatrix}\begin{pmatrix} \cos\theta & \sin\theta & 0 \\ -\sin\theta & \cos\theta & 0 \\ 0 & 0 & 1 \end{pmatrix} \qquad \text{(B.85)}$$

$$= \begin{pmatrix} \cos^2\theta + \sin^2\theta + 0 & \sin\theta\cos\theta - \sin\theta\cos\theta + 0 & 0 \\ \sin\theta\cos\theta - \sin\theta\cos\theta + 0 & \cos^2\theta + \sin^2\theta + 0 & 0 \\ 0 & 0 & 1 \end{pmatrix} = \begin{pmatrix} 1 & 0 & 0 \\ 0 & 1 & 0 \\ 0 & 0 & 1 \end{pmatrix}$$

Any matrix **B** that satisfies the relationship $\mathbf{AB} = \mathbf{BA} = \mathbf{I}$ is called the inverse matrix of **A** and is designated $\mathbf{A}^{-1}$. Inverse matrices play an important role in finding the energies of a set of molecular orbitals that is a linear combination of atomic orbitals.

Answers to Selected End-of-Chapter Problems

Numerical answers to problems are included here. Complete solutions to selected problems can be found in the *Student's Solutions Manual*.

Chapter 1

P1.1   723 K

P1.2   a. $P_{H_2} = 6.24 \times 10^5 \, \text{Pa}; \; P_{O_2} = 3.90 \times 10^4 \, \text{Pa}$

          $P_{total} = 6.57 \times 10^5 \, \text{Pa}$

          mol % H_2 = 94.1%; mol % O_2 = 5.9%

     b. $P_{N_2} = 4.45 \times 10^4 \, \text{Pa}; \; P_{O_2} = 3.90 \times 10^4 \, \text{Pa}$

          $P_{total} = 8.35 \times 10^4 \, \text{Pa}$

          mol % N_2 = 53.3%; mol % O_2 = 46.7%

     c. $P_{NH_3} = 7.32 \times 10^4 \, \text{Pa}; \; P_{CH_4} = 7.77 \times 10^4 \, \text{Pa}$

          $P_{total} = 1.51 \times 10^5 \, \text{Pa}$

          mol % NH_3 = 48.5%; mol % O_2 = 51.5%

P1.3   0.08200 atm mol^{-1} °C^{-1}; 280.2° C

P1.4 1.26×10^3 L

P1.5   26.8 L

P1.6   a.   68.8%, 18.5%, and 12.0% for N_2, O_2, Ar, and H_2O, respectively

     b.   12.2 L

     c.   0.992

P1.7   a. 2.88×10^{-2} bar

     b. $x_{O_2} = 0.0179$, $x_{N_2} = 0.803$, $x_{CO} = 0.178$, $P_{O_2} = 5.16 \times 10^{-4}$ bar,

          $P_{N_2} = 2.31 \times 10^{-2}$ bar, $P_{CO} = 5.10 \times 10^{-3}$ bar

P1.8   1.50 L

P1.9 4.84×10^5 Pa

P1.10 $x_{CO_2} = 0.176; \; x_{H_2O} = 0.235; \; x_{O_2} = 0.588$

P1.11   158 amu

P1.12 $x_{O_2} = 0.20; \; x_{H_2} = 0.80$

P1.13   59.9%

P1.14   54

P1.15 $x_{CO_2} = 0.028$; $x_{N_2} = 0.972$

P1.16 8.34×10^4 Pa

P1.17 17.3 bar

P1.18 41.6 bar

P1.20 $x_{H_2}^\circ = 0.103$; $x_{O_2}^\circ = 0.897$

Chapter 2

P2.1 a. -4.00×10^3 J

b. -8.22×10^3 J

P2.2 $w = 2.03 \times 10^4$ J; $\Delta U = 0$ and $\Delta H = 0$;
$q = -2.03 \times 10^4$ J

P2.3 312 K

P2.4 $\Delta U = -935$ J; $\Delta H = q_P = -1.56 \times 10^3$ J; $w = 624$ J

P2.5 -28.6×10^3 J, -15.1×10^3 J, 0

P2.6 $\Delta H = 1.19 \times 10^4$ J; $\Delta U = 9.41 \times 10^3$ J

P2.7 $w = -1.87 \times 10^3$ J; 0.944 bar

P2.8 $q = 0$; $\Delta U = w = -1200$ J; $\Delta H = -2.00 \times 10^3$ J

P2.10 235 K

P2.11 a. -21.1×10^3 J

b. -9.34×10^3 J

P2.12 step 1, 0; step 2, -23.0×10^3 J; step 3, 9.00×10^3 J;
cycle, -14.0×10^3 J

P2.13 4.25×10^3 J; 0.69 m

P2.14 110.5×10^3 Pa; 107.8×10^3 Pa

P2.15 -379 J

P2.16 $\Delta U = q = 5.72 \times 10^3$ J; $\Delta H = 8.01 \times 10^3$ J

P2.17 a. $P_2 = 0.500 \times 10^6$ Pa ; $w = -1.69 \times 10^3$ J; $\Delta U = 0$
and $\Delta H = 0$; $q = -w = 1.69 \times 10^3$ J

b. $P_2 = 6.02 \times 10^5$ Pa; $\Delta U = 748$ J; $w = 0$; $q = 748$ J;
$\Delta H = 1.25 \times 10^3$ J

Overall: $q = 2.44 \times 10^3$ J; $w = -1.69 \times 10^3$ J;
$\Delta U = 748$ J; $\Delta H = 1.25 \times 10^3$ J

P2.18 749 K; $q = 0$; $w = 0$; $w = \Delta U = 5.62 \times 10^3$ J;
$\Delta H = 9.37 \times 10^3$ J

P2.19 a. $w = -496$ J; ΔU and $\Delta H = 0$; $q = -w = 496$ J

b. $\Delta U = -623$ J; $w = 0$; $q = \Delta U = -623$ J;
$\Delta H = -1.04 \times 10^3$ J

$\Delta U_{total} = 623$ J; $w_{total} = -496$ J; $q_{total} = -127$ J ;
$\Delta H_{total} = -1.04 \times 10^3$ J

P2.20 $q = 0$; $w = \Delta U = 463$ J; $\Delta H = 771$ J

P2.21 a. $\Delta U = \Delta H = 0$; $w = -q = -1.25 \times 10^3$ J

b. $w = 0$; $q = \Delta U = 854$ J; $\Delta H = 1.42 \times 10^3$ J

For the overall process, $w = -1.25 \times 10^3$ J,
$q = 2.02 \times 10^3$ J, $\Delta U = 854$ J, and $\Delta H = 1.42 \times 10^3$ J

P2.22 a. 667 K, $w = 9.30 \times 10^3$ J

b. 3.80×10^3 K,, 67.3×10^3 J

P2.23 299 K

P2.24 -8.99×10^3 J

P2.25 a. 188 K

b. 217 K

P2.26 c. 1.27×10^3 kg, 2.54×10^3 kg

P2.27 $q = 0$; $\Delta U = w = -2.43 \times 10^3$ J; $\Delta H = -4.05 \times 10^3$ J

P2.28 475 K

P2.29 $q = 0$; $\Delta U = w = -1.43 \times 10^3$ J; $\Delta H = -2.39 \times 10^3$ J

P2.30 a. -5.54×10^3 J

b. -5.52×10^3 J; -0.4%

Chapter 3

P3.5 77.8 bar

P3.9 $\Delta H = q = 6.67 \times 10^4$ J; $\Delta U = 5.61 \times 10^4$ J;
$w = -1.06 \times 10^4$ J

P3.15 306 K

P3.16 303 K

P3.17 345 K

P3.18 $q = \Delta H = 4.35 \times 10^4$ J; $w = -3.74 \times 10^3$ J;
$\Delta U = 3.98 \times 10^4$ J

P3.19 3.06×10^3 J, 0

Chapter 4

P4.1 a. -1816 kJ mol^{-1}; -1814 kJ mol^{-1}

b. -116.2 kJ mol^{-1}; -113.7 kJ mol^{-1}

c. 62.6 kJ mol^{-1}; 52.7 kJ mol^{-1}

d. -111.6 kJ mol^{-1}; -111.6 kJ mol^{-1}

e. 205.9 kJ mol^{-1}; 200.9 kJ mol^{-1}

f. -172.8 kJ mol^{-1}; -167.8 kJ mol^{-1}

P4.2 $\Delta H_{combustion}^\circ = -3268$ kJ mol^{-1};
$\Delta U_{reaction}^\circ = -3264$ kJ mol^{-1}; 0.0122

P4.3 49.6 kJ mol^{-1}

P4.4 10.41 kJ mol^{-1}, -1.54%

P4.5 -59.8 kJ mol^{-1}

P4.6 -266.3 kJ mol^{-1}; -824.2 kJ mol^{-1}

P4.7 91.6 kJ mol^{-1}

P4.8 -1810 kJ mol^{-1}

P4.9 -20.6 kJ mol^{-1}; -178.2 kJ mol^{-1}

P4.10 415.8 kJ mol^{-1}; 1.2%

P4.11 -134 kJ mol^{-1}, $\approx 0\%$

P4.12 -180.0 kJ mol^{-1}

P4.13 $-91.96 \text{ kJ mol}^{-1}$

P4.14 $132.86 \text{ kJ mol}^{-1}$

P4.15 $-812.2 \text{ kJ mol}^{-1}$

P4.16   a. $-73.0 \text{ kJ mol}^{-1}$

      b. -804 kJ mol^{-1}

P4.17   a. $428.22 \text{ kJ mol}^{-1}$; $425.74 \text{ kJ mol}^{-1}$

      b. $926.98 \text{ kJ mol}^{-1}$; $922.02 \text{ kJ mol}^{-1}$

      c. $498.76 \text{ kJ mol}^{-1}$; $498.28 \text{ kJ mol}^{-1}$

P4.18   Si-F 596 kJ mol^{-1}; 593 kJ mol^{-1}

      Si-Cl 398 kJ mol^{-1}; 396 kJ mol^{-1}

      C-F 489 kJ mol^{-1}; 487 kJ mol^{-1}

      N-F 279 kJ mol^{-1}; 276 kJmol^{-1}

      O-F 215 kJ mol^{-1}; 213 kJ mol^{-1}

      H-F 568 kJ mol^{-1}; 565 kJ mol^{-1}

P4.19   a. 416 kJ mol^{-1}; 413 kJ mol^{-1}

      b. 329 kJ mol^{-1}; 329 kJ mol^{-1}

      c. 589 kJ mol^{-1}; 588 kJ mol^{-1}

P4.20 $\Delta U_f - 757 \text{ kJ mol}^{-1}$; $\Delta H_f^\circ = -756 \text{ kJ mol}^{-b1}$

P4.21 $5.16 \times 10^3 \text{ J} \,^\circ\text{C}^{-1}$

P4.22 $-2.86 \times 10^3 \text{ J mol}^{-1}$, 16%

Chapter 5

P5.2   a. $V_c = 29.6 \text{ L}$, $V_d = 10.4 \text{ L}$

      b. $w_{ab} = -7.62 \times 10^3 \text{ J}$

          $w_{bc} = -5.61 \times 10^3 \text{ J}$

          $w_{cd} = 3.68 \times 10^3 \text{ J}$

          $w_{da} = 5.61 \times 10^3 \text{ J}$

          $w_{total} = -3.94 \times 10^3 \text{ J}$

      c. 0.515; 1.94 kJ

P5.3 $a \rightarrow b$: $\Delta U = \Delta H = 0$; $q = -w = 7.62 \times 10^3 \text{ J}$

      $b \rightarrow c$: $\Delta U = w = -5.61 \times 10^3 \text{ J}$; $q = 0$

      $\Delta H = -9.35 \times 10^3 \text{ J}$

      $c \rightarrow d$: $\Delta U = \Delta H = 0$; $q = -w = 3.68 \times 10^3 \text{ J}$

      $d \rightarrow a$: $\Delta U = w = 5.61 \times 10^3 \text{ J}$; $q = 0$

      $\Delta H = 9.35 \times 10^3 \text{ J}$

      $q_{total} = 3.94 \times 10^3 \text{ J} = -w_{total}$

      $\Delta U_{total} = \Delta H_{total} = 0$

P5.4 $a \rightarrow b$: $\Delta S = -\Delta S_{surroundings} = 8.73 \text{ J K}^{-1}$

      $\Delta S_{total} = 0$

      $b \rightarrow c$: $\Delta S = -\Delta S_{surroundings} = 0$; $\Delta S_{total} = 0$

      $c \rightarrow d$: $\Delta S = -\Delta S_{surroundings} = -8.70 \text{ J K}^{-1}$

      $\Delta S_{total} = 0$

$d \rightarrow a$: $\Delta S = -\Delta S_{surroundings} = 0$. $\Delta S_{total} = 0$ to within the round-off error; for the cycle, $\Delta S = \Delta S_{surroundings} = \Delta S_{total} = 0$ to within the round-off error

P5.5   a. 17.6 J K^{-1}

      b. 10.6 J K^{-1}

P5.6 16.8 J K^{-1}

P5.7   a. $w = -1.25 \times 10^3 \text{ J}$; $\Delta U = 1.87 \times 10^3 \text{ J}$; $q = \Delta H = 3.12 \times 10^3 \text{ J}$; $\Delta S = 8.43 \text{ J K}^{-1}$

      b. $w = 0$; $\Delta U = q = 1.87 \times 10^3 \text{ J}$; $\Delta H = 3.12 \times 10^3 \text{ J}$; $\Delta S = 5.06 \text{ J K}^{-1}$

      c. $\Delta U = \Delta H = 0$; $w_{reversible} = -q = -1.73 \times 10^3 \text{ J}$; $\Delta S = 5.76 \text{ J K}^{-1}$

P5.8   a. $\Delta S_{surroundings} = -6.93 \text{ J K}^{-1}$

      $\Delta S_{total} = 1.50 \text{ J K}^{-1}$; spontaneous

      b. $\Delta S_{surroundings} = -4.16 \text{ J K}^{-1}$

      $\Delta S_{total} = 0.90 \text{ J K}^{-1}$; spontaneous

      c. $\Delta S_{surroundings} = -5.76 \text{ J K}^{-1}$

      $\Delta S_{total} = 0$; not spontaneous

P5.9   a. $1.03 \text{ J K}^{-1} \text{ mol}^{-1}$

      b. $3.14 \text{ J K}^{-1} \text{ mol}^{-1}$

      c. $\Delta S_{transition} = 8.24 \text{ J K}^{-1} \text{ mol}^{-1}$; $\Delta S_{fusion} = 25.12 \text{ J K}^{-1} \text{ mol}^{-1}$

P5.10   a. 23.49 J K^{-1}

      b. 154.4 J K^{-1}

P5.11   a. $q = 0$; $\Delta U = w = -935 \text{ J}$; $\Delta H = -1.31 \times 10^3 \text{ J}$; $\Delta S = 0$

      b. $q = 0$; $\Delta U = w = -748 \text{ J}$; $\Delta H = -1.05 \times 10^3 \text{ J}$; $\Delta S = 1.24 \text{ J K}^{-1}$

      c. $w = 0$; $\Delta U = \Delta H = 0$; $\Delta S = 5.76 \text{ J K}^{-1}$

P5.12   a. $100.8 \text{ J mol}^{-1} \text{ K}^{-1}$

      b. $18.94 \times 10^3 \text{ J mol}^{-1}$

P5.13 $\Delta H_m = 2.84 \times 10^3 \text{ J mol}^{-1}$; $\Delta S_m = 8.90 \text{ J K}^{-1} \text{ mol}^{-1}$

P5.14 $\Delta U = 18.5 \text{ J}$; $w = -2.73 \times 10^3 \text{ J}$

      $\Delta H = 32.1 \text{ J}$; $q \approx 2.73 \times 10^3 \text{ J}$

      $\Delta S = 9.10 \text{ J K}^{-1}$

P5.15 $21.88 \text{ J K}^{-1} \text{ mol}^{-1}$

P5.16   a. $\Delta S_{total} = \Delta S + \Delta S_{surroundings} = 0 + 0 = 0$; not spontaneous

      b. $\Delta S_{total} = \Delta S + \Delta S_{surroundings} = 1.24 \text{ J K}^{-1} + 0 = 1.24 \text{ J K}^{-1}$; spontaneous

P5.17   a. $\Delta S_{surroundings} = 0$; $\Delta S = 0$; $\Delta S_{total} = 0$; not spontaneous

      b. $\Delta S = 27.7 \text{ J K}^{-1}$; $\Delta S_{surroundings} = 0$; $\Delta S_{total} = 27.17 \text{ J K}^{-1}$; spontaneous

      c. $\Delta S = 27.7 \text{ J K}^{-1}$; $\Delta S_{surroundings} = -27.7 \text{ J K}^{-1}$; $\Delta S_{total} = 0$; not spontaneous

P5.18 $\Delta U = -4.36 \times 10^3$ J; $\Delta H = -7.27 \times 10^3$ J;
$\Delta S = -30.7$ J K^{-1}

P5.19
a. $q = 0$; $\Delta U = w = -5.21 \times 10^3$ J; $\Delta H = -8.68 \times 10^3$ J
$\Delta S = 0$; $\Delta S_{surroundings} = 0$; $\Delta S_{total} = 0$

b. $w = 0$; $\Delta U = q = 5.21 \times 10^3$ J; $\Delta H = 8.68 \times 10^3$ J
$\Delta S = 14.5$ J K^{-1}; $\Delta S_{surroundings} = -17.4$ J K^{-1};
$\Delta S_{total} = -2.90$ J K^{-1}

c. $\Delta H = \Delta U = 0$; $w = -q = 6.48 \times 10^3$
$\Delta S = -14.5$ J K^{-1}; $\Delta S_{surroundings} = 21.6$ J K^{-1};
$\Delta S_{total} = 7.1$ J K^{-1}

For the cycle:

$w_{total} = 1.27 \times 10^{-3}$ J; $q_{total} = -1.27 \times 10^{-3}$ J
$\Delta U_{total} = 0$; $\Delta H_{total} = 0$; $\Delta S_{total} = 0$
$\Delta S_{surroundings} = 0$; $\Delta S_{total} = 4.20$ J K^{-1}

P5.20 30.7 J K^{-1}

P5.21 9.0 J K^{-1}

P5.22 18.2 J K^{-1}

P5.23 0.564, 0.744

P5.24 $\Delta S = -21.7$ J K^{-1}; $\Delta S_{surroundings} = 21.9$ J K^{-1};
$\Delta S_{total} = 0.2$ J K^{-1}

P5.25 3.24×10^8 J

P5.26 2.5

P5.27
a. 0.627
b. 0.398
c. 110.7

P5.28 6.25 m^2

P5.29 640 J s^{-1}

P5.30 4.5×10^2 g

P5.31 30.69 J K^{-1} mol^{-1}

Chapter 6

P6.1 -40.96 kJ g^{-1}, -117.6 kJ g^{-1}

P6.2 5.30×10^3 J

P6.3 $-22.1 \times 10^3 - $ J

P6.4
a. -9.97×10^3 J, -9.97×10^3 J
b. same as part a

P6.5 216

P6.6 52.8 J; 11.4×10^3 J; -218.5×10^3 J mol^{-1}

P6.7 -257.2×10^3 J mol^{-1}; -226.8×10^3 J mol^{-1}

P6.8 $\Delta G^\circ_{combustion} = -818.6 \times 10^3$ J mol^{-1};
$\Delta A^\circ_{combustion} = -813.6 \times 10^3$ J mol^{-1}

P6.9
a. 0.1408
b. 2.00×10^{-18}
c. 101 kJ mol^{-1}

P6.10
a. 0.379; 1.284
b. $\Delta H^\circ_{reaction} = 56.8 \times 10^3$ J mol^{-1};
$\Delta G^\circ_{reaction}(298.15$ K$) = 35.0 \times 10^3$ J mol^{-1}

P6.11
a. 1.40; $\Delta G^\circ_{reaction} = -2.80 \times 10^3$ J mol^{-1}
b. -29.7 kJ mol^{-1}

P6.12
a. $\Delta H^\circ_{reaction} = -19.0$ kJ mol^{-1}
$\Delta G^\circ_{reaction}(600°C) = 765$ J mol^{-1}
$\Delta S^\circ_{reaction}(600°C) = -22.6$ J mol^{-1} K^{-1}
b. $x_{CO_2} = 0.47$; $x_{CO} = 0.53$

P6.13
a. $K_P(700$ K$) = 3.85$; $K_P(800$ K$) = 1.56$
b. $\Delta H^\circ_{reaction} = -42.1$ kJ mol^{-1}
$\Delta G^\circ_{reaction}(700$ K$) = -7.81$ kJ mol^{-1}
$\Delta G^\circ_{reaction}(800$ K$) = -2.91$ kJ mol^{-1}
$\Delta S^\circ_{reaction}(700$ K$) = 60.1$ J mol^{-1} K^{-1}
$\Delta S^\circ_{reaction}(800$ K$) = 52.6$ J mol^{-1} K^{-1}
c. -27.5 kJ mol^{-1}

P6.14
a. $\dfrac{x_F}{x_G} = 2.025 \times 10^{-4}$

$\dfrac{x_E}{x_G} = 4.581 \times 10^{-7}$; $\dfrac{x_D}{x_G} = 2.486 \times 10^{-5}$;

$\dfrac{x_C}{x_G} = 4.109 \times 10^{-6}$; $\dfrac{x_B}{x_G} = 1.497 \times 10^{-6}$;

$\dfrac{x_A}{x_G} = 9.803 \times 10^{-8}$

c. F 2.025×10^{-2} %; E 4.581×10^{-5} %
D 2.486×10^{-3} %; C 4.109×10^{-4} %
B 1.497×10^{-4} %; A 9.803×10^{-6} %

P6.15
b. 3.78×10^{-5} bar
c. 6.20×10^{-5} bar

P6.17 -65.2×10^3 J mol^{-1}

P6.18 $= -18.6 \times 10^3$ J; 62.5 J K^{-1}

P6.19
a. -34.4 kJ
b. -47.3 kJ
c. -12.9 kJ

P6.20 4; -32.7×10^3 J mol^{-1}

P6.21 468 K; 1.03×10^4

P6.22 9.95×10^5

P6.23 371

P6.24 4.68×10^{-2}

P6.25 1456 K; 9.12 Torr

P6.26
a. 1.11×10^{-2}
c. 1.76 moles of $N_2(g)$, 5.28 moles of $H_2(g)$, and
0.48 moles of $NH_3(g)$

P6.27	c.	5.13×10^{-35}
	d.	1.03×10^{-34}
P6.28	c.	8.68×10^{-2}, 0.045
	d.	2.2%
P6.29	a.	3.31×10^{-3}
	b.	0.0139 bar
P6.30	b.	0.55
	d.	0.72

Chapter 7

P7.1	vdW: 169 bar; R-K: 174 bar
P7.2	$\rho_{idealgas} = 395 \text{ g L}^{-1}$; $\rho_{vdW} = 369 \text{ g L}^{-1}$
P7.3	Ideal gas: 9.62 mol L^{-1}; vdW gas: 8.73 mol L^{-1}
P7.4	111 K, 426 K, 643K
P7.5	$b = 0.0431 \text{ dm}^3\text{mol}^{-1}$ $a = 2.303 \text{ dm}^6\text{bar mol}^{-2}$
P7.6	$a = 32.20 \text{ dm}^6 \text{ bar K}^{1/2}\text{mol}^{-2}$ $b = 0.02985 \text{ dm}^3\text{mol}^{-1}$
P7.7	0.105 L mol^{-1}
P7.8	51.2 K; 18.8 bar
P7.10	298 K, 297.6 K
P7.11	Ideal gas: -10.34×10^3 J; vdW gas: -10.41×10^3 J; 1.3%
P7.14	$b = 3.59 \times 10^{-5} \text{m}^3 \text{ mol}^{-1}$ $a = 3.73 \times 10^{-2} \text{m}^6 \text{ Pa mol}^{-2}$
P7.15	Ideal gas: 0.2438 L; vdW gas: 91.4×10^{-3} L, R-K gas: 81.2×10^{-3} L
P7.18	$V_m = 3.34 \times 10^{-2} \text{ L mol}^{-1}$; $z = 0.602$
P7.23	$\gamma = 0.497, 0.368, 0.406, 0.670,$ and 1.65 at 100, 200, 300, 400, and 500 bar, respectively
P7.24	$a = 1.15 \text{ L}^2 \text{ bar mol}^{-2}$; $b = 0.0630 \text{ L mol}^{-1}$; $V_m = 22.72$ L
P7.25	Ideal gas: 0.5211 L; vdW gas: 0.1784 L

Chapter 8

P8.1	a.	110 J mol^{-1}
	b.	594 J mol^{-1}
P8.5		$T_{b,normal} = 271.8$ K; $T_{b,standard} = 269.6$ K
P8.6		354.4 K
P8.7		6.17×10^3 Pa
P8.8		$30.58 \text{ kJ mol}^{-1}$
P8.9		$22.88 \text{ kJ mol}^{-1}$

P8.10		$20.32 \text{ kJ mol}^{-1}$
P8.11		$25.28 \text{ kJ mol}^{-1}$
P8.12		$50.99 \text{ kJ mol}^{-1}$
P8.13	a.	$\Delta H_m^{vaporization} = 32.1 \times 10^3 \text{J mol}^{-1}$; $\Delta H_m^{sublimation} = 37.4 \times 10^3 \text{J mol}^{-1}$
	b.	$5.3 \times 10^3 \text{ J mol}^{-1}$
	c.	349.5 K; $91.8 \text{ J mol}^{-1} \text{ K}^{-1}$
	d.	$T_{tp} = 264$ K; $P_{tp} = 2.84 \times 10^3$ Pa
P8.14	a.	720 bar
	b.	2.2×10^2 bar
	c.	$-1.5°C$
P8.15	a.	56.22 Torr
	b.	52.65 Torr
P8.16	a.	4.66 bar
	b.	4.10 bar
P8.17		$8.2°C$
P8.18		269 Pa
P8.19	a.	335.9 K
	b.	$38.19 \text{ kJ mol}^{-1}$ at 298 K; $37.20 \text{ kJ mol}^{-1}$ at 335.9 K
P8.21	a.	$\Delta H_{sublimation} = 16.92 \times 10^3 \text{ J mol}^{-1}$; $\Delta H_{vaporization} = 14.43 \times 10^3 \text{ J mol}^{-1}$
	b.	$2.49 \times 10^3 \text{ J mol}^{-1}$
	c.	$T_{tp} = 73.62$ K; $P_{tp} = 5.36 \times 10^{-3}$ Torr
P8.22		$38.4 \text{ J K}^{-1} \text{ mol}^{-1}$; $16.4 \times 10^3 \text{ J mol}^{-1}$
P8.23		0.061%
P8.24		467.7 K; 2.513×10^5 Pa
P8.25		7.806×10^4 Pa
P8.26		$\Delta H_{sublimation} = 231.7 \text{ kJ mol}^{-1}$ $\Delta H_{vaporization} 206.5 \text{ kJ mol}^{-1}$ $\Delta H_{fusion} = 25.2 \text{ kJ mol}^{-1}$ 1398 K; 128 Torr
P8.27		$\Delta H_{sublimation}° = 32.6 \text{ kJ mol}^{-1}$ $\Delta H_{vaporization}° = 26.9 \text{ kJ mol}^{-1}$ $\Delta H_{fusion}° = 5.6 \text{ kJ mol}^{-1}$ 240.3 K; 402 Torr
P8.28		142 K; 2984 Torr $\Delta H_{sublimation} = 10.07 \times 10^3 \text{ J mol}^{-1}$ $\Delta H_{vaporization} = 9.38 \times 10^3 \text{ J mol}^{-1}$ $\Delta H_{fusion} = 0.69 \times 10^3 \text{ J mol}^{-1}$
P8.29		-0.72 K at 100 bar and -3.62 K at 500 bar
P8.30		1.95 atm
P8.31		8.5 kJ mol^{-1}

P8.32 9.60×10^5 Pa

P8.33 1.068

P8.34 6.66×10^4 Pa

Chapter 9

P9.1 121 Torr

P9.2 0.116 bar

P9.3 0.272

P9.4 $P_a^* = 0.623$ bar; $P_B^* = 1.414$ bar

P9.5 a. $P_A = 63.0$ Torr; $P_B = 25.4$ Torr

   b. $P_A = 74.9$ Torr; $P_B = 18.2$ Torr

P9.6 $x_{bromo} = 0.67$; $y_{bromo} = 0.44$

P9.7 a. 2651 Pa

   b. 0.525

   c. $Z_{chloro} = 0.614$

P9.8 0.301

P9.9 a. 25.0 Torr, 0.50

   b. $Z_{EB} = (1 - Z_{EC}) = 0.387$

P9.10 a. 0.560

   b. 0.884

P9.11 0.337

P9.13 a. for ethanol: $a_1 = 0.9504$; $\gamma_1 = 1.055$

      for isooctane: $a_2 = 1.411$; $\gamma_2 = 14.20$

   b. 121.8 Torr

P9.16 413 Torr

P9.17 61.9 Torr

P9.18 -4.2 cm^3

P9.19 33.5 g mol^{-1}

P9.20 1.86 K kg mol^{-1}

P9.21 $M = 37.6$ g mol^{-1}; $\Delta T_f = -1.26$ K;

   $\dfrac{P_{benzene}}{P_{benzene}^*} = 0.981$

   $\pi = 5.37 \times 10^5$ Pa

P9.22 2.37 m; 2.32×10^4 Pa

P9.23 1400 kg mol^{-1}

P9.25 57.8 cm^3 mol^{-1}

P9.26 0.327 mol

P9.27 -0.034 L

P9.28 $a_A = 0.569$; $\gamma_A = 2.00$; $a_B = 0.986$; $\gamma_B = 1.38$

P9.29 $a_{CS_2}^R = 0.8723$; $\gamma_{CS_2}^R = 1.208$;

   $a_{CS_2}^H = 0.2223$; $\gamma_{CS_2}^H = 0.3079$

P9.30 7.14×10^{-3} g; 2.67×10^{-3} g

Chapter 10

P10.1 $\Delta H_{reaction}^\circ = -65.4$ kJ mol^{-1};
$\Delta G_{reaction}^\circ = -55.7$ kJ mol^{-1}

P10.2 $\Delta H_{reaction}^\circ = 17$ kJ mol^{-1}; $\Delta G_{reaction}^\circ - 16.5$ kJ mol^{-1}

P10.3 -32.9 J K^{-1} mol^{-1}

P10.4 1.1 J K^{-1} mol^{-1}

P10.5 $\Delta G_{solvation}^\circ = -379$ kJ mol^{-1}

P10.6 a. 5.0×10^{-4} mol kg^{-1}

   b. 7.9×10^{-4} mol kg^{-1}

   c. 5.0×10^{-4} mol kg^{-1}

P10.10 0.0285 mol kg^{-1}

P10.11 0.0111

P10.12 0.238 mol kg^{-1}; 0.0393

P10.13 43.0 nm

P10.14 304 nm

P10.15 0.736

P10.16 a. 0.92

   b. 0.77

   c. 0.52

P10.17 $I = 0.1500$ mol kg^{-1}

   $\gamma_\pm = 0.2559$

   $a_\pm = 0.0146$

P10.18 $I = 0.0750$ mol kg^{-1}

   $\gamma_\pm = 0.523$

   $a_\pm = 0.0209$

P10.19 $I = 0.325$ mol kg^{-1}

   $\gamma_\pm = 0.069$

   $a_\pm = 0.0068$

P10.20 a. 1.07×10^{-5} mol L^{-1}

   b. 1.21×10^{-5} mol kg^{-1}

P10.21 a. 49%

   b. 40%

P10.22 a. 13.6%

   b. 14.8%

P10.23 a. 6.89%

   b. 8.08%

P10.24 a. 0.0770

   b. 0.0422

   c. 0.0840

P10.25 a. 0.0794 mol kg^{-1}

   b. 0.0500 mol kg^{-1}

c. 0.0500 mol kg^{-1}

d. 0.1140 mol kg^{-1}

P10.26 a. 0.150 mol kg^{-1}

b. 0.0500 mol kg^{-1}

c. 0.200 mol kg^{-1}

d. 0.300 mol kg^{-1}

P10.27 Using limiting law 0.100m 1.37%;1.00m 0.453%; no ionic interactions 0.100m 1.31%;1.00m 0.418%

P10.28 a. 2.91%

b. 2.02%

c. 17.7%

Chapter 11

P11.1 210.4 kJ mol^{-1}; 1.21×10^{-37}

P11.2 713.2 kJ mol^{-1}; 9.06×10^{124}

P11.3 8.28×10^{-84}; -1.22869 V

P11.4 -103.8 kJ mol^{-1}

P11.5 -131.2 kJ mol^{-1}

P11.6 a. 1.30×10^8

b. 6.67×10^{-56}

P11.7 a. 1.52×10^{-82}

b. 3.34×10^{13}

P11.8 1.178 V; 1.49×10^{-36}; 204.5 kJ mol^{-1}

P11.9 -0.698 V; 1.63×10^{-23}

P11.10 -0.7910 V; 7.23×10^{-22}

P11.11 -2.340 V; 1.38×10^{-78}

P11.12 4.16×10^{-4}

P11.13 2.65×10^6; -36.7 kJ mol^{-1}

P11.14 -0.913 V

P11.15 c. -1108 kJ

P11.16 4.90×10^{-13}

P11.17 $\Delta G_R^\circ = -212.3$ kJ mol^{-1}

$\Delta S_R^\circ = -12.5$ J K^{-1}

$\Delta H_R^\circ = -216.0$ kJ mol^{-1}

P11.18 2.38 V; 1.81×10^{-4} V K^{-1}

P11.19 1.094 V; 1.097 V; 0.27%

P11.20 a. 1.0122 V

b. 1.0050 V; 0.72% or ~0 with correct number of significant figures

P11.22 0.769

P11.23 a. 1.110 V

b. 0.626

c. 1.106 V

P11.24 a. 9.94

b. 0.101

P11.26 -131.1 kJ mol^{-1}

P11.28 $\Delta G = -369.99$ kJ mol^{-1}; $\Delta S = 10.8$ J mol^{-1} K^{-1}; $\Delta H = -367.0$ kJ mol^{-1}

P11.29 $\Delta G = -33.4$ kJ mol^{-1}; $\Delta S = -29.9$ J mol^{-1} K^{-1}; $\Delta H = -43.1$ kJ mol^{-1}

P11.30 1.75×10^{-12}

Chapter 12

P12.1 a. 4/52

b. 1/52

c. 12/52 and 3/52, respectively

P12.2 a. 0.002

b. 1.52×10^{-6}

P12.3 a. 1/6

b. 1/18

c. 21/36

P12.4 a. 8/49

b. 6/49

c. 23/49

P12.5 a. 720

b. 360

c. 1

d. 3.73×10^{16}

P12.6 120

P12.7 a. 1

b. 15

c. 1

d. 1.03×10^{10}

P12.8 a. 4.57×10^5

b. 1.76×10^4

c. 3.59×10^5

P12.9 a. 9.54×10^{-7}

b. 9.54×10^{-7}

c. 1.27×10^{-6} and 3.77×10^{-7}

P12.10 0.004

P12.11 a. 4.52×10^{-8}

b. 1.04×10^{-6}

c. 8.66×10^{-9}

P12.12 a. bosons: 220; fermions: 45

b. bosons: 1.72×10^5; fermions: 1.62×10^5

P12.13 a. 7.41×10^{11}

b. 2.97×10^{10}

c. 2.04×10^6

P12.14 a. 9.77×10^{-4}

b. 0.044

c. 0.246

d. 0.044

P12.15 a. 1.69×10^{-5}

b. 3.05×10^{-3}

c. 0.137

d. 0.195

P12.16 a. 9.57×10^{-7}

b. 0.176

c. 0.015

P12.17 a. $(n)(n-1)$

b. $\dfrac{(n)(n-1)(n-2)(n/2+1)}{(n/2)!}$

P12.18 $1.91

P12.20 c. 182 J mol^{-1}

P12.21 c. case 1: 0.245; case 2: 0.618

P12.22 a. $2/a$

b. $a/2$

c. $a^2\left(\dfrac{1}{3} - \dfrac{1}{2\pi^2}\right)$

d. $a^2\left(\dfrac{1}{12} - \dfrac{1}{2\pi^2}\right)$

P12.23 a. $\sqrt{m/2\pi kT}$

b. 0

c. kT/m

d. kT/m

P12.24 a. 0

b. 1/3

Chapter 13

P13.1 b. $\exp(693)$

c. $\exp(673)$

P13.3 0.25

P13.4 a. 2.60×10^6

b. 5148

P13.7 $P_{N_2} = 0.230$ atm; $P_{O_2} = 0.052$ atm

P13.8 a. 248 K

b. 178 K

P13.9 a. 254 K

b. 179 K

P13.10 a. 6.07×10^{-20} J

b. Set C

P13.12 0.333

P13.13 4150 K

P13.14 0.999998

P13.15 $a_- = 0.333334$

$a_0 = 0.333333$

$a_+ = 0.333333$

P13.16 432 K

P13.17 1090 K

P13.18 At 300 K, p = 0.074; F_2 equivalent at 524 K

At 1000 K, p = 0.249; F_2 equivalent at 1742 K

P13.19 5.85×10^4 K

P13.20 At 100 K, $p = 0.149$

At 500 K, $p = 0.414$

At 200 K, $p = 0.479$

Chapter 14

P14.1 $q_T(H_2) = 2.74 \times 10^{26}; q_T(N_2) = 1.42 \times 10^{28}$

P14.2 $q_T(Ar) = 2.44 \times 10^{29}, T = 590$ K

P14.3 0.086 K

P14.4 3.91×10^{17}

P14.5 2.00×10^5

P14.6 a. 1

b. 2

c. 2

d. 12

e. 4

P14.7 Rotational: HD; translational: D_2

P14.8 H_2: 1.00; HD: 1.22

P14.9 a. no

b. no

c. yes

d. yes

P14.10 $q_R = 5832$

P14.11 $q_R = 3.78 \times 10^4$

P14.12 a. 616 K

b. $J = 5$

c. 615 K

P14.13 $q_R = 21.8$; by summation, $q_R = 22.0$

P14.14 a.

J	p_J	J	p_J
0	0.041	5	0.132
1	0.113	6	0.095
2	0.160	7	0.062
3	0.175	8	0.037
4	0.167	9	0.019

b.

J	p_J	J	p_J
0	0.043	5	0.131
1	0.117	6	0.093
2	0.165	7	0.059
3	0.179	8	0.034
4	0.163	9	0.018

P14.15 $\Theta_R = 7.58$ K

P14.16 At 300 K, $q = 1$, $p_0 = 1$
At 3000 K, $q = 1.32$, $p_0 = 0.762$

P14.17 IF at 300 K: $q = 1.06$, $p_0 = 0.943$, $p_1 = 0.051$,
$p_2 = 0.003$
IF at 3000 K: $q = 3.94$, $p_0 = 0.254$, $p_1 = 0.189$,
$p_2 = 0.141$
IBr at 300 K: $q = 1.38$, $p_0 = 0.725$, $p_1 = 0.199$,
$p_2 = 0.054$
IBr at 3000 K: $q = 8.26$, $p_0 = 0.121$, $p_1 = 0.106$,
$p_2 = 0.094$

P14.18 $q_V = 1.67$

P14.19 $q_V = 1.09$

P14.20 $q_V = 1.70$

P14.22 451 cm^{-1}

P14.24 a. $q = L/\Lambda$
b. $q = V/\Lambda^3$

P14.25 $q_E = 10.2$

P14.26 a. $q_E = 4.77$
b. 2138 K

P14.27 a. $q_E = 3.12$
b. 251 K

P14.28 $q = 1.71 \times 10^{34}$

P14.29 $q = 1.30 \times 10^{29}$

P14.30 $q_R = 265$

Chapter 15

P15.1 Ensemble B

P15.2 nRT

P15.3 $U = 1/2\,NkT$; $C_V = 1/2\,Nk$

P15.4

Molecule	θ_R (K)	High-T for R?	θ_V (K)	High-T for V?
H^{35}Cl	15.3	no	4153	no
^{12}C^{16}O	2.78	yes	3123	no
^{39}KI	0.088	yes	288	no
CsI	0.035	yes	173	no

P15.5 1.71 kJ mol^{-1}

P15.6 $U = \dfrac{Nm_1\varepsilon_1 e^{-\varepsilon_1/kT}}{m_0\left(1+\left(\dfrac{m_1}{m_0}\right)e^{-\varepsilon_1/kT}\right)}$

P15.7 3.72 kJ mol^{-1}

P15.8 6.19 kJ mol^{-1}

P15.10 C_V values are as follows:

	298 K	**500 K**	**1000 K**
2041 cm^{-1}	0.042	0.811	4.24
712 cm^{-1}	3.37	5.93	7.62
3369 cm^{-1}	0.000	0.048	1.56
Total	6.78	12.7	21.0

P15.11 $C_V = n(1.86 \text{ J mol}^{-1}\text{ K}^{-1})$

P15.12 c. 352 m s^{-1}

P15.15 S(Ar) at 200 K = 123 J mol^{-1} K^{-1}
S(Ar) at 300 K = 128 J mol^{-1} K^{-1}
S(Ar) at 500 K = 135 J mol^{-1} K^{-1}

P15.16 1.28×10^{-10} m

P15.17 219 J mol^{-1} K^{-1}

P15.18 256 J mol^{-1} K^{-1}

P15.19 260 J mol^{-1} K^{-1}

P15.20 186 J mol^{-1} K^{-1}

P15.22 11.2 J mol^{-1} K^{-1}

P15.23 a. $R \ln 2$
b. $R \ln 4$
c. $R \ln 2$
d. 0

P15.27 Ne: -40.0 kJ mol^{-1}; Kr: -40.3 kJ mol^{-1}

P15.29 -57.2 kJ mol^{-1}

P15.30 $G^\circ_{R,m} = -15.4$ kJ mol^{-1}; $G^\circ_{V,m} = -0.30$ kJ mol^{-1}

P15.31 2.25×10^{-9}

P15.33 a. 143.4 mJ mol^{-1}
b. 28.0

Chapter 16

P16.2 a. $\nu_{mp} = 495$ m s^{-1}, $\nu_{ave} = 559$ m s^{-1}, $\nu_{rms} = 607$ m s^{-1}

b. $\nu_{mp} = 243$ m s^{-1}, $\nu_{ave} = 274$ m s^{-1}, $\nu_{rms} = 298$ m s^{-1}

c. $\nu_{mp} = 555$ m s^{-1}, $\nu_{ave} = 626$ m s^{-1}, $\nu_{rms} = 680$ m s^{-1}

d. $\nu_{mp} = 406$ m s^{-1}, $\nu_{ave} = 458$ m s^{-1}, $\nu_{rms} = 497$ m s^{-1}

e. $\nu_{mp} = 82.9$ m s^{-1}, $\nu_{ave} = 93.6$ m s^{-1}, $\nu_{rms} = 102$ m s^{-1}

P16.3 300 K: $\nu_{mp} = 395$ m s^{-1}, $\nu_{ave} = 446$ m s^{-1}, $\nu_{rms} = 484$ m s^{-1}

500 K: $\nu_{mp} = 510$ m s^{-1}, $\nu_{ave} = 575$ m s^{-1}, $\nu_{rms} = 624$ m s^{-1}

$v_{H_2} = (3.98)v_{O_2}$

P16.4 $\nu_{ave,CCl_4} = 203$ m s^{-1}

$\nu_{ave,O_2} = 444$ m s^{-1}

$KE_{ave} = 6.17 \times 10^{-21}$ J (for both)

P16.5 a. 5.66×10^{-4} s

b. 2.11×10^{-3} s

c. 0.534

P16.6 a. $\dfrac{\nu_{avg}}{\nu_{mp}} = \dfrac{2}{\sqrt{\pi}}$, $\dfrac{\nu_{rms}}{\nu_{mp}} = \sqrt{\dfrac{3}{2}}$

P16.7 81.5 K (for both cases)

P16.8 0.843 and 0.157, respectively

P16.9 a. Ne: 829 m s^{-1}, Kr: 407 m s^{-1}, Ar: 589 m s^{-1}

b. 2100 K

P16.10 0.392

P16.11 a. 2.10×10^5 K

b. 3.00×10^4 K

P16.12 At 298 K: 0.132

At 500 K: 0.071

P16.19 1.70×10^{24} collisions s^{-1}

P16.20 a. 4.97 s

b. 6.03×10^{-4} kg

P16.21 a. 2.73×10^{23} collisions s^{-1}

b. 3.60×10^{14} collisions s^{-1}

P16.22 A $= 1.07 \times 10^{-5}$ m^2

P16.23 a. 7.11×10^9 collisions s^{-1}

b. 0.382 atm

c. 1.29×10^{-7} m

P16.24 a. 8.44×10^{34} m^{-3} s^{-1}

b. 227 K

P16.25 a. $z_{11} = 9.35 \times 10^3$ s^{-1}, $\lambda = 0.051$ m

b. $z_{11} = 9.35 \times 10^{-4}$ s^{-1}, $\lambda = 5.08 \times 10^5$ m

P16.26 a. 1.60×10^{-7} m

b. 1.60×10^{-5} m

c. 1.60×10^{-2} m

P16.27 Ne: 2.01×10^{-7} m, Kr: 9.26×10^{-8} m, CH$_4$: 1.05×10^{-7} m

P16.28 4.58×10^{-4} torr

P16.30 a. 9.60×10^7 s^{-1}

b. 1.80×10^{31} m^{-3} s^{-1}

c. 4.39×10^{-6} m

Chapter 17

P17.1 0.318 nm^2

P17.2 a. 0.368 nm^2

b. 0.265 nm^2

P17.3 a. 319 s

b. 6.13×10^{-10} s

P17.4 a. 1.60×10^{-3} s

c. 2.40×10^{-3} s

P17.5 b. 2.58×10^{-5} m

P17.6 a. -9.60 J s^{-1}

b. -33.6 J s^{-1}

c. -22.8 J s^{-1}

P17.7 -1.80×10^{-4} W cm^{-2}

P17.8 a. 0.00516 J K^{-1} m^{-1} s^{-1}

b. 0.00249 J K^{-1} m^{-1} s^{-1}

c. 0.0050 J K^{-1} m^{-1} s^{-1}

P17.9 1.50×10^{-19} m^2

P17.10 c. 1.50×10^{-19} m^2

P17.11 a. 1.14

b. 0.659

P17.12 1.33

P17.13 a. 6.23×10^{-19} m^2

b. 0.00389 J K^{-1} m^{-1} s^{-1}

P17.14 a. 2.05×10^{-3} m^3 s^{-1}

b. 1.88×10^{-3} m^3 s^{-1}

P17.15 a. 37.3 m s^{-1}

b. 0.893 m s^{-1}

P17.16 D_2: 118 μP, Hd: 103 μP

P17.17 1.89 cP

P17.18 21.9 s

P17.19 b. 1.34×10^{-5} m^2 s^{-1}

P17.20 b. 6.95×10^{-3} J K^{-1} m^{-1} s^{-1}

c. 1.47×10^{-2} J K^{-1} m^{-1} s^{-1}

P17.21 E $= 10.7$ kJ mol^{-1}, A $= 8.26 \times 10^{-3}$

P17.22 a. 1.89 nm

b. 16.8 kg mol^{-1}

P17.23 a. Catalase: 238 kg mol^{-1}, alcohol dehyd.: 74.2 kg mol^{-1}

b. 3.10×10^4 s

P17.24 a. 1.70×10^{-13} s

b. 1.89 nm

P17.25 3.75×10^{20} electrons

P17.26 0.0123 S m^2 mol^{-1}

P17.27 Strong electrolyte; $\Lambda_m^\circ = 0.0125$ S m^2 mol^{-1}

P17.28 $\Lambda_m^\circ = 0.00898$ S m^2 mol^{-1}

P17.30 10 steps: 4.39×10^{-2}, 20 steps: 7.39×10^{-2}, 100 steps: 6.66×10^{-2}

P17.31 1.17×10^{-10} m^2 s^{-1}

P17.32 a. 4.48×10^{-6} S m^{-1} ohm

b. 4.34×10^{-6} S m^{-1}

P17.33 b. K $= 1.00 \times 10^{-14}$

Chapter 18

P18.2 b. 2.79×10^3 s

c. 425 s

P18.3 First order, $k = 3.79 \times 10^{-5}$ s^{-1}

P18.5 a. First order with regard to (wrt) ClO

First order wrt BrO

Second order overall

Units of k: s^{-1} M^{-1}

b. Second order wrt NO

First order wrt O$_2$

Third order overall

Units of k: s^{-1} M^{-2}

c. Second order wrt HI

First order wrt O$_2$

$-1/2$ order wrt H$^+$

2.5 order over all

Units of k: s^{-1} M$^{-3/2}$

P18.6 a. Second

b. Third

c. 1.5

P18.7 Second order with respect to NO$_2$, first order with respect to H$_2$, $k = 6.43 \times 10^{-5}$ kPa^2 s^{-1}

P18.9 Second order; $k = 0.317$ M^{-1} s^{-1}

P18.11 a. $k = 2.08 \times 10^{-3}$ s^{-1}

b. $t = 1.11 \times 10^3$ s

P18.12 1.43×10^{24}

P18.13 1.50×10^4 days

P18.14 a. $k = 3.65 \times 10^{-4}$ day^{-1}

b. 1.90×10^3 days

P18.18 2.55×10^{-7} s

P18.21 a. $k = 0.0329$ M^{-1} s^{-1}

b. 120 s

P18.22 a. 1.08×10^{-12} s

P18.23 4.76×10^{-10} yr^{-1} and $k_2 = 5.70 \times 10^{-11}$ yr^{-1}

P18.24 a. Rate $= 2.08 \times 10^{-15}$ M s^{-1}

b. Rate $= 1.56 \times 10^{-15}$ M s^{-1}

c. [Cl] $= 1.10 \times 10^{-18}$ M, [O$_3$] $= 4.24 \times 10^{-11}$ M

P18.28 a. $E_a = 1.50 \times 10^5$ J mol^{-1}, $A = 1.02 \times 10^{10}$ M^{-1} s^{-1}

b. $k = 0.0234$ M^{-1} s^{-1}

P18.29 269 s

P18.30 $A = 7.38 \times 10^{13}$ s^{-1}, $k = 3.86 \times 10^{-5}$ s^{-1}

P18.32 1845 s^{-1} and 1291 M^{-1} s^{-1}

P18.33 a. $E_a = 108$ kJ mol^{-1}, $A = 1.05 \times 10^{10}$ s^{-1}

b. $\Delta S^\ddagger = -60.8$ J mol K^{-1}; $\Delta H^\ddagger = 105.2$ kJ mol^{-1}

P18.34 a. $E_a = 219$ kJ mol^{-1}, $A = 7.20 \times 10^{12}$ s^{-1}

b. $\Delta S^\ddagger = -14.0$ J mol K^{-1}; $\Delta H^\ddagger = 212.4$ kJ mol^{-1}

P18.35 a. $E_a = 790$ J mol^{-1}, $A = 4.88 \times 10^{10}$ M^{-1} s^{-1}

b. $k = 3.17 \times 10^{10}$ M^{-1} s^{-1}

c. $\Delta S^\ddagger = -57.8$ J mol K^{-1}; $\Delta H^\ddagger = -2.87$ kJ mol^{-1}

P18.36 $\Delta S^\ddagger = 62.0$ J mol K^{-1}; $\Delta H^\ddagger = 270$ kJ mol^{-1}

P18.37 $\Delta S^\ddagger = 7.33$ J mol K^{-1}; $\Delta H^\ddagger = 37$ kJ mol^{-1}

Chapter 19

P19.5 b. 170 kJ mol^{-1}

c. 3.98×10^7

P19.11 $k_1 = 1.19 \times 10^4$ M^{-1} s^{-1}; $k_1/k_2 = 1.05 \times 10^3$ M^{-1}

P19.12 2.45×10^{-3} M s^{-1}

P19.13 $Rate_{max} = 3.75 \times 10^{-2}$ M s^{-1}, $K_m = 2.63 \times 10^{-2}$ M, $k_2 = 1.08 \times 10^7$ s^{-1}

P19.14 a. 2.5 M

b. 2.71×10^{-5} M

P19.15 0.0431 M

P19.16 $K_m = 6.49$ μM, $rate_{max} = 4.74 \times 10^{-8}$ μM s^{-1}, $K_m^* = 24.9$ μM, $K_i = 70.4$ μM

P19.22

P (atm)	θ
20	0.595
50	0.754
100	0.853
200	0.932
300	0.952

P19.23 $r = 0.674$ mL, $s = 4.99$ torr^{-1}

P19.24 $V_m = 1.56$ cm^3 g^{-1}, $K = 221$ torr^{-1}

P19.26 a. At 290 nm: 7.29×10^{16} photons cm^{-2}; at 320 nm: 8.05×10^{16} photons cm^{-2}

b. 34.5 s

P19.28 0.95

P19.29 4.55×10^{18} photons, I $= 2.41 \times 10^{-3}$ J s^{-1}

P19.30 3.02×10^{19} photons s^{-1}

P19.31 $k_{ISC}^S/k_f = 7.33$, $k_P = 3.88 \times 10^{-2}$ s^{-1}, $k_{ics}^T = 0.260$ s^{-1}

P19.32 a. 0.91

b. 7.19×10^{18} photons cm^{-1} s^{-1}

c. 1.53 nW

P19.33 b. $A = 0.312$, $B = 0.697$

c. -0.172 atm

Index

Page numbers in italics indicate tables and page numbers in bold indicate figures.

Masses and Natural Abundances for Selected Isotopes

Nuclide	Symbol	Mass (amu)	Percent Abundance
H	^{1}H	1.0078	99.985
	^{2}H	2.0140	0.015
He	^{3}He	3.0160	0.00013
	^{4}He	4.0026	100
Li	^{6}Li	6.0151	7.42
	^{7}Li	7.0160	92.58
B	^{10}B	10.0129	19.78
	^{11}B	11.0093	80.22
C	^{12}C	12 (exact)	98.89
	^{13}C	13.0034	1.11
N	^{14}N	14.0031	99.63
	^{15}N	15.0001	0.37
O	^{16}O	15.9949	99.76
	^{17}O	16.9991	0.037
	^{18}O	17.9992	0.204
F	^{19}F	18.9984	100
P	^{31}P	30.9738	100
S	^{32}S	31.9721	95.0
	^{33}S	32.9715	0.76
	^{34}S	33.9679	4.22
Cl	^{35}Cl	34.9688	75.53
	^{37}Cl	36.9651	24.4
Br	^{79}Br	79.9183	50.54
	^{81}Br	80.9163	49.46
I	^{127}I	126.9045	100